Methods in Enzymology

Volume 352

REDOX CELL BIOLOGY AND GENETICS

Part A

METHODS IN ENZYMOLOGY

Methods in Enzymology

Volume 352

Redox Cell Biology and Genetics

Part A

EDITED BY

Chandan K. Sen

LABORATORY OF MOLECULAR MEDICINE
DEPARTMENTS OF SURGERY AND MOLECULAR AND CELLULAR BIOCHEMISTRY
DAVIS HEART AND LUNG RESEARCH INSTITUTE
THE OHIO STATE UNIVERSITY MEDICAL CENTER
COLUMBUS, OHIO

Lester Packer

DEPARTMENT OF MOLECULAR PHARMACOLOGY AND TOXICOLOGY
UNIVERSITY OF SOUTHERN CALIFORNIA
LOS ANGELES, CALIFORNIA

ACADEMIC PRESS

An imprint of Elsevier Science

Amsterdam Boston London New York Oxford Paris
San Diego San Francisco Singapore Sydney Tokyo

This book is printed on acid-free paper. ∞

Academic Press
An imprint of Elsevier Science.
525 B Street, Suite 1900, San Diego, California 92101-4495, USA
http://www.academicpress.com

Academic Press
84 Theobalds Road, London WC1X 8RR, UK
http://www.academicpress.com

International Standard Book Number: 0-12-182255-9

PRINTED IN THE UNITED STATES OF AMERICA
02 03 04 05 06 07 SB 9 8 7 6 5 4 3 2 1

Table of Contents

CONTRIBUTORS TO VOLUME 352 ix

PREFACE xv

VOLUMES IN SERIES xvii

Section I. Cellular Responses

1. Measurement of Absolute Oxygen Levels in Cells and Tissues Using Oxygen Sensors and 2-Nitroimidazole EF5	CAMERON J. KOCH	3
2. Detection of Oxygen-Sensing Properties of Mitochondria	NAVDEEP S. CHANDEL	31
3. Direct Detection of Singlet Oxygen via Its Phosphorescence from Cellular and Fungal Cultures	PIOTR BILSKI, MARGARET E. DAUB, AND COLIN F. CHIGNELL	41
4. Molecular Analysis of Mitogen-Activated Protein Kinase Signaling Pathways Induced by Reactive Oxygen Intermediates	M. LIENHARD SCHMITZ, SUSANNE BACHER, AND WULF DRÖGE	53
5. Detection of Intracellular Reactive Oxygen Species in Cultured Cells Using Fluorescent Probes	ANNE NEGRE-SALVAYRE, NATHALIE AUGÉ, CARINE DUVAL, FANNY ROBBESYN, JEAN-CLAUDE THIERS, DANI NAZZAL, HERVÉ BENOIST, AND ROBERT SALVAYRE	62
6. Flow Cytometric Determination of Cytoplasmic Oxidants and Mitochondrial Membrane Potential in Neuronal Cells	FRANCESC X. SUREDA, MERCÈ PALLÀS, AND ANTONI CAMINS	71
7. Flow Cytometric Determination of Lipid Peroxidation Using Fluoresceinated Phosphoethanolamine	GAUTAM MAULIK, RAVI SALGIA, AND G. MIKE MAKRIGIORGOS	80
8. Determination of Intracellular Reactive Oxygen Species as Function of Cell Density	GIOVANNI PANI, RENATA COLAVITTI, BARBARA BEDOGNI, ROSANNA ANZEVINO, SILVIA BORRELLO, AND TOMMASO GALEOTTI	91

9. Identification of Redox-Active Proteins on Cell Surface	NEIL DONOGHUE AND PHILIP J. HOGG	101
10. Probing Redox Activity of Human Breast Cells by Scanning Electrochemical Microscopy	MICHAEL V. MIRKIN, BIAO LIU, AND SUSAN A. ROTENBERG	112
11. Determining Influence of Oxidants on Nuclear Transport Using Digitonin-Permeabilized Cell Assay	RANDOLPH S. FAUSTINO, MICHAEL P. CZUBRYT, AND GRANT N. PIERCE	123
12. Functional Imaging of Mitochondrial Redox State	DAGMAR KUNZ, KIRSTIN WINKLER, CHRISTIAN E. ELGER, AND WOLFRAM S. KUNZ	135
13. Hydrogen Peroxide-Induced Apoptosis: Oxidative or Reductive Stress?	SHAZIB PERVAIZ AND MARIE-VÉRONIQUE CLÉMENT	150
14. Peroxidation of Phosphatidylserine in Mechanisms of Apoptotic Signaling	YULIA Y. TYURINA, VLADIMIR A. TYURIN, ANNA A. SHVEDOVA, JAMES P. FABISIAK, AND VALERIAN E. KAGAN	159
15. Quantitative High Throughput Endothelial Cell Migration and Invasion Assay System	JAMES C. MALIAKAL	175
16. *In Vitro* Model of Oxidative Stress in Cortical Neurons	RAJIV R. RATAN, HOON RYU, JUNGHEE LEE, AZIZA MWIDAU, AND RACHEL L. NEVE	183
17. Glutamate-Induced c-Src Activation in Neuronal Cells	SAVITA KHANNA, MIKA VENOJARVI, SASHWATI ROY, AND CHANDAN K. SEN	191
18. Measurement of Inflammatory Properties of Fatty Acids in Human Endothelial Cells	MICHAL TOBOREK, YONG WOO LEE, SIMONE KAISER, AND BERNHARD HENNIG	198
19. Redox Control of Tissue Factor Expression in Smooth Muscle Cells and Other Vascular Cells	OLAF HERKERT AND AGNES GÖRLACH	220
20. Redox Processes Regulate Intestinal Lamina Propria T Lymphocytes	BERND SIDO, RAOUL BREITKREUTZ, CORNELIA SEEL, CHRISTIAN HERFARTH, AND STEFAN MEUER	232

21. Linker for Activation of T Cells: Sensing Redox Imbalance — SONJA I. GRINGHUIS, FERDINAND C. BREEDVELD, AND CORNELIS L. VERWEIJ 248

22. Generation of Prooxidant Conditions in Intact Cells to Induce Modifications of Cell Cycle Regulatory Proteins — FRANCA ESPOSITO, TOMMASO RUSSO, AND FILIBERTO CIMINO 258

23. Analysis of Transmembrane Redox Reactions: Interaction of Intra- and Extracellular Ascorbate Species — MARTIJN M. VANDUIJN, JOLANDA VAN DER ZEE, AND PETER J. A. VAN DEN BROEK 268

24. Regulation of Endothelial Cell Proliferation by Nitric Oxide — CYNTHIA J. MEININGER AND GUOYAO WU 280

25. Fluorescent Imaging of Mitochondrial Nitric Oxide in Living Cells — MANUEL O. LÓPEZ-FIGUEROA, CLAUDIO A. CAAMAÑO, M. INÉS MORANO, HUDA AKIL, AND STANLEY J. WATSON 296

Section II. Tissues and Organs

26. Detection of Reactive Oxygen and Nitrogen Species in Tissues Using Redox-Sensitive Fluorescent Probes — LI ZUO AND THOMAS L. CLANTON 307

27. Simultaneous Detection of Tocopherols and Tocotrienols in Biological Samples Using HPLC-Coulometric Electrode Array — SASHWATI ROY, MIKA VENOJARVI, SAVITA KHANNA, AND CHANDAN K. SEN 326

28. *In Vivo* Measurement of Oxidative Stress Status in Human Skin — JÜRGEN FUCHS, NORBERT GROTH, AND THOMAS HERRLING 333

29. Localization of Oxidation-Specific Epitopes in Tissue — GREGORY D. SLOOP 340

30. Quantitation of *S*-Nitrosothiols in Cells and Biological Fluids — VLADIMIR A. TYURIN, YULIA Y. TYURINA, SHANG-XI LIU, HÜLYA BAYIR, CARL A. HUBEL, AND VALERIAN E. KAGAN 347

31. Peroxisomal Fatty Acid Oxidation and Cellular Redox — INDERJIT SINGH 361

32. Ultrastructural Localization and Relative Quantification of 4-Hydroxynonenal-Modified Proteins in Tissues and Cell Compartments — TERRY D. OBERLEY 373

33. Ultrastructural Localization of Light-Induced Lipid Peroxides — PETER KAYATZ, GABRIELE THUMANN, AND ULRICH SCHRAERMEYER 378

34. A Survival Model for Study of Myocardial Angiogenesis — NILANJANA MAULIK, SHOJI FUKUDA, AND HIROAKI SASAKI 391

35. Determination of Angiogenesis-Regulating Properties of NO — MARINA ZICHE AND LUCIA MORBIDELLI 407

36. Hemangioma Model for *in Vivo* Angiogenesis: Inducible Oxidative Stress and MCP-1 Expression in EOMA Cells — GAYLE M. GORDILLO, MUSTAFA ATALAY, SASHWATI ROY, AND CHANDAN K. SEN 422

37. Redox Aspects of Vascular Response to Injury — FRANCISCO R. M. LAURINDO, HERALDO P. DE SOUZA, MARCELO DE A. PEDRO, AND MARIANO JANISZEWSKI 432

38. Involvement of Superoxide in Pathogenic Action of Mutations That Cause Alzheimer's Disease — MARK P. MATTSON 455

39. Three-Dimensional Redox Imaging of Frozen-Quenched Brain and Other Organs — AKIHIKO SHIINO, MASAYUKI MATSUDA, AND BRITTON CHANCE 475

40. *In Vivo* Fluorescent Imaging of NADH Redox State in Brain — ROBERT E. ANDERSON AND FREDRIC B. MEYER 482

41. Nitroxyl Probes for Brain Research and Their Application to Brain Imaging — HIDEO UTSUMI, HIROAKI SANO, MASAICHI NARUSE, KEN-ICHIRO MATSUMOTO, KAZUHIRO ICHIKAWA, AND TETSUO OI 494

42. Analytical Implications of Iron Dithiocarbamates for Measurement of Nitric Oxide — ALEXANDRE SAMOUILOV AND JAY L. ZWEIER 506

AUTHOR INDEX 523

SUBJECT INDEX 561

Contributors to Volume 352

Article numbers are in parentheses following the names of contributors.
Affiliations listed are current.

HUDA AKIL (25), *Mental Health Research Institute, University of Michigan, Ann Arbor, Michigan 48109*

ROBERT E. ANDERSON (40), *Department of Neurosurgery, Thoralf M. Sundt, Jr. Neurosurgery Research Laboratory, Mayo Clinic, Rochester, Minnesota 55905*

ROSANNA ANZEVINO (8), *Institute of General Pathology, Catholic University, 00168 Rome, Italy*

MUSTAFA ATALAY (36), *Department of Physiology, University of Kuopio, FIN-70211 Kuopio, Finland*

NATHALIE AUGÉ (5), *INSERM U-466, Institut Louis Bugnard IFR 31, 31403 Toulouse Cedex 4, France*

SUSANNE BACHER (4), *Department of Immunochemistry, German Cancer Research Center, D-69120 Heidelberg, Germany*

HÜLYA BAYIR (30), *Departments of Anesthesiology and Critical Care Medicine, University of Pittsburgh, Pittsburgh, Pennsylvania 15260*

BARBARA BEDOGNI (8), *Institute of General Pathology, Catholic University, 00168 Rome, Italy*

HERVÉ BENOIST (5), *INSERM U-466, Institut Louis Bugnard IFR 31, 31403 Toulouse Cedex 4, France*

PIOTR BILSKI (3), *Laboratory of Pharmacology and Chemistry, National Institute of Environmental Health Sciences, National Institutes of Health, Research Triangle Park, North Carolina 27709*

SILVIA BORRELLO (8), *Institute of General Pathology, Catholic University, 00168 Rome, Italy*

FERDINAND C. BREEDVELD (21), *Department of Rheumatology, Leiden University Medical Center, 2300 RC Leiden, The Netherlands*

RAOUL BREITKREUTZ (20), *Department of Immunochemistry, Deutsches Krebsforschungszentrum, D-69120 Heidelberg, Germany*

CLAUDIO A. CAAMAÑO (25), *Mental Health Research Institute, University of Michigan, Ann Arbor, Michigan 48109*

ANTONI CAMINS (6), *Pharmacology and Pharmacognosy Unit, Universitat de Barcelona, 08028 Barcelona, Spain*

BRITTON CHANCE (39), *Department of Biochemistry and Biophysics, University of Pennsylvania School of Medicine, Philadelphia, Pennsylvania 19104*

NAVDEEP S. CHANDEL (2), *Department of Medicine, Northwestern University Medical School, Chicago, Illinois 60611*

COLIN F. CHIGNELL (3), *Laboratory of Pharmacology and Chemistry, National Institute of Environmental Health Sciences, National Institutes of Health, Research Triangle Park, North Carolina 27709*

FILIBERTO CIMINO (22), *Department of Biochemistry and Medical Biotechnology, Università di Napoli Federico II, 80131 Naples, Italy*

THOMAS L. CLANTON (26), *Department of Internal Medicine, Pulmonary and Critical Care Medicine, Davis Heart and Lung Research Institute, Ohio State University, Columbus, Ohio 43210*

MARIE-VÉRONIQUE CLÉMENT (13), *Department of Biochemistry, National University of Singapore and National University Medical Institutes, Singapore 117597*

RENATA COLAVITTI (8), *Institute of General Pathology, Catholic University, 00168 Rome, Italy*

MICHAEL P. CZUBRYT (11), *Division of Stroke and Vascular Disease, St. Boniface General Hospital Research Centre, Winnipeg, Manitoba, Canada R2H 2A6*

MARGARET E. DAUB (3), *Department of Botany, North Carolina State University, Raleigh, North Carolina 27695*

HERALDO P. DE SOUZA (37), *Emergency Medicine Department, University of São Paulo School of Medicine, CEP 05403-000 São Paulo, Brazil*

NEIL DONOGHUE (9), *Center for Thrombosis and Vascular Research, School of Pathology, University of New South Wales, 2052 Sydney, Australia*

WULF DRÖGE (4), *Department of Immunochemistry, German Cancer Research Center, D-69120 Heidelberg, Germany*

CARINE DUVAL (5), *INSERM U-466, Institut Louis Bugnard IFR 31, 31403 Toulouse Cedex 4, France*

CHRISTIAN E. ELGER (12), *Department of Epileptology, University of Bonn Medical Center, D-53105 Bonn, Germany*

FRANCA ESPOSITO (22), *Department of Biochemistry and Medical Biotechnology, Università di Napoli Federico II, 80131 Naples, Italy*

JAMES P. FABISIAK (14), *Department of Environmental and Occupational Health, University of Pittsburgh, Pittsburgh, Pennsylvania 15260*

RANDOLPH S. FAUSTINO (11), *Division of Stroke and Vascular Disease, St. Boniface General Hospital Research Centre, Winnipeg, Manitoba, Canada R2H 2A6*

JÜRGEN FUCHS (28), *Department of Dermatology, Johann Wolfgang Goethe University, 60590 Frankfurt am Main, Germany*

SHOJI FUKUDA (34), *Cardiovascular Research Center, University of Connecticut School of Medicine, Farmington, Connecticut 06030*

TOMMASO GALEOTTI (8), *Institute of General Pathology, Catholic University, 00168 Rome, Italy*

GAYLE M. GORDILLO (36), *Department of Surgery, Laboratory of Molecular Medicine, Davis Heart and Lung Research Institute, Ohio State University, Columbus, Ohio 43210*

AGNES GÖRLACH (19), *Department of Experimental Pediatric Cardiology, German Heart Center Munich, Technical University Munich, 80636 Munich, Germany*

SONJA I. GRINGHUIS (21), *Department of Rheumatology, Leiden University Medical Center, 2300 RC Leiden, The Netherlands*

NORBERT GROTH (28), *FOM Institute, TFH Berlin, University of Applied Science and Technology, 12489 Berlin, Germany*

BERNHARD HENNIG (18), *Cell Nutrition Group, College of Agriculture, University of Kentucky, Lexington, Kentucky 40546*

CHRISTIAN HERFARTH (20), *Department of Surgery, University of Heidelberg, D-69120 Heidelberg, Germany*

OLAF HERKERT (19), *Institute for Cardiovascular Physiology, Johann Wolfgang Goethe University Hospital, 60590 Frankfurt am Main, Germany*

THOMAS HERRLING (28), *FOM Institute, TFH Berlin, University of Applied Science and Technology, 12489 Berlin, Germany*

PHILIP J. HOGG (9), *Center for Thrombosis and Vascular Research, School of Pathology, University of New South Wales, 2052 Sydney, Australia*

CARL A. HUBEL (30), *Magee Womens Research Institute and Department of Obstetrics, Gynecology and Reproductive Sciences, University of Pittsburgh, Pittsburgh, Pennsylvania 15260*

KAZUHIRO ICHIKAWA (41), *Department of Biophysics, Graduate School of Pharmaceutical Sciences, Kyushu University, Higashi-ku, Fukuoka 812-8582, Japan*

MARIANO JANISZEWSKI (37), *Vascular Biology and Applied Physiology Laboratories, Heart Institute, University of São Paulo School of Medicine, CEP 05403-000 São Paulo, Brazil*

VALERIAN E. KAGAN (14, 30), *Department of Environmental and Occupational Health, University of Pittsburgh, Pittsburgh, Pennsylvania 15260*

SIMONE KAISER (18), *Department of Surgery, Division of Neurosurgery, University of Kentucky Medical Center, Lexington, Kentucky 40536*

PETER KAYATZ (33), *Laboratory of Experimental Ophthalmology, University Eye Clinic, University of Cologne, D-50931 Cologne, Germany*

SAVITA KHANNA (17, 27), *Departments of Surgery and Molecular and Cellular Biochemistry, Laboratory of Molecular Medicine, Davis Heart and Lung Research Institute, Ohio State University, Columbus, Ohio 43210*

CAMERON J. KOCH (1), *Department of Radiation Oncology, University of Pennsylvania, Philadelphia, Pennsylvania 19104*

DAGMAR KUNZ (12), *Institute of Clinical Chemistry and Pathobiochemistry, Medical Center of RWTH Aachen, 52072 Aachen, Germany*

WOLFRAM S. KUNZ (12), *Department of Epileptology, University of Bonn Medical Center, D-53105 Bonn, Germany*

FRANCISCO R. M. LAURINDO (37), *Vascular Biology and Applied Physiology Laboratories, Heart Institute, University of São Paulo School of Medicine, CEP 05403-000 São Paulo, Brazil*

JUNGHEE LEE (16), *Department of Neurology, Harvard Medical School, Boston, Massachusetts 02115*

YONG WOO LEE (18), *Department of Surgery, Division of Neurosurgery, University of Kentucky Medical Center, Lexington, Kentucky 40536*

BIAO LIU (10), *Department of Chemistry and Biochemistry, Queens College of the City University of New York, Flushing, New York 11367*

SHANG-XI LIU (30), *Department of Environmental and Occupational Health, University of Pittsburgh, Pittsburgh, Pennsylvania 15260*

MANUEL O. LÓPEZ-FIGUEROA (25), *Pritzker Neuropsychiatric Disorders Research Consortium, San Francisco, California 94111*

G. MIKE MAKRIGIORGOS (7), *Department of Adult Oncology, Dana-Farber Cancer Institute, Harvard Medical School, Boston, Massachusetts 02115*

JAMES C. MALIAKAL (15), *BD Biosciences Discovery Labware, Bedford, Massachusetts 01730*

MASAYUKI MATSUDA (39), *Department of Neurosurgery, Shiga University of Medical Science, Ohtsu, Shiga 520-2192, Japan*

KEN-ICHIRO MATSUMOTO (41), *Showa College of Pharmaceutical Sciences, Machida, Tokyo 194-8543, Japan*

MARK P. MATTSON (38), *Laboratory of Neurosciences, Gerontology Research Center, National Institute on Aging, Baltimore, Maryland 21224*

GAUTAM MAULIK (7), *Department of Adult Oncology, Dana-Farber Cancer Institute, Harvard Medical School, Boston, Massachusetts 02115*

NILANJANA MAULIK (34), *Cardiovascular Research Center, University of Connecticut School of Medicine, Farmington, Connecticut 06030*

CYNTHIA J. MEININGER (24), *Cardiovascular Research Institute and Department of Medical Physiology, Texas A&M University System Health Science Center, Temple, Texas 76504*

STEFAN MEUER (20), *Institute of Immunology, University of Heidelberg, D-69120 Heidelberg, Germany*

FREDRIC B. MEYER (40), *Department of Neurosurgery, Thoralf M. Sundt, Jr. Neurosurgery Research Laboratory, Mayo Clinic, Rochester, Minnesota 55905*

MICHAEL V. MIRKIN (10), *Department of Chemistry and Biochemistry, Queens College of the City University of New York, Flushing, New York 11367*

M. INÉS MORANO (25), *Mental Health Research Institute, University of Michigan, Ann Arbor, Michigan 48109*

LUCIA MORBIDELLI (35), *Institute of Pharmacological Sciences, University of Siena, 53100 Siena, Italy*

AZIZA MWIDAU (16), *Department of Neurology, Harvard Medical School, Boston, Massachusetts 02115*

MASAICHI NARUSE (41), *Imaging Research Laboratories I, Daiichi Radioisotope Laboratories Ltd., Matsuo-machi, Chiba 289-1592, Japan*

DANI NAZZAL (5), *INSERM U-466, Institut Louis Bugnard IFR 31, 31403 Toulouse Cedex 4, France*

ANNE NEGRE-SALVAYRE (5), *Department of Biochemistry, INSERM U-466, Institut Louis Bugnard IFR 31, 31403 Toulouse Cedex 4, France*

RACHEL L. NEVE (16), *Department of Psychiatry, Harvard Medical School, Boston, Massachusetts 02115*

TERRY D. OBERLEY (32), *Department of Pathology and Laboratory Medicine, University of Wisconsin Medical School, and William S. Middleton Memorial Veterans Administration Hospital, Madison, Wisconsin 53705*

TETSUO OI (41), *Imaging Research Laboratories I, Daiichi Radioisotope Laboratories, Ltd., Matsuo-machi, Chiba 289-1592, Japan*

MERCÈ PALLÀS (6), *Pharmacology and Pharmacognosy Unit, Universitat de Barcelona, 08028 Barcelona, Spain*

GIOVANNI PANI (8), *Institute of General Pathology, Catholic University, 00168 Rome, Italy*

MARCELO DE A. PEDRO (37), *Vascular Biology and Applied Physiology Laboratories, Heart Institute, University of São Paulo School of Medicine, CEP 05403-000 São Paulo, Brazil*

SHAZIB PERVAIZ (13), *Department of Physiology, National University of Singapore, Singapore 117597*

GRANT N. PIERCE (11), *Division of Stroke and Vascular Disease, St. Boniface General Hospital Research Centre, Winnipeg, Manitoba, Canada R2H 2A6*

RAJIV R. RATAN (16), *Department of Neurology, Harvard Medical School, Boston, Massachusetts 02115*

FANNY ROBBESYN (5), *INSERM U-466, Institut Louis Bugnard IFR 31, 31403 Toulouse Cedex 4, France*

SUSAN A. ROTENBERG (10), *Department of Chemistry and Biochemistry, Queens College of the City University of New York, Flushing, New York 11367*

SASHWATI ROY (17, 27, 36), *Departments of Surgery and Molecular and Cellular Biochemistry, Laboratory of Molecular Medicine, Davis Heart and Lung Research Institute, Ohio State University, Columbus, Ohio 43210*

TOMMASO RUSSO (22), *Department of Biochemistry and Medical Biotechnology, Università di Napoli Federico II, 80131 Naples, Italy*

HOON RYU (16), *Department of Neurology, Harvard Medical School, Boston, Massachusetts 02115*

RAVI SALGIA (7), *Department of Adult Oncology, Dana-Farber Cancer Institute, Harvard Medical School, Boston, Massachusetts 02115*

ROBERT SALVAYRE (5), *Department of Biochemistry, INSERM U-466, Institut Louis Bugnard IFR 131, 31403 Toulouse Cedex 4, France*

ALEXANDRE SAMOUILOV (42), *Department of Medicine, Johns Hopkins University School of Medicine, Baltimore, Maryland 21224*

HIROAKI SANO (41), *Imaging Research Laboratories I, Daiichi Radioisotope Laboratories, Ltd., Matsuo-machi, Chiba 289-1592, Japan*

HIROAKI SASAKI (34), *Cardiovascular Research Center, University of Connecticut School of Medicine, Farmington, Connecticut 06030*

M. LIENHARD SCHMITZ (4), *Department of Chemistry and Biochemistry, University of Bern, CH-3012 Bern, Switzerland*

ULRICH SCHRAERMEYER (33), *Laboratory of Experimental Ophthalmology, University Eye Clinic, University of Cologne, D-50931 Cologne, Germany*

CORNELIA SEEL (20), *Department of Surgery, University of Heidelberg, D-69120 Heidelberg, Germany*

CHANDAN K. SEN (17, 27, 36), *Departments of Surgery and Molecular and Cellular Biochemistry, Laboratory of Molecular Medicine, Davis Heart and Lung Research Institute, Ohio State University, Columbus, Ohio 43210*

AKIHIKO SHIINO (39), *Department of Neurosurgery, Shiga University of Medical Science, Ohtsu, Shiga 520-2192, Japan*

ANNA A. SHVEDOVA (14), *Department of Environmental and Occupational Health, University of Pittsburgh, Pittsburgh, Pennsylvania 15260*

BERND SIDO (20), *Department of Surgery, University of Heidelberg, D-69120 Heidelberg, Germany*

INDERJIT SINGH (31), *Medical University of South Carolina, Charleston, South Carolina 29425*

GREGORY D. SLOOP (29), *Department of Pathology, Louisiana State University School of Medicine, New Orleans, Louisiana 70112*

FRANCESC X. SUREDA (6), *Pharmacology Unit, Universitat Rovira i Virgili, 43201 Reus (Tarragona), Spain*

JEAN-CLAUDE THIERS (5), *INSERM U-466, Institut Louis Bugnard IFR 31, 31403 Toulouse Cedex 4, France*

GABRIELE THUMANN (33), *Laboratory of Experimental Ophthalmology, University Eye Clinic, University of Cologne, D-50931 Cologne, Germany*

MICHAL TOBOREK (18), *Department of Surgery, Division of Neurosurgery, University of Kentucky Medical Center, Lexington, Kentucky 40536*

VLADIMIR A. TYURIN (14, 30), *Department of Environmental and Occupational Health, University of Pittsburgh, Pittsburgh, Pennsylvania 15260*

YULIA Y. TYURINA (14, 30), *Department of Environmental and Occupational Health, University of Pittsburgh, Pittsburgh, Pennsylvania 15260*

HIDEO UTSUMI (41), *Laboratory of Biofunction Analysis, Graduate School of Pharmaceutical Sciences, Kyushu University, Higashi-ku, Fukuoka 812-8582, Japan*

PETER J. A. VAN DEN BROEK (23), *Department of Molecular Cell Biology, Sylvius Laboratory, Leiden University Medical Center, 2333 AL Leiden, The Netherlands*

JOLANDA VAN DER ZEE (23), *Department of Molecular Cell Biology, Sylvius Laboratory, Leiden University Medical Center, 2333 AL Leiden, The Netherlands*

MARTIJN M. VANDUIJN (23), *Institute for Atomic and Molecular Physics, Foundation for Fundamental Research on Matter, 1098 SJ Amsterdam, The Netherlands*

MIKA VENOJARVI (17, 27), *Departments of Surgery and Molecular and Cellular Biochemistry, Laboratory of Molecular Medicine, Davis Heart and Lung Research Institute, Ohio State University, Columbus, Ohio 43210*

CORNELIS L. VERWEIJ (21), *Department of Molecular Cell Biology, Vrije Universiteit Medical Center, 1081 BT Amsterdam, The Netherlands*

STANLEY J. WATSON (25), *Mental Health Research Institute, University of Michigan, Ann Arbor, Michigan 48109*

KIRSTIN WINKLER (12), *Department of Neurology, University of Magdeburg Medical Center, D-39120 Magdeburg, Germany*

GUOYAO WU (24), *Departments of Animal Science and Medical Physiology, Texas A&M University System Health Science Center, College Station, Texas 77843*

MARINA ZICHE (35), *Institute of Pharmacological Sciences, University of Siena, 53100 Siena, Italy*

LI ZUO (26), *Department of Internal Medicine, Pulmonary and Critical Care Medicine, Davis Heart and Lung Research Institute and Biophysics Program, Ohio State University, Columbus, Ohio 43210*

JAY L. ZWEIER (42), *Department of Medicine, Johns Hopkins University School of Medicine, Baltimore, Maryland 21224*

Preface

Oxidants may serve as cellular messengers. Changes in oxidoreductive or redox status in the cell regulate several signal transduction pathways. Redox-sensitive changes in signal transduction processes translate to functional changes at the cellular, tissue, as well as organ levels.

Redox changes in biological cells, tissues, and organs are often transient. For years, investigators have been challenged by the lack of reliable techniques to assess such changes in intact biological samples. Only recently have novel cell biology and genetic techniques to visualize and document redox changes in intact cells become available. Unlike biochemical methods that rely on the study of biological extracts, these cell biology- and genetics-related techniques arrest transient redox changes in the intact cell, tissues, and even organs. Technologies dependent on laser illumination, advanced spectroscopy, DNA microarray, and related approaches allow visualization of redox changes in the intact biological sample. Such approaches, including but not limited to redox imaging of intact organs, gene therapy, gene screening, flow cytometry and advanced microscopy, represent the "cutting-edge" technology currently available to only select laboratories.

Our objective was to compile detailed protocols describing and critiquing essential methods in the field of redox cell biology and genetics. Redox Cell Biology and Genetics, Parts A and B, Volumes 352 and 353 of *Methods in Enzymology,* feature a diverse collection of novel cell biology and genetic protocols authored by highly recognized leaders in the field. Part A covers cellular responses and tissues and organs; Part B covers structure and functions of proteins and nucleic acids and genes.

The excellent editorial assistance of Dr. Savita Khanna and the outstanding contributions of the authors are gratefully acknowledged. We hope that this volume will contribute to the further development of this important field of biomedical research.

CHANDAN K. SEN
LESTER PACKER

Preface

METHODS IN ENZYMOLOGY

VOLUME I. Preparation and Assay of Enzymes
Edited by SIDNEY P. COLOWICK AND NATHAN O. KAPLAN

VOLUME II. Preparation and Assay of Enzymes
Edited by SIDNEY P. COLOWICK AND NATHAN O. KAPLAN

VOLUME III. Preparation and Assay of Substrates
Edited by SIDNEY P. COLOWICK AND NATHAN O. KAPLAN

VOLUME IV. Special Techniques for the Enzymologist
Edited by SIDNEY P. COLOWICK AND NATHAN O. KAPLAN

VOLUME V. Preparation and Assay of Enzymes
Edited by SIDNEY P. COLOWICK AND NATHAN O. KAPLAN

VOLUME VI. Preparation and Assay of Enzymes (*Continued*)
Preparation and Assay of Substrates
Special Techniques
Edited by SIDNEY P. COLOWICK AND NATHAN O. KAPLAN

VOLUME VII. Cumulative Subject Index
Edited by SIDNEY P. COLOWICK AND NATHAN O. KAPLAN

VOLUME VIII. Complex Carbohydrates
Edited by ELIZABETH F. NEUFELD AND VICTOR GINSBURG

VOLUME IX. Carbohydrate Metabolism
Edited by WILLIS A. WOOD

VOLUME X. Oxidation and Phosphorylation
Edited by RONALD W. ESTABROOK AND MAYNARD E. PULLMAN

VOLUME XI. Enzyme Structure
Edited by C. H. W. HIRS

VOLUME XII. Nucleic Acids (Parts A and B)
Edited by LAWRENCE GROSSMAN AND KIVIE MOLDAVE

VOLUME XIII. Citric Acid Cycle
Edited by J. M. LOWENSTEIN

VOLUME XIV. Lipids
Edited by J. M. LOWENSTEIN

VOLUME XV. Steroids and Terpenoids
Edited by RAYMOND B. CLAYTON

VOLUME XVI. Fast Reactions
Edited by KENNETH KUSTIN

VOLUME XVII. Metabolism of Amino Acids and Amines (Parts A and B)
Edited by HERBERT TABOR AND CELIA WHITE TABOR

VOLUME XVIII. Vitamins and Coenzymes (Parts A, B, and C)
Edited by DONALD B. MCCORMICK AND LEMUEL D. WRIGHT

VOLUME XIX. Proteolytic Enzymes
Edited by GERTRUDE E. PERLMANN AND LASZLO LORAND

VOLUME XX. Nucleic Acids and Protein Synthesis (Part C)
Edited by KIVIE MOLDAVE AND LAWRENCE GROSSMAN

VOLUME XXI. Nucleic Acids (Part D)
Edited by LAWRENCE GROSSMAN AND KIVIE MOLDAVE

VOLUME XXII. Enzyme Purification and Related Techniques
Edited by WILLIAM B. JAKOBY

VOLUME XXIII. Photosynthesis (Part A)
Edited by ANTHONY SAN PIETRO

VOLUME XXIV. Photosynthesis and Nitrogen Fixation (Part B)
Edited by ANTHONY SAN PIETRO

VOLUME XXV. Enzyme Structure (Part B)
Edited by C. H. W. HIRS AND SERGE N. TIMASHEFF

VOLUME XXVI. Enzyme Structure (Part C)
Edited by C. H. W. HIRS AND SERGE N. TIMASHEFF

VOLUME XXVII. Enzyme Structure (Part D)
Edited by C. H. W. HIRS AND SERGE N. TIMASHEFF

VOLUME XXVIII. Complex Carbohydrates (Part B)
Edited by VICTOR GINSBURG

VOLUME XXIX. Nucleic Acids and Protein Synthesis (Part E)
Edited by LAWRENCE GROSSMAN AND KIVIE MOLDAVE

VOLUME XXX. Nucleic Acids and Protein Synthesis (Part F)
Edited by KIVIE MOLDAVE AND LAWRENCE GROSSMAN

VOLUME XXXI. Biomembranes (Part A)
Edited by SIDNEY FLEISCHER AND LESTER PACKER

VOLUME XXXII. Biomembranes (Part B)
Edited by SIDNEY FLEISCHER AND LESTER PACKER

VOLUME XXXIII. Cumulative Subject Index Volumes I-XXX
Edited by MARTHA G. DENNIS AND EDWARD A. DENNIS

VOLUME XXXIV. Affinity Techniques (Enzyme Purification: Part B)
Edited by WILLIAM B. JAKOBY AND MEIR WILCHEK

VOLUME XXXV. Lipids (Part B)
Edited by JOHN M. LOWENSTEIN

VOLUME LV. Biomembranes (Part F: Bioenergetics)
Edited by SIDNEY FLEISCHER AND LESTER PACKER

VOLUME LVI. Biomembranes (Part G: Bioenergetics)
Edited by SIDNEY FLEISCHER AND LESTER PACKER

VOLUME LVII. Bioluminescence and Chemiluminescence
Edited by MARLENE A. DELUCA

VOLUME LVIII. Cell Culture
Edited by WILLIAM B. JAKOBY AND IRA PASTAN

VOLUME LIX. Nucleic Acids and Protein Synthesis (Part G)
Edited by KIVIE MOLDAVE AND LAWRENCE GROSSMAN

VOLUME LX. Nucleic Acids and Protein Synthesis (Part H)
Edited by KIVIE MOLDAVE AND LAWRENCE GROSSMAN

VOLUME 61. Enzyme Structure (Part H)
Edited by C. H. W. HIRS AND SERGE N. TIMASHEFF

VOLUME 62. Vitamins and Coenzymes (Part D)
Edited by DONALD B. MCCORMICK AND LEMUEL D. WRIGHT

VOLUME 63. Enzyme Kinetics and Mechanism (Part A: Initial Rate and Inhibitor Methods)
Edited by DANIEL L. PURICH

VOLUME 64. Enzyme Kinetics and Mechanism (Part B: Isotopic Probes and Complex Enzyme Systems)
Edited by DANIEL L. PURICH

VOLUME 65. Nucleic Acids (Part I)
Edited by LAWRENCE GROSSMAN AND KIVIE MOLDAVE

VOLUME 66. Vitamins and Coenzymes (Part E)
Edited by DONALD B. MCCORMICK AND LEMUEL D. WRIGHT

VOLUME 67. Vitamins and Coenzymes (Part F)
Edited by DONALD B. MCCORMICK AND LEMUEL D. WRIGHT

VOLUME 68. Recombinant DNA
Edited by RAY WU

VOLUME 69. Photosynthesis and Nitrogen Fixation (Part C)
Edited by ANTHONY SAN PIETRO

VOLUME 70. Immunochemical Techniques (Part A)
Edited by HELEN VAN VUNAKIS AND JOHN J. LANGONE

VOLUME 71. Lipids (Part C)
Edited by JOHN M. LOWENSTEIN

VOLUME 72. Lipids (Part D)
Edited by JOHN M. LOWENSTEIN

VOLUME 91. Enzyme Structure (Part I)
Edited by C. H. W. HIRS AND SERGE N. TIMASHEFF

VOLUME 92. Immunochemical Techniques (Part E: Monoclonal Antibodies and General Immunoassay Methods)
Edited by JOHN J. LANGONE AND HELEN VAN VUNAKIS

VOLUME 93. Immunochemical Techniques (Part F: Conventional Antibodies, Fc Receptors, and Cytotoxicity)
Edited by JOHN J. LANGONE AND HELEN VAN VUNAKIS

VOLUME 94. Polyamines
Edited by HERBERT TABOR AND CELIA WHITE TABOR

VOLUME 95. Cumulative Subject Index Volumes 61–74, 76–80
Edited by EDWARD A. DENNIS AND MARTHA G. DENNIS

VOLUME 96. Biomembranes [Part J: Membrane Biogenesis: Assembly and Targeting (General Methods; Eukaryotes)]
Edited by SIDNEY FLEISCHER AND BECCA FLEISCHER

VOLUME 97. Biomembranes [Part K: Membrane Biogenesis: Assembly and Targeting (Prokaryotes, Mitochondria, and Chloroplasts)]
Edited by SIDNEY FLEISCHER AND BECCA FLEISCHER

VOLUME 98. Biomembranes (Part L: Membrane Biogenesis: Processing and Recycling)
Edited by SIDNEY FLEISCHER AND BECCA FLEISCHER

VOLUME 99. Hormone Action (Part F: Protein Kinases)
Edited by JACKIE D. CORBIN AND JOEL G. HARDMAN

VOLUME 100. Recombinant DNA (Part B)
Edited by RAY WU, LAWRENCE GROSSMAN, AND KIVIE MOLDAVE

VOLUME 101. Recombinant DNA (Part C)
Edited by RAY WU, LAWRENCE GROSSMAN, AND KIVIE MOLDAVE

VOLUME 102. Hormone Action (Part G: Calmodulin and Calcium-Binding Proteins)
Edited by ANTHONY R. MEANS AND BERT W. O'MALLEY

VOLUME 103. Hormone Action (Part H: Neuroendocrine Peptides)
Edited by P. MICHAEL CONN

VOLUME 104. Enzyme Purification and Related Techniques (Part C)
Edited by WILLIAM B. JAKOBY

VOLUME 105. Oxygen Radicals in Biological Systems
Edited by LESTER PACKER

VOLUME 106. Posttranslational Modifications (Part A)
Edited by FINN WOLD AND KIVIE MOLDAVE

VOLUME 107. Posttranslational Modifications (Part B)
Edited by FINN WOLD AND KIVIE MOLDAVE

VOLUME 126. Biomembranes (Part N: Transport in Bacteria, Mitochondria, and Chloroplasts: Protonmotive Force)
Edited by SIDNEY FLEISCHER AND BECCA FLEISCHER

VOLUME 127. Biomembranes (Part O: Protons and Water: Structure and Translocation)
Edited by LESTER PACKER

VOLUME 128. Plasma Lipoproteins (Part A: Preparation, Structure, and Molecular Biology)
Edited by JERE P. SEGREST AND JOHN J. ALBERS

VOLUME 129. Plasma Lipoproteins (Part B: Characterization, Cell Biology, and Metabolism)
Edited by JOHN J. ALBERS AND JERE P. SEGREST

VOLUME 130. Enzyme Structure (Part K)
Edited by C. H. W. HIRS AND SERGE N. TIMASHEFF

VOLUME 131. Enzyme Structure (Part L)
Edited by C. H. W. HIRS AND SERGE N. TIMASHEFF

VOLUME 132. Immunochemical Techniques (Part J: Phagocytosis and Cell-Mediated Cytotoxicity)
Edited by GIOVANNI DI SABATO AND JOHANNES EVERSE

VOLUME 133. Bioluminescence and Chemiluminescence (Part B)
Edited by MARLENE DELUCA AND WILLIAM D. MCELROY

VOLUME 134. Structural and Contractile Proteins (Part C: The Contractile Apparatus and the Cytoskeleton)
Edited by RICHARD B. VALLEE

VOLUME 135. Immobilized Enzymes and Cells (Part B)
Edited by KLAUS MOSBACH

VOLUME 136. Immobilized Enzymes and Cells (Part C)
Edited by KLAUS MOSBACH

VOLUME 137. Immobilized Enzymes and Cells (Part D)
Edited by KLAUS MOSBACH

VOLUME 138. Complex Carbohydrates (Part E)
Edited by VICTOR GINSBURG

VOLUME 139. Cellular Regulators (Part A: Calcium- and Calmodulin-Binding Proteins)
Edited by ANTHONY R. MEANS AND P. MICHAEL CONN

VOLUME 140. Cumulative Subject Index Volumes 102–119, 121–134

VOLUME 141. Cellular Regulators (Part B: Calcium and Lipids)
Edited by P. MICHAEL CONN AND ANTHONY R. MEANS

VOLUME 142. Metabolism of Aromatic Amino Acids and Amines
Edited by SEYMOUR KAUFMAN

VOLUME 143. Sulfur and Sulfur Amino Acids
Edited by WILLIAM B. JAKOBY AND OWEN GRIFFITH

VOLUME 144. Structural and Contractile Proteins (Part D: Extracellular Matrix)
Edited by LEON W. CUNNINGHAM

VOLUME 145. Structural and Contractile Proteins (Part E: Extracellular Matrix)
Edited by LEON W. CUNNINGHAM

VOLUME 146. Peptide Growth Factors (Part A)
Edited by DAVID BARNES AND DAVID A. SIRBASKU

VOLUME 147. Peptide Growth Factors (Part B)
Edited by DAVID BARNES AND DAVID A. SIRBASKU

VOLUME 148. Plant Cell Membranes
Edited by LESTER PACKER AND ROLAND DOUCE

VOLUME 149. Drug and Enzyme Targeting (Part B)
Edited by RALPH GREEN AND KENNETH J. WIDDER

VOLUME 150. Immunochemical Techniques (Part K: *In Vitro* Models of B and T Cell Functions and Lymphoid Cell Receptors)
Edited by GIOVANNI DI SABATO

VOLUME 151. Molecular Genetics of Mammalian Cells
Edited by MICHAEL M. GOTTESMAN

VOLUME 152. Guide to Molecular Cloning Techniques
Edited by SHELBY L. BERGER AND ALAN R. KIMMEL

VOLUME 153. Recombinant DNA (Part D)
Edited by RAY WU AND LAWRENCE GROSSMAN

VOLUME 154. Recombinant DNA (Part E)
Edited by RAY WU AND LAWRENCE GROSSMAN

VOLUME 155. Recombinant DNA (Part F)
Edited by RAY WU

VOLUME 156. Biomembranes (Part P: ATP-Driven Pumps and Related Transport: The Na, K-Pump)
Edited by SIDNEY FLEISCHER AND BECCA FLEISCHER

VOLUME 157. Biomembranes (Part Q: ATP-Driven Pumps and Related Transport: Calcium, Proton, and Potassium Pumps)
Edited by SIDNEY FLEISCHER AND BECCA FLEISCHER

VOLUME 158. Metalloproteins (Part A)
Edited by JAMES F. RIORDAN AND BERT L. VALLEE

VOLUME 159. Initiation and Termination of Cyclic Nucleotide Action
Edited by JACKIE D. CORBIN AND ROGER A. JOHNSON

VOLUME 160. Biomass (Part A: Cellulose and Hemicellulose)
Edited by WILLIS A. WOOD AND SCOTT T. KELLOGG

VOLUME 178. Antibodies, Antigens, and Molecular Mimicry
Edited by JOHN J. LANGONE

VOLUME 179. Complex Carbohydrates (Part F)
Edited by VICTOR GINSBURG

VOLUME 180. RNA Processing (Part A: General Methods)
Edited by JAMES E. DAHLBERG AND JOHN N. ABELSON

VOLUME 181. RNA Processing (Part B: Specific Methods)
Edited by JAMES E. DAHLBERG AND JOHN N. ABELSON

VOLUME 182. Guide to Protein Purification
Edited by MURRAY P. DEUTSCHER

VOLUME 183. Molecular Evolution: Computer Analysis of Protein and Nucleic Acid Sequences
Edited by RUSSELL F. DOOLITTLE

VOLUME 184. Avidin-Biotin Technology
Edited by MEIR WILCHEK AND EDWARD A. BAYER

VOLUME 185. Gene Expression Technology
Edited by DAVID V. GOEDDEL

VOLUME 186. Oxygen Radicals in Biological Systems (Part B: Oxygen Radicals and Antioxidants)
Edited by LESTER PACKER AND ALEXANDER N. GLAZER

VOLUME 187. Arachidonate Related Lipid Mediators
Edited by ROBERT C. MURPHY AND FRANK A. FITZPATRICK

VOLUME 188. Hydrocarbons and Methylotrophy
Edited by MARY E. LIDSTROM

VOLUME 189. Retinoids (Part A: Molecular and Metabolic Aspects)
Edited by LESTER PACKER

VOLUME 190. Retinoids (Part B: Cell Differentiation and Clinical Applications)
Edited by LESTER PACKER

VOLUME 191. Biomembranes (Part V: Cellular and Subcellular Transport: Epithelial Cells)
Edited by SIDNEY FLEISCHER AND BECCA FLEISCHER

VOLUME 192. Biomembranes (Part W: Cellular and Subcellular Transport: Epithelial Cells)
Edited by SIDNEY FLEISCHER AND BECCA FLEISCHER

VOLUME 193. Mass Spectrometry
Edited by JAMES A. MCCLOSKEY

VOLUME 194. Guide to Yeast Genetics and Molecular Biology (Part A)
Edited by CHRISTINE GUTHRIE AND GERALD R. FINK

VOLUME 195. Adenylyl Cyclase, G Proteins, and Guanylyl Cyclase
Edited by ROGER A. JOHNSON AND JACKIE D. CORBIN

VOLUME 196. Molecular Motors and the Cytoskeleton
Edited by RICHARD B. VALLEE

VOLUME 197. Phospholipases
Edited by EDWARD A. DENNIS

VOLUME 198. Peptide Growth Factors (Part C)
Edited by DAVID BARNES, J. P. MATHER, AND GORDON H. SATO

VOLUME 199. Cumulative Subject Index Volumes 168–174, 176–194

VOLUME 200. Protein Phosphorylation (Part A: Protein Kinases: Assays, Purification, Antibodies, Functional Analysis, Cloning, and Expression)
Edited by TONY HUNTER AND BARTHOLOMEW M. SEFTON

VOLUME 201. Protein Phosphorylation (Part B: Analysis of Protein Phosphorylation, Protein Kinase Inhibitors, and Protein Phosphatases)
Edited by TONY HUNTER AND BARTHOLOMEW M. SEFTON

VOLUME 202. Molecular Design and Modeling: Concepts and Applications (Part A: Proteins, Peptides, and Enzymes)
Edited by JOHN J. LANGONE

VOLUME 203. Molecular Design and Modeling: Concepts and Applications (Part B: Antibodies and Antigens, Nucleic Acids, Polysaccharides, and Drugs)
Edited by JOHN J. LANGONE

VOLUME 204. Bacterial Genetic Systems
Edited by JEFFREY H. MILLER

VOLUME 205. Metallobiochemistry (Part B: Metallothionein and Related Molecules)
Edited by JAMES F. RIORDAN AND BERT L. VALLEE

VOLUME 206. Cytochrome P450
Edited by MICHAEL R. WATERMAN AND ERIC F. JOHNSON

VOLUME 207. Ion Channels
Edited by BERNARDO RUDY AND LINDA E. IVERSON

VOLUME 208. Protein–DNA Interactions
Edited by ROBERT T. SAUER

VOLUME 209. Phospholipid Biosynthesis
Edited by EDWARD A. DENNIS AND DENNIS E. VANCE

VOLUME 210. Numerical Computer Methods
Edited by LUDWIG BRAND AND MICHAEL L. JOHNSON

VOLUME 211. DNA Structures (Part A: Synthesis and Physical Analysis of DNA)
Edited by DAVID M. J. LILLEY AND JAMES E. DAHLBERG

VOLUME 212. DNA Structures (Part B: Chemical and Electrophoretic Analysis of DNA)
Edited by DAVID M. J. LILLEY AND JAMES E. DAHLBERG

VOLUME 213. Carotenoids (Part A: Chemistry, Separation, Quantitation, and Antioxidation)
Edited by LESTER PACKER

VOLUME 214. Carotenoids (Part B: Metabolism, Genetics, and Biosynthesis)
Edited by LESTER PACKER

VOLUME 215. Platelets: Receptors, Adhesion, Secretion (Part B)
Edited by JACEK J. HAWIGER

VOLUME 216. Recombinant DNA (Part G)
Edited by RAY WU

VOLUME 217. Recombinant DNA (Part H)
Edited by RAY WU

VOLUME 218. Recombinant DNA (Part I)
Edited by RAY WU

VOLUME 219. Reconstitution of Intracellular Transport
Edited by JAMES E. ROTHMAN

VOLUME 220. Membrane Fusion Techniques (Part A)
Edited by NEJAT DÜZGUÜNES

VOLUME 221. Membrane Fusion Techniques (Part B)
Edited by NEJAT DÜZGÜNES

VOLUME 222. Proteolytic Enzymes in Coagulation, Fibrinolysis, and Complement Activation (Part A: Mammalian Blood Coagulation Factors and Inhibitors)
Edited by LASZLO LORAND AND KENNETH G. MANN

VOLUME 223. Proteolytic Enzymes in Coagulation, Fibrinolysis, and Complement Activation (Part B: Complement Activation, Fibrinolysis, and Nonmammalian Blood Coagulation Factors)
Edited by LASZLO LORAND AND KENNETH G. MANN

VOLUME 224. Molecular Evolution: Producing the Biochemical Data
Edited by ELIZABETH ANNE ZIMMER, THOMAS J. WHITE, REBECCA L. CANN, AND ALLAN C. WILSON

VOLUME 225. Guide to Techniques in Mouse Development
Edited by PAUL M. WASSARMAN AND MELVIN L. DEPAMPHILIS

VOLUME 226. Metallobiochemistry (Part C: Spectroscopic and Physical Methods for Probing Metal Ion Environments in Metalloenzymes and Metalloproteins)
Edited by JAMES F. RIORDAN AND BERT L. VALLEE

VOLUME 227. Metallobiochemistry (Part D: Physical and Spectroscopic Methods for Probing Metal Ion Environments in Metalloproteins)
Edited by JAMES F. RIORDAN AND BERT L. VALLEE

VOLUME 228. Aqueous Two-Phase Systems
Edited by HARRY WALTER AND GÖTE JOHANSSON

VOLUME 229. Cumulative Subject Index Volumes 195–198, 200–227

VOLUME 230. Guide to Techniques in Glycobiology
Edited by WILLIAM J. LENNARZ AND GERALD W. HART

VOLUME 231. Hemoglobins (Part B: Biochemical and Analytical Methods)
Edited by JOHANNES EVERSE, KIM D. VANDEGRIFF, AND ROBERT M. WINSLOW

VOLUME 232. Hemoglobins (Part C: Biophysical Methods)
Edited by JOHANNES EVERSE, KIM D. VANDEGRIFF, AND ROBERT M. WINSLOW

VOLUME 233. Oxygen Radicals in Biological Systems (Part C)
Edited by LESTER PACKER

VOLUME 234. Oxygen Radicals in Biological Systems (Part D)
Edited by LESTER PACKER

VOLUME 235. Bacterial Pathogenesis (Part A: Identification and Regulation of Virulence Factors)
Edited by VIRGINIA L. CLARK AND PATRIK M. BAVOIL

VOLUME 236. Bacterial Pathogenesis (Part B: Integration of Pathogenic Bacteria with Host Cells)
Edited by VIRGINIA L. CLARK AND PATRIK M. BAVOIL

VOLUME 237. Heterotrimeric G Proteins
Edited by RAVI IYENGAR

VOLUME 238. Heterotrimeric G-Protein Effectors
Edited by RAVI IYENGAR

VOLUME 239. Nuclear Magnetic Resonance (Part C)
Edited by THOMAS L. JAMES AND NORMAN J. OPPENHEIMER

VOLUME 240. Numerical Computer Methods (Part B)
Edited by MICHAEL L. JOHNSON AND LUDWIG BRAND

VOLUME 241. Retroviral Proteases
Edited by LAWRENCE C. KUO AND JULES A. SHAFER

VOLUME 242. Neoglycoconjugates (Part A)
Edited by Y. C. LEE AND REIKO T. LEE

VOLUME 243. Inorganic Microbial Sulfur Metabolism
Edited by HARRY D. PECK, JR., AND JEAN LEGALL

VOLUME 244. Proteolytic Enzymes: Serine and Cysteine Peptidases
Edited by ALAN J. BARRETT

VOLUME 245. Extracellular Matrix Components
Edited by E. RUOSLAHTI AND E. ENGVALL

VOLUME 246. Biochemical Spectroscopy
Edited by KENNETH SAUER

VOLUME 247. Neoglycoconjugates (Part B: Biomedical Applications)
Edited by Y. C. LEE AND REIKO T. LEE

VOLUME 248. Proteolytic Enzymes: Aspartic and Metallo Peptidases
Edited by ALAN J. BARRETT

VOLUME 249. Enzyme Kinetics and Mechanism (Part D: Developments in Enzyme Dynamics)
Edited by DANIEL L. PURICH

VOLUME 250. Lipid Modifications of Proteins
Edited by PATRICK J. CASEY AND JANICE E. BUSS

VOLUME 251. Biothiols (Part A: Monothiols and Dithiols, Protein Thiols, and Thiyl Radicals)
Edited by LESTER PACKER

VOLUME 252. Biothiols (Part B: Glutathione and Thioredoxin; Thiols in Signal Transduction and Gene Regulation)
Edited by LESTER PACKER

VOLUME 253. Adhesion of Microbial Pathogens
Edited by RON J. DOYLE AND ITZHAK OFEK

VOLUME 254. Oncogene Techniques
Edited by PETER K. VOGT AND INDER M. VERMA

VOLUME 255. Small GTPases and Their Regulators (Part A: Ras Family)
Edited by W. E. BALCH, CHANNING J. DER, AND ALAN HALL

VOLUME 256. Small GTPases and Their Regulators (Part B: Rho Family)
Edited by W. E. BALCH, CHANNING J. DER, AND ALAN HALL

VOLUME 257. Small GTPases and Their Regulators (Part C: Proteins Involved in Transport)
Edited by W. E. BALCH, CHANNING J. DER, AND ALAN HALL

VOLUME 258. Redox-Active Amino Acids in Biology
Edited by JUDITH P. KLINMAN

VOLUME 259. Energetics of Biological Macromolecules
Edited by MICHAEL L. JOHNSON AND GARY K. ACKERS

VOLUME 260. Mitochondrial Biogenesis and Genetics (Part A)
Edited by GIUSEPPE M. ATTARDI AND ANNE CHOMYN

VOLUME 261. Nuclear Magnetic Resonance and Nucleic Acids
Edited by THOMAS L. JAMES

VOLUME 262. DNA Replication
Edited by JUDITH L. CAMPBELL

VOLUME 263. Plasma Lipoproteins (Part C: Quantitation)
Edited by WILLIAM A. BRADLEY, SANDRA H. GIANTURCO, AND JERE P. SEGREST

VOLUME 264. Mitochondrial Biogenesis and Genetics (Part B)
Edited by GIUSEPPE M. ATTARDI AND ANNE CHOMYN

VOLUME 265. Cumulative Subject Index Volumes 228, 230–262

VOLUME 266. Computer Methods for Macromolecular Sequence Analysis
Edited by RUSSELL F. DOOLITTLE

VOLUME 267. Combinatorial Chemistry
Edited by JOHN N. ABELSON

VOLUME 268. Nitric Oxide (Part A: Sources and Detection of NO; NO Synthase)
Edited by LESTER PACKER

VOLUME 269. Nitric Oxide (Part B: Physiological and Pathological Processes)
Edited by LESTER PACKER

VOLUME 270. High Resolution Separation and Analysis of Biological Macromolecules (Part A: Fundamentals)
Edited by BARRY L. KARGER AND WILLIAM S. HANCOCK

VOLUME 271. High Resolution Separation and Analysis of Biological Macromolecules (Part B: Applications)
Edited by BARRY L. KARGER AND WILLIAM S. HANCOCK

VOLUME 272. Cytochrome P450 (Part B)
Edited by ERIC F. JOHNSON AND MICHAEL R. WATERMAN

VOLUME 273. RNA Polymerase and Associated Factors (Part A)
Edited by SANKAR ADHYA

VOLUME 274. RNA Polymerase and Associated Factors (Part B)
Edited by SANKAR ADHYA

VOLUME 275. Viral Polymerases and Related Proteins
Edited by LAWRENCE C. KUO, DAVID B. OLSEN, AND STEVEN S. CARROLL

VOLUME 276. Macromolecular Crystallography (Part A)
Edited by CHARLES W. CARTER, JR., AND ROBERT M. SWEET

VOLUME 277. Macromolecular Crystallography (Part B)
Edited by CHARLES W. CARTER, JR., AND ROBERT M. SWEET

VOLUME 278. Fluorescence Spectroscopy
Edited by LUDWIG BRAND AND MICHAEL L. JOHNSON

VOLUME 279. Vitamins and Coenzymes (Part I)
Edited by DONALD B. MCCORMICK, JOHN W. SUTTIE, AND CONRAD WAGNER

VOLUME 280. Vitamins and Coenzymes (Part J)
Edited by DONALD B. MCCORMICK, JOHN W. SUTTIE, AND CONRAD WAGNER

VOLUME 281. Vitamins and Coenzymes (Part K)
Edited by DONALD B. MCCORMICK, JOHN W. SUTTIE, AND CONRAD WAGNER

VOLUME 282. Vitamins and Coenzymes (Part L)
Edited by DONALD B. MCCORMICK, JOHN W. SUTTIE, AND CONRAD WAGNER

VOLUME 283. Cell Cycle Control
Edited by WILLIAM G. DUNPHY

VOLUME 284. Lipases (Part A: Biotechnology)
Edited by BYRON RUBIN AND EDWARD A. DENNIS

VOLUME 305. Bioluminescence and Chemiluminescence (Part C)
Edited by THOMAS O. BALDWIN AND MIRIAM M. ZIEGLER

VOLUME 306. Expression of Recombinant Genes in Eukaryotic Systems
Edited by JOSEPH C. GLORIOSO AND MARTIN C. SCHMIDT

VOLUME 307. Confocal Microscopy
Edited by P. MICHAEL CONN

VOLUME 308. Enzyme Kinetics and Mechanism (Part E: Energetics of Enzyme Catalysis)
Edited by DANIEL L. PURICH AND VERN L. SCHRAMM

VOLUME 309. Amyloid, Prions, and Other Protein Aggregates
Edited by RONALD WETZEL

VOLUME 310. Biofilms
Edited by RON J. DOYLE

VOLUME 311. Sphingolipid Metabolism and Cell Signaling (Part A)
Edited by ALFRED H. MERRILL, JR., AND YUSUF A. HANNUN

VOLUME 312. Sphingolipid Metabolism and Cell Signaling (Part B)
Edited by ALFRED H. MERRILL, JR., AND YUSUF A. HANNUN

VOLUME 313. Antisense Technology (Part A: General Methods, Methods of Delivery, and RNA Studies)
Edited by M. IAN PHILLIPS

VOLUME 314. Antisense Technology (Part B: Applications)
Edited by M. IAN PHILLIPS

VOLUME 315. Vertebrate Phototransduction and the Visual Cycle (Part A)
Edited by KRZYSZTOF PALCZEWSKI

VOLUME 316. Vertebrate Phototransduction and the Visual Cycle (Part B)
Edited by KRZYSZTOF PALCZEWSKI

VOLUME 317. RNA–Ligand Interactions (Part A: Structural Biology Methods)
Edited by DANIEL W. CELANDER AND JOHN N. ABELSON

VOLUME 318. RNA–Ligand Interactions (Part B: Molecular Biology Methods)
Edited by DANIEL W. CELANDER AND JOHN N. ABELSON

VOLUME 319. Singlet Oxygen, UV-A, and Ozone
Edited by LESTER PACKER AND HELMUT SIES

VOLUME 320. Cumulative Subject Index Volumes 290–319

VOLUME 321. Numerical Computer Methods (Part C)
Edited by MICHAEL L. JOHNSON AND LUDWIG BRAND

VOLUME 322. Apoptosis
Edited by JOHN C. REED

VOLUME 323. Energetics of Biological Macromolecules (Part C)
Edited by MICHAEL L. JOHNSON AND GARY K. ACKERS

VOLUME 324. Branched-Chain Amino Acids (Part B)
Edited by ROBERT A. HARRIS AND JOHN R. SOKATCH

VOLUME 325. Regulators and Effectors of Small GTPases (Part D: Rho Family)
Edited by W. E. BALCH, CHANNING J. DER, AND ALAN HALL

VOLUME 326. Applications of Chimeric Genes and Hybrid Proteins (Part A: Gene Expression and Protein Purification)
Edited by JEREMY THORNER, SCOTT D. EMR, AND JOHN N. ABELSON

VOLUME 327. Applications of Chimeric Genes and Hybrid Proteins (Part B: Cell Biology and Physiology)
Edited by JEREMY THORNER, SCOTT D. EMR, AND JOHN N. ABELSON

VOLUME 328. Applications of Chimeric Genes and Hybrid Proteins (Part C: Protein-Protein Interactions and Genomics)
Edited by JEREMY THORNER, SCOTT D. EMR, AND JOHN N. ABELSON

VOLUME 329. Regulators and Effectors of Small GTPases (Part E: GTPases Involved in Vesicular Traffic)
Edited by W. E. BALCH, CHANNING J. DER, AND ALAN HALL

VOLUME 330. Hyperthermophilic Enzymes (Part A)
Edited by MICHAEL W. W. ADAMS AND ROBERT M. KELLY

VOLUME 331. Hyperthermophilic Enzymes (Part B)
Edited by MICHAEL W. W. ADAMS AND ROBERT M. KELLY

VOLUME 332. Regulators and Effectors of Small GTPases (Part F: Ras Family I)
Edited by W. E. BALCH, CHANNING J. DER, AND ALAN HALL

VOLUME 333. Regulators and Effectors of Small GTPases (Part G: Ras Family II)
Edited by W. E. BALCH, CHANNING J. DER, AND ALAN HALL

VOLUME 334. Hyperthermophilic Enzymes (Part C)
Edited by MICHAEL W. W. ADAMS AND ROBERT M. KELLY

VOLUME 335. Flavonoids and Other Polyphenols
Edited by LESTER PACKER

VOLUME 336. Microbial Growth in Biofilms (Part A: Developmental and Molecular Biological Aspects)
Edited by RON J. DOYLE

VOLUME 337. Microbial Growth in Biofilms (Part B: Special Environments and Physicochemical Aspects)
Edited by RON J. DOYLE

VOLUME 338. Nuclear Magnetic Resonance of Biological Macromolecules (Part A)
Edited by THOMAS L. JAMES, VOLKER DÖTSCH, AND ULI SCHMITZ

VOLUME 339. Nuclear Magnetic Resonance of Biological Macromolecules (Part B)
Edited by THOMAS L. JAMES, VOLKER DÖTSCH, AND ULI SCHMITZ

VOLUME 340. Drug–Nucleic Acid Interactions
Edited by JONATHAN B. CHAIRES AND MICHAEL J. WARING

VOLUME 341. Ribonucleases (Part A)
Edited by ALLEN W. NICHOLSON

VOLUME 342. Ribonucleases (Part B)
Edited by ALLEN W. NICHOLSON

VOLUME 343. G Protein Pathways (Part A: Receptors)
Edited by RAVI IYENGAR AND JOHN D. HILDEBRANDT

VOLUME 344. G Protein Pathways (Part B: G Proteins and Their Regulators)
Edited by RAVI IYENGAR AND JOHN D. HILDEBRANDT

VOLUME 345. G Protein Pathways (Part C: Effector Mechanisms)
Edited by RAVI IYENGAR AND JOHN D. HILDEBRANDT

VOLUME 346. Gene Therapy Methods
Edited by M. IAN PHILLIPS

VOLUME 347. Protein Sensors and Reactive Oxygen Species (Part A: Selenoproteins and Thioredoxin)
Edited by HELMUT SIES AND LESTER PACKER

VOLUME 348. Protein Sensors and Reactive Oxygen Species (Part B: Thiol Enzymes and Proteins)
Edited by HELMUT SIES AND LESTER PACKER

VOLUME 349. Superoxide Dismutase
Edited by LESTER PACKER

VOLUME 350. Guide to Yeast Genetics and Molecular and Cell Biology (Part B)
Edited by CHRISTINE GUTHRIE AND GERALD R. FINK

VOLUME 351. Guide to Yeast Genetics and Molecular and Cell Biology (Part C)
Edited by CHRISTINE GUTHRIE AND GERALD R. FINK

VOLUME 352. Redox Cell Biology and Genetics (Part A)
Edited by CHANDAN K. SEN AND LESTER PACKER

VOLUME 353. Redox Cell Biology and Genetics (Part B) (in preparation)
Edited by CHANDAN K. SEN AND LESTER PACKER

VOLUME 354. Enzyme Kinetics and Mechanism (Part F: Detection and Characterization of Enzyme Reaction Intermediates) (in preparation)
Edited by DANIEL L. PURICH

VOLUME 355. Cumulative Subject Index Volumes 321–354 (in preparation)

VOLUME 356. Laser Capture Microscopy and Microdissection (in preparation)
Edited by P. MICHAEL CONN

Section I

Cellular Responses

[1] Measurement of Absolute Oxygen Levels in Cells and Tissues Using Oxygen Sensors and 2-Nitroimidazole EF5

By CAMERON J. KOCH

Despite the enormous range and variety of chemical, biochemical, and biological effects of oxygen, accurate control and assessment of the "oxygen concentration" remain extremely difficult experimental tasks in all but the simplest of systems. The difficulty of such control and assessment increases greatly as the oxygen concentration decreases below physiological levels, often the range of greatest biological significance. This chapter discusses possible approaches to these difficult tasks using quantitative measurements of binding of a metabolic marker of hypoxia, the 2-nitroimidazole EF5 [2-(2 nitro-1*H*-imidazol-1-yl)-*N*-(2,2,3,3,3-pentafluoropropyl)acetamide]. The accuracy, dynamic range, and specificity of the pO_2 measurement capability of EF5 is discussed in the context of an accurate polarographic pO_2 sensor used as a calibration source. Present results suggest that it is indeed possible to assess tissue pO_2 at high accuracy and spatial resolution.

Introduction

Low tumor tissue oxygen concentrations pose a therapeutically important problem. The importance of hypoxic cells in limiting radiation response in *animal* tumors is very well recognized (for review see Refs. 1 and 2). Because of their relative isolation from the blood supply, many scientists believe that hypoxic tumor cells could also be resistant to cytotoxic drugs.[3,4] The importance of hypoxia in the development, treatment, and cure of human cancers has, until recently, been less certain. This view was reinforced by the relatively modest success of various trials designed to allow the sensitization of hypoxic cells by hyperbaric oxygen[5] or 5- and 2-nitroimidazoles.[6–9] A fundamental problem in assessing the outcome

[1] J. D. Chapman, A. J. Franko, and C. J. Koch, *in* "Biological Bases and Clinical Implications of Tumor Radioresistance" (G. H. Fletcher, C. Nervi, and H. R. Withers, eds.), p. 61. Masson Pub., New York, NY, 1983.

[2] J. E. Moulder and S. C. Rockwell, *Int. J. Radiat. Oncol. Biol. Phys.* **10,** 695 (1984).

[3] G. E. Adams, *Cancer* **48,** 696 (1981).

[4] K. A. Kennedy, B. A. Teicher, S. C. Rockwell, and A. C. Sartorelli, *Biochem. Pharm.* **29,** 1 (1980).

[5] J. M. Henk, *Int. J. Radiat. Oncol. Biol. Phys.* **12,** 1339 (1986).

[6] J. Overgaard, H. Hansen, A. P. Sandersen, M. Hjelm-Hansen, K. Jorgensen, E. Sandberg, A. Berthelsen, R. Hammer, and M. Pedersen, *Int. J. Radiat. Oncol. Biol. Phys.* **16,** 1065 (1989).

[7] J. Overgaard, *Radiother. Oncol.* **24,** S64 (1992).

[8] T. L. Phillips, T. H. Wasserman, R. J. Johnson, V. A. Levin, and G. VanRaalte, *Cancer* **48,** 1687 (1981).

[9] R. C. Urtasun, C. N. Coleman, T. H. Wasserman, and T. L. Phillips, *Int. J. Radiat. Oncol. Biol. Phys.* **10,** 1691 (1984).

0076-6879/02 $35.00

of these trials was the lack of a clinically relevant oxygen-monitoring technique. Although one might expect that specific tumor types (e.g., soft tissue sarcomas) might be characterized by specific causes of resistance, such as hypoxia, tumors are instead "heterogeneous" with respect to such characteristics. This heterogeneity is particularly problematic in the design of experimental or clinical trials if relatively small improvements in therapeutic response are expected. Heterogeneity in the presence vs absence of an associated risk factor (e.g., hypoxia) can dramatically increase the statistically required number of experimental subjects while simultaneously decreasing the apparent therapeutic benefit.[10]

A new type of needle electrode measuring system developed by the Eppendorf company has been used to prove that tissue oxygenation is important in the biology and treatment of human tumors. Initial studies demonstrated tremendous heterogeneity in human tumors of a similar type.[11–13] These seminal results emphasized the importance of individual tumor monitoring and provided support for the statistical problems described earlier. More recent studies, also using the electrode technique, have shown a good correlation between tumor therapy outcome and the distribution of intratumoral oxygen concentration.[14–17] These results confirmed those of Gatenby and others in earlier, but much less comprehensive studies.[18–22] While recent electrode studies clearly demonstrate the importance of monitoring hypoxia in patients who will receive radiation therapy, the role of oxygen in determining the efficacy of other therapeutic modalities has also been shown to be important. This is illustrated by the finding that the presence of hypoxia is associated with poor outcome following surgery alone (local-regional recurrence)[16,17] and failure from distant metastasis.[14] It is now generally acknowledged that hypoxia

[10] H. B. Stone, J. M. Brown, T. L. Phillips, and R. M. Sutherland, *Radiat. Res.* **136,** 422 (1993).

[11] D. M. Brizel, G. L. Rosner, J. Harrelson, L. R. Prosnitz, and M. W. Dewhirst, *Int. J. Radiat. Oncol. Biol. Phys.* **30,** 635 (1994).

[12] P. Okunieff, M. Hockel, E. P. Dunphy, K. Schlenger, C. Knoop, and P. Vaupel, *Int. J. Radiat. Biol. Oncol. Phys.* **26,** 631 (1993).

[13] R. Rampling, G. Cruickshank, A. Lewis, S. A. Fitzsimmons, and P. Workman, *Int. J. Radiat. Oncol. Biol. Phys.* **29,** 427 (1994).

[14] D. M. Brizel, S. P. Scully, J. M. Harrelson, L. J. Layfield, J. M. Bean, L. R. Prosnitz, and M. W. Dewhirst, *Cancer Res.* **56,** 941 (1996).

[15] A. W. Fyles, M. Milosevic, R. Wong, M. C. Kavanagh, M. Pintilie, W. Chapman, W. Levin, L. Manchul, T. J. Keane, and R. P. Hill, *Radiother. Oncol.* **48,** 149 (1998).

[16] M. Hockel, C. Knoop, K. Schlenger, B. Vordran, E. Baussmann, M. Mitze, P. G. Knapstein, and P. Vaupel, *Radiother. Oncol.* **26,** 45 (1993).

[17] M. Nordsmark, M. Overgaard, and J. Overgaard, *Radiother. Oncol.* **41,** 31 (1996).

[18] D. B. Cater and I. A. Silver, *Acta Radiol.* **23,** 233 (1942).

[19] R. A. Gatenby, H. B. Kessler, J. S. Rosenblum, L. R. Coia, P. J. Moldofsky, W. H. Hartz, and G. J. Broder, *Radiology* **156,** 211 (1985).

[20] R. A. Gatenby, H. B. Kessler, J. S. Rosenblum, L. R. Coia, P. J. Moldofsky, W. H. Hartz, and G. J. Broder, *Int. J. Radiat. Oncol. Biol. Phys.* **14,** 831 (1988).

[21] P. Vaupel, P. Okunieff, F. Kallinowski, and L. J. Neuringer, *Radiat. Res.* **120,** 477 (1989).

[22] P. Wendling, R. Manz, G. Thews, and P. Vaupel, *Adv. Exp. Med. Biol.* **180,** 293 (1984).

(or cyclic changes in oxygen concentration such as seen in reperfusion injury) can lead to genetic changes, including selection of cells resistant to apoptosis, often a normal consequence of severe hypoxia.[23–25]

In normal tissue, hypoxia is more typically associated with vascular insufficiency as found in diabetes, wounding, stroke, and infarct. In some situations, however (embryonic development; closure of ductus arteriosus), tissue hypoxia plays an important role in normal developmental processes wherein the pathological condition may be hyperoxia.[26,27]

Why is oxygen measurement and control so difficult? Oxygen is the only key nutrient in gaseous form; the exquisitely complex distribution of oxygen via red cell hemoglobin and the vasculature of normal tissue is the subject of a distinct research area (e.g., see proceedings of the International Society for Oxygen Transport to Tissue[27a]). The nature of the oxygen measurement and control problem can be appreciated more easily by considering the consumption vs supply of oxygen. Tumor cells in tissue culture consume oxygen at a typical rate of the order of 5×10^{-17} mol/cell/sec. At tissue densities (4×10^{11} cells/liter), this would imply a total consumption rate of 20 μmol/liter/sec. Because the solubility of oxygen in physiological fluids at 37° is only about 210 μM[28] this implies that oxygen must be supplied by the vasculature at several times the solubility limit each minute. Furthermore, many of the most interesting effects of oxygen occur at much lower oxygen concentrations (micromolar range), making measurement and control even more challenging (Fig. 1).

Oxygen Sensors and Experimental Systems

In the 1970s, most oxygen-related studies emphasized experimental situations with substantially less complexity than that of organized tissue, e.g., cells in tissue culture. Even for tissue culture experiments, oxygen control and measurement

[23] T. G. Graeber, C. Osmanian, T. Jacks, D. E. Housman, C. J. Koch, S. W. Lowe, and A. J. Giaccia, *Nature* **379,** 88 (1996).

[24] T. Y. Reynolds, S. Rockwell, and P. M. Glazer, *Cancer Res.* **56,** 5754 (1990).

[25] S. D. Young, R. S. Marshall, and R. P. Hill, *Proc. Natl. Acad. Sci. U.S.A.* **85,** 9533 (1988).

[26] E. Y. Chen, M. Fujinaga, and A. J. Giaccia, *Teratology* **60,** 215 (1999).

[27] R. I. Clyman, C. Y. Chan, F. Mauray, Y. Q. Chen, W. Cox, S. R. Seidner, H. Weiss, E. M. Lord, N. Waleh, S. M. Evans, and C. J. Koch, *Pediatr. Res.* **45,** 19 (1999).

[27a] *Adv. Exp. Med. Biol.*, published annually.

[28] The solubility of gas is temperature dependent, as is the partial pressure due to water vapor. In this report, the following assumptions are made regarding oxygen solubility. At 37°, the partial pressure of water vapor is ~45 mm of Hg; thus, atmospheric pressure of 760 mm of Hg of air with saturated water vapor allows for oxygen (21%) + nitrogen (79%) partial pressures of 149 + 566 mm of Hg, respectively. The solubility of oxygen in physiological saline (or equivalent) equilibrated with water vapor-saturated air at 37° is about 0.20 m*M*. At 25° and 4°, the corresponding solubility is about 0.25 and 0.45 m*M*, respectively. In the text, we report absolute concentrations (in micromolar) and mm of Hg or percentage oxygen values assuming a temperature of 37°.

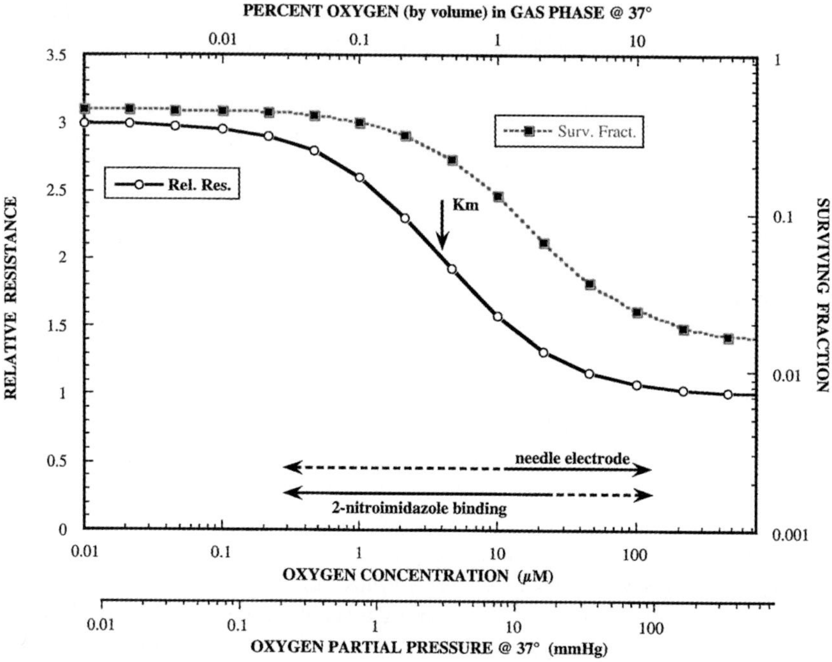

FIG. 1. The classical modification of radiation damage by oxygen is plotted against three separate measures of oxygen (micromolar, mm of Hg, and percentage oxygen). The second and third measures represent the partial pressure of oxygen for liquid in equilibrium with gas at 37°; the first measure is the actual concentration of oxygen in the liquid, in this case physiological saline or equivalent. The relative resistance (left vertical axis, open circles) is defined by the relative dose of radiation required for any effect of interest. Thus, hypoxic cells require three fold higher radiation doses than aerobic cells for the same amount of killing. On the right axis is plotted the actual change in radiation-induced colony-forming ability for a constant dose of radiation, in this case 10 Gy given to Chinese hamster fibroblasts. The base of the figure illustrates the measurement capability range of needle electrodes vs EF5 binding, with solid lines indicating best accuracy and dashed lines indicating less reliable measurements. Note that the two measuring systems have opposite or complementary accuracies.

were compromised by poor instrumentation. Even after the major advance of the membrane-covered polarographic oxygen sensor by Clark[29] the stability of commercial sensors at both high and zero oxygen values was problematic. Most sensors required high stirring speeds in liquids, preventing use in tissue culture (for discussion, see Koch[30]). One of the best designed oxygen sensors at that time was made by the Beckman Company. The Model 325814 included a fine platinum

[29] L. C. J. Clark, *Trans. Am. Soc. Artif. Int. Organs* **2,** 41 (1956).

[30] C. J. Koch, "Polarographic Oxygen Sensor with Glass Seal." Canada Patent, 1304449: June 30, 1992.

wire cathode, sealed within a glass tube. The cathode and a spiral silver anode wire were encapsulated within a plastic body. We found that about one sensor in five had much better stability than specified at low oxygen levels.[31] A subsequent quest for the basis of this superior response led the author to a greatly improved design where more emphasis was placed on a highly improved cathode and membrane seal. This new sensor (Controls Katharobic or CK sensor) had (a) a dynamic range extending from above 100% oxygen to well below the micromolar range, (b) a narrow tip with a tapered ceramic body, which could be inserted into completely sealed systems (standard glassware taper of 1 : 10), and (c) a composition of glass and ceramic rather than plastic.[30]

Elimination of plastics in the newly designed oxygen sensor was essential because plastics have the undesirable properties of both a high oxygen solubility and a low diffusion constant.[30,32–34] While their convenience, cost, and engineering advantages are obvious, plastics have caused, and continue to cause, problems far exceeding those limited to oxygen sensor design. All plastics act as high impedance oxygen reservoirs, delivering contaminating levels of oxygen for many hours to days.[33–35] Plastic connecting hoses and containers cause oxygen "leaks," which can only be minimized by using high gas flow rates. Teflon-coated spin bars provide a virtually inexhaustible source of oxygen for solutions flushed by oxygen-free gas (Fig. 2). Cells growing in monolayer cultures on plastic dishes pose a particularly challenging problem. At low cell densities, the plastic acts as an oxygen reservoir, inhibiting attempts at deoxygenation, whereas at high cell densities, cultures may become hypoxic, irrespective of the dish composition, due to respiration and the diffusion barrier posed by the medium.[36]

After the development of the CK oxygen sensor, it became possible to fabricate and test two all-glass/ceramic systems for control and monitoring of oxygen levels in tissue culture. The first was described as a "thin-film" system. Cells were plated[37] onto the central region of 50-mm-diameter glass dishes selected for flat bases.

[31] C. J. Koch and J. Kruuv, *Anal. Chem.* **44,** 1258 (1972).

[32] Polycarbonate is the most oxygen permeable of plastics made for tissue culture use (dishes are made by NUNC). The least toxic and water-impermeable tubing is made from silicone, but it is by far the worst in the sense of high oxygen permeability. It is not the present intention to make a blanket condemnation of experiments using plastics. Rather, it should be appreciated that many oxygen-dependent chemical and biological phenomena are very sensitive to the range of oxygen below 2% (some much lower still). Thus, use of plastics may prevent a proper understanding of the oxygen dependence of various biological phenomena.

[33] J. D. Chapman, J. Sturrock, J. W. Boag, and J. O. Crookall, *Int. J. Radiat. Biol.* **17,** 305 (1970).

[34] C. J. Koch and J. Kruuv, *Br. J. Radiol.* **45,** 787 (1972).

[35] A. J. Franko, H. I. Freedman, and C. J. Koch, *in* "Spheroids in Cancer Research," p. 162. Springer Verlag, Berlin, 1984.

[36] C. J. Koch, *in* "Proc. 6th L. H. Gray Conf. (1974)," p. 167. Institute of Physics, London, 1975.

When the medium volume on these dishes is reduced to 1 ml, the effective medium thickness covering the cells is only about 100 μm (because of the peripheral meniscus), allowing very rapid equilibration between gas and liquid (a few seconds).[38] The gas-phase oxygen concentration is controlled via sealed aluminum chambers, whose gas content, in turn, is varied using a brass and stainless-steel manifold. A gas inlet is provided by high-purity stainless-steel regulators and bellows tubing, and a gas outlet is provided by a vacuum pump. All gas flow is controlled by bellows-seal valves. The CK oxygen sensors respond only to oxygen partial pressure, not absolute pressure. In other words, the sensor output is the same for 1% oxygen plus 99% nitrogen at atmospheric pressure vs 100% oxygen at 1/100 atmospheric pressure. This allows great flexibility in monitoring leaks and/or checking absolute zero signals. It is also possible, through gas transfer techniques, to monitor the gas phase pO_2 of the chambers at both the beginning and the end of experiments.

The thin-film system provides highly controlled pO_2 in the gas and liquid volumes contained within the sealed chambers. Under conditions where cellular respiration would lower the cellular pO_2, despite the thin medium film (high cell density, 37°), the chambers are placed on a reciprocating shaker table (2.5-cm stroke, 1 Hz). Shaking improves oxygen exchange greatly while simultaneously promoting an exchange of nutrients with the bulk of the medium contained in the peripheral meniscus of the dishes. Despite these precautions, we have estimated that intracellular pO_2 may be as much as 1 μM lower than the equilibrium value predicted by the gas phase of the chambers.

Most cells can tolerate suspension culture for at least a few hours. Thus, a second experimental system was designed consisting of a glass vial with a spin bar and a precision tapered stopper made from ceramic—the "spinner-vial" system.[39,40] When the vials are completely filled and sealed, absolute rates of oxygen consumption for chemical or cellular oxidations can be measured. Because Clark-type oxygen sensors respond to oxygen partial pressure rather than concentration, it is important to be able to calibrate the oxygen concentration, which is affected by temperature and solution composition. For example, alcohols typically have oxygen solubility almost 10 times higher than water; salts reduce the equilibrium oxygen concentration. Such calibrations require a standard oxygen-consuming chemical or enzymatic reaction. Unfortunately, most published examples only work in very simple solvents because reactive and radical species produced

[37] To promote cell attachment, glass is cleaned, heated to 500°, and then conditioned by overnight treatment with water containing 50 mM sodium carbonate and 15% serum, followed by 1 hr with water containing tissue culture grade gelatin (0.1%). The gelatin is removed, but the thin remaining film is allowed to dry onto the glass under UV light in a sterile hood.

[38] C. J. Koch, *Radiat. Res.* **97,** 434 (1984).

[39] C. J. Koch, *Adv. Exp. Biol. Med.* **157,** 123 (1983).

[40] L.-W. Lo, C. J. Koch, and D. F. Wilson, *Anal. Biochem.* **236,** 153 (1996).

(e.g., hydrogen peroxide, superoxide radical, hydroxyl radical) may react with other components of the experimental solutions of interest. We found that a very reliable and rapid calibration could be achieved through the use of ascorbic acid (which is available in very pure form) and ascorbate oxidase.[40] This remarkable enzyme, like cytochrome oxidase, catalyzes the 4-electron reduction of oxygen while consuming two molecules of ascorbic acid, without release of radical intermediates. A further advantage of this enzymatic reaction is that its K_m for ascorbate is extremely low, allowing complete calibration curves in a relatively short time. In contrast, enzymes such as glucose oxidase have a high substrate K_m while releasing equal quantities of hydrogen peroxide in the process. Additionally, glucose is often a metabolically required solute and so cannot be used for calibration purposes (Fig. 3).

If an attempt is made to maintain a constant pO_2 in a spinner culture by use of a gas flow above the liquid, several difficulties arise: (1) the gas is often flowed at high rates to minimize the problems of oxygen contamination via plastic tubing; (2) this exacerbates the likelihood of solvent evaporation, which often requires complex humidification and flow control through another vessel; and (3) there will inevitably arise a large gradient of oxygen concentration from the supply (gas) to the oxygen-consuming solution.[41] The first two problems are solvable. By making an otherwise sealed and plastic-free vessel, as described earlier, and using fine-bore stainless-steel tubing with precision regulators, gas flow can be minimized to a few milliliters per minute, effectively eliminating the need for gas humidification.[42]

Stainless-steel tubing is quite difficult to bend and position. Thus, the present availability of 1/16-inch diameter PEEK tubing, for use in high-performance liquid chromatography, allows the elimination of the mechanically "difficult" stainless steel at the sacrifice of only a small amount of oxygen leakage through the tubing (see Fig. 2). Because the bore of PEEK tubing is available from 0.005 to 0.03 inch, the tubing presents a high resistance to flow, which can be tailored over a broad variety of (low) flow rates. Thus, it is possible to regulate flow via pressure, thereby eliminating flow meters, which are largely inaccurate at such low flow rates.

While eliminating the first two problems with spinner cultures, the third remains (lack of equilibrium between gas and liquid phase) and requires continuous monitoring of the liquid pO_2. Because the liquid itself is well stirred, this disequilibrium is modeled by assuming a thin unstirred layer between the gas and

[41] R. S. Marshall, C. J. Koch, and A. M. Rauth, *Radiat. Res.* **108,** 91 (1986).

[42] It is important to realize that the internal volume of gas regulators is several tens of milliliters. Thus, when a regulator is first connected to a gas cylinder (or pressurized after disuse), this internal oxygen-containing volume is highly pressurized by the cylinder gas (e.g., nitrogen) and can cause a long-lasting contamination of oxygen if used at low flow rates. To prevent this, the regulator must be purged by a series of on–off cycles of the high-pressure cylinder valve, each followed by release of the gas from the regulator.

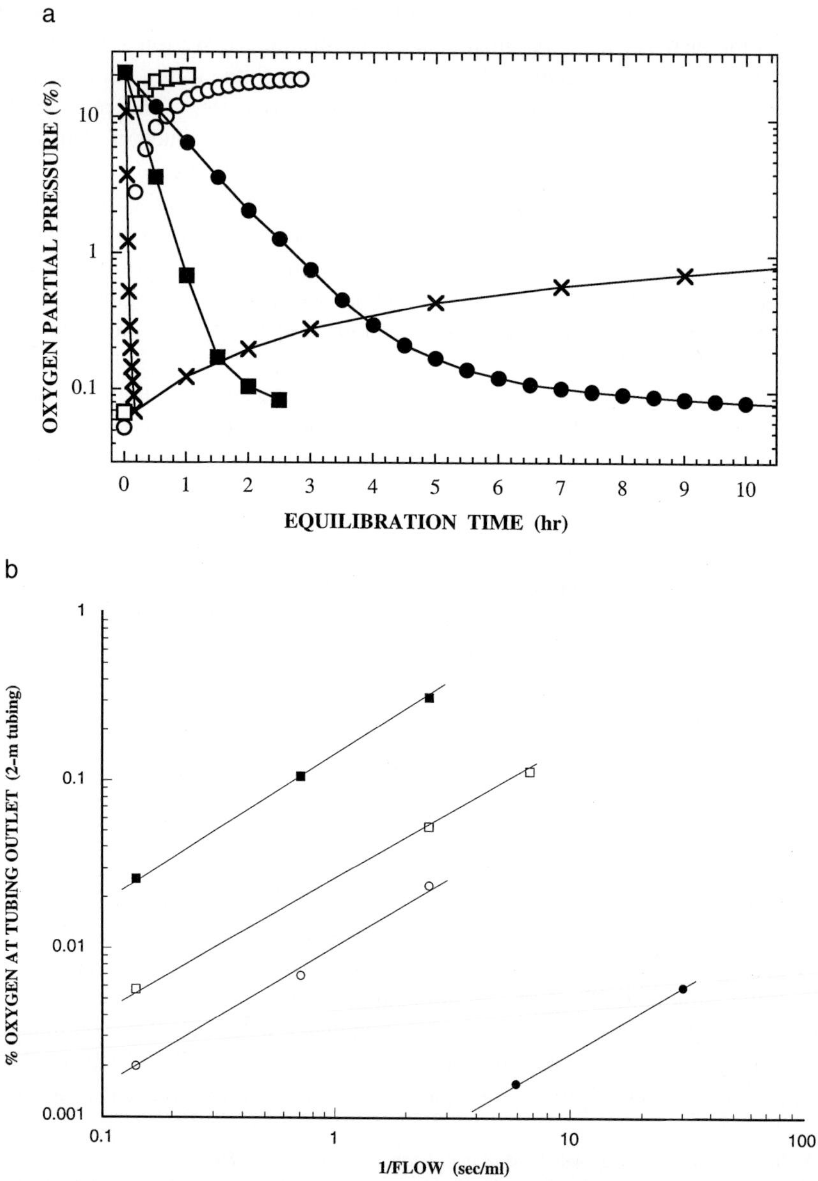

FIG. 2. (a) Examples of problems posed by plastics in culture systems. Closed circles and squares illustrate the temporal dependence of oxygen loss in a 1 liter Bellco spinner flask with 400 vs 200 ml of medium, respectively, stirred at 150 rpm under constant flushing by oxygen-free nitrogen. Note that equilibrium takes several hours to achieve, even with the smaller volume. Return to atmospheric oxygen levels has the same kinetics, but appears much more rapid (open symbols). The "X" symbols

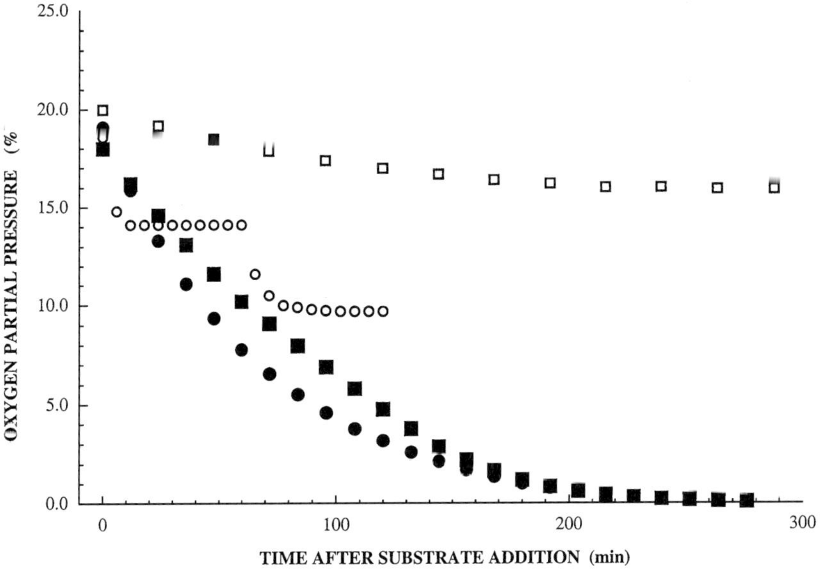

FIG. 3. Calculation of the oxygen solubility of physiological solutions at 25° using the oxygen consumption capabilities of glucose oxidase (GO, squares) vs ascorbate oxidase (AO, circles). A sealed 10-ml all-glass/ceramic vial with a CK oxygen sensor is filled with equilibrated physiological saline containing enzyme (for GO, catalase is also included). Typically, GO has other oxidizing contaminants, which reduce its specificity. The purest form available, to the author's knowledge, is insoluble *Aspergillus niger* GO attached to polyacrylamide beads (Sigma G-9255). In an excess of substrate (2 m*M* ascorbic acid or 5.5 m*M* glucose), both enzymes consume oxygen at a rate that is roughly first order in oxygen concentration (closed squares, 0.025 U/ml GO; closed circles 0.01 U/ml ascorbate oxidase, Sigma C-3515). However, at 10-fold higher enzyme concentrations, where a titrating amount of substrate is added (100 nmol/ml glucose or ascorbic acid), AO is at least 10-fold more efficient (open circles, two additions at 1-hr intervals) than GO (open squares). Because of the leak-proof characteristics of the spinner systems, it is possible to wait for many hours for the new equilibrium to be established. The oxygen concentration can be calculated as 227 ± 5 μM for either assay. In actual practice, 10-fold higher AO concentrations allow such calibrations in a few minutes. [L.-W. Lo, C. J. Koch, and D. F. Wilson, *Anal. Biochem.* **236,** 153 (1996)].

---

show the change in pO_2 in a Billups–Rothenberg chamber containing the usual plastic tray and twelve 100-mm plastic petri dishes. Gas flow is 10 liters/min for the first 10 min. Oxygen diffuses from the plastic components for many hours, resulting in a final oxygen content of more than 1%. (b) Oxygen content at outflow of 2 m lengths of various tubing flushed with oxygen-free nitrogen at several flow rates [closed squares, 6 × 3-mm (o.d. vs i.d.) silicone tubing; open squares, 6 × 3-mm amber rubber; open circles, 6 × 3-mm Tygon; closed circles 1/16 × 0.007-inch PEEK]. The leakage rate of oxygen into the PEEK tubing is below measurement capabilities at the higher flow rates. The take-home message is that silicone tubing is 10-fold more oxygen diffusive than Tygon, which is two orders of magnitude more diffusive than the PEEK microbore tubing. The PEEK tubing also has the significant "advantage" of a high flow resistance. Thus, flow can be precisely determined by pressure.

the liquid phase. This situation occurs whenever there is a discontinuity between oxygen delivery and consumption. When more complex tissue samples (spheroids or tissue cubes) are monitored in suspension, an even more complicated situation arises. An additional unstirred layer occurs at the surface of the tissue, and dramatic changes in pO_2 occur as a function of depth into the tissue from its surface. Some information can be gained from modeling this dynamic situation (Fig. 4), but the real hope is that it will be possible to measure tissue pO_2 under conditions of such steep oxygen gradients.

Oxygen Measurements under Conditions of Steep Oxygen Gradients

The problems with oxygen measurement in tissue parallel those described earlier, but with the added requirement of discriminating changes over very small dimensions. Consideration of this complex situation leads to two observations: (1) oxygen concentrations can change dramatically over dimensions of a few micrometers and (2) the absolute values of oxygen concentration in this same range are substantially below the limitations of many measurement techniques.[1]

At present, only two possibilities exist for measuring tissue pO_2: true microelectrodes and 2-nitroimidazole-binding techniques. Unlike the Eppendorf needle electrode discussed earlier (which has an unusual inset membrane and is 300 μm in diameter), true microelectrodes are only a few micrometers in diameter.[43] Although studies with these fragile devices have been accomplished in truly elegant fashion,[44] their invasive character, extreme fragility, and modest spatial range do not allow sufficient flexibility to monitor tissue pO_2 in a clinically relevant manner. Often, the overwhelming challenge is to mechanically fix in place the tissue of interest to allow controlled insertion of the microsensor.[45] A particularly interesting use of this method, where the exact position of the microneedle tip with respect to the vasculature was visualized via a window chamber, has been published by Li *et al.*[46] Recognizing the exacting requirements of high spatial resolution in *in situ* measurements, we investigated the potential of the 2-nitroimidazole-binding technique.

Oxygen Detection by Metabolism of Nitroheterocyclics

The requirement for hypoxia in the radiosensitizing, cytotoxic, and metabolism-perturbing properties of bioreductively activated drugs is well recognized

[43] D. G. Buerk, "Biosensors: Theory and Applications." Technomic, Lancaster, PA, 1993.

[44] O. Thews, D. K. Kelleher, and P. W. Vaupel, *in* "Tumor Oxygenation," p. 27. Gustav Fisher Verlag, Stuttgart, Germany, 1995.

[45] W. Mueller-Klieser, J. P. Freyer, and R. M. Sutherland, *Br. J. Cancer* **53,** 345 (1986).

[46] C. Y. Li, S. Shan, Q. Huang, R. D. Braun, J. Lanzen, K. Hu, P. Lin, and M. W. Dewhirst, *J. Nat. Cancer Inst.* **92,** 143 (2000).

(see Ref. 47 for review). Specifically, the reductive metabolism of nitroheterocyclics leads to their activation and subsequent formation of covalent bonds with cellular macromolecules; this process is inhibited greatly as a function of increasing oxygen concentration.[38,48–50] It is thought that the oxygen dependence of the process occurs after a single electron is transferred to the nitro compound. The resulting nitro-radical anion can then either be further reduced, leading to reactive species that form covalent bonds with cellular macromolecules, or be reoxidized by transfer of the electron to oxygen. Macromolecular adducts form very preferentially with protein thiols.[51,52] This process, referred to as "binding" throughout this chapter, was first studied with the 2-nitroimidazole, misonidazole. Binding of misonidazole has been shown to vary over oxygen concentrations known to affect radiation response and drug cytotoxicity in a number of cell and multicellular systems[38,53,54] and *in vivo*.[48,55] Problems associated with the use of misonidazole and its derivatives include dramatic variations in absolute binding rates, variations in the oxygen-dependent kinetics of the binding rates, and susceptibility to nonoxygen-dependent metabolism.[54,56–59] It is likely that such problems contributed to the suggestion that some normal tissues were severely hypoxic.[57] As discussed later, it is possible to design 2-nitroimidazoles that do not suffer from the problems just discussed.

The detection technique for bound adducts initially involved the use of a radioactive drug and either liquid scintillation methods to determine the average uptake *in vitro* and *in vivo* or autoradiography to determine the distribution of drug adducts in tissue sections.[48,60] The latter allows good spatial resolution but is extemely time-consuming and the use of a radioactive probe (e.g., tritium) imposes many restrictions, particularly for clinical application. Despite these restrictions, regions of high binding and, more importantly, *high binding gradients,* presumably detecting *oxygen gradients,* have been found in both animal and human tumors.[60,61]

[47] G. E. Adams, A. Breccia, E. M. Fielden, and P. Wardman (eds.), "Selective Activation of Drugs by Redox Processes." Plenum Press, New York, 1990.

[48] J. D. Chapman, K. A. Baer, and J. Lee, *Cancer Res.* **43,** 1523 (1983).

[49] Y. C. Taylor and A. M. Rauth, *Cancer Res.* **38,** 2745 (1978)

[50] A. J. Varghese, S. Gulyas, and J. K. Mohindra, *Cancer Res.* **36,** 3761 (1976).

[51] C. J. Koch and J. A. Raleigh, *Arch. Biochem. Biophys.* **287,** 75 (1991).

[52] J. A. Raleigh and C. J. Koch, *Biochem. Pharmacol.* **40,** 2457 (1990).

[53] A. J. Franko and C. J. Koch, *Int. J. Radiat. Oncol. Biol. Phys.* **10,** 1333 (1984).

[54] A. J. Franko, C. J. Koch, B. M. Garrecht, J. Sharplin, and J. Howorko, *Cancer Res.* **47,** 5367 (1987).

[55] D. G. Hirst, J. L. Hazlehurst, and J. M. Brown, *Int. J. Radiat. Biol. Oncol. Phys.* **11,** 1349 (1984).

[56] L. M. Cobb, J. Nolan, and P. O'Neill, *Br. J. Cancer* **59,** 12 (1989).

[57] L. M. Cobb, J. Nolan, and T. Hacker, *Int. J. Radiat. Oncol. Biol. Phys.* **22,** 655 (1992).

[58] C. J. Koch, *in* "Selective Activation of Drugs by Redox Processes" (G. E. Adams, A. Breccia, E. M. Fielden, and P. Wardman, eds.), p. 237. Plenum Press, NY, 1990.

[59] D. J. Van Os-Corby, C. J. Koch, and J. D. Chapman, *Biochem. Pharmacol.* **36,** 3487 (1987).

[60] B. M. Garrecht and J. D. Chapman, *Br. J. Radiol.* **56,** 745 (1983).

[61] R. C. Urtasun, C. J. Koch, A. J. Franko, J. A. Raleigh, and J. D. Chapman, *Br. J. Cancer* **54,** 453 (1985).

A breakthrough in eliminating the need for a radioactive drug was made by Raleigh and colleagues in 1987, when polyclonal antibodies to detect drug adducts of CCI-103F (a hexafluorinated derivative of misonidazole) were first made.[62] Shortly thereafter, existing antibodies to theophylline were used to detect adducts of NITP (a derivative of azomycin containing theophylline on its side chain)[63] in a flow cytometry assay. Unfortunately, there are many problems inherent in the use of polyclonal antibodies, and both drugs have limited solubility. For NITP, drug distribution and stability *in vivo* are extremely poor.[64]

To summarize, the 2-nitroimidazole-binding technique has the dual advantages of sensitivity at very low oxygen levels and resolution at the cell–cell level. These advantages, until recently, have been balanced by several problems: (1) variations in the absolute rates of binding appear to vary with several factors in addition to, but always including, oxygen; (2) suitable marker techniques that could be used for routine diagnosis have not been available; (3) drug pharmacology and stability have been highly problematic; and (4) binding techniques have not been shown previously to predict radiation resistance or therapeutic outcome in individual tumors.

We have made substantial progress in solving these problems. Briefly, a 2-nitroimidazole named EF5 has been developed. The drug contains multiple fluorines, yet is quite soluble, allowing detection by magnetic resonance imaging or spectroscopy (MRI/MRS) and positron emission tomography (PET). EF5 is much less susceptible to nonoxygen-dependent variations in binding found previously for misonidazole.[65] Highly specific monoclonal antibodies (MAbs) against the drug and its adducts have been made,[65–67] allowing a number of clinically relevant biopsy techniques. We have not identified any nonbioreductive metabolism *in vivo*. EF5 distributes evenly to all tissues, including brain, and has a simple exponential pharmacological decay, dominated by renal clearance of the unmodified drug.[68] EF5 binding can predict oxygen gradients corresponding to increasing radiation resistance in multicell tumor spheroids.[69] It has also been possible to predict the radiation response of cells from individual rat[66] and mouse[69a] tumors. This is the first predictive assay developed for measuring therapy-relevant hypoxia

[62] J. A. Raleigh, G. G. Miller, A. J. Franko, C. J. Koch, A. F. Fuciarelli, and D. A. Kelley, *Br. J. Cancer* **56,** 395 (1987).

[63] R. J. Hodgkiss, G. Jones, A. Long, J. Parrick, K. A. Smith, and M. R. L. Stratford, *Br. J. Cancer* **63,** 119 (1991).

[64] R. J. Hodgkiss, M. R. Stratford, M. F. Dennis, and S. A. Hill, *Br. J. Cancer* **72,** 1462 (1995).

[65] C. J. Koch, S. M. Evans, and E. M. Lord, *Br. J. Cancer* **72,** 869 (1995).

[66] S. M. Evans, W. T. Jenkins, B. Joiner, E. M. Lord, and C. J. Koch, *Cancer Res.* **56,** 405 (1996).

[67] E. M. Lord, L. W. Harwell, and C. J. Koch, *Cancer Res.* **53,** 5271 (1993).

[68] K. M. Laughlin, S. M. Evans, W. T. Jenkins, M. Tracy, C. Y. Chan, E. M. Lord, and C. J. Koch, *J. Pharmacol. Exp. Ther.* **277,** 1049 (1996).

[69] M. R. Woods, E. M. Lord, and C. J. Koch, *Int. J. Radiat. Oncol. Biol. Phys.* **34,** 93 (1996).

[69a] J. Lee, D. W. Siemann, C. J. Koch, and E. M. Lord, *Int. J. Cancer* **67,** 372 (1996).

in individual rodent tumors. Furthermore, EF5 has been used in several normal tissue animal models of pathological conditions involving low tissue pO_2, including patent ductus arteriosus,[27,70] cardiac infarct,[71] brain ischemia,[72] and others. An example of the type of colocalization analysis that can be achieved routinely in tumors is illustrated by three-color images of EF5 binding, blood perfusion (as measured by injecting the DNA-binding dye Hoechst 33324 IV just before tissue collection), and blood vessels (as measured by a monoclonal antibody to CD31) (Fig. 5).

Images such as that shown in Fig. 5 engender a "feel-good" impression, as they support reasonable physiological interpretation. It is clear that EF5 binding increases with distance from a perfused blood vessel, that overlap of EF5 binding with CD31 occurs almost exclusively in nonperfused regions, and that essentially all perfused regions are associated with CD31 staining (one would not expect universal colocalization of these two stains, as the perfusion marker may originate from vessels on either side of the two-dimensional section). The CD31 image is binary in nature, but obviously there is considerable variation in vessel size and multiple vessels may cooperate in the nutrient delivery to a specified region. The perfusion and EF5-binding signals are analog in nature (i.e., they are characterized by intensity as well as location) so the types of analysis that can be made are very complex, and this complexity may increase arbitrarily with other parameters (cells that are cycling, presence of cytokines, three-dimensional considerations, etc.). Because this is such a complex subject area, this chapter emphasizes a central question: "Can we abstract absolute tissue pO_2 information from the analog signal representing EF5 binding?" The answer to this question has been under detailed study in this laboratory for several years, and the remaining portion of this paper emphasizes the criteria required for such an analysis and the progress made to date.

Criteria for Absolute Oxygen Measurements *in Vivo*

We suggest that five critical requirements must be met to measure tissue pO_2 using a 2-nitroimidazole marker: (1) the detection system must be quantitative; (2) the dynamic range of the method must be large enough to include the physiological and pathological pO_2 range; (3) drug metabolism must be understood to exclude nonoxygen-dependent binding; (4) the pharmacokinetics and compound stability must be well understood; and (5) the oxygen dependence of binding must be quantified.

[70] H. Kajino, Y.-Q. Chen, S. Chemtob, N. Waleh, C. J. Koch, and R. I. Clyman, *Am. J. Physiol. Regul. Integr. Comp. Physiol.* **279,** 278 (2000).

[71] S. Bialik, D. L. Geenen, I. Sasson, R. Cheng, J. Horner, S. M. Evans, E. M. Lord, C. J. Koch, and R. N. Kitsis, *J. Clin. Invest.* **100,** 1363 (1997).

[72] M. Bergeron, S. M. Evans, F. R. Sharp, C. J. Koch, E. M. Lord, and D. M. Ferriero, *Neuroscience* **89,** 1357 (1999).

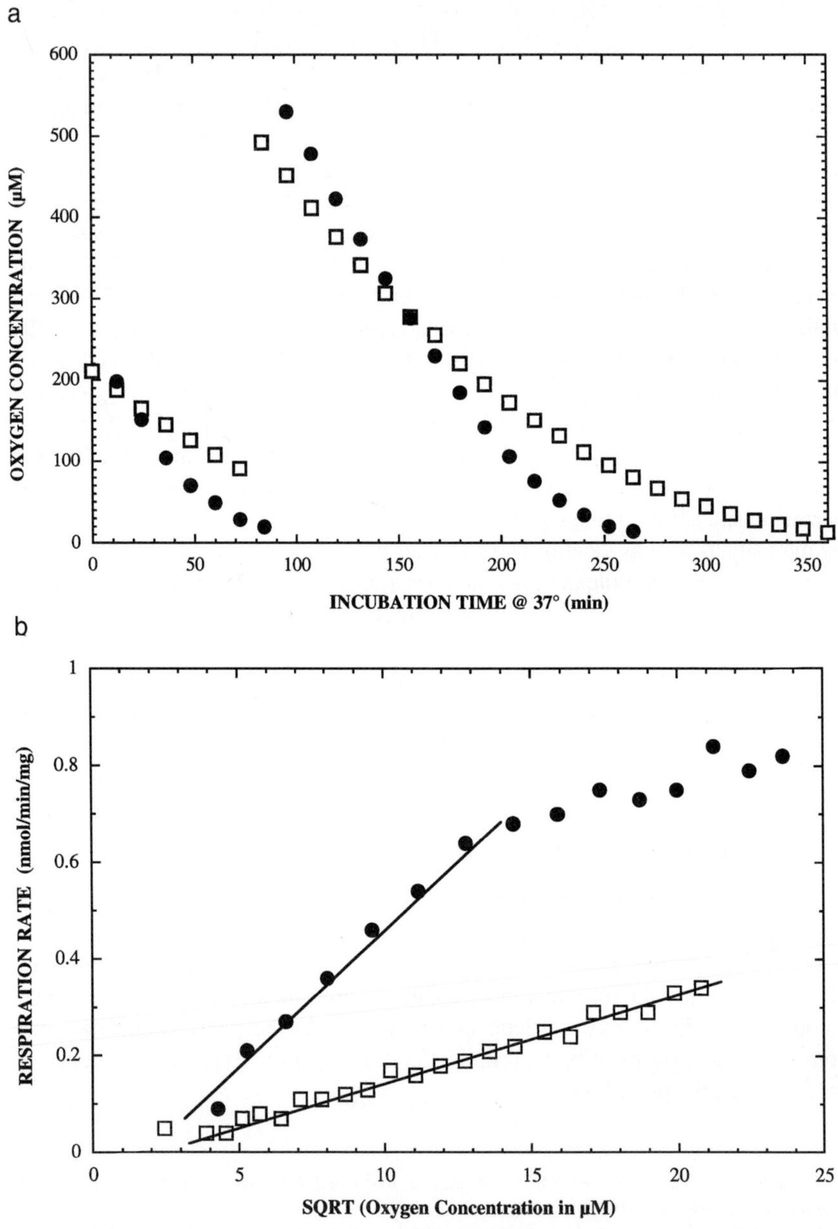

FIG. 4. (a) Using sealed spinner vials (described in the text and as used in Fig. 3), the respiration rate of tissue cubes can be measured as a function of oxygen concentration. HEPES-buffered medium is equilibrated with 100% oxygen at 1/5 atmosphere pressure (to eliminate nitrogen and allow "space" for

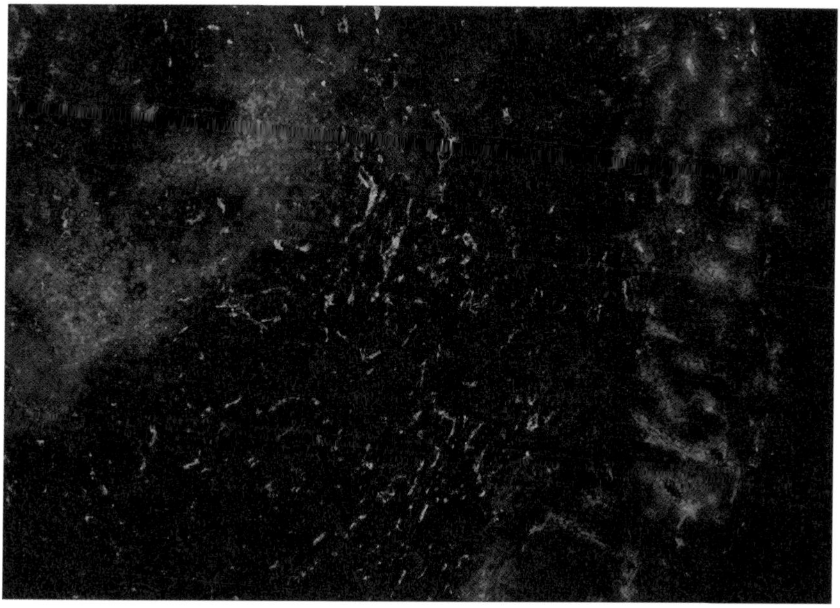

FIG. 5. Multicolor fluorescence immunohistochemical staining of a Morris 7777 hepatoma rat tumor. The tumor-bearing animal was treated with EF5, given as a tail-vein injection of 10 m*M* in 5% glucose. The injection volume (in ml) was equal to 1% of the animal's mass (in g). At 2.85 hr, the animal was injected with a similar volume of 2.5 m*M* Hoechst 33342. The animal was euthanized at 3 hr, and the tumor was removed with frozen sections analyzed as shown previously [S. M. Evans, S. Hahn, D. R. Pook, P. J. Zhang, W. T. Jenkins, D. Fraker, R. A. Hsi, W. G. McKenna, and C. J. Koch, *Int. J. Radiat. Oncol. Biol. Phys.* **49,** 587 (2000)]. EF5 binding is illustrated in red, blood vessels (as assayed by anti-CD31 Ab) in green, and the Hoechst assay of perfusion in blue. Note that most perfusion (Hoechst) is in the right periphery, is colocalized with CD31, and has no visible EF5 binding. Other vessels (central) have substantially less Hoechst staining, but maintain a lack of EF5 binding. The bright EF5-binding regions are characterized by low vessel density and lack of Hoechst perfusion. The total area of the image is 3 × 2 mm.

above-ambient oxygen addition via catalase plus hydrogen peroxide), open squares; 8 cubes of HCT116 human colon carcinoma weighing 102 mg, closed circles; 8 cubes of murine liver weighing 55 mg. The break in oxygen concentration at about 90 min results from the addition of catalase plus hydrogen peroxide. Note that the respiration rate after the catalase hydrogen peroxide addition is similar, over similar oxygen concentrations, over the 6-hr course of the experiment. (b) A semiinfinite plane of tissue would be expected to have a respiration rate proportional to the square root of oxygen concentration until all tissue is oxygenated, at which point the respiration rate should be independent of oxygen concentration. Such analysis of data from (a) shows that the smaller liver cubes reach a maximum rate of oxygen consumption of ~0.7 nmol/min/mg of tissue (closed circles), whereas the larger tumor cubes (open squares) do not reach a maximal respiration rate over the range of oxygen levels used in this experiment. However, the maximum rate of oxygen consumption for the tumor tissue is expected to be at least half of the value found for liver.

Quantitative Detection of EF5 Adducts by Fluorescence-Based Immunohistochemistry

Fluorescence detection has not previously been considered a quantitative method, particularly for imaging applications. The characteristics of optical systems and fluorescent dyes provide a litany of explanations underlying this view: instability of light sources; lack of standard optical filter design; variable optical characteristics arising from microscope design; temporal, chemical, and optical instability of dyes; and lack of consistency and uniformity of fluorescent dye labeling. Similarly, the quantitative aspects of antibody detection are seldom characterized because they can be influenced by a large number of factors involving access, sensitivity, and specificity of the antibody and microenvironment of the antigen. For detailed studies comparing different experimental conditions with a consistent cell type, flow cytometric assays offer some improvement, as calibrated beads are available. However, in our experience, the fluorescence of the beads varies from batch to batch. More problematic, the optical properties of the beads only approximate, or in some cases are greatly different than, the actual dye of interest. The beads are often much smaller than tumor cells, yet are usually much brighter than cells stained with typical procedures. Beads do not share light transmission and scattering properties common to mammalian cells. Consistency of antibody binding adds further complications, particularly since the time lag between staining and assay may vary. Finally, we are not aware of any antibody-based detection system that also requires a large dynamic range (see next section). This long list of problems has gradually been resolved, and a brief summary of the methods and results, developed over several years, is provided. The solutions described may not be unique, but the multiple variations of conditions tested cannot be detailed in this brief paper.

Hybridomas making monoclonal antibodies to EF5 and its adducts were first developed and characterized by Dr. Edith Lord.[67] These were the first MAbs ever developed against 2-nitroimidazole adducts. Antibodies are currently produced using a capillary culture system (CellCo CM-MAX). The hybridoma supernatant is concentrated using ultrafilters (Amicon YM-30), and the antibodies (mouse IgG_1) are purified to homogeneity using protein A columns (Pierce). The antibodies were found to be completely stable at -65° in phosphate-buffered saline (PBS) with 25% glycerol. The antibodies were characterized using precision microdialysis devices designed by the author (Fig. 6).

A single antibody detection system was developed because it was felt that dual antibodies or other methods (e.g., avidin–streptavidin) would add an additional level of complexity to the optimization procedures. Thus, MAbs were conjugated with the newly developed (at that time) Cy3 dye,[73] as this dye was reported to

[73] P. L. Southwick, L. A. Ernst, E. W. Tauriello, S. R. Parker, R. B. Mujumdar, S. R. Mujumdar, H. A. Clever, and A. S. Waggoner, *Cytometry* **11,** 418 (1990).

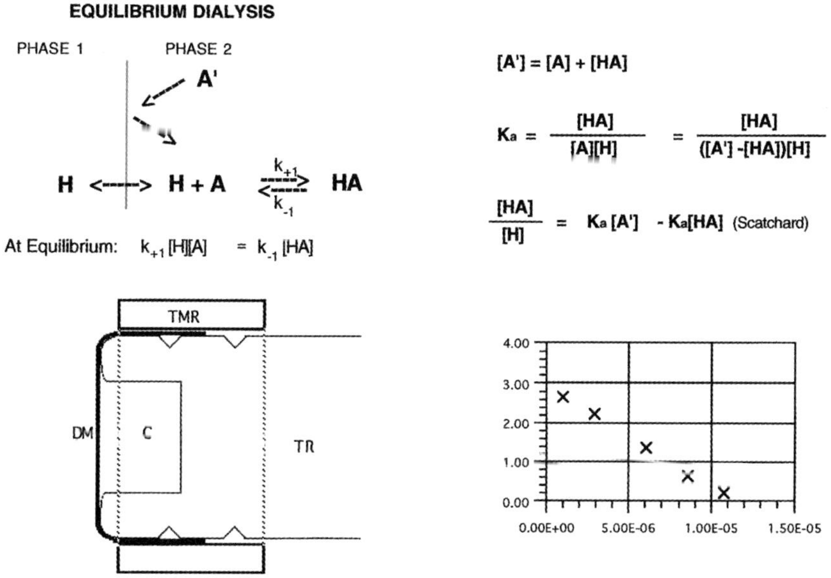

FIG. 6. Antibodies against EF5 were assessed by equilibrium dialysis using miniature devices designed by the author. (Lower left) An 8-mm-diameter Teflon rod (TR) with 40 μl cavity (C) was filled with Ab solution (typically 2 mg/ml) and covered by a dialysis membrane (DM, 10 kDa pore size) held in place by a Teflon membrane retainer (TMR). When immersed in a small volume of radioactive EF5, EF5 diffuses across the membrane, but Ab cannot escape. Thus, an equilibrium is set up, described in the upper left, where the EF5 hapten (H) is concentrated by the antibody (A) forming the complex HA. The Scatchard relationship is described in the upper right, where the ratio of complex to free hapten (HA/H) is proportional to the affinity constant multiplied by the complex concentration. Using various concentrations of EF5, a Scatchard plot for the original ELK2-4 Ab (lower right) shows that its affinity for EF5 is 300,000. The antibodies under current use (ELK3-51) have affinity constants that are more than 20-fold higher ($>5 \times 10^6$).

be free from photobleaching and other variability associated with classical dyes, such as FITC. We found that the manufacturers' suggested methods (adding dilute antibody to concentrated dye or dye powder) tended to produce heterogeneity in the number of dye molecules per antibody molecule (as determined by nonlinearity in the Scatchard plot of binding affinity) and also nonspecific binding (data not shown). When the dye was dissolved separately and added more slowly as a dilute solution, its efficiency of binding dropped but was very reproducible. Dye-conjugated antibodies (2–2.5 dye molecules per Ab molecule) are repurified using protein A columns as described previously.

Nonreacted Cy3 dye can be stored in the dark in a preservative (1% paraformaldehyde in PBS) and forms the basis of a fluorescence standard for microscopy. The dye solution is diluted to an absorbance at 549 nm of 1.25. This is loaded into a hemocytometer, which provides a standard thickness (100 μm) of fluorescent dye. We have not tested other dyes, but Cy3 fluorescence can be predicted accurately by absorbance for several years when stored this way (we determine this by comparing the fluorescence of freshly purchased dye with that stored; data not shown).

Detection of bound, radioactive drug adducts can be measured quantitatively (using a liquid scintillation counter with internal standards). Therefore, we first determined that EF5 adduct formation was consistent between experiments. Using the "thin-film" incubation method described earlier, we found this to be the case. This led to the adoption of a "standard," used for calibrations of the flow cytometer, consisting of V79 Chinese hamster fibroblasts incubated with 100 μM EF5 for 4 hr under conditions of severe hypoxia (defined herein as a pO_2 of less than 0.005%). Absolute EF5 binding is $2.85 \pm 0.4 \times 10^{-4}$ (SD) pmol/cell for this standard condition. Because V79 cells have a volume of about 1 μl per million cells, this corresponds to an adduct concentration of 285 μM (i.e., the binding rate per hour is about equal to the steady-state concentration of drug).

Using fluorescent MAbs, we found that the cellular morphology and contained adduct concentration were stable indefinitely if cells were trypsinized from the experimental dishes, fixed for 1 hr at 0° in freshly prepared 4% paraformaldehyde, rinsed twice and suspended in 25% glycerol, and frozen at −65° (all solutions contain Dulbecco's phosphate-buffered saline, pH 7.3). In 1995, we showed that *relative* fluorescence, as measured by flow cytometry, varied over the same range as radioactive drug uptake and radiation survival for cells treated by a constant EF5 exposure at various oxygen concentrations.[65] At that time, we were using commercially available beads as a calibration source. Having determined that the beads could not be used as a universal calibration source for microscopy, or even for other cytometers using different wavelengths, because of the problems mentioned earlier, i.e., the fluorescence properties of the beads were not the same as the actual dye, we switched to the new cellular standard. Using the new cellular calibration source, we have now repeated and extended these experiments to define a calibrated fluorescence scale. It can be seen that radioactive drug uptake parallels fluorescent antibody binding for all cells tested and over a large dynamic range (Fig. 7). These results suggest that the other concerns listed earlier (e.g., antibody-binding characteristics and access to antigen for different cell types) are relatively unimportant for the specified fixation, staining, and storage methods we have developed. Of course, these results can only be deemed applicable to the present drug/antibody system and would not necessarily apply to other antibodies and antigens. In addition to defining a calibrated fluorescence scale for the Cy3 dye, this method can be used for any dye of interest, irrespective of the optical characteristics of the measurement, as the experimental sample is always compared to the same standard and using the same antibodies. Most importantly, the same standard can be equally applied to microscopy.

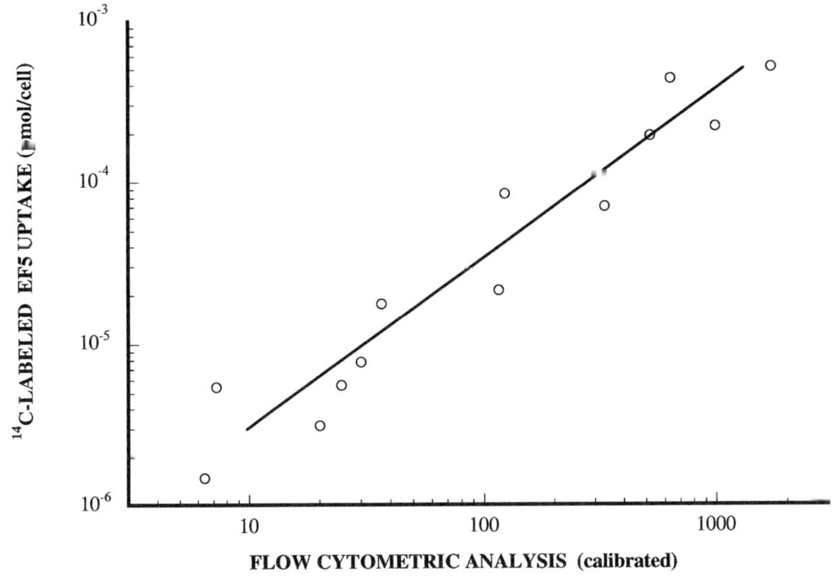

FIG. 7. Summary of EF5 binding, as determined using radioactive EF5 uptake,[65] vs antibody detection of EF5 adducts as determined by flow cytometry using a calibration system described in the text. Included in this summary are normal Chinese hamster fibroblasts and one cell line each derived from feline, murine, rat, and human tumor origin. All data points can be fit by a common line with slope 1.0.

A seldom-discussed problem with immunohistochemistry is nonspecific binding by the antibody (or its equivalent: antibody-binding proteins in the sample of interest). We have found that necrotic tumor areas are particularly problematic in this regard. Because of the high affinity of the ELK3-51 antibodies for authentic EF5, we have developed a method to assess nonspecific binding, which has been called the "competed antibody stain" (see Fig. 8a). In this method, EF5 is added to the antibody stain at substantial excess (0.5 m*M* EF5 in solutions containing 0.5 μM Ab, i.e., 75 μg/ml). Initial rinses also contain EF5 at 0.25 m*M*.[73a] We cannot rule out the possibility that nonspecific binding is modified by the presence of an authentic antigen in the binding site in some instances. However, this does not appear to be the case, as specific binding is typically reduced to levels seen for non-EF5-treated samples. The competed Ab stain innovation is particularly valuable for human tumor studies where it is logistically and ethically impossible

[73a] S. M. Evans, S. Hahn, D. R. Pook, W. T. Jenkins, A. A. Chalian, P. Zhang, C. Stevens, R. Weber, I. Benjamin, N. Mirza, M. Morgan, S. Rubin, W. G. McKenna, E. M. Lord, and C. J. Koch, *Cancer Res.* **60,** 2018 (2000).

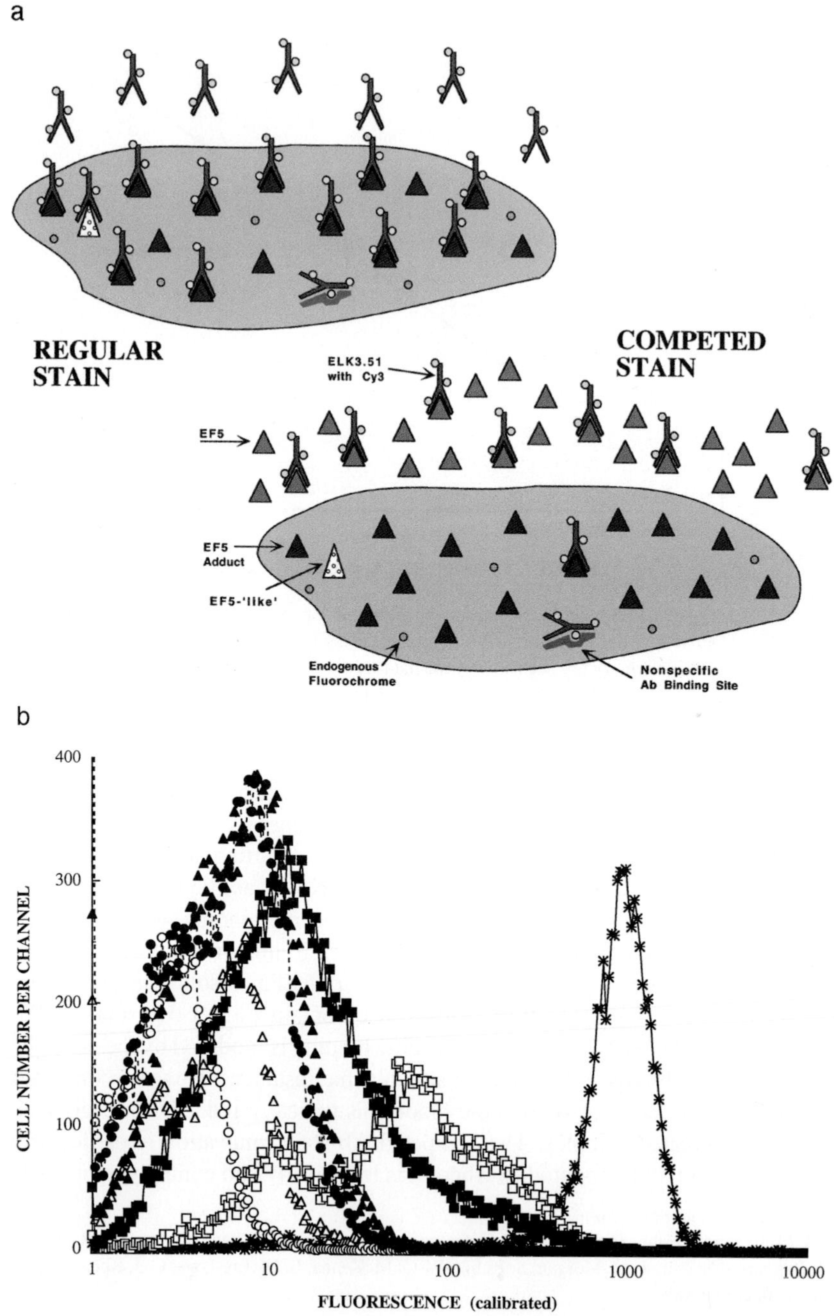
a
REGULAR STAIN
COMPETED STAIN
ELK3.51 with Cy3
EF5
EF5 Adduct
EF5-'like'
Endogenous Fluorochrome
Nonspecific Ab Binding Site
b
CELL NUMBER PER CHANNEL
FLUORESCENCE (calibrated)
400
300
200
100
0
1
10
100
1000
10000

to obtain tissue biopsies before the EF5 is given (for a no-drug control) and where each tumor is likely to have uniquely different cell types and characteristics. It is also extremely useful in experimental systems where the amount of nonspecific binding is variable or where there is endogenous tissue fluorescence. This problem is seen for Morris 7777 hepatoma tumors, which tend to have nonspecific binding by the antibody as well as variable background fluorescence in the Cy3 wavelength range, compared with 9L glioma tumors, which have neither (Fig. 8b). In general, much lower background fluorescence is seen at higher wavelengths (e.g., Cy5; data not shown), but nonspecific antibody binding is independent of conjugated dye in our hands.

The data and references given demonstrate that our monoclonal antibody detection system with self-contained fluorescent dye can quantitatively measure EF5 adducts over a broad dynamic range for cells *in vitro*. The dynamic range of signal detection is clearly sufficient to monitor tumor cell pO_2 over the complete physiological and pathological range, although the best resolution is obtained at oxygen levels below 1–2%. As suggested earlier, however, three additional steps are required to convert tissue EF5 binding to tissue pO_2. These steps are to quantify the drug metabolism and pharmacokinetics *in vivo* and the oxygen dependence of binding.

Importance of Drug Metabolism and Pharmacokinetics

Drug pharmacokinetics are of obvious importance because the absolute binding (at constant pO_2) depends directly on tissue exposure or "area under the curve" (AUC).[74] An exception has been noted for misonidazole where the binding kinetics are more complex, but only under severely hypoxic conditions.[58] Thus, any variation in pharmacokinetic properties of a drug will lead directly to variations in observed binding. Such variation includes consideration of not only serum or tissue

[74] T. M. Busch, S. M. Hahn, S. M. Evans, and C. J. Koch, *Cancer Res.* **60,** 2636 (2000).

FIG. 8. (a) The competed Ab method adds an excess of authentic EF5 (purple triangles) to the normal Ab staining solution (inverted "Y's") to tie up all specific binding sites. Nonspecific binding of the antibody is not affected by this control, allowing the assessment of nonspecific binding. (b) Flow cytometric analysis of cells from Morris 7777 hepatoma (closed symbols) vs 9L glioma (open symbols) rat tumors. The three conditions represented for each tumor cell type are regular stain (squares), competed stain (triangles), and no stain (circles). Elk3-51 Ab were conjugated with the green excited, orange-emitting Cy3 and analyzed on a Becton Dickinson FacsCaliber. It can be seen that significant autofluorescence and nonspecific binding occur for the Morris 7777 tumor (the amount varies from tumor to tumor), whereas neither problem is evident in the 9L glioma tumor. Similar variations have been observed in human tumor samples from individual patients, emphasizing the need for this type of control (data not shown). Also shown is the calibration signal provided by Chinese hamster cells exposed to 400 μM hr of EF5 under severe hypoxia (*).

half-life, but also drug stability, nonoxygen-dependent metabolism (often seen in excretory organs and liver), and drug biodistribution. These drug characteristics are by no means a trivial issue. An example posed by drug biodistribution in the absence of other problems is illustrated by etanidazole. This drug is a highly polar 2-nitroimidazole, which served as the nonfluorinated model for EF5. It shares the same *in vivo* stability as EF5, but its concentration in tumors varies from 0 (brain tumors) to 2.5 times the plasma level.[9,75] The only other 2-nitroimidazole in current use for hypoxia detection by immunohistochemical techniques is pimonidazole.[76] The percentage of authentic compound excreted renally has shown substantial variability.[77,78] This drug is relatively unstable *in vivo*. Its side chain is modified by a nonoxygen-dependent process unrelated to nitro reduction.[79] More importantly, its weakly basic side chain causes pH-dependent accumulation in tumors, varying from less than half to 12 times that of plasma.[78] A recent report of pharmacological aspects of fluoromisonidazole (a 2-nitroimidazole used for noninvasive measurement of hypoxia via ^{18}F-PET) has been made.[79a] This report showed multiple radioactive compounds in urine, with only a small fraction of parent drug. Although such pharmacological problems have not, to our knowledge, been discussed in the literature, we would predict a direct proportionality between tissue drug concentration and rate of drug metabolism. Therefore, substantial variations in drug stability and tissue access are bound to contribute to nonoxygen-dependent variations in drug adduct formation.

EF5 was designed specifically as a hypoxia marker, and a major aspect of this design was to optimize its pharmacological properties. Unlike etanidazole, EF5 is lipophilic enough to diffuse evenly to all tissues, including brain.[68] It is somewhat surprising that such a lipophilic drug would have dominant renal clearance, but this is indeed the case. This prevents problems expected for drugs that have predominant clearance by gut metabolism. EF5, like its design model (etanidazole), is remarkably stable *in vivo,* and we have seen no evidence for nonbioreductive metabolism. In other words, EF5 decays exponentially in plasma, with half-lives varying from 50 min in mice to 12 hr in humans.[68,80] The only compound found (as detected by absorbence at 325 n*M* by HPLC) in either blood or urine is authentic EF5.[80]

[75] C. N. Coleman, L. Noll, A. E. Howes, J. R. Harris, J. Zakar, and R. A. Kramer, *Int. J. Radiat. Oncol. Biol. Phys.* **16,** 1085 (1989).

[76] A. S. Kennedy, J. A. Raleigh, G. M. Perez, D. P. Calkins, D. E. Thrall, D. B. Novotny, and M. Varia, *Int. J. Radiat. Oncol. Biol. Phys.* **37,** 897 (1997).

[77] J. T. Roberts, N. M. Bleehen, M. I. Walton, and P. Workman, *Br. J. Radiol.* **59,** 107 (1986).

[78] M. I. Saunders, S. Dische, D. Fermont, A. Bishop, I. Lenox-Smith, J. G. Allen, and S. L. Malcolm, *Br. J. Cancer* **46,** 706 (1982).

[79] M. I. Walton, N. M. Bleehen, and P. Workman, *Biochem. Pharmacol.* **34,** 3939 (1985).

[79a] J. S. Rasey, P. D. Hofstrand, L. K. Chin, and T. K. Tewson, *J. Nucl. Med.* **40,** 1072 (1999).

[80] C. J. Koch, S. M. Hahn, K. J. Rockwell, J. M. Covey, W. K. McKenna, and S. M. Evans, *Cancer Chemother. Pharmacol.* **48,** 177 (2001).

Quantification of Binding vs Tissue pO_2

In order to make the final step of quantification of tissue pO_2, a method is required to calibrate the oxygen dependence of binding *in vivo*. A tremendous simplification would be possible if the absolute rate and oxygen dependence of EF5 binding were the same for all tissues and species. Because the assay depends on the biochemical rate of drug reduction, this seems an extremely unlikely possibility. Indeed, the range of rates for maximum *in vitro* EF5 binding for various tumor cells varies by a factor of about 5, with rodent tumor cells and human tumor cells in the upper and lower halves of this range, respectively. Similarly, the oxygen dependence of binding has subtle variations from cell to cell (Fig. 9). A good approximation to the oxygen dependence of EF5 binding is to assume an inverse relationship over the two decades of 0.1 to 10.0% oxygen, with constant values above and below this range.

The five fold variation in the maximum EF5-binding rate suggests a corresponding variation in the effective cellular "nitroreductase" levels. The variation in oxygen dependence of binding suggests that the critical oxygen-dependent step is not quite as simple as is implied by the chemical process of donating an electron from the nitro radical anion to oxygen. Such a simple chemical reaction should

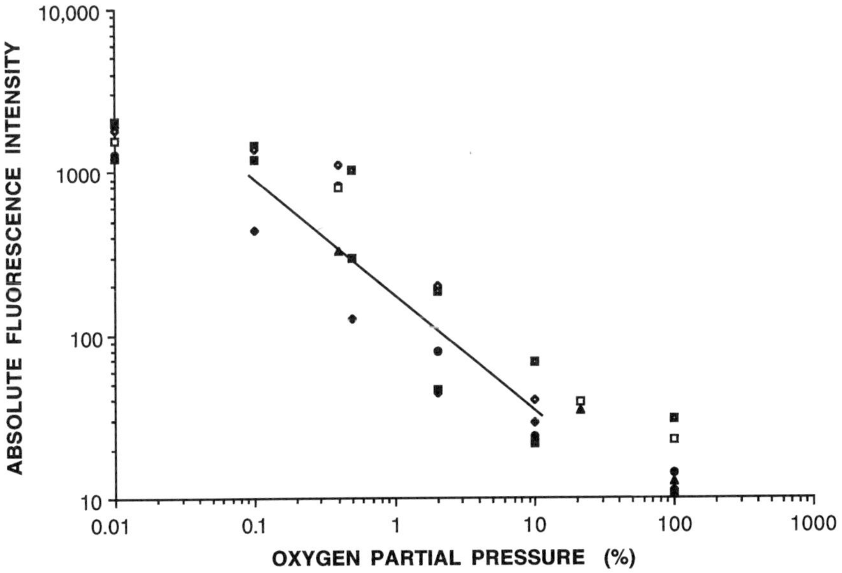

FIG. 9. Summary of oxygen dependence of EF5 binding, determined using uptake of radioactive EF5, for seven rodent tumor cell lines. The assumption of a linear relationship over the 0.1 to 10% oxygen range provides a good fit to all data.

be cell line independent for a given drug, although there could be cell-dependent variations in the physical–chemical environment of the electron transfer reaction. Despite these caveats, in all instances investigated to date, the maximum binding rate found *in vivo* appears to be limited by the drug exposure *in vivo* (AUC) and by the maximum binding rate determined *in vitro*. In other words, the maximum binding rate *in vivo* may be much less but is never more than the maximum predicted binding rate. We interpret this result to mean that the binding rates are the same between *in vivo* and *in vitro* conditions, so that decreases in observed binding *in vivo* reflect increases in tissue pO_2. To test this interpretation, we have devised a method to directly determine the maximum EF5-binding rate for intact tissue. We call this method "cube reference binding," wherein small tissue cubes (~1.5 mm on the side) are dissected from fresh tumor and then exposed to a low oxygen level (0.5%) in medium containing 200 μM EF3 or EF5. Identical results are obtained using either glass spinner vials or dishes in chamber methods. EF3 is an analog of EF5 wherein the C2 propyl carbon has two hydrogen rather than fluorine atoms. For these *in vitro* calibration studies, EF3 is used where tissue has already been exposed to EF5 *in vivo* (e.g., patient tissues). Anti-EF3 antibodies are much less active against EF5 adducts, allowing a discrimination against the endogenous EF5 signal. Because dissection of the tissue cubes is likely to cause surface tissue damage, we use a nonzero pO_2 for the incubation conditions. This allows respiration-induced severe hypoxia, and hence maximal binding, to occur at a depth beneath the cube surface. After the incubation period, the cubes are frozen, sectioned, and analyzed.[73,81] Using this method, we have shown that transplanted rodent tumors of the same type have a consistent maximal binding rate (C. J. Koch, in preparation). As expected from the geometry of the tissue cubes, the binding shows a characteristic reduction at the cube surface, followed by a steep gradient to maximal binding and then decreasing with further depth, probably due to nutrient and/or drug diffusion limitations. Two examples of this type of experiment illustrate its application but also occasional problems (Fig. 10).

For the 9L glioma cube, binding is completely "as expected," with very uniform binding from cell to cell and gradients toward the surface and cube interior. The Morris 7777 hepatoma cube shows a more spotty and variable binding. We occasionally see this pattern in both human and rodent tumor samples. There are several possible reasons for such heterogeneity. It could reflect true variability in the actual rate of EF5 metabolism—which would be problematic—or it could reflect other types of heterogeneity, which would not influence the purpose of the calibration: tumor tissue viability, physical aspects of the cube geometry, etc. Studies on the cause of such heterogeneity are underway.

[81] S. M. Evans, S. Hahn, D. R. Pook, P. J. Zhang, W. T. Jenkins, D. Fraker, R. A. Hsi, W. G. McKenna, and C. J. Koch, *Int. J. Radiat. Oncol. Biol. Phys.* **49,** 587 (2000).

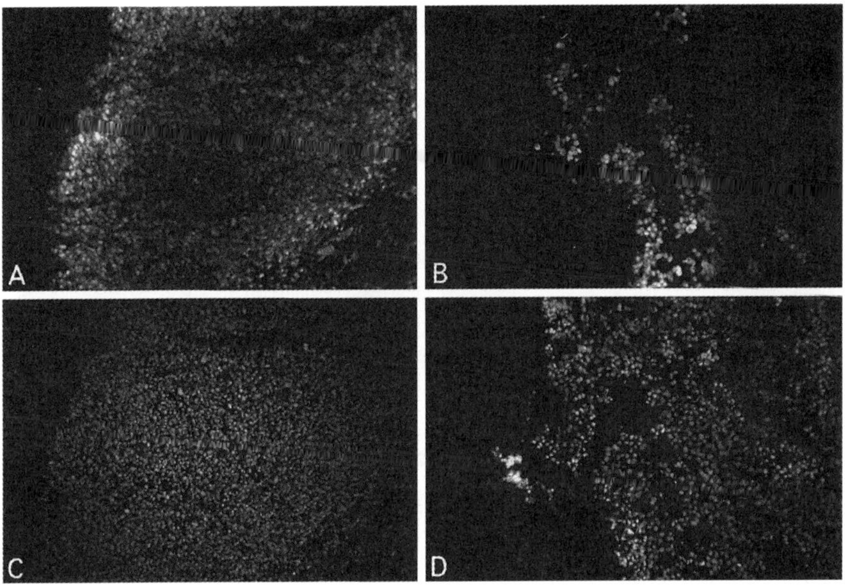

FIG. 10. Examples of cube calibrations for 9L glioma (A) vs Morris 7777 hepatoma (B) tumors. Staining of the sections with Hoechst 33342 (B and D, respectively) provides information on cellularity of the specimens. Thus, it can be seen that there are substantial acellular areas in the Morris 7777 tissue (upper right of D), whereas the 9L tissue has uniform cellularity. Despite "spotty" EF5 binding in the hepatoma tissue piece, the maximum binding rate is very similar for the two tissues, a result that is in keeping with *in vitro* results (data not shown).

For many tumors, we can perform an additional type of calibration using cells dissociated from the intact tissue. The cells are incubated with EF3 or EF5 under conditions of severe hypoxia using the spinner vial apparatus described earlier. We do not have as much experience with this method, but preliminary data (not shown) suggest that maximal binding rates are the same as for cells directly from tissue culture.[65,82]

The cube-binding method can often be applied directly to normal tissues, but usually it is not possible to obtain cultures of dissociated cells to perform definitive tests for the oxygen dependence of binding. In this case, we need to assume an "average" oxygen dependence of binding[68] because it is not possible to control the pO_2 of a tissue piece. Using such approximations, we estimated the median liver pO_2 in mice to be about 20 mm of Hg.[68] This result was quite different from the former estimates of severe hypoxia using radioactive misonidazole,[57] but appears more consistent with known physiological principles. Similarly, we tested the hypothesis

[82] T. M. Busch, S. M. Hahn, S. M. Evans, and C. J. Koch, *Cancer Res.* **60,** 2636 (2000).

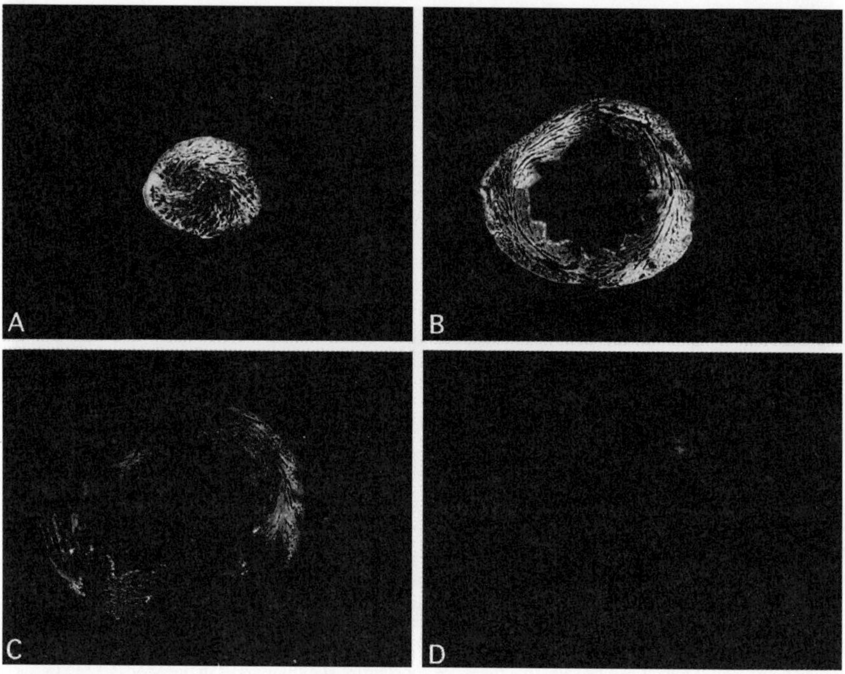

FIG. 11. Frozen sections of a murine myocardial infarct at 2-mm intervals from apex (A) to above the infarct (D). In our published studies, apoptosis of cardiac myocytes closely corresponded with regions of maximal EF5 binding [S. Bialik, D. L. Geenen, I. Sasson, R. Cheng, J. Horner, S. M. Evans, E. M. Lord, C. J. Koch, and R. N. Kitsis, *J. Clin. Invest.* **100,** 1363 (1997)].

that chondrocytes in the growth plate were severely hypoxic and found this to be incompatible with EF5-binding values. We are currently working on similar calibrations for ductus arteriosus tissue.[70] Some normal tissues are too heterogeneous or fragile to allow a simple interpretation of the cube-binding method. Thus, the maximal binding in the growth plate depended on the specific cell type.[83] The size of skeletal muscle cells has prevented their dissection without damage. In experimental model systems, it is sometimes possible to observe the maximum binding of infarcted tissue *in vivo* (Fig. 11). As with the cubes, diffusion of essential nutrients and drug may complicate such analysis, but under conditions of complete ischemia, we typically observe a penumbra of maximal binding downstream of the infarct.[71] A hallmark of severe hypoxia in normal tissue is apoptosis,[27,71] but complete ischemia again appears to cause rapid and uncontrolled cell death, as expected.

[83] I. M. Shapiro, K. D. Mansfield, S. M. Evans, E. M. Lord, and C. J. Koch, *Am. J. Physiol.* **272,** c1134 (1997).

Future Direction

EF5 has now been used in a broad variety of clinical and experimental situations to detect low pO_2s in normal and tumor tissue. Although it is somewhat difficult to predict its optimal utility, there is no question that noninvasive assays are required for many clinical situations. Noninvasive assays of metabolized compounds depend on MRI, single photon emission computed tomography (SPECT), or PET. In addition to the relative paucity of PET imaging centers, there is considerably greater flexibility in isotope characteristics for SPECT than PET. Additionally, the optimal choice of compound for hypoxia imaging is not clear at the present time; certainly our choice of ^{18}F and PET in a lipophilic compound runs completely counter to the present trend.[84,85] Lipophilic compounds tend to have longer serum half-lives (preventing rapid clearance of unmetabolized drug), and the half-life of most common PET isotopes is much shorter than the typical biological half-life of the drugs.[79] These characteristics tend to make detection of the metabolized signal difficult to distinguish from the nonmetabolized signal. However, drug pharmacology, drug biodistribution and stability *in vivo,* stability of the isotopic chemical linkage, and resolution of the imaging technique are also important considerations. The drug stability, biodistribution, and freedom from nonhypoxia-dependent metabolism of EF5 are all excellent. Similarly, in contrast to chelates or chemical linkages containing other halogens, the C–F bond strength prevents loss of isotope. Finally, the potential spatial resolution of PET far exceeds that of SPECT. Thus we are optimistic that detection of ^{18}F-labeled EF5 by PET holds significant promise for both cancer and other disease states (Fig. 12).[86]

Summary

We have established basic methods, using quantitative measures of EF5 binding, to estimate the actual pO_2 of cells and tissues. In situations where the tissue can be dissociated into single cells, or for cell cultures, we can measure the distribution of cellular binding rates using flow cytometry and these can be compared with cells treated under pO_2s controlled by the spinner vial or thin-film methods *in vitro*. The flow cytometer is calibrated by staining V79 cells treated with EF5 under "standard" conditions.

For intact tissues treated with EF5 *in vivo,* we need to correct for possible variations in drug exposure (AUC). Frozen sections are stained for EF5 binding and are analyzed by a sensitive (cooled) CCD camera with linear output vs fluorescence

[84] J. D. Chapman, E. L. Engelhardt, C. C. Stobbe, R. F. Schneider, and G. E. Hanks, *Radiother. Oncol.* **46,** 229 (1998).

[85] L. L. Wiebe and D. Stypinski, *Q. J. Nucl. Med.* **40,** 270 (1996).

[86] S. M. Evans, A. V. Kachur, C.-Y. Shiue, R. Hustinx, W. T. Jenkins, G. G. Shiue, J. S. Karp, A. Alavi, E. M. Lord, W. R. Dolbier, Jr., and C. J. Koch, *J. Nucl. Med.* **41,** 327 (1999).

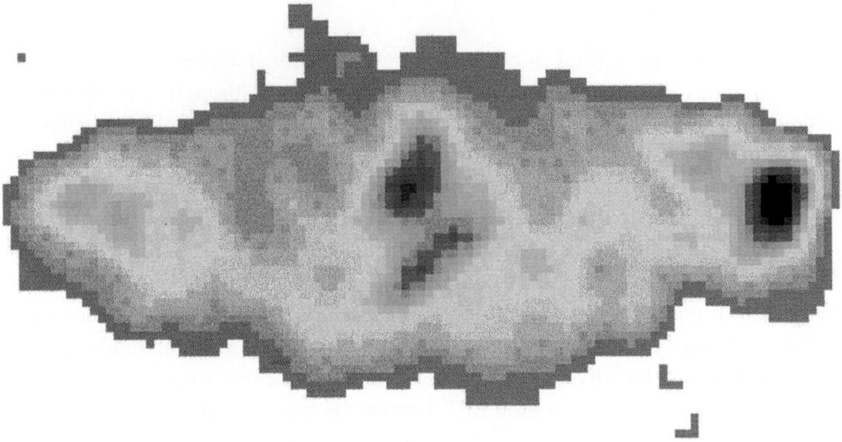

FIG. 12. First PET image of ^{18}F-labeled EF5 administered 2 hr prior to the image acquisition. Experiments were performed as described in S. M. Evans *et al., J. Nucl. Med.* **41,** 327 (1999). A cross-sectional view through the center of a Morris 7777 hepatoma shows prominent binding compared with same or opposite leg muscle (tumor : muscle ratio ~2.5). The high binding region at the midpoint indicates the route of renal excretion.

input. The camera has very consistent sensitivity, but the entire optical system, including the camera, can be calibrated by an absolute fluorescence standard (dye in hemocytometer). This system can also be used to measure the fluorescence of the flow cytometer standards, providing a direct link between the two assays. We can measure the maximum binding rate using the tissue cube method, but need to assume an "average" oxygen dependence of binding for intact tissues. The best-fit approximation for existing data is an inverse relationship between binding and pO_2, with binding decreasing 50-fold between 0.1 and 10% oxygen.

Using these methods, we routinely estimate the minimum pO_2 (maximum binding) in experimental rodent and human tumors. In normal tissue models, an excellent correlation is found between near-maximal binding (severe hypoxia) and apoptosis (heart infarct and ductus arteriosus). Some normal tissues (e.g., skeletal muscle) are refractory to both cellular disaggregation and cube calibration methods. To extend the tissue imaging measurements to a complete two- or three-dimensional analysis of the distribution of tissue pO_2s requires a substantial additional investment of imaging methods, which are currently being implemented.

Acknowledgments

At the beginning of the EF5 project (1988), 2-nitroimidazole development was rapidly waning, and it was generally agreed that monoclonal antibodies could not be made against their bioreductive adducts. Every chemist contacted acknowledged that a multiply fluorinated side chain could not be made in radioactive form using ^{18}F. Therefore, much of the research summarized in this chapter was unfunded, if not unfundable. The author thanks Dr. Edith Lord (University of Rochester) for making the antibodies that could not be made, Dr. Gillies McKenna (Chairman, Radiation Oncology, University of Pennsylvania) for supporting the unfundable research, and Drs. Bill Dolbier, Alex Kachur, and Kirsten Skov for performing the impossible chemistry. Finally, great thanks are due Dr. Sydney Evans for proving that it was indeed possible to predict radiation response in individual animal tumors. Dr. Evans and I wish, in turn, to thank the many investigators at the NCI (particularly Drs. Ed Sausville, Percy Ivy, Joe Covey, and Rao Vishnuvajjala) for their help in getting EF5 to the clinic. We also thank our many colleagues who have developed dozens of normal and tumor tissue models in which hypoxia has been assessed using EF5.

[2] Detection of Oxygen-Sensing Properties of Mitochondria

By NAVDEEP S. CHANDEL

Introduction

Oxygen is required for the survival and energetic needs of mammalian cells. Accordingly, mammalian cells have developed multiple adaptive strategies at both the organ and the cellular level to protect against the consequences of decreases in oxygen tension (*hypoxia,* $PO_2 < 20$ Torr). Adaptive responses to hypoxia at the organismal level include pulmonary vasoconstriction (to maintain pulmonary gas exchange), carotid body chemotransduction (to stimulate ventilation), and stimulating red blood cell production (to increase blood oxygen carrying capacity).[1–3] Cellular and molecular adaptive responses to hypoxia include the activation of various transcription factors, such as *hypoxia inducible factor 1* (HIF-1), and signaling pathways, such as MAP kinases.[4] Although much progress has been achieved in determining the signaling pathways, transcription factors, and genes that are activated during hypoxia, the mechanisms by which cells detect decreases in oxygen tension and initiate signaling events remain unknown. Currently, there are two redox-dependent models to explain oxygen sensing in mammalian cells.[5] One model assumes that an oxygen sensor located in the cytosol detects decreases in oxygen tension by decreasing the generation of reactive oxygen species (ROS).[6]

[1] E. D. Michelakis, S. L. Archer, and E. K. Weir, *Physiol. Res.* **44,** 361 (1995).
[2] N. R. Prabhakar, *J. Appl. Physiol.* **88,** 2287 (2000).
[3] B. L. Ebert and H. F. Bunn, *Blood* **94,** 1864 (1999).
[4] G. L. Semenza, *Annu. Rev. Cell Dev. Biol.* **15,** 551 (1999).
[5] G. L. Semenza, *Cell* **98,** 281 (1999).
[6] H. F. Bunn and R. O. Poyton, *Physiol. Rev.* **76,** 839 (1996).

0076-6879/01 $35.00

In an alternative model, we have proposed that mitochondria detect decreases in oxygen concentration and respond by *increasing* the generation of ROS, which are required and sufficient to activate HIF-1.[7,8] This chapter describes two techniques described to detect the role of mitochondria in regulating the hypoxic generation of ROS and activation of HIF-1. In the first method, the generation of ROS and the activation of HIF-1 are examined in cells depleted of mtDNA (ρ° cells). The second method examines the generation of ROS and the activation of HIF-1 in cells treated with mitochondrial inhibitors. Accompanying these methods are protocols for detecting nuclear HIF-1α protein levels by immunoblotting and detecting intracellular ROS by the oxidation of 2′,7′-dichlorofluorescin (DCFH). In theory, the same principles and methods can be applied in examining a different hypoxic response, such as activation of kinases or other transcription factors.

Detecting HIF-1α Protein Levels

Background

At the cellular level, one of the most crucial responses to hypoxia is the activation of the transcription factor HIF-1. HIF-1 is a heterodimer of two basic helix loop–helix/PAS proteins: HIF-1α and the aryl hydrocarbon nuclear translocator (ARNT or HIF-1β). The significance of HIF-1 in hypoxic transcriptional regulation has been demonstrated by a marked decrease in the induction of mRNA expression of vascular endothelial growth factor (VEGF) and glycolytic enzymes during hypoxia in HIF-1α- and ARNT-deficient murine embryonic stem cells.[9,10] Both HIF-1α and ARNT mRNAs are constitutively expressed under normoxic and hypoxic conditions, suggesting that functional activity of the HIF-1α/ARNT complex is regulated by a posttranscriptional mechanism(s). ARNT protein levels are constitutively expressed and are not significantly affected by oxygen, whereas the HIF-1α protein is present only in hypoxic cells. The HIF-1α protein is rapidly degraded under normoxic conditions by the ubiquitin–proteasome system.[11] Hypoxia significantly prolongs HIF-1α protein half-life, thereby allowing its accumulation and dimerization with ARNT. Thus, experimentally, functional HIF-1α/ARNT complex formation can be determined by assessing nuclear the HIF-1α protein by immunoblotting methods. Aside from physiological hypoxia, the iron chelator desferrioxamine (DFO) and divalent cations such as cobalt under normal oxygen conditions can be used to activate HIF-1.

[7] N. S. Chandel, E. Maltepe, E. Goldwasser, C. E. Mathieu, M. C. Simon, and P. T. Schumacker, *Proc. Natl. Acad. Sci. U.S.A.* **95,** 11715 (1998).

[8] F. H. Agani, P. Pichiule, J. C. Chavez, and J. C. LaManna, *J. Biol. Chem.* **275,** 35863 (2000).

[9] E. Maltepe, J. V. Schmidt, D. Baunoch, C. A. Bradfield, and M. C. Simon, *Nature* **386,** 403 (1997).

[10] N. V. Iyer, L. E. Kotch, F. Agani, S. W. Leung, E. Laughner, R. H. Wenger, M. Gassmann, J. D. Gearhart, A. M. Lawler, A. Y. Yu, and G. L. Semenza, *Genes Dev.* **12,** 149 (1998).

[11] L. E. Huang, J. Gu, M. Schau, and H. F. Bunn, *Proc. Natl. Acad. Sci. U.S.A.* **95,** 7987 (1998).

Materials

- Phosphate-buffered saline (PBS) (GIBCO, Gaithersburg, MD)
- Buffer A (cytosolic extract): 10 m*M* HEPES–KOH, 1.5 m*M* $MgCl_2$, 10 m*M* KCl, 50 μg/ml phenylmethylsulfonyl fluoride (PMSF), 1 m*M* dithiothreitol (DTT), and cocktail of protease inhibitors; sterilize by filtration and store at 4°
- Buffer C (nuclear extract): 20 m*M* HEPES–KOH, 1.5 m*M* $MgCl_2$, 420 m*M* KCl, 25% (v/v) glycerol, 0.2 m*M* EDTA, 50 μg/ml PMSF, 1 m*M* DTT, and a cocktail of protease inhibitors; sterilize by filtration and store at 4°
- Cocktail of protease inhibitors (Roche Molecular Biochemicals/Boehringer Mannheim, Indianapolis, IN)
- Bio-Rad protein assay (Bio-Rad, Hercules, CA)
- Sample loading buffer (2X): 125 m*M* Tris base (pH 6.8), 4% (w/v) sodium dodecyl sulfate (SDS), 20% (v/v) glycerol, 200 m*M* DTT, 0.02% (w/v) bromphenol blue; store in aliquots at −20°
- Running buffer: 25 m*M* Tris base (pH 8.3), 192 m*M* glycine, 0.1% (w/v) SDS
- Transfer buffer: 25 m*M* Tris base (pH 8.3), 192 m*M* glycine, 20% (v/v) methanol
- Ponceau S solution: 0.1% (w/v) Ponceau S in 5% (v/v) acetic acid
- Tris-buffered saline (TBS): 100 m*M* Tris base (pH 7.5), 0.9% (w/v) NaCl
- Blocking solution: 5% (w/v) nonfat milk powder, 0.1% (v/v) Tween 20 in TBS; prepare fresh each time
- Wash buffer: 0.1% (v/v) Tween 20 in TBS
- Antihypoxia-inducible factor 1α monoclonal antibody (clone H1α67) (Novus Biologicals, Littleton, CO)
- Sheep anti-mouse Ig horseradish peroxidase antibody (Amersham, Buckinghamshire, England)
- Hybond-ECL nitrocellulose (0.45 μm) (Amersham)
- ECL Western blotting detection reagents (Amersham)
- Hyperfilm ECL (Amersham)

Equipment

- Minielectrophoresis and semidry transfer unit (Bio-Rad)
- INVIVO O_2 hypoxic workstation (Ruskin Technologies, Leeds, UK)

Methods

Nuclear Extract Preparation. Prior to experimentation, 100-mm petri dishes containing 10 ml of cell culture media are placed either in a cell culture incubator (21% O_2, 5% CO_2, 74% N_2) or in the hypoxic workstation (1.5% O_2, 5% CO_2, 93.5% N_2) for 4 hr. Subsequently, cells cultured on 100-mm petri dishes at approximately 70% confluence are incubated in 10 ml of either hypoxic or normoxic equilibrated media for 4 hr. Afterward, media are removed and adherent cells are scraped off the culture dish in 10 ml of PBS. The cells are centrifuged at 500*g* for 5 min at 4°, and the resulting pellet is resuspended in 1 ml cold buffer A. Incubate the cell/buffer A suspension for a minimum of 10 min on ice, gently flicking the tubes every 2–3 min. The cell suspension is then transferred to a 1.5-ml Eppendorf tube and centrifuged at 12,000*g* for 1 min. Aspirate and discard the buffer A supernatant from the resulting nuclear pellet and resuspend in 25–100 μl cold buffer C (25 μl of buffer C per 1×10^6 cells). Incubate the nuclear suspension for 15 min on ice, gently flicking the tubes every 2–3 min. The suspension is then centrifuged at 12,000*g* for 1 min, and the resulting supernatants are saved as nuclear extracts at −70° until they can be analyzed by gel electrophoresis.

Immunoblotting Using Antihypoxia-Inducible Factor 1α Antibody. After determining the protein concentration of the nuclear extracts (Bio-Rad protein assay), 20–40 μg of protein is mixed with 2× sample loading buffer and heated for 5 min at 85°. The samples are loaded onto a 7.5% SDS–polyacrylamide gel (minigel electrophoresis unit) and run at 200 V for approximately 40 min. After electrophoresis, the proteins are transferred and bound to a nitrocellulose membrane in transfer buffer at 25 V for 30 min (semidry transfer unit). To prevent nonspecific protein binding, after transfer the membrane is incubated in blocking solution for 1 hr at room temperature on a rocking plate. The blocking solution is then discarded and the membrane is incubated in 10 ml of monoclonal antihypoxia-inducible factor 1α antibody H1α67 (diluted 1 : 500 in blocking solution) on a rocking plate overnight at 4°. The next day the membrane is washed three times with 15 ml of blocking solution for 10 min. Then, 15 ml of anti-mouse horseradish peroxidase antibody (diluted 1 : 1000 in blocking solution) is added, and the membrane is incubated on a rocking plate for 1.5 hr at room temperature. Finally, the membrane is washed three times with 15 ml of wash buffer, and the protein–antibody complexes are visualized with chemiluminescence reagents on autoradiography film.

Note: To control for equal loading, after transfer the membrane can be stained with Ponceau S solution for 5 min on a rocking plate and then rinsed with ddH_2O until clear protein bands can be seen. The stained membrane is further rinsed with ddH_2O until most of the Ponceau S solution is gone and is then incubated in blocking solution and treated as described earlier.

Detecting Intracellular ROS Levels

Background

Currently there are many different qualitative and quantitative methods to determine oxidant generation within cells. The high reactivity and relative instability of ROS make them extremely difficult to detect or measure in biological systems. The assessments of ROS generation have largely been made by indirect measurement from the interaction of ROS with various "detector" molecules that are oxidatively modified to elicit luminescent or fluorescent signals. A variety of methods detect only extracellular ROS and are thus reflective of ROS generators located on the plasma membrane. Because ROS are likely to be important intracellular signaling molecules, we have employed the oxidation of the nonfluorescent substrate DCFH to a fluorescent product 2′,7′-dichlorofluorescein (DCF) as a marker of intracellular oxidant generation in cells.[12] Cells are incubated with dichlorofluorescin diacetate (DCFH-DA), an esterified form of DCFH that readily crosses cell membranes and then deacetylates to be trapped intracellularly as DCFH. The oxidation of DCFH is more sensitive for the detection of hydrogen peroxide than it is for the detection of superoxide. The dye is also sensitive to reactive nitrogen species (RNS) such as nitric oxide and peroxynitrite. During hypoxia, an increase in the oxidation of DCFH occurs that indicates an increase in oxidant production. Hydrogen peroxide scavengers suppress the oxidation of DCFH. In contrast, agents that scavenge superoxide, hydroxyl radicals, or RNS do not affect the hypoxia-induced oxidation of DCFH. These results indicate that the oxidation of DCFH during hypoxia is primarily due to the production of hydrogen peroxide.

Materials

2′,7′-Dichlorofluorescin diacetate (Molecular Probes, Eugene, OR)
Minimum essential media α (MEMα)

Equipment

INVIVO O_2 hypoxic workstation
96-well fluorescence spectrometer (Molecular Devices)

Methods

Prior to experimentation, 100-mm petri dishes containing 10 ml of MEMα supplemented with DCFH-DA (10 μM) are placed either in a cell culture incubator (21% O_2, 5% CO_2, 74% N_2) or in the hypoxic workstation (1.5% O_2, 5% CO_2, 93.5% N_2) for 4 hr. Subsequently, cells cultured on 60-mm petri dishes at approximately 50% confluence are incubated in 4 ml of either hypoxic- or normoxic-equilibrated MEMα containing DCFH-DA (10 μM) for 4 hr. Afterward, media are removed and cells are lysed with 1 ml of 0.1% Triton X-100 within the normoxic cell culture incubator or the hypoxic workstation. Lysed cells are then centrifuged to remove the debris, and a 200-μl aliquot of the lysed cells is measured using a fluorescence spectrometer at an excitation wavelength of 500 nm and an emission

[12] D. A. Bass, J. W. Parce, L. R. Dechatelet, P. Szejda, M. C. Seeds, and M. Thomas, *J. Immunol.* **130,** 1910 (1983).

wavelength of 530 nm. A 200-μl aliquot of 0.1% Triton X-100 is used a blank. Data are normalized to values obtained from normoxic-untreated samples.

Note: Regardless of the media that cells are cultured in, MEMα works best for the DCFH oxidation assay.

Method 1: Use of ρ° Cells to Examine Hypoxic Accumulation of HIF-1α Protein in the Nucleus

Background

To investigate whether respiratory-competent mitochondria are required for the hypoxic activation of HIF-1, the accumulation of HIF-1α protein levels in the nucleus is examined in ρ° cells derived from human cell lines. Because ρ° cell lines have depleted mtDNA, they do not contain an active electron transport chain and are respiratory incompetent. The mtDNA genome encodes 13 polypeptides, including the critical catalytic subunits for complex III (bcl complex) and complex IV (cytochrome *c* oxidase). The isolation of ρ° cells is based on the use of ethidium bromide to deplete cells of their mtDNA. In mammalian cells, ethidium bromide at low concentrations (50–400 ng/ml) results in inhibition of mtDNA replication, but has no effect on the replication of nuclear DNA.[13] Because mtDNA replication is inhibited, the initial number of mtDNA molecules present in the cells is decreased by a factor of two each time the cell divides. The survival of ρ° cells in culture is dependent on high glucose, pyruvate, and pyrimidines, such as uridine.

Materials

Human lung carcinoma A549 cell line or the human hepatoma Hep3B cell line (ATCC, Rockville, MD)
Dulbecco's modified essential medium (DMEM) with high glucose (4500 mg/liter)
Minimum essential medium α
Penicillin–streptomycin liquid (5000 units/ml and 5000 μg/ml, respectively, GIBCO)
Fetal bovine serum (FBS)
10 mg/ml solution of ethidium bromide (Sigma)
Uridine (Sigma)
Antimycin A (Sigma)
Rotenone (Sigma)
Desferrioxamine (Sigma)
Cytochrome *c* oxidase subunit II primers for PCR:
  Forward primer A: 5′-GCT CAG GAA ATA GAA ACC GTC-3′

[13] M. P. King and G. Attardi, *Methods Enzymol.* **264,** 304 (1996).

Reverse primer B: 5′-GGT TTG CTC CAC AGA TTT CAG-3′

PCR SuperMix (GIBCO)

Agarose, ultrapure DNA grade (Bio-Rad)

TBE buffer: 89 m*M* Tris base, 89 m*M* boric acid, and 2 m*M* EDTA (pH 8.3)

Methods

Wild-type A549 cells and wild-type Hep3B cells are grown in DMEM and MEMα, respectively, on either 100-mm petri dishes to detect HIF-1 protein levels or 60-mm petri dishes to detect ROS. Both media are supplemented with 10% FBS, penicillin–streptomycin (1%). To generate ρ°-A549 cells or ρ°-Hep3B cells, wild-type A549 cells or Hep3B cells are grown in filter-sterilized DMEM or MEMα containing 10% FBS, ethidium bromide (50 ng/ml), sodium pyruvate (1 m*M*), and uridine (100 μg/ml) for a minimum of 2–3 weeks. Because ethidium bromide can be a potential mutagen, we routinely make new ρ° cells for each set of experiments. Thus if a mutation has occurred in the genes involved in the oxygen-sensing pathway due to ethidium bromide treatment, the probability of mutating the same set of genes is reduced with different batches of ρ° cells. Prior to experimentation, the cells are treated for 2–3 days with mitochondrial electron transport chain inhibitors antimycin A (1 μg/ml) and rotenone (1 μg/ml) to eliminate any cells relying on a functional electron transport chain for survival. It is important to validate that 1 μg/ml is a sufficient dose of rotenone or antimycin A to induce cell death in wild-type cells. Furthermore, two independent tests are performed to validate whether cells are ρ° cells. First, the growth of ρ° cells is monitored in media with or without 100 μg/ml of uridine. If cells without uridine stop growing and begin to die while cells with uridine continue to grow, then these cells are likely to be ρ° cells. The second test to verify the status of ρ° cells is to employ a polymerase chain reaction (PCR)-based method to detect any residual mtDNA sequences in ρ° cells. Mitochondrial DNA from wild-type or ρ° cells is isolated by using the Qiagen blood and cell culture DNA minikit according to the manufacturer's protocol. The isolated DNA undergoes PCR reaction using cytochrome *c* oxidase subunit II primers. The PCR reaction is conducted using the GIBCO PCR SuperMix protocol. The PCR reaction undergoes a 5-min hold at 94° followed by 25 cycles of the following: 30 sec 94° $\rightarrow$ 30 sec 55° $\rightarrow$ 30 sec 72°. Subsequently, the reaction is held for 7 min at 72° followed by storage at 4°. The PCR product is run on a 1% agarose gel in TBE buffer.

Wild-type cells and ρ° cells are both exposed to normoxia (21% O_2), hypoxia (1.5% O_2), or DFO (100 μM) for 4 hr. Subsequently, nuclear accumulation of HIF-1α protein levels and DCFH oxidation are measured in both cell types as indicated in the protocols described earlier. Our experimental findings demonstrate that wild-type cells increase ROS levels and the accumulation of HIF-1α protein levels while ρ° cells fail to increase ROS levels or HIF-1α protein levels. In contrast, DFO induces the accumulation of HIF-1 protein levels in both wild-type or ρ° cells. Moreover, the oxidation of DCFH was not observed in either wild-type or ρ° cells treated with DFO. The ability of ρ° cells to accumulate HIF-1 protein levels in response to DFO indicates that only sensitivity to hypoxia is lost in ρ° cells and that most of the signaling elements responsible for HIF-1 accumulation are intact.

Method 2: Use of Mitochondrial Inhibitors to Examine Hypoxic Accumulation of HIF-1α Protein in the Nucleus

Background

The potential for mitochondrial complex III-generated ROS to mediate cell signaling has gained significant attention. Mitochondrial-generated ROS can account for ~1–2% of total O_2 consumption under reducing conditions.[14] The superoxide generated during the Q cycle within the mitochondrial electron complex III (bcl complex) is considered the main site of ROS generation within the electron transport chain (Fig. 1).[15,16] The Q cycle involves the transfer of two electrons to ubiquinone from complex I or complex II, resulting in the reduction of ubiquinone to ubiquinol. Subsequently, ubiquinol oxidation requires the donation of two electrons: the first electron transfer is to the iron–sulfur centers and cytochrome *cl*, resulting in the oxidation of ubiquinol to ubisemiquinone. This reaction is inhibited by myxothiazol. The second electron is transferred to cytochrome *b* and results in the oxidation of ubisemiquinone to ubiquinone. This step is inhibited by antimycin A. The oxidation of ubisemiquinone to ubiquinone is the main site of ROS generation during the Q cycle (Fig. 1). We propose that hypoxia increases superoxide generation during the Q cycle within complex III. The increase in mitochondrial-derived ROS is required for hypoxic stabilization of the HIF-1α protein. Mitochondrial inhibitors such as myxothiazol, rotenone (complex I inhibitor), or thenyltrifluoroacetone (TTFA, complex II inhibitor) that prevent the transfer of electrons to ubisemiquinone interfere with the production of ROS and the stabilization of HIF-1α protein levels during hypoxia. In contrast, antimycin A increases the lifetime of ubisemiquinone, resulting in the maintenance of the production of ROS and the stabilization of HIF-1α protein levels during hypoxia.

Materials

Human lung carcinoma A549 cell line or the human hepatoma Hep3B cell line (ATCC)

Dulbecco's modified essential medium with high glucose (4500 mg/liter)

[14] B. A. Freeman and J. D. Crapo, *Lab. Invest.* **47,** 412 (1982).

[15] J. F. Turrens, A. Alexandre, and A. L. Lehninger, *Arch. Biochem. Biophys.* **237,** 408 (1985).

[16] A. Boveris, N. Oshino, and B. Chance, *Biochem. J.* **128,** 617 (1972).

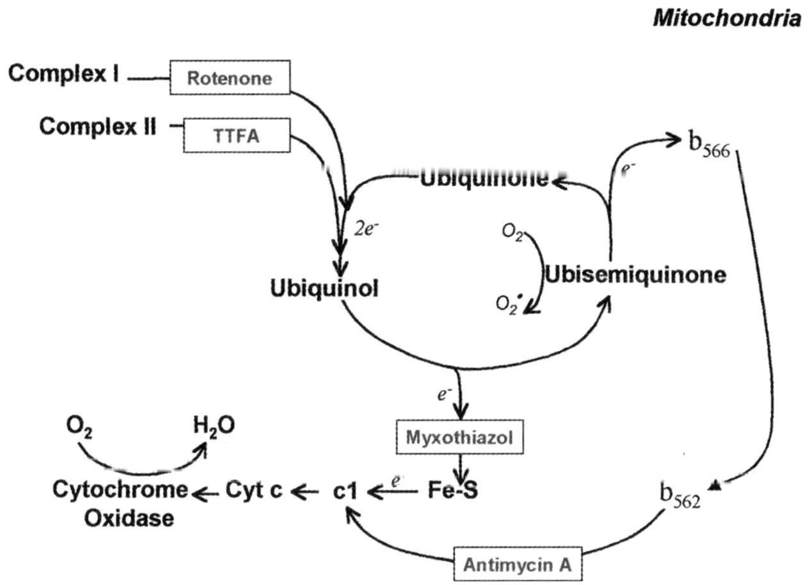

FIG. 1. In the electron transport chain, ubisemiquinone appears to be a major site of superoxide generation because of its predisposition for univalent electron transfer to O_2. Electron transport inhibition with rotenone (complex I) plus TTFA (complex II) or myxothiazol limits the generation of superoxide by attenuating the formation of ubisemiquinone, whereas antimycin A, an inhibitor of complex III, augments superoxide generation by increasing the lifetime of ubisemiquinone.

Minimum essential medium α
Penicillin–streptomycin liquid (5000 units/ml and 5000 μg/ml, respectively, GIBCO)
Fetal bovine serum
Rotenone (Sigma)
Antimycin A (Sigma)
Myxothiazol (Sigma)
Thenyltrifluoroacetone (TTFA, Sigma)

Methods

Wild-type A549 cells and wild-type Hep3B cells are grown in DMEM and MEMα, respectively, on either 100-mm petri dishes to detect HIF-1 protein levels or 60-mm petri dishes to detect ROS. Both media are supplemented with 10% FBS, penicillin–streptomycin (1%). Cells on 60-mm petri dishes placed in MEMα media supplemented with 10 μM DCFH-DA are exposed to normoxia or treated with rotenone (0, 0.1, or 1.0 μg/ml) under hypoxic conditions. The oxidation of DCFH as a marker of ROS production is assessed as described in the protocol described earlier. Rotenone should diminish the oxidation of DCFH in hypoxic cells down to levels observed in normoxic cells. If rotenone at the highest dose fails to completely diminish DCFH oxidation to normoxic control levels, then examine DCFH oxidation under hypoxic conditions in the presence of the highest dose of rotenone and TTFA (10 or 20 μM). This protocol ensures the inhibition of electron transport into complex III from both complexes I and II and prevents the formation of free radicals within the Q cycle. The attenuation of DCFH oxidation under hypoxic conditions can also be achieved using myxothiazol (100, 200, or 400 ng/ml). In contrast, DCFH oxidation in hypoxic cells can be maintained in the presence of antimycin A (1 μg/ml). The doses of rotenone/TTFA or myxothiazol most effective in decreasing the oxidation of DCFH under hypoxia should be used to test whether these doses are effective in abolishing the hypoxia-induced accumulation of HIF-1α protein levels in the nucleus. Alternatively, antimycin A should maintain HIF-1α protein levels in the nucleus during hypoxia. HIF-1α protein levels should be detected using the protocol described earlier.

Note: Rotenone/TTFA and myxothiazol abolish DCFH oxidation and the accumulation of the HIF-1 protein in the nucleus during hypoxia, whereas antimycin A preserves these responses. These results indicate that mitochondrial ATP during hypoxia is not required for the response because all of the mitochondrial inhibitors block electron transport and abolish oxidative phosphorylation.

Concluding Remarks

Currently there is much interest in deciphering the oxygen-sensing pathway. The role of the mitochondrion as an oxygen sensor has relied on the use of ρ° cells and mitochondrial inhibitors to perturb the oxygen-sensing pathway. As other techniques are developed to examine the role of mitochondria in oxygen sensing, it is important to examine both the status of ROS production and to correlate that with the hypoxic response such as HIF-1 activation. To date, in our experimental observations, any technique that diminishes the increase in mitochondrial ROS generation during hypoxia has prevented the activation of a variety of hypoxia-induced responses, such as HIF-1-mediated gene expression.

Acknowledgments

This work was supported by the National Institutes of Health Grant GM60472-03 to NSC. I thank David McClintock, Emin Maltepe, and Scott Budinger for their insightful comments on this chapter.

[3] Direct Detection of Singlet Oxygen via Its Phosphorescence from Cellular and Fungal Cultures

By PIOTR BILSKI, MARGARET E. DAUB, and COLIN F. CHIGNELL

Introduction

Singlet molecular oxygen ($^1O_2$), an excited form of oxygen, is an important oxidative transient showing much higher reactivity than ground state (triplet) oxygen. Singlet oxygen is involved in numerous biological processes, including cell signaling during oxidative stress. While $^1O_2$ is commonly formed via energy transfer from excited (generally triplet) sensitizer states to molecular oxygen during photosensitization processes[1-3] (which can occur in the eye and skin[4]), $^1O_2$ may also be generated during some enzymatic oxidation processes[5] and in many other reactions not requiring illumination.[6]

Singlet oxygen can be detected directly by its emission (phosphorescence), although this emission is extremely weak due to spin restrictions. In an aqueous environment, phosphorescence is additionally attenuated by resonance energy transfer from $^1O_2$ to the O–H vibrations in water molecules and by partial reabsorption of the emitted phosphorescence by water. In cells and tissues, however, the major problem is that the presence of biological quenchers/antioxidants may shorten $^1O_2$ lifetime to below or close to the limit of detection by spectral means.[7,8] Thus, although the spectroscopic detection of singlet oxygen in biological systems has been a challenge,[7,8] it is also an active issue because improved techniques always bring new information that would not be available otherwise.

The spectral detection of $^1O_2$ is superior to other detection methods, such as chemical trapping, quenching by specific quenchers, or enhancement by D_2O, because it can be virtually artifact free and gives instant results. Singlet oxygen is directly observed by its simple, but characteristic infrared phosphorescence spectrum with a maximum at 1268 nm. For spectral detection of $^1O_2$ in biological systems, we initially used cultured keratinocytes.[9] We then adapted the technique to

[1] F. Wilkinson, D. J. McGarvey, and A. F. Olea, *J. Phys. Chem.* **98,** 3762 (1994).
[2] A. P. Darmanyan and C. S. Foot, *J. Phys. Chem.* **97,** 5032 (1993).
[3] A. J. McLean and M. A. J. Rodgers, *J. Am. Chem. Soc.* **115,** 4786 (1993).
[4] T. Hikichi, N. Ueno, C. L. Trempe, and B. Chakrabari, *Biochem. Mol. Biol. Inter.* **33,** 497 (1994).
[5] R. D. Hall, W. Chamulitrat, H. Takahashi, C. F. Chignell, and R. P. Mason, *J. Biol. Chem.* **264,** 7900 (1989).
[6] Q. Jason Niu and G. D. Mendenhall, *J. Am. Chem. Soc.* **114,** 165 (1992).
[7] A. A. Gorman and M. A. J. Rodgers, *J. Photochem. Photobiol. B Biol.* **14,** 159 (1992) and references therein.
[8] D. Gal, *Biochem. Biophys. Res. Commun.* **202,** 10 (1994).
[9] P. Bilski, B. M. Kukielczak, and C. F. Chignell, *Photochem. Photobiol.* **68,** 675 (1998).

investigate the *in vivo* production of $^1O_2$ from cercosporin, a highly toxic photoactivated toxin produced by plant pathogenic fungi in the genus *Cercospora*.[10] These fungi use cercosporin as a means to parasitize plants via peroxidative damage of the host cell membranes presumably through the formation of $^1O_2$. Cercosporin generates both $^1O_2$ and radical forms of active oxygen *in vitro,* but direct production of $^1O_2$ by endogenously produced cercosporin *in vivo* had never been demonstrated. In addition, these fungi are immune to cercosporin, making them an ideal model system for understanding the basis of cellular resistance to photosensitizers and, if present, $^1O_2$ itself.

This chapter provides a description of the apparatus design and procedural issues associated with the spectral detection of $^1O_2$ in animal cell and fungal cultures. We limit our discussion to processes initiated by light, and not due to natural physiological processes. Our methods allow the detection of $^1O_2$ from living keratinocytes and fungal mycelial cells both treated with externally applied photosensitizers and endogenously produced cercosporin. Our results suggest that the *Cercospora* fungus appears to be a superior model in which it is easier than in keratinocytes to study cellular defense mechanisms against severe oxidative stress, as these fungi have evolved to tolerate high concentrations of $^1O_2$-generating compounds.

Singlet Oxygen Spectrophotometers

Singlet oxygen phosphorescence, like other luminescence, can be measured either in a time-resolved or a steady-state manner, each requiring a different technical approach. To obtain accurate kinetic information, speed is an essential factor in time-resolved $^1O_2$ detection, which is usually achieved at the expense of detection sensitivity. In contrast, high sensitivity and wavelength specificity are the main concerns in the steady-state technique, which is a better choice to measure weak $^1O_2$ phosphorescence and to distinguish it from other (artifactual) emissions in biological systems. While we employed a steady-state spectrophotometer for our main experiments, our time-resolved $^1O_2$ spectrophotometer was used to select and optimize cell and fungal growth media and to identify $^1O_2$ quenchers and antioxidants.

Extremely weak $^1O_2$ phosphorescence occurs in a near-infrared spectral region, where, until recently, photomultipliers were not available for its detection. Nitrogen-cooled germanium diodes have proved to be useful $^1O_2$ detectors. Detection sensitivity requires efficient transmission of $^1O_2$ phosphorescence from the production site (cuvette) to the detection element (crystal). It is particularly vital to optimize the optical system when limited irradiation times are employed to measure $^1O_2$ production *in vitro,* when sample degradation usually precludes multiple exposures necessary for signal averaging. Both of our singlet oxygen

[10] M. E. Daub and M. Ehrenshaft, *Annu. Rev. Phytopathol.* **38,** 461 (2000).

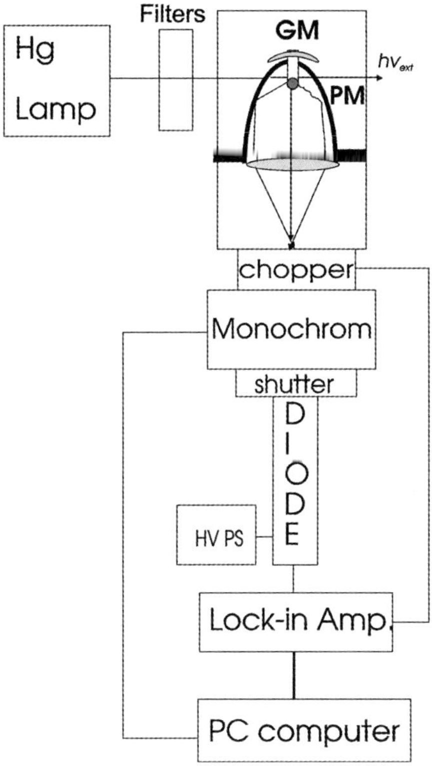

FIG. 1. Block diagram of the steady-state $^1O_2$ spectrometer. Hg, mercury lamp, excitation filters, gold concave mirror (GM), parabolic mirror (PM), light chopper, monochromator, germanium diode, high-voltage power supply (HV-PS), lock-in amplifier, and a PC computer.

spectrophotometers share a similar efficient optical design. Details of our pulse singlet oxygen spectrophotometer have been published elsewhere.[11]

A schematic of the steady-state $^1O_2$ spectrophotometer is shown in Fig. 1. Briefly, its optical configuration consists of a parabolic mirror (PM) whose focal point accommodates a hanging fluorescence (micro) cuvette. A small concave gold mirror (GM) is mounted at the rear of the PM to redirect $^1O_2$ emission toward the monochromator. Excitation light from a mercury lamp reaches the cuvette through one of two small vertical holes in the PM body: one to permit sample excitation and the other to allow nonabsorbed light to escape. The PM is capped with a concave lens prefocusing the emitted light into a cylindrical lens matching the focal length of the monochromator. Such an arrangement improves

[11] P. Bilski and C. F. Chignell, *J. Biochem. Biophys. Methods.* **33,** 73 (1996).

light collection, especially in cell suspensions that are opaque and scatter the emitted phosphorescence in all directions. Singlet oxygen phosphorescence is focused through these lenses onto a scanning monochromator grating blazed at 1200 nm that redirects the inspected wavelength to the exit of the monochromator where the $^1O_2$ detector is mounted. Thus, most of the phosphorescence is collected and directed to a germanium diode. The excitation path accommodates a heat-absorbing filter, while the excitation side contains a filter wheel with a shutter. All cuvette holder components are placed in a light-tight compartment. Other parts of our spectrophotometer are standard data acquisition and signal processing components, such as a light chopper, a digital lock-in amplifier, and a PC computer. The high sensitivity of our $^1O_2$ spectrometer allows us to conveniently record the $^1O_2$ spectrum in one 20-sec scan over the 1200- to 1350-nm range, usually without resorting to multiscan averaging. Measurements are fast and conserve the sample material, which has been especially practical in highly reactive systems such as cell and tissue samples.

Spectral Detection of Singlet Oxygen in Keratinocytes Stained with Rose Bengal

We chose keratinocytes as a model cell system because they are exposed in the skin to photooxidative processes that can be mediated by $^1O_2$. Keratinocytes adhere well to surfaces and grow naturally as a monolayer, thereby eliminating centrifugation and cell damage during sample preparation and handling. We used Rose Bengal (RB) to stain keratinocytes because RB is a very efficient $^1O_2$ photosensitizer. As evidenced by fluorescence, RB localizes exclusively in the form of ion pairs[12,13] inside keratinocyte hydrophobic regions such as the membrane. The higher viscosity of the membrane, as compared to the cellular aqueous milieu, may not only slow $^1O_2$ quenching, but $^1O_2$ molecules may also escape from the monolayer to the cell exterior (air) and not be subject to intracellular quenching agents. Thus, such a system appeared to be a promising choice to detect $^1O_2$ phosphorescence in living cells.

Sample Preparation

A detailed description of keratinocyte preparation has been published elsewhere[9]; we will focus only on the elements specific to our photochemical experiments. For tissue culture, we selected plastic coverslips from Nalge Nunc International (Naperville, IL), as they were easy to cut to the proper size and keratinocytes grew quickly on them. Keratinocytes, grown to cover at least 85% of the coverslip surface (about 2400 cells/cm^2), are treated topically with ca. 50 μl

[12] P. Bilski, R. Dabestani, and C. F. Chignell, *J. Photochem. Photobiol. A Chem.* **79,** 121 (1994).
[13] P. Bilski, R. Dabestani, and C. F. Chignell, *J. Phys. Chem.* **95,** 5784 (1991).

of Rose Bengal solution (2–30 μM) in phosphate-buffered saline (PBS). After incubation in the dark for 2–4 min, the coverslips are washed several times with a 10% dimethyl sulfoxide (DMSO)/H_2O buffer mixture to completely remove traces of nonabsorbed RB. If required, the washed samples are then placed in an incubator for various times to allow better penetration of RB molecules into the cell organelles. Such stained keratinocytes feature very faint but detectable RB absorption, and very clear RB fluorescence is observable using a fluorescence microscope. After incubation, each coverslip is dried gently with blotting paper before measurement.

For irradiation, coverslips with attached cells are fitted diagonally into a fluorescence cuvette (1-cm path length), which positions them at 45° to both excitation and emission beams inside the spectrophotometer. The same procedure is performed on coverslips without the keratinocytes and with unstained keratinocytes. All operations between staining and spectrum acquisition are performed in less than 2 min in dim room light, plus incubation time in the dark, if any.

Problems and Artifacts

Detection of extremely weak emission in nonhomogeneous (biological) systems creates many experimental problems and is inherently prone to artifacts. There are many other types of weak emissions that do not belong to $^1O_2$. The most frequent and dominant one is the tail fluorescence/phosphorescence from a photosensitizer. These emissions are more difficult to discern in the pulse detection technique, where on laser excitation, even impurities trapped in glass and quartz may emit weak radiation with a lifetime similar to singlet oxygen.[11] Recording a complete spectrum makes it easier to determine whether the recorded emission belongs to $^1O_2$ phosphorescence. Thus, the main advantage of the steady-state technique is that, if properly executed, it provides the best specificity for $^1O_2$ emission and superior sensitivity.

Additional artifacts were associated with $^1O_2$ phosphorescence originating from unexpected sources. For example, in our experiments, we found that cutting polystyrene produced edges that were usually rough and cracked and trapped RB molecules that could not be removed easily by washing, which produced $^1O_2$ during irradiation. Also, when polystyrene absorbs light, $^1O_2$ and its subsequent phosphoresence are produced inside the polymer. These problems disqualified polystyrene as a support for cellular cultures in photochemical experiments. RB adsorbed on plastic coverslips, and even on the edges of quartz plates, also produced a clear spectrum of $^1O_2$. Appropriate experimental protocols and tedious washing control experiments were the only remedy to distinguish genuine $^1O_2$ phosphorescence from cells from artifactual $^1O_2$ emissions associated with photosensitizer adsorption to supporting materials.

We also attempted to use cell suspensions to detect $^1O_2$ phosphorescence. For experiments in cell suspension, keratinocytes are trypsinized, incubated for a few minutes with RB (30 μM) solution, centrifuged, and washed several times until the solution is colorless. They are resuspended either in water or D_2O buffer, transferred to a fluorescence cuvette, and used for experiments. However, residual release of RB from keratinocyte debris is a major problem, which could not be fully eliminated, as washing and centrifugation would always further damage some cells. Even though we could detect a clear $^1O_2$ signal from the keratinocyte suspension, we considered it too susceptible to artifacts. Thus our experiments used coverslips with intact cell monolayers.

Another problem associated with cell systems is that growth media contain $^1O_2$ quenchers that hinder phosphorescence detection. While we largely eliminated this problem for keratinocytes due to their ability to attach to plastic coverslips, fungal mycelium does not adhere well and could not be washed as thoroughly. Thus some residual medium was present. To solve this problem, we used the pulse technique to first test various fungal growth media to identify one that had limited quenching activity (vide infra). Following these optimization studies, we satisfactorily resolved most of the issues mentioned earlier and were able to detect $^1O_2$ from both keratinocytes and *Cercospora* fungi.

Detection of Singlet Oxygen Phosphorescence

A coverslip with RB-stained keratinocytes is placed in a 1-cm quartz cuvette suspended inside the parabolic mirror cell compartment and irradiated using a Hg 250-W lamp through a water filter and a combination of 475-nm cutoff and infrared-blocking filters. After one 20-sec scan, most cells do not take up Trypan blue, indicating minimal direct cellular damage. A clear $^1O_2$ spectrum is observed when the RB-stained keratinocyte coverslip is simply irradiated in an empty cuvette in air (Fig. 2). The spectral intensity decreases when the coverslip is immersed in D_2O during irradiation and is almost completely quenched when the coverslip is irradiated while immersed in water, showing that an aqueous environment outside the keratinocytes severely compromises $^1O_2$ detection. Control spectra from unstained keratinocytes or from a bare coverslip do not show any $^1O_2$ emission. These results suggest that most of the detectable $^1O_2$ phosphorescence originates from those $^1O_2$ molecules that escape from the cell through its membrane into D_2O or into the air, where $^1O_2$ has a longer lifetime. However, we cannot rule out some contribution(s) from other organelles, such as nuclei, that were also stained with RB and provide a hydrophobic environment. Because the RB was inside the keratinocytes, $^1O_2$ must also be produced inside the keratinocytes. Thus, independent of its precise origin, we have detected a clear steady-state emission spectrum of $^1O_2$ phosphorescence from HaCaT keratinocytes stained with RB, which confirms that $^1O_2$ can indeed contribute to oxidative stress in photosensitizer-exposed skin. Information gathered from keratinocytes in a monolayer can be used as a basis for *in vivo* investigations of other photosensitizers and biological organisms, and

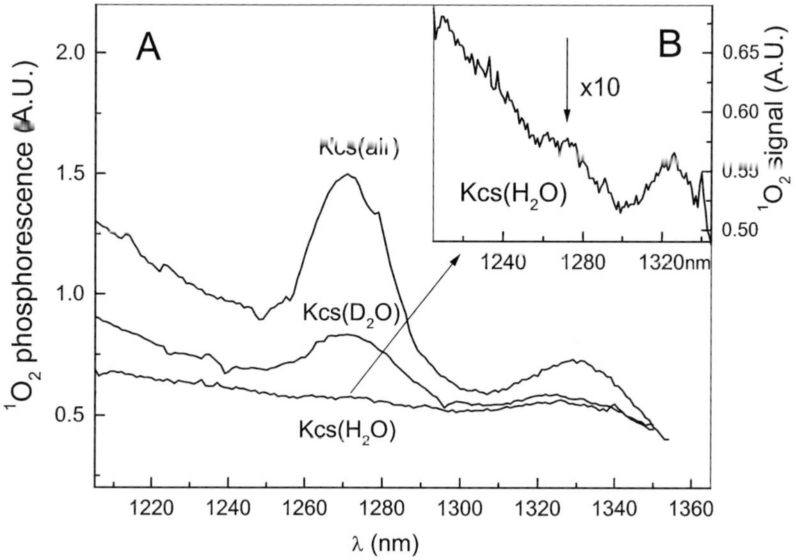

FIG. 2. Single scan spectra of $^1O_2$ phosphorescence from RB-stained keratinocytes from a coverslip irradiated at 45° to the keratinocyte layer in air, Kcs(air); from a coverslip immersed in D_2O during spectrum acquisition, Kcs(D_2O); and from a coverslip immersed in H_2O during spectrum acquisition, Kcs(H_2O). This spectrum is magnified 10 times in B, which shows residual $^1O_2$ production that mostly originated from the cell interior. Control spectra from keratinocytes not stained with RB and from the bare coverslips incubated with RB did not show any $^1O_2$ production and are omitted for presentation clarity. Keratinocytes were incubated with RB (30 μM) in PBS buffer for about 3 min. $^1O_2$ spectra were obtained using a 475-nm cutoff filter combination for RB excitation. Each spectrum was acquired during one 20-sec scan and represents nonfiltered raw data that were only shifted for presentation clarity. Figure adapted from data in P. Bilski, B. M. Kukielczak, and C. F. Chignell, *Photochem. Photobiol.* **68,** 675 (1998).

these studies were applied to investigations of *Cercospora* fungi and the phototoxin cercosporin.

Spectral Detection of $^1O_2$ in *Cercospora* Fungus Containing Endogenously Produced Cercosporin

Fungi in the genus *Cercospora* produce a potent perylenequinone photosensitizer, cercosporin (Fig. 3), a phototoxin that plays an important role in the ability of these pathogens to parasitize plants.[10] On activation by light, cercosporin generates $^1O_2$ with a high quantum yield (ca. 0.8).[14,15] Consistent with its high $^1O_2$ production, cercosporin is toxic to plants, mice, bacteria, and many fungi.

[14] D. C. Dobrowolski and C. S. Foote, *Angewante Chem.* **95,** 729 (1983).

[15] G. B. Leisman and M. E. Daub, *Photochem. Photobiol.* **55,** 373 (1992).

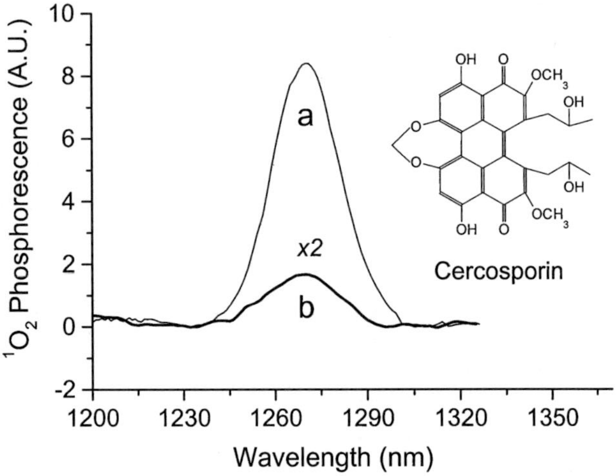

FIG. 3. Production of $^1O_2$ by cercosporin *in vivo* in contact with fungal hyphae. The $^1O_2$ phosphorescence spectrum from 25 μM cercosporin (a) in D_2O solution and from endogenous cercosporin in *Cercospora nicotianae* mycelium grown in malt medium (b) is shown. The *in vivo* signal was acquired under the same conditions as in Fig. 2 and is magnified two times for presentation clarity. Figure adapted from data in M. E. Daub, M. Y. Li, P. Bilski, and C. F. Chignell, *Photochem. Photobiol.* **71,** 135 (2000).

In plant cells, cercosporin induces the peroxidation of cell membranes, leading to leakage of nutrients from the cells into the leaf intercellular spaces where the pathogen colonizes. Thus cercosporin is used by these pathogens to obtain nutrients needed for growth and sporulation in the host and is hypothesized to be responsible for the wide host range found in this group of pathogens. Although cercosporin appears to have almost universal toxicity to cells of most organisms, *Cercospora* fungi and other plant pathogenic fungi that produce similar perylenequinone toxins can tolerate millimolar concentrations of these potent toxins in culture. *Cercospora* species also tolerate high concentrations of other potent $^1O_2$-generating photosensitizers with no measurable decrease in growth.

The ability of *Cercospora* species to grow in the presence of cercosporin and other generators of $^1O_2$ has made these organisms a model for the study of cellular photosensitizer resistance. Studies utilizing confocal microscopy, along with ones on the solvent dependence of $^1O_2$ generation from reduced cercosporin, have suggested that the ability of the fungus to biochemically reduce cercosporin and to localize it within an aqueous environment in the cell (such as cytoplasm of the hyphal cells) is a major mechanism of detoxification.[16] Other protection mechanisms also contribute to this remarkable immunity against oxidative stress in *Cercospora*. The most notable one appears to be a new antioxidant role for

[16] M. E. Daub, M. Y. Li, P. Bilski, and C. F. Chignell, *Photochem. Photobiol.* **71,** 135 (2000).

pyridoxine (vitamin B_6).[17,18] Pyridoxine and its vitamers, pyridoxal, pyridoxal phosphate, and pyridoxamine, were shown to have strong $^1O_2$-quenching activity, and fungal mutants lacking a gene encoding pyridoxine biosynthetic production become sensitive to cercosporin phototoxicity.

Studies on the resistance of *Cercospora* species to cercosporin and other photosensitizers are shedding light on cellular resistance to photosensitizers. Whether such studies also provide insight into mechanisms used by cells to tolerate $^1O_2$ itself depends on whether $^1O_2$ is actually produced in cells. Although the production of $^1O_2$ by cercosporin has been confirmed *in vitro,* the production of $^1O_2$ by cercosporin in fungal cells *in vivo* had never been definitively shown. It is possible that fungal resistance is mainly due to the prevention of $^1O_2$ production *in vivo* rather than protection against $^1O_2$ itself. Thus we applied our model detection of $^1O_2$ in keratinocytes to assay for *in vivo* $^1O_2$ production in cultures of *Cercospora nicotianae* producing cercosporin.

Preparation of Fungal Samples

Our assay of *in vivo* $^1O_2$ production in fungi utilized *C. nicotianae* strain ATCC#18366 and procedures that were developed and used to detect $^1O_2$ phosphorescence from keratinocytes. By testing different inoculation and culturing conditions, we developed a culture system that allowed for growth of a thin mycelial mat coating the plastic coverslips that we utilized in the keratinocyte studies. The fungus is grown on solid malt medium[19] at 24° in the dark. Inoculum for the *in vivo* $^1O_2$ emission studies is produced by harvesting 4- to 5-mm agar plugs from the margins of 2-week-old mycelial cultures and grinding in water (6 plugs/5 ml) with glass beads. The sterile plastic coverslips are placed in 60-mm petri dishes to which 4 ml of medium (minimal medium, malt broth, or potato dextrose broth)[19] is added. The medium is inoculated with 100 μl of the macerated mycelial inoculum, and cultures are incubated in a 24° growth chamber and illuminated with constant fluorescent light (30 $\mu E\ m^{-2}\ sec^{-1}$) for 4–5 days until a confluent mat of hyphae has grown across the surface. Illumination is necessary to induce production of cercosporin by the fungus.[10] Dark-incubated cultures are unable to synthesize cercosporin, and these cultures are used as controls and for staining with exogenous RB or cercosporin. Mycelial mats on coverslips from nonproducing cultures treated with RB or cercosporin are washed with a 10% DMSO aqueous solution to remove external RB or cercosporin. Samples obtained from cultures endogenously producing cercosporin are often used unwashed. Even though the fungus adheres

[17] P. Bilski, M. Y. Li, M. Ehrenshaft, M. E. Daub, and C. F. Chignell, *Photochem. Photobiol.* **71,** 129 (2000).

[18] M. Ehrenshaft, P. Bilski, M. Y. Li, C. F. Chignell, and M. E. Daub, *Proc. Natl. Acad. Sci. U.S.A.* **96,** 9374 (1999).

[19] A. E. Jenns, M. E. Daub, and R. G. Upchurch, *Phytopathology* **79,** 213 (1989).

weakly to the coverslips (as compared to keratinocytes), it is possible to stain and wash them to obtain decent $^1O_2$ signals.

$^1O_2$ Quenchers in the Media

An essential component contributing to the successful detection of $^1O_2$ in biological samples is information on the quenching activity of nutrient media and

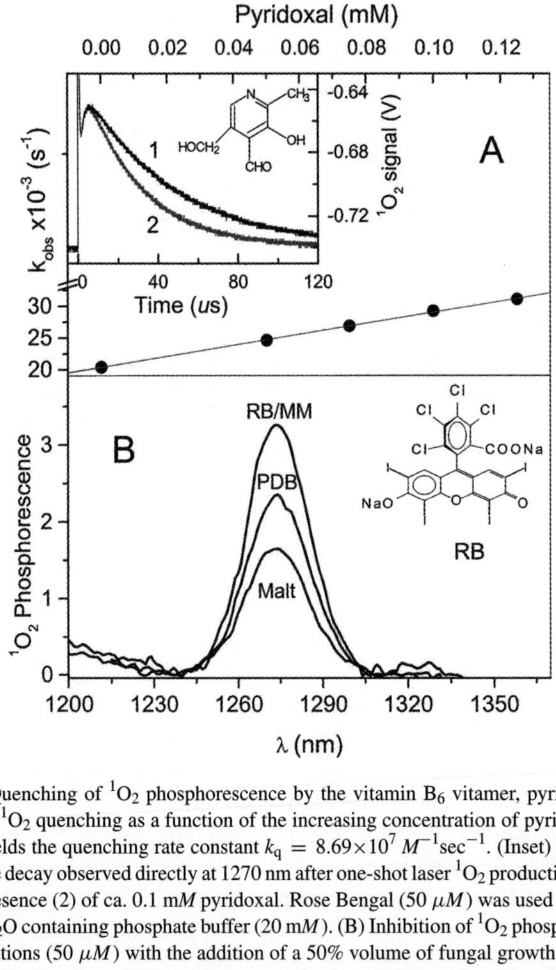

FIG. 4. (A) Quenching of $^1O_2$ phosphorescence by the vitamin B_6 vitamer, pyridoxal. Observed rate constant for $^1O_2$ quenching as a function of the increasing concentration of pyridoxial at pD 7.4; the line slope yields the quenching rate constant $k_q = 8.69 \times 10^7\ M^{-1}\text{sec}^{-1}$. (Inset) Examples of $^1O_2$ phosphorescence decay observed directly at 1270 nm after one-shot laser $^1O_2$ production in the absence (1) and in the presence (2) of ca. 0.1 m*M* pyridoxal. Rose Bengal (50 μM) was used to photosensitize $^1O_2$ in aerobic D_2O containing phosphate buffer (20 m*M*). (B) Inhibition of $^1O_2$ phosphorescence from aqueous RB solutions (50 μM) with the addition of a 50% volume of fungal growth media: RB/MM, RB alone and minimal medium; PDB, potato dextrose broth; and Malt, malt medium. RB was irradiated through a 550-nm interference filter, and $^1O_2$ spectra were normalized to the same number of absorbed excitation photons at this wavelength. The $^1O_2$ phosphorescence signal was decreased significantly in the presence of potato dextrose broth and malt medium, the two media supporting cercosporin production by the fungus. Figure adapted from data in M. E. Daub, M. Y. Li, P. Bilski, and C. F. Chignell, *Photochem. Photobiol.* **71,** 135 (2000) and P. Bilski, M. Y. Li, M. Ehrenshaft, M. E. Daub, and C. F. Chignell, *Photochem. Photobiol.* **71,** 129 (2000).

TABLE I
QUENCHING RATE CONSTANTS, k_q, FOR $^1O_2$ QUENCHING BY SELECTED BIOMOLECULES[a]

Quencher	Solvent/pH	k_q ($M^{-1}sec^{-1}$)	Source
Pyridoxine	D_2O, pD = 6.3	6.3×10^7	Ref. 17
Pyridoxal	D_2O, pD = 6.3	7.7×10^7	Ref. 17
Pyridoxamine	D_2O, pD = 6.3	8.9×10^7	Ref. 17
Pyridoxal 5-phosphate	D_2O, pD = 6.3	5×10^7	Ref. 17
Cysteine	D_2O, pD = 6	5.1×10^5	This work
Cysteine	D_2O, pD = 7.4	1.3×10^7	This work
Urocanic acid	D_2O, pD = 7.4	1.3×10^8	This work
Ascorbic acid	D_2O, pD = 6	1.8×10^8	This work
Uric acid	D_2O, pD = 8	3.6×10^8	This work
Imidazole	D_2O, pD = 6	1.2×10^7	This work

[a] Some of the listed compounds may not be explicitly present in the growth media; however, they contain representative quenching functionalities, such as the imidazole or SH moieties, which are very active in $^1O_2$ quenching.

selected nutrients. Therefore, we provide here abbreviated procedures on time-resolved methods for detecting $^1O_2$ quenching by selected nutrients and by whole media. To determine separate contributions of selected nutrients, samples containing the RB photosensitizer and the nutrient of interest are excited using a 532-nm laser pulse, and $^1O_2$ phosphorescence intensity is recorded as a function of time on a digital oscilloscope (Fig. 4A, inset). For known nutrient (pyridoxal) concentrations, the quenching rate constant is calculated (Fig. 4A) as described previously,[17] and the rate constant values for selected biomolecules are shown in Table I. It is worth mentioning that such a screening procedure led to the identification of a new role of vitamin B_6 in fungal biochemistry.[17,18]

For multi-ingredient medium samples in which many components may contribute to $^1O_2$ quenching, only the overall attenuation of $^1O_2$ lifetime could be measured in the time-resolved $^1O_2$ spectrophotometer. Therefore, for more illustrative visualization, we determined the effect of the fungal nutrient medium composition on the $^1O_2$ phosphorescence spectrum using RB as the photosensitizer. When the same volumes of malt and PDB medium were mixed with aqueous RB solutions (ca. 50 μM), the steady-state intensity of $^1O_2$ phosphorescence decreased significantly (Fig. 4B). In contrast, minimal medium did not appreciably inhibit $^1O_2$ phosphorescence. However, minimal medium does not support the production of cercosporin by the fungal cultures. Therefore, for the experiments on cultures endogenously producing cercosporin, malt and PDB media were used to grow the fungus, despite their high quenching activity. For the experiments involving cultures stained with RB, minimal medium was used.

In Vivo Phosphorescence of $^1O_2$ from Cultures of Cercospora nicotianae

To measure the $^1O_2$ phosphorescence spectrum from *C. nicotianae,* the coverslip covered with mycelium is fitted diagonally into a 1-cm^2 quartz cuvette and observed in the $^1O_2$ steady-state spectrophotometer, using a 418-nm cutoff filter combination for excitation. First we use fungal hyphae that are stained with RB, as described for the keratinocytes. When slides are treated with RB and then washed thoroughly, strong red RB fluorescence is observed by fluorescence microscopy in the fungal nucleus and membranes. A clear $^1O_2$ phosphorescence signal is obtained from these mycelial cultures,[16] confirming that it is possible to detect $^1O_2$ in the living fungus.

We next measured $^1O_2$ phosphorescence in undisturbed wild-type *C. nicotianae* mycelium actively producing cercosporin. During growth, most cercosporin is bioreduced inside the wild-type fungus, which is evidenced by its green fluorescence.[16] When reduced cercosporin is excreted from the mycelium, it becomes oxidized and exhibits a red fluorescence. Bioreduction of cercosporin makes it a very poor $^1O_2$ generator, especially in aqueous environments.[16] In contrast, oxidized cercosporin in an even stronger $^1O_2$ photosensitizer than RB.[16] Despite the inhibitory effect of numerous nutrients in malt medium, we detected a clear signal of $^1O_2$ phosphorescence from the fungal cultures producing endogenous cercosporin (Fig. 3, spectrum b). This result confirms that $^1O_2$ is generated from cercosporin *in vivo* and that *Cercospora* uses $^1O_2$ from this endogenous photosensitizer to kill plant leaves. Because reduced cercosporin present inside the hyphae is a poor $^1O_2$ photosensitizer, most $^1O_2$ phosphorescence we recorded is likely produced from excreted cercosporin adhering to the cell walls. While this suggests that the fungal cell must be heavily exposed to $^1O_2$ when producing cercosporin, the fungus itself is resistant to the (photo)toxicity of $^1O_2$. One component of this resistance is the antioxidant role of cellular pyridoxine in $^1O_2$ quenching. Other mechanisms by which these organisms can tolerate such an incredibly high level of oxidative stress are under investigation, following our key observation of the *in vivo* $^1O_2$ phosphorescence spectrum from fungal and cellular cultures.

Summary

In summary, we have developed methods to detect $^1O_2$ spectrally from viable keratinocytes and fungal mycelial cultures. Both time-resolved and steady-state spectrophotometers were used, providing complementary information on optimal conditions for the unambiguous detection of $^1O_2$ phosphorescence. By using our techniques, we were able to confirm that photosensitizers in contact with living cells indeed generate $^1O_2$. The model systems and the procedures described can be adopted for other studies involving $^1O_2$ and oxidative stress in living cells.

[4] Molecular Analysis of Mitogen-Activated Protein Kinase Signaling Pathways Induced by Reactive Oxygen Intermediates

By M. LIENHARD SCHMITZ, SUSANNE BACHER, and WULF DRÖGE

Introduction

Reactive oxygen intermediates (ROIs) have many different functions and are widely used as signal transmitting molecules in many different physiological processes, such as the control of protein kinase activity,[1] but kinase activity can also be modulated by changes in the intracellular thiol/disulfide redox state, which trigger the same redox responsive signaling proteins and pathways as those activated by hydrogen peroxide.[2–4] A prominent example of redox- and thiol/disulfide-regulated kinases is the family of mitogen-activated protein kinases (MAPKs), which translate extracellular signals into intracellular responses. The family of MAPKs consists of three groups: cJun N-terminal kinases (JNKs), p38 MAPKs, and extracellular signal-regulated kinases (ERKs). While the ERK kinases are at the heart of signaling pathways governing proliferation and differentiation,[5] JNKs and p38 kinases are induced by stress responses and cytokines, thus modulating inflammation, innate immunity, and cell death.[6] MAPKs are activated by dual phosphorylation at conserved threonine and tyrosine residues by MAPK kinases (MAPKKs), which in turn are substrates for MAPK kinase kinases (MAPKKKs). While the direct activators for each MAPK are relatively well defined, the highly complex and interconnected signaling pathways triggering the activation of MAPKKs are less completely understood. ROIs can affect all three MAPK signaling pathways, but the concentration and chemical structure of reactive oxygen species, the cell type, and the physiological context determine which MAPK is activated.[7]

Originally identified as an ultraviolet (UV)-responsive protein kinase phosphorylating the c-Jun transactivation domain,[8] there is ample evidence that elevated levels of ROIs are necessary for the activation of JNK. Treatment of cells with

[1] W. Dröge, *Physiol. Rev.*, in press (2001).
[2] D. Galter, S. Mihm, and W. Dröge, *Eur. J. Biochem.* **221,** 639 (1994).
[3] S. Kuge and N. Jones, *EMBO J.* **13,** 655 (1994).
[4] S. P. Hehner, R. Breitkreutz, G. Shubinsky, H. Unsoeld, K. Schulze-Osthoff, M. L. Schmitz, and W. Dröge, *J. Immunol.* **165,** 4319 (2000).
[5] W. Kölch, *Biochem. J.* **351,** 289 (2000).
[6] M. Rincon, R. A. Flavell, and R. A. Davis, *Free. Radic. Biol. Med.* **28,** 1328 (2000).
[7] V. Adler, Z. Yin, K. D. Tew, and Z. Ronai, *Oncogene* **18,** 6104 (1999).
[8] M. Hibi, A. Lin, T. Smeal, A. Minden, and M. Karin, *Genes Dev.* **7,** 2135 (1993).

0076-6879/02 $35.00

H_2O_2 strongly induces JNK activity. Conversely, antioxidant treatment antagonizes interleukin (IL)-1 and tumor necrosis factor (TNF) α-mediated activation of JNK.[9] These early findings can now be explained on a mechanistic basis by the constitutive association of JNK with glutathione *S*-transferase Pi (GSTp). H_2O_2 treatment or UV irradiation causes oligomerization of GSTp and dissociation of the GSTp–JNK complex, indicating that the monomeric form of GSTp mediates JNK inhibition. GSTp exerts its inhibitory effect only on JNK and is independent of the upstream kinases MEKK1 and MKK4,[10] but ROIs also trigger the activity of a kinase, which is located more upstream in the JNK signaling cascade: apoptosis signal-regulating kinase 1 (ASK1), a MAPKKK.[11] ASK1 becomes activated on association with the TNF receptor-associated factor 2 (TRAF2) protein, an adapter protein that couples tumor necrosis factor receptor signals to JNK, but ASK can also bind to thioredoxin (Trx), an endogenous inhibitor of ASK1 in non-stressed cells. Trx prevents the interaction between ASK1 and TRAF2, but this inhibitory effect is reversed by oxidants. Inhibition of ASK1 is lost on diminished expression of thioredoxin. Furthermore, ROIs induce the homooligomerization of ASK1, thus triggering its activity.[12] Activation by multimerization is also seen by coumermycin-induced dimerization of an ASK1-gyrase B fusion protein, which activates the kinase function of ASK1.[11] JNK activation is mediated by singlet oxygen, as ultraviolet A (UV-A)-induced activation of JNK still occurs in the presence of the hydroxyl radical scavengers mannitol or dimethyl sulfoxide, but can be inhibited by the singlet oxygen quenchers azide and imidazole.[13] Further experiments showed that intracellularly generated $^1O_2$ is mediating the activation of JNK and p38.

p38 is also activated by oxidative stress, as H_2O_2 is a powerful inducer of p38 in the perfused heart.[14] Generally, JNK and p38 show a strongly overlapping activation pattern by ROIs.[4,13,14,15,16] This might be taken as an indication that p38 and JNK share a redox-sensitive activator. One candidate is ASK1, which is known to trigger both pathways. The inhibitory effect of herbimycin A on redox-induced activation of JNK and p38 in T cells points to the involvement of protein tyrosine kinases (PTKs) for this activation pathway, at least in T lymphocytes.[4]

[9] Y. Y. C. Lo, J. M. S. Wong, and T. F. Cruz, *J. Biol. Chem.* **271,** 15703 (1996).

[10] V. Adler, Z. Yin, S. Y. Fuchs, M. Benezra, L. Rosario, K. D. Tew, M. R. Pincus, M. Sardana, C. J. Henderson, C. R. Wolf, R. J. Davis, and Z. Ronai, *EMBO J.* **18,** 1321 (1999).

[11] Y. Gotoh and J. A. Cooper, *J. Biol. Chem.* **273,** 17477 (1998).

[12] H. Liu, H. Nishitoh, H. Ichijo, and J. M. Kyriakis, *Mol. Cell. Biol.* **20,** 2198 (2000).

[13] L. O. Klotz, C. Pellieux, K. Briviba, C. Pierlot, J. M. Aubry, and H. Sies, *Eur. J. Biochem.* **260,** 917 (1999).

[14] A. Clerk, S. J. Fuller, A. Michael, and P. H. Sugden, *J. Biol. Chem.* **273,** 7228 (1998).

[15] J. Tao, J. S. Sanghera, S. L. Pelech, G. Wong, and J. G. Levy, *J. Biol. Chem.* **271,** 27107 (1996).

[16] L. O. Klotz, C. Fritsch, K. Briviba, N. Tsacmacidis, F. Schliess, and H. Sies, *Cancer Res.* **58,** 4297 (1998).

The activity of ERK-1 and ERK-2 (also called p44 and p42 MAPK) can be activated by ROIs. ERKs are generally less responsive to oxidative stress than JNK and p38 kinases, as the activation of ERK-1/2 requires the addition of H_2O_2 in millimolar concentrations. Evidence shows that the H_2O_2-induced Ras/Raf/ERK pathway employs the kinases Fyn and JAK2. H_2O_2-stimulated JAK2 activity is completely inhibited, and Shc tyrosine phosphorylation and Ras activation are strongly impaired in $Fyn^{-/-}$ cells, but not in $Src^{-/-}$ cells.[17] These experiments also point to the importance of early activated Src kinase Fyn for the initiation of redox signals.

Detection of Redox-Mediated MAP Kinase Activity Using Phospho-specific Antibodies

This chapter describes methods to measure MAPK phosphorylation and activity induced by changes in the thiol state or by ROIs. Depending on the cell type, approximately 1–20× 10^6 cells are stimulated. The stimulation time required to see optimal induction of the MAPK under investigation may vary—dependent on the oxidant—between 5 min and several hours. Cells are washed twice with cold TBS buffer (25 m*M* Tris–HCl, pH 7.4, 137 m*M* NaCl, 5 m*M* KCl, 0.7 m*M* $CaCl_2$, 0.1 m*M* $MgCl_2$). After centrifugation at 3000*g* in a cooled table centrifuge, the cell pellet is lysed by the addition of 60 μl NP-40 lysis buffer [50 m*M* Tris–HCl, pH 7.5, 150 m*M* NaCl, 1 m*M* phenylmethylsulfonyl fluoride (PMSF), 10 m*M* NaF, 0.5 m*M* sodium vanadate, leupeptine (10 μg/ml), aprotinin (10 μg/ml), 1% (v/v) NP-40, and 10% (v/v) glycerol] on ice for 20 min. NP-40 lysis buffer is freshly prepared by adding PMSF, NaF, sodium vanadate, leupeptine, and aprotinin just prior to use. During the incubation time, the lysate is carefully vortexed twice. Following centrifugation in a cooled centrifuge for 10 min at 14,000*g*, the supernatant is transferred in a new Eppendorf tube. Protein concentration is determined colorimetrically with the Bradford reagent.[18] Because the lysis buffer contains 1% (v/v) NP-40, the Bradford reagent is supplemented with 4% (w/v) SDS in order to prevent precipitation of the Folin reagent. Equal amounts of protein are mixed with 5× SDS sample buffer [250 m*M* Tris–HCl, pH 6.8, 10% (w/v) SDS, 40% (v/v) glycerol, and 15% (v/v) 2-mercaptoethanol] and heated for 5 min at 95°. The samples can now be frozen until use or used directly for denaturing SDS–PAGE on 12% polyacrylamide gels.[19] Samples typically containing 50 μg of protein are separated on the SDS gel and are subsequently blotted onto a polyvinylidene difluoride (PVDF) Immobilon-P membrane (Millipore). The stacking gel is removed from the SDS gel and a PVDF membrane and Whatman 3MM paper is cut to fit

[17] J. Abe and B. C. Berk, *J. Biol. Chem.* **274,** 21003 (1999).
[18] M. M. Bradford, *Anal. Biochem.* **72,** 248 (1970).
[19] U. K. Laemmli, *Nature* **227,** 680 (1970).

the size of the gel. The PVDF membrane is activated by incubation for 15 sec in methanol. Subsequently the PVDF membrane is incubated for 5 min together with the 3MM paper in semidry transfer buffer [25 m*M* Tris, 40 m*M* glycine, and 10% (v/v) methanol, pH 9.4]. The Western blot is assembled by placing wet 3MM papers in a semidry blotting apparatus (Bio-Rad Laboratories), followed by the PVDF membrane, the SDS gel, and another layer of 3MM paper. Air bubbles prevent the transfer of proteins to the membrane and are therefore removed carefully by rolling a pipette over the surface of each layer in the stack prior to closing the gel chamber. A current is applied in order to allow the transfer of proteins to the PVDF membrane. In order to avoid excessive heat development, the current should not exceed 1.5 mA/cm^2 of membrane. After 1 hr, the transfer of proteins to the membrane is complete.

The membrane is then blocked for 1 hr in TBST buffer [25 m*M* Tris–HCl, pH 7.4, 137 m*M* NaCl, 5 m*M* KCl, 0.7 m*M* $CaCl_2$, 0.1 m*M* $MgCl_2$, 0.05% (v/v) Tween 20] containing 5% (w/v) skim milk powder. Blocking can alternatively be done with bovine serum albumin (BSA). The membrane is then incubated for several hours in a small volume of TBST containing 0.5% (w/v) skim milk powder in TBST and the phospho-specific antibody of interest in an appropriate dilution according to the suggestion of the manufacturer. These antibodies are available from various commercial suppliers (e.g., New England Biolabs, Upstate Biotechnology, Santa Cruz Biotechnology, Zymed Laboratories Inc.) and recognize the phosphorylated and thus activated forms of MAPKs: α-phospho-p38 antibodies recognize p38 phosphorylated at Thr180/Tyr182, α-phospho- JNK antibodies bind to Thr183/Tyr185-phosphorylated JNK, and α-phospho-ERK-1/2 antibodies are specific for ERK-1/2 phosphorylated at Thr202/Tyr204. The PVDF membrane is incubated with the phospho-specific antibodies for 2 hr. Because many phospho-specific antibodies are generated in rabbits, the quality and the signal/background ratio of the antibodies may display some variability. Subsequently, the diluted primary antibody is removed and can be stored at 4° after the addition of sodium azide [final concentration 0.02% (w/v)]. This solution is stable for several months and can be reused more than 50 times. The PVDF membrane is washed extensively in TBST buffer and incubated for another hour in TBST/0.5% (w/v) skim milk powder containing a 1 : 3000 dilution of an appropriate secondary antibody coupled to horseradish peroxidase. After washing the membrane at least five times in TBST, the immunoreactive bands are visualized by enhanced chemiluminescence according to the instructions of the manufacturers (e.g., Amersham Pharmacia Biotech, NEN Life Science Products). An example for the analysis of redox-induced MAPK activation using phospho-specific antibodies is shown in Fig. 1.[20]

The membrane can be reprobed after stripping for α-JNK, α-p38, or α-ERK antibodies, which serve as gel-loading and protein controls. The membrane is stripped

[20] O. W. Griffith and A. Meister, *J. Biol. Chem.* **254,** 7558 (1979).

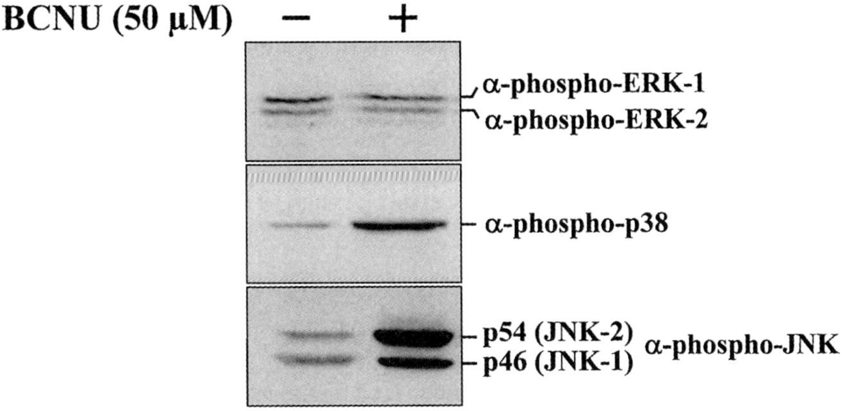

FIG. 1. Jurkat T cells were incubated for 90 min with 50 μM 1,3-bis(2-chloroethyl)-1-nitrosourea (BCNU), an inhibitor of glutathione reductase [O. W. Griffith and A. Meister, *J. Biol. Chem.* **254,** 7558 (1979)] that induces oxidative stress. Whole cell extracts were analyzed by Western blotting for phosphorylated and activated forms of endogenous ERK-1/2, p38, and JNK-1/2.

by incubation for 30 min at 55° in stripping buffer [100 mM 2-mercaptoethanol, 2% (w/v) SDS, 62.5 mM Tris, pH 6.7]. In most cases, the use of stripped PVDF membranes for another round of incubation with phospho-specific antibodies is not recommended. A used membrane can be reprobed with phospho-specific antibodies by washing the membrane in TBST, followed by the inactivation of horseradish peroxidase for 15 min in TBST containing 0.5% (w/v) sodium azide. The membrane can then be incubated with primary and secondary antibodies as described earlier. It will be important to reprobe the blot for the phosphorylation of a protein having a molecular weight clearly distinguishable from the phosphoprotein previously detected on the membrane.

Detection of Oxidant-Induced MAP Kinase Activity Using Immune Complex Kinase Assays

While phospho-specific antibodies determine the phosphorylation of the MAPKs, immune complex kinase assays determine the activity of MAPKs. Following oxidant treatment, 2–20 × 10^6 cells are washed twice with cold TBS buffer and harvested by centrifugation. The cell pellet is resuspended on addition of NP-40 lysis buffer and cell extracts are prepared as described earlier. In order to remove proteins binding nonspecifically to protein A/G Sepharose, the lysate is incubated for 1 hr with protein A/G Sepharose [20 μl of a 50 : 50 (v/v) slurry] and rotated on a spinning wheel at 4°. After centrifugation for 1 min at 13,000g, equal amounts of protein (approximately 100 μg) contained in the supernatant containing the endogenous kinase or an epitope-tagged kinase produced by transient transfection are immunoprecipitated. Because all MAPKs contain several family members,[21] it may be useful to immunoprecipitate with antibodies that recognize several family members or to mix antibodies recognizing various individual family members. One to 2 μg of antibodies recognizing the kinase is added with 25 μl protein A/G Sepharose, and binding of the interaction partners occurs for 4 hr at 4° in Eppendorf safe-lock test tubes rotating on a spinning wheel. The tubes are centrifugated for 1 min at 8000*g* and the supernatant is carefully aspirated off. The removal of the supernatant should be performed with caution, as the Sepharose beads form only relatively loose pellets. The pellets are washed three more times in NP-40 lysis buffer (1 ml buffer/tube). After removal of the buffer, the pellet is washed in kinase buffer [typically 20 m*M* HEPES–KOH, pH 7,4, 25 m*M* 2-glycerophosphate, 2 m*M* dithiothreitol (DTT), 20 m*M* $MgCl_2$). Dependent on the specific requirements of the kinase under investigation, other kinase buffers may also be used. The protein A/G Sepharose pellet is washed with 1 ml of kinase buffer, and 20% of the volume is transferred into new tubes and centrifuged. The immunoprecipitated kinases will be eluted with 1× SDS loading buffer and used later as loading controls. After a final centrifugation for 1 min at 8000*g*, the supernatant is completely removed and the tubes are kept on ice. Subsequently, 20 μl of the kinase reaction mix [20 μl kinase buffer containing 2 μg of substrate protein, 20 μM ATP, and 5 μCi [γ-^{32}P]ATP (5000 Ci/mmol)] is added to the beads, and the reaction is simultaneously started on incubation of the reactions at 30°. JNK activity is measured using GST-cJun (5-89),[8] and p38 activity is determined with GST-ATF-2 (1-109)[22] as a substrate. ERK activity is measured using the myelin basic protein (MBP) or GST-Elk (307-428)[23] substrate proteins. The kinase reaction is stopped by the addition of 8 μl 5× SDS sample buffer and denaturation at 95° for 5 min. Following centrifugation for 1 min at 13,000 rpm in a table centrifuge, the supernatant is loaded on a denaturing SDS gel. After electrophoresis, the gel is fixed for 15 min by shaking in fixation solution [10% (v/v) acetic acid, 30% (v/v) methanol], rinsed in water, and dried for 90 min at 65° on a vacuum dryer. The gel is subsequently autoradiographed or quantitatively evaluated by phosphorimaging.

Determination of Redox-Mediated Kinase Function by In-Gel Kinase Assays

In-gel kinase assays are employed when specific antibodies for the kinase are unavailable or when the identity of the kinase is unknown. This protocol involves

[21] H. Ichijo, *Oncogene* **18,** 6087 (1999).

[22] J. Raingeaud, S. Gupta, J. S. Rogers, M. Dickens, J. Han, R. J. Ulevitch, and R. J. Davis, *J. Biol. Chem.* **270,** 7420 (1995).

[23] H. Gille, M. Kortenjann, O. Thomae, C. Moomaw, C. Slaughter, M. H. Cobb, and P. E. Shaw, *EMBO J.* **14,** 951 (1995).

copolymerization of a substrate protein with SDS and polyacrylamide, followed by SDS gel electrophoresis of a protein sample containing the kinase(s) under investigation. Following several denaturing/renaturing steps, the copolymerized substrate protein is phosphorylated by the renatured kinase within the gel, and the phosphorylated bands are visualized by phosphorimaging or autoradiography. This method reveals the molecular weight of the protein kinase and allows the determination of its kinase activity. However, not all kinases can be detected with this method, as not all protein kinases can be successfully renatured and also multisubunit kinases remain inactive.

Cells are lysed in NP-40 lysis buffer as described previously, and 15 μg of total cell lysate is loaded together with ^{14}C-labeled molecular weight markers (Amersham Pharmacia Biotech) on two SDS gels. The separating gel contains either copolymerized GST protein (5 μg/ml) as a control or 5 μg/ml of GST fused to the substate protein adequate for the kinase under investigation (see earlier discussion). Gel electrophoresis is done with 10 V/cm, and the separating gel is washed twice for 10 min in buffer A [50 m*M* HEPES, pH 7.4, 10 m*M* DTT, 20% (v/v) isopropanol] and then equilibrated for 1 hr in buffer B (50 m*M* HEPES, pH 7.4, 10 m*M* DTT). Proteins are denatured for 2 $\times$ 30 min in buffer C (50 m*M* HEPES, pH 7.4, 10 m*M* DTT, 6 *M* guanidine–HCl). Renaturing of proteins is done overnight by adding sequentially buffer D [50 m*M* HEPES, pH 7.4, 10 m*M* DTT, 0.04% (v/v) Tween 20] containing 6, 3, 1.5, and 0.75 *M* urea. Subsequently, the gel is equilibrated 2$\times$ for 1 hr with reaction buffer (25 m*M* HEPES, pH 7.4, 10 m*M* DTT, 10 m*M* $MgCl_2$, 90 μM Na_3VO_4). The reaction buffer is decanted and replaced by a small volume of reaction buffer supplemented with 250 μCi [γ-^{32}P]ATP (5000 Ci/mmol) for 1 hr. The gel is placed on a shaker in order to prevent dehydration of the gel. After removal of the radioactivity, the gel is washed extensively in washing buffer [1% (w/v) sodium pyrophosphate and 5% trichloroacetic acid] until the radioactivity is at background level. The gel can then be dried and analyzed by autoradiography.

Analysis of Redox-Induced MAPK Signaling Pathways

ROIs can target MAPK cascades at different levels (see Introduction). In order to identify the redox-activated part of the MAPK signaling cascade, redox-activated MAPK activity is studied in the presence of either pathway-specific inhibitors or dominant-negative members of MAPK signaling cascades.[24] This experimental approach is based on the assumption that inhibition of components located upstream from the redox-activated kinase does not interfere with signaling events triggered by the activated kinase, as shown schematically in Fig. 2.

[24] R. Seger and E. G. Krebs, *FASEB J.* **9,** 726 (1995).

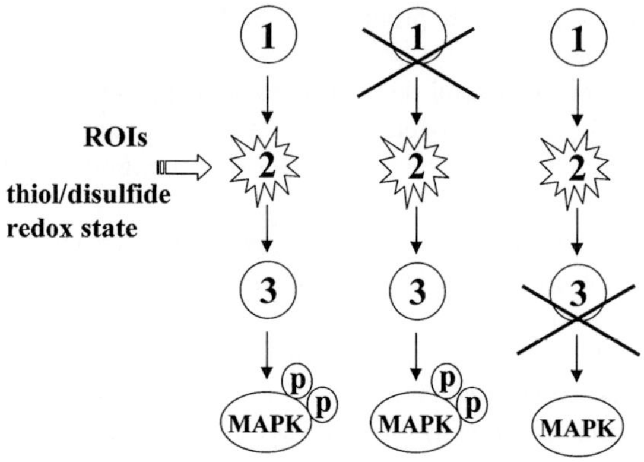

FIG. 2. Schematic representation of a directional signaling cascade leading to MAPK phosphorylation and activation. In the scenario depicted here, ROIs target signaling component 2 so that the expression of dominant-negative (represented by the X) component 1 remains without impact on MAPK activation induced by ROIs or changes in the thiol/disulfide state. In contrast, dominant-negative component 3 inhibits activation, revealing the level of redox regulation within the signaling cascade.

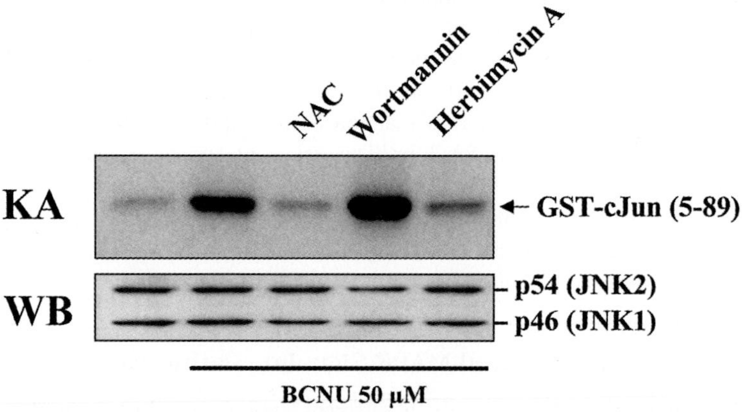

FIG. 3. Jurkat T cells were preincubated for 20 min with the antioxidant *N* acetyl cysteine (NAC, 20 m*M*), the PI3 kinase antagonist wortmannin (100 n*M*), or the PTK inhibitor herbimycin A (0.5 μg/ml) prior to the administration of 50 μM BCNU for 90 min. JNK-1 and JNK-2 were immunoprecipitated from cellular extracts and analyzed either by immune complex kinase assays (KA) using recombinant GST-c-Jun (5-89) as the substrate (upper) or by Western blotting (WB) for the expression of JNK-1 and JNK-2 (lower). This experiment shows that PI3 kinase-derived pathways are not important for BCNU-induced JNK activation, while it reveals the importance of PTKs and oxidants for this signaling cascade.

Cells are incubated with the inhibitory compound followed by the investigation of redox-induced MAPK activity either by Western blotting using phospho-specific antibodies or by kinase assays, as exemplified in the experiment displayed in Fig. 3.

It may be necessary to include positive controls in order to ensure that the inhibitory compound is active in the experimental setting chosen. The redox-regulated part of the MAPK signaling cascade can be determined on expression of dominant-negative versions of signaling molecules. Dominant-negative proteins usually lack their enzymatic function or their ability to be recruited to their site of action. However, the endogenous MAPK can only be inhibited when the dominant-negative signal transmitter is expressed in the majority of cells. If such a transfection efficiency cannot be achieved, the dominant-negative signaling molecule should be cotransfected with a tagged form of the MAPK, which allows isolation of the MAPK from transfected cells by immunoprecipitation. Please note that because the inhibitory activity of dominant-negative proteins is concentration dependent, it may be required to express the dominant-negative forms of proteins to rather high levels.

However, signaling cascades can also be investigated by recently developed phospho-specific antibodies recognizing the upstream components of the various MAPK pathways. These antibodies are available from various commercial suppliers (e.g., New England Biolabs) and can be used to determine the redox-regulated level within a regulatory network.

Acknowledgments

We thank Ingrid Fryson for critically reading the manuscript. Work from our laboratory is supported by grants from the EU (QLK3-CT-2000-00463), Deutsche Krebshilfe, Deutsche Forschungsgemeinschaft (Schm 1417/3-1), and Fonds der Chemischen Industrie.

[5] Detection of Intracellular Reactive Oxygen Species in Cultured Cells Using Fluorescent Probes

By ANNE NEGRE-SALVAYRE, NATHALIE AUGÉ, CARINE DUVAL, FANNY ROBBESYN, JEAN-CLAUDE THIERS, DANI NAZZAL, HERVÉ BENOIST, and ROBERT SALVAYRE

Introduction

Reactive oxygen species (ROS) are small chemically reactive molecules, traditionally considered undesirable by products of aerobic metabolism or environment, involved in oxidative damage of cell components, which are potentially responsible for cellular injury, cellular dysfunction, and cell death. ROS are thought to be implicated in the pathogenesis of various diseases, including cardiovascular diseases, neurodegenerative disorders, and inflammatory and infectious events. In addition to these deleterious effects, evidence suggests that ROS also function as second messengers in signal transduction and are involved in essential cellular functions, such as proliferation or death.[1–3] ROS, such as superoxide anion ($O_2^{\cdot -}$), hydroxyl radical ($HO^{\cdot}$), and hydrogen peroxide (H_2O_2), are ubiquitously generated through diverse enzymatic pathways, e.g., NADPH oxidases, mitochondrial electron transport chain, xanthine–xanthine oxidase, and oxygenases. Nitric oxide ($NO^{\cdot}$) is generated by NO synthases.[3] Cellular generation of ROS may occur during the oxidative metabolism of various substrates (arachidonic acid and polyunsaturated fatty acids, cholesterol, amino acids), in response to various agonists (cytokines, growth factors), or as a stress response to various exogenous agents, such as chemicals, drugs, radiation, or infectious agents.[2,3]

Regarding the increasing importance of ROS in cellular biology and pathophysiology, several fluorescent probes and new methods have been developed to monitor ROS generation in intact living cells.[4]

This chapter presents simple techniques utilizing fluorescent probes suitable for intracellular ROS detection. These probes do not discriminate directly between the various ROS produced by living cells nor indicate the cellular source of ROS. This can be accomplished by exploiting the differential reactivities of the probes and with the use of inhibitors and scavengers.[4,5]

[1] K. Irani, *Circ. Res.* **87,** 179 (2000).

[2] K. J. Davies, *IUBMB Life* **48,** 41 (1999).

[3] M. S. Wolin, *Arteriosc. Thromb. Vasc. Biol.* **20,** 1430 (2000).

[4] J. A. Royall and H. Ischiropoulos, *Arch. Biochem. Biophys.* **302,** 348 (1993).

[5] H. Ischiropoulos, A. Gow, S. R. Thom, N. W. Kooy, J. A. Royall, and J. Crow, *Methods Enzymol.* **301,** 367 (1999).

0076-6879/02 $35.00

Fluorescent Probes Used for Intracellular ROS Detection

General Principles

Fluorescent probes used for determining intracellular ROS in intact living cells must fulfill specific requirements[4,5]: (1) effective cell loading with the cell-permeant fluorogenic probe, (2) oxidation of the fluorogenic probe by cellular ROS (but not by spontaneous autooxidation); (3) stability and retention (low leakage) of the fluorescent probe inside the cell; (4) easy detection of the fluorescent probe with commonly used fluorescent microscopes; and (5) lack of self-toxicity.

The commonly used fluorescent probes, derived from fluorescein, rhodamine, ethidium, or other fluorescent dyes, are dihydrogenated into colorless and nonfluorescent dihydro derivatives (Fig. 1). To improve cell loading, the probes are rendered more cell permeant by esterification of their hydrophilic groups (OH and COOH). Inside the living cell, cellular esterases hydrolyze ester bonds, thereby liberating hydrophilic groups, and favoring intracellular sequestration of the probe. After entering the cell, fluorogenic (dihydrogenated, nonfluorescent probes) probes can be converted back into fluorescent compounds through intracellular oxidation (Fig. 2). The oxidized fluorescent probes are retained inside the cell, either in the cytoplasm (e.g., dichlorofluorescein derivatives) or in subcellular compartments (e.g., rhodamine 123 in mitochondria and ethidium in the nucleus), as visualized by fluorescence microscopy or determined by fluorometry. It may be noted that, after oxidation, the distinct fluorophores end up in distinct subcellular compartments (Fig. 2), but do not sense ROS in distinct compartments.

Fluorogenic Compounds

Dihydrofluorescein (or fluorescin, H_2F), dihydrodichlorofluorescein (or dichlorofluorescin, H_2DCF), chloromethyl-dihydrodichlorofluorescein (chloromethyl-H_2DCF), and carboxydihydrodichlorofluorescein (carboxy-H_2DCF) are largely used (Molecular Probes, Eugene, OR). The esterified derivatives dihydrofluorescein diacetate (or H_2FDA), dihydrodichlorofluorescein diacetate (or H_2DCFDA), carboxydihydrodichlorofluorescein diacetate diacetoxy-methylester (carboxy-H_2DCFDA diacetoxymethylester), or chloromethyl-dihydrodichlorofluorescein diacetate (chloromethyl H_2DCFDA) (see formula in Fig. 1) are loaded more effectively within the cells because they are more cell permeant and can enter intact living cells more easily. In the cytoplasm, the ester groups are hydrolyzed by cellular esterases, thereby releasing dihydro derivatives that may be oxidized back to the fluorescent parent dye (fluorescein derivatives) by intracellular ROS. The fluorescence resulting from oxidation can be observed by fluorescence microscopy[6,7]

[6] E. L. Greene, V. Velarde, and A. A. Jaffa, *Hypertension* **35,** 942 (2000).

[7] A. Görlach, R. P. Brandes, K. Nguyen, M. Amidi, F. Dehghani, and R. Busse, *Circ. Res.* **87,** 26 (2000).

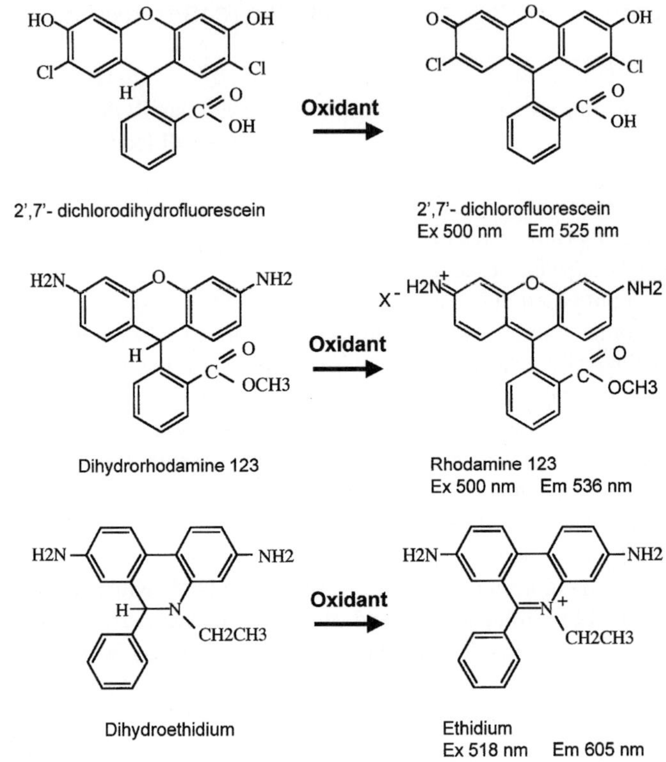

FIG. 1. Chemical structures of ROS-sensitive fluorogenic probes: 2′,7′-dichlorodihydrofluorescein diacetate (or dihydrodichlorofluorescein diacetate or H_2DCFDA), dihydrorhodamine 123 (or DHR), and dihydroethidium (or hydroethidine or HEt).

or measured by spectrofluorometry[4,8,9] or flow cytometry.[10,11] It may be noted that a nonnegligible leakage of the probe may render it difficult to monitor slow oxidation rates. This leakage may be reduced for more hydrophilic derivatives, such as carboxy-DCF or chloromethyl-DCF, which are negatively charged at physiological intracellular pH.

Dihydrorhodamine 123 (DHR) is the nonfluorescent dihydroderivative of rhodamine 123, a fluorescent, positively charged compound used as a mitochondria-specific dye. When reacting with an oxidant, dihydrohodamine 123 is oxidized

[8] P. A. Hyslop and L. A. Sklar, *Anal. Biochem.* **141,** 280 (1984).

[9] W. O. Carter, P. K. Narayanan, and J. P. Robinson, *J. Leukocyte Biol.* **55,** 253 (1994).

[10] S. L. Hempel, G. R. Buettner, Y. Q. O'Malley, D. A. Wessels, and D. M. Flaherty, *Free Radic Biol. Med.* **27,** 146 (1999).

[11] S. J. Vowells, S. Sekhsaria, H. L. Malech, M. Shalit, and T. A. Fleisher, *J. Immunol. Methods* **178,** 89 (1995).

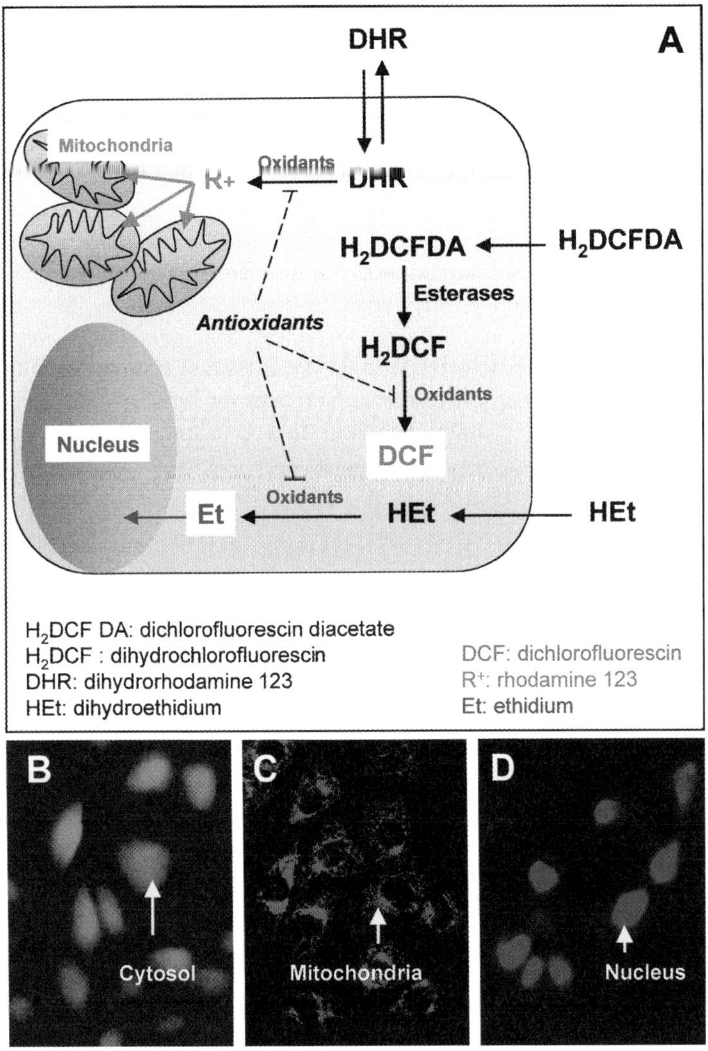

FIG. 2. Metabolism and subcellular localization of three probes commonly used for ROS determination. (A) Schemes of metabolism and subcellular localization of the probes. After cellular uptake, H_2DCFDA (dihydrodichlorofluorescein diacetate) is hydrolyzed to H_2DCF by cellular esterases and is converted by oxidation to DCF localized in the cytosol. After cellular uptake, DHR (dihydrorhodamine 123) is oxidized to rhodamine 123 (R+), which then enters the mitochondria. After cellular uptake, dihydroethidium (or HEt) is oxidized to Et, which then intercalates into nuclear DNA. Note that the distinct dihydrogenated fluorogenic compounds do not sense ROS in distinct compartments and end up, only after oxidation, in distinct subcellular compartments only (i.e., DCF in the cytosol, R+ in mitochondria, or Et into nuclear DNA). Fluorescence microscopy of cells loaded with H_2DCFDA (A), DHR (B), and HEt (C) and treated with H_2O_2 (100 μM) for 20 min.

back to rhodamine 123, which moves (after oxidation) to the inside, negative (−180 mV) mitochondrial environment, where it is sequestered.[12] DHR, like H_2DCF, is not supposed to detect $O_2^{\cdot -}$, but reacts with hydrogen peroxide produced in the cell.[4,13]

Dihydroethidium [or Hydroethidine (HEt); Prescott Labs] is a dihydro derivative of ethidium, which has blue fluorescence in the cytoplasm (dihydroethidium does not intercalate into DNA). Once dihydroethidium is oxidized into ethidium, it binds DNA and stains the nucleus a bright fluorescent red. This red fluorescence results from (at least) two parameters: the oxidation rate (which converts dihydroethidium into ethidium) and the subsequent intercalation of ethidium into DNA (which enhances its bright red fluorescence). This rise of fluorescence on intercalation increases the sensitivity of the methods and also suggests that the apparent fluorescence does not necessarily reflect the oxidation of the probe. This artifactual risk can be eliminated with the use of low HEt concentrations, as reported.[14,15]

Experimental Procedures

The oxidative activity of living cells can be determined by several technical approaches utililizing fluorogenic probes and detection by spectrofluorometry, confocal microscopy, or flow cytometry. The choice of the technique depends on the cell type. For instance, flow cytometry is largely used for cells in suspension, whereas confocal microscopy or spectrofluorometry has been used for adherent vascular cells.

This chapter gives examples of ROS production measured by spectrofluorometry in adherent vascular cells and by flow cytometry in blood phytohemagglutinin-stimulated lymphocytes.

Determination of ROS by Spectrofluorometry

Materials. Stock solution (50 mM) of H_2DCFDA and DHR (Molecular Probes) are prepared in ethanol and stored under nitrogen at −20° in the dark for up to 6 months. Working solutions of H_2DCFDA or DHR are diluted to 500 μM in phosphate-buffered saline (PBS) (10 μl of stock solution + 1 ml PBS). Fresh dilute solutions are prepared just before use (do not store). A slight opalescence of the working solution is sometimes observed, which has no effect on ROS determination.

[12] L. V. Johnson, M. L. Walsh, and J. B. Chen, *Proc. Natl. Sci. U.S.A.* **77,** 990 (1980).
[13] R. K. Emaus, R. Grunwald, and J. LeMasters, *Biochim. Biophys. Acta* **850,** 436 (1986).
[14] L. Benov, L. Sztejnberg, and J. Fridovich, *Free Radic. Biol. Med.* **25,** 826 (1998).
[15] S. C. Budd, R. F. Castillo, and D. L. Nichols, *FEBS Lett.* **415,** 21 (1997).

H_2O_2 (30%, Sigma, St. Louis, MO) is diluted to 10 m*M* in distilled water just before use. A stock solution of antimycin (Sigma) is prepared in ethanol (stock 5 m*M*). Other reagents are from Sigma.

Low-density lipoproteins are prepared from human blood serum by sequential ultracentrifugation,[16] and mild oxidation is induced by UV-C radiation in combination with Cu_2+/EDTA, as indicated previously.[17]

Spectrofluorometric Determination of ROS. The delay in the increase of ROS after stimulation by the agonist leads to the choice of the respective time for cell loading (with the fluorogenic probe) and treatment with the agonist. For example, H_2O_2 induces a relatively rapid increase in intracellular ROS (in this case, cell loading with the fluorogenic probe is performed before the addition of H_2O_2). In contrast, oxidized LDL evokes a 2-hr-delayed ROS generation (here cell loading is performed during the pulse with oxidized LDL, 30 min before fluorescence reading).

1. Cells are seeded in 6- or 24-multiwell plates (Nunc) at 10^5 cells/ml.
2. Cells are loaded in serum-free culture medium containing H_2DCFDA or DHR (5 μM, i.e., 10 μl of H_2DCFDA or DHR working solutions added to 1 ml of culture medium) for 30 min at 37°. Thirty-minute loading is generally enough to get a good intracellular distribution of the probe (this concentration is around 1.5–2.2 nmol/mg protein according to Royall and Ischiropoulos[4] and remains stable for at least 1 hr in the continuous presence of the probe). The continuous presence of the probe in the culture medium does not impair ROS determination. Alternatively, when ROS generation is relatively intense (fast), the probe can be removed from the culture medium, at the end of the loading period, just before starting the stimulation. However, it should be noted that diffusion of the probe out of the cell can occur relatively rapidly (in endothelial cells, 90% of the intracellular probe is lost after 1 hr of incubation after removing the probe from the culture medium[4]). The outward diffusion rate is highly variable depending on the cell type; for example, rhodamine 123 (sequestered in mitochondria) is only slightly diffusible in some cells, but is lost rapidly in other cell types.[18]
3. Addition of agonists: In cases of rapid increases of ROS (e.g., H_2O_2 or antimycin), cells are loaded with the fluorogenic probe just before agonist addition. For instance, cells are loaded as indicated earlier for 30 min and then H_2O_2

[16] R. I. Havel, H. A. Eder, and J. H. Braigon, *J. Clin. Invest.* **39,** 1345 (1955).

[17] A. Negre-Salvayre, N. Paillous, N. Dousset, J. Bascoul, and R. Salvayre, *Photochem. Photobiol.* **55,** 197 (1992).

[18] R. P. Haughland, "Handbook of Fluorescent Probes and Research Chemicals," 6th Ed. Molecular Probes, 1996.

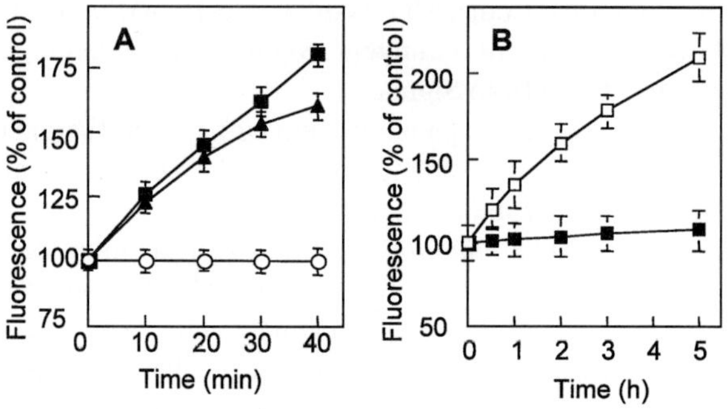

FIG. 3. Fluorometric determination of the time course of intracellular ROS production induced by H_2O_2 or antimycin (A) and by oxidized LDL (B) in ECV-304 cells. (A) Cells are loaded with DHR (5 μM) for 30 min and stimulated for various times with H_2O_2 (100 μM final concentration, ■) or antimycin (10 μM final concentration, ▲) (antimycin generates mitochondrial-free radical production through inhibition of complex III) or nonstimulated (control, ○). (B) Cells are incubated for various periods of time with 100 μg apoB/ml mildly UV-oxidized LDL (obtained by irradiation of human LDL by UV 234 nm 0.5 mW/cm^2 for 2 hr in PBS containing 0.3 mM EDTA and 5 μM CuSO4[17]) in the absence (□) or presence of antioxidant PDTC (pyrrolidinedithiocarbamate 100 μM, ■). ROS were determined at the indicated time, after cellular loading with of H_2DCFDA (5 μM final concentration, added 30 min before reading the fluorescence). Data are expressed as a percentage of the unstimulated control.

or antimycin is added to the culture medium and the fluorescence is read at the indicated time (Fig. 3A). In cases of slow increases of ROS (e.g., induced by oxidized LDL), cells are incubated with oxidized LDL and then cells are loaded with the fluorogenic probe 30 min before the time of fluorescence reading (Fig. 3B).

4. Read the fluorescence. The fluorescence of the oxidized probe is read either on cell extracts using a classical fluorometer or directly on intact cells using a fluorescence multiwell plate reader.

When using a classical cuvette spectrofluorometer, cells are washed three times with PBS, scraped off in 1 ml PBS, and sonicated. The fluorescence of the cell lysates is read using a JY3C Jobin-Yvon spectrofluorometer (excitation/emission wavelengths are 498/525 nm for dichlorofluorescein and 500/536 nm for rhodamine 123, respectively).

When using a fluorescence multiwell plate reader, cells are washed three times with phenol red free medium and the fluorescence (of intact live cells) is read directly. This method is much more rapid and permits the exclusion of possible artifacts due to scrapping and lysis of cells.

5. An aliquot of the cell lysate is used for protein determination.

6. Data are usually expressed as arbitrary fluorescence units per milligram cell protein or per well or as percentage of the unstimulated control. Quantitative evaluation of oxidized DHR can be performed using a standard prepared, as described by Royall and Ischiropoulos[4] by measuring the fluorescence intensity of 5 μM of DHR oxidized by increasing concentrations of H_2O_2 in the presence of 10 μM cytochrome *c*. The reaction is linear up to 100 μM H_2O_2.[4]

Comments. This technique is relatively simple, easy to use, and reliable, but H_2DCFDA or DHR is poorly specific because it can be oxidized by $O_2^{\cdot -}$ and H_2O_2 (in the presence of peroxidases, transition metal ions or cytochrome *c*) and also by $ONOO^-$ or HOCl.[5] Identification of ROS can be achieved by taking advantage of the differential susceptibility of the probes to various ROS and of the specificity of antioxidants, scavengers, or enzyme inhibitors[5] (Fig. 3B).

Fluorescence Microscopy of Intracellular ROS

Cells are prepared as indicated earlier (*Spectrofluorometric determination*). Cellular fluorescence of oxidized probes generated by intracellular ROS may be visualized by fluorescence microscopy. As discussed previously and as shown in Fig. 2, the fluorescent (oxidized) probes may be localized in different compartments: DCF exhibits a diffuse cytoplasmic fluorescence and rhodamine 123 and ethidium localize in the mitochondria and in the nucleus, respectively.

It may be noted that more precise subcellular localization and quantitative data may be obtained by analyzing the cellular fluorescence by confocal microscopy (according to the technique described by Bkaily *et al.*[19]). As documented by Hempel *et al.*,[9] HF-DA seems to be the most suitable probe for this kind of analysis because it generates brighter confocal images than the other probes.

Flow Cytometry

Cell Preparation and Culture. This technique is useful chiefly for cells grown in suspension. In the experiments reported here, as an example, intracellular ROS was determined in peripheral blood lymphocytes (PBL) treated with H_2O_2 (Fig. 4).

PBL is prepared from the buffy coat of healthy donors by centrifugation over Ficoll (GIBCO) and removal of adherent cells after incubation for 2 hr at 37° in 60-mm plastic petri dishes. PBL (5×10^5/ml), in RPMI 1640 supplemented with glutamine and 2% human delipidated serum, is activated by phytohemagglutinin P (PHA-P, 1 μg/ml; Sigma), as described previously.[20]

[19] G. Bkaily, D. Jacques, and P. Pothier, *Methods Enzymol.* **307,** 119 (1999).

[20] S. Caspar-Bauguil, S Saadawi, A. Negre-Salvayre, M. Thomsen, R. Salvayre, and H. Benoist, *Biochem. J.* **330,** 659 (1998).

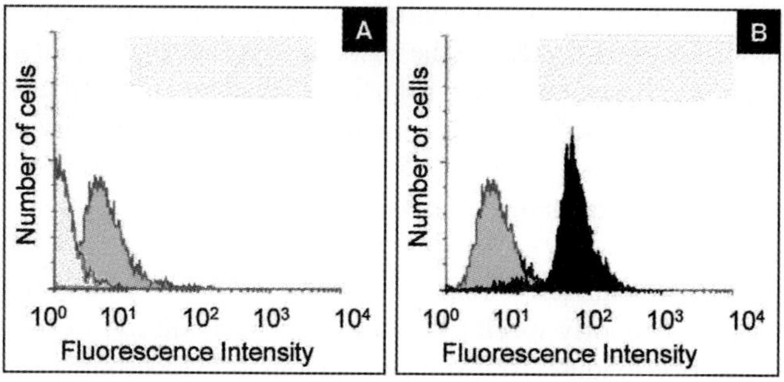

FIG. 4. Flow cytometry analysis of intracellular ROS in blood PHA-stimulated PBL incubated without (A) or with H_2O_2 (100 μM) (B). PBL are preloaded with 5 μM H_2DCFDA for 30 min and stimulated for 20 min with H_2O_2. Cells are analyzed by flow cytometry (green fluorescence, excitation 490 nm, emission 530 nm). Open histograms (A), control unloaded cells (autofluorescence); gray-filled histograms (A,B), control H_2DCFDA-loaded (H_2O_2 untreated) cells; and black-filled histograms (B), cells loaded with H_2DCFDA and treated with H_2O_2.

ROS Determination. For staining, add 10 μl of H_2DCFDA working solution (500 μM in PBS) to 1 ml of cell suspension (0.5×10^6/ml) (final concentration 5 μM) and incubate at 37° for 30 min. Then add H_2O_2 (100 μM) for an additional 20 min, wash three times in PBS (at 37°), and count immediately by flow cytometry[21] (Beckman Coulter XL4). Ten thousand events were collected and analyzed.

As shown in Fig. 4, unloaded and untreated cells exhibit a slight spontaneous autofluorescence; the fluorescence intensity of cells loaded with H_2DCFDA before the addition of H_2O_2 and after treatment with H_2O_2 (under the conditions indicated earlier was 10 and 150 times higher than the autofluorescence background, respectively (Figs. 4A and 4B).

Concluding Remarks

Various ROS and oxidants, among them H_2O_2, NOO^-, and to a lesser extent, $O_2^{\cdot -}$, are able to react with these fluorogenic probes. Intracellular oxidation of the probe is a complex process depending on a wide range of cellular parameters, such as the nature of ROS and the relative rate of ROS generation and degradation (the latter depending on cellular antioxidant systems). Moreover, intracellular oxidation and fluorescence of the probe are largely dependent on and influenced by the presence of cofactors, catalyzers, or enzymes (Fe^{2+}, ascorbate, peroxidases, Cu,ZnSOD, catalase, but not glutathione peroxidase).[5,9] The choice of a suitable

[21] G. Rothe and G. Valet, *Methods Enzymol.* **233,** 539 (1994).

probe for investigating ROS production will consequently depend on the nature of the stimulus, the cell type, and the technique of ROS determination.

Finally, the identity of ROS generated and involved in the oxidation of the probes may be investigated by using the most appropriate probe (the sensitivity and selectivity of the probes, H_2DCF or DHR, for a particular ROS are different) in combination with selective scavengers, antioxidants, or enzymes inhibitors.[5]

[6] Flow Cytometric Determination of Cytoplasmic Oxidants and Mitochondrial Membrane Potential in Neuronal Cells

By FRANCESC X. SUREDA, MERCÈ PALLÀS, and ANTONI CAMINS

Introduction

In recent years, new technologies have emerged that have given us greater knowledge of various aspects of research in the life sciences. Although many assays are routinely used to ascertain changes in intracellular processes, the study of these processes in the brain has been hampered by the number of different cell types and the difficulty in separating and maintaining the viability of a particular cell population. Even though a number of *in vitro* culture methodologies have been developed for neurons and glia, the purity of the cell preparation and other factors make it difficult to study their physiology and pharmacology. For this reason, new techniques are needed to identify and determine changes in intracellular processes in neurons.

Flow cytometry is a technique that has been used mainly in hematology and immunology. The application of flow cytometry in neuroscience dates from the early 1980s, when it was used to separate and study neurons and other types of cells from cows and rats.[1] Its main advantages are that it can identify and/or separate one particular cell type from a pool of mixed cells and measure intracellular signals by using previously loaded fluorescent probes. Although it has been used extensively to study protein expression or to quantify DNA, until the last decade little effort was made to adapt flow cytometry to the study of intracellular processes in neurons.[2–4]

The study of intracellular signaling in neurons is especially interesting because little is known about the participation of macromolecular components in several

[1] R. A. Meyer, M. E. Zaruba, and G. M. McKhann, *Anal. Quant. Cytol.* **2,** 66 (1980).

[2] Y. Oyama, T. Ueha, A. Hayashi, L. Chikahisa, and K. Noda, *Japan. J. Pharmacol.* **60,** 385 (1992).

[3] Y. Oyama, A. Hayashi, T. Ueha, and K. Maekawa, *Brain Res.* **635,** 113 (1994).

[4] K. Furukawa, Y. Oyama, L. Chikahisa, Y. Hatakeyama, and N. Akaike, *Brain Res.* **662,** 259 (1994).

0076-6879/02 $35.00

neurologic and psychiatric disorders. For instance, oxidative stress is one of the processes that leads to neuronal death.[5] It is well known that reactive oxygen species, also known as "free radicals," are harmful to cells. These highly reactive compounds oxidize a number of essential constituents of the cell, such as proteins or membrane phospholipids. In fact, free radicals are involved in both acute (ischemia) and chronic disorders (Alzheimer's disease, Parkinson's disease). Other aspects of the functionalism of the neuron might be involved in neurodegenerative disorders, such as calcium homeostasis[6] or dysfunction of important organelle such as mitochondria.[7] Mitochondria not only participate in the generation of free radicals, but also have a role in regulating the intracellular calcium concentration. Complexes I, III, and IV of the mitochondrial respiratory chain pump protons from the mitochondria to the cytoplasm, thus allowing for a mitochondrial membrane potential (MMP). The activation of certain cytoplasmic membrane receptors that are involved in neurodegenerative disorders (i.e., glutamate receptors) increases cytoplasmic calcium. If too much calcium enters the mitochondria, the membrane potential can be impaired. In fact, MMP has been shown to decrease after the glutamate receptors are activated.[8,9] Associated to this decrease in MMP, there is a decrease in ATP production, which leads to ATPases (which pump calcium outside the cell) and other ATP-dependent processes to malfunction. For all these reasons, oxidative stress and MMP in neuronal populations need to studied if the underlying mechanisms in neurogenerative disorders are to be understood. These studies have a clear application in the development of new neuroprotective drugs and in the evaluation of the toxic effects of chemical compounds.[10]

Changes in oxidative stress and MMP can be detected using other fluorimetric techniques, such as spectrofluorimetry or fluorescence microscopy. Like these methods, flow cytometry is specific and has a wide noise-to-signal ratio, but it also has other advantages: it is fast, sensitive, and can separate a particular cell type for further examination or purification. Moreover, cell suspensions can be obtained either from freshly dissected tissue or from cell cultures.

This study describes the method used to measure changes in cytoplasmic oxidative stress and in MMP in viable freshly dissociated cerebellar granule cells obtained from 9- to 11-day-old rat pups or in granule cerebellar cells from *in vitro* cultures. We have also used the same methodology (with little modification) to

[5] N. A. Simonian and J. T. Coyle, *Annu. Rev. Pharmacol. Toxicol.* **36,** 83 (1996).

[6] D. W. Choi, *Trends Neurosci.* **18,** 58 (1995).

[7] A. F. Schinder, E. C. Olson, N. C. Spitzer, and M. Montal, *J. Neurosci.* **16,** 6125 (1996).

[8] F. X. Sureda, E. Escubedo, C. Gabriel, J. Camarasa, and A. Camins, *Naunyn-Schmiedeberg's Arch. Pharmacol.* **354,** 420 (1996).

[9] A. Camins, C. Gabriel, L. Aguirre, F. X. Sureda, M. Pallàs, E. Escubedo, and J. Camarasa, *J. Neurosci. Res.* **52,** 684 (1998).

[10] K. M. Savolainen, J. Loikkanen, and J. Naarala, *Toxicol. Lett.* **82–83,** 399 (1995).

study other processes, such as changes in cytoplasmic calcium,[11] cytoplasmic membrane potential,[12] and apoptosis.

Experimental Procedure

Isolation of Dissociated Cerebellar Cells

Cells are obtained from six rat pups (Sprague–Dawley) aged between 9 and 11 days old. The pups are anesthetized with ether and killed by decapitation. The cerebella are quickly dissected out and placed on an ice-cooled surface. The meninges are removed, and the cerebella are cut with a McIlwain tissue chopper (400 μm thick). The slices are then incubated in a shaking water bath for 30 min (37°) in a solution of 10 ml of Tyrode–HEPES Mg^{2+}-free buffer (THB) containing 0.33 mg/ml collagenase A (from *Clostridium histolyticum*, EC 3.4.24.3, Boehringer-Mannheim, Germany). THB is composed of 137 m*M* NaCl, 2.6 m*M* KCl, 0.42 m*M* NaH_2PO_4, 11.9 m*M* $NaHCO_3$, 1.5 m*M* $CaCl_2$, 5.6 m*M* D-glucose, and 10 m*M* HEPES. THB is previously gassed with CO_2/O_2 (5%/95%) and adjusted to pH 7.4 with NaOH. All the inorganic salts and the glucose were provided by Panreac Química (Barcelona, Spain) and HEPES by Sigma-Aldrich (St. Louis, MO). After incubation, the slices are gently pipetted with a glass Pasteur pipette and filtered through a 63-μm nylon mesh. Then, dissociated cells are centrifuged in polycarbonate tubes for 3 min at 300*g*. Finally, the pellet is resuspended in new THB to yield a final density of 10^6 cells/ml. Aliquots of 1 ml of the cell suspension are prepared in borosilicate glass tubes and kept in ice before the dyes are loaded.

Preparation of Cerebellar Granule Cell Cultures

Primary cultures of 90–95% enriched cerebellar granule cells are prepared from 10 rat pups aged 7 or 8 days old using the method described by Gallo *et al.*[13] Briefly, the cerebella are quickly dissected out, and the meninges are removed under aseptic conditions. Then, the cerebella are cut with a razor blade and incubated in a (gently shaking) water bath for 15 min at 37° in a Krebs solution (free of Ca^{2+} and Mg^{2+}) containing trypsin (from bovine pancreas, EC 3.4.21.4, Sigma-Aldrich). After dissociation with a glass Pasteur pipette, the cells are centrifuged, and the pellet is resuspended in basal medium with Eagle's salts (GIBCO, Life Technologies, UK) supplemented with 10% fetal bovine serum (GIBCO), 25 m*M* KCl, 2 m*M* L-glutamine, and 0.3 mg/ml gentamicin (all from Sigma-Aldrich). The cells are plated on poly-L-lysine-coated, 24-well plates at

[11] F. X. Sureda, A. Camins, R. Trullas, J. Camarasa, and E. Escubedo, *Brain Res.* **723,** 110 (1996).
[12] F. X. Sureda, C. Gabriel, M. Pallas, J. Adan, J. M. Martínez, E. Escubedo, J. Camarasa, and A. Camins, *Neuropharmacology* **38,** 671 (1999).
[13] V. Gallo, A. Kingsbury, R. Balazs, and O. S. Jorgensen, *J. Neurosci.* **7,** 2203 (1987).

a density of 10^6 cells/ml. After 16–18 hr of plating, 10 μM of cytosine arabinoside (Sigma-Aldrich) is added to the wells to prevent nonneuronal cells from proliferating.

Loading of 2′,7′-Dichlorofluorescin Diacetate (DCFH-DA)

Oxidative stress is measured using the intracellular probe DCFH-DA (Fluka A.G., Germany), dissolved in anhydrous dimethyl sulfoxide. Viable cells incorporated and deacetylated DCFH-DA to 2′,7′-dichlorofluorescin (DCFH), which is not fluorescent. DCFH reacts quantitatively with oxygen species to produce the fluorescent dye 2′,7′-dichlorofluorescein (DCF). Fluorescence emission at 525 nm provides an index of the intracellular oxidative metabolism.

In experiments using dissociated cerebellar cells, 1 ml of the cell suspension is loaded for 1 hr under an atmosphere of CO_2/O_2 (5%/95%, 37° in a shaking water bath) with a final concentration of 100 μM of DCFH-DA. After loading, drugs are added and their effect on oxidative stress is evaluated.

In assays using cultured cells, DCFH-DA is added to the wells (final concentration: 10 μM) and loading is performed for 30 min in the cell culture incubator.

Loading of Rhodamine 123 (Rh123)

Rh123 (Sigma-Aldrich) is a cationic, fluorescent dye that is sequestered readily by active mitochondria without inducing cytotoxic effects. The uptake by mitochondria is dependent on their transmembrane potential. The loss of fluorescence emission at 525 nm is indicative of decay in the mitochondrial membrane potential.

Rh123 is loaded in cerebellar cells in the same way as DCFH-DA. In accordance with previosly published results,[14] we use a final concentration for Rh123 of 1 μM, both for dissociated and for cultured cells.

Loading of Propidium Iodide (PI)

PI (Sigma-Aldrich) is used routinely to assess cell viability. Dead cells are permeable to PI, which stains DNA and shows red fluorescence (detected at 675 nm). Viable cells do not load PI, and no fluorescent signal is detected at this wavelength. A final concentration of 10 μg/ml is used in all the assays. The dye is added (from an aqueous solution) 5 min before analysis.

Flow Cytometric Experiments: General Settings

We use a Coulter Epics Elite flow cytometer, equipped with an air-cooled, argon-ion laser tuned at 488 nm. The cell suspension is sipped up by a capillary tube,

[14] F. X. Sureda, E. Escubedo, C. Gabriel, J. Comas, J. Camarasa, and A. Camins, *Cytometry* **28,** 74 (1996).

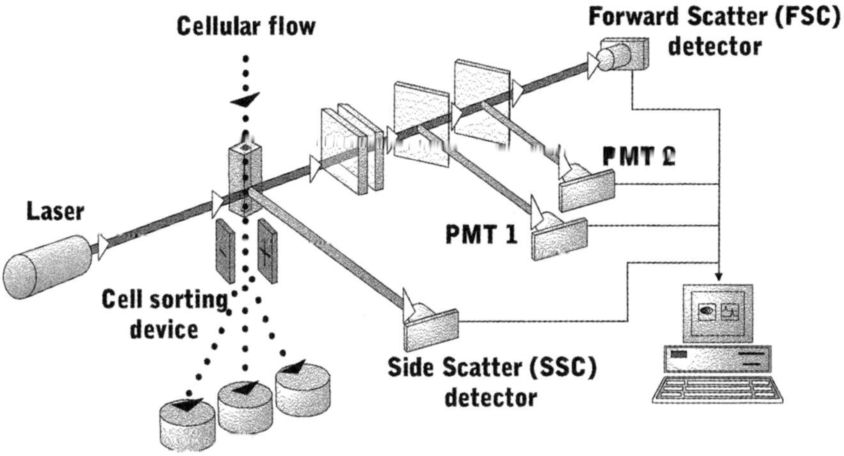

FIG. 1. Diagram of the flow cytometer used in the assays.

forming a laminar flow. The laser beam excites every single cell. The light emitted is scattered by the cell and is detected by two different photomultipliers. One is located in front of the laser beam [forward scatter detector (FSC)] and the other one receives the side-angle scattered light [side-scatter detector (SSC)] as shown in Fig. 1. Fluorescence emission of loaded indicators is selected by appropriate band-pass filters and is detected by photomultipliers PMT1 and PMT2. Signals from all these detectors are analyzed in real time by a computer using proprietary software.

The scattered radiation enables the various cell populations present in the cell suspension to be characterized. Whereas the FSC signal tends to correlate with cell size, SSC measures cell granularity. Both parameters are analyzed on a bidimensional density plot (Fig. 2), and a "gate" that contains the selected cell population is defined. In previous studies with dissociated cerebellar cells, we separated the cell population defined by this gate with the cell sorter feature of the apparatus. Using transmission electronic microscopy, we verified that only one cell type was present: cerebellar granule cells.[11] To measure the fluorescence emission from previously loaded probes, we used one optical filter to select Rh123 or DCF emission (band-pass filter of 525 nm) and another to select PI fluorescence (band-pass filter of 625).

Assays were performed only on PI-negative cells. The percentage of PI-positive cells was checked regularly. Usually, 10–15% of cells died, mainly due to the dissociation process. An increase in the number of PI-positive cells meant that the drug assayed had a toxic effect. There is also the possibility that drugs or conditions can change the size and shape of neurons. For instance, cells may swell or shrink

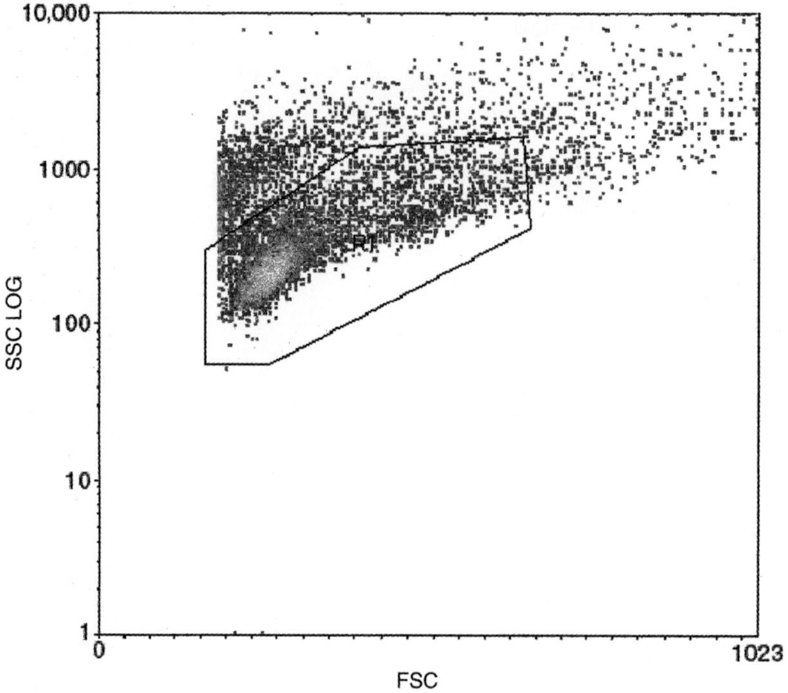

FIG. 2. Bidimensional density plot of a sample of dissociated cerebellar cells. The gate defines the cell population used in the assays, corresponding to cerebellar granule cells.

if ions enter or exit neurons en masse. For this reason, the percentage of cells in the gate is also checked and compared to control values.

Measurement of Oxidative Stress and MMP in Dissociated Cerebellar Cells

The fluorescence emission of each PI-negative cell of the gate was recorded and plotted as a histogram of frequencies. The fluorescence of 10,000–20,000 cells was analyzed and the peak in Fig. 3 was plotted. The peak shifted to the left or right if the emission fluorescence of the probe decreased or increased, respectively. To calculate the shift in the emission, we compared the mean fluorescence of two samples. Figure 3 (white peak) shows how L-glutamate (300 μM) affects the oxidative stress of the neurons. L-Glutamate, acting on its receptors, increases the oxidative metabolism and, subsequently, the oxidation of DCFH to DCF. The fluorescence emission of most of the cells increased and the peak shifted to the right. However, L-glutamate (300 μM) decreased the mitochondrial membrane potential (Fig. 4). In this case, the function of the mitochondria

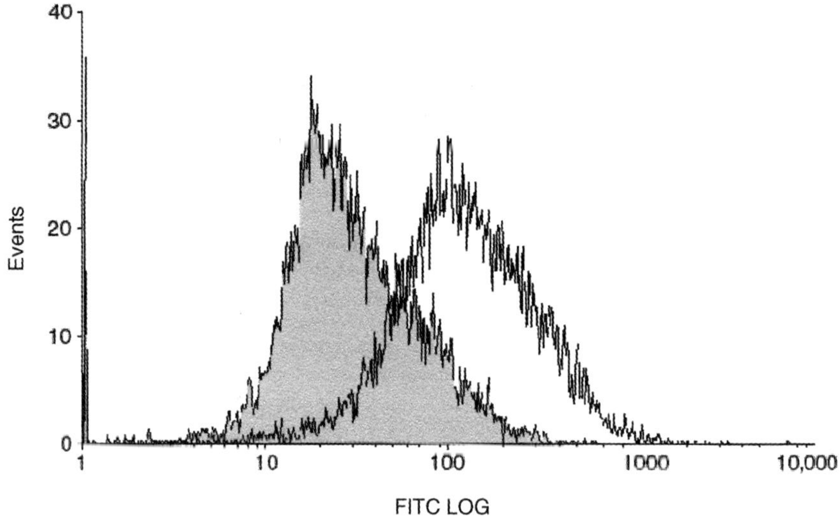

FIG. 3. Single parameter histograms of DCF fluorescence from viable cells. Cells incubated in the presence of L-glutamate show greater DCF fluorescence (white peak) than control cells (gray peak).

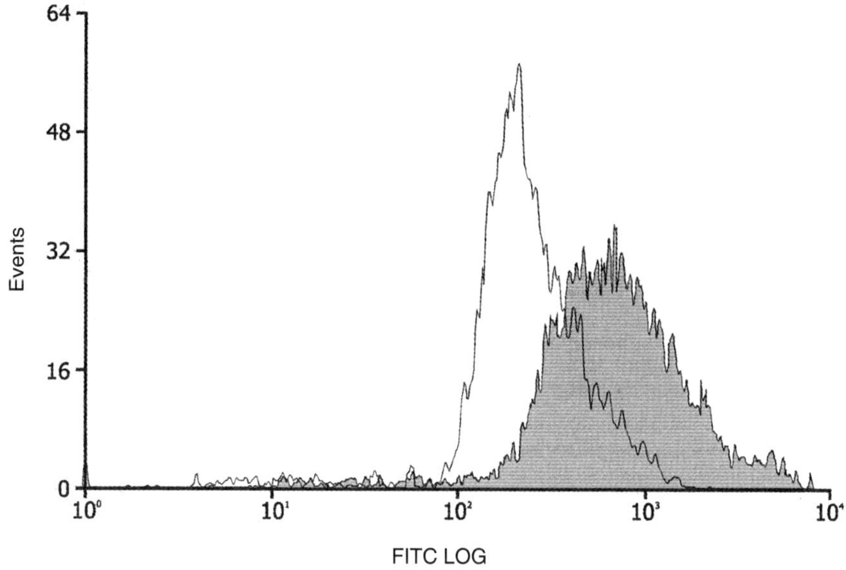

FIG. 4. Single parameter histograms of Rh123 fluorescence from viable cells. Incubation of cells with L-glutamate (white peak) decreased the fluorescence of control cells (gray peak).

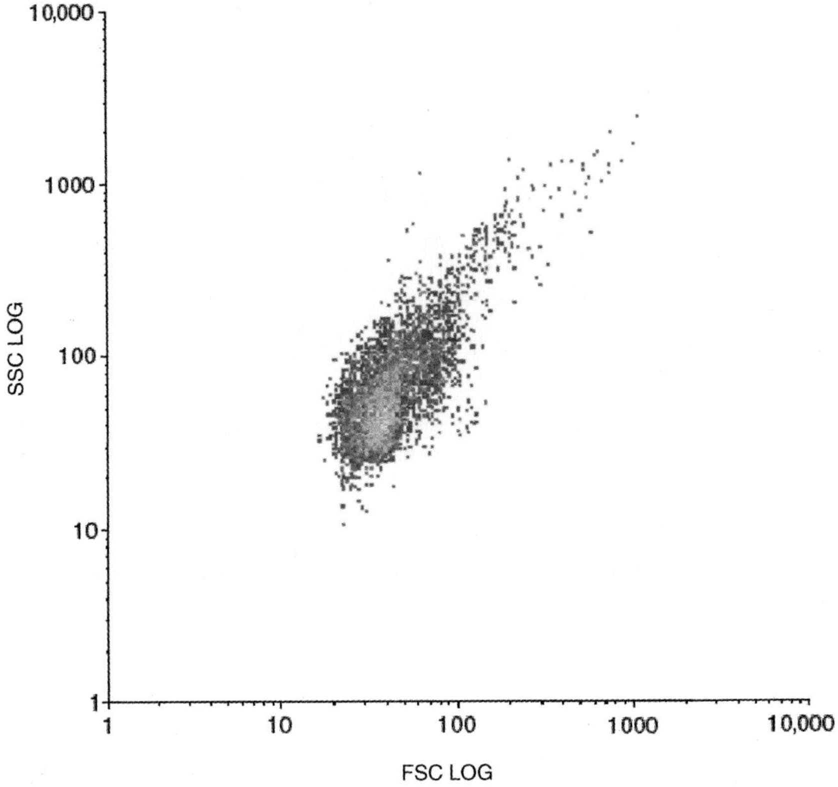

FIG. 5. Bidimensional density plot of a sample of cerebellar granule cells 8 days after plating.

was impaired, and the MMP was lost or substantially decreased. The peak, therefore, shifted to the left.

Measurement of Oxidative Stress in Cerebellar Granule Cell Cultures

In this case and as we have stated earlier, fluorescent probes are added directly to the culture well. After the cells had been incubated with drugs and the PI had been loaded, they are detached by gentle pipetting with a 1-ml automatic pipette. Figure 5 shows the bivariate density plot resulting from the cytometric analysis of the suspension of cerebellar granule cells. The effect of drugs or treatments on the oxidative metabolism is assessed in the same way as for dissociated cerebellar cells. Again, an increase in oxidative metabolism (in this case, induced by a lack of fetal bovine serum and potassium) was indicated by the peak shifting to higher emission values (Fig. 6).

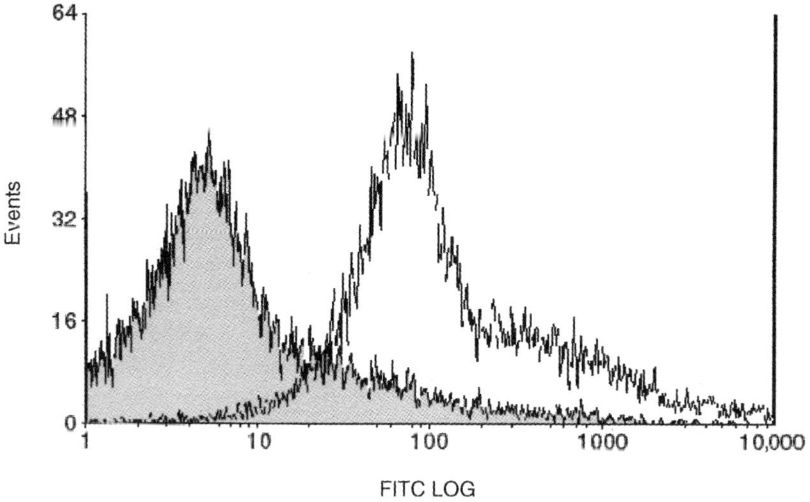

FIG. 6. Single parameter histograms of DCF fluorescence from viable cells. Potassium and serum deprivation increased DCF fluorescence (white peak) compared to control cells (gray peak).

Concluding Remarks

In recent years, a considerable number of cytometric methods involving cells from the brain have been developed. From the study of neuronal or glial markers to the purification and separation of particular cell types, flow cytometry has demonstrated its usefulness to neuroscience. In recent years, our group has also proved that it is useful for studying processes resulting from the activation of glutamate receptors on dissociated cerebellar cells and cultured cerebellar granule cells. Undoubtedly, the next few years will see the increasing application of flow cytometry in the study of receptor-mediated intracellular processes.

Acknowledgments

We are indebted to Mr. J. Comas and R. González from the Scientific Services of the University of Barcelona for their excellent technical assistance. We also thank Mr. J. Bates from the Language Service of the Rovira i Virgili University for revising the language of the typescript. This work was supported by a grant from Fundació "La Marató de TV3."

[7] Flow Cytometric Determination of Lipid Peroxidation Using Fluoresceinated Phosphoethanolamine

By GAUTAM MAULIK, RAVI SALGIA, and G. MIKE MAKRIGIORGOS

Introduction

Reactive oxygen species (ROS) are important intermediates in cell growth as well as apoptosis. ROS can affect proteins, DNA, and lipids. The formation of peroxidation products is often used as a presumptive marker for oxidative damage. It is important to study the effects of free radicals because they have been closely linked to the pathogenesis and/or progression of diseases such as cancer, cardiovascular disease, and diabetes.[1–4] Blood cells are considered prime targets for free radical attack, and red blood cells (RBC), for example, are highly susceptible to oxidative damage due to the presence of both high membrane concentrations of polyunsaturated fatty acids (PUFA) and oxygen-carrying hemoglobin, a potent promoter of oxidative processes.[5] The ability to reliably quantitate the interaction of free radicals with such cells is essential in understanding the pathophysiology of such diseases.

Because of the difficulty in directly measuring short-lived, reactive-free radicals, the formation of lipid, protein, or DNA breakdown products is often used as a presumptive marker for oxidative damage.[6,7] Lipid peroxidation is considered to be a significant stage in the pathogenic processes related to oxidative stress.[8] When lipid peroxidation is initiated, polyunsaturated fatty acids (PUFA) are subject to oxidization by lipid peroxyl radicals and by oxygen-derived free radicals, resulting in the formation of lipid hydroperoxides. In biological tissues, these lipid hydroperoxides further degrade into a variety of products, including aldehydes and ketones.[9] Methods available for the detection of lipid peroxidation products include the determination of diene conjugation,[10] lipid hydroperoxides,[11] hydroxy

[1] J. R. Totter, *Proc. Natl. Acad. Sci. U.S.A.* **77,** 1763 (1980).

[2] B. Halliwell, *FASEB J.* **1,** 358 (1987).

[3] G. Witz, *Proc. Soc. Exp. Biol. Med.* **198,** 675 (1991).

[4] D. K. Das and N. Maulik, *in* "Exercise and Oxygen Toxicity" (C. K. Sen, L. Packer, and O. Hannimen, eds.). Elsevier, New York, 1994.

[5] S. M. H. Sadrzadeh, E. Graf, S. S. Panter, P. E. Hallaway, and J. W. Eaton, *J. Biol. Chem.* **259,** 14354 (1984).

[6] S. P. Wolff, A. Garner, and R. T. Dean, *Trends Biol. Sci.* **11,** 27 (1986).

[7] G. Maulik, G. A. Cordis, and D. K. Das, *Ann. N. Y. Acad. Sci.* **793,** 431 (1996).

[8] R. P. Bird and H. H. Draper, *Methods Enzymol.* **105,** 299 (1984).

[9] R. O. Recknagel and A. K. Ghoshal, *Exp. Mol. Pathol.* **5,** 413 (1966).

[10] W. A. Pryor and L. Castle, *Methods Enzymol.* **105,** 293 (1984).

0076-6879/02 $35.00

acids,[12] ethane,[13] and thiobarbituric acid (TBA)-reactive materials.[14] Estimation of TBA-reactive products is the most widely used in biological samples. The success of this method depends on the accuracy of the determination of malondialdehyde (MDA).[15] However, TBA reacts with many other compounds in addition to MDA, and as a result it often overestimates MDA content.[16] This method has been modified to detect MDA–TBA complexes and the dinitrophenylhydrazine derivative of MDA with high-precision liquid chromatography (HPLC).[17]

A limitation of all currently used methods, however, is that they only provide the average lipid peroxidation over the whole cell population. Consequently, these methods fail to detect cell subpopulations with increased lipid peroxidation or cell type-specific lipid peroxidation.

This methodology chapter reports a novel fluorescein-based flow cytometric method to monitor lipid peroxidation on a cell-by-cell basis in red blood cells undergoing oxidative stress.[18] This approach relies on our previous observation that the fluorescence of fluorescein is diminished upon reaction with peroxyl radicals,[18] and the hypothesis that labeling erythrocyte membranes with lipophilic fluoresceins would provide a fluorescent signal that could be used to monitor lipid peroxidation on a cell-by-cell basis with flow cytometry. In this study, fluoresceinated phosphoethanolamine (fluor-DHPE) is found to be a nonexchangeable flow cytometric probe among several lipophilic fluoresceins tested. The utility of fluor-DHPE as a lipid peroxidation probe for erythrocytes is also demonstrated in RBC of varying vitamin E content.

The results demonstrate that fluor-DHPE, a head group fluoresceinated lipophilic phosphoethanolamine, can be used to monitor lipid peroxidation in red blood cells exposed to oxidative stress. It must be noted, however, that the cell type tested (RBC) is unique in some sense because erythrocytes are loaded with iron-containing hemoglobin, which participates and facilitates the lipid peroxidation process; also RBC contain higher than usual amounts of polyunsaturated fatty acids. Therefore it is recommended that prior to using fluor-DHPE for other cell lines, a similar set of preliminary experiments be conducted.

[11] J. R. Wright, R. C. Rumbaugh, H. D. Colby, and P. R. Miles, *Arch. Biochem. Biophys.* **192,** 344 (1979).

[12] C. A. Riely, G. Cohen, and M. Lieberman, *Science* **183,** 20810 (1974).

[13] J. J. M. Van den Berg, F. A. Kuypers, B. H. Lubin, B. Roelofsen, and J. A. F. Op den Kamp, *Free Radic. Biol. Med.* **11,** 255 (1991).

[14] J. J. M. Van den Berg, F. A. Kuypers, J. H. Qju, D. Chiu, B. Lubin, B. Roelofsen, and J. A. F. Op den Kamp, *Biochim. Biophys. Acta* **944,** 29 (1988).

[15] T. F. Slater, *Methods Enzymol.* **105,** 283 (1984).

[16] J. M. C. Gutteridge and T. R. Tickner, *Anal. Biochem.* **91,** 250 (1978).

[17] G. A. Cordis, N. Maulik, and D. K. Das, *J. Mol. Cell. Cardiol.* **27,** 1645 (1995).

[18] G. M. Makrigiorgos, A. I. Kassis, A. Mahmood, E. A. Bump, and P. Savvides, *Free Radic. Biol. Med.* **22,** 93 (1997).

Methods

Chemicals

Fluorescein-derivatized dihexadecanoylglycerophosphoethanolamine (fluor-DHPE), 5-dodecanoylaminofluorescein (C_{11}-fluor), 5-hexadecanoylaminofluorescein (C_{16}-fluor), and 5-octadecanoylaminofluorescein (C_{18}-fluor) are from Molecular Probes, Incorporated (Eugene, OR). Cumene hydroperoxide, Dulbecco's phosphate-buffered saline (PBS) without calcium and magnesium chloride, and benzoyl peroxide are from Sigma Chemical Company (St. Louis, MO) and Trolox (water-soluble vitamin E) is from Aldrich Chemical Company (Milwaukee, WI). Solutions are prepared in ultrapure water (Millipore Corporation, Bedford, MA).

Isolation of RBC

RBC is isolated from male rats. Twenty Fisher 344 male rats weighing 175–200 g are purchased from Taconic Farms (Germantown, NY). Animals are housed two per cage, with approved IRB protocols at our institution, and allowed to acclimatize for 2 days prior to experimental use. Animals are housed in an environmentally controlled animal facility operating on a 12-hr dark–light cycle at 22–24°. Over a period of 1 month, one group (10 rats) is fed a normal diet containing vitamin E (55 IU/kg diet, Purina rodent lab chow #5001) and water *ad libitum,* whereas the other group (10 rats) is fed a vitamin E-deficient diet (Purina Mills, St. Louis, MO, #5827C) and water.

Rats are lightly anesthetized using Metofane (methoxyflurane, Mallinckrodt Veterinary, Incorporated, Mundelein, IL). Blood is collected in tubes containing ethylenediaminetetraacetate (EDTA) by retroorbital puncture with a capillary tube. EDTA–blood (0.1–0.15 ml) is diluted to 50 ml with PBS in a 50-ml polypropylene test tube and centrifuged at 700 rpm for 20 min at room temperature. The supernatant is discarded. Cells are washed three times with PBS and then suspended in PBS (2×10^7 cells/ml, counted using a hemocytometer).

Labeling of RBC

Aliquots of RBC (2×10^7 cells/ml) in PBS are labeled with fluorescein-DHPE, C_{11}-fluor, C_{16}-fluor, and C_{18}-fluor by adding the probes dissolved in ethanol (less than 0.5% final ethanol concentration and 0.1–1 m*M* final probe concentration). The cells are incubated for 1 hr at 37° with continuous agitation, centrifuged once to remove unbound label, and then examined by flow cytometry or fluorometry.

Labeling and Exchangeability of C_{11}-, C_{16}, and C_{18}-Fluor Probes in RBC

Labeling of RBC with the lipophilic fluorescein derivatives C_{11}-, C_{16}-, and C_{18}-fluor results in an intense fluorescence, which is measurable either by conventional fluorometry or by flow cytometry. In Fig. 1,[18a] the flow cytometric profiles of

[18a] G. Maulik, A. I. Kassis, P. Savvides, and G. M. Makrigiorgos, *Free Radic. Biol. Med.* **25,** 645(1998).

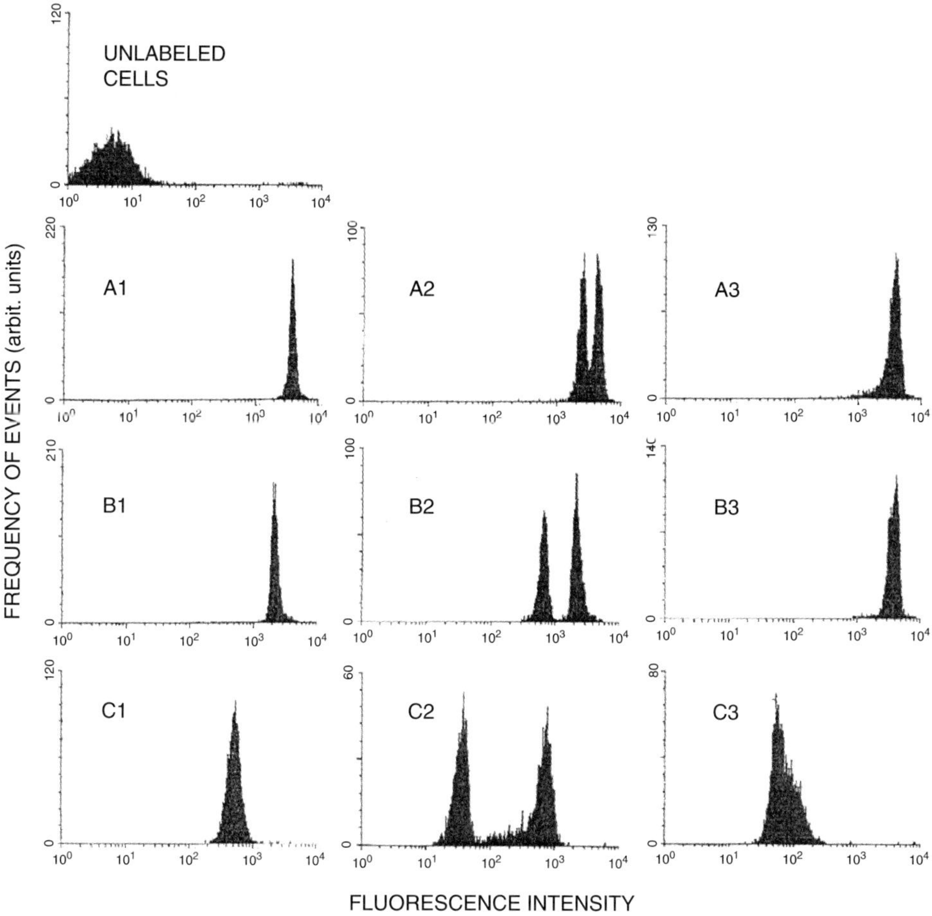

FIG. 1. Flow cytometric profiles of RBC labeled with (A) C_{16}-fluor, (B) C_{18}-fluor, and (C) C_{11}-fluor. Unlabeled RBC are also shown. (A1) C_{16}-fluor-labeled RBC; (A2) C_{16}-fluor-labeled RBC mixed for 2 min with an equal number of unlabeled RBC; (A3) C_{16}-fluor-labeled RBC mixed for 60 min with an equal number of unlabeled RBC; (B1) C_{18}-fluor-labeled RBC; (B2) C_{18}-fluor-labeled RBC mixed for 2 min with an equal number of unlabeled RBC; (B3) C_{18}-fluor-labeled RBC mixed for 60 min with an equal number of unlabeled RBC; (C1) C_{11}-fluor-labeled RBC; (C2) C_{11}-fluor-labeled RBC mixed for 2 min with an equal number of unlabeled RBC; and (C3) C_{11}-fluor-labeled RBC mixed for 60 min with an equal number of unlabeled RBC. Adapted with permission from G. Maulik, A. I. Kassis, P. Savvides, and G. M. Makrigiorgos, *Free Radic. Biol. Med.* **25,** 645 (1998).

labeled (A1, B1, and C1) and unlabeled RBC are depicted. When these labeled cells are mixed with an equal number of unlabeled RBC for 2 min (A2, B2, C2) or 60 min (A3, B3, C3), a rapid mixing of fluorescent and nonfluorescent cell populations is observed. For example, unlabeled RBC mixed with C_{16}-fluor-labeled RBC within 2 min acquires about half the fluorescence per cell of the labeled cells (A2). The mixing is complete by 60 min, indicating that all three lipophilic fluoresceins are readily exchangeable among cells.

Labeling and Exchangeability of Fluor-DHPE Probe in RBC

In Fig. 2, A and B depict the fluorescence of RBC unlabeled and labeled with fluor-DHPE, respectively. Fluorescent microscopy examination of these cells reveals a diffuse but inhomogeneous distribution of the probes throughout the cells, with fluorescence being more evident in the outer leaflet (data not shown). When these cells are mixed with unlabeled cells at a ratio of 10 : 90 or 1 : 99, there is no apparent exchange of fluor-DHPE among cells after 1hr of mixing (C1 and D1,

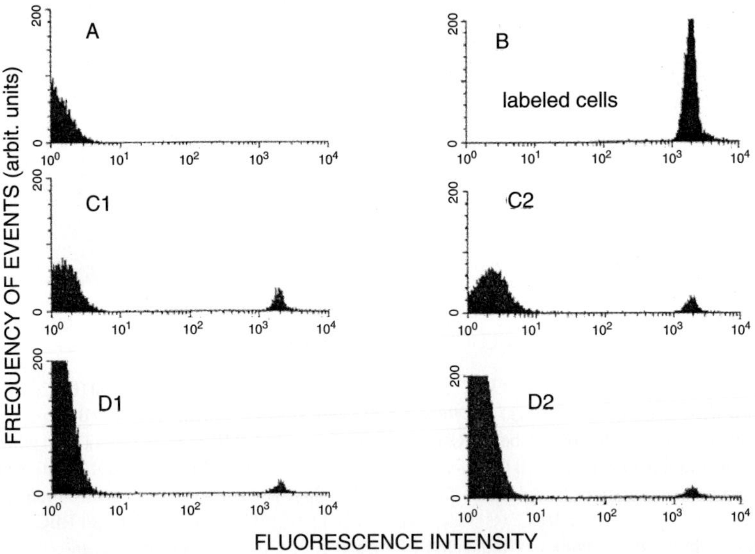

FIG. 2. Flow cytometric profiles of RBC labeled with fluor-DHPE (structure depicted) 1 and 48 hr after mixing with unlabeled cells. (A) Unlabeled RBC; (B) fluor-DHPE-labeled RBC 1 hr after labeling; (C1) 90% unlabeled RBC plus 10% labeled RBC 1 hr after mixing; (C2) 90% unlabeled RBC plus 10% labeled RBC 48 hr after mixing; (D1) 99% unlabeled RBC plus 1% labeled RBC 1 hr after mixing; and (D2) 99% unlabeled RBC plus 1% labeled RBC 48 hr after mixing. Adapted with permission from G. Maulik, A. I. Kassis, P. Savvides, and G. M. Makrigiorgos, *Free Radic. Biol. Med.* **25,** 645 (1998).

respectively). Even when the cells are mixed and stirred for 48 hr at 4°, there is little apparent change in the flow cytometric profile of the populations (C2 and D2). These data indicate that, unlike the three single acyl chain probes (C_{11}-fluor, C_{16}-fluor, and C_{18}-fluor), fluor-DHPE can be considered, for most practical purposes, nonexchangeable among cells. All further experiments are thus conducted with fluor-DHPE.

Stability of Fluor-DHPE Probe in RBC

The stability of the fluorescence per RBC in cells labeled with fluor-DHPE and maintained at 37° under stirring for up to 48 hr is shown in Fig. 3A. The fluorescence is stable for several hours; however, after the first 24 hr, a slow decline in the fluorescence per cell is observed. Figure 3B shows the dependence of the fluorescence per cell (1×10^8 RBC/ml) on the concentration of fluor-DHPE (0.6–10 μM). A linear increase in fluorescence is depicted for up to 1.2 μM fluor-DHPE, whereas the increase is nonlinear above this concentration. The change in slope above 1.2 μM may reflect the self-quenching among fluorescein molecules localized in close proximity in the membrane. All subsequent experiments are conducted using 0.1 μM per 2×10^7 RBC/ml.

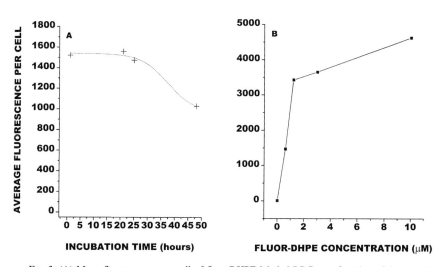

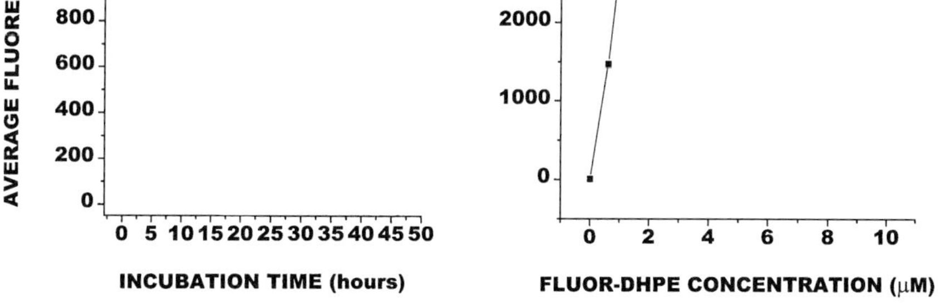

FIG. 3. (A) Mean fluorescence per cell of fluor-DHPE-labeled RBC as a function of time (cells incubated in 0.6 μM fluor-DHPE). (B) Mean fluorescence per cell of fluor-DHPE-labeled RBC as a function of fluor-DHPE concentration in labeling solution. Adapted with permission from G. Maulik, A. I. Kassis, P. Savvides, and G. M. Makrigiorgos, *Free Radic. Biol. Med.* **25,** 645 (1998).

Usefulness of Fluor-DHPE Probe in Measuring Lipid Peroxidation

In this report, we explore the efficacy of lipophilic fluorescein-based probes to overcome some of the problems. Four fluorescein-based lipophilic probes were studied, C_{11}-fluor, C_{16}-fluor, C_{18}-fluor, and fluor-DHPE, for their exchangeability among labeled and unlabeled cells as described previously. All probes produced an intense labeling of RBC as measured by flow cytometry. Among the four probes, fluor-DHPE was found to be the only one that is nonexchangeable among labeled and unlabeled RBC and the only one, therefore, that could identify differences that potentially exist in lipid peroxidation among cell subpopulations. While C_{11}-fluor was also found in an earlier investigation[18] to be suitable as a flow cytometric probe to evaluate lipid peroxidation in red blood cells, it exchanges readily among labeled and unlabeled cells (Fig. 1C) and this property limits its usefulness. An additional problem of C_{11}-fluor is that it tends to stick to the test tube unless polymethacrylate cuvettes are used. Fluor-DHPE, however, was found to be stable for at least 24 hr in polymethacrylate, polystyrene, or glass containers. Similar to earlier results with C_{11}-fluor,[18,19] benzoyl peroxide and cumene hydroperoxide both decreased the fluorescence of fluor-DHPE-labeled RBC. This decrease in the lipophilic fluorescein fluorescence was effectively arrested by addition of a lipid-soluble antioxidant, vitamin E.

Flow Cytometric and Fluorometric Evaluation of Susceptibility of Fluor-DHPE-Labeled Cells to in Vitro-Applied Lipid Peroxidation

In order to examine the response of the system to lipid peroxidation applied *in vitro*, fluor-DHPE-labeled RBC are exposed to 0–50 μM cumene hydroperoxide or 0–20 μM benzoyl peroxide. The peroxide dissolved in ethanol [0.5–1% (v/v), final volume] is added directly to the cell suspension, and cell-associated fluorescence is then determined by fluorometry (PerkinElmer LS50B fluorometer, excitation and emission wavelengths of 488 and 520 nm, respectively, slit width 5 nm) as a function of time with or without vitamin E (10 μM). Cells are analyzed using a flow cytometer (FACScan, Becton Dickinson, San Jose, CA) equipped with a laser-emitting visible blue light (488 nm) at 15 mW (Coherent, Incorporated, Palo Alto, CA). Acquisition and analysis are performed using Lysys II software (Becton Dickinson Immunocytometry Systems, San Jose, CA). Forward and side scatter is collected on a linear scale. During analysis, a gate is set on the dot plot of forward and side scatter to include red blood cells and to exclude nucleated cells and debris. Approximately 10,000 events are collected from each sample. Mean fluorescence of each cell is determined and expressed as the percentage reduction

[19] G. M. Makrigiorgos, *J. Biochem. Biophys. Methods* **35,** 23 (1997).

of mean fluorescence relative to control (labeled cells without stress). Data are compiled and computed, and statistical significance is evaluated using Student's *t*-test.

Detection of Lipid Peroxidation Caused by Benzoyl Peroxide on Fluor-DHPE-labeled RBC

To examine the utility of fluor-DHPE for monitoring lipid peroxidation on RBC, benzoyl peroxide, an oxidant known to generate lipid peroxidation on erythrocytes,[19] is used. Cells labeled with fluor-DHPE are exposed for 75 min to benzoyl peroxide (0–30 μM) and then their fluorescence is quantified by flow cytometry. A decrease in fluorescence per cell (Fig. 4), similar to that reported earlier with C_{11}-fluor,[18] is recorded. This decrease is consistent with the expected loss of fluorescence following oxidation of the fluorescein moiety by lipid peroxidation radicals.[18] To further assess the utility of fluor-DHPE for scoring lipid peroxidation in RBC subpopulations, RBC are first labeled with fluor-DHPE, then a fraction of these labeled cells is exposed for 15 min to 20 μM benzoyl peroxide and mixed with an equal number of RBC unexposed to the oxidant, and the flow cytometric profile of the mixtures is taken immediately and 50 min later. The results (Fig. 5) indicate that lipid peroxidation proceeds in the benzoyl peroxide-exposed cells independently of the presence of unexposed RBC (compare the positions of each population in C and D) and that both populations can be monitored independently and simultaneously by the present method.

Detection of Lipid Peroxidation on Fluor-DHPE-Labeled RBC Following in Vitro and in Vivo Modulation of Vitamin E Levels

Tests were performed to examine whether fluor-DHPE can monitor differences in the susceptibility of RBC to lipid peroxidation when these cells are deprived of or supplemented with normal levels of vitamin E. When fluor-DHPE-labeled RBC are exposed to cumene hydroperoxide, another oxidant known to initiate lipid peroxidation on RBC,[20] a gradual decrease in fluorescence is observed, similar to that reported with C_{11}-fluor.[18] Figure 6 shows the time-dependent (over 75 min), gradual decline of fluorescence following the addition of 50 μM cumene hydroperoxide, as measured in a spectrofluorometer. When cells are presupplemented *in vitro* with 10 μM α-tocopherol (vitamin E) prior to the addition of cumene hydroperoxide, the decrease in fluorescence is arrested (Fig. 6).

[20] J. J. M. Van den Berg, J. A. F. Op den Kamp, B. H. Lubin, B. Roelofsen, and F. A. Kuypers, *Free Radic. Biol. Med.* **12,** 487 (1992).

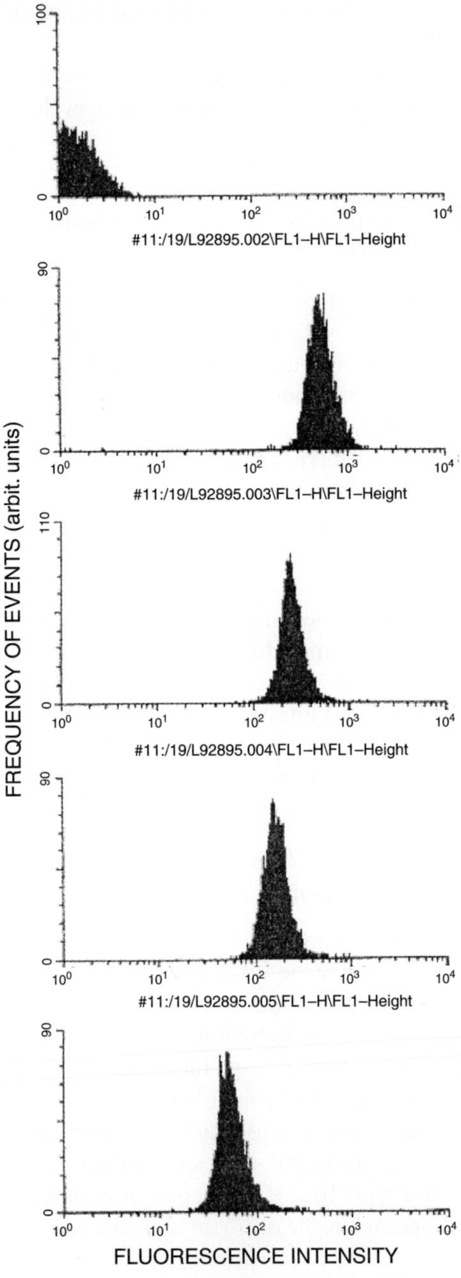

FIG. 4. Flow cytometric profile of RBC labeled with fluor-DHPE (0.1 μM) and exposed to 0, 5, 10, or 30 μM benzoyl peroxide (top to bottom, respectively) for 75 min. Adapted with permission from G. Maulik, A. I. Kassis, P. Savvides, and G. M. Makrigiorgos, *Free Radic. Biol. Med.* **25,** 645 (1998).

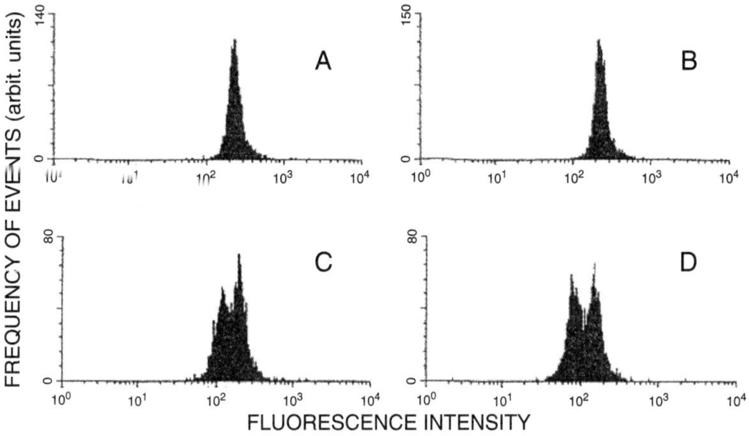

FIG. 5. Flow cytometric profile of RBC labeled with fluor-DHPE (0.1 μM) and exposed to 20 μM benzoyl peroxide. (A) RBC not exposed to benzoyl peroxide; (B) same as A 50 min later; (C) RBC exposed to benzoyl peroxide for 15 min, mixed with an equal number of nonexposed RBC, and measured immediately after mixing; (D) RBC exposed to benzoyl peroxide for 15 min, mixed with an equal number of nonexposed RBC, and measured 50 min later. Adapted with permission from G. Maulik, A. I. Kassis, P. Savvides, and G. M. Makrigiorgos, *Free Radic. Biol. Med.* **25,** 645 (1998).

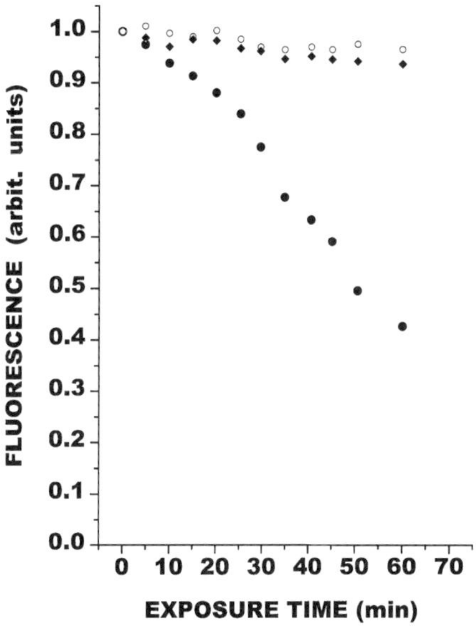

FIG. 6. Fluorometric detection of lipid peroxidation caused by cumene hydroperoxide in fluor-DHPE-labeled RBC. ○, no cumene hydroperoxide; ●, 50 μM cumene hydroperoxide; ◆, 50 μM cumene hydroperoxide plus 10 μM vitamin E. Adapted with permission from G. Maulik, A. I. Kassis, P. Savvides, and G. M. Makrigiorgos, *Free Radic. Biol. Med.* **25,** 645 (1998).

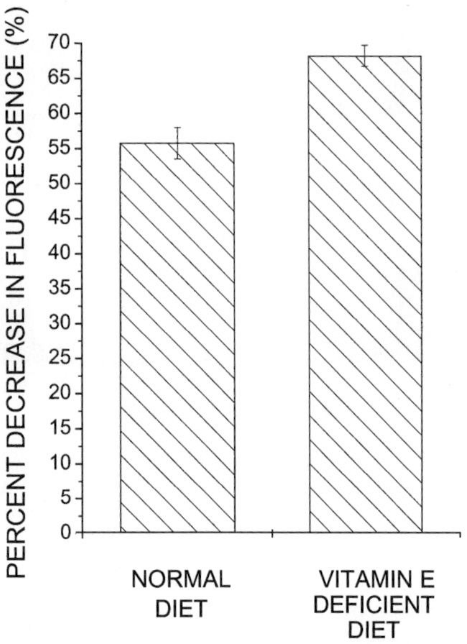

FIG. 7. Effect of cumene hydroperoxide on fluorescence exhibited by fluor-DHPE-labeled RBC from vitamin E-deficient or normal rats. Results expressed as percentage reduction of fluorescence ($\pm$SE, 10 rats per group) when RBC are exposed for 75 min to 50 μM hydroperoxide. Adapted with permission from G. Maulik, A. I. Kassis, P. Savvides, and G. M. Makrigiorgos, *Free Radic. Biol. Med.* **25,** 645 (1998).

The reprodubility of these data is about 5%, as infered by performing three independent experiments.

RBC obtained from rats fed a normal diet containing vitamin E or a vitamin E-deficient diet for 4 weeks are labeled with fluor-DHPE and examined using flow cytometry for their susceptibility to applied lipid peroxidation by 50 μM cumene hydroperoxide. The results (Fig. 7) demonstrate a statistically significant difference ($P < 0.0005$, Student's t test) in the susceptibility of erythrocytes to lipid peroxidation among the two groups of rats.

Summary

In conclusion, we describe the use of fluor-DHPE as a flow cytometric probe to assess lipid peroxidation in erythrocytes. The stability and nonexchangeability of this probe make it suitable for monitoring lipid peroxidation in a particular cell type via flow cytometry, which can be done in the presence of other cells.

Application of the probe in erythrocytes from rats fed a vitamin E-deficient diet demonstrated a higher susceptibility to lipid peroxidation among these cells than among erythrocytes from rats fed a normal diet. The current flow cytometric lipid peroxidation detection method can be interfaced directly with several standard techniques that are available to measure specific blood cell populations via flow cytometry.

Acknowledgment

Parts of this work were supported by Grant SBIR HL 57150-01.

[8] Determination of Intracellular Reactive Oxygen Species as Function of Cell Density

By GIOVANNI PANI, RENATA COLAVITTI, BARBARA BEDOGNI, ROSANNA ANZEVINO, SILVIA BORRELLO, and TOMMASO GALEOTTI

Introduction

Reactive species generated in biological systems by the incomplete reduction of molecular oxygen to water have potentially deleterious effects on almost any cellular component and represent important mediators of cellular damage in an endless series of human and experimental diseases.[1]

Nevertheless, oxygen species have also important physiological functions, and it is now believed that their formation in living matter represents much more than just a biochemical trade-off for aerobic metabolism. For instance, reactive oxygen species (ROS) such as superoxide ($O_2^{\cdot -}$) hydrogen peroxide (H_2O_2), and nitric oxide (NO), generated by activated phagocytes, are well recognized as crucial for the immune response against pathogens.

In the last decade, evidence has accumulated on the potential role of oxygen species as endogenous mediator molecules involved in signal transduction by inflammatory, cytotoxic, and mitogenic stimuli.[2] It has become evident that ROS, while responsible for indiscriminate oxidative cell damage when released in excess amounts, can also, at low concentrations and in very limited time frames, behave as

[1] J. Kehrer, *Crit. Rev. Toxicol.* **23,** 21 (1993).
[2] T. Finkel, *Curr. Opin. Cell Biol.* **10,** 248 (1998).

0076-6879/02 $35.00

fine modulators of important intracellular events, such as protein phosphorylation[3] and gene transcription,[4] through which the cell response to environmental stimuli is triggered. In particular, while nitric oxide has a major role in the regulation of endothelial cell functions and in the orchestration of learning circuits in the central nervous system,[5] hydrogen peroxide and superoxide have been directly involved in the inflammatory and mitogenic response to, respectively, cytokines[6] and growth factors.[7–9] In these physiological settings, oxygen species are generated on ligand–receptor interaction and contribute, through biochemical mechanisms still poorly understood, to the intracellular propagation of the signal from the cell surface to the nucleus.

The cellular response to several biochemical stimuli, as those delivered by mitogens, differentiating agents, and cytotoxic drugs, is largely affected by cell–cell contact and cell density. Normal fibroblasts stop growing on reaching confluence ("contact inhibition") even in the presence of optimal amounts of growth factors,[10] and myoblasts differentiate into myotubes when plated at high density in low serum. However, cancer cells are more susceptible to killing by antinflammatory cytokines, antibiotics, and anticancer drugs when seeded sparsely to prevent cell–cell contact.[11,12] Given the growing importance of intracellular ROS, and of hydrogen peroxide in particular, as putative mediators of signals involved in the regulation of both cell proliferation and cell death, the possibility of evaluating the generation of these species as a function of cell density appears to be potentially relevant.

Among the many procedures commonly utilized for measuring ROS production in cultured cells, some assess the release of oxygen species in the culture medium,[13] whereas others utilize cell-permeant, oxidant-sensitive fluorescent probes to indirectly evaluate ROS concentration in the intracellular compartment.[14] The latter are in general more sensitive, allowing the detection of even small and transient variations in the content of oxidants (mainly peroxides) inside the cells,

[3] H. P. Monteiro and A. Stern, *Free Radic. Biol. Med.* **21,** 323 (1996).

[4] Y. Sun and L. W. Oberley, *Free Radic. Biol. Med.* **21,** 335 (1996).

[5] C. J. Lowenstein and S. H. Snyder, *Cell* **70,** 705 (1992).

[6] Y. Y. C. Lo and T. Cruz, *J. Biol. Chem.* **270,** 11727 (1995).

[7] M. Sundaresan, Z. X. Yu, V. J. Ferrans, K. Irani, and T. Finkel, *Science* **270,** 296 (1995).

[8] Y. S. Bae, S. W. Kang, M. S. Seo, I. C. Baines, E. Tekle, P. B. Chock, and S. G. Rhee, *J. Biol. Chem.* **272,** 217 (1997).

[9] G. Pani, R. Colavitti, S. Borrello, and T. Galeotti, *Biochem. J.* **347,** 173 (1999).

[10] R. J. Wieser, D. Renauer, A. Schafer, R. Heck, R. Engel, S. Schutz, and F. Oesch, *Environ. Health Perspect.* **88,** 251 (1990).

[11] M. Brielmeier, J. M. Bechet, M. H. Falk, M. Pawlita, A. Polack, and G. W. Bornkamm, *Nucleic Acids Res.* **26,** 2082 (1998).

[12] J. W. Kao and J. L. Collins, *Cancer Invest.* **7,** 303 (1989).

[13] W. Ruch, P. H. Cooper, and M. Baggiolini, *J. Immunol. Methods* **63,** 347 (1983).

[14] J. A. Royall and H. Ischiropoulos, *Arch. Biochem. Biophys.* **302,** 348 (1993).

and are therefore more suitable for signal transduction studies. Another aspect to be taken into account when trying to correlate the production of oxygen species to cell density is that a procedure which allows ROS determination at a single cell level should be preferred to others that read bulk ROS production and therefore require further normalization for the number of cells plated on a given surface area.

In view of these considerations, we have elaborated a protocol in which the intracellular production of oxygen species grown at different plating densities is evaluated in single cells by flow cytometry after cell loading with the peroxide-sensitive probe dichlorofluorescein diacetate (DCF-DA) and following rapid detachment from the substrate.

Basic Procedure for ROS Measurement in Sparse and Dense Cells

An SV40-immortalized human fibroblastoid line and its "rho zero" (respiration-deficient) derivative[15] (Gm701 and Gm701 "rho zero"; kind gift of Dr. M. Jacobson, MRC London, UK) are utilized for most experiments. These cells, although transformed, retain a significant growth inhibition by cell density. In some experiments, Swiss 3T3 mouse fibroblasts are also used. All cell lines are routinely maintained in DMEM to which is added 10% fetal bovine serum, L-glutamine, nonessential amino acids, 1 m*M* pyruvate and antibiotics, and split at 80% confluence. Rho zero cells also require 50 μM uridine in the growth medium.

In most experiments, different cell culture densities are obtained by plating different cell numbers on an equal surface area (24-well plate, 2 cm^2/well, Corning) in an equal volume of medium (2 ml). In these conditions, dense and sparse cultures are prepared on the same plate and are identical with respect to substrate for adhesion, areation, and volume/surface ratio. These conditions are therefore optimal to assess the effect of cell confluence on the production of ROS, provided that the cell number is not limiting for the availability of oxygen and nutrients and that the fluorescent probe is in large excess with respect to cells.

The basic procedure for ROS determination at high and low cell density is outlined in Fig. 1. Exponentially growing cells are trypsinized, counted, and replated in a 24-well plate at a density of 6×10^5/well (i.e., 300,000/cm^2; dense), 180,000/well (subconfluent), and 60,000/well (sparse) and incubated for 1 to 16 hr at 37°/5% CO_2 in a humidified incubator. DCF-DA (Molecular Probes) is prepared in ethanol at a concentration of 5 mg/ml and added 1 : 1000 (5 μg/ml final concentration) in fresh medium to sparse and dense cultures for 1 hr. At the end of the incubation, cells are washed with phosphate-buffered saline (PBS), trypsinized, resuspended in 1 ml of PBS, and immediately analyzed by flow cytometry using

[15] M. D. Jacobson, J. F. Burne, M. P. King, T. Miyashita, J. C. Reed, and M. C. Raff, *Nature* **361,** 365 (1993).

Exponentially growing cells

↓

Detached (EDTA)

↓

Plated at different densities (24-well plate)

↓

Overnight incubation in complete medium

↓

Medium replaced with fresh complete DMEM containing 5 μg/ml DCF-DA

↓

1-hour incubation

↓

Detachment and flow cytometry (green fluorescence)

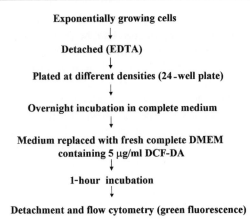

FIG. 1. Flow chart summarizing the basic protocol for determination of intracellular ROS at different plating densities. See text for details.

a Coulter-Epics cytometer equipped with an argon laser lamp (emission 488 n*M*, 530 n*M* band-pass filter for green fluorescence). At least 5000 cells per sample are analyzed, and the mean cell fluorescence values are reported in column histograms.

A typical experiment is reported in Fig. 2; as shown, the intracellular content of ROS, indirectly indicated by DCF oxidation, is higher in sparse than in dense cells and increases progressively in inverse relation with cell density (white columns).

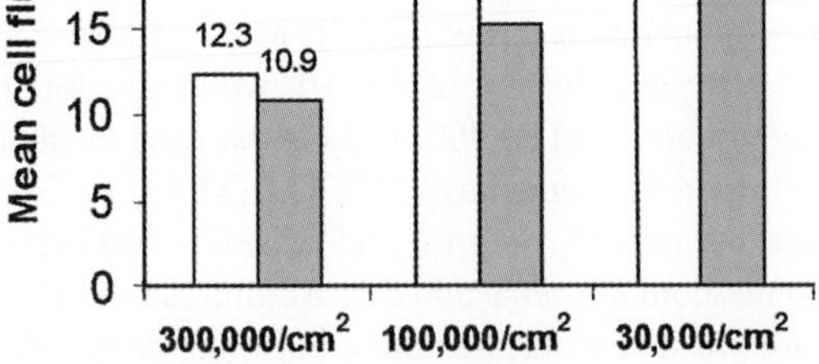

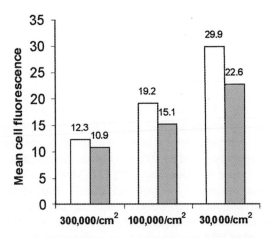

FIG. 2. Determination of intracellular ROS as a function of cell density in Gm701 and Gm701 "rho zero" cells. Parental (white columns) and "rho zero" (gray columns) Gm701 cells were assayed for ROS production after overnight incubation at the indicated cell densities in 2 ml of complete medium. Numbers indicate the mean cell fluorescence value for each sample. The figure is representative of several independent experiments.

Such an effect of cell number on intracellular ROS is unlikely due to reduced oxygen levels in crowded wells, as analogous differences between sparse and dense cultures are also observed with "rho zero" cells, which have a very low oxygen consumption due to the inactivation of the mitochondrial respiratory chain (Fig. 2, gray columns). However, it should be noted that overall ROS generation is, as expected, slightly lower in respiration deficient cells.

To confirm that the observed differences in cell fluorescence reflect differences in DCF oxidation and are not due to reduced cellular uptake of the fluorescent probe at high plating density, a number of controls were designed. First, several concentrations of of DCF-DA were utilized to exclude that the amount of DCF-DA is limiting in dense cultures; similar differences between sparse and confluent cells were observed at all the DCF concentration tested (data not shown), suggesting that the amount of DCF available per cell is not critical in dense cultures. Second, to exclude an effect of plating density on cell permeability to DCF, the basic protocol was modified as follows. After trypsinization, cells are preloaded while kept in suspension (300,000 cells/ml) with 5 μg/ml DCF for 30 min and subsequently plated at high or low density in 2 ml of complete medium containing 5 μg/ml DCF and incubated for an additional 60 min. As controls, some cells are also plated immediately at high or low density and exposed to DCF-DA, or left unlabeled. As shown in Fig. 3, preloading of DCF does not increase the fluorescence of dense cells, as one would expect if DCF loading were impaired by cell–cell contact in

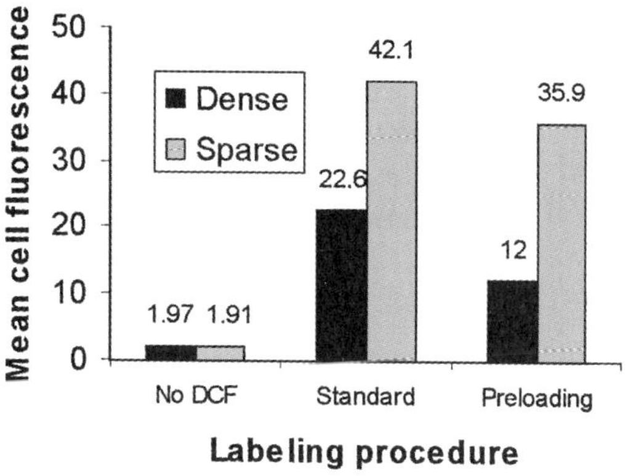

FIG. 3. Differences in DCF oxidation between sparse and dense cells are independent from the DCF-loading procedure. Cells were treated as described in the text. For the standard procedure, cells were plated at high (300,000/cm^2, dense) or low (30,000/cm^2, sparse) density and incubated for 90 min at 37° in humidified atmosphere/5% CO_2. Alternatively (preloading), cells were loaded while in suspension with 5 μg/ml DCF-DA for 30 min and subsequently plated at high or low density and incubated for an additional 60 min in complete medium containing the same concentration of DCF. As a negative control, cells were processed as in the standard procedure, except that DCF was not added (no DCF). Mean cell fluorescence values, determined by flow cytometry, are indicated. The figure is representative of two independent experiments.

confluent monolayers, nor attenuates the difference with sparse cultures, consistent with the hypothesis that the oxidation of DCF is intrinsically reduced in cells plated at high density. The fluorescence of unlabeled cells is very low and is unaffected by cell density (Fig. 3). It should be noted that in this experiment 60 min was sufficient time for cell adhesion and spreading on the substrate.

Because fibroblasts attach rapidly to the bottom of the plate, the most evident consequences of increasing the number of cells seeded per well are an increase in cell–cell contacts and spatial limitation to cell spreading. However, in our experimental model, not only the cell-to-surface ratio, but also the cell-to-volume ratio is different between sparse and dense cultures. Experiments were therefore conducted to address whether the production of oxygen species is a function of the number of cells per unit of surface, per unit of volume, or both. Cells are seeded in a 24-well plate at the following densities: 300,000 cells/cm^2/2 ml; 300,000 cells/cm^2/0.2 ml; 30,000 cells/cm^2/2 ml; and 30,000 cells/cm^2/0.2 ml. After 3 hr of incubation, cells are labeled with 5 μg/ml DCF for 1 hr. Following trypsinization, cells are analyzed by flow cytometry as described earlier.

As shown in Fig. 4, changes in the cell-to-volume ratio have marginal effects on DCF oxidation (compare column 1 with column 2 and column 3 with column 4; mean fluorescence values are indicated). Conversely, ROS production is increased significantly upon reduction of the cell-to-surface ratio, independently of the volume of medium (compare in particular columns 1 and 4, in which the cell number/volume ratio is identical). Note that while the values reported in Fig. 4 refer to single determinations and have therefore no error bars, this result has been confirmed in several independent experiments.

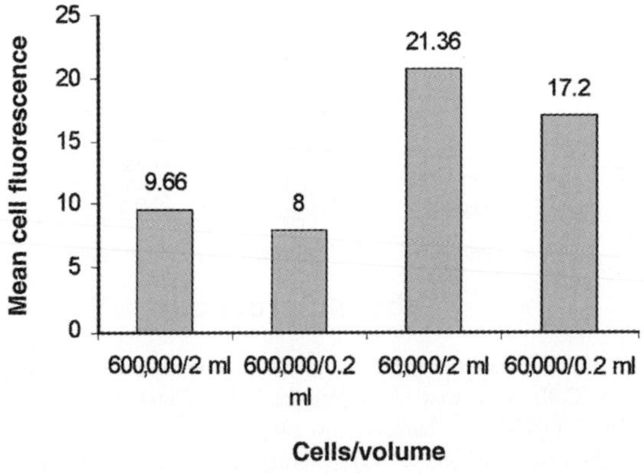

FIG. 4. Relative effect of culture area and culture volume on ROS production in Gm701 cells. Six hundred thousand or 60,000 cells were plated on 2-cm^2 culture wells in the indicated volume of complete medium. After 2 hr, DCF-DA was added at the final concentration of 5 μg/ml. Sixty minutes later, cells were detached and analyzed. Mean cell fluorescence values are indicated by numbers. The figure is representative of several independent experiments.

Identification of a Potential Source of Oxygen Radicals Modulated by Cell Density

Oxygen radicals involved in signal transduction can be generated by several intracellular sources, including NADPH oxidases, lipoxygenases, and the mitochondrial respiratory chain.[2] While mitochondria are unlikely to contribute to redox changes induced by cell confluence in view of the results obtained with Gm701 "rho zero" cells (Fig. 2), which lack this potential source of ROS, both lipoxygenase and NADPH oxidase represent good candidates, as both are controlled by the small GTPase Rac-1, a transducer known to be modulated by cell adhesion and cell shape changes.[16,17]

We have used the NADPH oxidase inhibitor DPI and two structurally unrelated inhibitors of arachidonic acid metabolism, the phospholipase A_2 inhibitor 4BPB and the lipoxygenase inhibitor NDGA, to verify the relative contribution of these radical sources to the intracellular redox modifications induced by cell density. For this purpose the basic protocol was slightly modified. After overnight incubation at high (600,000/well) or low (60,000/well) density, cells are washed with PBS and refed with 2 ml of fresh medium containing either DMSO 1 : 500 or the following inhibitors in the same volume of DMSO: 10 μM diphenyleneiodonium (DPI), 20 μM 4BPB, or 5 μM NDGA. After a 1-hr incubation, cells are labeled with 5 μg/ml DCF for an additional hour. Finally, cells are washed, detached, and analyzed as described earlier. As shown in Fig. 5, inhibitors of the arachidonate metabolism, but not the NADPH oxidase inhibitor DPI, are effective in decreasing DCF oxidation in sparse cells to a level comparable to dense cultures [Fig. 5; the fluorescence of dense cells + dimethyl sulfoxide (DMSO) is indicated by a horizontal dotted line]. These treatments had marginal or no effect on confluent cells (not shown).

It thus appears that lipoxygenases represent important sources of intracellular ROS in sparsely growing GM701 cells and, by extension, that their inhibition may be involved in the decreased generation of oxygen radicals in confluent Gm701 cells.

Modulation of Intracellular ROS by Cell Density Is Mediated by a Cell Autonomous, Nonautocrine Mechanism

Modulation of intracellular ROS by cell density could be a direct consequence of cell–cell contact (maybe through integrin signaling or cell shape changes) or be determined by an autocrine mechanism, involving, for instance, the release of a ROS-inhibiting factor or the down-regulation of a ROS-inducing soluble ligand in confluent cultures. To address this point, special cocultures are prepared in which

[16] G. M. Bokoch, *Curr. Opin. Cell. Biol.* **6,** 212 (1994).

[17] M. P. Peppelembosh, R.-G. Qiu, A. M. M. de Vries-Smits, L. G. J. Tertoolen, S. W. de Laat, F. McCormick, A. Hall, M. H. Symons, and J. L. Bos, *Cell* **81,** 849 (1995).

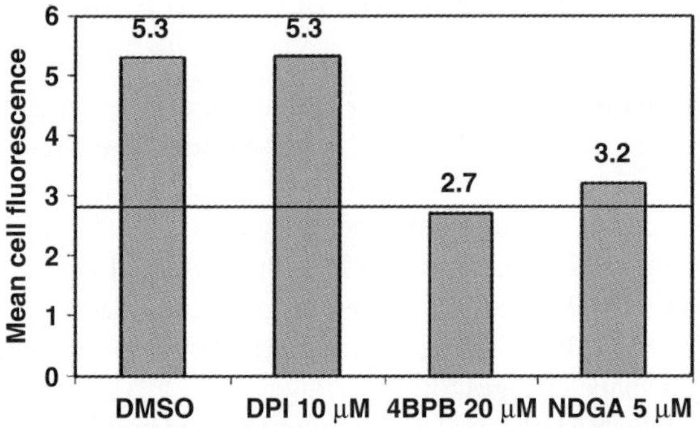

FIG. 5. Involvement of arachidonate metabolism in redox changes induced by cell density. Gm701 cells were incubated overnight at high or low density in 24-well plates. Cells were then washed and refed with fresh medium (2 ml) containing 1 : 500 DMSO or the indicated inhibitors in the same vehicle. After 1 hr, DCF-DA was added and cells were were incubated for another hour before analysis. 4BPB and NDGA, but not DPI, were effective in reducing ROS of sparse cells to the level observed in dense cultures (indicated by the dotted horizontal line). The figure is representative of two independent experiments. Mean fluorescence values are indicated. The fluorescence of dense cells + DMSO was 2.8.

sparse and dense cells share the same medium, and DCF oxidation is evaluated as a function of cell density. A standard 24-well plate is used in which the well surface is divided into two compartments by gluing a small plastic cylinder of 6 mm diameter onto the bottom of the well. Two separate areas are therefore obtained of about 1.7 and 0.3 cm^2, respectively. The two compartments communicate with each other, provided that enough medium (2 ml) is poured into the well (Fig. 6). Cells [Swiss 3T3 in complete DMEM+ 10% fetal calf serum (FCS)] are seeded as indicated in Fig. 6 in such a way that 60,000 cells are put in either of the two compartments (only the large one, B, or only the small one, A) or in both (Fig. 6). After 2 hr of incubation, cells are labeled with 2.5 μg/ml DCF-DA for an additional hour. Finally, cells are detached from the well and analyzed by flow cytometry. During incubations the plates are shaken occasionally to favor the exchange of medium between the two compartments.

As indicated in Fig. 6, sparse (60,000/1.7 cm^2) 3T3 had significantly higher levels of intracellular ROS in comparison to dense (60,000/0.3 cm^2) cells. This difference was maintained when sparse and dense cells were incubated in the same well (the increase in DCF oxidation in cocultured sparse cells was only observed occasionally), consistent with the idea that redox differences between sparse and dense cells are not mediated by soluble factors but are rather due to a cell autonomous mechanism. It should be noted that in these experiments, unlike the ones described earlier, differences in cell density are obtained by seeding an equal number of cells (in equal volumes of medium) on different surface areas.

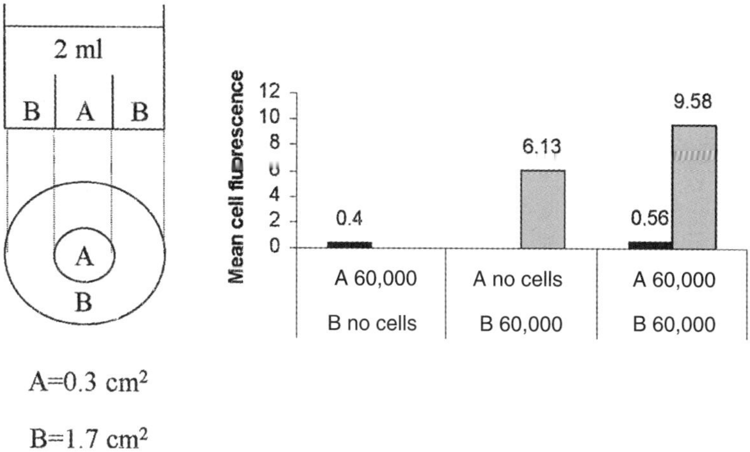

FIG. 6. Redox changes induced by cell density are not mediated by soluble factors. A special coculture of sparse and dense cells was set as described in the text. The profile and the section of the modified wells are shown. Swiss 3T3 cells plated in sections A and B shared the same medium, but could be incubated, detached, and analyzed separately. The figure is representative of two independent experiments. Numbers are mean cell fluorescence values.

Moreover, because dense and sparse cells share the same medium, the results further confirm that differences in DCF oxidation are unlikely to be due to artifacts such as insufficient oxygenation, limited availability of the fluorescent probe, or abnormal pH in crowded cultures.

Conclusions

We have developed a method to evaluate the intracellular content of oxidants as a function of cell density, i.e., in cells plated at different degree of confluence. This method is based on the cytofluorimetric analysis of the oxidative modification of a widely used probe, dichlorofluorescein diacetate, which gains a bright green fluorescence on interaction with hydrogen peroxide and organic peroxides. Importantly, DCF can be oxidized only in the intracellular environment and has proved to be an excellent tool to study transient intracellular modifications of the redox balance involved in signal transduction. The same procedure can be conceivably extended to other fluorescent markers for oxygen species, as, for instance dihydroethidium (sensitive to superoxide)[18] and DHR-123 (sensitive to peroxides and nitric oxide).[19] Our observation that the generation of ROS (most likely hydrogen

[18] M. A. Model, M. A. Kukuruga, and R. F. Todd, *J. Immunol. Methods* **202,** 105 (1997).
[19] J. P. Crow, *Nitric Oxide* **1,** 145 (1997).

peroxide) is decreased at high cell density needs to be strengthened by further biochemical and functional studies; however, these data, if confirmed and developed, could have important physiological implications.

Cell density, for instance, exerts, in normal cells a negative control on cell proliferation ("contact inhibition"), and ROS are currently considered important vehicles of mitogenic signals downstream of growth factor receptors and oncogenic ras. It is therefore tempting to hypothesize that contact inhibition of cell growth could depend, at least in part, to a decrease in intracellular oxygen species.[20]

Moreover, because the intensity of the bactericidal respiratory burst in neutrophils has been reported to decrease at increasing cell density, it is conceivable that the modulation of ligand-induced generation of ROS by the degree of cell–cell contacts represents a conserved mechanism in mammalian cells of different lineages.[21,22]

Finally, from a technical point of view, the evidence that intracellular redox balance is affected by cell density and cell–cell contact should be taken into account when designing experiments aimed at evaluating the cellular response to prooxidant stimuli in terms of cell proliferation or apoptosis.

Acknowledgment

This work was supported by Consiglio Nazionale delle Ricerche Target Project on Biotechnology (Grant 99.00302.PF49).

[20] G. Pani, R. Colavitti, B. Bedogni, R. Anzevino, S. Borrello, and T. Galeotti, *J. Biol. Chem.* **275,** 38891 (2000).

[21] G. F. Sud'ina, S. I. Galkina, L. B. Margolis, and V. Ullrich, *Cell Adhes. Commun.* **5,** 27 (1998).

[22] S. P. Peters, F. Cerasoli, Jr., K. H. Albertine, M. H. Gee, D. Berd, and Y. Ishihara, *J. Leukocyte Biol.* **47,** 457 (1990).

[9] Identification of Redox-Active Proteins on Cell Surface

By NEIL DONOGHUE and PHILIP J. HOGG

Introduction

Redox-active proteins are defined herein as those having one or more disulfide bonds that can be reversibly reduced to a pair of closely spaced thiols under physiological conditions. This process requires a reductant, usually another redox-active protein such as thioredoxin[1] or protein disulfide isomerase (PDI),[2] which itself becomes oxidized to the disulfide form. The thiol groups of the cysteine residues are close enough in space to form an intramolecular disulfide bond, but this disulfide bond is not so stable as to be nonreducible. Any pair of thiols that can be brought into close proximity by protein folding can be considered to be potentially a redox-active pair. If the disulfide bond, however, cannot be reduced back to the dithiol form through physiological processes, then it is not considered to be redox active. In these situations the disulfide bond may be thermodynamically too stable or buried in the protein and therefore inaccessible to suitable reductants.

The intracellular region is maintained in a reducing state by the glutathione redox buffer, in which the overall proportion of reduced glutathione (GSH) to oxidized glutathione is fixed at 30–100 : 1, with the total concentration of glutathione being 2 m*M*.[3] This environment facilitates redox reactions between redox-active proteins. In the intercellular space, no redox buffer is maintained, and the presence of reactive oxygen species makes this region an oxidizing environment. Hence, the conventional view is that all thiols exist in the intercellular space as the oxidized disulfide form, either as intramolecular disulfides or as mixed disulfides with low molecular weight thiols such as GSH. We have suggested that reduction, formation, or rearrangement of disulfide bonds in secreted proteins is not an uncommon event and that the different disulfide-bonded forms of the proteins can function in different ways.[4–7]

[1] A. Holmgren, *J. Biol. Chem.* **264,** 13963 (1989).
[2] H. F. Gilbert, *J. Biol. Chem.* **272,** 29399 (1997).
[3] J. B. Huppa and H. L. Ploegh, *Cell* **92,** 145 (1998).
[4] X.-M. Jiang, M. Fitzgerald, C. M. Grant, and P. J. Hogg, *J. Biol. Chem.* **274,** 2416 (1999).
[5] N. Donoghue, P. T. W. Yam, X.-M. Jiang, and P. J. Hogg, *Protein Sci.* **9,** 2436 (2000).
[6] J. K. Burgess, K. A. Hotchkiss, C. Suter, N. P. B. Dudman, J. Szollosi, C. N. Chesterman, B. H. Chong, and P. J. Hogg, *J. Biol. Chem.* **275,** 9758 (2000).
[7] A. J. Lay, X.-M. Jiang, O. Kisker, E. Flynn, A. Underwood, R. Condron, and P. J. Hogg, *Nature* **408,** 869 (2000).

0076-6879/02 $35.00

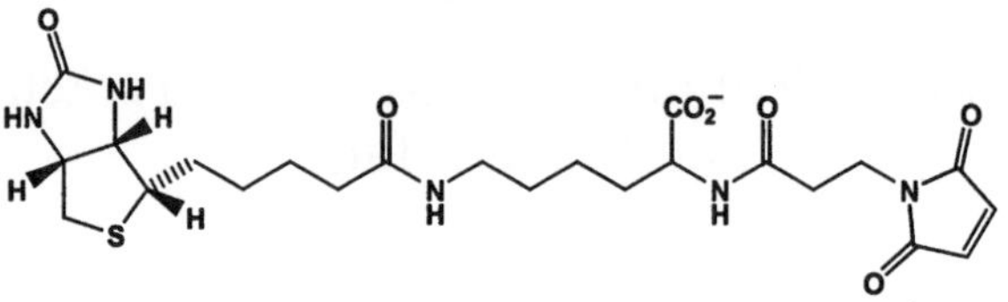

FIG. 1. The structure of MPB. MPB is expected to possess a single negative charge at physiological pH.

Redox events in secreted proteins have been probed using the substantially cell membrane-impermeable biotin-linked maleimide, *N*-[3-(*N*-maleimidyl)propionyl] biocytin (MPB),[4,6,7] shown in Fig. 1. MPB reacts with reduced protein thiols, whether or not they are part of a redox-active pair, but will not label disulfides (Fig. 2). Protein thiols labeled by MPB on the cell surface are detected using streptavidin peroxidase.

Trivalent arsenicals,* such as arsenoxides or dichloroarsines, bind tightly to closely spaced dithiols, forming cyclic adducts. Such arsenicals bind only weakly to monothiols and do not bind at all to disulfides (Fig. 3). An example is the adduct formed between *p*-tolyldichloroarsine and 2,3-dimercaptopropanol (DMP), where the crystal structure reveals a five-membered dithioarsonite ring.[8] These cyclic adducts are markedly more stable than the noncyclic adducts formed from trivalent arsenicals and monothiols due to entropic reasons.[9] Because of this, trivalent arsenicals such as phenylarsenoxide (PAO) have been used to investigate proteins that contain closely spaced thiols.[10–12] We have prepared the substatially cell membrane-impermeable trivalent arsenical, 4-[*N*-(*S*-glutathionylacetyl)amino] phenylarsenoxide (GSAO).[5] GSAO binds tightly to closely spaced synthetic, peptide and protein dithiols, but not to monothiols.[5] In order to identify cell surface proteins that contain closely spaced thiols, a biotin moiety was attached through a spacer arm to the primary amino group of the γ-glutamyl residue of GSAO, giving GSAO-B[5] (Fig. 4).

This chapter describes the synthesis of GSAO-B and the use of MPB and GSAO-B to identify and characterize redox-active proteins on the cell surface.

* Trivalent arsenoxides, such as PAO, are usually written with an arsenic-oxygen double bond, i.e., RAs=O. This has been shown not to be the case, the actual structure most probably is either a cyclic polymer $[RAsO]_n$ containing As-O-As bonds or the hydrate, $RAs(OH)_2$, an organoarsonous acid (see Ref. 14). In dilute aqueous solution, the hydrate is the form most likely to be adopted by an arsenoxide, and we have chosen to reflect this fact in our diagrams.

[8] E. Adams, D. Jeter, A. W. Cordes, and J. W. Kolis, *Inorg. Chem.* **29,** 1500 (1990).

[9] L. A. Stocken and R. H. S. Thompson, *Biochem. J.* **40,** 535 (1946).

[10] C. Gitler, M. Mogyoros, and E. Kalef, *Methods Enzymol.* **233,** 403 (1994).

[11] E. Kalef and C. Gitler, *Methods Enzymol.* **233,** 395 (1994).

[12] B. A. Griffin, S. R. Adams, and R. Y. Tsien, *Science* **281,** 269 (1998).

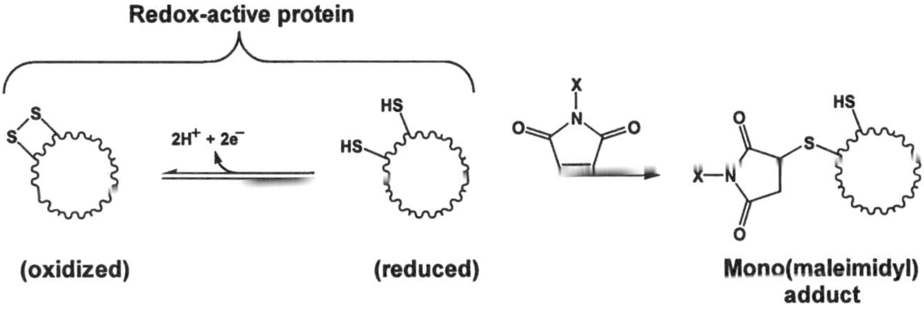

FIG. 2. The reaction of maleimide reagents with mono- and dithiols. A redox-active protein can be labeled with a maleimide reagent such as MPB only when it is in the reduced dithiol form. Excess maleimide reagent will also label the second thiol if it is sterically accessible (maleimide reagents will also label monothiols in the same manner).

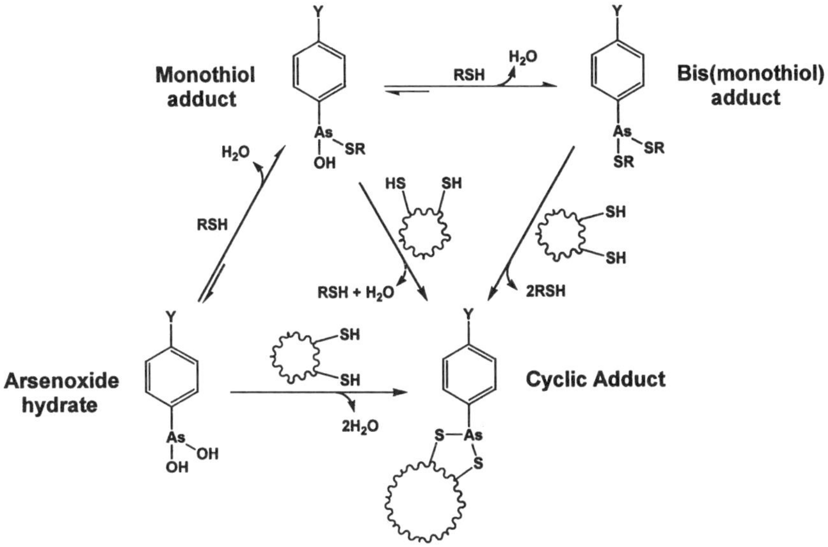

FIG. 3. The reaction of trivalent arsenicals with mono-and dithiols. Reaction of an arsenoxide hydrate with either one or two monothiols (RSH) at physiological pH will give the mono- or bis(monothiol) adduct, respectively. For steric reasons, these monothiols are most likely to be low molecular weight thiols, such as GSH. For each thiol sulfur binding to arsenic, one molecule of water is produced. Reaction of the arsenical with a monothiol results in an equilibrium, and reaction of either the arsenoxide hydrate itself or either of the monothiol adducts with a dithiol such as a reduced redox-active protein will drive the equilibria toward the cyclic protein–arsenical adduct.

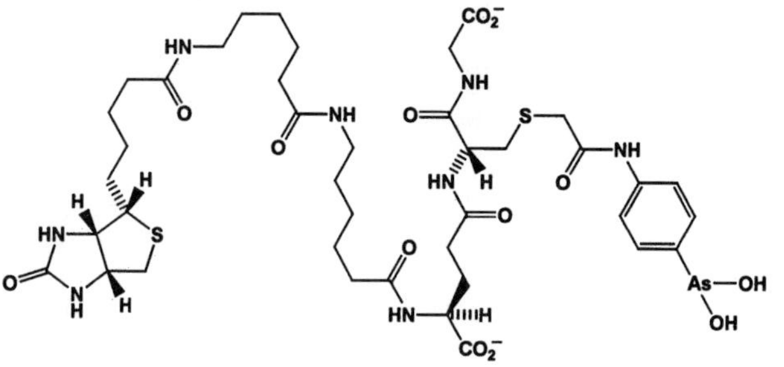

FIG. 4. Structure of GSAO-B. At physiological pH, GSAO-B is expected to exist as a doubly charged dianion due to ionization of both carboxylic acid groups on the glutathione pendant.

Synthesis of GSAO

The general scheme for the three-step synthesis of GSAO is shown in Fig. 5. The method produces two intermediates: BRAA and BRAO. Each step of the reaction is described in turn. It should be noted that compounds of arsenic are toxic and should be handled in a fume hood whenever there is a possibility of exposure either to the dust or to the fumes when dissolved in dimethyl sulfoxide (DMSO).

Synthesis of 4-[N-(Bromoacetyl)amino]phenylarsonic acid (BRAA)

Principle. Amino groups can be acylated with either acyl halides or carboxylic anhydrides to give carboxylic amides. One molecule of acid is produced for every amine acylated, and sodium carbonate is used as a base to remove this acid, thereby inhibiting the reverse reaction. Bromoacetyl bromide is corrosive and a lachrymator and should be handled with care in a fume hood.

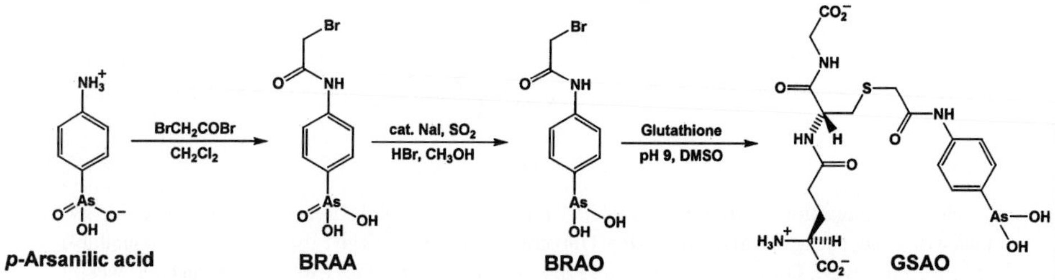

FIG. 5. The three-step synthesis of GSAO from *p*-arsanilic acid. Acylation of *p*-arsanilic acid with bromoacetyl bromide gives BRAA. The pentavalent arsonic acid in BRAA is reduced to the trivalent arsenoxide hydrate BRAO, which is then coupled to GSH under mildly alkaline conditions to give GSAO. Adapted with permission from N. Donoghue, P. T. W. Yam, X.-M. Jiang, and P. J. Hogg, *Protein Sci.* **9,** 2436 (2000).

Materials

4-Aminobenzenearsonic acid (*p*-arsanilic acid) was obtained from Tokyo Kasei Kogyo (Tokyo, Japan)
Bromoacetyl bromide (fume hood)
A solution of 16 g bromoacetic acid in 25 ml dichloromethane (fume hood)
Sodium carbonate
Sulfuric acid, 98% (w/w)

Procedure. p-Arsanilic acid (10 g) is added in portions to a well-stirred solution of sodium carbonate (30 g) in water (150 ml). When all the solids dissolve, 200 ml of water is added; the solution is about pH 10–11 at this point. Bromoacetyl bromide (10 ml) is added to the solution of bromoacetic acid in dichloromethane, and the solution is transferred to a pressure-equalizing dropping funnel fitted to the flask containing the alkaline arsonate solution. The acylating solution is allowed to run into the aqueous solution at a rate of about 2 drops per second, the whole time maintaining efficient stirring. It should take 30 to 45 min to add the acylating solution, and the reaction mixture should be about pH 8. After stirring at ambient room temperature for 2 hr, the organic layer is separated and discarded, and the aqueous solution is carefully acidified with 98% sulfuric acid to pH 4. A white precipitate forms and is collected by suction filtration on a water pump.

Reduction of BRAA to 4-[N-(Bromoacetyl)amino]phenylarsenoxide (BRAO)

Principle. To be able to bind dithiols, the arsenic atom must be in the trivalent state. Because the arsenic atom in an arsonic acid is in the pentavalent state, it must be reduced. Reduction involves bubbling sulfur dioxide through a strongly acidic solution of the arsonic acid (BRAA) containing a catalytic amount of sodium iodide. Hydrobromic acid is used in place of hydrochloric acid to prevent halogen exchange in the product. The actual reducing agent is hydroiodic acid, which is converted into iodine on reaction with pentavalent arsenic. The iodine is then reduced by sulfur dioxide, regenerating the hydroiodic acid.

Materials

Hydrobromic acid, 48% (w/w) (fume hood)
Methanol (Ajax, Auburn, Australia)
Sodium iodide
Cylinder of sulfur dioxide (fume hood)

Procedure. Suspend about 10 g of BRAA in 75 ml of methanol with constant stirring in a two-necked 250-ml round-bottomed flask and add 75 ml of hydrobromic acid (48%). All of the solids should dissolve within a few minutes (any solids that do not dissolve should be removed by filtration). The flask is placed in a cold water bath to prevent any increase in temperature above normal. A crystal of sodium iodide is added, with an immediate production of a brown color due to the formation of iodine. The supply of sulfur dioxide is connected to the tube in the side arm, and the gas is allowed to bubble through the rapidly stirred solution at a rate of about two bubbles per second. After some time (usually 2–3 hr), a white precipitate will form. Allow the reaction to continue for about 6–9 hr overall. The reaction mixture is then poured into 150 ml of cold water, and the product is collected by suction filtration.

Synthesis of 4-[N-(S-glutathionylacetyl)amino]phenylarsenoxide (GSAO)

Principle. Buffering the GSH to around pH 9–10 ensures that a significant percentage of the thiol group is deprotonated to the thiolate anion. The thiolate can then attack the bromoacetyl group of BRAO, displacing bromide to form the product, GSAO. Because arsenic in the trivalent state is oxidized easily by dissolved oxygen when buffered in a neutral or alkaline solution, all oxygen must be removed from the water and DMSO to be used prior to carrying out the reaction.

Materials

Glutathione, reduced (GSH)

Deoxygenated DMSO is prepared by bubbling nitrogen gas through an excess of DMSO for about 10 min in a fume hood

Deoxygenated 0.5 *M* sodium bicarbonate buffer, pH 9.6, prepared using deoxygenated water: about 300 ml of Milli-Q water is brought to boil for 2–3 min and then allowed to cool to room temperature under a flow of nitrogen gas. The buffer should be made up with sodium bicarbonate and sodium carbonate in the bottle it is to be kept in (a Pyrex bottle with a screw cap is recommended). Nitrogen gas should be bubbled through the buffer each time it is used.

Procedure. Dissolve about 1.0 g of BRAO in 10 ml of deoxygenated DMSO and determine the concentration of active arsenic using the activity assay described later. From the volume of the solution, the number of moles of active arsenical is calculated using $n = cv$ (for 1.0 g of perfectly dry BRAO hydrate in exactly 10 ml of solution, $n = 3.1$ mmol, but both the possibility of oxidation by molecular oxygen and the presence of water in the sample of BRAO will produce a smaller value for n, typically about 2.5 mmol). Dissolve an identical number of moles of GSH in the deoxygenated bicarbonate buffer (use about 5 ml of buffer per millimole of GSH) and add to the BRAO solution, mixing thoroughly. The solution will warm up somewhat from the mixing of DMSO with water and should be left to stand overnight (about 9 hr) before being neutralized by the dropwise addition of 32% hydrochloric acid. Lyophilization gives GSAO as a hygroscopic white solid. GSAO is stable toward oxidation as a dry solid, whereas oxidation of trivalent arsenoxides such as GSAO and GSAO-B is inhibited in solution either by buffering to pH 5.5 or by adding a twofold molar excess of glycine.

Synthesis of GSAO-B

The conversion of GSAO into GSAO-B is a single step, involving reaction with the amine-reactive biotinylating reagent 6-{*N*-[6-(*N*-biotinoylamino)hexanoyl]-amino}hexanoic acid, succinimidyl ester (biotin-XX, SE), as shown in Fig. 6.

Principle. The biotinylating reagent biotin-XX, SE contains an activated succinimidyl ester, which can couple to amino groups, in this case the amino group on the GSH pendant of GSAO. The amino group needs to be in the form of the free base to react with the ester, and therefore coupling must be carried out under mildly alkaline conditions (pH 8.5); having the conditions too alkaline (pH > 9.5) will result in the succinimidyl ester being hydrolyzed by hydroxide. The by-product from the coupling reaction is *N*-hydroxysuccinimide, which is soluble in both water and DMSO. Because it does not interfere with any subsequent reactions, removal by further purification is unnecessary.

Materials

6-{*N*-[6-(*N*-biotinoylamino)hexanoyl]amino}hexanoic acid, succinimidyl ester (biotin-XX, SE) was obtained from Molecular Probes (Eugene, OR)
Glycine

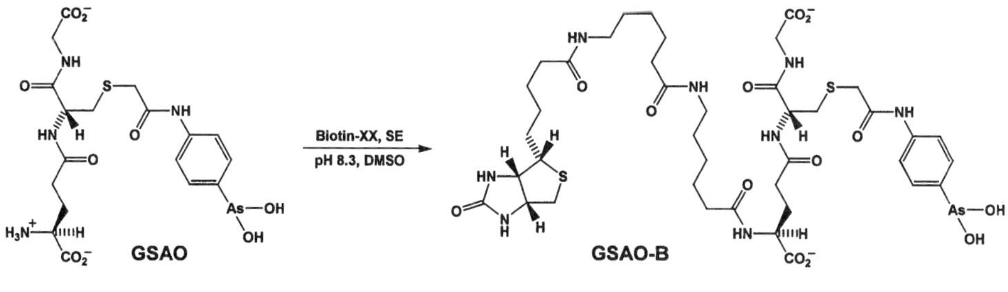

FIG. 6. The conversion of GSAO into GSAO-B. The amine-reactive biotin-XX, SE is coupled to GSAO under mildly alkaline conditions, giving GSAO-B.

Deoxygenated 0.5 *M* sodium bicarbonate buffer, pH 8.5, prepared using deoxygenated water in the same way as described for the deoxygenated 0.5 *M* sodium bicarbonate buffer, pH 9.6.

Procedure. Dissolve about 0.2 g GSAO in 5 ml of 0.5 *M* sodium bicarbonate buffer (pH 8.5) and determine the active concentration using the activity assay described later. A solution of 1.1 equivalents of biotin-XX, SE in DMSO is prepared at a concentration of 100 mg/ml. Mix the two solutions thoroughly and incubate at 4° for 4 hr. About 2 equivalents of glycine is then added, and the solution is kept at 4° overnight.

Determination of the Activity of Trivalent Arsenicals Such as BRAO, GSAO, and GSAO-B

Principle. 5,5′-Dithiobis(2-nitrobenzoic acid) (DTNB) reacts with free thiols at neutral pH, producing the strongly yellow 5-thio-2-nitrobenzoate (TNB) dianion. DMP will react with DTNB to produce the yellow color, but no reaction will occur with DMP complexed to trivalent arsenic. Hence, the active concentration of the trivalent arsenical can be determined by a titration between increasing amounts of the arsenical and a constant amount of DMP (Fig. 7).

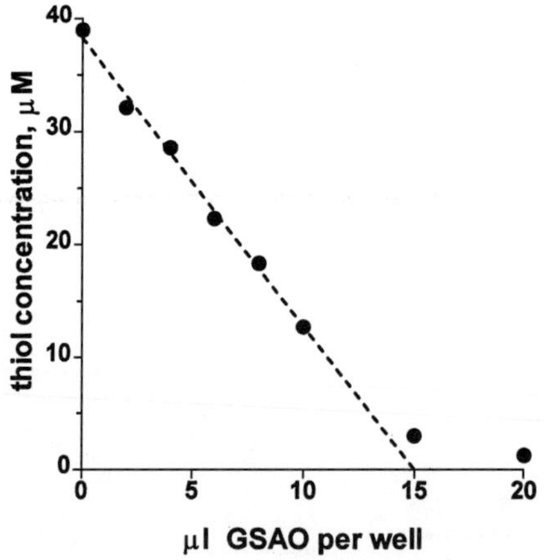

FIG. 7. Titration of GSAO with DMP. DMP (20 μM) was incubated with increasing volumes of a stock solution of GSAO (~270 μM) for 20 min. DTNB (950 μM) was then added, and the reactions was incubated for a further 20 min. The concentration of thiol in the reactions was determined from the absorbance of the TNB dianion at 412 nm. The dotted line represents the linear least-squares fit of first six data points. The equivalence point was 15 μl of GSAO per well.

Materials

Reaction buffer: 0.1 *M* HEPES, 0.3 *M* NaCl, 1 m*M* EDTA, pH 7.0
Working solution of ~40 m*M* DTNB in DMSO
Stock solution of 50 m*M* DMP in DMSO (fume hood)
Working solution of 500 μ*M* DMP: dissolve 10 μl stock solution into 1 ml reaction buffer
Working solution of the arsenical, approximately 500 μ*M*

Procedure. The activity of the arsenical is determined by titrating varying amounts of arsenical against the DMP working solution (10 μl) in a 96-well microtitre plate, with the total volume made up to 195 μl by the addition of reaction buffer. The reaction is incubated for 20 min at room temperature with constant agitation on a plate shaker, and then 5 μl of the working solution of DTNB is added and the plate is agitated for a further 20 min. The absorbance at 412 nm is measured, and the concentration of thiols is calculated using the extinction coefficient for the TNB dianion at pH 7.0 of 14,150 M^{-1} cm^{-1}.

Labeling of Cell Surfaces with MPB or GSAO-B

Principle. Cultured cells are incubated with either MPB or GSAO-B in the absence or presence of GSH or DMP, respectively. The labeled lysate is resolved on SDS–polyacrylamide gel electrophoresis (SDS–PAGE), transferred to a polyvinyldiethylene fluoride (PVDF) membrane, and the biotin-labeled proteins detected by blotting with streptavidin peroxidase (Fig. 8).

Materials

Working solution of 10 m*M* MPB in *N,N*-dimethylformamide (DMF)
Working solution of 10 m*M* GSAO-B in phosphate-buffered saline (PBS) containing 100 m*M* glycine
Working solution of 40 m*M* DMP in DMSO
Working solution of 40 m*M* GSH (Milli-Q water)
Sonication buffer: 50 m*M* Tris–HCl, 0.5 *M* NaCl, 1% (v/v) Triton X-100, 10 μ*M* leupeptin, 10 μ*M* aprotinin, 2 m*M* phenylmethylsulfonyl fluoride, 5 m*M* EDTA, pH 8.0
8–16% SDS–PAGE gel (Gradipore, Sydney, Australia)
PVDF membrane (Millipore, Bedford, MA)
Streptavidin horseradish peroxidase (Amersham, Sydney, Australia)
Chemiluminescence (NEM Life Science Products, Boston, MA)

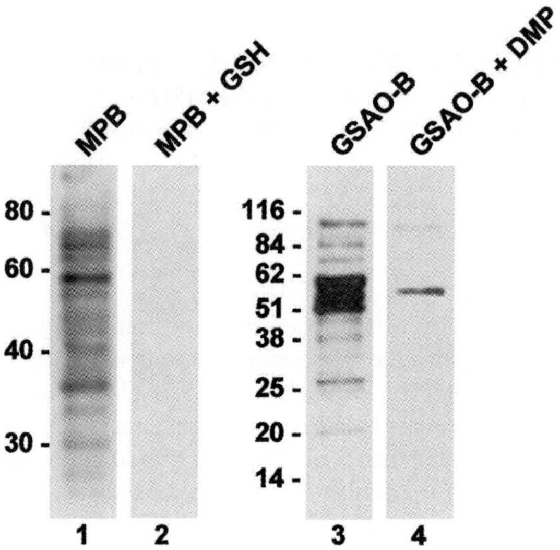

FIG. 8. Identification of redox-active proteins on the fibrosarcoma cell surface using MPB or GSAO-B. Human fibrosarcoma HT1080 cells (3×10^6 cells in 0.5 ml) were labeled with either MPB or GSAO-B (100 μM) for 30 min. Labeling with MPB was in the absence (lane 1) or presence (lane 2) of GSH (400 μM), whereas labeling with GSAO-B was in the absence (lane 3) or presence (lane 4) of DMP (400 μM). The cells were lysed, and the lysate (2.5 μg, $\sim 1 \times 10^4$ cells for MPB and 50 μg, $\sim 2 \times 10^5$ cells for GSAO-B) was resolved on 4–15% SDS–PAGE, transferred to a PVDF membrane, and blotted with streptavidin peroxidase to detect the biotin label. The positions of M_r markers are shown at the left.

Procedure. Cultured mammalian cells (1 ml of 5–10 $\times 10^6$ ml in Hank's balanced salt solution) are incubated with 10 μl of the working solutions of MPB or GSAO-B (final concentrations of 100 μM) for 30 min at room temperature. On some occasions, cells were labeled with MPB in the presence of 10 μl of the working solution of GSH (final concentration of 400 μM DMP) or labeled with GSAO-B in the presence of 10 μl of the working solution of DMP (final concentration of 400 μM DMP). Labeled cells are then washed three times with PBS and sonicated in 1 ml of sonication buffer at 4°. Samples of the cell lysates (2.5 μg, $\sim 1 \times 10^4$ cells for MPB and 50 μg, $\sim 2 \times 10^5$ cells for GSAO-B) are resolved on 4–15% SDS–PAGE, transferred to a PVDF membrane, blotted with a 1 : 2000 dilution of streptavidin horseradish peroxidase, and then developed and visualized using chemiluminescence according to the manufacturer's instructions.

Identification of MPB- and GSAO-B-Labeled Proteins

Principle. Cultured cells are labeled with either MPB or GSAO-B in the absence or presence of GSH or DMP, respectively. Biotin-labeled proteins are collected of streptavidin agarose, resolved on SDS–PAGE, transferred to a PVDF membrane, and detected by Western blotting.

Materials

Washing buffer: 50 m*M* Tris–HCl, 0.15 *M* NaCl, 0.05% Triton X-100, pH 8.0
8–16% SDS–PAGE gel (Gradipore, Sydney, Australia)
PVDF membrane (Millipore)
Streptavidin horseradish peroxidase (Amersham)
Chemiluminescence (NEM Life Science Products)

Procedure. Streptavidin-agarose beads (50 μl of packed beads) are incubated with the MPB- or GSAO-B-labeled cell lysates for 1 hr at 4° on a rotating wheel to isolate the biotin-labeled proteins. The streptavidin-agarose beads are washed five times with washing buffer, and the biotin-labeled proteins are released from the beads by boiling in 30 μl of SDS–Laemmli buffer for 2 min. Labeled proteins are resolved on 4–15% SDS–PAGE, transferred to a PVDF membrane, and detected by Western blotting.

Conclusion

We describe two reagents that can be used to detect and identify redox-active proteins on the cell surface. MPB and GSAO-B do not cross the plasma membrane to any appreciable extent. GSAO, in particular, has a GSH pendant, which is triply charged at physiological pH, making it hydrophilic and unlikely to dissolve in the predominantly nonpolar membrance of cells. Moreover, GSH is constituitively secreted by mammalian cells[13] but is not actively taken up by these cells. These factors make GSH an ideal pendant to limit entry of the trivalent arsenical into the cell. Any uptake of the reagents into cells is minimized further by restricting the labeling time to 30 min.

MPB labels any accessible thiol, whereas GSAO-B labels closely spaced dithiols on the cell surface. GSAO-B, therefore, will bind to a subset of the MPB-labelled proteins. The rationale is that if the thiols of the dithiol are close enough to complex with the arsenical of GSAO, then the thiols are close enough to form a disulfide bond. GSAO-B can identify, therefore, those proteins that contain one or more redox-active disulfide bonds by binding to the reduced dithiol form of the disulfide.

Several proteins on the human fibrosarcoma cell surface were labeled with MPB.[4] One of these proteins was shown to be PDI.[4] Both an increase in the labeling

[13] S. Bannai and H. Tsukeda, *J. Biol. Chem.* **254,** 3444 (1979).

[14] G. O. Doak and L. D. Freedman, "Organometallic Compounds of Arsenic, Antimony, and Bismuth." Wiley-Interscience, New York, 1970.

intensity of 11 of the proteins upon overexpression of PDI and a clear decrease in the labeling intensity of 3 of the 11 proteins on underexpression of PDI indicated that secreted PDI was controlling the redox state of existing exofacial redox-active proteins. By comparison, 12 distinct redox-active proteins were labeled with GSAO-B on the surface of human fibrosarcoma cells and 10 distinct proteins on the surface of bovine aortic endothelial cells.[5] PDI was labeled with GSAO-B on both cell surfaces.[5]

It is possible that plasma membrane dipeptidases cleave the GSH pendant of GSAO-B, although there is currently no evidence for this. The biotin moiety, however, will most likely prevent cleavage of the γ-glutamyl residue by membrane-bound γ-glutamyl transpeptidases (R. Hughey, personal communication). Cleavage of the Gly from GSAO-B would be expected to enhance the membrane permeability of the resulting fragment. This may change the biological effects of the arsenical. Cleavage of the Gly, however, should not change the reactivity of the resulting arsenical with cell surface proteins.

The overall aim is to identify all the cell surface proteins that contain closely spaced dithiols and to understand how redox events in these proteins affect function. Individual proteins can be identified by Western blotting if the identity of the protein is suspected, as was the case for PDI. Ultimately, however, GSAO-B-labeled proteins will need to be purified in sufficient quantity for peptide fingerprinting and microsequencing. We are currently isolating human platelet surface proteins[6] that bind GSAO-B. Biotin-labeled proteins are being collected on streptavidin-agarose beads and eluted with DMP. Elution with DMP excludes those proteins that have bound nonspecifically to the agarose beads. We anticipate that the function of many, if not most, of the GSAO-B-labeled platelet proteins will be influenced by reversible reduction/oxidation of one or more disulfide bonds.

[10] Probing Redox Activity of Human Breast Cells by Scanning Electrochemical Microscopy

By MICHAEL V. MIRKIN, BIAO LIU, and SUSAN A. ROTENBERG

Introduction

Developments in the design and miniaturization of electrochemical probes are creating new possibilities for investigating reduction–oxidation ("redox") processes occurring in living cells.[1–3] Single cell voltammetry has traditionally monitored and quantified the dynamic release of biologically important molecules such as catecholamines, insulin, and anticancer drugs[4–9] from living cells. To conduct such experiments, a micrometer-sized ultramicroelectrode is positioned in close

0076-6879/02 $35.00

proximity to a cell membrane, whereupon molecules ejected from the cell undergo oxidation (or reduction) at the electrode, resulting in an electrochemical signal. Measurements with an ultramicroelectrode can be further enhanced in combination with the scanning capability afforded by the scanning electrochemical microscope (SECM).[10] The ultramicroelectrode is moved in a horizontal plane above an immobilized sample, such as a cell, to obtain topographic images or maps of redox reactivity across the cell surface. Cellular redox activity can be characterized quantitatively by measuring the rate of transmembrane charge transfer (CT) with the SECM. The fundamentals of such measurements were previously developed in model studies conducted at solid/liquid and liquid/liquid interfaces.[11–14]

The microscopic size of the electrode tip (<5-μm radius) makes the SECM uniquely suited to probe redox processes occurring within single cells. Single cell SECM measurements provide valuable information that is hard to extract from the averaged signal produced by a large number of cells. Moreover, measurements can be accomplished on a milli- to microsecond time scale, which permits detection of short-lived radicals such as reactive oxygen species, and is therefore superior to the minute-long measurements required to analyze a large cell population.[15]

In one application of the SECM, redox reactivity of a cell can be measured by using a redox-sensitive chemical mediator (Fig. 1A[10]). In this experiment, the ultramicroelectrode (called the "tip") and cells that are immobilized in a plastic culture dish are immersed in buffered solution containing the oxidized (or reduced) form of a membrane-permeable redox mediator. Prior to entering the cell, the mediator is converted to its reduced (or oxidized) form at the ultramicroelectrode surface. In this form, the mediator crosses the membrane and engages with a suitably reactive

[1] R. N. Adams, *Anal. Chem.* **48,** 1128A (1976).

[2] R. M. Wightman, R. T. Kennedy, D. J. Wiedemann, K. T. Kawagoe, J. B. Zimmerman, and D. J. Leszczyszyn, *in* "Microelectrodes: Theory and Applications" (M. Montenegro, M. A. Queirós, and J. L. Daschbach, eds.), p. 453. Kluwer Academic, Dordrecht, 1991.

[3] R. D. O'Neill, *Analyst* **119,** 767 (1994).

[4] R. M. Wightman, J. A. Jankowski, R. T. Kennedy, K. T. Kawagoe, T. J. Schroeder, D. J. Leszczyszyn, J. A. Near, E. M. Diliberto, Jr., and O. H. Viveros, *Proc. Natl. Acad. Sci. U.S.A.* **88,** 10754 (1991).

[5] A. G. Ewing, T. S. Strein, and Y. Y. Lau, *Acc. Chem. Res.* **25,** 440 (1992).

[6] R. T. Kennedy, L. Huang, M. A. Atkinson, and P. Dush, *Anal. Chem.* **65,** 1882 (1993).

[7] W. G. Kuhr and P. Pantano, *Electroanalysis* **7,** 405 (1995).

[8] H. Lu and M. Gratzl, *Anal. Chem.* **71,** 2821 (1999).

[9] C. Yi and M. Gratzl, *Biophys. J.* **75,** 2255 (1998).

[10] B. Liu, S. A. Rotenberg, and M. V. Mirkin, *Proc. Natl. Acad. Sci. U.S.A.* **97,** 9855 (2000).

[11] A. J. Bard and M. V. Mirkin, eds., "Scanning Electrochemical Microscopy." Dekker, New York, 2001.

[12] C. Wei, A. J. Bard, and M. V. Mirkin, *J. Phys. Chem.* **99,** 16033 (1995).

[13] M. Tsionsky, A. J. Bard, and M. V. Mirkin, *J. Phys. Chem.* **100,** 17881 (1996).

[14] A. L. Barker, J. V. Macpherson, C. J. Slevin, and P. R. Unwin, *J. Phys. Chem. B* **102,** 1586 (1998).

[15] J. D. Rabinowitz, J. F. Vacchino, C. Beeson, and H. M. McConnell, *J. Am. Chem. Soc.* **120,** 2464 (1998).

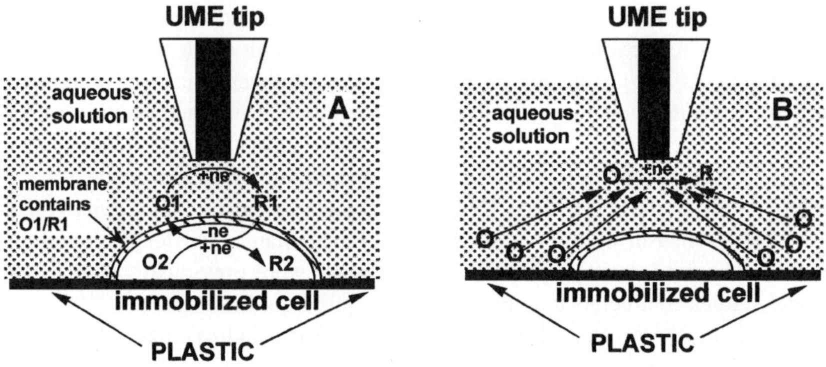

FIG. 1. Schematic diagrams of SECM experiments with two different types of mediator regeneration. The tip of the ultramicroelectrode (UME) is positioned in the solution close to the cell surface. (A) Bimolecular electron transfer between hydrophobic redox mediator (O1/R1) and cell-bound redox moieties (O2/R2). (B) The lipid cell membrane is impermeable for a hydrophilic redox mediator. Negative feedback is due to the hindered diffusion of redox species to the tip electrode. Reprinted from B. Liu, S. A. Rotenberg, and M. V. Mirkin, *Proc. Natl. Acad. Sci. U.S.A.* **97,** 9855 (2000), with permission from The National Academy of Sciences, U.S.A.

intracellular redox center that results in its oxidation (or reduction). On return to the extracellular compartment, the change in oxidation state caused by the intracellular redox center is reversed by the reaction at the tip. When the tip is positioned at a distance (d) from the cell membrane such that d is less than or comparable to the tip radius, this process produces an enhancement (positive feedback) in the faradaic current (i_T) at the tip electrode in a manner that depends on the tip-to-membrane distance. The electrochemical reaction therefore reports on the extent to which the mediator is oxidized (or reduced) by the cell (redox reactivity).

Alternatively, if the cell membrane is impermeable to a mediator, it will serve as an electrical insulator (Fig. 1B). As a result, no reactions of the tip-generated species (e.g., O) will occur at the surface. The closer the tip to the cell surface, the smaller the i_T (negative feedback) because the insulator blocks diffusion of species R from the bulk solution to the tip.[11] Thus, by scanning the tip closely over the cell surface, variations in i_T can be related to changes in cell surface topography.

As a specific application of SECM, this chapter compares the intrinsic redox reactivity and topographies of metastatic and nonmetastatic human breast cell lines. One of the many properties that distinguish these cell lines is the expression level of protein kinase Cα (PKCα), a redox-sensitive enzyme[16] that has been linked with motility and metastasis of various cell types.[17,18] MCF-10A

[16] R. Gopalakrishna and W. B. Anderson, *Proc. Natl. Acad. Sci. U.S.A.* **86,** 6758 (1989).

[17] R. Gopalakrishna and S. H. Barsky, *Proc. Natl. Acad. Sci. U.S.A.* **85,** 612 (1988).

[18] E. Batlle, J. Verdu, D. Dominguez, M. del Mont Llosas, V. Diaz, N. Loukili, R. Paciucci, F. Alameda, and A. Garcia de Herreros, *J. Biol. Chem.* **273,** 15091 (1998).

cells are nontransformed, nonmetastatic cells that express very low levels of PKCα. When MCF-10A cells are genetically engineered to overproduce PKCα, these normally nonmotile cells exhibit a high degree of motility (renamed as 11α cells), as described previously.[19] The SECM is used to measure and compare the redox reactivity of MCF-10A cells, 11α cells, and overtly metastatic MDA-MB-231 human breast cells (which also express high levels of PKCα). With respect to the kinetics of transmembrane charge transfer by a variety of redox mediators, nonmotile and motile cells exhibit distinct and reproducible differences.

Materials and Methods

Chemicals

All aqueous solutions are prepared from deionized water (Milli-Q, Millipore Corp.). Menadione (General Biochemicals, Chagrin Falls, OH), 1,2-naphthoquinone (Aldrich, Milwaukee, WI), $Ru(NH_3)_6Cl_3$ (Strem Chemicals, Newburyport, MA), and $Na_4Fe(CN)_6$ (Fisher Scientific, Fair Lawn, NJ) are used as received. All other chemicals are reagent grade.

SECM Probes

A two-electrode setup is employed. AgCl-coated Ag wire (0.25 mm) serves as a reference electrode. Pt wires (12.5-, 5-, and 1-μm radius) and carbon fibers (5.5- and 3.5-μm radius) are heat-sealed in glass tubes under vacuum and then beveled to produce SECM tips, as described previously.[11,20]

Cell Culture

Midpassage MCF-10A cells, a human breast epithelial cell line, are cultured in DMEM/F12 media (1 : 1) supplemented with 15% equine serum, insulin (10 μg/ml), epidermal growth factor (20 ng/ml), cholera toxin (100 ng/ml), and hydrocortisone (0.5 μg/ml) and are maintained with penicillin (100 units/ml), streptomycin (100 μg/ml), and fungizone (0.5 μg/ml). Cells are passaged at 1 : 3 to 1 : 6 every 3 to 4 days. A stable transfectant clone (11α cells) that constitutively expresses bovine protein kinase Cα (PKCα) is isolated by neomycin resistance as described previously.[19] Transfectants are maintained in complete growth medium containing 125 μg/ml G418 and are cultured up to 20 passages. MDA-MB-231 cells are cultured in Iscove's modified Dulbecco's medium with L-glutamine, 10% fetal bovine serum, and 1% penicillin/streptomycin. Culture media, serum, and antibiotics (fungizone, penicillin, streptomycin) are from (Life Technologies, Gaithersburg, MD).

[19] X.-G. Sun and S. A. Rotenberg, *Cell Growth Diff.* **10,** 343 (1999).

[20] A. J. Bard, F.-R. F. Fan, J. Kwak, and O. Lev, *Anal. Chem.* **61,** 1794 (1989).

Instrumentation and Procedures

All measurements are performed at ambient temperature in a plastic culture dish mounted on a horizontal stage of the SECM and placed inside a Faraday cage. The SECM apparatus and procedures have been described previously.[12,13,21]

All measurements are conducted with live biological cells that are plated at low density in a 60-mm plastic culture dish and immersed in phosphate-buffered saline (PBS). The ultramicroelectrode is carefully positioned by use of a micromanipulator. To measure rate constants, the i_T is recorded as a function of the tip position as the tip is scanned vertically above the cell surface. One-dimensional current profiles and gray-scale constant-height images of cells are obtained by scanning the tip horizontally in the x–y plane above the cell surface.

According to previously developed models,[12–14] the rate of interfacial charge transfer should appear immeasurably low if the concentration of redox species in solution is much higher than the intracellular concentration of redox centers. We have found that quantitative kinetic measurements can be made with concentrations of redox species as low as 10 μM, as confirmed by fitting experimental current–distance curves obtained at both conductive and insulating substrates to theory.

Due to the negative standard potentials of the menadione and 1,2-naphthoquinone redox couples used in our studies, electron transfer can occur between these mediators and oxygen in the medium. As a result, deoxygenation of the solution with argon gas is necessary in experiments with these mediators. To prevent potential damage to the cells, oxygen removal from solution is for only the short period of time (<1 hr) required for an entire experiment. Thus, after positioning the tip above an immobilized cell, a flow of argon is passed quickly through a small volume (~5 ml) of PBS that covers the cells. The nearly complete removal of O_2 can be verified by cyclic voltammetry.

An additional caveat is to avoid touching the cell surface with the ultramicroelectrode, as this could lead to penetration and puncture of the cell membrane. The destruction of the cell in this manner results in pure negative feedback.

Applications

Topographic Imaging of Cultured Human Breast Cells with Hydrophilic Mediators

When mediator species are not soluble in lipid phase and cannot permeate the membrane, no mediator regeneration by the cell is possible because the lipid bilayer is too thick for efficient electron tunneling.[22–24] As a result, SECM current

[21] B. Liu and M. V. Mirkin, *J. Am. Chem. Soc.* **121,** 8352 (1999).

[22] H. Yamada, T. Matsue, and I. Uchida, *Biochem. Biophys. Res. Commun.* **180,** 1330 (1991).

[23] M. Tsionsky, A. J. Bard, and M. V. Mirkin, *J. Am. Chem. Soc.* **119,** 10785 (1997).

[24] M. Tsionsky, J. Zhou, S. Amemiya, F.-R. F. Fan, A. J. Bard, and R. A. W. Dryfe, *Anal. Chem.* **71,** 4300 (1999).

vs distance (i_T-d) curves display pure negative feedback when the tip approaches a mammalian cell in the presence of different hydrophilic redox mediators. When the tip is scanned in a horizontal plane parallel to the bottom of the cell culture dish, the tip current above the cell is significantly lower than that above the plastic surface (Fig. 2A). The cell surface thus impedes diffusion of the mediator to the tip (Fig. 1B). Similar experiments can be performed with various hydrophilic mediators [e.g., ferrocene carboxylate, $Ru(NH_3)_6^{3/2+}$, $Fe(CN)_6^{3/4-}$, and $Ru(CN)_6^{3/4-}$]. Despite very different standard potentials of those redox couples, pure negative feedback is always observed, and no unmediated CT across the cell membrane can be detected.

A typical gray-scale image of a nontransformed, nonmetastatic (MCF-10A) breast cell (Fig. 3A) contains only topographic information. In agreement with a one-dimensional scan in Fig. 2A, it shows an irregularly shaped cell with a nucleus near the center and surrounded by a much thinner cytoplasm. As nonmotile cells, MCF-10A cells spread over the surface of the plate.

A topographic image of a metastatic (MDA-MB-231) breast cell (Fig. 3B) yields a diameter of approximately 50 μm, i.e., close to the size of a typical cell when observed by light microscopy. Unlike a MCF-10A breast cell (Fig. 3A), the MDA-MB-231 cell appears to be essentially hemispherical. This feature is in part due to higher motility that is accompanied by weaker adherence to the plastic surface of the culture dish. We note that based on *in vitro* migration assays with MDA-MB-231 cells (unpublished data), the motility of these cells is likely to be much slower than the time interval required for data collection during a typical scan (approximately 10 min). During this brief period, motility-related changes in cell geometry are too slow to affect the resulting redox image.

Mapping Redox Activity in the Cell

In contrast with hydrophilic mediators, lipid-soluble mediators, such as menadione or 1,2-naphthoquinone and their corresponding diols, can travel back and forth across the membrane and thus produce an increase in i_T when the tip is scanned horizontally above the cell (Fig. 1A). The oxidized and reduced forms of menadione

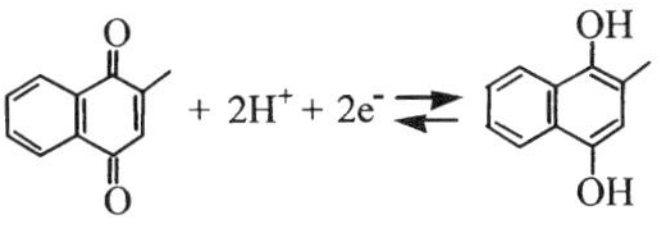

and 1,2-naphthoquinone

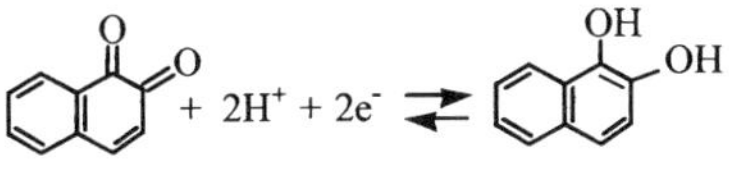

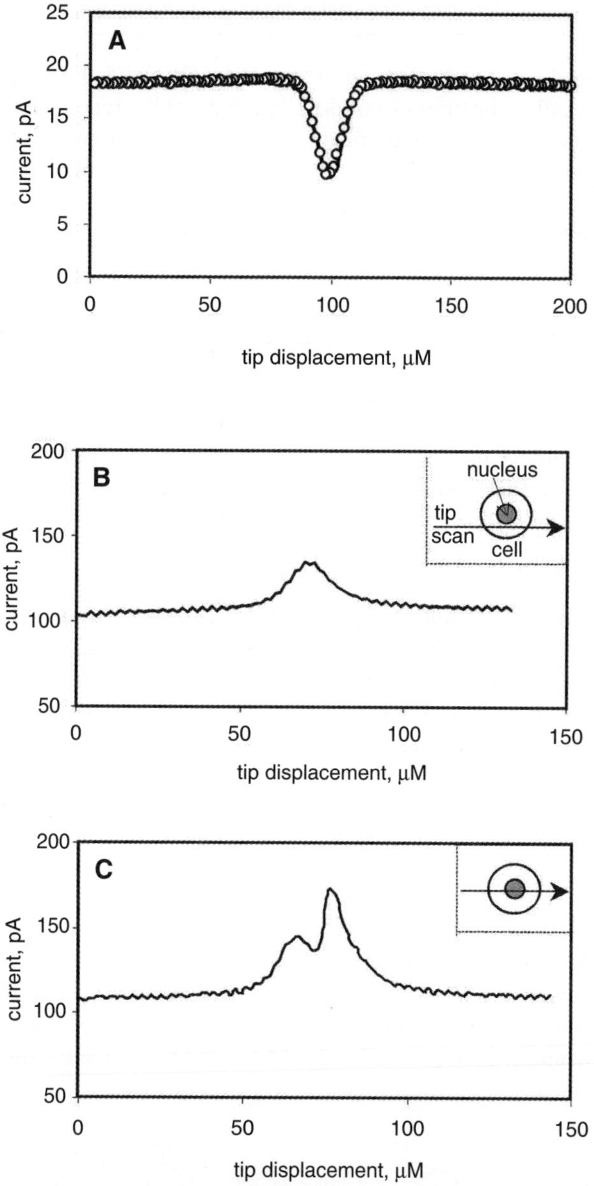

FIG. 2. Current vs tip position dependencies for an 11-μm (A) and 7-μm (B and C) carbon tip scanned laterally over a MCF-10A cell in phosphate buffer solution containing 30 μM FcCOONa (A) and 40 μM 1,2-naphthoquinone (B and C). The scan rate was 20 μm/sec. The tip potential was 0.35 V (A) and −0.36 V (B and C) vs Ag/AgCl reference. The inserts in B and C show schematically the trajectory of the tip scan line relative to the cell nucleus. (A) The solid line and symbols are obtained by scanning the probe in the forward and reverse directions above the same cell. Reprinted from B. Liu, W. Cheng, S. A. Rotenberg, and M. V. Mirkin, *J. Electroanal. Chem.* **500,** 590 (2001), with permission from Elsevier Science.

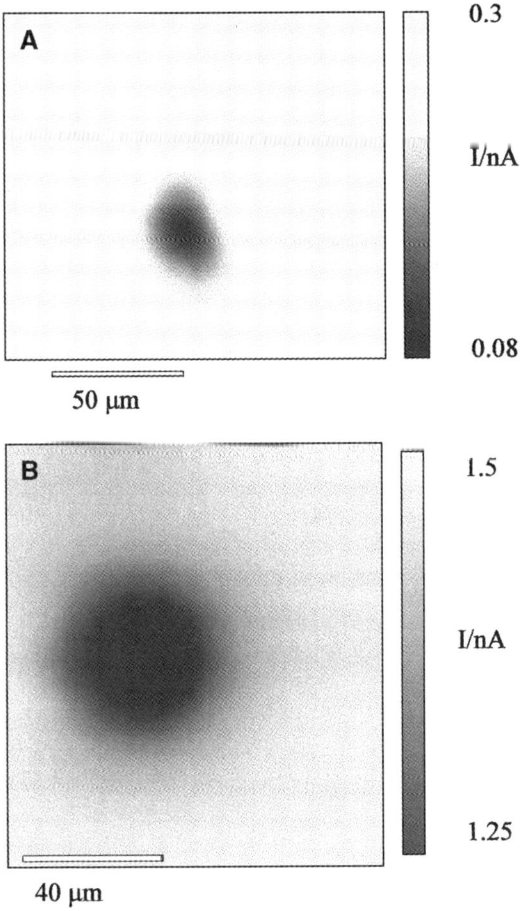

FIG. 3. Normal human breast (A) and metastatic (B) cells imaged by the SECM in solutions containing hydrophilic redox mediators [1 m*M* $Na_4Ru(CN)_6$ (A) and 1 m*M* $Ru(NH)_6Cl_3$ (B)]. The tip was an 5.5-μm radius carbon (A) and a 12.5-μm radius Pt UME (B). The tip potential was 0.84 V (A) and −0.3 V (B). The scan rate was 10 μm/sec. Reprinted from B. Liu, W. Cheng, S. A. Rotenberg, and M. V. Mirkin, *J. Electroanal. Chem.* **500,** 590 (2001), with permission from Elsevier Science.

are neutral and cross the cell membrane readily. The increase in the i_T above the cell shown in Fig. 2B is attributed to the regeneration of the reduced form (menadiol) by the ultramicroelectrode following the release of menadione that had been formed by intracellular oxidation. The tip current variations in this case represent the distribution of redox reactivity over the cell surface. The SECM measurement requires approximately 1 min and is reproducible in repetitive recordings for a single cell. An interesting feature in Fig. 2C is the presence of two closely spaced current maxima corresponding to the same immobilized cell. Such dual peaks are

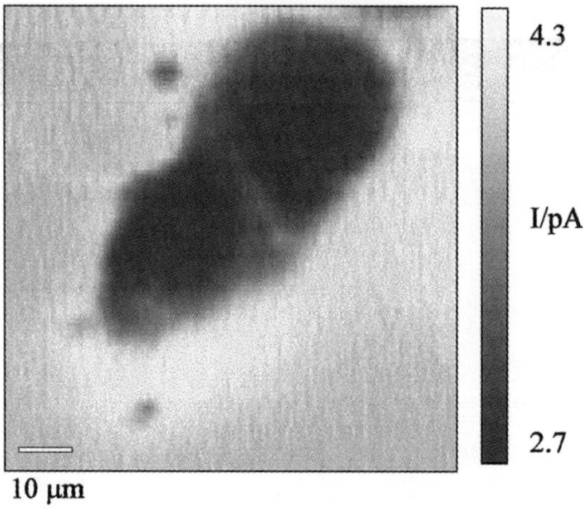

FIG. 4. Map of redox activity in normal human breast (MCF-10A) cells imaged by the SECM with a 1-μm radius Pt tip and 40 μM 1,2-naphthoquinone as mediator. Reprinted from B. Liu, S. A. Rotenberg, and M. V. Mirkin, *Proc. Natl. Acad. Sci. U.S.A.* **97,** 9855 (2000), with permission from The National Academy of Sciences, U.S.A.

observed when the tip is scanned over the center of the cell rather than over its periphery (compare inserts in Figs. 2B and 2C). Apparently, the mediator regeneration rate is higher at the periphery of the cell than around the center. This deduction is confirmed by the gray-scale image of two aggregated MCF-10A cells in Fig. 4 obtained by using 40 μM 1,2-naphthoquinone as mediator. The bright halo over the cell cytoplasm signifies the region of highest redox activity, whereas the dark, redox-inactive area in the center of the cell is a nucleus that is apparently impenetrable to the mediator species. The spatial resolution of this figure is significantly higher because of a smaller tip size (1-μm radius).

Current–distance curves (approach curves) were obtained to calculate rate constants (Fig. 5). In conducting this type of experiment with human breast cells, the kinetic rate constants are calculated as described previously.[12] As given in Table I, the apparent cellular redox activity is higher in MCF-10A breast cells, somewhat lower in 11α cells, and significantly lower in metastatic (MDA-MB-231) cells. With either 1,2-napthoquinone or menadione as mediator, the average feedback current for a MCF-10A cell is significantly higher than that obtained for an MDA-MB-231 cell. A MCF-10A breast cell regnerates 1,2-naphthoquinone at a substantially higher rate than a metastatic cancer cell. The slower mediator regeneration rates observed for 11α and MDA-MB-231 cells may be related to PKCα overexpression, which is a common factor in these cell lines. Although untested in these cells, a direct redox interaction of menadione and napthoquinone with intracellular PKC (which converts them to their reduced forms) has been reported.[25]

[25] G. E. Kass, S. K. Duddy, and S. Orrenius, *Biochem. J.* **260,** 499 (1989).

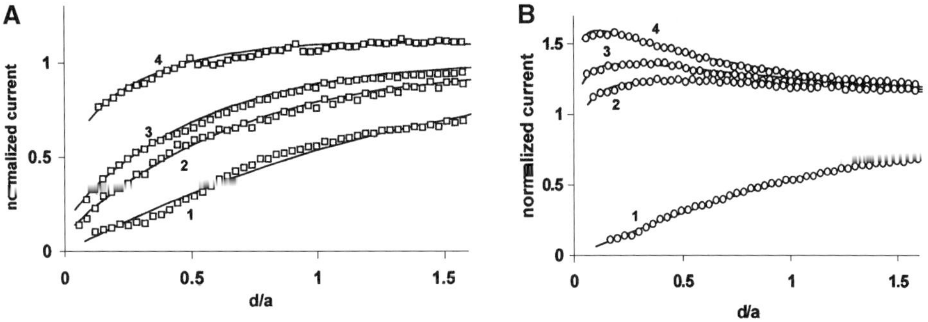

FIG. 5. An ultramicroelectrode tip approaches the (1) plastic surface, (2) MDA-MB-231 cell, (3) 11α cell, and (4) MCF-10A cell. Solid lines are the theoretical curves for an insulating substrate (curve 1) and finite heterogeneous kinetics (curves 2, 3, and 4). Phosphate buffer (pH 7.4) contained (A) 20 μM menadione and (B) 30 μM 1,2-naphthoquinone. The tip current is normalized by the current value measured in the bulk solution. The tip was a 5.5-μm radius carbon fiber. Reprinted from B. Liu, S. A. Rotenberg, and M. V. Mirkin, *Proc. Natl. Acad. Sci. U.S.A.* **97,** 9855 (2000), with permission from The National Academy of Sciences, U.S.A.

This possibility may explain in part the diminished signal produced by cells expressing high levels of PKC (Fig. 5).

The magnitude of the difference between the redox responses given by a normal breast cell and a metastatic breast cell depends on the nature of the redox mediator whose ability to donate electrons inside the cell is a function of the available redox centers and their reduction potentials. Undoubtedly, there are numerous

TABLE I
VARIABILITY OF HETEROGENEOUS RATE CONSTANTS MEASURED FOR DIFFERENT CELL LINES WITH MENADIONE AND 1,2-NAPHTHOQUINONE MEDIATORS[a]

	Rate constant (10^{-3} × cm/sec)	
Cell type	Menadione	Naphthoquinone
MCF-10A	3.8 ± 0.4 (12)	14 ± 1.1 (15)
11α	2.0 ± 0.3 (11)	11 ± 1.0 (13)
MDA-MB-231	1.5 ± 0.15 (9)	8 ± 1.2 (11)

[a] Measurements were made with either 20 μM menadione or 30 μM 1,2-naphthoquinone. The shown uncertainties are 95% confidence intervals. The number of experiments with different cells of the same type is indicated in parentheses. [Reprinted from B. Liu, S. A. Rotenberg, and M. V. Mirkin, *Proc. Natl. Acad. Sci. U.S.A.* **97,** 9855 (2000), with permission from The National Academy of Sciences, U.S.A.]

active redox couples within the cell, including those couples that reside in different subcellular compartments. The total amount of those redox centers initially present in the cell can be evaluated by integrating tip current vs time curves. By performing similar measurements using different redox mediators with different standard potentials, one can evaluate the concentrations of different redox components in the cell. An ongoing focus of our studies is to identify additional lipophilic compounds that may improve on the differential observed between normal and metastatic human breast cells with menadione or 1,2-napthoquinone (Table I).

Conclusions

Topography and redox reactivity of human breast cells have been characterized and imaged by the scanning electrochemical microscope. SECM imaging of biological cells is challenging because they are soft, possess variable thickness, and therefore can be damaged easily by the tip. This is especially true for feedback mode experiments in which the feedback from the cell is neither purely positive nor purely negative. Nevertheless, conventional, constant-height imaging is feasible and can provide useful information on both topography and reactivity. Although the SECM instrumentation used in our studies was assembled from laboratory equipment, the complete system may be purchased from CH Instruments, Inc. (Austin, TX).

The measurements of redox reactivity may help define a property of human breast cells that distinguishes nonmetastatic and metastatic cells. In addition to redox reactivity, other chemical processes occurring in living cells (e.g., diffusion of oxygen or pumping of H^+ across the membrane) can be detected with potentiometric or amperometric electrodes and imaged with submicrometer spatial resolution.[26] Mapping by SECM of such chemical reactions in subcellular organelles or localized regions of the cell can provide valuable spatial and activity information about metabolism, respiration, photosynthesis, and other important biochemical processes.

Acknowledgments

Funding for this work was provided by the Donors of the Petroleum Research Fund administered by the American Chemical Society (MVM), the Gustavus and Louise Pfeiffer Foundation (SAR), and the Research Foundation of PSC-CUNY (MVM).

[26] B. Liu, W. Cheng, S. A. Rotenberg, and M. V. Mirkin, *J. Electroanal. Chem.* **500,** 590 (2001).

[11] Determining Influence of Oxidants on Nuclear Transport Using Digitonin-Permeabilized Cell Assay

By RANDOLPH S. FAUSTINO, MICHAEL P. CZUBRYT, and GRANT N. PIERCE

Introduction

Nuclear transport is a fundamental physiological process that occurs in eukaryotic cells. It is essential for the transport of proteins, such as helicases and histones, into the nucleus[1–4] and is also responsible for the movement of molecules, such as mRNA and tRNA, from the nucleus to the cytoplasm.[1,5–9]

Nucleocytoplasmic trafficking is a highly controlled process with regulatory mechanisms at each stage of nuclear transport. Nuclear import occurs in two distinct stages: an energy-independent stage and an energy-dependent translocation step.[10] Substrates to be imported into the nucleus possess a stretch of amino acids termed the nuclear localization signal (NLS) (Table I). This NLS can be masked or unmasked by the phosphorylation[11–15] or dephosphorylation of adjacent amino acids. Exposure of the NLS may also occur after proteolytic cleavage of the protein.[16,17] An exposed, "activated" NLS may then bind to an NLS receptor

[1] H. Tang, D. McDonald, T. Middlesworth, T. J. Hope, and F. Wong-Staal, *Mol. Cell Biol.* **19,** 3540 (1999).

[2] Y. Miyamoto, N. Imamoto, T. Sekimoto, T. Tachibana, T. Seki, S. Tada, T. Enomoto, and Y. Yoneda, *J. Biol. Chem.* **272,** 26375 (1997).

[3] M. Johnson-Saliba, N. A. Siddon, M. J. Clarkson, D. J. Tremethick, and D. A. Jans, *FEBS Lett.* **467,** 169 (2000).

[4] S. Jakel, W. Albig, U. Kutay, F. R. Bischoff, K. Schwamborn, D. Doenecke, and D. Gorlich, *EMBO J.* **18,** 2411 (1999).

[5] A. H. Corbett, D. M. Koepp, G. Schlenstedt, M. S. Lee, A. K. Hopper, and P. A. Silver, *J. Cell. Biol.* **130,** 1017 (1995).

[6] H. Großhans, G. Simos, and E. Hurt, *J. Struct. Biol.* **129,** 288 (2000).

[7] S. Nakielny and G. Dreyfuss, *Curr. Opin. Cell Biol.* **9,** 420 (1997).

[8] S. Nakielny and G. Dreyfuss, *Cell* **99,** 677 (1999).

[9] J. C. Politz, *J. Struct. Biol.* **129,** 252 (2000).

[10] M. S. Moore and G. Blobel, *Cell* **69,** 939 (1992).

[11] C. P. Mattison and I. M. Ota, *Genes Dev.* **14,** 1229 (2000).

[12] E. K. Heist, M. Srinivasan, and H. Schulman, *J. Biol. Chem.* **273,** 19763 (1998).

[13] D. A. Jans and S. Hübner, *Physiol. Rev.* **76,** 651 (1996).

[14] K. Mishra and V. K. Parnaik, *Exp. Cell Res.* **216,** 124 (1995).

[15] T. Tagawa, T. Kuroki, P. K. Vogt, and K. Chida, *J. Cell. Biol.* **130,** 255 (1995).

[16] R. T. Hay, L. Vuillard, J. M. Desterro, and M. S. Rodriguez, *Phil. Trans. R. S. Lond. B Biol. Sci.* **354,** 1601 (1999).

[17] L. Ghoda, X. Lin, and W. C. Greene, *J. Biol. Chem.* **272,** 21281 (1997).

0076-6879/02 $35.00

TABLE I
SELECTION OF IDENTIFIED NUCLEAR LOCALIZATION SIGNALS AND THEIR COGNATE RECEPTORS

Protein	Sequence (length)	Receptor
SV40 large T antigen	PKKKRKV (7)	Importin-α1
Nucleoplasmin	KRPAATKKAGQAKKKK (16)	Importin-α1
HRNPA1 M9	NQSSNFGPMKGGNFGGRSSGPYG GGGQYFAKPRNQGGY (38)	Importin-β2
HIV-1 Rev	RQARRNRRRRWR (12)	Importin-β1
U snRNA	m3G (trimethylguanosine) cap (35)	Snurportin

and is translocated in an energy-independent step to the nuclear pore complex. This cargo–receptor complex associates with a host of accessory proteins en route to the nuclear envelope, and any one of these other proteins may be conformationally altered such that nuclear transport may be up- or downregulated.[18–20] Finally, at the nuclear envelope itself, cargoes are actively transported into the nucleus through the nuclear pore complex,[21,22] which in turn may be regulated by altering local calcium levels,[23–25] as well as by phosphorylation, glycosylation, or acetylation of nuclear pore components[26,27] (Fig. 1). Although possessing a few different regulatory mechanisms, nuclear export occurs in much the same manner[7] in that an amino acid signal sequence is recognized by its cognate receptor and transported to the cytoplasmic compartment.

Control of nuclear trafficking, therefore, is a multivariate affair involving the concerted action of kinases, phosphatases, proteolytic enzymes,[16] associated nuclear transport factors, and calcium.[11,14,28] In one study, our laboratory used a cell-based assay to examine the regulation of transport and demonstrated that oxidative species such as hydrogen peroxide may participate in the control of nuclear import through an effect on at least one phosphorylation pathway.[29] This

[18] S. A. Adam, R. Sterne-Marr, and L. Gerace, *J. Cell. Biol.* **111,** 807 (1990).
[19] S. A. Adam, *Curr. Opin. Cell Biol.* **11,** 402 (1999).
[20] S. M. Fujihara and S. G. Nadler, *Biochem. Pharmacol.* **56,** 157 (1998).
[21] L. F. Pemberton, G. Blobel, and J. S. Rosenblum, *Curr. Opin. Cell Biol.* **10,** 392 (1998).
[22] D. Görlich, *Curr. Opin. Cell Biol.* **9,** 412 (1997).
[23] C. Perez-Terzic, J. Pyle, M. Jaconi, L. Stehno-Bittel, and D. E. Clapham, *Science* **273,** 1875 (1996).
[24] C. Perez-Terzic, M. Jaconi, and D. E. Clapham, *Bioessays* **19,** 787 (1997).
[25] C. Perez-Terzic, A. M. Gacy, R. Bortolon, P. P. Dzeja, M. Puceat, M. Jaconi, F. G. Prendergast, and A. Terzic, *Circ. Res.* **84,** 1292 (1999).
[26] M. W. Miller, M. R. Caracciolo, W. K. Berlin, and J. A. Hanover, *Arch. Biochem. Biophys.* **367,** 51 (1999).
[27] T. Kouzarides, *EMBO J.* **19,** 1176 (2000).
[28] P. Gallant, A. M. Fry, and E. A. Nigg, *J. Cell Sci.* **21** (1995).
[29] M. P. Czubryt, J. A. Austria, and G. N. Pierce, *J. Cell. Biol.* **148,** 7 (2000).

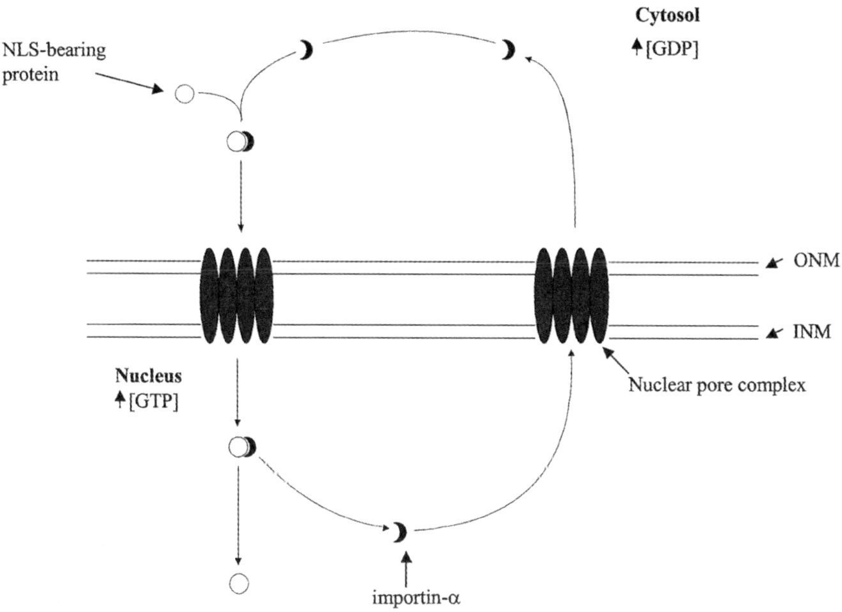

FIG. 1. Schematic of nuclear import. A protein bound for the nucleus possesses a nuclear localization signal that is recognized by a receptor. In this example, the receptor is importin-α. The NLS-bearing protein and the receptor form a stable complex in a high [GDP] environment. This complex moves to the nuclear pore and is then translocated into the nucleoplasm in an energy-requiring step. Once inside, high concentrations of GTP dissociate the complex within the nuclear interior. This releases the protein and allows importin-α to cycle back to the cytoplasm where it can participate in another round of nuclear import.

chapter provides a protocol that directs an investigator in the methodology required to study the effects of oxidative species on nuclear transport in cultured cells.

Methodology

Preparation of Cultured Smooth Muscle Cells

Smooth muscle cells are isolated from the thoracic aorta of New Zealand white rabbits. The aortic tissue is removed from the animal and placed in 1× phosphate-buffered saline (PBS) to wash it clean of red blood cells. Extraneous fat and connective tissue are removed, and the aorta is cut into ~3-mm-wide rings, which are then placed in Dulbecco's modified Eagle's medium (DMEM) containing 20% fetal bovine serum (FBS) and 10% fungizone. These rings are incubated for 10–12 days at 37° in 95% : 5% $O_2 : CO_2$ to allow the migration of fibroblasts and endothelial cells out of the tissue into the media. The rings are allowed to incubate for another 7 days before being transferred to new media (same as described earlier) for 7 more days to allow the migration of smooth muscle cells (SMC). After SMC migration, the rings are removed and the cells are grown to confluency in DMEM supplemented with 10% FBS and 1% fungizone. Cells are passaged with 0.05% trypsin and 0.53 m*M* EDTA onto glass coverslips in a dish containing new media. At this time, cells are either maintained in starvation (STV) media (DMEM, 5 μg/ml holotransferrin, 1 m*M* sodium selenite, 200 μM ascorbate, 10 m*M* insulin, 2.5 μM sodium pyruvate, 1% fungizone) or in DMEM containing 10% FBS and 1% fungizone. Only cells from the first passage are used and always no more than 2 weeks old. To prepare for the nuclear import assay, cells are fed with DMEM + 10% FBS + 1% fungizone 24–48 hr prior to use.

Preparation of $ALEXA_{488}$-BSA-NLS Conjugate for Use in the Nuclear Import Assay

The import substrate used is a fluorescently tagged $ALEXA_{488}$-BSA molecule fused to the "classical" SV40 large T antigen nuclear localization signal, i.e., -PKKKRKV. In order to generate the substrate, 4 mg of $ALEXA_{488}$-BSA is suspended in 500 μl PBS, and 1 mg of a cross-linking agent [sulfosuccinimidyl 4-(*N*-maleimidomethyl)cyclohexane-1-carboxylate] is added (Fig. 2). The solution is incubated at 37° for 30 min and is then desalted on a 5-ml Excellulose column that has been equilibrated with 5 column volumes of PBS prior to use. Using PBS as an eluant, aliquots are collected in 500-μl fractions and those containing the fluorescent tag are identified by their bright yellow/orange color. Two milligrams of NLS is suspended in 500 μl of coupling buffer [50 m*M* MES, 0.4 m*M* tris-(2-carboxyethyl) phosphine HCl, pH 5.0] and incubated for 30 min at 37°. Once incubation is complete, the solution containing the NLS is combined with the previously isolated $ALEXA_{488}$-BSA fractions. The resulting solution is placed in a dark bottle and allowed to conjugate overnight at 4° with gentle stirring in a darkened environment. The following day, the conjugated solution is passed through another 5-ml Excellulose column equilibrated with a second elution buffer (10 m*M* HEPES, 110 m*M* potassium acetate, pH 7.3). Eight fractions of 500 μl each are collected, using the second elution buffer as eluant. To prevent overloading of the column, smaller volumes of the conjugated solution are added, and the elution step is done once on each of the smaller volumes. Once again, fractions containing the fluorophore are identified by their bright yellow/orange color (typically the second to the fifth fractions) and are pooled together. To determine whether the conjugation has been successful, the original $ALEXA_{488}$-BSA and the NLS–$ALEXA_{488}$-BSA conjugate are run on a 9% SDS–PAGE at 30 mA for 45 min and stained with Coomassie Brilliant Blue R-250. A gel shift of the conjugated

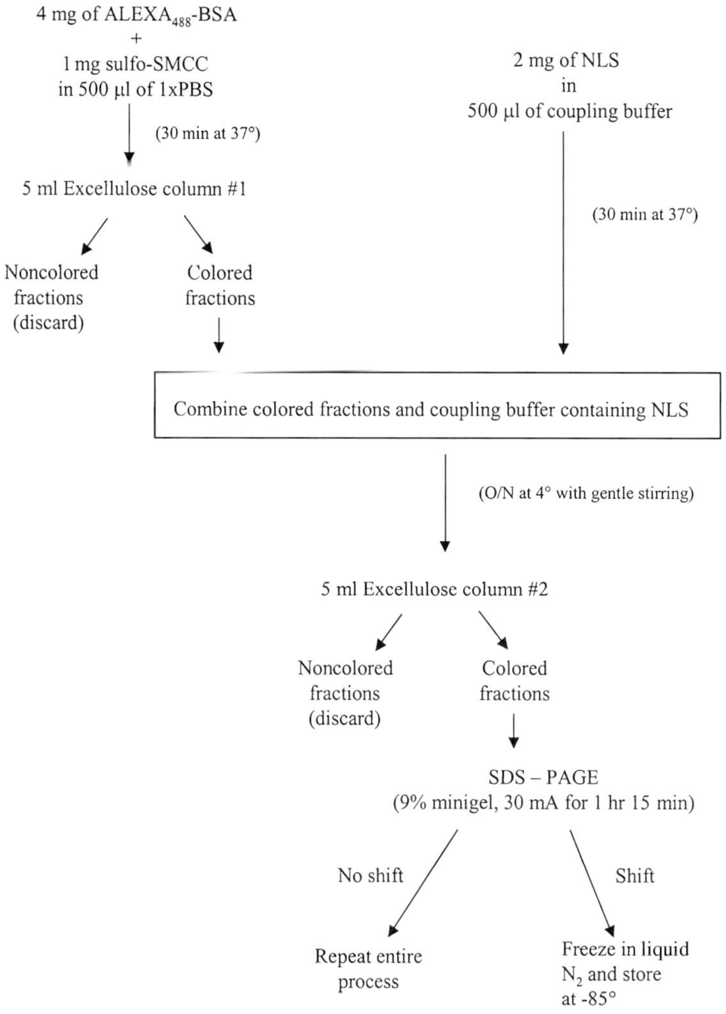

FIG. 2. Synthesis of the ALEXA488-BSA–NLS conjugate. Flow chart outlining steps involved in the preparation of the fluorescent BSA–NLS molecule.

fluorophore, when compared to the original, allows for estimation of the number of NLS molecules linked to the fluorescently tagged protein (Fig. 3). Using the ratio of 1.5 kDa : 1 NLS, approximately 10–15 NLSs are calculated to be attached to each ALEXA488-BSA molecule. The final conjugate is aliquoted as necessary, frozen in liquid N_2, and stored at −85°.

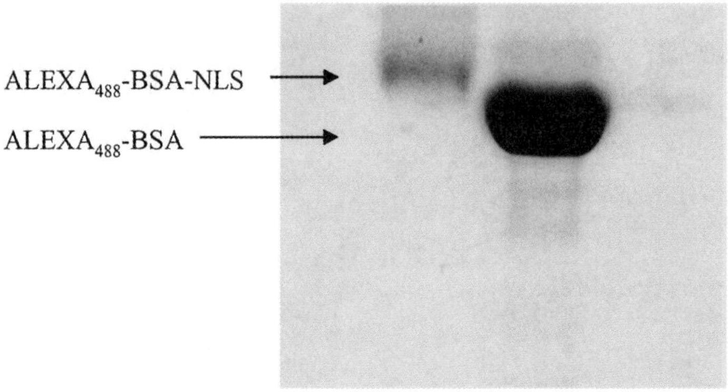

FIG. 3. SDS–PAGE of conjugated and nonconjugated $ALEXA_{488}$-BSA. To determine whether conjugation of the fluorescent BSA to the NLS had occurred, samples were run on a 9% gel. An upward shift of the BSA molecule is indicative of successful conjugation with the added NLS, as shown.

Nuclear Import Assay

Smooth muscle cells cultured from rabbit thoracic aorta are permeabilized by incubating the coverslip with the cells in nuclear import buffer (20 m*M* HEPES, 110 m*M* potassium acetate, 5 m*M* sodium acetate · $3H_2O$, 2 m*M* magnesium acetate · $4H_2O$, 0.5 m*M* EGTA, pH 7.3) with 40 μg/ml digitonin for 5.0 min. After permeabilization, cells are washed gently three times with import buffer and inverted on 50 μl of treated or untreated import mix (50% rat liver cytosol, 1 m*M* ATP, 5 m*M* phosphocreatine, 20 units/ml creatine phosphokinase, 1 μg/ml each of leupeptin, aprotinin, and pepstatin A) in a humidified, parafilm-lined box, which is then placed in a 37° water bath for 30 min. After incubation, coverslips are lifted off by the addition of 1.0 ml of import buffer, removed, and placed in a quartered petri dish. Care must be taken when lifting coverslips off the parafilm, as the cells are loosely attached. Coverslips are washed once with 1.0 ml of import buffer and fixed by adding 1.0 ml of import buffer containing 3.7% paraformaldehyde for 10 min. Fixative is removed after the incubation period, and 2.0 ml of import buffer is added to the coverslips. Cells are visualized using confocal microscopy, and images are analyzed using Molecular Dynamics Imagespace software on a SGI (Silicon Graphics Indigo O_2) workstation (Fig. 4).

Treatment of Cells or Cytoplasmic Mix with Hydrogen Peroxide

Prior to incubation of the coverslips with import substrate, the permeabilized cells or cytosolic mix may be treated with the oxidant being studied. We looked at the effects of hydrogen peroxide in the cytoplasm or the nuclear envelope. In order to treat the cytosol, the import mix is incubated with 0.1, 0.5, and 1.0 m*M*

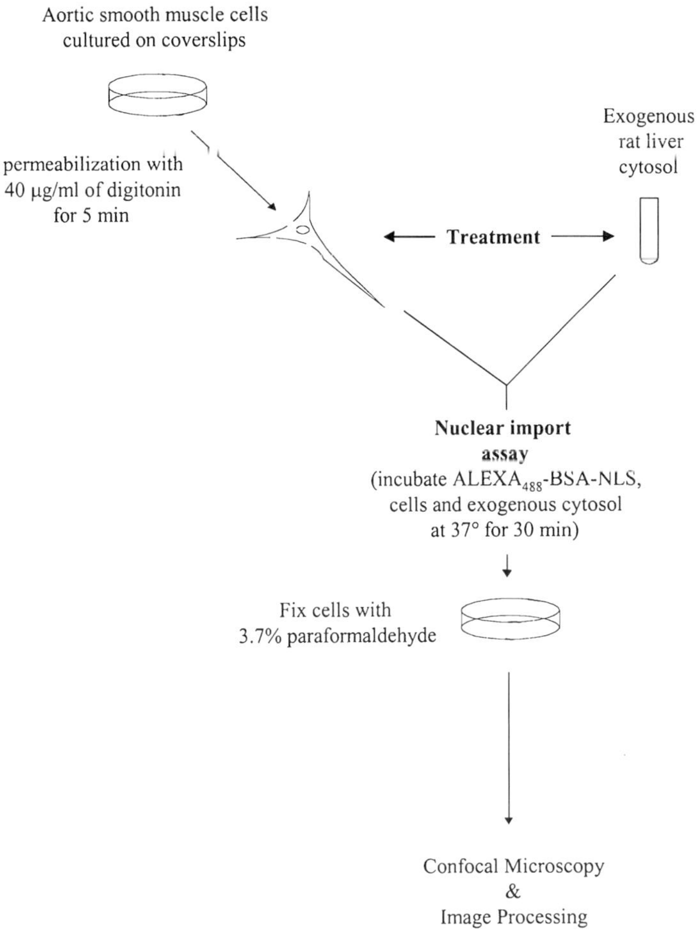

FIG. 4. The nuclear import assay. Cells were grown on coverslips and permeabilized using 40 μg/ml digitonin. At this point, either cells or exogenously added cytosol was treated with hydrogen peroxide. Nuclear import was carried out for 30 min at 37°, and cells were fixed with 3.7% paraformaldehyde. Confocal microscopy was used to visualize nuclear fluorescence.

of H_2O_2 for 30, 60, or 120 min at 37°. Alternately, permeabilized cells may be treated directly with the oxidant being studied to determine if there are any effects on the nuclear envelope itself. In this instance, aortic smooth muscle cells are permeabilized with digitonin (described earlier) and then treated with 1.0 mM H_2O_2 for 30 min. Coverslips are then rinsed briefly before proceeding with the nuclear import assay.

Confocal Microscopy and Image Analysis

After fixing the cells, coverslips are placed in a Leyden chamber, and 1.0 ml of nuclear import buffer is added to prevent the samples from drying out. Using a Bio-Rad MRC-600 confocal microscope, the 488 laser (at 10% transmission) and a small pinhole size (~3.0–5.0 units) are used to visualize the localization of the import substrate. Each of the images acquired was Kalman filtered three times per field to remove background noise originating from the photomultiplier and initially obtained using a 40×, N.A. 1.3 oil immersion objective lens. The images are processed on an SGI workstation using the Molecular Dynamics Imagespace image analysis software.

The SGI software assigns a numerical value to the fluorescence in control and treated cells. Changes in the rate of nuclear transport are determined by comparing the fluorescence of treated to nontreated cells and are reported as a percentage of the control. In this manner, the results of three to four different experiments are averaged and reported as mean value plus or minus standard error. Typically, each of these experiments contained numerical values of fluorescence intensity from 75 to 150 cells.

Nuclear Import Assay (Digitonin-Permeabilized Cell Assay)

The technique we used was performed according to the method of Adam *et al.*[18] In studying nuclear trafficking, the digitonin-permeabilized cell assay has become invaluable in determining the mechanisms and components involved in nuclear import. By varying the different conditions in which nuclear import takes place, an investigator can look at the effects of treatments at both the level of the cytosol and the nuclear envelope. The role of many cytosolic factors participating in nuclear import has been elucidated in this manner.[19,30–33]

We successfully used this method to investigate the influence of H_2O_2 on nuclear import. Using this protocol, we demonstrated that a cytosolic component of the transport machinery was modified by hydrogen peroxide such that nuclear import was attenuated in the presence of H_2O_2 and was significantly reversed by the addition of catalase (Fig. 5). Negligible effects were seen when permeabilized cells were treated similarly. Using this as a model, one can study the effects of different oxidants as well as different free radical-generating systems on nuclear transport within cultured cells. For example, a xanthine/xanthine oxidase-free radical-generating system was used and found to have no effects on nuclear import (unless nuclear integrity was compromised).[29] We concluded that H_2O_2

[30] U. Kutay, F. R. Bischoff, S. Kostka, R. Kraft, and D. Görlich, *Cell* **90,** 1061 (1997).

[31] R. Mahajan, C. Delphin, T. Guan, L. Gerace, and F. Melchior, *Cell* **88,** 97 (1997).

[32] M. W. Miller and J. A. Hanover, *J. Biol. Chem.* **269,** 9289 (1994).

[33] M. V. Nachury and K. Weis, *Proc. Natl. Acad. Sci. U.S.A.* **96,** 9622 (1999).

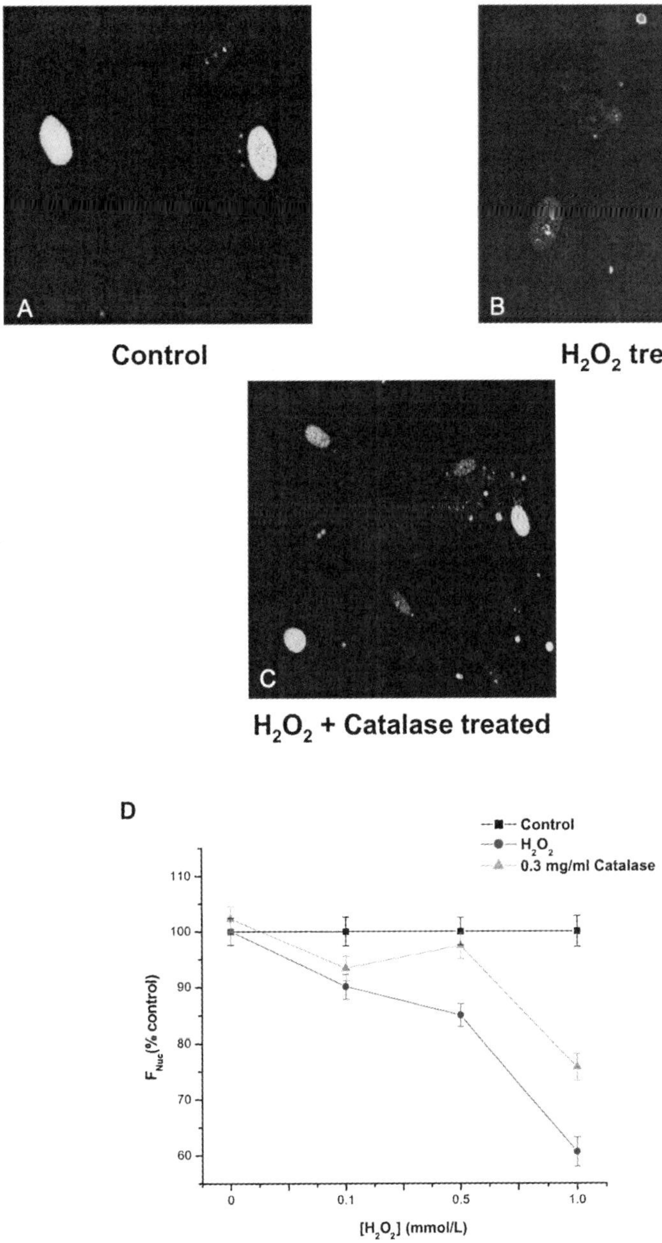

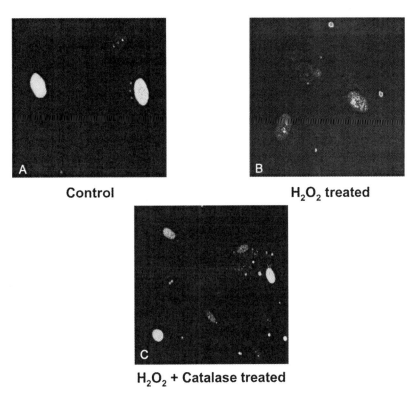

FIG. 5. Inhibition of nuclear import by hydrogen peroxide. The addition of 1.0 m*M* H_2O_2 to permeablized smooth muscle cells decreased nuclear import. Treatment with catalase reversed the effects, whereas treatment with catalase alone did not affect basal levels of nuclear import (data not shown) (A–C). (D) H_2O_2 inhibition of nuclear import is significantly reversed by the addition of catalase. Modified from M. P. Czubryt, J. A. Austria, and G. N. Pierce, *J.Cell. Biol.* **148,** 7 (2000).

modifies transport by inducing specific MAP kinase signaling pathways, and other free radicals do not act in a similar manner.

Conclusions: Advantages and Limitations of the Technique

The protocol described here was used to determine the role, if any, that hydrogen peroxide plays in regulating nuclear import. While the approach we used investigated the role of the "classical" transport pathway, i.e., the importin-α pathway, other modes of transport can also be studied by varying the type of localization signal in the import substrate. For example, one can change the nuclear localization signal so that the M9 "transportin" pathway[34] is used to import the fluorophore. In addition to changing the type of NLS that can be used, an investigator may also try using substrates other than a bovine serum albumin (BSA)–fluorophore conjugate. Using fluorescently tagged, small nuclear ribonucleoproteins would make use of the snurportin pathway,[35] another nuclear transport mechanism distinct from importin-α and transportin pathways. The ease and simplicity of the assay, combined with its flexibility, allow the researcher many options when deciding which element(s) of nuclear transport to investigate. Although the focus of this chapter has been on using hydrogen peroxide, there are many other significant substances that can be tested.

The use of confocal microscopy over conventional epifluorescence provides several advantages to the investigator when studying nuclear import. One is that it allows fluorescent imaging of live cells. Using the confocal microscope to observe nuclear transport of an NLS-tagged substrate that has been microinjected into a cell allows one to track its movements *in vivo* over a period of time. Another significant advantage is that it removes blurring from light sources outside the plane of focus. A stack of images, or optical sections, can be taken of a sample that can be reconstructed to form a three-dimensional representation of the sample and show the distribution and movements of a fluorescent molecule in the x, y, and z planes. In addition to looking at the movement and distribution of a single molecule within a cell, dual labeling of a sample provides the opportunity to look at the colocalization of two molecules in the third dimension when used in conjunction with confocal microscopy. In relation to nuclear transport studies, three-dimensional colocalization is a useful tool in determining whether a transport intermediate is at the nuclear pore complex at a particular time or after a particular treatment.

Although many advantages come from using the permeabilized cell assay together with confocal techniques, there are a few caveats of which to take note.

[34] S. Nakielny, M. C. Siomi, H. Siomi, W. M. Michael, V. Pollard, and G. Dreyfuss, *Exp. Cell Res.* **229,** 261 (1996).

[35] J. Huber, U. Cronshagen, M. Kadokura, C. Marshallsay, T. Wada, M. Sekine, and R. Luhrmann, *EMBO J.* **17,** 4114 (1998).

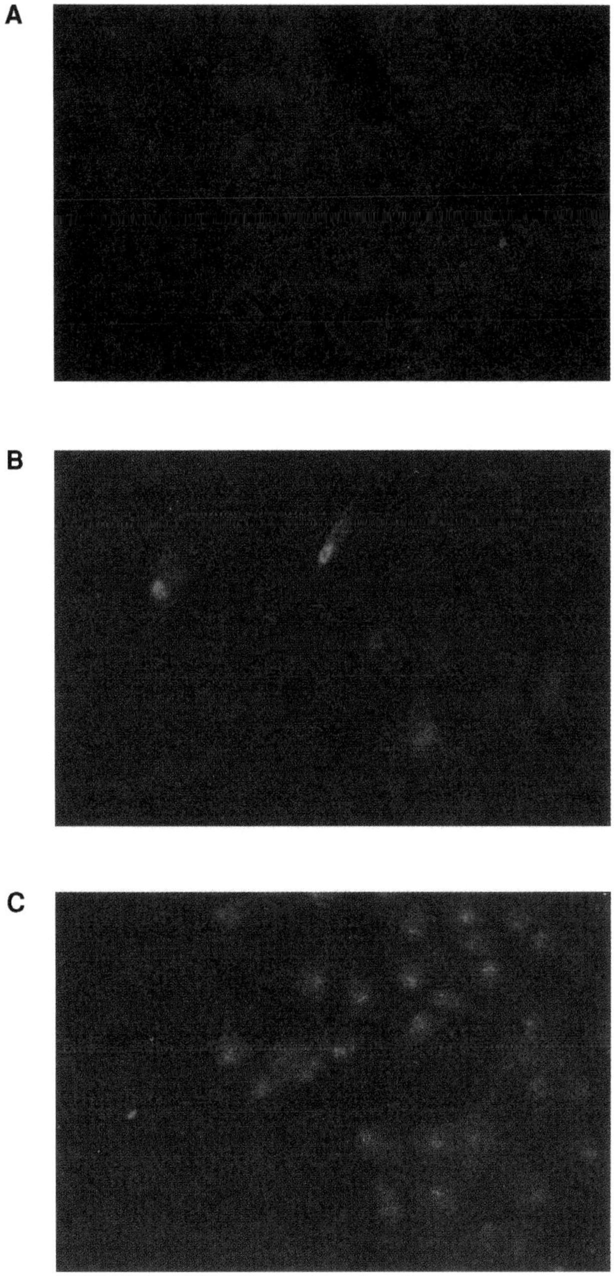

FIG. 6. Treatment of smooth muscle cells with various concentrations of digitonin. Aortic smooth muscle cells were permeabilized with 40(A), 80(B), or 800(C) μg/ml digitonin and were stained with anti-DNA antibodies to determine their degree of permeabilization. For this particular cell type, no nuclear staining is seen in A, but nuclear fluorescence seen in B and C indicates perforation of the nuclear envelope.

When permeabilizing smooth muscle cells, 40 μg/ml digitonin used for 5.0 min is sufficient to disrupt the cell membrane without damaging the nuclear envelope. This may not be the case for other cell types and one must determine the optimum concentration and time for the particular cell being studied. As illustrated in Fig. 6, excessive amounts of digitonin (or prolonged exposure) can lead to permeabilization of the nuclear envelope, which may lead to false positive or negative results when performing the nuclear import assay. The former would occur due to entry of the fluorophore through a permeabilized nuclear envelope, whereas the latter would be the result of an inability of the nucleus to retain an NLS-tagged substrate.

It was mentioned earlier that one of the strengths of confocal microscopy is that it allows imaging of live cells. However, because the digitonin-permeabilized assay fixes the cells, the confocal could not be used in this context for this experiment. Therein lies one significant limitation of the permeabilized cell assay—*in vivo* conditions cannot be assayed.

With regard to the study mentioned here, it has been shown that bovine serum albumin is a potential H_2O_2 scavenger.[36,37] In our experiments, the concentration of BSA is not high enough to scavenge the H_2O_2 that we have added.[36,37] However, it remains noteworthy for other researchers wishing to use the same experimental procedure.

Despite these cautions, the cell permeabilization assay, together with confocal microscopy, remains a powerful tool in identifying elements of the nucleocytoplasmic transport process that may be influenced or regulated by reactive oxidative species.

[36] M. J. Dumoulin, R. Chahine, R. Atanasiu, R. Nadeau, and M. A. Mateescu, *Arzneim.-Forsch.* **46,** 855 (1996).

[37] T. Ogino and S. Okada, *Biochim. Biophys. Acta* **1245,** 359 (1995).

[12] Functional Imaging of Mitochondrial Redox State

By DAGMAR KUNZ, KIRSTIN WINKLER, CHRISTIAN E. ELGER, and WOLFRAM S. KUNZ

Introduction

The main source of reducing equivalents for the mitochondrial respiratory chain is mitochondrial NADH. Therefore, the redox state of the NAD system reflects the rate of mitochondrial oxidative phosphorylation at a constant supply of substrates and allows a monitoring of mitochondrial metabolism. In the classical literature, determination of the NAD redox state in cellular systems has been performed by NAD(P)H fluorescence measurements[1-3] or by the detection of the fluorescence of α-lipoamide dehydrogenase, a mitochondrial flavoprotein in close redox equilibrium with the mitochondrial NAD system.[4-6] These investigations have been performed with perfused tissues or cellular suspensions. However, until now, much less information is available on the distribution of the mitochondrial redox states at the single cell level. Considerable technical progress in fluorescence microscopy (adequate laser light sources and highly sensitive CCD cameras are available) allows one to address this problem. For example, in postmitotic cells such as skeletal muscle fibers there seems to be considerable heterogeneity of mitochondria in respect to the fluorescence properties of flavoproteins.[7] Additionally, the use of intrinsic fluorophores allows evaluation of the function of mitochondrial oxidative phosphorylation without having the phototoxic side effects of the various fluorescent mitochondrial dyes.[8]

Methods

Preparation of Muscle Fibers

About 50 mg of human diagnostic biopsy tissue (m. vastus lateralis) is used for the isolation of saponin-permeabilized fibers. Approximately 10-mm-long bundles of muscle fibers containing, usually, two to four single fibers are isolated at 4°

[1] R. W. Estrabrook, *Anal. Biochem.* **4,** 231 (1962).
[2] H. Franke, C. H. Barlow, and B. Chance, *Am. J. Physiol.* **231,** 1082 (1979).
[3] L. A. Katz, A. P. Koretsky, and R. S. Balaban, *FEBS Lett.* **221,** 270 (1987).
[4] R. Scholz, R. G. Thurman, J. R. Williamson, B. Chance, and T. Bücher, *J. Biol. Chem.* **244,** 2317 (1969).
[5] A. Mayevsky and B. Chance, *Science* **217,** 537 (1982).
[6] K. H. Vourinen, A. Ala-Rämi, Y. Yan, P. Ingman, and I. Hassinen, *J. Mol. Cell. Cardiol.* **27,** 1581 (1995).

0076-6879/02 $35.00

by mechanical dissection in a relaxing solution (for composition, see later). The saponin treatment is performed by a 30-min incubation of the fiber bundles in ice-cold relaxing solution, which contains 50 μg/ml saponin as described by Kunz *et al.*[9] To remove the saponin, fibers are incubated for 10 min at 4° in the medium for measurements (for composition, see later) which contains 2 mg/ml bovine serum albumin. The relaxing solution contains 10 m*M* Ca/CaEGTA buffer, a free concentration of 0.1 μM calcium, 20 m*M* imidazole, 20 m*M* taurine, 49 m*M* K–MES, 0.5 m*M* dithiothreitol (DTT), 3 m*M* KH_2PO_4, 9.5 m*M* $MgCl_2$, 5 m*M* ATP, and 15 m*M* phosphocreatine, pH 7.1. All measurements are performed in a medium consisting of 110 m*M* mannitol, 60 m*M* KCl, 10 m*M* KH_2PO_4, 5 m*M* $MgCl_2$, 0.5 m*M* Na_2EDTA, and 60 m*M* Tris–HCl, pH 7.4.

Fibroblast Culture Conditions

Fibroblasts from a 5 × 5-mm human diagnostic skin biopsy are cultivated in Dulbecco's modified Eagle's medium (DMEM) supplemented with 10% fetal calf serum, 100 U/ml penicillin, 100 U/ml streptomycin, 10 μg/ml tylosin, and 2 m*M* glutamine in a 6.5% CO_2 atmosphere at 37°. After reaching confluence, cells are harvested between passages 6 or 7 and are transferred to special chambered coverslips (borosilicate glass, 4 well) from Nalge Nunc International (Naperville, IL). After 24–36 hr, the attached cells are rinsed carefully with HBSS (PAA Laboratories GmbH, Colbe, Germany). To avoid cell detachment during microscopic experiments, measurements are performed in a medium containing 49 m*M* MES, 3 m*M* KH_2PO_4, 20 m*M* taurine, 0.5 m*M* DTT, 20 m*M* imidazole, 9.5 m*M* $MgCl_2$, 35 m*M* KCl, and 10 m*M* EGTA–CaEGTA (pH 7.4). Fibroblasts are permeabilized with 10 μg/ml digitonin, and the NAD(P)H autofluorescence images are acquired after the addition of 1 m*M* octanoylcarnitine, 5 m*M* malate, 1 m*M* ADP, and 4 m*M* KCN.

Confocal Microscopy

The skeletal muscle fibers are fixed at both ends on a coverslip in a Heraeus flexiperm chamber (Hanau, Germany) and incubated in 0.5 ml of the medium for measurements. The confocal autofluorescence images of muscle fibers are acquired with a Noran Instrument Odyssey XL CLSM (Middleton, WI) supported

[7] A. V. Kuznetsov, O. Mayboroda, D. Kunz, K. Winkler, W. Schubert, and W. S. Kunz, *J. Cell. Biol.* **140,** 1091 (1998).

[8] M. Reers, S. T. Smiley, C. Mottola-Hartshorn, A. Chen, M. Lin, and L. B. Chen, *Methods Enzymol.* **260,** 406 (1995).

[9] W. S. Kunz, A. V. Kuznetsov, W. Schulze, K. Eichhorn, L. Schild, F. Striggow, R. Bohnensack, S. Neuhof, H. Grasshoff, H. W. Neumann, and F. N. Gellerich, *Biochim. Biophys. Acta* **1144,** 46 (1993).

with Intervision 1.4.1 software at argon-ion laser excitation at 488 nm and an emission barrier filter at 514 nm (long path). To obtain a sufficient z axis resolution, between 80 and 100 images (step size 0.72 μm) are acquired using an UApo 340 20× objective (Olympus) having a numerical aperture of 0.75.

Fluorescence Microscopy

In experiments with isolated skeletal muscle fibers, samples are fixed at both ends on a coverslip in a Heraeus flexiperm chamber (Hanau, Germany) and incubated in 0.5 ml of the medium for measurements. Experiments with fibroblasts are performed in the special chamber coverslips as described earlier. Digital video images are acquired with an inverse fluorescence microscope (Model IX70; Olympus, Tokyo, Japan) equipped with a 100-W extra-high-pressure mercury lamp USH-102D (Ushio Inc., Tokyo, Japan) and a CCD camera (Model CF 8/1 DXC; Kappa, Gleichen, Germany). The NAD(P)H fluorescence image is obtained using 360–370 nm excitation and >450-nm long path emission [U excitation cube (narrow band), Olympus, Tokyo, Japan], and the flavoprotein fluorescence image is obtained using 470–490 nm excitation and 515–550 nm narrow-band emission [IB excitation cube (narrow band), Olympus, Tokyo, Japan]. To obtain reproducible fluorescence images on different days, it is highly recommended to check the excitation intensity of the fluorescence microscope with a standard luminescent glass (e.g., GG 17, Jenaer Glaswerke, Jena, Germany). Digital ratio images are calculated from unprocessed eight-bit images using the Metaview software 4.5 (Universal Imaging Corporation, West Chester, PA), which is also used for gray value determinations. For the detection of NAD(P)H fluorescence in fibroblasts, the UApo 340 40× objective from Olympus (Tokyo, Japan) is used.

Isolation of Mononuclear Cells and Flow Cytometry

The mononuclear cell (MNC) preparation is performed from 5 to 10 ml heparinized blood of healthy donors by density gradient centrifugation.[10] For one determination of mitochondrial NAD redox state, between 0.5 and 1.5×10^6 MNC are necessary. The MNC cell suspension contains about 20% monocytes, about 80% lymphocytes, and below 2% granulocytes. The individual composition and cell counts are determined with a Cell Dyn 4000 (Abbott Diagnostika). Autofluorescence changes in subpopulations of human mononuclear cells are recorded using a FACSCalibur flow cytometer (Becton Dickinson) using 488 nm excitation and a 530 ± 15-nm fluorescence emission filter. Reproducibility of the fluorescence readings of the instrument should be checked before all measurements using FITC-labeled polymethylmethacrylate microspheres (CaliBRITE, Becton Dickinson). Flow cytometric measurements are performed in 0.5-ml suspensions of MNC in phosphate-buffered saline (PBS) containing between 0.2 and 0.5×10^6 cells. The

[10] A. Boyum, *Scand. J. Clin. Lab. Invest.* **21** (Suppl. 27), 77 (1968).

sample flow rate is adjusted to about 1000 cells/sec. During the measurement, simultaneous gating for lymphocytes and monocytes is performed. The respective gates are defined using the distinctive forward scatter and side scatter properties of the individual cell populations. The measurement is routinely finished if the fluorescence properties of about 2000 monocytes are acquired. For redox state determinations, two separate incubations are necessary: (i) MNC suspension +1 m*M* octanoic acid (mean of the autofluorescence distribution-F_{oc}) and the subsequent addition of 4 m*M* KCN (mean of the autofluorescence distribution-F_{KCN}) and (ii) MNC suspension +2 μM TTFB (mean of the autofluorescence distribution-F_{TTFB}). At the end of experiments (to avoid an initial contamination of the flow cytometer with fluorescent substances) the viability of cells is determined by propidium iodide staining, which should exceed 96%. Because the flavoprotein autofluorescence signals are very low, the instrument has to be washed carefully with azide-free PBS as sheath fluid to remove completely residual fluorochroms and colored buffers have to be avoided. Data are analyzed using the CELLQuest software for Macintosh.

Results and Discussion

Measurements of α-Lipoamide Dehydrogenase Redox State Changes in Muscle Fibers by Confocal Microscopy

To detect the changes of the mitochondrial redox state in response to alterations of oxidative phosphorylation (OxPhos) at the single cell level, we investigated the fluorescence changes of the α-lipoamide dehydrogenase flavin moiety by confocal microscopy. This flavoprotein is in tight redox equilibrium with the mitochondrial NAD system and contributes in skeletal muscle mainly to the respiratory chain-dependent green autofluorescence signal if the sample is excited at 488 nm (argon-ion laser excitation) and the fluorescence emission is detected above 520 nm (Scheme 1).[11,12]

This is due to the selection of excitation and emission wavelengths and the low content of electron transfer flavoprotein in skeletal muscle having a 436-nm excitation maximum and a 490-nm emission maximum and an about 2.3-fold lower content than in liver.[12] Typical fluorescence changes of this flavoprotein in response to changes in the rate of oxidative phosphorylation detected by confocal microscopy in a human skeletal muscle fiber are shown in Fig. 1. In the endogenous oxidized state, a bright fluorescence image can be detected showing the distribution of mitochondria along the myofibrils in one confocal plane across the center of the fiber (Fig. 1A). The addition of mitochondrial substrates octanoylcarnitine and malate causes a reduction of the mitochondrial NAD system and a quenching of the α-lipoamide dehydrogenase fluorescence (Fig. 1B).

[11] I. Hassinen and B. Chance, *Biochem. Biophys. Res. Commun.* **31,** 895 (1968).
[12] W. S. Kunz and F. N. Gellerich, *Biochem. Med. Metab. Biol.* **50,** 103 (1993).

$$
\begin{array}{ccccc}
SH_2 & \xrightarrow{DH} & (NADH / NAD^+)_{mit} & \xrightarrow{RC} & O_2 \\
 & & \downarrow\uparrow & & \\
 & & (FpH_2 / Fp)_{\alpha\text{-lip DH}} & &
\end{array}
$$

SCHEME 1. Redox equilibrium of the mitochondrial NAD system and the flavin moiety of α-lipoamide dehydrogenase. DH, dehydrogenases; RC, respiratory chain.

It can be seen that the entire fiber gets darker, indicating no redox state differences among all mitochondria. Next, we added ADP to the skeletal muscle fiber to stimulate the electron flow through the mitochondrial respiratory chain (Fig. 1C). Very clearly, the fiber became uniformly brighter, indicating a reoxidation of the mitochondrial NAD system in all mitochondria. It can be noted, however, that the longitudinal striated pattern changed slightly, most likely due to fiber relaxation by the adenine nucleotide addition. To obtain a fully reduced reference state of the mitochondrial NAD system, we added cyanide to block electron flow through the respiratory chain. Due to the reduction of α-lipoamide dehydrogenase, a dark image is seen in Fig. 1D. Since not only the confocal planes but also the reconstructed cross sections (Figs. 1A′–1D′) show that the expected redox state changes of α-lipoamide dehydrogenase occur in the whole mitochondrial population of the fiber, this experiment shows the absence of diffusion restrictions for substrates, ADP, or respiratory chain inhibitors. Thus, confocal microscopy allows detection of redox state changes of the mitochondrial NAD system on the single cell level and visualization of possible heterogeneities of the mitochondrial redox state due to diffusion limitations or due to the heterogeneity of mitochondrial subpopulations.[7] It should be noted, however, that this method is not very sensitive and, due to the high laser excitation light requirements, is prone to bleaching problems. This can cause problems in the quantification of mitochondrial redox states.

Detection of Heterogeneously Distributed Defects of Mitochondrial Oxidative Phosphorylation: Quantitative Redox State Imaging by Fp/NAD(P)H Ratioing in Muscle Fibers

In contrast to confocal microscopy, conventional fluorescence microscopy offers a much higher sensitivity and flexibility in the choice of excitation wavelengths but has, of course, a much greater limitation in respect to the subcellular resolution. With this method, simultaneously with flavoprotein fluorescence measurements, the UV excitation at about 340 nm can be used safely to detect the blue NAD(P)H fluorescence at 450 nm emission [the bleaching of UV lasers is very large; only two photon techniques allow measurement of NAD(P)H fluorescence with a confocal microscope[13]]. In muscle tissue, this signal originates mainly

[13] S. Vielhaber, K. Winkler, E. Kirches, D. Kunz, M. Büchner, H. Feistner, C. E. Elger, A. C. Ludolph, M. W. Riepe, and W. S. Kunz, *J. Neurol. Sci.* **169,** 133 (1999).

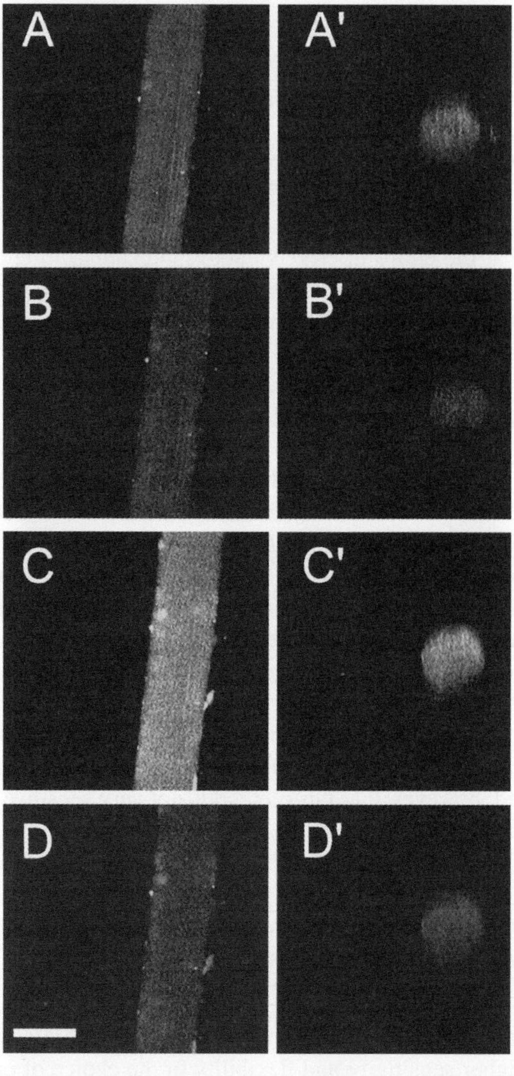

FIG. 1. Confocal laser-scanning micrographs showing the autofluorescence of flavoproteins within a single saponin-permeabilized human skeletal muscle fiber. (A–D) One single confocal plane across the fiber is illustrated. (A′–D′) One cross section, reconstructed from a stack of 90 different confocal planes (*z* stack), is shown A,A′, endogenous state; B,B′, addition of 1 m*M* octanoylcarnitine and 5 m*M* malate; C,C′, addition of 1 m*M* ADP; and D,D′, addition of 4 m*M* potassium cyanide. Bar: 50 μm.

from mitochondrial NADH.[1,3] Because pyridine nucleotides are fluorescent in the reduced state and flavoproteins in the oxidized state, the ratio of flavoprotein fluorescence and NAD(P)H fluorescence (Fp/NAD(P)H) can be used as an extremely sensitive indicator of the mitochondrial redox state as introduced by Mayevsky and Chance.[5] We used this ratioing approach to detect putative differences of mitochondrial redox states between individual skeletal muscle fibers.

In Fig. 2, characteristic Fp/NAD(P)H ratio images of two human muscle fibers containing different amounts of mitochondria (the upper fiber is glycolytic, having only 60% of the flavoprotein fluorescence intensity of the lower oxidative fiber) are shown. In the oxidized state, the ratio image of the glycolytic fiber is darker than of the oxidative fiber (Fig. 2B). However, on addition of octanoylcarnitine + malate (Fig. 2C), ADP (Fig. 2D), and cyanide (Fig. 2E), comparable intensity changes of the ratio image are detected. Because the ratio image is also sensitive to changes in the mitochondrial content, the uneven intensity of the ratio image within the individual fibers (cf. upper glycolytic fiber) should be attributed to differences in the mitochondrial distribution. To investigate putative differences between oxidative and glycolytic fibers, we have determined the redox states of α-lipoamide dehydrogenase and NAD(P)H in individual human muscle fibers and compared the results with laser fluorimetric determinations of the redox states of the α-lipoamide dehydrogenase and NAD(P)H in fiber bundles (according to Ref. 14) of the same biopsy samples. As shown in Table I, the determined single fiber redox states were slightly lower than the redox states in fiber bundles, but no differences between glycolytic and oxidative fibers were detected. This result is in accordance with data obtained from mice muscle fibers from m. quadriceps.[7] Furthermore, it can be seen that at comparable intensities of the initial Fp and PN signals, determination of the "redox state" of the Fp/PN ratio signal yields a value close to both individual redox states. It should be mentioned, however, that this value is not a linear function of both redox states and is dependent on the individual contribution of each signal. This technique has been applied to detect skeletal muscle fibers, which contain mitochondria with defective respiratory chain function.[13,15] This type of heterogeneous muscle pathology is known for different mitochondrial DNA diseases,[16] but can be also found in motor neuron disease (amyotrophic lateral sclerosis, ALS).[13,15] A typical experiment with muscle fibers of a patient suffering from ALS is shown in Fig. 3. In the fully oxidized state (Fig. 3B), the brightness of the ratio image of both fibers is nearly identical,

[14] K. Winkler, A. V. Kuznetsov, H. Lins, E. Kirches, P. von Bossanyi, K. Dietzmann, B. Frank, H. Feistner, and W. S. Kunz, *Biochim. Biophys. Acta* **1272,** 181 (1995).

[15] S. Vielhaber, D. Kunz, K. Winkler, F. R. Wiedemann, E. Kirches, H. Feistner, H.-J. Heinze, C. E. Elger, W. Schubert, and W. S. Kunz, *Brain* **123,** 1339 (2000).

[16] R. Schröder, S. Vielhaber, F. R. Wiedemann, C. Kornblum, A. Papassotiropoulos, P. Broich, S. Zierz, C. E. Elger, H. Reichmann, P. Seibel, T. Klockgether, and W. S. Kunz, *J. Neuropathol. Exp. Neurol.* **59,** 353 (2000).

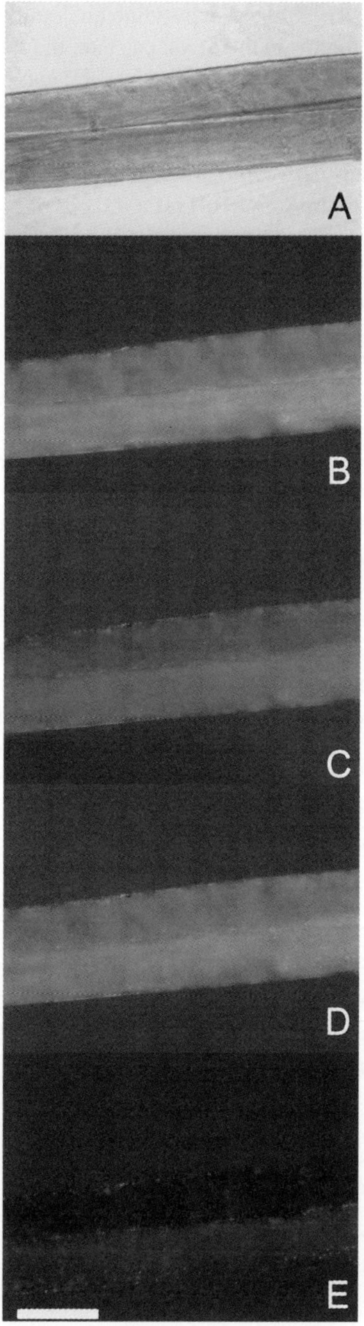
A
B
C
D
E

TABLE I
STEADY-STATE REDOX STATES OF NAD(P)H AND FLUORESCENT FLAVOPROTEINS IN HUMAN SAPONIN-PERMEABILIZED MUSCLE FIBERS DETERMINED BY REDOX STATE IMAGING[a]

	Oxidative fibers	Glycolytic fibers	Fiber bundles[b]
NAD(P) redox state ($R_{\mathrm{NAD(P)H}}$, %)	14.8 ± 8.9	14.1 ± 9.2	27.0 ± 4.2
NAD(P)H signal (arb. u.)	17.4 ± 3.4	9.2 ± 4.9	—
Fp redox state (R_{Fp}, %)	30.6 ± 13.8	32.2 ± 13.4	40.7 ± 4.8
Fp signal (arb. u.)	23.9 ± 4.0	12.3 ± 3.6	—
Fp/NAD(P)H redox state ($R_{\mathrm{Fp/NAD(P)H}}$, %)	29.1 ± 11.8	29.6 ± 13.7	—
Fp/NAD(P)H signal (arb. u. × 1000)	714 ± 165	427 ± 138	—

[a] Redox states are determined applying the following equations:

$$R_{\mathrm{NAD(P)H}} = (F_{\mathrm{ADP}} - F_{\mathrm{endo}})/(F_{\mathrm{KCN}} - F_{\mathrm{endo}}) \times 100$$
$$R_{\mathrm{Fp}} = (F_{\mathrm{endo}} - F_{\mathrm{ADP}})/(F_{\mathrm{endo}} - F_{\mathrm{KCN}}) \times 100$$
$$R_{\mathrm{Fp/NAD(P)H}} = (Q_{\mathrm{endo}} - Q_{\mathrm{ADP}})/(Q_{\mathrm{endo}} - Q_{\mathrm{KCN}}) \times 100.$$

F are the intensities of the fluorescence signals, Q are the intensities of the ratio signals in the endogenous state, the state in the presence of substrate and ADP, or substrate and KCN. For comparison, maximal individual intensities (maximal gray value differences) of the individual signals are given.

[b] Redox states of fiber bundles are determined by laser fluorimetry.[14]

indicating similar mitochondrial content. On reduction of the mitochondrial NAD system by octanoylcarnitine + malate addition, both fibers get darker, but already a difference between the pathological hypotrophic fiber and the normal fiber can be seen (Fig. 3C). Addition of ADP, which stimulates the electron flow through the respiratory chain, leading to the reoxidation of the mitochondrial NAD system, causes an increase in the brightness of both fibers, but a clear difference in the steady-state redox state of both fibers can be detected (Fig. 3D). The thinner fiber has a largely elevated ratio signal "redox state" of 79%, whereas the redox state

FIG. 2. Phase-contrast and video autofluorescence ratio images [flavoprotein/NAD(P)H] of two saponin-permeabilized fibers from human control skeletal muscle. Digital video images of NAD(P)H and flavoprotein autofluorescence, respectively, were obtained as described in the methods section. To visualize metabolic alteration of the mitochondrial NAD redox state, the flavoprotein image was divided by the NAD(P)H image. A, phase contrast; B, endogenous oxidized state; C, addition of 1 mM octanoylcarnitine and 5 mM malate; D, addition of 1 mM ADP; and E, addition of 4 mM potassium cyanide. Bar: 60 μm.

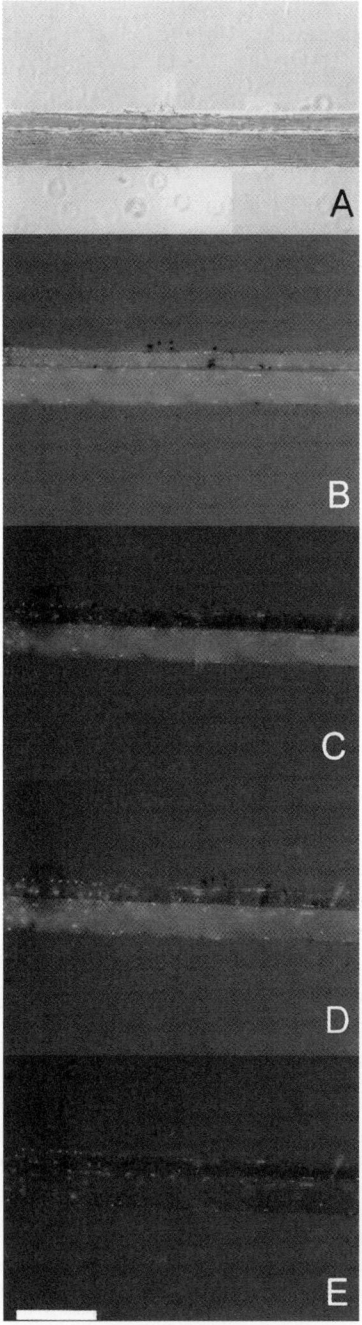
A
B
C
D
E

of the normal fiber is, with 51%, somewhat higher than normal (cf. Table I). On addition of the respiratory chain inhibitor cyanide, both fibers become comparably dark (Fig. 3E). This control experiment clearly demonstrates that the differences in fiber brightness under steady-state conditions (Figs. 3C and 3D) can be attributed to differences in the mitochondrial NAD redox state caused by a different degree of enzyme deficiencies of the mitochondrial respiratory chain (cf. discussion in Ref. 15).

Visualization of Mitochondrial Defects in Digitonin-Treated Cultured Skin Fibroblasts by Mitochondrial NAD(P) Redox State Determinations

In cultured cells, such as human skin fibroblasts, which have a rather low mitochondrial content, the respiratory chain-linked flavoprotein fluorescence signal is very low. Therefore, changes in the mitochondrial redox state can be detected only by pyridine nucleotide fluorescence measurements. Two problems have to be considered in these measurements. (i) Because the cells are usually extremely sensitive to intense UV illumination, high excitation intensities should be avoided. (ii) The contribution of extramitochondrial pyridine mucleotides to the NAD(P)H fluorescence signal can vary considerably. Therefore, we determined the mitochondrial redox state in permeabilized cells applying low digitonin concentrations (10 μg/ml). Under these conditions, cytosolic pyridine nucleotides are lost. A typical experiment with control fibroblasts and fibroblasts of a CPEO patient harboring the 4977-bp "common deletion" of mitochondrial DNA is shown in Figs. 4 and 5. Figures 4B and 5B show pyridine nucleotide autofluorescence images of the permeabilized cells. Addition of substrates (octanoylcarnitine and malate) and ADP causes increased reduction of the mitochondrial NAD system, resulting in a brighter fluorescence image (Figs. 4C and 5C). In control fibroblasts the maximal NAD(P) reduction is obtained by cyanide addition (Fig. 4D). In contrast, CPEO fibroblasts failed to show an additional fluorescence increase on addition of the respiratory chain inhibitor (Fig. 5D). This behavior is a clear indication of the presence of a respiratory chain defect in CPEO fibroblasts.

FIG. 3. Phase-contrast and video autofluorescence ratio images [flavoprotein/NAD(P)H] of two saponin-permeabilized fibers from ALS skeletal muscle. Digital video images of NAD(P)H and flavoprotein autofluorescence, respectively, were obtained as described in the methods section. To visualize metabolic alteration of the mitochondrial NAD redox state, the flavoprotein image was divided by the NAD(P)H image. A, phase contrast; B, endogenous oxidized state; C, addition of 1 m*M* octanoylcarnitine and 5 m*M* malate; D, addition of 1 m*M* ADP; and E, addition of 4 m*M* potassium cyanide. Bar; 60 μm. The steady-state redox state of the two fibers in the presence of substrates and ADP (D) is 72 and 57% for flavoproteins and 53 and 32%, for NAD(P)H, respectively. The "redox state" of the ratio signal was 79% for the upper fiber and 51% for the lower fiber.

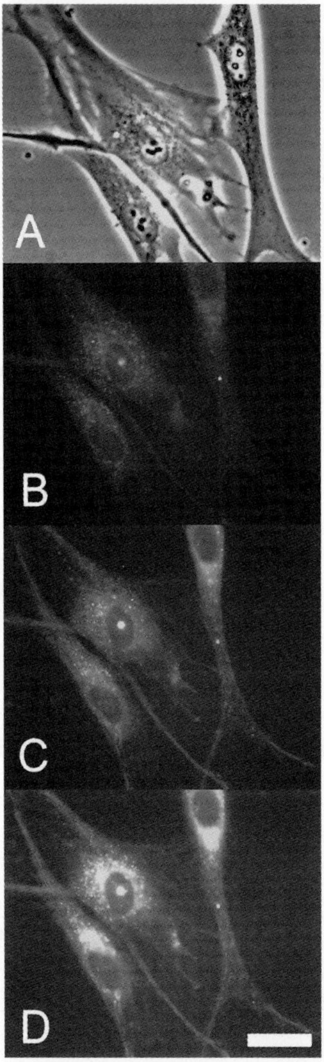

FIG. 4. Phase-contrast and video autofluorescence images of NAD(P)H fluorescence of digitonin-treated control fibroblasts. A, phase contrast; B, endogenous state in the presence of 10 μg/ml digitonin; C, addition of 1 m*M* octanoylcarnitine, 5 m*M* malate, and 1 m*M* ADP; and D, addition of 4 m*M* potassium cyanide. Bar: 15 μm.

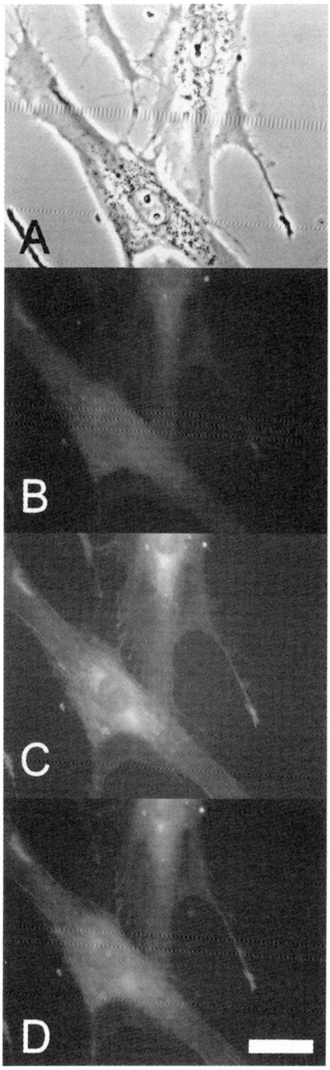

FIG. 5. Phase-contrast and video autofluorescence images of NAD(P)H fluorescence of digitonin-treated fibroblasts of a CPEO patient harboring the 4977-bp "common deletion" of mitochondrial DNA. A, phase contrast; B, endogenous state in the presence of 10 μg/ml digitonin; C, addition of 1 mM octanoylcarnitine, 5 mM malate, and 1 mM ADP; and D, addition of 4 mM potassium cyanide. Bar: 15 μm.

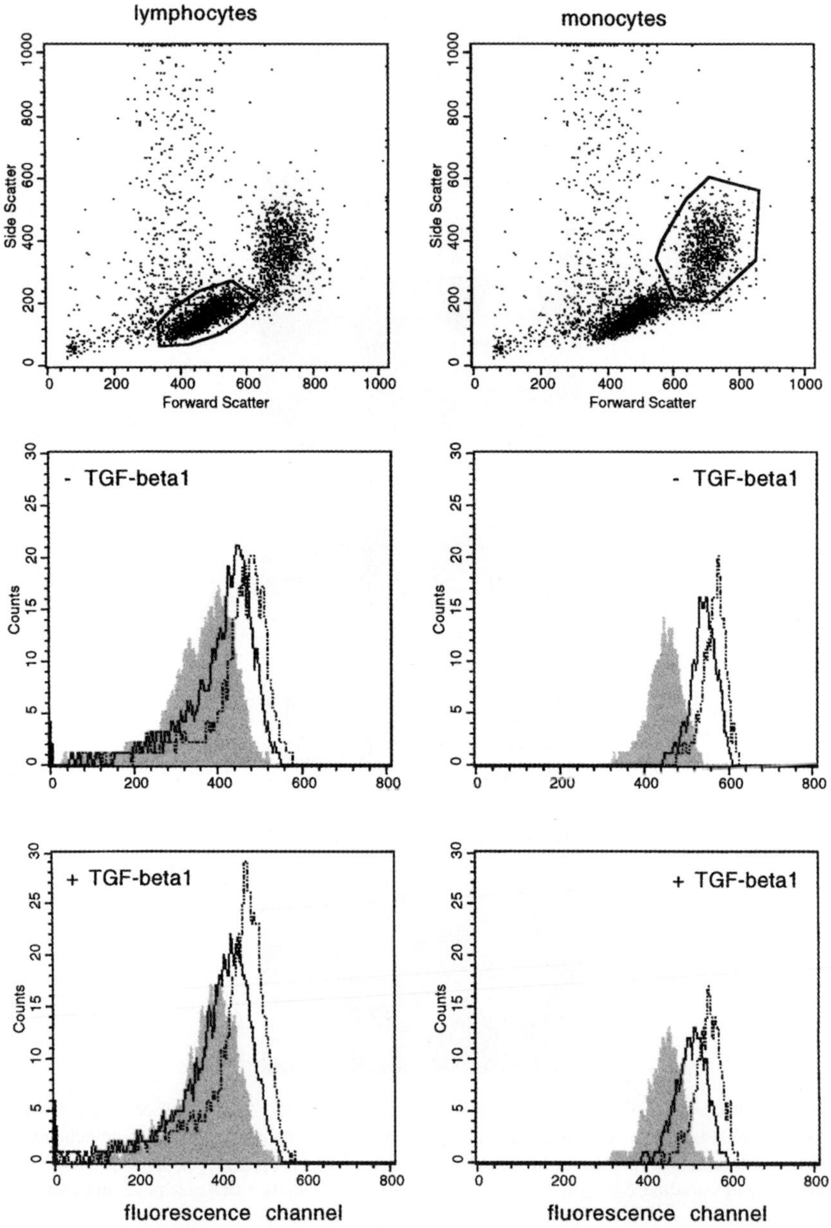

FIG. 6. Flow cytometric analysis of flavoprotein autofluorescence distribution of a suspension of mononuclear cells. For one single analysis the fluorescence properties of 10,000 mononuclear cells within the gates for lymphocytes (upper left) and monocytes (upper right) are collected.

Mitochondrial Redox State Determination in Heterogeneous Cell Suspensions by Flow Cytometry

Imaging techniques are not suitable for obtaining information on the mitochondrial redox state in heterogeneous suspensions of cells that cannot be attached to a coverslip. Under these circumstances, flow cytometry offers the possibility of observing mitochondrial flavoprotein redox state changes, at least on the level of individual cell populations, if they have distinct light-scattering properties. To detect mitochondrial flavoprotein fluorescence with flow cytometry, argon-ion laser excitation at 488 nm can be used.[17] Typical recordings of a mononuclear cell suspension are shown in Fig. 6. The upper panels show the light scatter gates for the lymphocyte and monocyte subpopulation. In the presence of an uncoupler and the absence of substrates (oxidized state), the fluorescence properties of the individual cell populations are distributed as shown in the dotted histogram plots (middle panels). The addition of octanoic acid to the cell suspension shifts the fluorescence distribution to the right (to lower fluorescence intensities), indicating flavoprotein reduction (solid lines). For the complete reduction of the mitochondrial NAD system, we added potassium cyanide, causing a further shift of the flavoprotein fluorescence distribution to lower intensities (gray histogram plot). With this approach, the putative effects of various substances on respiratory chain function can be tested. As an example, the effects of treatment with the cytokine transforming growth factor-beta 1 (TGF-β1) on both the lymphocyte and the monocyte subpopulation of mononuclear cells are shown in the lower panels of Fig. 6. Whereas no influence on fluorescence distribution in the oxidized state was observed, fluorescence distribution in the presence of octanoic acid appears to be shifted to lower fluorescence intensities in both subpopulations (solid lines lower panels, compare with solid lines in middle panels). This is an indication for a higher steady-state redox state of the mitochondrial NAD system on TGF-β1 treatment. This effect could be caused either by an improved supply of reducing equivalents or by a

[17] D. Kunz, C. Luley, K. Winkler, H. Lins, and W. S. Kunz, *Anal. Biochem.* **246,** 218 (1997).

(Middle) Flavoprotein autofluorescence distribution in the presence of 5 μM TTFB (4,5,6,7-tetrachloro-2-trifluoromethylbenzimidazole), dotted curves, in the presence of 1 mM octanoic acid, solid line curves; and 1 mM octanoic acid and 4 mM KCN, gray histogram plot. (Bottom) Prior to the flow cytometric analysis, the mononuclear cell suspension is incubated for 30 min at 37° with 100 ng/ml TGF-β1 (R&D systems). Autofluorescence distribution in the presence of 5 μM TTFB, dotted curves; in the presence of 1 mM octanoic acid, solid line curves, and 1 mM octanoic acid and 4 mM KCN, gray histogram plot. Redox states were calculated from the means of the individual histograms according to the following equation: $R_{Fp} = (F_{TTFB} - F_{OC})/(F_{TTFB} - F_{KCN}) \times 100\%$. Flavoprotein redox state in control mononuclear cells: lymphocytes, 35.0 $\pm$ 4.9%; monocytes, 25.2 $\pm$ 5.7%; and after preincubation with 100 ng/ml TGF-β1: lymphocytes, 55.8 $\pm$ 15.2%; monocytes, 33.4 $\pm$ 5.1% (four independent cell preparations).

partial inhibition of the mitochondrial respiratory chain. Because TGF-β1 elicits a proapoptotic action on these cells,[18] it is reasonable to assume that its action could be mediated by respiratory chain inhibition due to increased concentrations of oxygen radicals.[19] This example clearly shows that the determination of the mitochondrial redox state in heterogeneous cell suspensions by flow cytometry allows one to obtain information about a possible (even indirect) action of a compound of interest on oxidative phosphorylation.

Concluding Remarks

Functional imaging of the mitochondrial redox state can be performed by microscopic detection of the flavoprotein fluorescence of the mitochondrially localized α-lipoamide dehydrogenase, which is in tight redox equilibrium with the mitochondrial NAD system. Direct detection of the NAD(P)H fluorescence signal for mitochondrial NAD redox state determinations can be performed in digitonin-permeabilized cells or saponin-permeabilized muscle fibers at considerably higher sensitivity. This procedure is applicable for cultured cells having low mitochondrial content. For the determination of putative heterogeneities of the mitochondrial redox state in permeabilized muscle fibers, the application of Fp/NAD(P)H ratio imaging is advantageous because it offers an increased sensitivity with respect to metabolic alterations of the fluorescence signals. In sum, the monitoring of intrisic fluorophores allows one to obtain insights into the mitochondrial function of single living cells, avoiding the problems of phototoxicity and photobleaching.

[18] C. M. Rodrigues, G. Fan, X. Ma, B. T. Kren, and C. J. Steer, *J. Clin. Invest.* **101,** 2790 (1998).
[19] V. J. Thannickal and B. L. Fanburg, *J. Biol. Chem.* **270,** 30334 (1995).

[13] Hydrogen Peroxide-Induced Apoptosis: Oxidative or Reductive Stress?

By SHAZIB PERVAIZ and MARIE-VÉRONIQUE CLÉMENT

Based on morphologic and pathologic criteria, cell death is currently subdivided into two categories: necrosis (accidental) and apoptosis (suicidal). Apoptosis is an active process that involves cross talk between caspase proteases and apoptogenic factors released from the mitochondria.[1] This interaction amplifies the death signal and triggers a cascade of events leading to the acquisition of the apoptotic

[1] M. O. Hengartner, *Nature* **407,** 770 (2000).

0076-6879/02 $35.00

phenotype via proteolytic degradation of a number of cellular proteins. Necrosis results from a passive disruption of the cell membrane with a rapid cessation of cellular function and death.[2,3] Whereas it has long been accepted that an overwhelming increase in intracellular radical oxygen intermediates (ROI), such as superoxide anion (O_2^-), hydrogen peroxide (H_2O_2), and hydroxyl radical ($OH^·$), induces necrotic death by generating oxidative stress,[4,5] the role of ROI in apoptosis is more controversial. Due to the fact that many of the known triggers of apoptosis are oxidants or stimulators of intracellular ROI production, in addition to their role as inducers of necrosis, ROI have been regarded as mediators of apoptotic cell death.[6] However, studies on the effects of hypoxia on cell survival have suggested that ROI are not necessary to induce apoptosis.[7,8] Consistent with the latter, an increase in the intracellular O_2^- concentration through inhibition of the principal intracellular O_2^- scavenger, Cu, ZnSOD, or by stimulating cells with phorbol esters and chemicals that directly induce O_2^- production inhibits apoptosis in mammalian cell lines. Conversely, inhibition of cellular O_2^- production enhances the sensitivity of tumor cells to apoptotic triggers.[9,10] These results suggest that alterations in the intracellular redox state could trigger or block the apoptotic death program. Indeed, consistent with this hypothesis, we have shown that exposure of cells to apoptotic concentrations (<0.5 m*M*) of H_2O_2 induces a decrease in intracellular O_2^- via an inhibitory effect on the NADH/NADPH oxidase system.[5] Interestingly, at these concentrations of H_2O_2, the intracellular pH (pH_i) drops significantly prior to the activation of caspase proteases. At higher concentrations of H_2O_2 (0.5–2 m*M*) the cells undergo necrosis with a concomitant increase in pH_i and a drop in the GSH/GSSG ratio, consistent with oxidative stress. Furthermore, anticancer drugs that activate the apoptotic pathway via the production of intracellular H_2O_2 also trigger an early acidification of the intracellular milieu, which could be reversed on scavenging H_2O_2.[11] In view of these data, it appears that concentrations of O_2^-/H_2O_2 that activate the apoptotic cell death pathway favor the reduction of intracellular milieu, thereby creating an environment permissive for apoptotic execution as opposed to very high concentrations that kill cells by oxidative stress-induced necrosis. We therefore refer to

[2] A. H. Wyllie, J. F. Kerr, and A. R. Currie, *Int. Rev. Cytol.* **68,** 251 (1980).
[3] G. Majno and I. Joris, *Am. J. Pathol.* **146,** 3 (1995).
[4] M. B. Hampton and S. Orrenius, *FEBS Lett.* **414,** 552 (1997).
[5] M. V. Clement, A. Ponton, and S. Pervaiz, *FEBS Lett.* **440,** 13 (1998).
[6] T. M. Buttke and P. A. Sandstrom, *Immunol. Today* **15,** 7 (1994).
[7] S. Shimizu, Y. Eguchi, H. Kosaka, W. Kamiike, H. Matsuda, and Y. Tsujimoto, *Nature* **374,** 811 (1995).
[8] M. D. Jacobson and M. C. Raff, *Nature* **374,** 814 (1995).
[9] M. V. Clement and I. Stamenkovic, *EMBO J.* **15,** 216 (1996).
[10] S. Pervaiz, J. Ramalingan, J. Hirpara, and M.-V. Clement, *FEBS Lett.* **459,** 343 (1999).
[11] J. L. Hirpara, M. V. Clement, and S. Pervaiz, *J. Biol. Chem.* **276,** 514 (2001).

the apoptotic signaling triggered by exogenous or endogenous H_2O_2 as "reductive stress-induced apoptosis" as opposed to the oxidative damage induced by necrotic concentrations.[5]

General Methods

Cell Lines and Apoptotic Triggers

The human promyelocytic leukemia cell line HL60 (ATCC, Rockville, MD) is maintained in culture in RPMI 1640 supplemented with 10% fetal bovine serum (FBS; GIBCO-BRL, Gaithersburg, MD). The M14 human melanoma cell line is a generous gift from Dr. Armando Bartolazi (Oncologia Clinica e Sperimentale, Rome, Italy) and is cultured in Dulbecco's modified Eagle's medium (DMEM)/5% FBS. Apoptosis is induced by the exogenous addition of H_2O_2 (0.1 to 1 m*M*) or 50 μg/ml of C2 (merocil)[12] to 1×10^6 cells.

Assessment of H_2O_2-Induced Apoptosis by Poly(ADP-Ribose) Polymerase (PARP) Cleavage

One of the classical hallmarks of apoptosis is the activation of intracellular caspase proteases. Caspase activation can be determined either by direct measurement of the enzymatic activity using specific substrates conjugated to fluorophores, or by evaluating cleavage of substrates specific for the executioner caspases. The nuclear repair enzyme PARP is a specific substrate of caspase 3 and is cleaved during apoptotic execution.[13] PARP cleavage can be detected by Western blot analysis of lysates following triggering of apoptosis from the 116-kDa band to fragments of 89 and 27 kDa. In order to determine the apoptotic concentration range of H_2O_2, M14 cells (2×10^6) are exposed to increasing concentrations of H_2O_2 in the presence or absence of 500 μM general inhibitor of caspases (ZVAD-fmk; Bio-Rad Laboratories, Hercules, CA) for 18 hr. Cells are then lysed directly in 500 μl of sample lysis buffer [62.5 m*M* Tris–HCl, pH 6.8; 6 *M* urea; 10% glycerol; 2% sodium dodecyl sulfate (SDS); 0.00125% bromphenol blue; 5% 2-mercaptoethanol], boiled, subjected to 10% polyacrylamide gel electrophoresis (PAGE), and transferred to a nitrocellulose membrane. Keeping the lysate at 4° for overnight before performing SDS–PAGE usually gives a better detection of the cleaved fragments. The membrane is then exposed to a 1 : 5000 dilution of anti-PARP antibody C-2-10 (Pharmingen, San Diego, CA) at room temperature for 2 hr made in Tris-buffered saline (TBS: 50 mmol/liter Tris/HCl, pH 7.4; 150 mmol/liter NaCl) + 0.1% Tween 20 and 1% bovine serum albumin. Following three washes with TBST

[12] S. Pervaiz, M. A. Seyed, J. L. Hirpara, M. V. Clement, and K. W. Loh, *Blood* **93,** 4096 (1999).
[13] P. J. Duriez and G. M. Shah, *Biochem. Cell Biol.* **75,** 337 (1997).

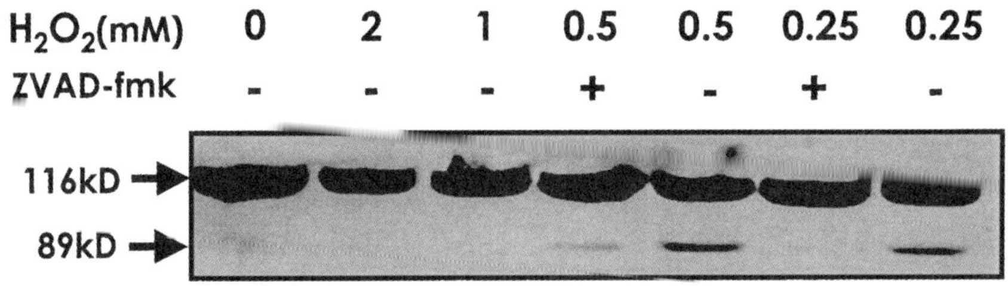

FIG. 1. M14 cells (1×10^6) are exposed to increasing concentrations of H_2O_2 for 18 hr in the presence or absence of the general caspase inhibitor ZVAD-fmk (500 μM). PARP cleavage is analyzed by Western blot analysis as described in the text.

(TBS + 0.1% Tween 20), the membrane is exposed to a 1 : 10,000 dilution of goat anti-mouse IgG–HRP conjugate (Pierce, Rockford, IL) for 1 hr and washed three times with TBST. Chemiluminescence is detected using the SuperSignal substrate Western blotting kit (Pierce, Rockford, IL). Figure 1 shows that H_2O_2 induces cleavage of PARP at concentrations of 0.5 mM and below, which is inhibitable by the general caspase inhibitor ZVAD. Conversely, concentrations above 0.5 mM result in no detectable cleavage of PARP; however, these cells still undergo death (data not shown), which is indicative of necrosis. The concentration range of H_2O_2 may differ from cell line to cell line; however, the same effects (apoptosis at submicromolar concentrations and necrosis at higher concentrations) are universally observed.

Measurement of Intracellular O_2^- and H_2O_2

Accurate assessment of intracellular levels of small miscible molecules, such as O_2^- and H_2O_2, is problematic. Out of the many assays described,[14] in our experience, a modified version of the lucigenin-based chemiluminescence assay originally described by Porter *et al.*[15] and staining of the cells with dichlorohydrofluorescine diacetate (DCHF-DA) are efficient and reproducible assays for measuring intracellular ROI.

Lucigenin-Based Assay to Detect Level of Intracellular O_2^-. Despite some controversy with respect to the specificity of the lucigenin-based chemiluminescence assay for the detection of O_2^-,[16,17] we have used this assay in the presence

[14] B. Halliwell and J. M. C. Gutteridge, *Free Radic. Biol. Med.* **351** (1999).

[15] C. D. Porter, M. H. Parka, M. K. L. Collins, R. J. Levinsky, and C. Kinnon, *J. Immunol. Methods* **155,** 151 (1992).

[16] S. I. Liochev and I. Fridovich, *Proc. Natl. Acad. Sci. U.S.A.* **94,** 2891 (1997).

[17] Y. Li, H. Zhu, P. Kuppusamy, V. Roubaud, J. L. Zweier, and M. A. Trush, *J. Biol. Chem.* **273,** 2015 (1998).

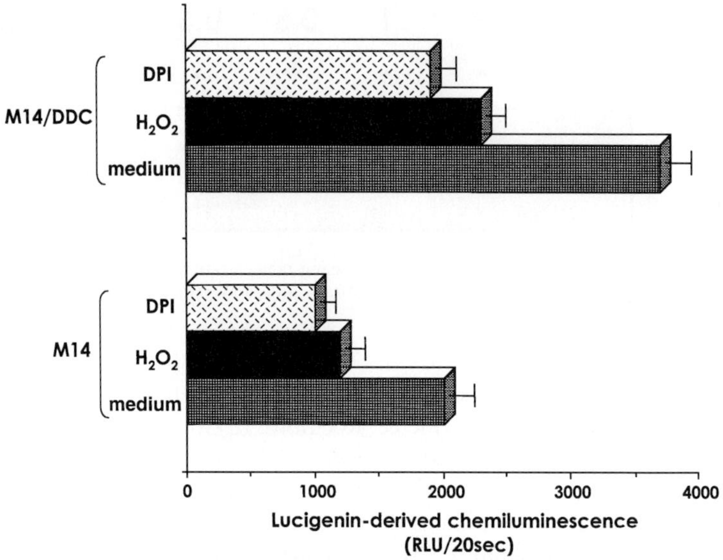

FIG. 2. Detection of intracellular O_2^- using lucigenin-derived chemiluminescence. M14 cells (1×10^6) are treated with 1 m*M* DDC (M14/DDC) for 1 hr followed by 250 μM H_2O_2 or 25 μM DPI for 1 hr (top). In a separate experiment, M14 cells are directly exposed to DPI and H_2O_2 for 1 hr before assaying for lucigenin-derived chemiluminescence (bottom). Chemiluminescence is detected as described in the text.

of a mild detergent. Using inducers and inhibitors of intracellular O_2^-, we consistently obtain a strong correlation between lucigenin-derived chemiluminescence (LDC) and intracellular O_2^- concentration (Fig. 2). For adherent cells, 1×10^6 cells are seeded in 100-mm tissue culture plates 24 hr before performing the experiment. We recommend 2×10^6 cells for nonadherent cell lines. The chemiluminescence falls off the linear scale if $<0.5 \times 10^6$ cells are used per assay. Cells are washed once with 1× PBS, trypsinized (Trypsin, Hyclone, Logan, UT), transferred to a sample cuvette, and centrifuged at 2000 rpm at 25° for 1 min. The supernatant is removed, and the cell pellet is resuspended in 400 μl of 1× ATP-releasing agent (Sigma Chemical Co., St. Louis, MO) at room temperature. The ATP-releasing agent was originally used for the bioluminescence determination of ATP in somatic cells. However, we find that using this reagent to permeabilize cells allows a reproducible measurement of intracellular O_2^- by LDC. A stock solution (850 μM) of Lucigenin (bis-*N*-methylacridinium nitrate; Sigma Chemical) is prepared in ddH_2O and filtered before use in the assay. One hundred microliters of Lucigenin stock solution is automatically injected into the cell lysate before measuring chemiluminescence. Chemiluminescence is immediately monitored for 20 sec to 3 min in a Lumat LB 9501 (Wallac Inc., Gaithersburg, MD). We have also used the TD-20/20 luminometer (Turner Designs, Sunnyvale, CA) with similar results. In order to avoid variation in cell numbers, LDC can be standardized per microgram of total protein. Protein concentration is determined using the Coomassie Plus protein assay reagent (Pierce Chemical Company, Rockford, IL). Figure 2 shows an increase in LDC in M14 cells following a 1 hr incubation with diethyl dithiocarbamate (DDC; Sigma Chemical Co.), an inhibitor of the Cu,Zn superoxide dismutase. Similar treatment of M14 cells with 25 μM diphenylene iodonium (DPI), an inhibitor of O_2^- production, or 250 μM H_2O_2 induces a 50% decrease in LDC. However, incubation with 1 mM H_2O_2 does not result in any change in chemiluminescence compared to control cells (data not shown). These results show that apoptotic concentrations of H_2O_2 induce a detectable decrease in intracellular level of O_2^-. Similar results are obtained with other human tumor cell lines, such as HL60, Jurkat, U2-OS osteosarcoma, and T24 bladder carcinoma.

Flow Cytometric Detection of Intracellular H_2O_2 Using DCHF-DA Staining. A stock solution of DCHF-DA (5 mM; Molecular Probes Inc.) is made in dimethyl sulfoxide (DMSO) and stored in the dark at 4°. In a typical experiment, 1×10^6/ml cells are pelleted, washed with 1× PBS, and loaded with 5 μM DCHF-DA for 15–30 min at 37° in RPMI (100 μl), washed gently once with RPMI, resuspended in 0.5 ml of RPMI, and DCF fluorescence ($\lambda_{emission}$ = 525 nm) is analyzed using a flow cytometer (Coulter EPICS Elite ESP). At least 10,000 events are analyzed by WinMDI software. Higher concentrations of DCHF-DA do not essentially give better results, but the sensitivity of the assay could drop if the cells are loaded with higher than 10 μM chromophore. Incubation of cells with increasing concentrations of H_2O_2 (200–800 μM) for 15–30 min followed by DCHF-DA loading and flow cytometry analysis shows that the minimum concentration of H_2O_2 in the culture medium required to detect a significant increase in the intracellular DCF fluorescence indicative of intracellular H_2O_2 is ~200 μM. A newer derivative of DCHF-DA (CM-DCHF-DA) gives a significantly better retention of the oxidized derivative, and therefore enhances the sensitivity of the assay (Fig. 3b). In order to detect the drug-induced production of intracellular H_2O_2, cells (1×10^6) are treated with the apoptotic inducer for different incubation periods, followed by loading with 5 μM DCHF-DA and flow analysis (Fig. 4a). To ascertain that a drug-induced increase in DCF flourescence is specific for H_2O_2, preincubation of cells with scavengers of H_2O_2, such as catalase, should result in inhibition of a fluorescence shift detected by flow cytometry (Fig. 4b). We do not find a significant difference in the fluorescence yield on preloading the cells with DCHF-DA and subsequent exposure to apoptotic agents. It should be pointed out that because of the H_2O_2-scavenging ability of serum, the concentration of serum in the culture medium could decrease the sensitivity of detection; the sensitivity of this assay does enhance if the serum content of the culture medium is reduced from 10% to between 5 and 2.5%.

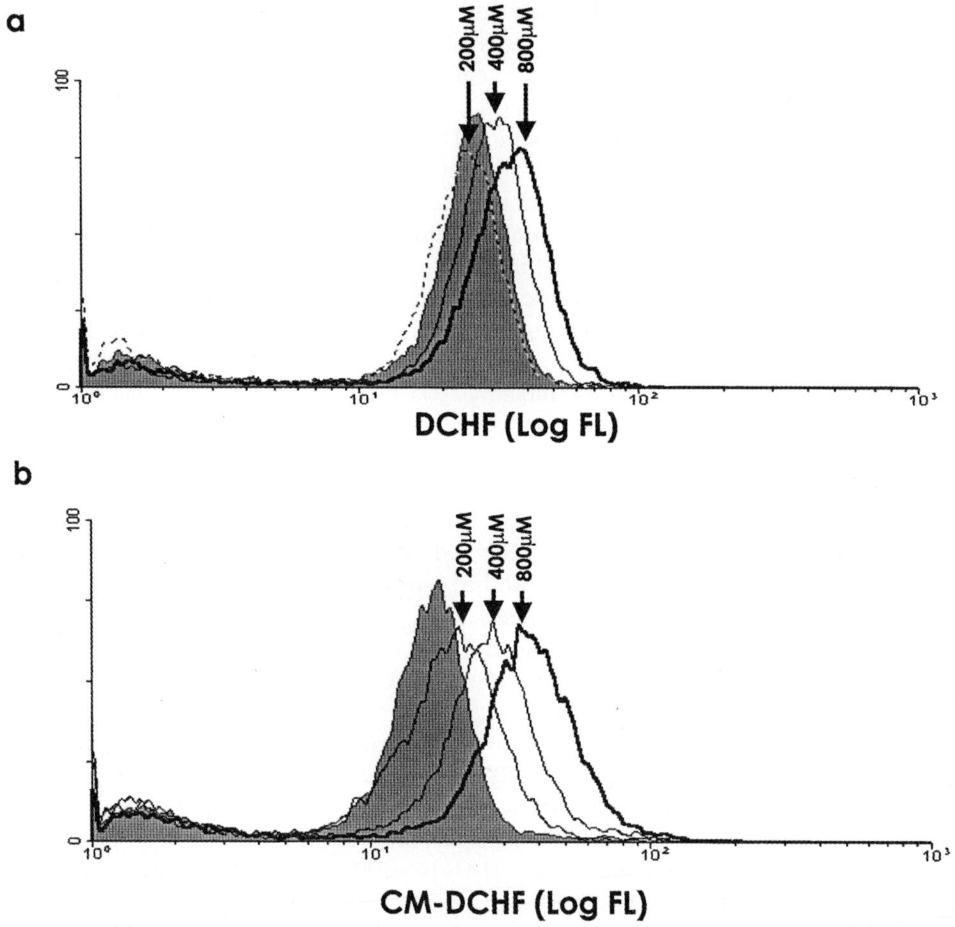

FIG. 3. Flow cytometric detection of intracellular H_2O_2 using DCHF-DA and CM-DCHF-DA. HL60 (1×10^6) cells are incubated with increasing concentrations of H_2O_2 for 30 min followed by loading with 5 μM DCHF-DA (a) or 5 μM CM-DCHF-DA (b) for 30 min at 37°. Cells are then analyzed by flow cytometry as described in the text.

Measurement of Intracellular pH

Intracellular pH (pH_i) is measured by loading cells with membrane-impermeant dye BCECF [2′, 7′-bis(2-carboxyethyl)-5,6-carboxyfluorescein; Sigma] following the protocol described by Musgrove and Hedley.[18] Before proceeding with the measurement of pH_i in experimental samples, a pH calibration curve is obtained

[18] E. A. Musgrove and D. W. Hedley, *Methods Cell Biol.* **33,** 59 (1990).

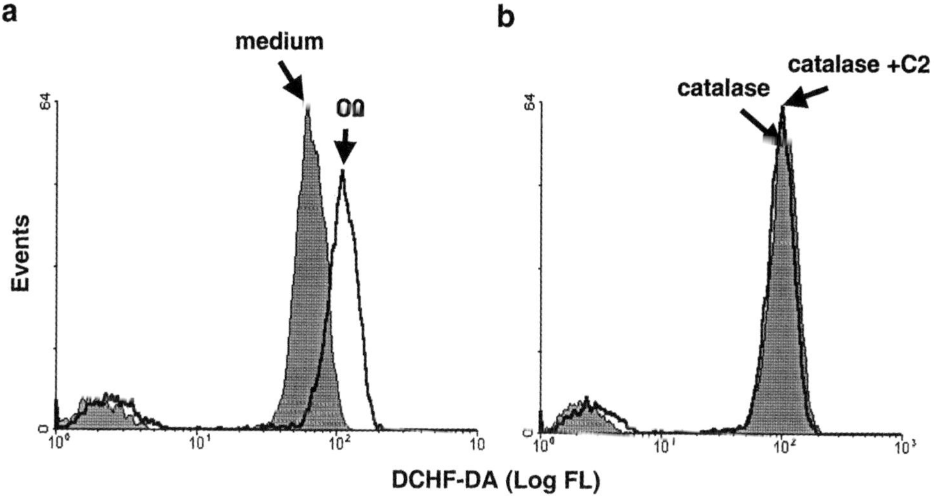

FIG. 4. HL60 cells (1×10^6) are exposed to 50 μg/ml of C2 (merocil) for 4 hr in the presence or absence of catalase (1000 U/ml), loaded with 5 μM DCHF-DA for 30 min at 37°, and subjected to flow cytometry as described in the text.

for the cell line to be assayed (Fig. 5) each time the experiment is performed. In order to generate a pH calibration curve, 1×10^6 cells/ml in Hanks' balanced salt solution (HBSS; Sigma) are loaded with BCECF (10 μl/ml from 1 mM stock) at 37° for 30 min in the dark, washed once with HBSS, and resuspended in high K^+ buffer. The high K^+ buffer is prepared by mixing 135 mM KH_2PO_4/20 mM NaCl (pH 8.6) with 110 mM K_2HPO_4/20 mM NaCl (pH 4.4). Buffers with a range of pH (6.0 to 8.0) are obtained by adjusting the volume of one or the other high K^+ buffer in the solution. Immediately before flow cytometry, cells (in 0.5 ml of various pH buffers) are loaded with 20 μM nigericin (Sigma, 10 mM stock in absolute alcohol), an electroneutral K^+/H^+ exchanger, and fluorescence ratio measurements (525/610 nm) are obtained using a Coulter EPICS Elite ESP (Coulter, Hialeah, FL) flow cytometer with excitation set at 488 nm. For the measurement of pH_i of experimental samples, cells are first exposed to the apoptotic triggers and then loaded with BCECF, resuspended in 0.5 ml of HBSS, and subjected to flow cytometry analysis as described earlier. Fluorescence ratios obtained for untreated and treated cells are then plotted on the calibration curve to determine pH_i (Fig. 5). To evaluate the role of H_2O_2, added exogenously or produced on drug exposure of cells, in the drop in pH_i, cells are preincubated with catalase (1000 U/ml) for 1 hr before triggering apoptosis. Pretreatment with catalase salvages cells from undergoing acidification, suggesting an involvement of H_2O_2 in the induction of intracellular acidification (Fig. 5). It should also be pointed out that a necrotic

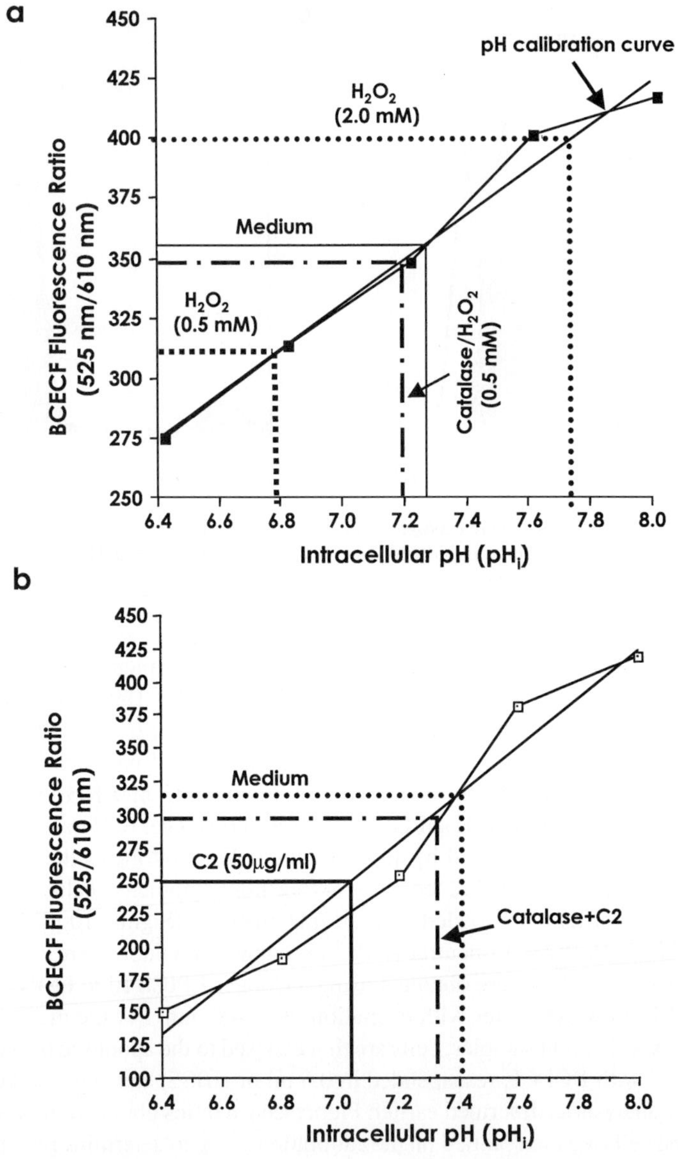

FIG. 5. Effect of H_2O_2 and C2 (merocil) on intracellular pH (pH_i) of M14 and HL60 cells, respectively. (a) M14 (1×10^6) cells are exposed to 0.5 or 2 mM H_2O_2 for 2 hr in the presence or absence of catalase (1000 U/ml) followed by loading with BCECF for 30 min at 37° as described in the text. (b) HL60 (1×10^6) cells are exposed to 50 μg/ml of C2 for 4 hr in the presence or absence of catalase, and pH_i is measured as described. pH_i is determined from the calibration curve obtained (BCECF fluorescence ratio) for each cell line using buffers with a range of pH (6.4 to 8).

concentration of H_2O_2 (2 m*M*) does not induce an intracellular drop in pH, but conversely the pH_i increases, as shown in Fig. 5a.

Conclusions

The intracellular concentration of ROI must be tightly regulated during homeostasis, and a significant deviation from this may tilt the balance between oxidative or reductive stress-induced death. We have shown that H_2O_2-induced apoptosis results in a significant drop in the intracellular pH and O_2^- concentration, irrespective of whether H_2O_2 is added exogenously or induced intracellularly using drugs.[5,11] The reduced intracellular milieu then provides an environment permissive for the execution of the death signal.[11] Therefore, it is important to differentiate between oxidative stress, induced by high concentrations of ROI that cause necrosis of the cells, from reductive stress-induced cell death that involves ROI production associated with a drop in the pH_i and subsequent activation of the apoptotic cell death pathway.

Acknowledgments

The authors thank J.L. Hirpara for technical assistance. This work was supported by Grants R-185-000-019-213 and R-185-000-009-112 from the NRMC, Singapore, and ARF, National University of Singapore, respectively, to S.P. and Grant R-364-000-008-213 from the NMRC, Singapore to M-V.C.

[14] Peroxidation of Phosphatidylserine in Mechanisms of Apoptotic Signaling

By YULIA Y. TYURINA, VLADIMIR A. TYURIN, ANNA A. SHVEDOVA, JAMES P. FABISIAK, and VALERIAN E. KAGAN

Introduction

Asymmetric distribution of major phospholipid classes across membranes is a fundamental characteristic of all cells whose disturbance is incompatible with physiological functions of membranes and with cell viability.[1,2] Under normal conditions, phosphatidylcholine (PC) and sphingomyelin (SPM) are located primarily in the outer leaflet of plasma membrane, whereas aminophospholipids—phosphatidylethanolamine (PE) and phosphatidylserine (PS)—are found almost

[1] J. A. F. Op den Kamp, *Annu. Rev. Biochem.* **48,** 47 (1979).
[2] R. F. A. Zwaal and A. J. Schroit, *Blood* **89,** 1121 (1997).

0076-6879/02 $35.00

entirely in the inner leaflet.[1–3] Transmembrane migration and externalization of PS by apoptotic cells are considered to be the triggering events for their recognition by cognate "scavenger" receptors of macrophages.[4] This is followed by phagocytosis and safe digestion of apoptotic cells, thus preventing development of an inflammatory response.[5]

Normal maintenance of PS asymmetry is mainly due to the constitutive activity of aninophospholipid translocase (APT), an ATP-dependent enzyme that transports aminophospholipids from the external to the internal surface of the plasma membrane lipid bilayer.[6] Activation of a Ca^{2+}-dependent phospholipid scramblase, which promotes random redistribution of all phospholipids in a bidirectional manner, is involved in the initiation of PS externalization during apoptosis. It is also clear, however, that downregulation of the surveillance function of APT is required to maintain PS externalization.[7] Previous work has demonstrated that apoptosis is associated with selective oxidation of specific phospholipid classes, most notably PS.[8,9] Site-specific oxidation of PS in plasma membrane may be an important early step in the mechanisms leading to its externalization[8,10,11] and binding of (oxidized) PS with scavenger receptors.[12,13] We speculated that externalization of (oxidized) PS might be due to the failure of APT to internalize it via either direct enzyme inhibition or the inability of the enzyme to recognize oxidized PS. It is also noteworthy that oxidized phospholipids undergo spontaneous "flip-flop" more readily than their nonoxidized counterparts.[14]

Oxidative stress is a frequent trigger of apoptosis in a variety of cells and is also thought to be involved as a component of the common pathway in the execution

[3] E. M. Bevers, E. M. Comfurius, D. W. C. Dekkers, and R. F. A. Zwaal, *Biochim. Biophys. Acta* **1439,** 317 (1999).

[4] V. Fadok, D. R. Voelker, P. A. Campbell, J. J. Cohen, D. L. Bratton, and P. M. Henson, *J. Immunol.* **148,** 2207 (1992).

[5] V. A. Fadok, D. L. Bratton, S. C. Frasch, M. L. Warner, and P. M. Henson, *Cell Death Differ.* **5,** 551 (1998).

[6] D. Daleke and J. V. Lyles, *Biochim. Biophys. Acta* **1486,** 108 (2000).

[7] B. Fadeel, B. Gleiss, K. Hogstrand, J. Chandra, T. Wiedmer, P. J. Sims, J. I. Henter, S. Orrenius, and A. Samali, *Biochem. Biophys. Res. Commun.* **266,** 504 (1999).

[8] J. P. Fabisiak, V. E. Kagan, V. B. Ritov, D. E. Johnson, and J. S. Lazo, *Am. J. Physiol. (Cell Physiol.)* **272,** C675 (1997).

[9] N. F. Schor, Y. Y. Tyurina, J. P. Fabisiak, V. A. Tyurin, J. S. Lazo, and V. E. Kagan, *Brain Res.* **831,** 125 (1999).

[10] J. P. Fabisiak, Y. Y. Tyurina, V. A. Tyurin, U. S. Lazo, and V. E. Kagan, *Biochemistry* **37,** 13781 (1998).

[11] J. P. Fabisiak, V. E. Kagan, Y. Y. Tyurina, V. A. Tyurin, and J. S. Lazo, *Am. J. Physiol. (Lung Cell. Mol. Physiol.)* **274,** L793 (1998).

[12] J. Tait and C. Smith, *J. Biol. Chem.* **274,** 3048 (1999).

[13] A. Boullier, K. L. Gillotte, S. Horkko, S. R. Green, P. Friedman, E. A. Dennis, J. L. Witztum, D. Steinberg, and O. Quehenberger, *J. Biol. Chem.* **275,** 9163 (2000).

[14] L. I. Barsukov, A. V. Victorov, I. A. Vasilenko, R. P. Evistigneeva, and L. D. Bergelson, *Biochim. Biophys. Acta* **598,** 153 (1980).

of apoptosis.[10,11,15–17] While effects of oxidative stress on protein components of apoptotic machinery, such as caspases,[18,19] have been well characterized, information on selective oxidation of specific classes of phospholipids in live cells is limited. This results mainly from the paucity of sensitive and specific quantitative assays for measuring the oxidation of different classes of phospholipids. The direct measurement of oxidized lipid products is problematic, as cells possess a very effective system for remodeling and repairing oxidatively modified phospholipids[20] that interferes with their accurate measurement. We have developed a sensitive, specific, and reliable procedure for the assessment of oxidative stress in different classes of membrane phospholipids in intact live cells based on their metabolic acylation with an oxidation-sensitive and fluorescent fatty acid, *cis*-parinaric acid (*cis*-PnA), as a reporter molecule. Basic procedures and some applications of the technique are described in this chapter.

Reagents and Cells

cis-Parinaric acid [(9Z, 11E, 13E, 15Z)-octadecatetraenoic acid] is from Molecular Probes (Eugene, OR). Chloroform, methanol, glacial acetic acid, hexane, 2-propanol (HPLC grade), Tween 20, butylated hydroxytoluene (BHT), malachite base green, phospholipase A_2 from bee venom, and mellitin from bee venom are from Sigma (St. Louis, MO). Ammonium hydroxide is from Fisher Scientific (Pittsburgh, PA). α-Tocopherol acetate is from Aldrich Chemical (Milwaukee, WI). KGM-2 medium is from Clonetics (San Diego, CA). AMVN, 2.2′-azobis (2,4-dimethylisovaleronitrile) is from Wako Chemicals (Richmond, VA). Silica G HPTLC plates (5 × 5 cm) are from Whatman (Clifton, NJ).

We have, to date, utilized a number of diverse cell lines in conjunction with the metabolic incorporation of *cis*-PnA to measure lipid peroxidation in various phospholipid classes following oxidative stress (see Tables I and II). Our collective experience with multiple cell types includes those that grow as suspension cultures, as well as attached monolayers. The following detailed methods describe the use of HL-60 cells and normal human epidermal keratinocytes (NHEK) as representative example of suspension and monolayer cells, respectively. NHEK from adults are from Clonetics (San Diego, CA). Cells are plated at a density of 6.25×10^4 cells per 75-ml tissue culture flask (Greiner Laboratories, GmbH, Germany) and grown in KGM-2 medium until confluent monolayers are obtained. Repeated counts

[15] D. M. Hockenberry, Z. N. Oltavi, X.-M. Yin, C. L. Milliman, and S. J. Korsmeyer, *Cell* **75,** 241 (1993).

[16] J. Cai and D. P. Jones, *J. Biol. Chem.* **273,** 11401 (1998).

[17] T. M. Buttke and P. A. Sandstrom, *Immunol. Today* **15,** 7 (1994).

[18] S. Dimmeler, J. Haendeler, M. Nehls, and A. M. Zeiher, *J. Exp. Med.* **185,** 601 (1997).

[19] M. B. Hampton and S. Orrenius, *FEBS Lett.* **414,** 552 (1997).

[20] E. H. Pacifici, L. L. McLeod, and A. Sevanian, *Free Radic. Biol. Med.* **17,** 297 (1994).

TABLE I
INTEGRATION OF *cis*-PARINARIC ACID INTO PHOSPHOLIPIDS OF DIFFERENT CELL TYPES[a]

		cis-Parinaric acid integrated into cell phospholipids (ng/μg total lipid phosphorus)			
Cell line	*N*	PI	PE	PS	PC
Rat vascular smooth muscle cells	7	13.4 ± 0.4	66.2 ± 2.4	11.6 ± 1.1	237.0 ± 4.8
Rat cardiomyocytes	3	N.D.	73.7 ± 4.8	6.5 ± 0.7	384.5 ± 27.4
Rat myoblasts	3	5.6 ± 0.6	46.3 ± 4.6	4.2 ± 0.7	165.5 ± 16.6
Rat hepatocytes	3	0.9 ± 0.6	6.8 ± 0.4	2.2 ± 0.7	43.8 ± 2.1
Rat spinal cord neurons	3	9.4 ± 0.4	85.8 ± 1.6	8.8 ± 0.6	361.9 ± 4.2
Rat pheochromocytoma PC-12 Cells	7	0.7 ± 0.1	8.4 ± 2.0	1.7 ± 0.3	105.5 ± 15.0
Murine myeloid 32D cells	7	12.7 ± 2.0	35.4 ± 1.8	5.2 ± 0.4	180.2 ± 9.2
Mouse lung fibroblasts	5	N.D.	1.6 ± 0.1	0.3 ± 0.1	9.4 ± 0.6
Sheep pulmonary artery endothelial cells	3	1.8 ± 0.3	33.7 ± 4.0	13.3 ± 2.6	141.2 ± 10.0
Bovine aorta endothelial cells	6	1.4 ± 0.2	7.7 ± 0.4	3.7 ± 0.3	118.8 ± 4.8
Normal human epidermal keratinocytes	10	2.0 ± 0.2	32.4 ± 1.9	3.9 ± 0.3	50.8 ± 2.9
MCF-7 breast cancer cells	13	4.4 ± 1.8	29.8 ± 3.1	4.6 ± 0.8	135.9 ± 14.4
Human leukemia HL-60 cells	5	12.8 ± 3.5	62.7 ± 12.0	6.4 ± 2.3	212.0 ± 39.6

[a] All data are means ±SEM.

of NHEK reveal a concentration of 1–2 × 10^6 cells per flask at the time of harvest. HL-60 human promyelocytic leukemia cells from American Type Culture Collection (Manassas, VA) are cultured in RPMI 1640 medium supplemented with 12.5% fetal bovine serum. Routine passages (1 : 5–1 : 10 splits) are performed when cells reach a density of between 1 and 1.5 × 10^6 cells/ml.

Integration of *cis*-PnA Into Cell Phospholipids

cis-PnA has been used as a fluorescent probe in physical–chemical studies of model membranes and as a reporter for the assessment of peroxidation in chemical systems, lipoproteins, and simple membrane systems[29] and in total cell lipids.[30] *cis*-PnA is a natural 18 carbon fatty acid with four conjugated double bonds. The four conjugated double bonds confer highly fluorescent properties to *cis*-PnA and render it highly susceptible to peroxidation. Upon peroxidation, fluorescence is irreversibly lost as mammalian cells do not synthesize fatty acids with conjugated double bond systems. *cis*-PnA can be metabolically incorporated into phospholipids similar to endogenous free fatty acids. If care is taken to remove free *cis*-PnA after metabolic integration so that it is no longer available for

TABLE II
EFFECT OF OXIDATIVE STRESS ON PHOSPHOLIPID COMPOSITION AND OXIDATION OF *cis*-PNA-LABELED PHOSPHOLIPIDS IN CELLS

Cell line	Oxidants	Changes in phospholipid composition	Oxidation of *cis*-parinaric acid-labeled phospholipids	References
Rat vascular smooth muscle cells	*tert*-BuOOH	−	+	21
Rat vascular smooth muscle cells	AMVN	−	+	22
Rat cardiomyocytes	*tert*-BuOOH	−	+	23
Rat spinal cord neurons	AMVN	−	+	
Sheep pulmonary artery endothelial cells	AMVN	−	+	
Normal human epidermal keratinocytes	AMVN	−	+	24
	Phenol	−	+	24
	Cumene-OOH	−	+	25
MCF-7 breast cancer cells	AMVN	−	+	26
Leukemia HL-60 cells	AMVN	−	+	11
	Cu-NTA	−	+	27
	tert-BuOOH	−	+	
Murine myeloid 32D cells	Paraquat	−	+	8, 10
Pheochromocytoma PC-12 cells	AMVN	−	+	28
	Neocarzinostatin	−	+	9

[21] A. A. Shvedova, Y. Y. Tyurina, N. V. Gorbunov, V. A. Tyurin, V. Castranova, J. Ojimba, R. Gandley, M. K. McLaughlin, and V. E. Kagan, *Biochem. Pharm.* **57,** 989 (1999).

[22] R. K. Dubey, Y. Y. Tyurina, V. A. Tyurin, D. Gillespio, R. A. Branch, E. K. Jackson, and V. E. Kagan, *Circ. Res.* **84,** 229 (1999).

[23] N. V. Gorbunov, Y. Y. Tyurina, G. Salama, B. W. Day, H. G. Claycamp, G. Argyros, N. M. Elsayed, and V. E. Kagan, *Biochem. Biophys. Res. Commun.* **244,** 647 (1998).

[24] A. A. Shvedova, C. Kommineni, B. A. Jeffries, V. Castranova, Y. Y. Tyurina, V. A. Tyurin, E. A. Serbinova, J. P. Fabisiak, and V. E. Kagan, *J. Invest. Dermatol.* **114,** 354 (2000).

[25] V. E. Kagan, Y. Y. Tyurina, V. A. Tyurin, K. Kawai, J. P. Fabisiak, C. Kommineni, V. Castranova, and A. A. Shvedova, *Toxicologist* **54,** 113 (2000).

[26] N. F. Schor, Y. Y. Tyurina, V. A. Tyurin, and V. E. Kagan, *Biochem. Biophys. Res. Commun.* **260,** 410 (1999).

[27] K. Kawai, S. L. Liu, V. A. Tyurin, Y. Y. Tyurina, G. G. Borisenko, J. P. Fabisiak, B. R. Pitt, and V. E. Kagan, *Chem. Res. Toxicol.* **13,** 1275 (2000).

[28] Y. Y. Tyurina, V. E. Tyurin, G. Carta, P. J. Quinn, N. F. Schor, and V. E. Kagan, *Arch. Biochem. Biophys.* **344,** 413 (1997).

phospholipid repair, then resolution of major phospholipid classes by fluorescence HPLC can be used to quantify their oxidative damage (as a decreased content of fluorescent *cis*-PnA residues in respective phospholipid classes). Importantly, the *cis*-PnA-based assay can identify the selectivity of phospholipid oxidation based on polar head groups and is independent of the fatty acid composition of phospholipids.[30,31]

cis-PnA is incorporated into HL-60 cell suspensions or NHEK monolayers by addition of its complex with human serum albumin (hSA). The purity of each lot of *cis*-PnA is determined by UV spectrophotometry using the molar extinction $\varepsilon_{304\text{ nm ethanol}} = 80 \times 10^3\ M^{-1}\text{cm}^{-1}$. The complex is prepared by adding *cis*-PnA (500 μg, 1.8 μmol) in 25 μl of dimethyl sulfoxide to hSA (50 mg, 750 nmol) in 1 ml of phosphate-buffered serum (PBS).

HL-60 cells (10^6/ml) or NHEK (80–90% confluent monolayers) are incubated in the presence of *cis*-PnA/hSA complex (2 or 5 μg *cis*-PnA/10^6 cells/ml medium, respectively) in serum-free RPMI 1640 medium or KGM-2 medium, respectively, at 37° in 5% CO_2 atmosphere for 2 hr. Preliminary studies reveal that the integration of *cis*-PnA into cellular phospholipids under these conditions is time dependent and reaches maximum after 1–2 hr of incubation. At the end of incubation, cells are washed once with PBS containing fatty acid-free hSA (0.5 mg/ml) and again without hSA. This step is necessary to remove excess free *cis*-PnA and to exclude its potential interference by a reacylation reaction that might mask the degree of actual oxidation of *cis*-PnA-labeled phospholipids. *cis*-PnA-labeled HL-60 cells or NHEK are then exposed to various oxidative stimuli: *tert*-butyl hydroperoxide (*tert*-BuOOH) (150 μM) for 20 min in the case of HL-60 cells or cumene hydroperoxide (cumene-OOH) (200 μM) for 2 hr (NHEK cells) at 37°. At the end of treatment, NHEK are scraped and HL-60 cells are collected by centrifugation. Lipids are extracted by the Folch procedure.[32]

HPLC Analysis of *cis*-PnA-Labeled Phospholipids

An HPLC procedure is used to separate and detect *cis*-PnA integrated into various cellular phospholipid classes. The lipid extracts are applied to a 5 μm Microsorb-MV column (4.5 × 250 mm, Rainin, Woburn, MA) equilibrated with a mixture of one part of solvent A [2-propanol : hexane : water (56 : 42 : 2, by volume)] and nine parts of solvent B [2-propanol : hexane : 40 mM aqueous ammonium acetate (54 : 41 : 10, by volume)] pH 6.7. The column is eluted during the first 3 min with a linear gradient from 10% solvent B to 37% solvent B, for 3–15 min

[29] F. A. Kuypers, J. J. M. van den Berg, C. Schalkwijk, B. Roelofsen, and J. A. F. Op den Kamp, *Biochim. Biophys. Acta* **921,** 266 (1987).

[30] G. P. Drummen, J. A. Op den Kamp, and J. A. Post, *Biochim. Biophys. Acta* **1436,** 370 (1999).

[31] V. B. Ritov, S. Banni, J. C. Yalowich, B. W. Day, H. G. Claycamp, F. P. Corongiu, and V. E. Kagan, *Biochim. Biophys. Acta* **1283,** 127 (1996).

[32] J. Folch, M. Less, and G. H. Sloan-Stanley, *J. Biol. Chem.* **226,** 497 (1959).

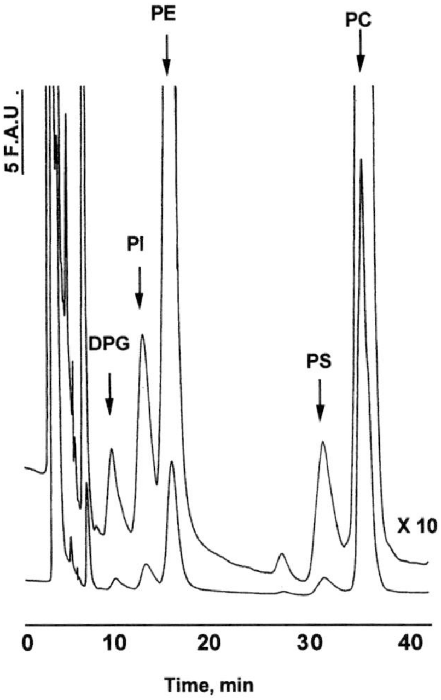

FIG. 1. A normal phase HPLC chromatogram of total *cis*-PnA-labeled phospholipids extracted from HL-60 cells. Fluorescence emission intensity, excitation at 324 nm, emission at 420 nm. PI, phosphatidylinositol; PE, phosphatidylethanolamine, PS, phosphatidylserine, PC, phosphatidylcholine; DPG, diphosphatidylgycerol.

with an isocratic gradient at 37% solvent B, for 15–23 min with a linear gradient to 100% solvent B, and for 23–45 min with an isocratic gradient at 100% solvent B. The solvent flow rate is maintained at 1 ml/min. Separations are performed using a high-performance liquid chromatograph (Shimadzu Model LC-600) equipped with an in-line configuration of fluorescence (Model RF-551) and UV-VIS (Model DPD-10AV) detectors. The effluent is monitored by absorbance at 205 nm to detect lipids; the fluorescence of *cis*-PnA is measured by its emission at 420 nm after excitation at 324 nm.

A typical fluorescence emission profile of the HPLC column eluate of total lipids extracted from HL-60 cells is shown in Fig. 1. Major fluorescence peaks were identified using authentic phospholipid standards and include diphosphatidylgycerol (DPG), phosphatidylinositol (PI), PE, PS, and phosphatidylcholine (PC). Under the indicated conditions, retention times for DPG, PI, PE, PS, and PC were determined as 11.3, 15.1, 17.3, 32.1, and 36.2 min, respectively. The identity of the fluorescence peaks was also confirmed by HPTLC of the individual collected HPLC fractions. Lipid phosphorus was determined using a micro method.[33] Data

[33] A. Chavardjian and E. Rubbnicki, *Anal. Biochem.* **36,** 225 (1970).

are expressed as nanograms of *cis*-PnA per microgram of total lipid phosphorus. Values for relative incorporation of *cis*-PnA into membrane phospholipids of examined cells are presented in Table I. This shows that incorporation of *cis*-PnA in the various phospholipids was differential, and the amount of *cis*-PnA incorporated was in the following order PC > PE > PI ≥ PS. In general, this parallels the relative abundance of each of these phospholipid classes within cells.

HPTLC of Cell Phospholipids during Apoptosis Induced by Oxidative Stimuli

Table II shows our collective experience at measuring *cis*-PnA oxidation following exposure of a variety of cell types to numerous oxidants. In all cases, *cis*-PnA oxidation (assessed by comparing *cis*-PnA fluorescent content in oxidant-treated cells compared to untreated controls) could be measured in individual classes of phospholipids following oxidative stress. Importantly, these changes were observed early following exposure to low level oxidants at a time when overall cell viability was unchanged by oxidant treatment. In live cells, oxidative modifications of only a relatively small fraction of membrane phospholipids are compatible with the preservation of cellular functions. Therefore, lipid oxidation as measured here represents relatively small levels of oxidation, and it is doubtful that significant changes of phospholipid composition would be detectable on oxidative challenge to cells within their limits of survival. It was important to determine if changes in phospholipid composition arose as a result of these oxidative challenges. In addition, information on phospholipid composition, i.e., distribution of phospholipids between their different classes, may be useful for determinations of changes in phospholipid peroxidation on a specific basis (per unit of a given class of phospholipids). To these ends, HPTLC separation and quantification of phospholipid distribution were performed as follows.

Cells after exposure to oxidative stimuli are collected by centrifugation and resuspended in 2 ml of 0.1 *M* NaCl. To extract lipids, methanol (3 ml) containing butylated hydroxytoluene (100 μM, as an antioxidant blocking peroxidation during work-up of samples) and chloroform (2 ml) are added to cell suspensions. After phase separation, the lower phase is collected, and the solvent is evaporated under nitrogen. The film of lipids is dissolved in hexane : propanol (4 : 3, by volume) and used for phospholipid analysis. Individual phospholid classes in lipid extracts are separated by two-dimensional HPTLC on silica G plates (5 × 5 cm, Whatman). The plates are first developed with a solvent system consisting of chloroform : methanol : 28% ammonium hydroxide (65 : 25 : 5, by volume). After drying the plates with a forced air blower to remove the solvent, plates are developed in the second dimension with a solvent system consisting of chloroform : methanol : glacial acetic acid : water (50 : 20 : 10 : 10 : 5, by volume). The phospholipid spots are visualized by exposure to iodine vapor and identified by comparison with the migration of authentic phospholipid standards. The spots identified by iodine staining are scraped and transferred on silica to glass tubes. Lipid phosphorus is determined by a micro method as described by Bottcher *et al.*[34] In all cell lines studied (Table II), PC and PE are the two major phospholipids, which represent 43.4–60.9% and 18.1–29.9% of total phospholipid, respectively, SPH, PI, DPG, and PS are also detectable on HPTLC plates, and their amount is dependent on the cell line (data not shown).

Expectedly, no significant difference in phospholipid distribution was detected after exposure of any examined cells to different oxidative stimuli (Table II). The lack of HPTLC-detectable changes in the phospholipid composition of different cells exposed to oxidative stimuli might be due to the effective repair of phospholipids via deacylation/reacylation pathways.[35,36] Oxidatively modified phospholipids are known to undergo rapid and effective remodeling that involves phospholipase A_2-catalyzed hydrolysis with subsequent acyltransferase-catalyzed reacylation of peroxided phospholipids.[37] Thus, in live cells, direct analysis of phospholipid composition may not be used for purposes of detection or quantitation of oxidatively modified phospholipids. The information from such determinations, however, may be utilized for more complete characterization of specific oxidative stress in different classes of membrane phospholipids using *cis*-PnA as a reporting molecule (see later).

Positional Distribution of *cis*-PnA Integrated in Cell Phospholipids

In mammalian cells, the *sn*-2 position in phospholipid molecules is usually occupied by a polyunsaturated fatty acid residue. We were anxious to determine whether *cis*-PnA, with its four conjugated double bonds, was similarly metabolically integrated into phospholipids as well, thus representing a positionally "normal" target for oxidative stress. To analyze the positional distribution of *cis*-PnA in membrane phospholipids, we treated homogenates of *cis*-PnA-prelabeled HL-60 cells with phospholipase A_2 whose catalytic action is enhanced in the presence of mellitin. Because phospholipase A_2 specifically hydrolyzes phospholipids in *sn*-2 position fluorescence HPLC, as well as HPTLC, analysis of the products permits detrmination of the relative amounts of *cis*-PnA esterified in the *sn*-1 and *sn*-2 positions of membrane phospholipids.

Homogenates (2.5×10^7 cells/ml) are prepared by freezing ($-80°$) and thawing *cis*-PnA-loaded cells treated with phospholipase A_2 from bee venom (20 U/ml)

[34] C. J. F. Bottcher, C. M. Van Gent, and C. Pries, *Anal. Chim. Acta* **24,** 203 (1961).

[35] A. Van der Vliet and A. Bast, *Chem. Biol. Interact.* **85,** 95 (1992).

[36] L. R. McLean, K. A. Hagaman, and W. S. Davidson, *Lipids* **28,** 505 (1993).

[37] J. Rashba-Step, A. Tatoyan, R. Duncan, D. Ann, T. R. Pushpa-Rehka, and A. Sevanian, *Arch. Biochem. Biophys.* **343,** 44 (1997).

and mellitin (10 μM) in 50 mM Tris–HCl buffer, pH 8.0, containing 2 mM $CaCl_2$ at 37° for 30 min. The reaction is terminated by the extraction of lipids by the Folch procedure.[32] HPTLC results demonstrate that >95% phospholipids underwent hydrolysis under the conditions used (data no shown). Our HPLC data showed that >99% of *cis*-PnA was confined to the *sn*-2 position in all major classes of phospholipids (99.3, 99.5, 99.2, and 95.6% for PC, PE, PS, and PI, respectively) (Table III). Thus, *cis*-PnA containing four conjugated double bonds was predominantly integrated into the *sn*-2 position of phospholipids in HL-60 cells in line with the positional distribution of endogenous polyunsaturated fatty acid residues in mammalian phospholipids.[38] This indicates that an oxidative attack on *cis*-PnA-labeled cells would also be occurring at the *sn*-2 position. In fact, our fluorescence HPLC measurements of HL-60 cell phospholipids after exposure to *tert*-BuOOH revealed no peaks corresponding to fluorescently labeled lysophospholipids (see later). This confirms that *cis*-PnA esterified in the *sn*-2 position was by far the major substrate for peroxidation in cells following oxidant exposure.

Thus our developed and optimized protocol yields cells containing the major classes of membrane phospholipids—PC, PE, PS, PI—fluorescently labeled with *cis*-PnA (Table I) and extremely low intracellular concentration of free *cis*-PnA. The level of *cis*-PnA labeling of endogenous phospholipids (≈1–3 mol%) was low enough to have minimal effects on cell viability and functions, yet sufficient to permit quantitative detection of oxidative stress. Using the assay we were able to reliably and sensitively detect phospholipid peroxidation in different cell lines induced by a variety of oxidants at sublethal levels of oxidative stress (Table II). Importantly, in none of these cases was conventional analysis of phospholipids sensitive enough to detect oxidation-induced changes in phospholipid composition.

Site-Selective Oxidation of PS Induced by Apoptotic Oxidative Stimuli

Because the *cis*-PnA-based assay permits quantification of the amount of oxidative stress in different classes of phospholipids in live cells, it can be utilized for revealing the roles that oxidative modification of specific phospholipids may play in cell function and signaling. In particular, oxidative stress-induced apoptosis represents an exciting area of research to discover specific mechanisms and pathways through which peroxidation of different classes of phospholipids participates in execution of the apoptotic program. Externalization of phosphatidylserine has been identified as a critical event in macrophage recognition of apoptotic cells. Therefore, we were interested in determining whether oxidant-induced apoptosis was also associated with selective oxidation of PS.

[38] A. L. Lehninger, D. L. Nelson, and M. M. Cox, *in* "Principles of Biochemistry," 2nd Ed. Worth Publishers Inc., 1993.

TABLE III

EFFECT OF PHOSPHOLIPASE A_2 ON PHOSPHOLIPID COMPOSITION AND CONTENT OF *cis*-PNA-LABELED PHOSPHOLIPIDS IN HL-60 CELLS[a]

Phospholipid	Content of phospholipids, % of total phospholipids,				*cis*-PnA-labeled phospholipids, ng PnA/μg total lipid Pi,			
	without phospholipase A_2		with phospholipase A_2		without phospholipase A_2		with phospholipase A_2	
	Control	*tert*-BuOOH	Control	*tert*-BuOOH	Control	*tert*-BuOOH	Control	*tert*-BuOOH
Phosphatidylcholine	45.9 ± 3.9	46.4 ± 1.1	0.5 ± 0.1	0.4 ± 0.4	314.2 ± 57.4	244.1 ± 27.5	2.2 ± 0.1	1.3 ± 0.1
Phosphatidylethanolamine	31.4 ± 5.4	27.9 ± 1.4	0.3 ± 0.2	0.3 ± 0.1	56.2 ± 14.1	40.4 ± 4.9	0.3 ± 0.1	0.2 ± 0.1
Phosphatidylserine	6.5 ± 0.5	5.8 ± 0.5	Tr.	Tr.	13.3 ± 5.1	4.1 ± 0.8	0.1 ± 0.1	0.1 ± 0.1
Spingomyelin	6.1 ± 0.5	7.9 ± 0.7	8.6 ± 1.2	8.4 ± 1.6	N.D.	N.D.	N.D.	N.D.
Phosphatidylinositol	7.7 ± 0.5	8.3 ± 0.8	Tr.	Tr.	17.3 ± 5.9	11.3 ± 1.8	0.8 ± 0.2	0.7 ± 0.1
Diphosphatidylglycerol	1.9 ± 0.9	3.2 ± 0.8	3.8 ± 0.4	3.7 ± 0.9	0.6 ± 0.1	0.5 ± 0.1	0.4 ± 0.1	0.4 ± 0.1
Lysophospholipids[b]	0.5 ± 0.2	0.5 ± 0.2	86.8 ± 1.9	87.2 ± 4.0	N.D.	N.D.	78.7 ± 6.8	70.5 ± 1.5
Free *cis*-parinaric acid	N.D.	N.D.	N.D.	N.D.	1.6 ± 0.7	1.2 ± 0.1	318.2 ± 10.9	245.5 ± 3.1

[a] All values are means ± SD ($n = 3$). N.D., not detectable. Tr., trace—less than 0.1%.

[b] Lysophospholipids: lysophosphatidylcholine, lysophosphatidylethanolamine, lysophosphatidylserine, lysophosphatidylinositol.

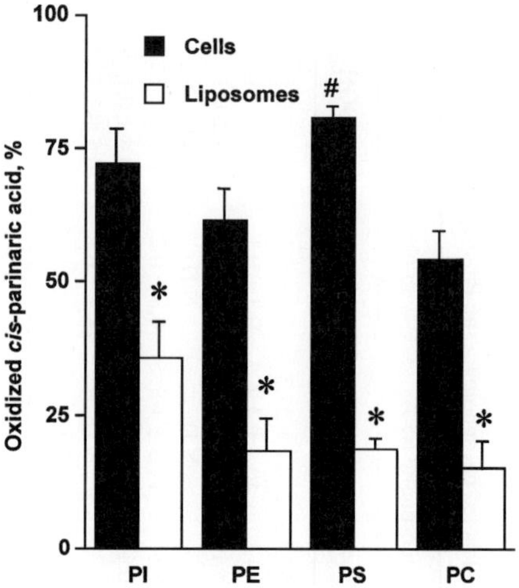

FIG. 2. Cumene hydroperoxide-induced oxidation of *cis*-PnA-labeled phospholipids in intact live normal human epidermal keratinocytes and cell-free *cis*-PnA-labeled liposomes derived from normal human epidermal keratinocytes. Intact living *cis*-PnA-labeled NHEK were exposed to cumene hydroperoxide (200 μM) for 1 hr at 37°. *cis*-PnA-labeled liposomes were prepared from *cis*-PnA-loaded NHEK and treated similarly with 200 μM cumeme hydroperoxide at 37° for 1 hr. At the end of the incubations, total lipids were extracted and resolved by HPLC. PI, phosphatidylinositol; PE, phosphatidylethanolamine; PS, phosphatidylserine; PC, phosphatidylcholine. Data represent means $\pm$ SEM, $n = 3$, $^{*}p < 0.01$.

Similar to our earlier studies observing the selective oxidation of PS as an early biomarker of apoptosis,[8,11] we documented that treatment of NHEK and HL-60 cells with organic hydroperoxides results in selective oxidation of PS that precedes PS externalization and activation of caspases in these cells.[25,39] NHEK are treated with cumene-OOH (200 μM) in phenol red-free KGM-2 medium for 1 hr at 37° in the dark. Data presented in Fig. 2 show that cumene-OOH (200 μM) causes substantial oxidation of all *cis*-PnA-labeled phospholipids in NHEK. At this concentration, however, cumene-OOH-induced oxidation of PS is significantly greater than oxidation of the two major phospholipids PC and PE.

To reveal whether this oxidation of PS was specific for the execution of apoptosis or simply reflected PS as a preferential target of cumene-OOH-induced oxidation of phospholipids, we performed studies to measure cumene-OOH-dependent oxidation of *cis*-PnA-labeled phospholipids in a cell-free system containing *cis*-PnA-labeled liposomes. Lipids are extracted from *cis*-PnA-loaded cells, the solvent is evaporated under N_2, and the film of lipids is suspended in 20 m*M* HEPES buffer,

[39] K. Kawai, Y. Y. Tyurina, V. A. Tyurin, V. E. Kagan, and J. P. Fabisiak, *Toxicologist* **54,** 165 (2000).

pH 7.4, to achieve the lipid concentration equivalent to that used in the experiments with live cells. Liposomes are prepared by sonication of lipid suspension (four 15-sec pulses on ice) using a tip sonicator (Ultrasonic Homogenizer 4710 series, Cole-Palmer Instrument Co., Chicago, IL). Liposomes containing *cis*-PnA-labeled phospholipids are incubated for 1 hr in the presence of cumene-OOH (200 μM) at 37° in the dark. At the end of incubation, lipids are extracted and resolved by HPLC. In liposomes, the oxidation of all classes of phospholipids was significantly less pronounced than that observed in live NHEK (Fig. 2). Importantly, preferential oxidation of PS observed in live NHEK cells was not detectable in the phospholipid-containing cell-free system. This suggests that nonrandom preferential oxidation of PS is characteristic of oxidant-induced apoptosis in NHEK cells.

The plasma membrane is the site where major events associated with PS signaling during apoptosis, including its externalization and subsequent recognition by macrophages, take place. This implies that selective oxidation of PS—if it is a part of apoptotic signaling—should occur within the plasma membrane compartment.[40,41] To establish whether plasma membrane PS undergoes peroxidation during apoptosis, we performed experiments in which we isolated different subcellular fractions and determined the amounts of oxidative stress in different classes of phospholipids in these organelles.[39] Importantly, the sensitivity of our PnA-based assay for phospholipid peroxidation permits us to conduct the study using practical amounts of cell material. Figure 3 compares the oxidation of *cis*-PnA-labeled PS, PE, and PC derived from whole HL-60 cells treated with *tert*-BuOOH to that observed within the plasma membrane of similarly treated cells. After a 20-min incubation of HL-60 cells with *tert*-BuOOH (150 μM) in serum-free RPMI medium 1640 without phenol red, oxidation of all phospholipids is observed. PS is remarkably more sensitive to oxidation than PC, PE, and PI (Fig. 3, left). We next determined the amounts of oxidative stress in plasma membrane phospholipids. Plasma membranes are isolated from *tert*-BuOOH-treated HL-60 cells as described by Storrie and Madden.[42] Figure 3 (right) shows that in HL-60 cells challenged with *tert*-BuOOH, plasma membrane PS was the largest source of oxidized PS (58%) and it was almost completely accountable for *tert*-BuOOH-induced PS oxidation in HL-60 cells. Notably, lower rates of PS oxidation were detected in other subcellular fractions, such as mitochondria, microsomes, lysosomes, and nuclei[39] (data not shown).

[40] V. E. Kagan, J. P. Fabisiak, A. A. Shvedova, Y. Y. Tyurina, V. A. Tyurin, N. F. Schor, and K. Kawai, *FEBS Lett.* **477,** 1 (2000).

[41] Y. Y. Tyurin, A. A. Shvedova, K. Kawai, V. A. Tyurin, C. Kommineni, P. J. Quinn, N. F. Schor, J. P. Fabisiak, and V. E. Kagan, *Toxicology* **148,** 93 (2000).

[42] B. Storrie and E. A. Madden, *Methods Enzymol.* **182,** 203 (1990).

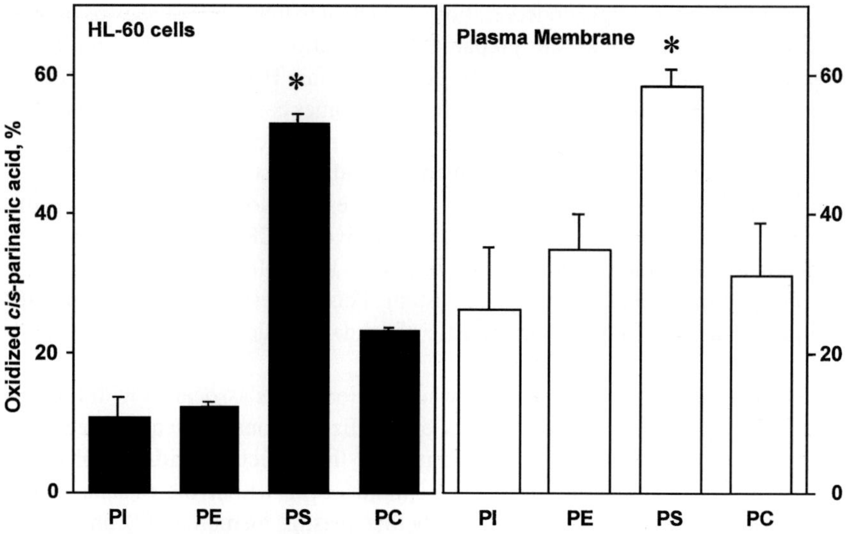

FIG. 3. tert-Butyl hydroperoxide-induced oxidation of *cis*-PnA-labeled phospholipids in intact live HL-60 cells and plasma membrane of HL-60 cells. *cis*-PnA-labeled HL-60 cells (left) were exposed to *tert*-butyl hydroperoxide (150 μM) for 20 min at 37°. The plasma membrane (right) was isolated from *cis*-PnA-labeled HL-60 cells treated with *tert*-butyl hydroperoxide (150 μM) at 37° for 20 min. PI, phosphatidylinositol; PE, phosphatidylethanolamine; PS, phosphatidylserine; PC, phosphatidylcholine. Data represent means $\pm$ SEM, $n = 3$, $^*p < 0.02$.

Note: The levels of *cis*-PnA-labeled phospholipids after incubation of cells without oxidants were used as controls for comparisons. During incubation in the absence of oxidants, the content of *cis*-PnA-labeled phospholipids was essentially unchanged (within 10% of the initial levels).

Antioxidants Protect against Phospholipid Peroxidation

To further prove that *cis*-PnA-labeled cells represent a good model for the quantitative assay of oxidative stress, we investigated the protective effects of vitamin E on the peroxidation of *cis*-PnA-labeled phospholipids induced by AMVN in NHEK and HL-60 cells. *cis*-PnA-loaded cells are incubated in the presence or in the absence of a lipid-soluble azoinitiator, AMVN [2,2′-azobis(2,4-dimethylisovaleronitrile], which generates peroxyl radicals within the lipid bilayer at a constant rate[43] and induces apoptosis in cells.[8,28,44] After incubation, NHEK

[43] E. Niki, *Methods Enzymol.* **186,** 100 (1990).

[44] J. P. Fabisiak, V. A. Tyurin, Y. Y. Tyurina, A. Sedlov, J. S. Lazo, and V. E. Kagan, *Biochemistry* **39,** 127 (2000).

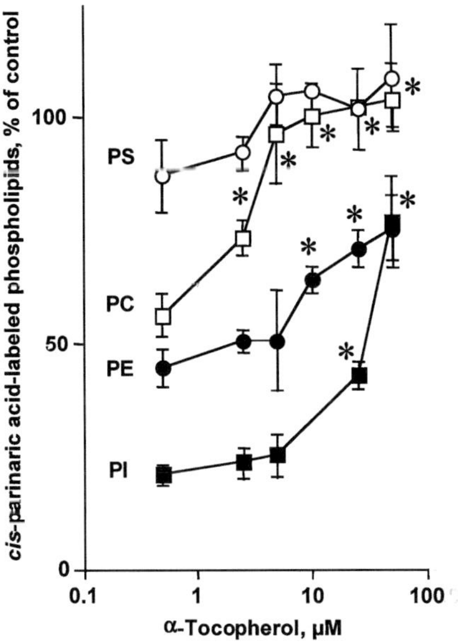

FIG. 4. Effect of α-tocopherol on oxidation of *cis*-PnA-labeled phospholipids induced by AMVN in normal human epidermal keratinocytes. NHEK (≈100% confluency) were cultured in the presence or in the absence of α-tocopherol (2.5–50 μ*M*) for 24 hr at 37° in KGM-2 medium. Unloaded NHEK and NHEK loaded with α-tocopherol were exposed to AMVN (500 μ*M*, 1 hr). At the end of incubation, cells were scraped and lipids were extracted and resolved by HPLC. PI, phosphatidylinositol; PE, phosphatidylethanolamine, PS, phosphatidylserine, PC, phosphatidylcholine. $^*p < 0.03$ vs AMVN. Data represent means ± SEM.

and HL-60 cells are collected and washed twice with PBS, and total lipids are extracted according to Folch *et al.*[32] in the presence of BHT (100 μ*M*) to prevent subsequent oxidation. The acyl chains of four classes of phospholipids in NHEK and HL-60 cells, namely PE, PC, PS, and PI, were the major targets for AMVN-induced peroxidation (Fig. 4 for NHEK and Fig. 5 for HL-60).

The sensitivity of this oxidation to the lipo-protective antioxidant, vitamin E, is then assessed. NHEK (≈100% confluence) and HL-60 cells (0.5×10^6) are cultured in the presence of α-tocopherol acetate (2.5–50 μ*M*) for 24 hr before *cis*-PnA labeling and exposure to oxidants. Vitamin E (α-tocopherol acetate) is added to the growth medium. Excess vitamin E that has not been integrated into cells is removed by washing cells with medium. The effect of α-tocopherol on the AMVN-induced oxidation of *cis*-PnA-labeled phospholipids is presented for NHEK in Fig. 4 and for HL-60 cells in Fig. 5. The protective effect of α-tocopherol on AMVN-induced oxidation of phospholipids is concentration dependent. As an effective radical scavenger, α-tocopherol (50 μ*M*) is able to completely protect all phospholipids against oxidation.

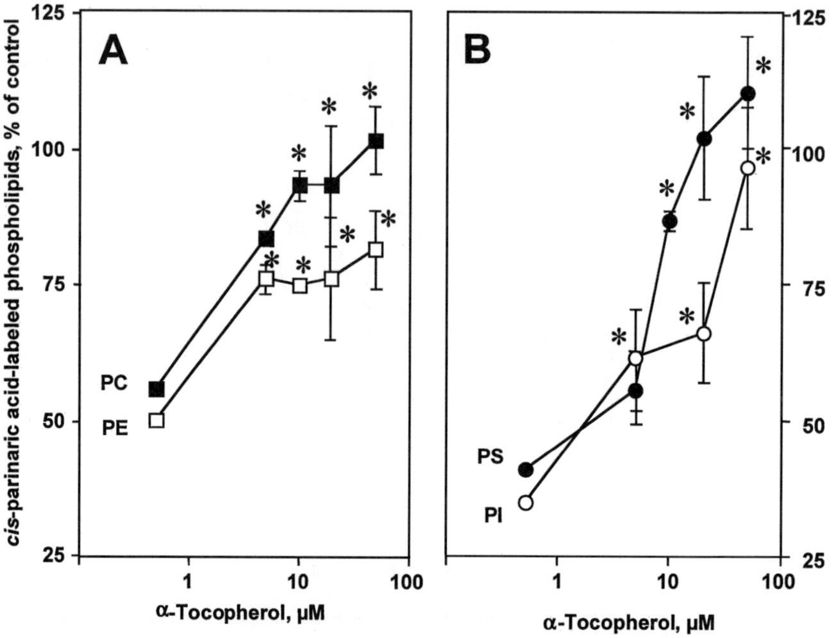

FIG. 5. Effect of α-tocopherol on oxidation of *cis*-PnA-labeled phospholipids induced by AMVN in HL-60 cells. HL-60 cells (0.5×10^6) were cultured in the presence or in the absence of α-tocopherol (5–50 M) for 24 hr at 37° in RPMI medium 1640. Unloaded HL-60 and HL-60 cells loaded with α-tocopherol were exposed to AMVN (500 μM). At the end of incubation, cells were collected by centrifugation, and lipids were extracted and resolved by HPLC. (A) PC, phosphatidylcholine; PE, phosphatidylethanolamine. (B) PS, phosphatidylserine; PI, phosphatidylinositol. $^*p < 0.01$ vs AMVN. Data represent means ± SEM.

Conclusion

In conclusion, our results clearly indicate that our model of *cis*-PnA-labeled cells offers a unique model system for quantitative studies of oxidative stress and selective oxidation of specific classes of phospholipids under normal physiological conditions, as well as during cell injury and apoptotic death.

Acknowledgments

Supported by grants from NIH 1RO1HL64145-01A1, EPA STAR Grant R827151, the NCI Oncology Research Faculty Development Program and Magee-Womens Reseach Institute (V.A.T.), and Leukemia Research Foundation and the International Neurological Science Fellowship Program (F05 NS 10669) administered by NIH/NINDS in collaboration with WHO, Unit of Neuroscience, Division of Mental Health and Prevention of Substance Abuse (Y.Y.T.).

[15] Quantitative High Throughput Endothelial Cell Migration and Invasion Assay System

By JAMES C. MALIAKAL

Introduction

Angiogenesis is the process that results in the formation of new blood vessels from preexisting capillaries. This process involves a cascade of highly regulated events that occur in many normal physiological processes, such as embryonic development, the menstrual cycle, and wound healing. In addition, angiogenesis is prominent during pathological processes, such as diabetes, chronic inflammation, cardiovascular diseases, and cancer. A spatiotemporal sequence of events regulates the net balance of angiogenic versus angiostatic (inhibition of angiogenesis) factors within the local microenvironment, which results in neovessel formation or inhibition. Angiogenic factors interact with extracellular matrix (ECM) proteins and activated endothelial cells to induce endothelial cell proliferation, migration, invasion and tube formation. Some of these dynamic cellular events can be reproduced *in vitro* by allowing the endothelial cells to migrate and invade toward an angiogenic factor through ECM on occluded membrane pores. In recent years, a great deal of work has focused on delineating various pathways and molecular mechanisms that dictate this regulation. In the private sector, researchers are striving to develop therapeutic molecules that either stimulate or inhibit new blood vessel growth. Such molecules are useful in treating pathologies such as tissue damage after ischemia reperfusion or in inhibiting new blood vessel formation in diseases such as diabetec retinopathy, cancer, inflammation, and psoriasis.[1] A major bottleneck restraining the compound screening process is the lack of an efficient and reproducible high throughput assay for evaluating directional cell motility toward chemotactic stimuli. A popular platform that is used to study directional cell motility is based on the Boyden chambers.[2] In this assay, two chambers are separated by a microporous membrane that is typically coated with collagen or other extracellular matrix molecules. The cells are seeded in the apical chamber and the chemotactic agent is added to the bottom chamber. Test or control substances are then placed in the apical chamber or to the bottom chamber. The cells that migrate through the pores and reach the bottom side of the membrane are fixed, stained, and then manually counted. In a modification of this assay, cells that have migrated to the underside of the membrane are treated with nuclear stain. The stain is then extracted with 0.1 *N* HCl

[1] A. W. Griffioen and G. Molema, *Pharmacol. Rev.* **52,** 237 (2000).
[2] S. Boyden, *J. Exp. Med.* **115,** 453 (1962).

0076-6879/02 $35.00

and transferred to a 96-well plate for spectrophotometric analysis of absorbance to quantitate the cells that have migrated.[3] These assays are very time-consuming and labor intensive. Therefore, the overall process is inefficient for high throughput compound screening. Furthermore, because the assay has to be terminated to fix the cells for quantitation, real-time kinetics studies are difficult to accomplish. Our goal was to develop an efficient high throughput cell motility assay using multiwell inserts with a fluorescence blocking microporous polyethylene terephthalate (PET) membrane (FluoroBlok) coated with ECM proteins. This invasion system consists of 24-multiwell insert plate with a 3-μm pore size PET membrane. This membrane has been uniformly coated with either the basement membrane protein BD Matrigel for the invasion chambers or thin coating of ECM proteins for migration assays. The optimum amount of Matrigel coating occludes the pores on the membrane, still allowing the invasive cells to digest through the membrane. Quantitation of migrated or invaded cells, which come to the underside of the membrane, is achieved by either pre- or postlabeling cells with a fluorescent dye and measuring the fluorescent signal from the invaded cells. The FluoroBlok membrane effectively blocks the passage of light from 490 to 700 nm at 99% efficiency. This virtually blocks the background fluorescence from the nonmigratory or noninvasive cells on the top of the membrane, while allowing reading of the fluorescent signal from the invaded or migrated cells on the underside of the membrane using a bottom-reading fluorescent reader. Using this assay, we have quantitated endothelial cell migration and invasion toward known angiogenic growth factors. We also tested the inhibition of angiostatic molecules in this assay system. This assay system allows fast and efficient quantitation of endothelial cell migration and invasion, which provides a valuable tool for angiogenic or angiostatic drug discovery efforts.

Materials and Methods

Human primary microvascular endothelial cells (HMVEC) and umbilical vein endothelial cells (HUVEC) are from BioWhittaker (Walkersville, MD). HMEC-1, the endothelial cell line, are licensed from the Center for Disease Control (Atlanta, GA). Cells are cultured in the recommended endothelial cell medium (EGM-2, BioWhittaker). Hanks' balanced salt solution (HBSS) and phosphate-buffered saline (PBS) are from Life Science Technologies (Rockville, MD). Growth factors are from R&D systems (Minneapolis, MN) and BD BioSciences (Bedford, MA). Basement membrane protein BD Matrigel and multiwell plates with inserts are from BD Biosciences.

Matrigel is diluted to 225–280 μg/ml, and 100 μl is added to the 3-μm membrane, polymerized for 2 hr, and air dried overnight to a thin even layer. Throughout

[3] G. R. Grotendorst, *Methods Enzymol.* **147,** 144 (1987).

the coating process, Matrigel and the coating buffer are kept on ice to prevent gelation of the Matrigel. Postcoating, insert plates are sealed in foil bags and stored at -20° until use.

Assay

Prior to initiating the assay, the membrane inserts are rehydrated for up to 2 hr in a 37° humidified incubator with 5% CO_2. In our studies, we found that endothelial cell monolayers greater than 50–60% in confluence migrated/invaded less than the less confluent cultures. Endothelial cells are trypsinized and made into a single cell suspension using the trypinization reagents recommended by BioWhittaker. Endothelial cells (5×10^4) are added to the inside of the Matrigel-coated inserts suspended in 250 μl of endothelial cell basal medium containing 0.1% bovine serum albumin (BSA). In some cases, when growth factors are analyzed, 0.4% fetal bovine serum (FBS) is also added to this medium. Various angiogenic compounds,

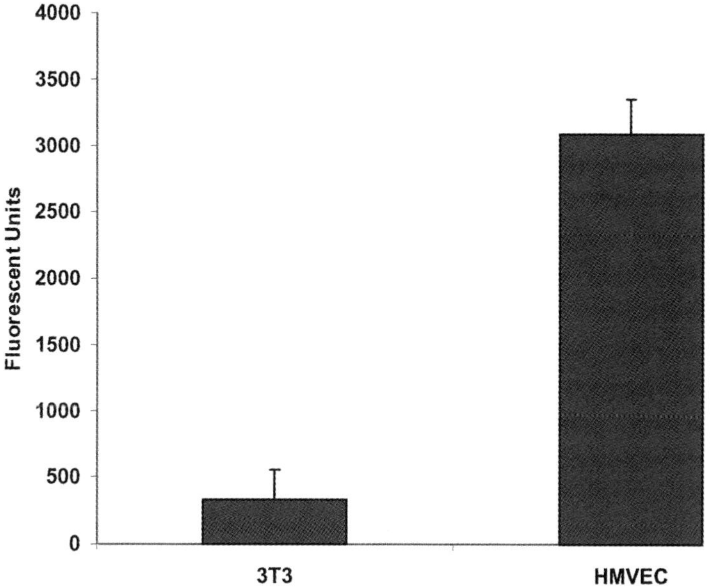

FIG. 1. Comparison of invasive endothelial cells and noninvasive NIH/3T3 cells on the BD Matrigel invasion system. HMVEC and 3T3 cells were seeded on Matrigel-coated, 3-μm FluoroBlok membrane inserts in 250 μl of medium. Seven hundred and fifty microliters of 5% FBS containing medium was added in the bottom wells. Cells were incubated for 22 hr, and the cells on the underside of the membrane were postlabeled with 4 μg/ml of calcein AM. The fluorescent signal was read on a bottom-reading fluorescent plate reader. Data presented are means of three inserts $\pm$ SD.

such as FBS, VEGF, and bFGF, are added to the bottom well of the multiwell insert assembly in a volume of 750 μl of the same medium in the presence or absence of angiostatic compounds. The cells are incubated for 22 hr in a humidified, vibration-free incubator in the presence of 5% CO_2.

For the invasion assay, the BD Matrigel coating of the microporous membrane occludes the pores on the membrane. In the assay, activated endothelial cells invade through the matrix and migrate to the bottom side of the membrane. In order to show that the vast majority of the pores are occluded, we assay noninvasive NIH/3T3 fibroblasts in the same coated inserts as negative controls. 3T3 monolayers cultured in Dulbecco's modified Eagle's medium (DMEM) containing 10% FBS medium are trypsinized the same way as the endothelial cells and made into a single cell suspension. Cells (5×10^4) suspended in DMEM medium in a volume of 250 μl are added to the inside of the insert. Seven hundred and fifty microliters of DMEM containing 5% FBS is added into the bottom well as a chemoattractant. These cells, along with endothelial cells, are incubated in a 37° incubator for 22 hr (see Fig. 1).

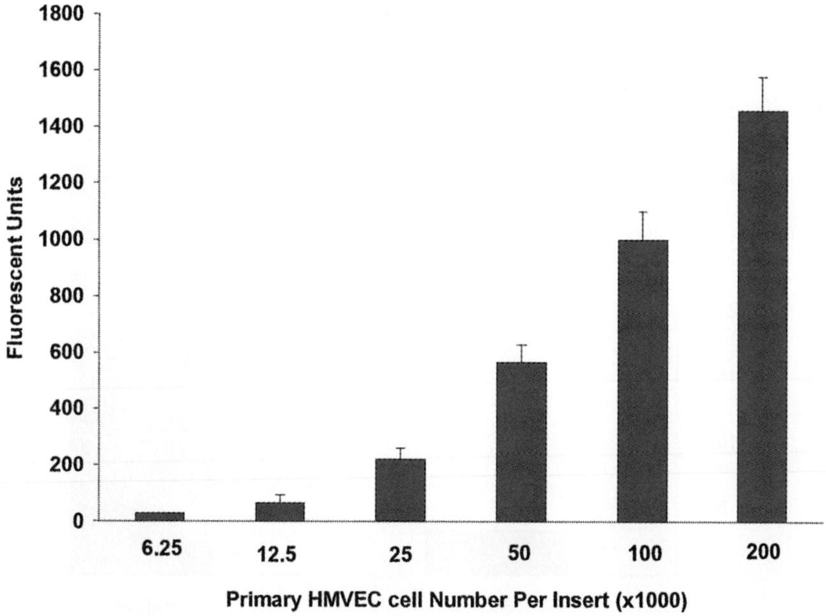

FIG. 2. Correlation of endothelial cell number vs invasion through the BD Matrigel FluoroBlok invasion system. Various numbers of HMVEC cells were seeded inside the membrane in 250 μl volume. Seven hundred and fifty microliters of 5% FBS containing medium was added to the bottom well. Invaded cells were postlabeled with 4 μg/ml calcein AM, and signals were read on a bottom-reading fluorescent plate reader. Data presented are means of three inserts ± SD.

Migration Assay

For migration assays, a thin uniform layer of extracellular matrix fibronectin collagen is adsorbed onto the topside of the membrane. Due to the thin coating, the pores are not occluded. Therefore, both endothelial cells and 3T3 fibroblasts migrate through the pores toward the chemoattractant.

Cell Labeling

Endothelial cells are either prelabled with fluorophore DiI (Molecular Probes, excitation/emission 530/590) or postlabeled with calcein AM or carboxyfluorescein CFDASE (Molecular Probes, excitation/emission 485/530) according to the manufacturer's recommendations. For prelabeling, endothelial cells growing in the flasks are labeled with 10 μg/ml of Dil diluted in endothelial growth medium for 2 hr at 37°. Prelabeling of the cells before the assay enables the kinetic analysis of the cell migration or invasion. When cells are prelabeled, high concentrations

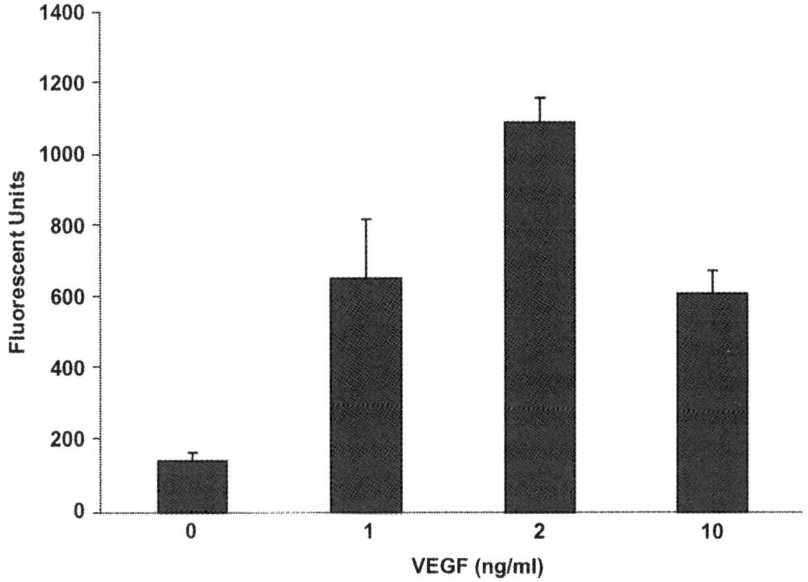

FIG. 3. Endothelial cell invasion toward VEGF. HMVEC cells (5×10^4) were grown on the Matrigel-coated FluoroBlok invasion system in 250 μl volume. Various concentrations of VEGF were added to the bottom well in 750 μl volume. After a 22-hr incubation, invaded cells were postlabeled, and fluorescent signals were read on a bottom-reading fluorescent plate reader. Data presented are means of three inserts $\pm$ SD.

of the label may have adverse effects on cell migration. During postlabeling, because the cells are not labeled before migration, labeling will not have any adverse effect on invasion or migration. The invaded/migrated cells are quantitated by reading the fluorescence signal derived from the bottom (underside) of the inserts in a fluorescent plate reader (CytoFluor 4000, Perseptive BioSystems).

Both calcein AM and CFDA SE are used for postlabeling. For calcein, 4 μg/ml of solution is prepared in HBSS. Following the migration or invasion assay, the multiwell inserts are placed in another 24-multiwell plate (Falcon) containing 0.5 ml/well of calcein solution. The cells are stained for 90 min at 37°, and the fluorescent signal is then measured using the plate reader.

For postlabeling with CFDASE fluorophore, a 20 m*M* solution is prepared in HBSS, and the insert is then placed in a 24-multiwell plate containing 0.5 ml/well of dye and incubated at 37° for 1 hr. After incubation, the insert is transferred to another multiwell plate containing 0.5 ml of HBSS. This wash step eliminates the high background generated from the hydrolysis of the fluorescent dye. The fluorescent signal is then read as described earlier.

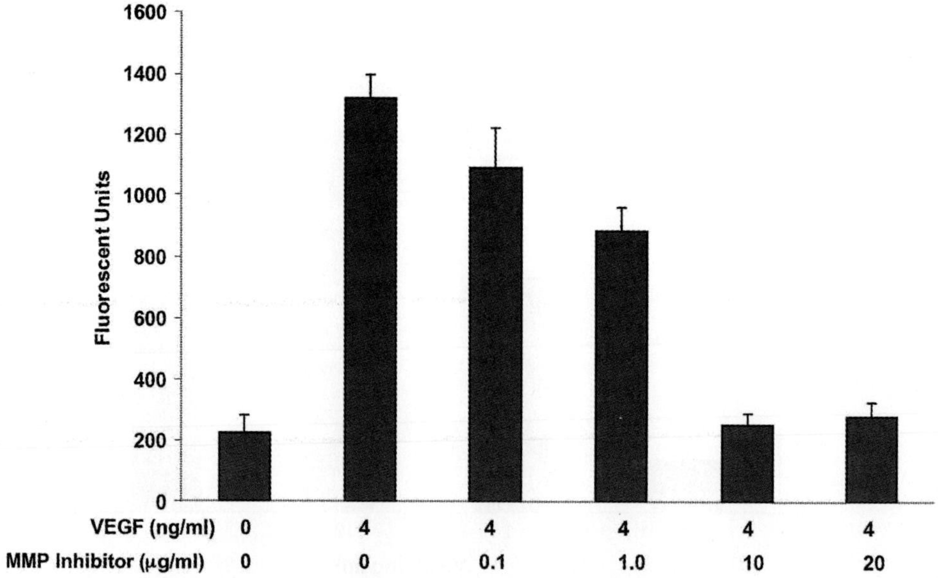

FIG. 4. Endothelial cell invasion inhibiton by the MMP inhibitor. Human microvascular endothelial cells (HMVEC) were grown on BD Matrigel-coated, 3-μm FluoroBlok cell culture inserts. Various concentrations of the MMP inhibitor, 1,10-phenanthroline, were added in the presence of VEGF (4 ng/ml) on the bottom well. The invaded cells on the bottom of the inserts were quantitated in a fluorescent reader. Data presented are means of three inserts $\pm$ SD.

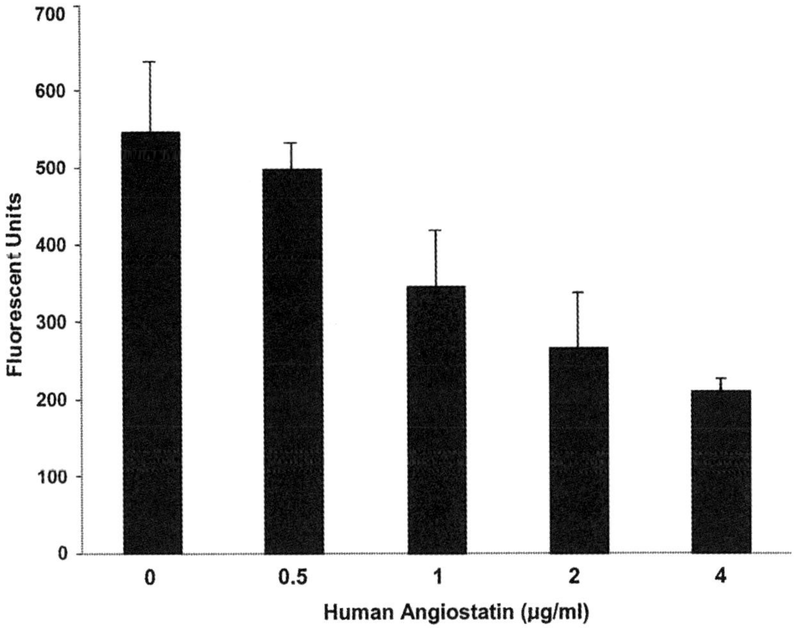

FIG. 5. Inhibition of endothelial cell invasion by angiostatin. HMEC-1 cells were grown on BD Matrigel-coated, 3-μm FluoroBlok cell culture inserts. Various concentrations of angiostatin were added in the presence of 5% FBS in the bottom well. The invaded cells on the bottom of the inserts were quantitated in a fluorescent plate reader. Data presented are means of three inserts ± SD.

Results

This assay evaluated pore occulsion by comparing the invasion of NIH/3T3 and endothelial cell line HMEC-1. As the signal reflects NIH/3T3 it was only less than 10% of the signal of the endothelial cells (Fig. 1). This demonstrates that the vast majority of the pores on the microporous membrane on the inserts are occluded by the basement membrane protein BD Matrigel. Only the activated endothelial cells were able to digest through the matrix and emerge to the bottom side of the membrane. We also examined the relationship between the cell number of HMVEC seeded on the top of the insert and their ability to invade through the invasion chambers toward 5% FBS. We found a direct correlation between cell number and cell invasion (Fig. 2). Using our novel assay system, we examined endothelial cell chemoattractants, such as FBS and VEGF, for their potential to stimulate migration and invasion across a microporous membrane. Both chemoattractants stimulated endothelial cell invasion and migration in a dose-dependent manner with a maximum stimulation of 10-fold for VEGF (Fig. 3). We examined the inhibitory effect of angiostatic compounds such as the MMP inhibitor 1,10-phenanthroline and angiostatin on the invasion of endothelial cells. In the invasion assay, 10 μg/ml of 1,10-phenanthroline inhibited a maximum of 71% of the VEGF-induced stimulation of invasion (Fig. 4). We also analyzed the angiogenesis inhibitor angiostatin in our assay system. Angiostatin inhibited the FBS-stimulated invasion of endothelial cells in a dose-dependent manner, with a maximum of 56% inhibition of the control at 4 μg/ml (Fig. 5).

Summary

We have developed a novel assay system combining fluorescent signal-blocking microporous PET membrane inserts and ECM to study dynamic endothelial cell migration and invasion. The assay described here may be applicable to other cell types for analogous assay. The currently used method for analyzing migration and invasion requires the removal of nonmigratory cells from the top of a clear microporous membrane. This laborious step is required to quantitate the invaded or migrated cells on the bottom of the membrane. When the fluorescent signal blocking the microporous membrane is used, the fluorescent signal from noninvaded or nonmigratory cells from the top of the membrane is virtually eliminated and allows for rapid and efficient measurement of the migrated or invaded cells on the underside of the insert. Unlike conventional low throughput migration assays, this assay is easy to perform and provides a quantitative invasion/migration profile of the chemoattractant of choice. This robust automation compatible assay will serve as a powerful tool to screen potential antiangiogenic compounds for drug discovery.

Acknowledgments

I thank Dr. Frank Mannuzza and Steve Ilsley for their advice and encouragement. Special thanks go out to Dr. Marshall Kosovsky for reviewing the manuscript.

[16] *In Vitro* Model of Oxidative Stress in Cortical Neurons

By RAJIV R. RATAN, HOON RYU, JUNGHEE LEE, AZIZA MWIDAU, and RACHEL L. NEVE

Oxidative Stress and Neurodegeneration: Cause or Consequence

Oxidative stress has been postulated to be a common mediator of a host of neurodegenerative diseases, including Alzheimer's disease, Parkinson's disease, Fredreich's ataxia, and stroke.[1] However, evidence that oxidative stress is an initiator or propagator of any of these diseases is incomplete. Indeed, clinical trials of antioxidants in a host of neurological conditions have not resulted in dramatic clinical improvement. The limited efficacy of antioxidant agents results, in part, from our limited understanding of the pathways activated in neurons by oxidative stress that affect cell viability.

Cortical neurons, grown in primary culture, provide a convenient *in vitro* environment for examining the mechanism(s) by which neuronal degeneration is induced by oxidative stress.[2] Early in their development in culture embryonic rat cortical neurons exposed continuously to elevated concentrations of extracellular glutamate or homocysteate (HCA) degenerate over 24 hr. Degeneration occurs subsequent to the depletion of intracellular glutathione, an important antioxidant. Under these conditions, glutathione depletion by glutamate occurs as a result of competitive inhibition of cystine uptake at its plasma membrane transporter rather than through the activation of ionotropic glutamate receptors (Fig. 1). Although neuronal death associated with a decreased presence of glutathione cannot be blocked by competitive or noncompetitive glutamate receptor antagonists, it can be effectively circumvented by treatment with the antioxidants vitamin E and idebenone.[2,3]

Our laboratory and others[4–6] have used glutamate-induced oxidative death as a model to elucidate the precise mechanisms by which oxidative stress can impact neuronal viability. The advantages of this model are numerous: (i) it involves primary postmitotic neurons; (ii) at least 100–150 million cells can be obtained

[1] R. R. Ratan, *in* "Cell Death and Diseases of the Nervous System" (V. E. Koliatsos and R. R. Ratan, eds.), p. 649. Humana Press, Totowa, 1999.

[2] T. H. Murphy, R. L. Schnaar, and J. T. Coyle, *FASEB J.* **4,** 1624 (1990).

[3] R. R. Ratan, T. H. Murphy, and J. M. Baraban, *J. Neurochem.* **62,** 376 (1994).

[4] Y. Li, Y., P. Maher, and D. Schubert. *Neuron* **19,** 453 (1997).

[5] C. K. Sen, S. Khanna, S. Roy, and L. J. Packer, *J. Biol. Chem.* **275,** 13049 (2000).

[6] M. Stanciu, Y. Wang, R. Kentor, N. Burke, S. Watkins, G. Kress, I. Reynolds, E. Klann, M. R. Angiolieri, J. W. Johnson, and D. B. DeFranco, *J. Biol. Chem.* **275,** 12200 (2000).

0076-6879/02 $35.00

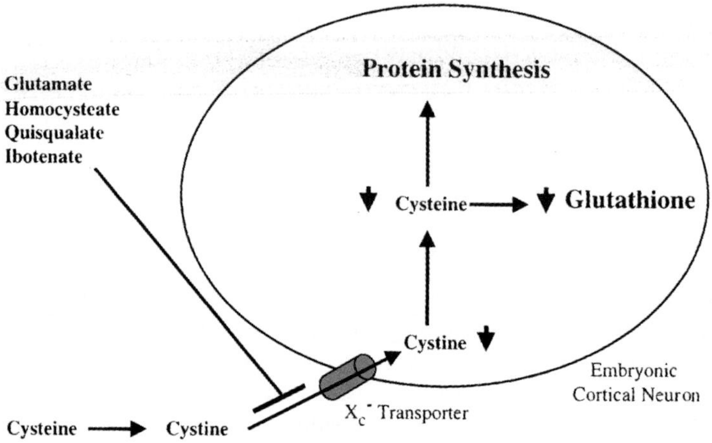

FIG. 1. An *in vitro* model of neuronal oxidative stress. Immature cortical neurons undergo apoptotic cell death in response to chronic exposure (>10 hr) of high concentrations of the excitatory amino acid glutamate (1–10 m*M*). Toxicity in this paradigm results not from the ability of glutamate to interact with a cell surface receptor (excitotoxcity), but rather through the ability of glutamate to inhibit the uptake of the amino acid cystine at its plasma membrane transporter. This transporting agency (designated X_c^-) is chloride dependent, nonelectrogenic (cystine is exchanged for glutamate), nonenergy requiring (facilitated diffusion is utilized), and is expressed in all tissues, including neurons. Only glutamate analogs (e.g., HCA) that can inhibit radioactive cystine uptake are toxic to cortical neurons. Inhibition of cystine transport leads to intracellular depletion of intracellular cysteine, the rate-limiting precursor in the synthesis of cellular glutathione, an important antioxidant. Depletion of glutathione results in oxidative stress and consequent cell death.

from each culture, facilitating biochemical and cell biological measurements; (iii) the cells are harvested from brain at a stage in development when the cultures are predominantly neuronal (>85%); (iv) oxidative stress is induced by the depletion of an endogenous antioxidant rather than the addition of high and possibly nonphysiological concentrations of an oxidant; and (v) after addition of glutamate, primary neurons do not commit to die until 8–12 hr later (depending on the antioxidant used), permitting primary oxidative events to be easily separated from those that occur as a consequence of cell death.

Methods for Culturing Primary Cortical Neurons for Glutathione Depletion Studies

The importance of performing the steps of the dissection in a timely fashion cannot be overstated. The procedure is generally broken down into three main segments: (i) culturing cortical neurons; (ii) exposing cortical neurons to glutamate or glutamate analogs, including HCA; and (iii) assessing cell viability. Timed-pregnant Sprague–Dawley rats are sacrificed at 17 days of gestation. At this stage of development, the brain tissue is still easy to dissociate, the meninges are easily removed, and the number of glial cells is still relatively modest.[7] The rat is euthanized by standard methods and placed on a diaper in a dissection hood. With the abdomen facing upward, 70% ethanol is liberally poured over the rat's abdomen to minimize contamination by aerosolized dander and to sterilize the field. It is preferable to hold the scissors (Fine Science Tools, Germany) at an inclined position at an angle of about 60° and a pair of forceps (Adumont and Fils, Switzerland) in the other hand. The pointed tips of the scissors are used to poke the abdomen, thus providing a route to begin opening it laterally. At this point, focus should primarily be on removing the amniotic sacs without disturbing the other organs and the areas around them.

Once the abdomen is open, the forceps are used to pick up the uterus, and a smaller pair of scissors (Fine Science Tools, Germany) in the other hand is used to free the multiple, connected amniotic sacs. The amniotic sacs are placed into a 100-mm cell culture dish (Corning Costar, polystyrene treated), with cold phosphate-buffered saline (PBS) where the fetuses are completely freed from the sac and placenta and then dropped into another 100-mm cell culture dish of cold PBS. Next, the head of each fetus is isolated from each body. The head is then placed in a saggital posture inside one of the 100-mm dishes with cold PBS. Forceps are placed into the eyes to hold the head steady. With another pair of forceps (Adumont and Fils, Switzerland) in the other hand, the skull plate (two layers composed of the nascent skull and the dura) is found. This should really be an effortless tug. Should one feel that excessive force is being applied to peel away the skull plate, stop, because the brain could be damaged. Once an opening in the skull and dura is made, the brain can be gently expelled from the skull and placed in a fresh plate of PBS. This is done with the aid of a dissecting microscope.

The next step is to remove the meninges from the rest of the brain. Place the brain in a coronal position, hold the brain stem with one set of forceps, and peel away the meninges with the other. Care should be taken to remove all connective tissue debris from the area. It is much easier to peel off the connective tissue and membranes if the brain has been sitting in very cold PBS. Once the meninges are completely removed, the cortex can be easily separated from the striatum by placing the forceps underneath the cortex and pinching. The cortex is thus removed much like a mushroom from its stalk.

We usually dissect cortices from 6 to 15 pups. Rats often have fewer pups during the summer months. The dissociated cortices are placed in sterile PBS (GIBCO BRL) and are ready for transfer to a 50-ml conical vial for washing, followed by dissociation with papain solution (30 ml Earle's basic salt solution, GIBCO; 300 μl of 50 mM EDTA, pH 8.0; 300 ml of papain suspension from Worthington Biochemical Corp.; (4 mg cysteine from Sigma) according to a

[7] K. Goslin and G. Banker, *in* "Culturing Nerve Cells" (G. Banker and K. Goslin, eds.), p. 251. MIT Press, Boston, 1991.

well-established method outlined by Baughman *et al.*[8] After the addition of warm papain solution, mix the contents of the vial well. Triturate several times to facilitate dissociation of the tissue and place the vial in a 37° water bath for 5–7 min, gently shaking it periodically to ensure good mixing. After pelleting the cells in a desktop centrifuge, remove the papain solution and add DNase (1 μg/ml, Sigma). After 5 min in DNase solution at 37°, layer the cells with a combination of bovine serum albumin (10 mg/ml) and trypsin inhibitor (10 mg/ml; Type I-S, Sigma) to inhibit papain. Remove the DNase/BSA/trypsin inhibitor supernatant and rinse the pelleted cells once with plating medium before resuspension in 10 ml of the same plating medium. The cells are counted in a hemocytometer. A good yield is 10^7 cells/ml.

Plating Out Cells and Performing Toxicity Studies

For cytotoxicity studies, we customarily plate cells at a density of 1×10^6 cells/ml in 12-well (1 ml/well; Corning Costar) or 24-well (500 μl/well; Corning Costar) dishes. The dishes are coated the night before with a solution of 10 μg/ml poly-L-lysine hydrobromide (Sigma, molecular weight 30,000 to 70,000) diluted in sterile PBS. Cells are coated at 37° in the incubator and are gently rinsed once with sterile PBS to remove insoluble aggregates of poly-L-lysine. Cells are plated in a standard plating medium [MEM with glutamax (GIBCO/BRL), 10% fetal calf serum (GIBCO/BRL), penicillin/streptomycin (Sigma), and 200 μM cystine (Sigma)]. After 16 to 18 hr of plating, the medium is changed to plating medium minus the added cystine and/or BHA, and the neurons are ready for treatment with glutamate or a glutamate analog. The LD_{50} of glutamate and various glutamate analogs is documented in an elegant study by Murphy and co-workers.[2] As a result of this analysis, we have chosen to do the majority of our studies with the glutamate analog homocysteate (HCA; Sigma), in addition to glutamate. HCA has a lower LD_{50} than glutamate, likely a result of its higher affinity for the cystine transporter. HCA (or glutamate) is prepared as a 100 mM stock normalized to pH 7.4 by the precise addition of dilute base (do not overshoot). We find that neurons exposed continuously to 1 mM HCA reproducibly undergo 60–80% loss in cell viability after 48 hr. Glial cells are not killed under these conditions, and thus the percentage of cell death will vary depending on the initial number of glial cells in the cultures.

The density of the cells is critical in the proper execution of these experiments. Low-density cultures (<600,000/ml) have significant spontaneous cell death. We suspect that this results from a deficiency in antioxidant defenses; addition of the antioxidant butylated hydroxyanisole (BHA, 10 μM) at the time of plating significantly reduces the basal cell death in low-density cultures. However, BHA

[8] R. W. Baughman, J. E. Huettner, K. A. Jones, and A. A. Khan, *in* "Culturing Nerve Cells" (G. Banker and K. Goslin, eds.), p. 227. MIT Press, Boston, 1991.

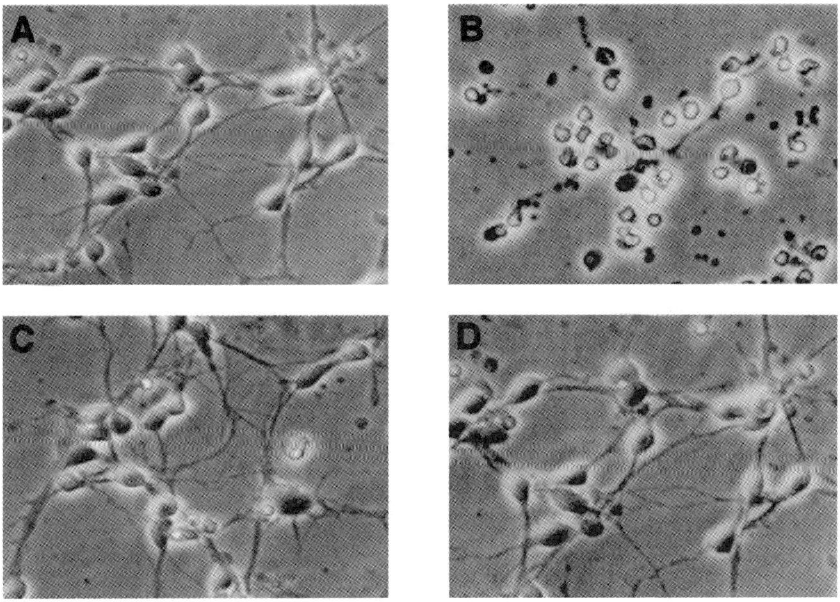

FIG. 2. Phase-contrast micrograph of cortical neurons (2 DIV). (A) Control; (B) 1 mM homocysteate (HCA), a glutamate analog that inhibits cystine uptake; (C) 1 mM HCA + the antioxidant iron chelator deferoxamine mesylate (DFO, 10 μM); and (D) DFO (10 μM).

must be rinsed from the cultures at the time of initiation of toxicity studies in order for them to be sensitive to oxidative stress-induced death. High-density cultures (>1,200,000/ml) are also problematic as these cultures are more resistant to oxidative death during the 48-hr experimental period. The precise reason for resistance to cystine deprivation in high-density cultures remains unclear, but it is worthy of further investigation. Nonreceptor-mediated glutamate toxicity *in vitro* is best observed in the first 2 to 3 days. After this time frame, the development of metabotropic and ionotropic glutamate receptors, as well as the proliferation of glial cells, can make interpretations of the biological effects of glutamate difficult.

The addition of glutamate (or HCA) leads to the depletion of total glutathione to 30% of control 5–6 hr later,[9] morphological evidence of apoptosis (as evidenced by DAPI or propidium iodide staining) by 12–14 hr,[3] and secondary necrosis as measured by lactate dehydrogenase (LDH) release by 24–48 hr. As expected, higher concentrations of glutamate that deplete glutathione more quickly can lead cells to necrosis than to apoptosis. Cell death can be completely blocked by a number of antioxidants, including *N*-acetylcysteine (100 μM), deferoxamine mesylate (10 μM, Fig. 2), and vitamin E (100 μM), as well as several inhibitors of the

[9] R. R. Ratan, T. H. Murphy, and J. M. Baraban, *J. Neurosci.* **14,** 4385 (1994).

lipoxygenase pathway.[2–4] These findings provide indirect evidence that oxidative stress is a primary mediator of cell death in the model described herein. As with other paradigms of apoptosis in neurons, cell death can be inhibited by macromolecular synthesis inhibitors[9] and caspase inhibitors,[10] but in contrast to other apoptotic paradigms, glutamate-induced oxidative death is bax independent.[10]

Measurements of Cell Viability in the Glutathione Depletion Model

We usually screen for inhibitors of oxidative stress-induced cell death using three assays. One assay is a morphological assay, which simply involves observing the cells by phase-contrast microscopy. A standard tissue culture microscope equipped with phase-contrast optics and a 20× objective is adequate. Photomicrographs (Fig. 2) of a field of cells can easily be taken using a standard Nikon camera that is mounted above the eyepiece on the scope.

The second assay is cell-associated LDH activity (an indirect measure of cells that survive). LDH is a cytosolic enzyme that is released into the bathing media in cells whose plasma membrane is disrupted. Cell-associated LDH activity can be measured rapidly and accurately in the lysates of remaining live cells using a standard, commercially available spectrophotometric assay (Promega Cytotox 96 nonradioactive assay), and percentage viability is determined by normalizing to control, untreated neurons. The Promega LDH assay kit is based on a coupled enzymatic assay involving the conversion of a tetrazolium salt (INT) into a formazan product. This reation is catalyzed by LDH, as well as the diaphorase supplemented in the substrate mixture. The absorbance of the formazan product (directly proportional to LDH activity) can be measured at 492 nm on a standard spectrophotometer. It is imperative to verify that the LDH activity measurements vary linearly with the number of cells within the dynamic range of maximal and minimal viability in the culture. This can be done by taking known numbers of cells and determining the LDH activities on these cells. If nonlinearity is observed, then the lysates can be appropriately diluted until a linear range is found. Once established, assessment of the linearity of the cell death assay need not be performed on a weekly basis.

Another quick assay for measuring cell viability utilizes 3-(4,5-dimethylthiazol-3-yl)-2,5-dipheyltetrazolum bromide (MTT) and can be performed using a commercial kit (Promega nonradioactive cell proliferation assay). When added to the culture well, soluble MTT is converted to an insoluble blue formazan by mitochondrial/glycolytic reductases in "live cells." In cells that are dead, these activities are commonly, albeit not uniformly, diminished. The insoluble blue formazan can be solublized using detergent, and the absorbance of the product can be read in a spectrophotometer at 570 nm and divided by the value at 690 nm. As with

[10] S. Tan, Y. Sagara, Y. Liu, P. Maher, and D. Schubert, *J. Cell Biol.* **141,** 1423 (1998).

the LDH assay, MTT incubation periods must be determined that result in a linear correlation between the amount of blue formazan produced and the number of live cells. We have found that with 1×10^6 neurons, 2–4 hr of incubation with MTT results in a linear relationship between the cells and the blue formazan product of MTT reduction.

A common problem with both LDH and MTT assays is that there are many compounds that will spuriously increase or decrease these activities in the absence of a corresponding change in cell number.[11] Morphological observations can often give a quick indication as to whether LDH or MTT assays are accurately reflecting the amount of cell death or cell protection that is actually occurring. When screening viability assays (e.g., phase microscopy, LDH activity, or MTT reduction) uniformly indicates that a compound can prevent oxidative death, additional markers of cell damage or cell death can be performed, such as Tdt-mediated dUTP digoxigenin nick end labeling (TUNEL) labeling (not specific for apoptosis or necrosis), DAPI staining (a DNA-intercalating dye that allows rapid assessment of nuclear morphology), or immunofluorescence staining of the activated form of one of the caspases.

Delivering Foreign Genes into Immature Cortical Neurons

Transfection of plasmids into immature cortical neurons is a difficult task. In the relatively low-density cultures required for the described studies, calcium phosphate seems to induce significant background levels of cell death. Lipofectamine and DMRIE-C (Life Technologies) are less toxic, but the transfection efficiencies and level of expression (using CMV promoters) are highly variable and are never consistently greater than 0.5–1%. Viral vectors such as adenovirus have been used successfully for mature neuronal cultures, but in our experience, the efficiency of infection and the level of expression in immature cultures have been disappointing, and glial cells are infected preferentially. It has been reported that herpes vectors are relatively nontoxic and allow high levels of rapid expression in immature cortical neurons.[12] We have confirmed these observations.

We infect cortical neurons 16–18 hr after plating with varying multiplicities of infection (MOI) in Opti-MEM (Life Technologies) for 3 hr. At this time, the cells are rinsed and fresh medium (with or without HCA) is added. We have found that at the appropriate MOIs, herpes vectors provide 20–40% transduction with virtually no toxicity and negligible affects on the sensitivity of cortical neurons to glutathione depletion-induced death. A herpes vector that expresses β-galactosidase (β-gal) is used as a protein control (Fig. 3). Cell death is quantitated by assessing the nuclear

[11] K. Zaman, H. Ryu, D. Hall, K. O'Donovan, K. I. Lin, M. P. Miller, J. C. Marquis, J. M. Baraban, G. L. Semenza, and R. R. Ratan, *J. Neurosci.* **19,** 9821 (1999).

[12] S. Bursztajn, R. DeSouza, D. L. McPhie, S. A. Berman, J. Shioi, N. K. Robakis, and R. L. Neve, *J. Neurosci.* **18,** 9790 (1998).

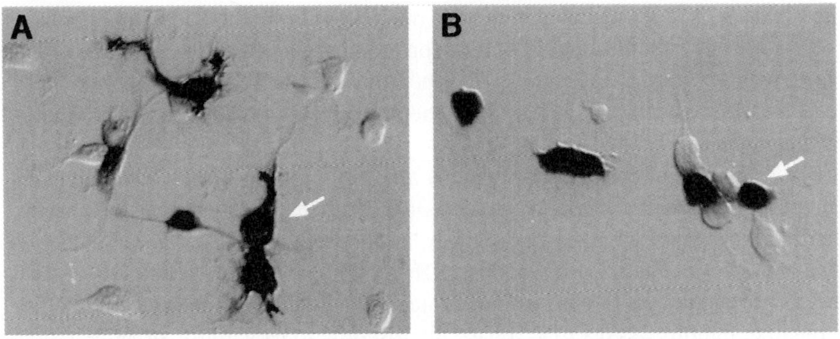

FIG. 3. Nomarski image of cortical neuronal cultures infected with HSV-lacz, fixed and incubated with X-gal. (A) Control. Arrow points to infected cell that is "blue" due to enforced expression of β-galactosidase. Note the ramifying neurites and normal morphology of the cell body. (B) 1 m*M* HCA. Arrow points to "blue"-infected cells. Note the absence of neurites and the shrinkage of the cell body.

morphology (DAPI fluorescence) in cells expressing the heterologous protein of interest or β-gal (as a control) in glutathione-replete and -depleted neurons. Cells that express the heterologous protein or β-gal are identified using immunofluorescence. The feasibility of using viral vectors to achieve enforced expression of foreign genes should facilitate the molecular elucidation of pathways involved in oxidative death in primary, postmitotic cortical neurons.

Conclusion

Glutamate or its analogs, such as HCA, induce nonreceptor-mediated toxicity in immature cortical neurons. Nonreceptor-mediated toxicity dominates, as functional glutamate receptors do not develop for up to a week after plating neurons from E17 rat embyros. Specifically, glutamate inhibits plasma membrane cystine transport, leading to depletion of glutathione and oxidative stress-induced cell death. The model is the result of pioneering studies initially performed by Bannai and Kitamura[13] and subsequently adapted to neurons by Murphy *et al.*[2] and has been used to probe mechanisms of oxidative death in neurons by a number of laboratories.

Acknowledgments

The authors acknowledge Tim Murphy, Jay Baraban, Khalequz Zaman, Paul Lee, and Kuo-I Lin for their intellectual and practical contributions to the methods outlined herein. This work was supported by the National Institutes of Health Grants K08 NS019151, R29 NS34943, and R01 NS39170 to R.R.R.

[13] S. Bannai and E. Kitamura, *J. Biol. Chem.* **255,** 2372 (1980).

[17] Glutamate-Induced c-Src Activation in Neuronal Cells

By SAVITA KHANNA, MIKA VENOJARVI, SASHWATI ROY, and CHANDAN K. SEN

Introduction

Glutamate toxicity is a major contributor to pathological cell death within the nervous system and appears to be mediated by reactive oxygen species.[1] Two pathways of glutamate toxicity have been defined: receptor-initiated excitotoxicity[2] and nonreceptor-mediated ROS-dependent toxicity.[3] One model used to study oxidative stress-related neuronal death is to inhibit cystine uptake by exposing cells to high levels of glutamate.[4] The induction of oxidative stress by glutamate in this model is a primary cytotoxic mechanism in C6 glial cells,[5,6] PC-12 neuronal cells,[7,8] immature cortical neurons cells,[4] and oligodendoglia cells.[9] Murine HT hippocampal neuronal cells, lacking an intrinsic excitotoxicity pathway, have been used as a model to characterize the oxidant-dependent component of glutamate.[10–12]

In neurons and astrocytes, c-Src is present at 15–20 times higher levels than that found in fibroblasts. The specific activity of the c-Src protein from neuronal cultures is 6–12 times higher than that from astrocyte cultures, suggesting a key function of this protein in neurons.[13] Src family kinases are able to induce caspase-independent cytoplasmic events leading to cell death.[14] We have reported that activation of c-Src kinase is a key event in glutamate-induced death of HT4 neurons.[10] A subsequent study showed that Src kinase-dependent neuronal damage plays a key role in stroke

[1] J. T. Coyle and P. Puttfarcken, *Science* **262,** 689 (1993).

[2] D. W. Choi, *Cerebrovasc. Brain Metab. Rev.* **2,** 105 (1990).

[3] S. Tan, Y. Sagara, Y. Liu, P. Maher, and D. Schubert, *J. Cell Biol.* **141,** 1423 (1998).

[4] T. H. Murphy, R. L. Schnaar, and J. T. Coyle, *FASEB J.* **4,** 1624 (1990).

[5] D. Han, C. K. Sen, S. Roy, M. S. Kobayashi, H. J. Tritschler, and L. Packer, *Am. J. Physiol.* **273,** R1771 (1997).

[6] S. Kato, K. Negishi, K. Mawatari, and C. H. Kuo, *Neuroscience* **48,** 903 (1992).

[7] P. Froissard, H. Monrocq, and D. Duval, *Eur. J. Pharmacol.* **326,** 93 (1997).

[8] C. M. Pereira and C. R. Oliveira, *Free Radic. Biol. Med.* **23,** 637 (1997).

[9] A. Oka, M. J. Belliveau, P. A. Rosenberg, and J. J. Volpe, *J. Neurosci.* **13,** 1441 (1993).

[10] C. K. Sen, S. Khanna, S. Roy, and L. Packer, *J. Biol. Chem.* **275,** 13049 (2000).

[11] Y. Li, P. Maher, and D. Schubert, *J. Cell Biol.* **139,** 1317 (1997).

[12] J. B. Davis and P. Maher, *Brain Res.* **652,** 169 (1994).

[13] J. S. Brugge, P. C. Cotton, A. E. Queral, J. N. Barrett, D. Nonner, and R. W. Keane, *Nature* **316,** 554 (1985).

[14] J. N. Lavoie, C. Champagne, M. C. Gingras, and A. Robert, *J. Cell Biol.* **150,** 1037 (2000).

0076-6879/02 $35.00

disorder.[15] Thus, Src represents a key intermediate and novel therapeutic target in the pathophysiology of neurodegenerative disorders.

Src family proteins are regulated through reversible phosphorylation and dephosphorylation events that alter the conformation of the kinase. A tyrosine kinase termed C-terminal Src kinase (Csk), expressed ubiquitously but predominantly in lymphoid tissues and neonatal brain, has been implicated to be the upstream regulatory tyrosine kinase by virtue of its ability to inactivate several Src family kinases.[16–20] Csk phosphorylates Tyr-527 in the C-terminal tail of c-Src and thus creates a binding site for the Src homology 2 (SH2) domain, locking the molecule in an inactive state. Dephosphorylation of Src Tyr-527 increases Src kinase activity up to 10- to 20-fold. The goal of this chapter is to present a detailed description of the methods that may be used to determine glutamate-induced c-Src activation in neuronal cells.

I. Determination of Glutamate-Induced Global Protein Tyrosine Phosphorylation Profile Using L-[^{35}S]Methionine Labeling of Proteins

Reagents

Lysis buffer: phosphate-buffered saline (PBS), pH 7.4, 1% (v/v) Nonidet P-40, 0.5% (w/v) sodium deoxycholate, 0.1% (v/v) sodium dodecyl sulfate, 0.25 m*M* sodium orthovanadate (Na_3VO_4), 10 m*M* phenylmethylsulfonyl fluoride (PMSF), 10 μg/ml aprotinin, and 10 μg/ml pepstatin

Wash buffer: phosphate buffer saline, pH 7.4, 1 *M* sodium chloride

Antibody: monoclonal protein phosphotyrosine antibody (PY99; Santa Cruz Biotech, Santa Cruz, CA)

Agarose beads: Protein A-agarose (Santa Cruz Biotech)

Other reagents and supplies: L-[^{35}S]methionine (NEN, Boston, MA); sodium orthovanadate (Sigma, St. Louis, MO); PBS, pH 7.4; cell lifter (Costar, Corning, NY); microfuge tubes; 2× Laemmli sample buffer

Procedure

Mouse hippocampal HT4 cells, kindly provided by D. E. Koshland, Jr., University of California at Berkeley, are grown in Dulbecco's modified Eagle's medium

[15] R. Paul, Z. G. Zhang, B. P. Eliceiri, Q. Jiang, A. D. Boccia, R. L. Zhang, M. Chopp, and D. A. Cheresh, *Nature Med.* **7,** 222 (2001).

[16] H. C. Cheng, J. D. Bjorge, R. Aebersold, D. J. Fujita, and J. H. Wang, *Biochemistry* **35,** 11874 (1996).

[17] B. W. Howell and J. A. Cooper, *Mol. Cell Biol.* **14,** 5402 (1994).

[18] M. Okada, S. Nada, Y. Yamanashi, T. Yamamoto, and H. Nakagawa, *J. Biol. Chem.* **266,** 24249 (1991).

[19] D. Sondhi and P. A. Cole, *Biochemistry* **38,** 11147 (1999).

[20] H. Yamashita, S. Avraham, S. Jiang, I. Dikic, and H. Avraham, *J. Biol. Chem.* **274,** 15059 (1999).

supplemented with 10% fetal calf serum, penicillin (100 U/ml), and streptomycin (100 μg/ml) at 37° in a humidified atmosphere containing 95% air and 5% CO_2.

HT4 cells (0.7×10^6) are seeded in a 100×20-mm plate using 10 ml growth medium.

1. Following 6 hr of seeding of cells, culture medium is changed and 5 ml fresh medium is added to the plate.
2. To radiolabel the cells, 36 μCi/ml L-[^{35}S]methionine (NEN) is added to cells. Cells are incubated at 37° in a humidified incubator containing 95% air and 5% CO_2 for 12 hr.
3. After 12 hr of such incubation, cells are challenged with 10 m*M* glutamate for 30 min under standard culture conditions. To inhibit protein tyrosine phosphatase activity, cells are treated with 0.25 m*M* sodium orthovanadate (Sigma) for 15 min prior to any treatment.
4. After 30 min (or any other desired duration) incubation of cells with glutamate, cells are washed twice with 10 ml of ice-cold PBS, pH 7.4.
5. Lysis buffer (1 ml) is added to cells and cells are kept on ice for 15 min for lysis.
6. After 15 min of lysis, cells are scraped using a cell lifter and collected in 2-ml microfuge tubes.
7. Cell lysates are centrifuged at 12,000*g* for 10 min at 4° and supernatants are used for immunoprecipitation.
8. Tyrosine-phosphorylated proteins are immunoprecipitated by adding 2 μg of monoclonal protein phosphotyrosine antibody to the cell extract contained in microfuge tubes.
9. Tubes are put on a rotating shaker at 4° for 4 hr followed by another 12 hr with 40 μl [25% (v/v) stock] protein A-agarose.
10. After 12 hr of incubation with antibody and agarose beads, the samples are washed once with lysis buffer and thrice with wash buffer.
11. The immunoprecipitated material is resuspended in 40 μl of 2× sample buffer and boiled for 10 min. After centrifugation (14,000*g*, 3 min), immunoprecipitated proteins are separated on a 10% SDS–PAGE gel, and the protein tyrosine phosphorylation profile is detected by autoradiography (Fig. 1).

Notes

1. To obtain reproducible data, it is important that attention be paid to having the same cell density in repeated experiments.
2. Treatment with 0.25 m*M* sodium orthovanadate for 15 min before challenging the cells to inhibit protein tyrosine phosphatase activity is critical. It is difficult to find phosporylated proteins without inhibiting phosphatase activity.
3. After extracting the cell, the whole procedure should be performed at 4°.
4. This radioisotope-dependent method suffers from the limitation that it is a

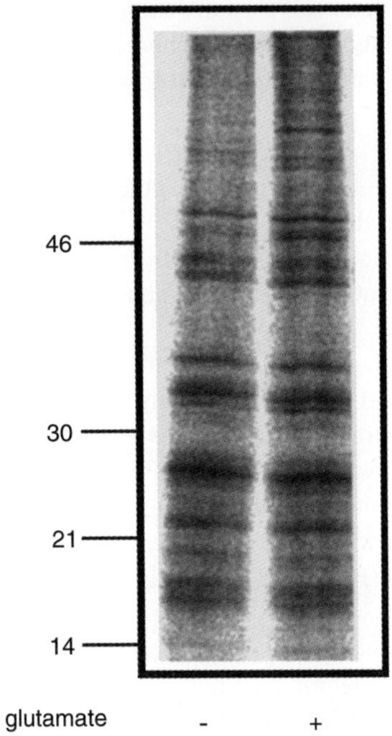

FIG. 1. Protein tyrosine phosphorylation profile in HT4 cells as determined by the method described in Section I. Glutamate treatment resulted in tyrosine phosphorylation of specific proteins.

"hot" procedure. The advantage is having sharp bands that are often difficult to get using standard Western blot detection.

II. Determination of Src Kinase Activity

Reagents

Lysis buffer: 20 m*M* HEPES–NaOH, pH 7.5; 3 m*M* $MgCl_2$, 100 m*M* NaCl, 1 m*M* dithiothreitol, 1 m*M* PMSF, 1 μg/ml leupeptin, 1 m*M* EGTA, 1 m*M* sodium orthovanadate, 10 m*M* NaF, 20 m*M* glycerophosphate, and 0.5% Nonidet P-40

Antibody: A-agarose-conjugated anti-Src family kinase antibody (Santa Cruz Biotechnology)

Reaction buffer: 40 m*M* HEPES–NaOH, pH 7.5, 10 m*M* $MgCl_2$, 3 m*M* $MnCl_2$, 0.5 m*M* dithiothreitol, 0.1 m*M* PMSF, 0.1 μg/ml leupeptin, 0.1 m*M* sodium orthovanadate, 1 m*M* NaF, and 2 m*M* glycerophosphate

Kinase activity buffer: acid-denatured enolase (Boehringer Mannheim, Germany), 10 μM ATP, and 10 μCi of [γ-^{32}P]ATP (NEN)

1. Enolase processing. Weigh enolase (5 μg/sample), add ice cold 10% (500 μl) trichloroacetic acid (TCA), keep on ice for 5 minutes. Add 500 μl reaction buffer and spin at maximum speed for 5 min. Repeat this step two more times. If the assay is to be performed on the same day, keep the pellet on ice. Otherwise it may be stored at -70° for weeks.

2. Preparation of cell extracts. To determine Src kinase activity, 1.6×10^6 HT4 cells are seeded in 140 × 20-mm plates containing 20 ml growth media. Following 12 hr of seeding, cells are activated with glutamate for 30 min as described in Section I. Cells are harvested and lysed in 1 ml of lysis buffer. Cell lysates are centrifuged at 12,000*g* for 10 min at 4°. The protein content of the sample is determined using the Pierce BCA protein assay kit (Rockford, IL).

Other reagents and supplies: cell lifter (Costar, Corning, NY) and microfuge tubes.

Kinase Assay

1. The cell extract (750 μg protein) is placed in a 1.5-ml microfuge tube.
2. The total volume of the cell extract is made to 1 ml using lysis buffer.
3. Src kinase is immunoprecipitated from the extract by adding 2 μg of protein A-agarose-conjugated anti-Src family kinase antibody.
4. The microfuge tubes are placed on a rotating shaker at 4° for 4 hr.
5. The immune complex is separated by centrifugation at 6000 rpm for 5 min in a refrigerated centrifuge.
6. The beads are washed twice with the lysis buffer (500 μl/wash) followed twice with reaction buffer (500 μl/wash).
7. Substrate mix is prepared by adding 20 μl of reaction buffer to the enolase pellet. Mix by vortexing and make sure that the pellet is dissolved.
8. Add 1 μl of 10 m*M* ATP (final 10 μM) and 10 μCi of [γ-^{32}P]ATP to each sample.
9. The kinase assay is performed for 10 min at 22° with 5 μg of acid-denatured enolase as substrate in 30 μl of the reaction buffer containing 10 μM ATP and 10 μCi of [γ-^{32}P]ATP per sample.
10. The kinase reaction is stopped by adding 10 μl of 4× Laemmli sample buffer. The mixture is boiled for 5 min and subjected to 10% SDS–PAGE.
11. The gel is dried and the radioactivity incorporated into enolase is determined using a phosphoimager (Molecular Dynamics, Sunnyvale, CA) as shown in Fig. 2.

Notes

1. Denatured enolase serves as a good substrate for this kinase assay. If the phosphorylation of any other specific physiological substrate is to be studied, enolase may be replaced.

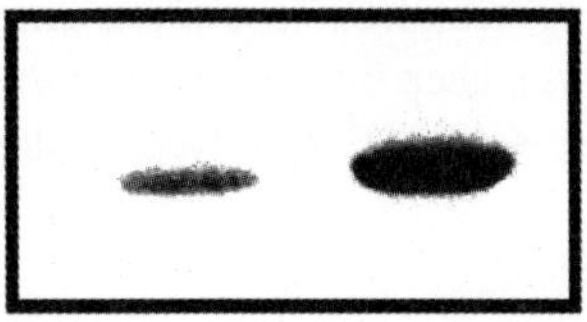

FIG. 2. Activity of Src kinase immunoprecipitated from HT4 cells. Glutamate treatment increased Src kinase activity.

III. Immunolocalization of Src and Phospho-Src

Reagents

Fixing solution: 3.7% formaldehyde in PBS pH 7.4, PBS-T: PBS containing 0.2% Triton X-100

Blocking buffer: PBS containing 0.2% (v/v) Triton X-100 and 1% (w/v) bovine serum albumin

Mounting media: GEL/MOUNT biomeda (Fisher Scientific)

Primary antibodies: anti-Src (c-Src B-12; Santa Cruz Biotech), anti-phospho-Src (Upstate Biotechnology, Lake Placid, NY)

Secondary antibodies: FITC-conjugated donkey anti-mouse IgG antibody and RhodamineRed-conjugated donkey anti-rabbit IgG antibody (Jackson Immunoresearch, West Grove, PA)

Microscope: Nikon E800, MetaMorph version 4.5 software (Universal Imaging Corp., West Chester, PA)

Other reagents and supplies: poly-L-lysine (Sigma); coverslips (Fisher Scientific); glass slides (Fisher Scientific); parafilm; and aluminum foil

Cell Culture

HT4 cells (0.1×10^6) are seeded on 22×22-mm (0.13 to 0.16 mm thick) autoclaved poly-L-lysine (0.01%)-coated coverslips at 37° in a humidified atmosphere containing 95% air and 5% CO_2 for 12 hr before challenging with glutamate.

Indirect Immunofluorescence Staining

1. Wash cells three times with PBS. Make sure that there is no cell culture medium left and check under microscope that cells are still attached to coverslips.
2. To fix cells, add 500 μl of fixing solution and keep cells at room temperature for 10 min.
3. Remove fixing solution and wash cells three time with PBS.

4. Add 1 ml of PBS-T for 20 min at room temperature to permeabilize cells.
5. Wash cells twice with 1 ml PBS-T and add 1 ml blocking buffer for 45 min at room temperature to block nonspecific antibody binding.
6. Wash cells with 1 ml PBS-T three times.
7. Take a 100 × 20-mm plate, place a piece of parafilm, and add 30 μl (1.5 μg/ sample or 1 : 50 dilution) primary antibody for anti-src and anti-phospho-src.
8. Place the coverslip (cells face down) on the antibody solution placed on the parafilm.
9. Place some wet tissue paper by the side of the 100 × 20-mm plates to have some humidity.
10. Incubate the coverslip under humid conditions overnight at 4°.
11. The next day, place the plates containing the coverslips at room temperature for 45 min.
12. Gently flip the coverslip (cells face up) and wash three times with 1 ml PBS-T.
13. Incubate the coverslip with secondary antibody (FITC-conjugated donkey anti-mouse IgG antibody, 1 : 100 dilution; and RhodamineRed-conjugated donkey anti-rabbit IgG antibody, 1 : 100 dilution) for 45 min at room temperature in the dark.
14. Wash cells with 2 ml PBS three times.
15. Add one drop of mounting medium containing antifade reagent on the glass slide. Put coverslips (cells face down) carefully on mounting media.

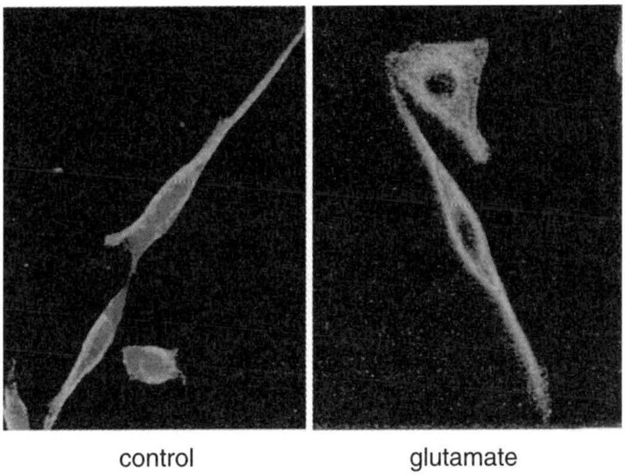

FIG. 3. HT4 cells immunostained with anti-Src or antiphospho-Src. Images are taken with 60× magnification (CFI Plan Apochoromat 60× oil, N.A. 1.40). Fluorescence analysis is performed using Metamorph Imaging software. FITC (representing Src) and rhodamine (representing phospho-Src) images are digitally overlayed. Less phospho-Src (Rhodamine; red) signal is visible in glutamate-treated cells (B versus A), suggesting glutamate-induced dephosphorylation (activation) of c-Src.

16. Keep for at least 30 min to dry mounting media before observing under a microscope.

Note

Avoid bubbles during antibody incubation and mounting of coverslips.

Microscopy

We recommend performing fluorescence microscopy using a Nikon E800 microscope with a 0.5 to 100× objective. Its differential interference contrast (DIC, Nomarski) optics allows excellent three-dimensional imaging of cells. Imaging is performed using a Photometrics Sen Sys CCD digital camera and MetaMorph 4.5 software. Simultaneous images of FITC/rhodamine fluorescence via epifluorescence and cell morphology via DIC are obtained.

Notes

1. A mercury lamp allows high magnification imaging with DIC and Plan-apochoromatic objectives provide the best correction for all aberrations.
2. Sensitivity and resolution with a Photometrics Syn CCD digital camera are exceptional, especially for fluorescence imaging.
3. The MetaMorph software allows image processing, contrast enhancement, color overlays, and intensity measurement. Dual color and overlay images are shown in Fig. 3.

[18] Measurement of Inflammatory Properties of Fatty Acids in Human Endothelial Cells

By MICHAL TOBOREK, YONG WOO LEE, SIMONE KAISER, and BERNHARD HENNIG

Introduction

Fatty acids can modulate inflammatory responses in numerous tissues, including the vascular endothelium. At least two different independent pathways can be responsible for these effects. These pathways are linked to either (1) eicosanoid production or (2) redox-regulated gene expression. Traditionally, lipid-mediated cellular inflammatory reactions have been linked to the release of arachidonic acid from the cellular membranes, activation of cyclooxygenases, and lipoxygenases

0076-6879/02 $35.00

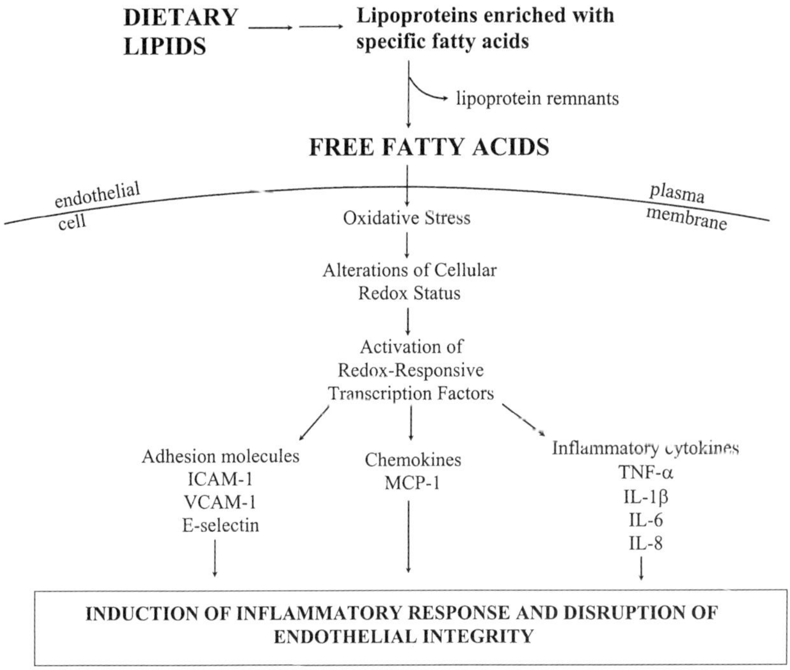

FIG. 1. Schematic diagram of proinflammatory pathways induced by dietary fatty acids in vascular endothelial cells. ICAM-1, intercellular adhesion molecule-1; IL, interleukin; MCP-1, monocyte chemoattractant protein-1; TNF-α, tumor necrosis factor-α; VCAM-1, vascular cell adhesion molecule-1.

with the subsequent overproduction of eicosanoids. These reactions have been relatively well studied and have been characterized in detail in several excellent reviews.[1-3] Therefore, this chapter focuses on dietary fatty acid-induced proinflammatory pathways mediated by the induction of oxidative stress, activation of redox-regulated transcription factors, and the inflammatory genes (Fig. 1).

Dietary Sources of Fatty Acids

Evidence indicates that selected fatty acids can stimulate inflammatory reactions through the transcriptional regulation of inflammatory genes, i.e., genes encoding for adhesion molecules and inflammatory cytokines.[4] However, it appears

[1] M. J. James, R. A. Gibson, and L. G. Cleland, *Am. J. Clin. Nutr.* **71**(1 Suppl.), 343S (2000).
[2] K. F. Scott, K. J. Bryant, and M. J. Bidgood, *J. Leukocyte. Biol.* **66,** 535 (1999).
[3] E. G. Spokas, J. Rokach, and P. Y. Wong, *Methods Mol. Biol.* **120,** 213 (1999).
[4] M. Toborek and B. Hennig, *Subcell. Biochem.* **30,** 415 (1998).

that the effects mediated by individual fatty acids are very specific and are influenced by diet and types of dietary fat. It is generally accepted that dietary profiles of fatty acids can influence lipoprotein lipid composition significantly. Thus, diets enriched in specific fatty acids result in high concentrations of these fatty acids in lipoprotein fractions. Nutritional analyses indicate that the typical Western diet contains 20- to 25-fold more n-6 (or omega-6) than n-3 (or omega-3) fatty acids. In addition, among n-6 fatty acids, linoleic acid (18 : 2, n-6) is the major dietary fatty acid present in high concentrations in corn, soy, sunflower, or safflower oils. It is estimated that linoleic acid provides approximately 7–8% of the average dietary energy intake.[1] In addition, linoleic acid is thought to be a predominant substrate for lipid peroxidation processes both in lipoproteins, such as low-density lipoproteins (LDL), and in tissues.[5] In contrast to linoleic acid, the dietary intake of α-linolenic acid (18 : 3, n-3), an essential fatty acid of the n-3 family, is relatively low. For example, α-linolenic acid, present in leafy, green vegetables, as well as in flaxseed and canola oils, constitutes only 0.3–0.4% of the average dietary energy intake. Oleic acid (18 : 1, n-9) is another main dietary fatty acid, present in high amounts in olive or sunola oils, as well as in meat. It is responsible for approximately 8–15% of the average dietary energy intake.[1] In contrast to linoleic acid or α-linolenic acid, oleic acid is not an essential fatty acid and can be synthesized from stearic acid by Δ-9 desaturation.[6]

Fatty acids can be hydrolyzed from lipoproteins in a reaction catalyzed by lipoprotein lipase, an enzyme associated with the vascular endothelium. Thus, endothelial cells can be directly exposed to high concentrations of free fatty acids.[7] Evidence indicates that specific free fatty acids can directly affect endothelial cell metabolism and induce potent proinflammatory reactions.[8] Because of their dietary significance and biological potential, our research has concentrated on the effects of 18 carbon fatty acids (such as oleic, linoleic, and linolenic acid) on endothelial cell metabolism in relationship to the development of atherosclerosis. Specifically, we study the roles of these fatty acids in the regulation of proinflammatory pathways in human endothelial cells.

Human Endothelial Cell Cultures and Preparation of Fatty Acid-Enriched Media

Background

Human umbilical vein endothelial cells (HUVEC) are the most common primary human endothelial cells available for routine cell culture research. Although

[5] G. Spiteller, *Chem. Phys. Lipids* **95,** 105 (1998).

[6] B. A. Watkins, B. Hennig, and M. Toborek, *in* "Bailey's Industrial Oil and Fat Products" (Y. H. Hui, ed.), p. 159. Wiley, New York, 1996.

[7] D. B. Zilversmit, *Circ. Res.* **33,** 633 (1973).

[8] B. Hennig, M. Toborek, and C. J. McClain, *J. Am. Coll. Nutr.* **20,** 97 (2001).

these cells are of vein origin, they appear to be well suited for research related to different aspects of vascular biology, including studies on inflammatory responses. For example, HUVEC express all mediators of inflammatory responses, such as genes encoding for adhesion molecules, inflammatory cytokines, and chemokines.[9] In addition, HUVEC are susceptible to the development of apoptosis.[10]

Human endothelial cells can also be isolated from other vessels, such as the aorta or the femoral artery. Moreover, a variety of different types of primary endothelial cells are available commercially (e.g., from Clonetics Corp., Walkersville, MD, or Cascade Biologics, Portland, OR). Several immortal cell lines of human endothelial cells are also available, such as human microvascular endothelial cells (HMEC-1), which originated from dermal microvascular endothelial cells transfected with the SV-40 large T promoter,[11] or the EA.hy926 cell line, produced by fusion of HUVEC with human A549 carcinoma cells.[12]

In most experiments, endothelial cells are exposed to fatty acids at concentrations of 60 or 90 μM, with experimental media albumin concentrations of about 60 μM. Normal plasma free fatty acid concentrations can range from approximately 90 to 1200 μM; however, the majority of free fatty acids are bound to plasma components, mostly albumin.[13,14] In fact, the main factor in the availability of fatty acids for cellular uptake is determined by the free fatty acid to albumin ratio. Normally, this ratio can range from 0.15 to 4 in response to various conditions, with an average of approximately 1.[13,14] Thus, the experimental conditions employed in our studies, which result in a free fatty acid to albumin ratio of 1 or 1.5, are within a physiological range.

Solutions for HUVEC Isolation and Culture

Dispase solution: Dispase (2 mg/ml) in M199 enriched with penicillin/streptomycin (400 U/ml) and 3% fetal bovine serum (FBS)

Growth medium: M199 with added $NaHCO_3$, pH 7.4, and enriched with heparin, 54.3 U/ml; HEPES, 25 mM; L-glutamine, 2 mM; sodium pyruvate, 1 μM; penicillin, 200 U/ml; streptomycin, 200 μg/ml; amphotericin B, 0.25 μg/ml; endothelial cell growth supplement (ECGS), 0.04 mg/ml; FBS, 20%

Experimental medium: composition is similar as that of growth medium, except for the serum content. FBS is added to the experimental medium at the final concentration of 10%

[9] B. Hennig, P. Meerarani, P. Ramadass, B. A. Watkins, and M. Toborek, *Metabolism* **49,** 1006 (2000).

[10] Y. W. Lee, H. Kühn, B. Hennig, and M. Toborek, *FEBS Lett.* **485,** 122 (2000).

[11] E. W. Ades, F. J. Candal, R. A. Swerlick, V. G. George, S. Summers, D. C. Bosse, and T. J. Lawley, *J. Invest. Dermatol.* **99,** 683 (1992).

[12] C. J. S. Edgell, C. C. McDonald, and J. B. Graham, *Proc. Natl. Acad. Sci. U.S.A.* **80,** 3734, (1983).

[13] A. A. Spector, *J. Lipid Res.* **16,** 165 (1975).

[14] B. Potter, J. D. Sorentino, and P. D. Berk, *Annu. Rev. Nutr.* **9,** 253 (1989).

Hanks' balanced salt solution: NaCl, 0.14 *M*; KCl, 5.36 m*M*; KH_2PO_4, 0.44 m*M*; $Na_2HPO_4 \cdot 7H_2O$, 0.63 m*M*; $NaHCO_3$, 4.16 m*M*; D-glucose, 5.55 m*M*; phenol red sodium salt, 0.001%

Procedure

Umbilical cords are collected in sterile beakers containing M199 and penicillin and streptomycin at concentrations of 400 U/ml. HUVEC are isolated under aseptic conditions as follows: umbilical cord is placed on sterile gauze and both ends are cut cleanly prior to locating the umbilical vein. Canuli with attached tubings are inserted into the vein from both ends of the cord, and umbilical tape is knotted tightly to uphold the canuli in the vein. Then, blood clots are rinsed from the inside of the umbilical vein by injection of Hanks' solution through the canula. After cleaning of the vein and clamping one end of the umbilical cord, the dispase solution is injected into the vein, allowing the cord to become fully distended. Tubings attached to each canula are clamped and the cords are placed in a sterile beaker containing Hanks' solution and refrigerated overnight for 15 to 18 hr to allow dislodging of endothelial cells. The following day, dispase-containing cells are collected by rinsing the lumen of the vein with Hanks' solution to further dislodge weakly attached cells. The cell suspension is centrifuged for 10 min at 250*g* at room temperature. Then, the pelleted cells are resuspended in growth medium and seeded in a cell culture flask. Two to 4 hr later, when endothelial cells are fully attached to the surface of the flask, the medium is removed and cells are rinsed gently with Hanks' solution to remove any remaining blood cells, and fresh growth medium is added to the flask. Endothelial cells are cultured at 37° in a humid atmosphere of 5% CO_2. Cell cultures are identified as endothelial by their cobblestone morphology and by the uptake of acetylated low-density lipoproteins labeled with 1,1′-dioctadecy1-3,3,3′,3′-tetramethylindocabocyanine perchlorate (Dil-Ac-LDL). Dil-Ac-LDL binds to the scavenger receptor present on endothelial cells, as well as on other cell types, such as macrophages or microglia. However, isolation of endothelial cells from umbilical veins and the subsequent cell culture procedures eliminate the possibility of macrophage or microglia contamination. Thus, under the described conditions, Dil-Ac-LDL uptake, combined with the morphological appearance of cells, can specifically identify endothelial cells. The passage of endothelial cells is performed by washing the cells with Hanks' solution and adding trypsin/EDTA at a 1 : 3 split ratio. All our experiments are conducted with cells from passage two.

Measurement of Dil-Ac-LDL Uptake by HUVEC

Dil-Ac-LDL (Molecular Probes, Eugene, OR) is diluted to 10 μg/ml in growth medium and added to cell cultures. Following a 4-hr incubation at 37°, medium containing Dil-Ac-LDL is removed from the cultures, and cells are washed three times with Hanks' solution. Then, cells are trypsinized, centrifuged at 500*g*, washed once with phosphate-buffered saline (PBS), and resuspended in PBS to obtain a final concentration of 1×10^6 cells/ml. The percentage of fluorescent-labeled cells is measured using an activated cell sorter, FACScan, with the wavelength for excitation and emission set at 514 and 500 nm, respectively. Cells from unlabeled cultures serve as negative controls.

Preparation of Fatty Acid-Enriched Media

Stock solutions of high purity (≥99%) fatty acids (Nu-Chek-Prep, Elysian, MN) are prepared in hexane. NaOH (6 *M*, or 30× molarity of fatty acid) is used for saponification to convert the fatty acids into a water-soluble form. The desired amount of fatty acid is aliquoted, mixed with 6 *M* NaOH, and dried under high purity nitrogen gas. The residue is dissolved in 1.0 ml of hot, distilled water, and the solution is immediately transferred to a beaker containing experimental medium. Then, the pH is adjusted to 7.4 with 1.2 *M* HCl and the medium is sterilized through a syringe-driven filter unit.

Fatty Acid-Induced Oxidative Stress in Endothelial Cells

Background

Among different methods to assess oxidative stress, 2′,7′-dichlorofluorescein (DCF) fluorescence appears to be very sensitive and especially useful in experimental settings that include cell cultures.[15] This method allows measurement of cellular oxidation in individual and viable cells directly on cell culture dishes. Cells are loaded with 2′,7′-dichlorofluorescin diacetate (DCF-DA), a stable, nonpolar compound that diffuses readily into the cells and is converted to a nonfluorescent polar derivative 2′,7′-dichlorofluorescin (DCF-H) by intracellular esterases. DCF-H can be oxidized to the highly fluorescent compound DCF by hydrogen peroxide or other peroxides produced by the cells. The intensity of cellular fluorescence can be assessed by a confocal laser-scanning microscope coupled to an inverted microscope.

We demonstrated that DCF fluorescence is a sensitive marker of cellular oxidation induced by fatty acids.[16] Figure 2 shows photomicrographs visualizing intracellular DCF fluorescence in control endothelial cells, as well as in endothelial cells exposed to 90 μM linoleic acid for 6 hr. The pseudocolor scale that reflects the levels of the intracellular peroxide tone is arranged in such a way that white color reflects the highest peroxide concentrations, red color high levels,

[15] M. P. Mattson, S. W. Barger, J. G. Begley, and R. J. Mark, *Methods Cell Biol.* **46,** 187 (1995).

[16] M. Toborek, S. W. Barger, M. P. Mattson, S. Barve, C. J. McClain, and B. Hennig, *J. Lipid Res.* **37,** 123 (1996).

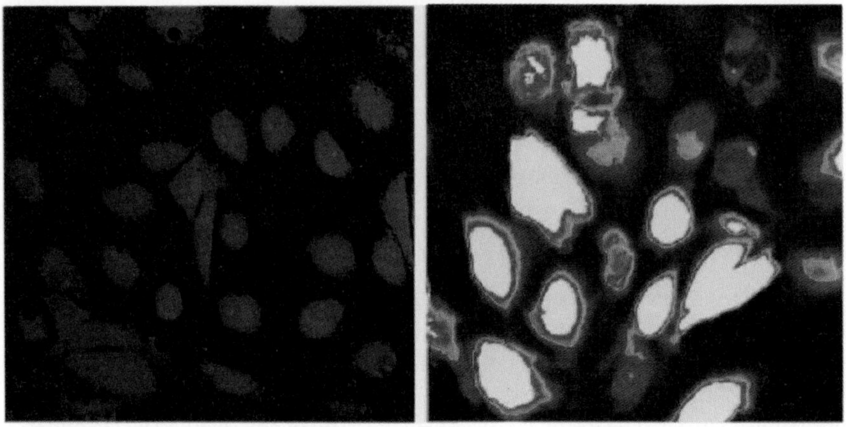

FIG. 2. Photomicrographs from confocal laser-scanning microscopy visualizing oxidative stress as DCF fluorescence emission. Endothelial cells were either untreated (left, control cells) or treated with 90 μM linoleic acid for 6 hr (right). Blue color on the pseudocolor scale reflects a low level of cellular oxidation, yellow intermediate, red high, and white the highest level of cellular oxidative stress. The intensity of fluorescence can be quantitated using "ImageSpace" software (Molecular Dynamics) and is expressed in relative units of DCF fluorescence.

yellow color intermediate, and blue color reflects the lowest levels of intracellular oxidizing compounds. The average pixel intensity is also measured within each field using the "ImageSpace" software supplied by the manufacturer (Molecular Dynamics) and is expressed in relative units of DCF fluorescence. DCF fluorescence can be utilized to measure cellular oxidation in different model systems. In fact, this method has been used successfully in a variety of cell types (e.g., neurons[15]) or treatments (e.g., amyloid β-peptide[17] or interleukin-4[18]).

Although DCF fluorescence is an excellent method to study cellular oxidation, it is variable throughout the cell. Therefore, we perform our measurements constantly 1 μm below the cell surface. In addition, loading of the dye may not be uniform across cells cultured in one dish. Thus, it is important to measure DCF fluorescent in a large number of cells in several independent cultures. A standard procedure in our laboratory involves measurements of up to 300 individual cells per culture in four independent cultures. In addition, DCF fluorescence is sensitive to pH changes. Therefore, it is important that the pH of each fatty acid-enriched medium is neutralized carefully to normal values before adding such a medium to endothelial cell cultures. Moreover, measurements of DCF fluorescence are performed in the presence of Hank's solution to buffer pH changes, which can occur during the procedure.

[17] Y. Goodman and M. P. Mattson, *Exp. Neurol.* **128,** 1 (1994).

[18] Y. W. Lee, H. Kühn, B. Hennig, A. S. Neish, and M. Toborek, *J. Mol. Cell. Cardiol.* **33,** 83 (2001).

Procedure

Endothelial cells (3.0×10^5 cells/dish) are plated on polyethylenimine-coated glass bottom 35-mm dishes (Mat-Tek, Inc., Ashland, MA), cultured for 3–4 days until confluent, and treated with fatty acids. The cells are loaded with 50 μM DCF-DA during the remaining 50 min of the experiment. At the end of the incubation period, cells are washed three times with Hanks' solution. Then, 1 ml of Hanks' solution is added to cell culture dishes and DCF fluorescence is measured using a confocal laser-scanning microscope (Molecular Dynamics, Sunnyvale, CA) coupled with a Nikon Diaphot inverted microscope (Nikon, Inc., Melville, NY) using 488-nm excitation and 510-nm emission filters. Operating conditions are as follows: objective, 60×; pinhole aperture, 50 μm; image size, 1024 × 1024 pixels; and pixel size, 0.21 μm. Average pixel intensity is measured within each individual cell and is expressed in the relative units of DCF fluorescence. Values are expressed as mean ± SEM of individual cells from three or four separate plates.

Other Methods Used to Assess Cellular Oxidation Status in Fatty Acid-Treated Endothelial Cells

Popular methods to assess cellular oxidative stress include measurements of thiobarbituric acid-reactive substances (TBARS), lipid hydroperoxides, conjugated dienes, and 4-hydroxynonenal (HNE).[19,20] Among them, measurement of TBARS is still used most frequently. The principle of the method is based on the reaction between malondialdehyde (MDA), an aldehyde product of lipid peroxidation, with thiobarbituric acid (TBA) at a high temperature, typically 100°. This method is criticized as a marker of lipid peroxidation because normally MDA is only a minor product of lipid peroxidation and the majority of detectable MDA is formed from hydroperoxides during heating in the reaction with TBA. In addition, MDA is formed from fatty acids, which contain a minimum of three double bonds. Thus, MDA is not generated from linoleic acid, which appears to be the main fatty acid involved in lipid peroxidation. Finally, several compounds can react with TBA in addition to MDA. A partial list of these compounds includes sialic acid, prostaglandins, thromboxanes, deoxyribose, and other carbohydrates.[19,20]

Lipid hydroperoxides are formed as intermediates of lipid peroxidation. A popular method to assess the lipid hydroperoxide level is based on the peroxide-mediated oxidation of ferrous (Fe^{2+}) to ferric (Fe^{3+}) iron. Ferric iron can bind to xylenol orange to produce a chromophore that can be quantitated at 560 nm (FOX assay).[21] We used this method successfully in our studies on linoleic acid-induced oxidation of cultured endothelial cells.[22]

[19] K. Moore and L. J. Roberts, *Free Radic. Res.* **28,** 659 (1998).

[20] H. Esterbauer, *Pathol. Biol.* **44,** 25 (1996).

[21] Z. Y. Jiang, J. V. Hunt, and S. P. Wolff, *Anal. Biochem.* **202,** 384 (1992).

[22] B. Hennig, M. Toborek, S. Joshi-Barve, S. Barve, M. P. Mattson, and C. J. McClain, *Am. J. Clin. Nutr.* **63,** 322 (1996).

Lipid peroxidation may also be simply assessed by accumulation of the conjugated dienes. This method is based on the principle that during the initiation of lipid peroxidation, isolated double bonds in fatty acid molecules are shifted to conjugated double bonds, which are detectable at 234 nm. Because of the simplicity of detection, measurements of conjugated dienes are useful for the continuous monitoring of lipid peroxidation to detect the susceptibility of biological samples to oxidation. This approach requires the stimulation of cellular oxidation, e.g., by adding copper or iron ions.[23]

HNE is another aldehyde product of lipid peroxidation.[20] In contrast to MDA, it can be generated during the peroxidation of linoleic acid. In fact, we indicated that HNE is formed in HUVEC exposed to this fatty acid for 24 hr. However, it should be noted that arachidonic acid, which is more unsaturated than linoleic acid, appears to be a better substrate for HNE. Levels of HNE in endothelial cells can be determined semiquantitatively by immunocytochemistry or Western blot and quantitatively by HPLC.[24]

In general, FOX method, TBARS levels, formation of conjugated dienes, and, to a lesser extent, production of HNE are much less sensitive than DCF fluorescence and can be performed only on large numbers of endothelial cells. In addition, they require extensive manipulations with cells, such as harvesting or sonication, which is inevitably connected with the creation of artificial oxidation in a test tube. Therefore, from our experience, DCF fluorescence appears to be the method of choice to study fatty acid-induced cellular oxidation in cultured endothelial cells.

Fatty Acid-Induced Alterations of Glutathione Levels and Cellular Redox Status

Background

Increased cellular oxidation results in alteration of the cellular redox status, which can be detected by the ratio of oxidative to reduced glutathione (GSSG and GSH, respectively). It is well known that oxidative stress results in decreased levels of total glutathione and increased concentrations of GSSG. Thus, the ratio of GSSG/GSH is recognized as a sensitive marker of cellular oxidative stress. In addition, because glutathione is the major nonprotein sulfhydryl compound, it plays a critical role in the maintenance of the cellular redox status.[25] In fact, the equilibrium between GSSG and GSH can regulate the activation of redox-regulated transcription factors, such as nuclear factor-κB (NF-κB) or activator protein-1 (AP-1). Our research indicated that treatment of endothelial cells with specific

[23] A. Chait, *Curr. Opin. Lipidol.* **3,** 389 (1992).

[24] U. Herbst, M. Toborek, S. Kaiser, M. P. Mattson, and B. Hennig, *J. Cell. Physiol.* **181,** 295 (1999).

[25] H. Sies, *Free Radic. Biol. Med.* **27,** 916 (1999).

fatty acids can lead to decreased levels of total glutathione, as well as alterations of the GSSG/GSH ratio. We also found a direct correlation between changes in cellular glutathione levels in fatty acid-treated endothelial cells and other methods of assessing cellular oxidative status which were described earlier.[26]

Among several methods to determine cellular glutathione content, an enzymatic recycling assay first described by Tietze appears to be simple and reliable.[27] There are several versions of this method, with a recent modification that allows one to perform the measurements using a microtiter plate reader.[28] Specificity of this method for glutathione assessment is ensured by highly specific glutathione reductase, which is added to the reaction mixture. This method also can be adapted to assay for GSSG by prior derivatization of GSH by adding 4-vinylpyridine. Then, levels of GSH can be calculated by a simple subtraction of GSSG concentration from total glutathione content. The detailed procedure to measure total cellular glutathione using a plate reader is given.[28]

Another popular method to detect the cellular glutathione level is based on the reaction of GSH and/or GSSG with *o*-phthalaldehyde (OPT) followed by fluorescence detection. The original method describing this approach was criticized because it overestimated levels of GSSG markedly.[29] However, a recently developed modification, in which GSSG is separated by HPLC and where the reaction with OPT is performed at high pH, appears to avoid these limitations.[30]

Procedure

Endothelial cells are cultured until confluence and treated with fatty acids at the concentration of 60–90 μM for up to 24 hr. At the end of the incubation period, cells are washed with PBS and scraped into 2.25% 5-sulfosalicylic acid. After centrifugation at 14,000*g* for 20 min at 4°, the supernatant is used for the determination of total glutathione, whereas the pellet is dissolved in 0.2 *M* NaOH containing 0.1% SDS for protein concentration analysis. Levels of total glutathione in acid-soluble fractions are determined by the enzymatic recycling assay in the presence of 0.15 m*M* 5,5′-dithiobis-2-nitrobenzoic acid (DTNB), 0.2 m*M* NADPH, and 1.0 unit of glutathione reductase/ml of assay mixture in 30 m*M* sodium phosphate buffer (pH 7.5) containing 0.3 m*M* EDTA. Total glutathione is estimated by monitoring the rate of formation of the chromophoric product 2-nitro-5-thiobenzoic acid at 405 nm. The glutathione content in samples is calculated on the basis of the standard curve obtained with known amounts of glutathione and expressed in nanomoles of glutathione per milligram of cellular protein.

[26] M. Toborek and B. Hennig, *Am. J. Clin. Nutr.* **59,** 60 (1994).
[27] F. Tietze, *Anal. Biochem.* **27,** 502 (1969).
[28] M. A. Baker, G. J. Cerniglia, and A. Zaman, *Anal. Biochem.* **190,** 360 (1990).
[29] P. J. Hissin and R. Hilf, *Anal. Biochem.* **74,** 214 (1976).
[30] K. J. Lenton, H. Therriault, and J. R. Wagner, *Anal. Biochem.* **274,** 125 (1999).

Fatty Acid-Induced Activation of Redox-Regulated Transcription Factors in Human Endothelial Cells

Background

Induction of cellular oxidative stress and/or changes of intracellular glutathione levels can trigger signal transduction pathways via activation of redox-responsive transcription factors and, hence, the transcription of specific genes. Among oxidative stress-responsive transcription factors, NF-κB and AP-1 appear to be most important. Indeed, increased activities of AP-1 and NF-κB are considered to be a part of a general regulation of gene expression by oxidative stress. NF-κB is composed of homo- or heterodimeric complexes of at least five distinct subunits, such as p50, p52, p65 (RelA), c-Rel, and Rel-B; however, the p50/p65 heterodimer is the predominant form of this transcription factor.[31] NF-κB-binding sites were identified in the promoter regions of genes encoding for adhesion molecules (intercellular adhesion molecule-1, ICAM-1; vascular cell adhesion molecule-1, VCAM-1; or E-selectin) and inflammatory cytokines (such as tumor necrosis factor-α, TNF-α; IL-1β, IL-6, or IL-8), growth factors, and chemokines. Although other transcription factors are also required for expression of these genes, NF-κB constitutes an important component of their transcriptional regulation. It is interesting that the expression of inflammatory cytokines is dependent on activated NF-κB and, in turn, these cytokines can stimulate activation of this transcription factor. Thus, it appears that inflammatory cytokines use NF-κB to amplify their own signals.[32]

Activation of AP-1 also can be implicated in the induction of inflammatory genes. AP-1 is a family of basic domain/leucine zipper transcription factors that have been characterized for the specific binding to and transactivation through a *cis*-acting 12-*O*-tetradecanoyl phorbol-13-acetate (TPA) response element. AP-1 is composed of Jun and Fos gene products, which can form heterodimers (Jun/Fos) or homodimers (Jun/Jun). It was shown that c-Fos/c-Jun-binding activity toward AP-1 sites is regulated by the oxidative status of cysteine residues of c-Fos and c-Jun proteins (Fos Cys-154 and Jun Cys-272, respectively). Oxidation of cysteine residues can convert c-Fos and/or c-Jun into inactive forms. In contrast, a reduction of these residues can reactivate c-Fos/c-Jun-binding activity.[33] AP-1-binding sites were identified in the promoter regions of genes encoding for inflammatory cytokines such as IL-6[34] and adhesion molecules such as ICAM-1,[35] VCAM-1,[36] and E-selectin.

[31] P. A. Baeuerle, *Cell* **95,** 729 (1998).

[32] J. A. Berliner, M. Navab, A. M. Fogelman, J. S. Frank, L. L. Demer, P. A. Edwards, A. D. Watson, and A. J. Lusis, *Circulation* **91,** 2488 (1995).

[33] D. Gius, A. Botero, S. Shah, and H. A. Curry, *Toxicol. Lett.* **106,** 93 (1999).

[34] U. Dendorfer, P. Oettgen, and T. A. Libermann, *Mol. Cell. Biol.* **14,** 4443 (1994).

[35] B. G. Stade, G. Messer, G. Riethmuller, and J. P. Johnson, *Immunobiology* **182,** 79 (1990).

[36] M. Ahmad, P. Theofanidis, and R. M. Medford, *J. Biol. Chem.* **273,** 4616 (1998).

Evidence indicates that not only NF-κB and AP-1 but also other transcription factors may belong to the family of the transcription factors whose activity is regulated by the cellular redox status. For example, our studies indicate that SP-1 and STAT1α may be regulated by cellular oxidative status.[18,37]

The electrophoretic mobility shift assay (EMSA) is utilized to determine the binding interaction of transcription factors with their specific DNA sequences. This assay is a simple, relatively rapid, and very sensitive method to perform. EMSA is based on the principle that specific protein–DNA-binding complexes have higher molecular weight than unbound oligonucleotide probes and migrate slower during a nondenaturing polyacrylamide gel electrophoresis (PAGE). The binding specificity of the bands corresponding to the specific transcription factors is established using at least three different experimental approaches: (i) competition binding with the molar excess of unlabeled oligonucleotide probes, (ii) binding with mutant oligonucleotides, and (iii) supershift with antibodies against specific subunits of individual transcription factors.

Using EMSA, we demonstrated that treatment of cultured endothelial cells with specific free fatty acids resulted in an increase in NF-κB- or AP-1-binding activity as well as transactivation of these transcription factors. Figure 3A depicts the effects of linoleic acid on the binding activity of NF-κB in human endothelial cells. A slight endogenous activity of NF-κB is observed in untreated control cell cultures (Fig. 3A, lane 2). However, when cells are stimulated with linoleic acid or lipopolysaccharide (LPS; positive control), a significant increase of binding activity is detected (Fig. 3A, lanes 3–6). The DNA binding is specifically inhibited by an unlabeled competitor DNA containing the consensus NF-κB sequence (Fig. 3A, lane 7). Identities of the bands also can be confirmed by antibodies against specific NF-κB subunits, i.e., anti-p50 and anti-p65 (Fig. 3B, lanes 3 and 4).

Reagents, Buffers, and Equipment Required for EMSA

- PBS: 137 m*M* NaCl, 2.7 m*M* KCl, 8 m*M* Na_2HPO_4, 1.5 m*M* KH_2PO_4, pH 7.4
- Lysis buffer: 10 m*M* Tris–HCl, pH 8.0, 60 m*M* KCl, 1 m*M* EDTA, 1 m*M* dithiothreitol (DTT), 100 μM phenylmethylsulfonyl fluoride (PMSF), 0.1% NP-40
- Nuclear extract buffer: 20 m*M* Tris–HCl, pH 8.0, 420 m*M* NaCl, 1.5 m*M* $MgCl_2$, 0.2 m*M* EDTA, 25% glycerol
- Redivue adenosine 5′-[γ-^{32}P]triphosphate, triethylammonium salt, 10 mCi/ml (Amersham Pharmacia Biotech, Piscataway, NJ)
- 0.25 × TBE buffer: 50 m*M* Tris–Cl, 45 m*M* boric acid, 0.5 m*M* EDTA, pH 8.4
- V16 vertical gel electrophoresis apparatus, 0.8-mm-thick gel (Life Technologies, Gaithersburg, MD) or similar instrument
- X-OMAT AR Kodak autoradiography film (Kodak, Rochester, NY)

[37] Y. W. Lee and M. Toborek, unpublished observation.

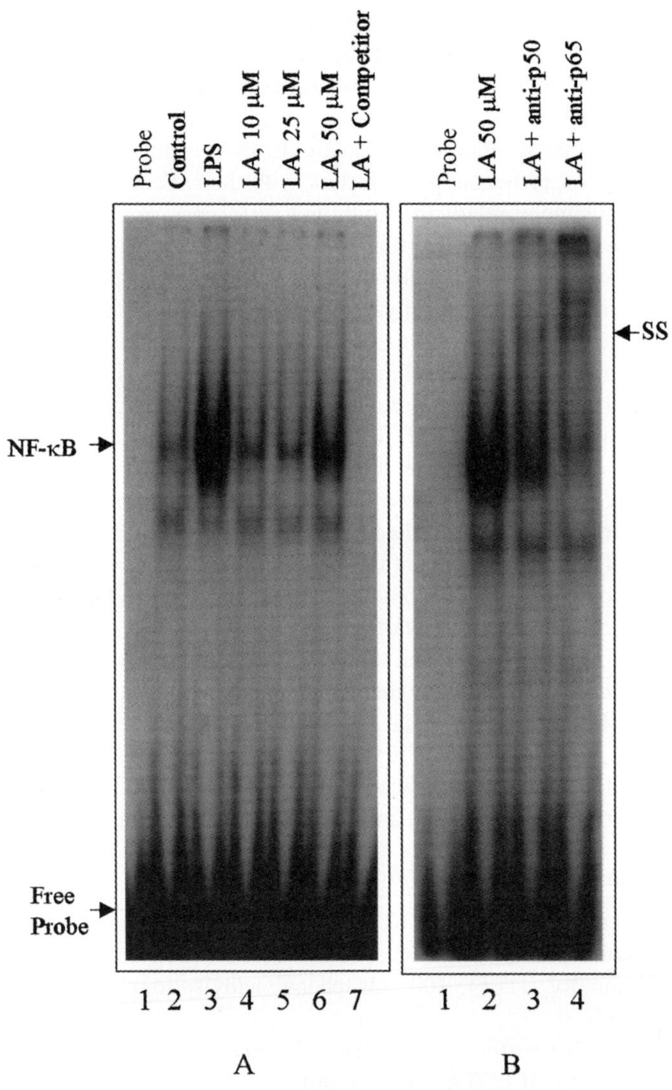

FIG. 3. (A) Linoleic acid (LA) treatment enhances NF-κB binding in human endothelial cells as analyzed by EMSA. Endothelial cells were either untreated (lane 2) or treated for 2 hr with increasing doses of linoleic acid (lanes 4–6). A competition study was performed by the addition of excess unlabeled oligonucleotide (lane 7) using nuclear extracts from cells treated with 50 μM linoleic acid. Lane 1, probe alone; lane 3, LPS (1 μg/ml, positive control). (B) Supershift analysis of LA-induced NF-κB-binding activity in human endothelial cells. Nuclear extracts were prepared from cells treated with 50 μM linoleic acid for 2 hr (lanes 2–4) and incubated with anti-p50 antibody (lane 3) or anti-p65 antibody (lane 4) for 25 min before the addition of the ^{32}P-labeled probe. Lane 1, probe alone. SS indicates bands shifted by specific antibodies.

Procedure

Isolation of Nuclear Extracts. All steps are performed on ice unless otherwise specified.[38]

1. Endothelial cell cultures (at least 6.0×10^6 cells per group) are treated with fatty acids for 0.5–6 hr, trypsinized, and collected by centrifugation at 2,500 rpm for 4 min at 4°. The pellet is washed once with PBS.
2. The cells are lysed in 1 ml of lysis buffer for 5 min on ice and centrifuged at 2500 rpm for 4 min at 4° to collect nuclei. Then, the nuclear pellets are washed with 1 ml of lysis buffer without NP-40.
3. The nuclear pellets are lysed in 100 μl of nuclear extract buffer for 10 min on ice and centrifuged at 14,000 rpm for 15 min at 4°.
4. Supernatants that contain nuclear extracts are frozen immediately in liquid nitrogen. Then, they can be transferred to a −80° freezer and stored for 2 weeks.

5′-End Labeling of Oligonucleotides with [γ-^{32}P]ATP

1. Double-stranded oligonucleotide probes are labeled with [γ-^{32}P]ATP using bacteriophage T4 polynucleotide kinase. The reaction mixture consists of 70 m*M* Tris–HCl, pH 7.6, 10 m*M* $MgCl_2$, 5 m*M* DTT, 5 pmol of double-stranded oligonucleotides, 30 μCi of [γ-^{32}P]ATP, and 20 units of T4 polynucleotide kinase (Promega, Madison, WI) in a total volume of 20 μl. The reaction mixture is incubated for 1 hr at 37°.
2. Following incubation, T4 polynucleotide kinase is inactivated by placing the tube on a heat block for 10 min at 68°.
3. Unincorporated nucleotides are removed by gel-filtration chromatography using mini Quick Spin Oligo columns (Boehringer Manheim Corporation, Indianapolis, IN).

Binding Reaction and Electrophoresis

1. Binding reactions are performed in a 20-μl volume containing 4–10 μg of nuclear protein extracts, 10 m*M* Tris–Cl, pH 7.5, 50 m*M* NaCl, 1 m*M* EDTA, 0.1 m*M* DTT, 10% glycerol, and 2 μg of poly[dI-dC], which is used as a nonspecific competitor. After adding the reagents, the mixture is incubated for 25 min at room temperature.
2. Then, 40,000 cpm of the ^{32}P-labeled specific oligonucleotide probe is added, and the binding mixture is incubated for 25 min at room temperature. Competition studies and supershift experiments are performed by the addition of a molar excess

[38] A. A. Beg, T. S. Finco, P. V. Nantermet, and A. S. Baldwin, Jr., *Mol. Cell. Biol.* **13,** 3301 (1993).

of unlabeled oligonucleotide probes or antibodies against specific transcription factors to the binding reaction.

3. Resultant protein–DNA complexes are electrophoresed on a nondenaturing 5% polyacrylamide gel (prerun for 2 hr at 150 V) using $0.25 \times$ TBE buffer for 2 hr at 150 V.

4. The gel is transferred to Whatman 3MM paper, dried on a gel dryer, and exposed to X-ray film overnight at $-70°$ with an intensifying screen.

Fatty Acid-Induced Inflammatory Genes in Endothelial Cells

Background

Fatty acid-induced inflammatory reactions in endothelial cells are mediated by the production of chemokines (e.g., monocyte chemoattractant protein-1; MCP-1), inflammatory cytokines (e.g., TNF-α), and adhesion molecules (e.g., ICAM-1 or VCAM-1).[39] Expression of these inflammatory mediators and their effects is closely interrelated. For example, ICAM-1 and VCAM-1 facilitate leukocyte adhesion to the vascular endothelium, and both MCP-1 and, to a lesser extent, TNF-α are potent chemoattractive factors, which play a significant role in recruiting lymphocytes and monocytes into the vessel wall.[40,41] In addition, TNF-α is a strong inducer of inflammatory reactions and can stimulate overexpression of MCP-1 and inflammatory cytokines, as well as ICAM-1 and VCAM-1.[42] We have obtained evidence that selective dietary fatty acids can induce expression of the inflammatory genes, such as ICAM-1, VCAM-1, MCP-1, or TNF-α in endothelial cells.[39] Reverse transcriptase polymerase chain reaction (RT-PCR) with specific primer pairs (Table I) is a very suitable experimental technique to perform these analyses.

Figure 4 depicts RT-PCR analysis of the effects of treatment with selected unsaturated fatty acids for 3 hr on MCP-1 gene expression in HUVEC. Among tested fatty acids, linoleic acid stimulated the most pronounced overexpression of the MCP-1 gene. Indeed, expression of this gene in endothelial cells treated with 90 μM linoleic acid was in the range of that observed in cells exposed to 20 ng/ml of TNF-α, which was used as a positive control. Expression of the MCP-1 gene was also increased in endothelial cells treated with linolenic acid. In contrast, expression of this gene in endothelial cells exposed to oleic acid appeared to be within or even below the control range observed in nonstimulated endothelial cells.

[39] M. Toborek, Y. W. Lee, R. Garrido, S. Kaiser, and B. Hennig, *Am. J. Clin. Nutr.* **75,** 119 (2002).

[40] N. W. Lukacs, R. M. Strieter, V. Elner, H. L. Evanoff, M. D. Burdick, and S. L. Kunkel, *Blood* **86,** 2767 (1995).

[41] R. M. Strieter, R. Wiggins, S. H. Phan, B. L. Wharram, H. J. Showell, D. G. Remick, S. W. Chensue, and S. L. Kunkel, *Biochem. Biophys. Res. Commun.* **162,** 694 (1989).

[42] J. S. Pober, *Pathol. Biol.* **46,** 159 (1998).

TABLE I
SEQUENCES OF PRIMER PAIRS EMPLOYED IN RT-PCR REACTIONS

Studied inflammatory gene[a]	Sequences of the primer pairs (5′-3′)
MCP-1[b]	Forward: CAG CCA GAT GCA ATC AAT GC
	Reverse: GTG GTC CAT GGA ATC CTG AA
TNF-α[b]	Forward: 5′-GTG ACA AGC CTG TAG CCC A-3′
	Reverse: 5′-ACT CGG CAA AGT CGA GAT AG-3′
ICAM-1	Forward: GGT GAC GCT GAA TGG GGT TCC
	Reverse: GTC CTC ATG GTG GGG CTA TGT CTC
VCAM-1[c]	Forward: ATG ACA TGC TTG AGC CAG G
	Reverse: GTG TCT CCT TCT TTG ACA CT
β-Actin (a housekeeping gene)[d]	Forward: AGC ACA ATG AAG ATC AAG AT
	Reverse: TGT AAC GCA ACT AAG TCA TA

[a] ICAM-1, intercellular adhesion molecule-1; MCP-1, monocyte chemoattractant protein-1; TNF-α, tumor necrosis factor-α; VCAM-1, vascular cell adhesion molecule-1.
[b] Primer pairs purchased from R&D Systems (Minneapolis, MN).
[c] L. Meagher, D. Mahiouz, K. Sugars, N. Burrows, P. Norris, H. Yarwood, M. Becker-Andre, and D. O. Haskard, *J. Immunol. Methods* **175,** 237 (1994).
[d] A. Ballester, A. Velasco, R. Tobena, and S. Alemany, *J. Biol. Chem.* **273,** 14099 (1998).

Fatty acid-mediated alterations of other inflammatory genes result in the same or similar pattern of changes.

The RT-PCR technique can be divided into three steps: (A) isolation of total RNA, (B) reverse transcription, and (C) polymerase chain reaction. For the isolation of high purity total RNA from cell cultures, TRI REAGENT is utilized. TRI REAGENT is a mixture of guanidine thiocyanate and phenol in a monophase solution, which effectively dissolves DNA, RNA, and protein in cell lysates. Then, isolated total RNA is utilized for the reverse transcription and polymerase chain reaction amplification of a specific target RNA. AMV reverse transcriptase, RNA-dependent DNA polymerase from the avian myeloblastosis virus, is utilized to synthesize the first single-stranded cDNA from isolated RNA. In addition, *Taq* DNA polymerase, a thermostable DNA-dependent DNA polymerase from *Thermus aquaticus,* is used to synthesize second strand cDNA and for DNA amplification.

Reagents and Equipment

TRI REAGENT (Sigma, St. Louis, MO)
Nuclease-free water (Promega, Madison, WI)
Reverse transcription system (Promega)
Taq PCR master mix kit (Qiagen, Valencia, CA)
GeneAmp PCR System 9700 (The PerkinElmer Corporation, Norwalk, CT) or similar instrument

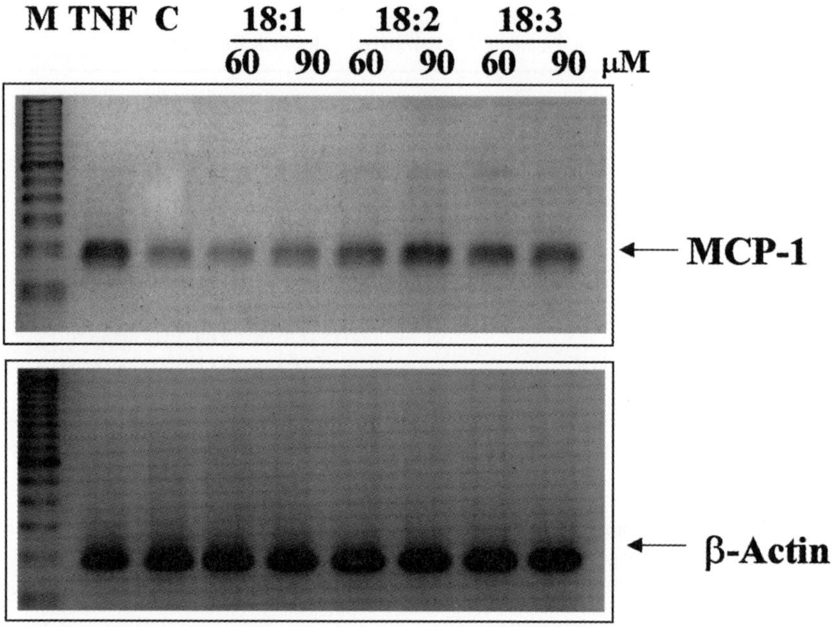

FIG. 4. Concentration-dependent upregulation of MCP-1 mRNA expression in human endothelial cells exposed to specific dietary fatty acids. Total cDNA was synthesized from 1 μg of cellular RNA isolated from HUVEC stimulated with 60 and 90 μM fatty acids or TNF-α (20 ng/ml, positive control) for 3 hr. Amplified PCR products were electrophoresed on a 2% TBE agarose gel, stained with SYBR Green I (Molecular Probes, Eugene, OR), and visualized using phosphoimaging technology (FLA-2000, Fuji, Stamford, CN). M, molecular weight markers (100-bp DNA ladder); 18 : 1, oleic acid; 18 : 2, linoleic acid; 18 : 3, linolenic acid.

Agarose, ultrapure (Life Technologies, Gaithersburg, MD)
SYBR Green I (Molecular Probes, Inc., Eugene, OR)
Horizontal gel electrophoresis system (e.g., Horizon 11.14, Life Technologies) or similar instrument
Phosphorimager (Fuji FLA-5000, Stamford, CN) or similar system

Procedure

Isolation of Total RNA for RT-PCR

1. Lysis of cultured monolayer cells. Endothelial cell cultures (3.0×10^6 cells per 100-mm culture dish) are treated with fatty acids, washed with PBS, and lysed by the direct addition of 1 ml of TRI REAGENT to culture dishes. The cell lysate is passed several times through a pipette to form a homogeneous lysate. The homogenate is centrifuged at 12,000g for 10 min at 4° to remove insoluble

material such as extracellular membranes, polysaccharides, and high molecular weight DNA. After centrifugation, the clear supernatant is transferred to a fresh tube and allowed to stand for 5 min at room temperature.

2. Phase separation. Chloroform (0.2 ml) is added to the cell lysate, shaken vigorously for 15 sec, and allowed to stand for 10 min at room temperature. Then, the resulting mixture is centrifuged at 12,000*g* for 15 min at 4°. Centrifugation separates the mixture into the following three phases: a red lower organic phase (protein part), a white interphase (DNA part), and a colorless upper aqueous phase (RNA part). The upper phase is transferred to a fresh tube. This is done very carefully in order not to disrupt the interphase or the lower phase.

3. Isolation of total RNA. Isopropanol (0.5 ml) is added to the tube containing the transferred upper aqueous phase (RNA part), mixed gently by inverting several times, and allowed to stand for 10 min at room temperature. The mixture is centrifuged at 12,000*g* for 10 min at 4° to precipitate RNA to the side and bottom of the tube. The supernatant is discarded and the white RNA pellet is washed with 1 ml of 75% ethanol. The mixture is vortexed briefly and centrifuged at 7500*g* for 5 min at 4°. Then, the supernatant is very carefully discarded and the RNA pellet is air dried for 10 min. Approximately 20–40 μl of nuclease-free water is added to the RNA pellet and mixed by repeated pipetting at 55° for 10 min to facilitate dissolution. Finally, the sample is incubated for 5 min at 70° and cooled quickly on ice. This procedure prevents RNA from possibly forming secondary structures, which may interfere with reverse transcription. The concentration and purity of total RNA are determined spectrophotometrically by measuring the absorbance at 260 nm (one absorbance unit at 260 nm equals 40 μg RNA/ml) and 280 nm. The ratio between the absorbance at 260 and 280 nm reflects RNA purity and should be $\geq$1.7. The sample concentration of RNA is adjusted to the final level of 1 μg RNA/μl.

Reverse Transcription (RT). One microgram of RNA, isolated from endothelial cells as described earlier, is reverse transcribed at 42° for 60 min, followed by a 5-min incubation at 99° and immediately cooled on ice. The complete reaction mixture for reverse transcription consists of 1 μg of isolated total RNA, 5 m*M* $MgCl_2$, 10 m*M* Tris–HCl, pH 9.0, 50 m*M* KCl, 0.1% Triton X-100, 1 m*M* dNTP, 1 unit/μl of recombinant RNasin ribonuclease inhibitor, 15 units/μg of AMV reverse transcriptase, and 0.5 μg of oligo$(dT)_{15}$ primer. The total volume of this mixture is 20 μl, adjusted with distilled water.

Polymerase Chain Reaction and Agarose Gel Electrophoresis. To perform PCR amplification, 2 μl of the reverse transcriptase reaction is mixed with a *Taq* PCR master mix kit and 20 pmol of primer pairs in a total volume of 50 μl, adjusted with distilled water. Table I depicts sequences of the primer pairs used for PCR amplification of most common human inflammatory genes in endothelial cells. Expression of β-actin is determined as a housekeeping gene. Expression of the studied genes is determined in the linear range of PCR amplification, specific for an individual PCR reaction. PCR products are separated by 2% agarose gel electrophoresis, stained with SYBR Green I and visualized using phosphoimaging technology (FLA-2000). For quantitation, intensity of the band corresponding to the specific inflammatory gene is related to the intensity of the band, which reflects expression of the β-actin mRNA.

Fatty Acid-Induced Protein Expression of Inflammatory Mediators in Endothelial Cells

Protein expression of the inflammatory mediators is determined using either flow cytometry or ELISA. Flow cytometry is employed to assay the expression of adhesion molecules, such as ICAM-1 or VCAM-1, which are present on the endothelial cell surface. FITC-labeled specific monoclonal antibodies (e.g., from R&D, Minneapolis, MN, or BD PharMingen, San Diego, CA) appear to be most suitable for this method. A protocol to determine the expression of adhesion molecules in fatty acid-treated endothelial cells by flow cytometry is given later.

ELISA kits, available commercially from several companies (e.g., R&D, Minneapolis, MN, or Amersham Pharmacia Biotech, Piscataway, NJ), allow for quantitative determination of soluble inflammatory mediators, such as chemokines (MCP-1) or inflammatory cytokines (e.g., TNF-α, IL-1β, IL-6, or IL-8) in culture media. Adhesion molecules shed from the surface of endothelial cells can also be present in cell culture media and can be assayed by commercially avaliable ELISA kits. However, determinations of soluble ICAM-1 or soluble VCAM-1 are more frequent in clinical studies, where serum levels of these molecules can serve as a marker of adhesion molecule expression.

Flow Cytometry Procedure to Detect Expression of Adhesion Molecules

Endothelial cells are cultured on six-well plates, grown to confluence, and treated with fatty acids for 12- or 24 hr. Cells are washed with Hanks' solution and harvested gently by trypsin/EDTA. It is important to note that excess and repeated exposure to trypsin can damage the endothelial cell surface and interfere with the results. Then, endothelial cells are washed twice with ice-cold PBS and incubated for 1 hr on ice with saturating amounts of specific monoclonal antihuman antibody labeled with FITC. FITC-labeled anti-human IgG is used as the isotype control. After incubation with antibodies, samples are washed twice with ice-cold PBS, suspended in 200 μl PBS, and analyzed with 10,000 cells per sample in a fluorescence-activated cell sorter (Becton Dickinson, San Jose, CA). Following correction for unspecific binding (isotype control), the intensity of fluorescence or the percentage of positively stained cells can be utilized as the indicator of adhesion molecule protein expression.

Transient Transfection of Endothelial Cells and Dual Luciferase Reporter Gene Assay

Transfections (physical or chemical methods of introducing foreign DNA into eukaryotic cells) and reporter gene assays provide powerful experimental tools to study mechanisms of gene regulation. During transient transfections, plasmid DNA is introduced into a cell population, and expression of the reporter gene is studied shortly after the transfection procedure, usually within 24–72 hr. Transfection methods include calcium–phosphate precipitation, electroporation, detergent–DNA complexes, DNA–DEAE complexes, microinjection, virus-mediated transfection, introduction of DNA via particle bombardment, and lipid-mediated transfection. In transfections performed *in vitro* in cultured cells, cationic lipids have become standard carriers of plasmid DNA. This method takes advantage of the associations of negatively charged DNA with positively charged liposomes to form a lipid–DNA complex, which can be introduced into cells relatively easily.

Endothelial cells, in general, are difficult to transfect. This may be related to the fact that these cells represent a physiologic barrier against invasion of the vessels and underlying tissues by exogenous substances. However, under carefully controlled experimental conditions, liposome-mediated transfection can be suitable to study the transactivation of redox-regulated transcription factors and mechanisms of expression of the inflammatory genes in fatty acid-treated human endothelial cells. We have optimized transfection conditions in human endothelial cells to achieve high-efficiency transient transfections.[43]

Firefly luciferase has been recognized to be the reporter gene of choice for transfection studies in cells resistant to the uptake of foreign DNA, such as endothelial cells.[44] The transgene is simple to measure and has no background levels in human tissues. Determination of luciferase activity also has the advantage of being several orders of magnitude more sensitive than other common reporter gene assays, such as activities of chloramphenicol acetyltransferase, β-galactosidase, or alkaline phosphatase. To correct for variations in transfection efficiency, a cotransfection with an internal control plasmid should be performed. β-Galactosidase or *Renilla* luciferase expression vectors are examples of plasmids that can be used as internal controls. When cells are transfected with constructs encoding for firefly luciferse as a targeted reporter gene and cotransfected with the *Renilla* luciferase expression vector, both a dual luciferase reporter assay system and a luminometer with double injector are required to perform transgene measurements. This system takes advantage of the fact that firefly luciferase and *Renilla* luciferase have distinct enzyme structures and substrate specificity. Thus, their activities can be sequentially measured in the same sample.

[43] S. Kaiser and M. Toborek, *J. Vasc. Res.* **38,** 133 (2001).

[44] L. Alam and J. L. Cook, *Anal. Biochem.* **188,** 245 (1990).

Reagents and Equipment

Renilla luciferase expression plasmid used as internal control (pRL-SV40 control vector, Promega, Madison, WI)
Dual-Luciferase reporter assay system (Promega, Madison, WI)
Luminometer fitted with two reagent injectors (TD-20/20, Turner Designs, Sunnyvale, CA, or similar model)

Procedure

Endothelial Cell Cultures. Endothelial cells are seeded on six-well plates (1.0–2.0 $\times$ 10^5 cells per well) and cultured for 24–48 hr until the cultures reach ~60% confluence.

Preparation of the Transfection Solution (1 ml per each well)

1. Specific firefly luciferase reporter plasmid (5–10 μg/ml) and *Renilla* luciferase expression plasmid (0.25–0.5 μg/ml, pRL-SV40 control vector) are mixed thoroughly with serum-free medium in a sterile tube in a volume of 0.5 ml/well.
2. In a separate sterile tube, a cationic liposome, such as pFx-7 (36 μg/ml) or DMRIE-C (40 μg/ml), is mixed thoroughly with serum-free medium in a volume of 0.5 ml/well.
3. The plasmid and liposome solutions (prepared in steps 1 and 2, respectively) are combined in a single tube, mixed thoroughly, and incubated for 30 min at 37° to allow the formation of liposome/DNA complexes.

Transient Transfection Procedure

1. Growth medium is removed from endothelial cell cultures and the monolayers are washed three times with serum-free medium.
2. The transfection solution (prepared as described earlier) is added to each well in a volume of 1 ml/well. Plates are returned to the cell culture incubator, and endothelial cells are transfected for 1.5 hr.
3. The transfection solution is removed carefully from each well, the cells are overlaid gently with 2 ml of normal growth medium, and returned to the incubator for 24 hr at 37°.

Treatment of Transfected Cells with Fatty Acids and Determination of Dual Luciferase Activity

1. Cultures are washed with Hanks' solution, and 2 ml of experimental media enriched with fatty acids is added into each well of the six-well plate. Cells are treated with fatty acids for 16–24 hr.
2. After incubation time, the cells are washed twice with PBS and lysed in 500 μl passive lysis buffer while shaking the plates for 15 min at room temperature.

3. Cell lysates are centrifuged at 12,000 rpm for 1 min to remove cell debris, and cell lysates are transferred to fresh tubes.

4. Firefly and *Renilla* luciferase activities are determined in 10–20 μl of cell lysates using a luminometer with a dual injector system. Injector #1 is set up to deliver 100 μl/tube of luciferase assay reagent II for determination of firefly luciferase activity, and injector #2 is set up to deliver 100 μl/tube of Stop & Glo reagent for determination of *Renilla* luciferase activity.

5. Relative luciferase activity is calculated as the ratio of firefly luciferase activity to *Renilla* luciferase activity.

Conclusions

Using the described methods, we determined that treatment of human endothelial cells with selected dietary fatty acids can induce oxidative stress, decrease cellular glutathione content, activate redox-responsive transcription factors, and induce expression of the inflammatory mediators, such as MCP-1, inflammatory cytokines (IL-6, IL-8, and TNF-α), and adhesion molecules (ICAM-1 and VCAM-1). However, the effects exerted by dietary fatty acids were highly specific. Among studied fatty acids, treatment with linoleic acid induced the most significant oxidative stress, alterations of cellular redox status, and induction of inflammatory genes.[9,24,38] The proinflammatory effects of linolenic acid were less pronounced. In contrast, exposure of human endothelial cells to oleic acid diminished expression of the inflammatory genes. Those results demonstrate that specific unsaturated dietary fatty acids, such as linoleic acid, the parent omega-6 fatty acid, which also is a major fatty acid in common vegetable oils, can stimulate inflammatory responses in vascular endothelial cells. These proinflammatory effects of selected fatty acids illustrate the significance of dietary lipids in the development, progression, or prevention of chronic vascular diseases, such as atherosclerosis.

Acknowledgments

This work was supported in part by grants from NRICGP/USDA, DOD, NIH/NINDS, NIH/NIEHS, and AHA, Ohio Valley Affiliate.

[19] Redox Control of Tissue Factor Expression in Smooth Muscle Cells and Other Vascular Cells

By OLAF HERKERT and AGNES GÖRLACH

Reactive oxygen species (ROS) have been demonstrated as novel signaling molecules in a variety of cell types. It has become clear that many cells, including vascular cells, are able to generate ROS at a basal level and in an inducible manner. Increased ROS production has been related to various diseases. Thus, the understanding of ROS-generating enzymes and the modulation of gene expression by redox-sensitive signaling pathways have gained increasing interest. Several enzymes, in addition to mitochondria, can produce ROS, including isoforms of the NADPH oxidase, a superoxide-generating enzyme originally identified in phagocytes. Many growth factors and cytokines, as well as physicochemical stress, stimulate ROS production and subsequently activate redox-sensitive signaling pathways. A role for ROS in the control of coagulation has been suggested, and the key activator of the extrinsic coagulation cascade, tissue factor (TF), has been shown to be modulated by ROS in several cell types, including vascular cells. This chapter presents some evidence for redox-sensitive signaling cascades involved in the regulation of TF in smooth muscle cells and other vascular cells.

Introduction

Reactive oxygen species have long been considered as unwanted by products of aerobic mitochondrial metabolism or other electron transfer reactions when dioxygen is not completely reduced to H_2O. Superoxide, H_2O_2, and hydroxyl radicals are widespread ROS, which can give rise to more toxic species. To limit cellular damage, counteracting mechanisms have been evolved, including antioxidative enzymes such as superoxide dismutase (SOD) and catalase.

In recent years, however, it has become evident that most cell types possess the ability to generate ROS at a basal level and in an inducible manner. Particular interest has been centered toward the role of ROS in the vasculature, as many cardiovascular diseases, among them atherosclerosis and hypertension, have been associated with increased ROS production.[1]

A number of enzymes, including NADPH oxidases, cyclooxygenases, xanthine oxidase, and lipoxygenases, have been identified in vascular cells to contribute to inducible ROS production in response to a variety of agonists and/or to

[1] T. Chakraborti, S. K. Ghosh, J. R. Michael, S. K. Batabyal, and S. Chakraborti, *Mol. Cell. Biochem.* **187,** 1 (1998).

0076-6879/02 $35.00

physicochemical stress, including shear stress and hypoxia.[2] This ROS production appears to be involved in signaling pathways, and there is evidence that ROS serve as second messengers. Modulation of ROS production has been shown to regulate a variety of genes, most of them linked to proliferation, migration, growth, and development, as well as to inflammation and chemotaxis.[3] ROS have also been shown to play a role in the control of coagulation by promoting a procoagulant state. *In vivo* models indicated the potent effects of ROS on platelet aggregation and thrombosis.[4] The key activator of the extrinsic coagulation cascade, tissue factor, is sensitive to increased levels of ROS, thus possibly explaining ROS-sensitive activation of the coagulation cascade.[4]

This chapter discusses current knowledge of redox-sensitive signaling cascades which may be involved in the regulation of TF.

Activation of Tissue Factor

Tissue factor is a 47-kDa single transmembrane glycoprotein that acts as the primary connecting link between vascular cells or mononuclear cells and the hemostatic system by initiating the extrinsic pathway of the blood coagulation cascade.[5] TF is a key determinant of hemostatic and thrombotic responses and plays a pivotal role in the procoagulant activity of disrupted atherosclerotic plaques or of acutely injured arteries as a consequence of balloon angioplasty or coronary atherectomy.[6–9]

TF initiates blood coagulation by binding factor VII/VIIa with high affinity. The resulting complex promotes the activation of factors IX and X with subsequent thrombin generation. Thrombin catalyzes the conversion of fibrinogen to fibrin, thus triggering rapid fibrin deposition and clot formation.[6]

An increased expression of TF is not inevitably associated with increased biological activity of TF. Functional TF is dependent on the expression of a biologically active form on the cell surface. In smooth muscle cells (SMC) and monocytes, only 10–20% of total cellular TF is available on the surface and reflects

[2] K. K. Griendling, D. Sorescu, and M. Ushio-Fukai, *Circ. Res.* **86,** 494 (2000).

[3] C. Kunsch and R. M. Medford, *Circ. Res.* **85,** 753 (1999).

[4] J. Ruef, Z. Y. Hu, L. Y. Yin, Y. Wu, S. R. Hanson, A. B. Kelly, L. A. Harker, G. N. Rao, M. S. Runge, and C. Patterson, *Circ. Res.* **81,** 24 (1997).

[5] E. W. Davie, K. Fujikawa, and W. Kisiel, *Biochemistry* **30,** 10363 (1991).

[6] Y. Nemerson, *Blood* **71,** 1 (1988).

[7] J. N. Wilcox, K. M. Smith, S. M. Schwartz, and D. Gordon, *Proc. Natl. Acad. Sci. U.S.A.* **86,** 2839 (1989).

[8] J. D. Marmur, S. V. Thiruvikraman, B. S. Fyfe, A. Guha, S. K. Sharma, J. A. Ambrose, J. T. Fallon, Y. Nemerson, and M. B. Taubman, *Circulation* **94,** 1226 (1996).

[9] B. H. Annex, S. M. Denning, K. M. Channon, M. H. Sketch, R. S. Stack, J. H. Morrissey, and K. G. Peters, *Circulation* **91,** 619 (1995).

the biological activity, whereas the remainder is contained in intracellular pools (~30%) and as latent surface TF (50–60%).[10,11]

The transmembrane human TF protein consists of 263 amino acids, 219 of which form the large extracellular domain and 23 are integrated membraneously; the cytoplasmic tail is only 21 amino acid residues in length. The extracellular part of TF is responsible for factor VIIa-mediated proteolytic signaling. On zymogen activation, factor VIIa remains in a state of very low catalytic activity. TF acts as a cofactor and supports specific conformational transitions of factor VIIa that "switch on" the protease domain to become catalytically active. Thus, factor VIIa cannot be considered an active enzyme, unless bound to TF as cofactor, and factor VIIa-mediated signaling is strictly dependent on TF to maintain its catalytic function.[12,13]

Expression of Tissue Factor

In normal adult blood vessels, little or no TF is constitutively expressed in the intima or media, whereas TF is abundant in adventitial fibroblasts. Quiescent endothelial cells *in vivo* do not express TF, possibly due to promoter elements in the TF gene that repress its transcription under basal conditions.[14,15] In addition, TF is not constitutively expressed in peripheral blood monocytes and macrophages.[11,16,17] Thus, cellular initiation of TF-dependent blood coagulation seems to require induction of TF expression in cells, which are under normal or pathophysiological conditions in contact with plasmatic factors of hemostasis.

Induction of TF has been shown to play a central role in several diseases. In sepsis, TF expression by monocytes and endothelial cells leads to life-threatening disseminated intravascular coagulation.[18–20] On the surface of monocytes/macrophages and endothelial cells, TF expression is highly inducible by a variety of agents, including phorbol esters, which are activators of protein kinase C, stimuli or mediators of inflammation such as bacterial lipopolysaccharide/endotoxin,

[10] K. T. Preissner, P. P. Nawroth, and S. M. Kanse, *J. Pathol.* **190,** 360 (2000).

[11] A. D. Schecter, P. L. Giesen, O. Taby, C. L. Rosenfield, M. Rossikhina, B. S. Fyfe, D. S. Kohtz, J. T. Fallon, Y. Nemerson, and M. B. Taubman, *J. Clin. Invest.* **100,** 2276 (1997).

[12] W. Ruf and B. M. Mueller, *Thromb. Haemost.* **82,** 175 (1999).

[13] C. D. Dickinson and W. Ruf, *J. Biol. Chem.* **272,** 19875 (1997).

[14] T. A. Drake, J. H. Morrissey, and T. S. Edgington, *Am. J. Pathol.* **134,** 1087 (1989).

[15] H. Holzmuller, T. Moll, R. Hofer-Warbinek, D. Mechtcheriakova, B. R. Binder, and E. Hofer, *Arterioscler. Thromb. Vasc. Biol.* **19,** 1804 (1999).

[16] T. S. Edgington, N. Mackman, K. Brand, and W. Ruf, *Thromb. Haemos.* **66,** 67 (1991).

[17] P. Oeth, G. C. Parry, and N. Mackman, *Arterioscler. Thromb. Vasc. Biol.* **17,** 365 (1997).

[18] C. E. Hack, *Crit. Care Med.* **28,** S25 (2000).

[19] C. T. Esmon, K. Fukudome, T. Mather, W. Bode, L. M. Regan, D. J. Stearns-Kurosawa, and S. Kurosawa, *Haematologica* **84,** 254 (1999).

[20] T. A. Drake, J. Cheng, A. Chang, and F. B. Taylor, *Am. J. Pathol.* **142,** 1458 (1993).

interleukin-1β (IL-1β), and tumor necrosis factor-α (TNF-α), as well as clotting factors, such as thrombin.[20–24]

In monocytes/macrophages and endothelial cells of human atherosclerotic plaques, as well as in the subendothelium at sides of vascular injury, following balloon angioplasty for example, TF expression and activity are induced in abundance.[10,12,25,26]

In animal models of balloon injury, TF expression increases rapidly in medial SMC after endothelial denudation and is markedly accumulated in SMC of the developing neointima. Thus, induction of TF in SMC appears to be primarily responsible for the prolonged vascular procoagulant activity after endothelial denudation *in vivo*.[27] These effects may be mediated by the increased availability of mitogens, growth factors, vasoactive agonists, and clotting factors under these conditions. It has been shown that TF expression is rapidly and markedly induced in cultured SMC by platelet-derived growth factor (PDGF), basic fibroblast growth factor (bFGF), transforming growth factor-β (TGF-β), epidermal growth factor (EGF), angiotensin II, thrombin, and activated platelets.[28–30] Moreover, the TF gene is also induced by serum, phorbol esters, and the Ca^{2+} ionophore ionomycin in SMC.[28]

Similarly, increased TF expression is found in foam cells as well as in SMC in atherosclerotic plaques and plays an important role in determining their thrombogenicity.[31,32]

Mediation and Regulation of Tissue Factor Expression

An immediate response to many TF-stimulating agonists is the activation of phospholipase C resulting in the generation of the second messengers,

[21] E. M. Scarpati and J. E. Sadler, *J. Biol. Chem.* **264,** 20705 (1989).

[22] G. C. Parry, J. H. Erlich, P. Carmeliet, T. Luther, and N. Mackman, *J. Clin. Invest.* **101,** 560 (1998).

[23] M. P. Bevilacqua, J. S. Pober, G. R. Majeau, R. S. Cotran, and M. A. Gimbrone, *J. Exp. Med.* **160,** 618 (1984).

[24] A. Bierhaus, Y. Zhang, Y. Deng, N. Mackman, P. Quehenberger, M. Haase, T. Luther, M. Müller, H. Böhrer, J. Greten, E. Martin, P. A. Baeuerle, R. Waldherr, W. Kisiel, R. Ziegler, D. M. Stern, and P. P. Nawroth, *J. Biol. Chem.* **270,** 26419 (1995).

[25] J. D. Marmur, M. Rossikhina, A. Guha, B. Fyfe, V. Friedrich, M. Mendlowitz, Y. Nemerson, and M. B. Taubman, *J. Clin. Invest.* **91,** 2253 (1993).

[26] J. D. Marmur, M. Poon, M. Rossikhina, and M. B. Taubman, *Circulation* **86,** 11153 (1992).

[27] C. M. Speidel, P. R. Eisenberg, W. Ruf, T. S. Edgington, and D. R. Abendschein, *Circulation* **92,** 3323 (1995).

[28] M. B. Taubman, J. D. Marmur, C. L. Rosenfield, A. Guha, S. Nichtberger, and Y. Nemerson, *J. Clin. Invest.* **91,** 547 (1993).

[29] N. Mackman, *Thromb. Haemost.* **78,** 747 (1997).

[30] C. A. McNamara, I. J. Sarembock, L. W. Gimple, J. W. Fenton, S. R. Coughlin, and G. K. Owens, *J. Clin. Invest.* **91,** 94 (1993).

[31] M. B. Taubman, J. T. Fallon, A. D. Schecter, P. Giesen, M. Mendlowitz, B. S. Fyfe, J. D. Marmur, and Y. Nemerson, *Thromb. Haemost.* **78,** 200 (1997).

[32] N. Mackman, *Front. Biosci.* **6,** D208 (2001).

diacylglycerol (DAG) and inositol triphosphate (IP_3). DAG activates protein kinase C, whereas IP_3 mobilizes Ca^{2+} from intracellular stores. The induction of TF expression and activity in SMC is dependent on the mobilization of intracellular Ca^{2+}. Moreover, the increased TF activity also seems to be attributed in part to a process of deencryption through changing of TF accessibility on the cell surface caused by increased intracellular Ca^{2+}, exposing previously encrypted and inactive TF molecules.[33]

At the transcriptional level, basal TF expression is controlled by the transcription factor Sp1, whereas inducible expression is regulated by c-Fos/c-Jun, c-Rel/p65, and Egr-1.[29] These transcription factors have been shown to be activated by ROS.[34,35]

Redox-Regulated Tissue Factor Expression

TF has been recognized as a redox-sensitive gene in several cell types, including monocytes, endothelial cells, and SMC, by responding to externally applied oxidant stress generated by H_2O_2 or xanthine/xanthine oxidase (X/XO) (Table I).

Application of H_2O_2 at an optimal concentration of 500 μM or exposure to X/XO stimulated TF activity in monocytes measured after 20 hr.[36] In endothelial cells, exposure to X/XO for only 5 min resulted in enhanced TF mRNA expression and TF activity.[37] Similarly, a flux of X/XO for 2.5 min induced TF activity in isolated rabbit hearts. TF activity was significantly upregulated in rabbit hearts subjected to 20 min of ischemia followed by 2 hr of reperfusion,[37] whereas administration of SOD at the moment of reperfusion decreased TF activity in these hearts, indicating that ROS contributed to TF activation in response to reperfusion.[37] In contrast, exposure of SMC to 1 mM H_2O_2 stimulated TF activity but had no effect on TF mRNA and protein expression.[38] Because TF can exist in a latent form, this finding has been related to enhanced activation of a preexisting inactive pool of cell surface TF by H_2O_2.[37] However, we found that H_2O_2 induced TF mRNA in a concentration-dependent manner at substantially lower doses in human SMC (A. Görlach *et al.,* unpublished observations, 2001), suggesting that mRNA expression and activation of TF exhibit different sensitivities on exposure to oxidant stress in these cells.

The involvement of ROS in agonist-induced TF activity, mRNA, and/or protein expression has been mainly related to the inhibitory action of various

[33] R. Bach and D. B. Rifkin, *Proc. Natl. Acad. Sci. U.S.A.* **87,** 6995 (1990).

[34] H. M. Lander, *FASEB J.* **11,** 118 (1997).

[35] R. G. Allen and M. Tresini, *Free Radic. Biol. Med.* **28,** 463 (2000).

[36] Y. Cadroy, D. Dupouy, B. Boneu, and H. Plaisancie, *J. Immunol.* **164,** 3822 (2000).

[37] P. Golino, M. Ragni, P. Cirillo, V. E. Avvedimento, A. Feliciello, N. Esposito, A. Scognamiglio, B. Trimarco, G. Iaccarino, M. Condorelli, M. Chiariello, and G. Ambrosio, *Nature Med.* **2,** 35 (1996).

[38] M. S. Penn, C. V. Patel, M. Z. Cui, P. E. DiCorleto, and G. M. Chisolm, *Circulation* **99,** 1753 (1999).

TABLE I
REACTIVE OXYGEN SPECIES AND TISSUE FACTOR EXPRESSION[a]

Cell type	Stimulus	Antioxidant	Tissue factor mRNA(R), protein (P), activity (A),	Inhibition (I)/ stimulation (S)	Ref.
Monocytes	AGE albumin	NAC	R, P, A	I	39
		TU	P, A, R(−)	I	
		DMTU	P, A	I	
Monocytes	Homocysteine	Catalase	A(−)		40
		SOD	A(−)		
		2-Me	A(−)		
Monocytes	Activated PMN	NAC	A	I	36
		PDTC	A	I	
	H_2O_2		A	S	
	X/XO		A	S	
EC	LPS	PDTC	A, R	I	42
	TNF-α	PDTC	A, R	I	
	IL-1β	PDTC	A, R	I	
	PMA	PDTC	A, R	I	
EC	Lipoprotein	NAC	R, P	I	44
		Vitamin E	R, P	I	
EC	TNF-α	PDTC	P	I	43
	Activated platelets	PDTC	P	I	
EC	X/XO		R, A	S	37
SMC	H_2O_2		A, R(−)	S	38
SMC	oxLDL	Ebselen	A, R(−)	I	45
		DFO, Tiron	A, R(−)	I	
		DPPD	A, R(−)	I	
		Vitamin E	A, R(−)	I	
SMC	LDL	Ebselen	R(−)		46
		DPPD	R(−)		
SMC	Activated platelets	PDTC	R	I	47
		NAC	R	I	
		DPI	R	I	
		o-Phen.	R	I	
		p22phoxAS	R	I	
Myocard	X/XO		A	S	37
	Ischemia/reperfusion	SOD	A	I	

[a] NAC, *N*-acetylcysteine; TU, thiourea; DMTU, dimethylurea; SOD, superoxide dismutase; 2-ME, 2-mercaptoethanol; PDTC, pyrrolidine dithiocarbamate; DFO, desferrioxamine; DPPD, *N*,*N*′-diphenyl-1,4-phenylenediamine; DPI, diphenylene iodonium; *o*-Phen, 11,12-*o*-phenanthroline; p22phoxAS, p22phox antisense oligonucleotides; PMN, polymorphonuclear neutrophils; X/XO, xanthine/xanthine oxidase; LPS, lipopolysaccharide; TNF-α, tumor necrosis factor-α; oxLDL, oxidized low-density lipoprotein; −, no effect.

antioxidants. Among the commonly used antioxidants are *N*-acetylcysteine (NAC), pyrrolidine dithiocarbamate (PDTC), and vitamins C and E, as well as *N,N′*-diphenyl-1,4-phenylenediamine (DPPD), thiolurea (TU), dimethlyurea (DMTU), 2-mercaptoethanol, and the iron chelators ebselen, tiron, desferrioxamine (DFO), and 11,12-*o*-phenanthroline (Table I).

In monocytes, TF mRNA and protein expression, as well as TF activity, was enhanced in response to AGE albumin, a factor increased in diabetes.[39] This response was inhibited in the presence of NAC, TU, and DMTU, suggesting a role for ROS in diabetes-induced coagulopathies.[39] Redox-sensitive TF activity has also been observed in monocytes in response to activated polymorphonuclear neutrophils (PMN), suggesting an interaction between these blood cells in promoting coagulation during sepsis.[36] However, antioxidant treatment had no effect on TF activity in response to homocysteine in these cells,[40] although ROS production by this agent has been reported.[41]

In endothelial cells, TF mRNA and protein expression, as well as TF activity in response to various cytokines, including IL-1β and TNF-α, to LPS, an important mediator of endotoxic shock and to proatherosclerotic factors such as remnant lipoprotein, as well as to phorbol ester (PMA), was sensitive to antioxidant treatment.[42–44] In addition, TF protein expression, enhanced by activated platelets, was inhibited by treatment with PDTC.[43]

Moreover, the stimulated TF activity in SMC by oxidized LDL (oxLDL), but not the increased mRNA or protein expression, was caused by a redox-sensitive mechanism, supporting the idea that oxidative stress is required to mediate the availability of biologically active TF molecules on the cell surface.[45,46]

However, exposure of SMC to activated platelets induced TF mRNA expression, and this response was abolished by a variety of antioxidants, including PDTC, NAC, and 11,12-*o*-phenanthroline.[47] Moreover, whereas inhibitors of cyclooxygenase (diclofenac) and xanthine oxidase (allopurinol) had no effect on TF mRNA induction by activated platelets, the flavin inhibitor diphenylene iodonium (DPI)

[39] F. Khechai, V. Ollivier, F. Bridey, M. Amar, J. Hakim, and D. de Prost, *Arterioscler. Thromb. Vasc. Biol.* **17,** 2885 (1997).

[40] A. Khajuria and D. S. Houston, *Blood* **96,** 966 (2000).

[41] S. Taha, A. Azzi, and N. K. Ozer, *Antioxid. Redox Signal* **1,** 365 (1999).

[42] C. L. Orthner, G. M. Rodgers, and L. A. Fitzgerald, *Blood* **86,** 436 (1995).

[43] J. R. Slupsky, M. Kalbas, A. Willuweit, V. Henn, R. A. Kroczek, and G. Muller-Berghaus, *Thromb. Haemost.* **80,** 1008 (1998).

[44] H. Doi, K. Kugiyama, H. Oka, S. Sugiyama, N. Ogata, S. I. Koide, S. I. Nakamura, and H. Yasue, *Circulation* **102,** 670 (2000).

[45] M. S. Penn, M. Z. Cui, A. L. Winokur, J. Bethea, T. A. Hamilton, P. E. DiCorleto, and G. M. Chisolm, *Blood* **96,** 3056 (2000).

[46] M. Z. Cui, M. S. Penn, and G. M. Chisolm, *J. Biol. Chem.* **274,** 32795 (1999).

[47] A. Gorlach, R. P. Brandes, S. Bassus, N. Kronemann, C. M. Kirchmaier, R. Busse, and V. B. Schini-Kerth, *FASEB J.* **14,** 1518 (2000).

abolished platelet-induced TF mRNA expression, suggesting that an NADPH oxidase was involved in this response.[47] In cells transfected with antisense plasmids or oligonucleotides against the NADPH oxidase subunit p22phox, platelet-stimulated TF mRNA expression, as well as TF reporter gene activity, was abolished.[47] These results not only demonstrated that TF is a redox-regulated gene in SMC but also characterized for the first time the direct involvement of an ROS-generating enzyme in the regulation of TF.

Methods

Analysis of Tissue Factor Expression and Activity in Smooth Muscle Cells

Tissue Culture. To investigate the redox regulation of tissue factor expression, we use cultured human or rat SMC in media that do not contain transition metals such as iron or copper (e.g., MEM) in order to prevent ROS formation by the Fenton or Haber–Weiss reaction.

For studying redox-mediated gene expression, we routinely use quiescent SMC that have been grown to confluency, washed with phosphate-buffered saline (PBS), and further incubated for 24 to 48 hr with medium without serum, which has been supplemented with 0.1% fatty acid-free bovine serum albumin and nonessential amino acids.

Northern Blot Analysis. Total cellular RNA from SMC cells is prepared according to standard protocols.[48] Total RNA (25 μg) is separated by electrophoresis through a 1.2% agarose gel containing 6% formaldehyde dissolved in 0.04 *M* morpholinopropanesulfonic acid, 0.01 *M* sodium acetate, and 1 m*M* EDTA, pH 7.0, visualized by ethidium bromide staining, transferred to a nylon membrane (Porablot NY amp, Macherey-Nagel, Düren, Germany), and UV cross-linked. Hybridization is performed at 42° for 16 hr with a human or mouse TF cDNA probe labeled with ^{32}P-dCTP at 2×10^6 cpm/ml using the Ready to Go DNA-labeling kit from Amersham Pharmacia Biotech Inc. (Freiburg, Germany). Subsequently, the blots are washed twice with 6× SSPE and 0.1% sodium dodecyl sulfate (SDS) at room temperature and at 42° and twice with 2× SSPE and 0.1% SDS at 42° and 54° for 30 min each time.

Tissue Factor Antigen Measurements. For tissue factor antigen measurements, confluent human SMC, seeded in 3.5-cm wells, are made quiescent by incubation for 48 hr in serum-free medium. After stimulation, the cells are washed once with PBS and lysed on ice with 400 μl lysis buffer containing 0.5% Triton X and 50 m*M* TEA in PBS. Cells are frozen, thawed three times, and then sonicated. Cell lysates are stored at −80° until assayed. TF antigen expression is determined using the Imubind Tissue factor ELISA kit according to the manufacturer's instructions (American Diagnostica, Greenwich, CT).

[48] P. Chomczynski and N. Sacchi, *Anal. Biochem.* **162,** 156 (1987).

Procoagulant Activity. The surface procoagulant activity of human SMC is determined by measurement of thrombin formation during the clotting process in recalcified human platelet poor plasma (PPP).[49] When SMC reach confluence (24-well microplate), they are serum deprived for 24 hr prior to stimulation. Following treatment, SMC are washed three times with a HEPES-Tyrode solution and then incubated with human PPP. The formation of thrombin is initiated by the addition of $CaCl_2$ (16.7 m*M*) to the incubation medium. Aliquots (20 μl) are removed at intervals of 1–2 min and the formation of thrombin is determined using the chromogenic substrate S-2238 (Haemochrom Diagnostica). Optical densities are measured in a spectrophotometer (Uvikon, Kontron Instruments) at 405 nm. The dependency of the surface procoagulant activity of SMC on the availability of membrane bound TF is demonstrated by using a neutralizing antibody directed specifically against human TF (Mab#4508; American Diagnostica; 10 μg/ml, added to SMC 20 min prior to recalcification of PPP).

Analysis of Redox Control of Tissue Factor Expression and Activity

Treatment with Prooxidants. Several prooxidant conditions have been tested in tissue culture in order to investigate the role of ROS in gene expression. Application of enzyme systems such as xanthine/xanthine oxidase or glucose/glucose oxidase leads to continuous generation of ROS. Other sources of ROS are redox-cycling compounds such as 2,3-dimethoxy-1,4-naphthoquinone (DMNQ). However, because H_2O_2 is probably one of the most important ROS acting as a signaling molecule in SMC and can enter the cell readily, we prefer exposure of SMC to H_2O_2 prepared from dilutions of a 30% stock solution in a concentration range of between 1 and 500 μM. A single application of H_2O_2, however, leads to a decrease in concentration over time due to antioxidant properties of the medium and the cells. The half-life of H_2O_2 in tissue culture depends on cell type, medium, serum, and the percentage of lysed cells that can release catalase and peroxidases. Elevated levels of H_2O_2 for a longer time period can be obtained by consecutive bolus applications.

Treatment with Antioxidants. The role of ROS in gene expression can be studied by preincubating cells with antioxidant compounds prior to stimulation. They can be grouped in several classes with respect to their mode of action, including chelators of transition metals, direct scavengers of ROS, enhancers of GSH level, modification of antioxidant enzymes, or inhibitors of ROS producing enzymes.

Iron chelators such as 1,10-phenanthroline (10 μM), tiron (10 m*M*), or desferrioxamine (100 μM) decrease the cellular levels of ROS by preventing hydroxyl radical formation from H_2O_2 and ferrous iron. Vitamins E (α-tocopherol, 20 m*M*)

[49] S. Beguin, T. Lindhout, and H. C. Hemker, *Thromb. Haemost.* **61,** 25 (1989).

and C (ascorbic acid, 100 μM) can act as direct scavengers of ROS. Whereas vitamin E is lipophilic, thus acting at the plasma membrane, vitamin C is hydrophilic and can be taken up by cells via an energy-dependent mechanism.[50] Pyrrolidine dithiocarbamate (PDTC, 100 μM) is a widely used hydrophilic compound that can act as a metal chelator and a direct ROS scavenger. Moreover, at low concentrations, PDTC (10 μM) can induce an increase in GSH levels, whereas at higher concentrations, PDTC can also have prooxidant effects, possibly through oxidation of GSH.[51]

N-Acetylcysteine (NAC, 10 mM) is another frequently used antioxidant that acts mainly by increasing levels of the endogenous antioxidant, reduced glutathione, and consequently decreases H_2O_2 formation by the glutathione peroxidase system.[52] Scavengers of hydroxyl radicals include thiourea and dimethylurea, dimethyl sulfoxide (DMSO), 2-mercaptoethanol, or *N,N'*-diphenyl-1,4-phenylenediamine (DPPD), whereas ebselen is a compound that reduces complex lipid hydroperoxides. In general, cells were treated for 30 min with these compounds prior to stimulation.

Finally, the ROS metabolizing enzymes, superoxide dismutase (SOD), which catalyzes the reaction form O_2^- to H_2O_2, and catalase, which converts H_2O_2 to O_2 and H_2O, can be applied to determine the involvement of superoxide anion or H_2O_2. However, both enzymes are not taken up readily by cells, thus requiring, for example, liposomal preparations or the application of modified forms such as the heme-like SOD mimetic manganese(III) tetrakis(benzoic acid)porphyrin chloride (MnTBAP). Alternatively, overexpression of these enzymes is a good method to obtain relatively reliable data on the influence of ROS on gene expression.

Determination of Sources of ROS Production. A variety of potential sources of ROS exist in SMC that can contribute to redox-dependent modulation of gene expression in SMC, including mitochondria, xanthine oxidase, cyclooxygenase, lipoxygenase, NO synthase, heme oxygenases, other peroxidases, and NADPH oxidases.

To block ROS production from the mitochondrial respiratory chain, we used sodium cyanide in a concentration of 10 μM. Rotenone or antimycin A can be used as an inhibitor of mitochondrial complex I and complex III, respectively. Allopurinol or oxypurinol (100 μM) is a frequently used blocker of xanthine oxidase, whereas diclofenac (100 μM) or indomethacin inhibits cyclooxygenase. The contribution of NO synthase can be determined using inhibitors such as $N(\omega)$-nitro-L-arginine methyl ester (L-NAME, 300 μM). Heme oxygenases can be inhibited by zinc protoporphyrin IX.

[50] V. Kagan, E. Witt, R. Goldman, G. Scita, and L. Packer, *Free Radic. Res. Commun.* **16,** 51 (1992).

[51] D. Moellering, J. McAndrew, H. Jo, and V. M. Darley-Usmar, *Free Radic. Biol. Med.* **26,** 1138 (1999).

[52] S. M. Deneke, *Curr. Topics Cell. Regul.* **36,** 151 (2000).

Several methods exist to determine the capacity of SMC to produce ROS and to identify potential sources of ROS generation, including fluorescence and chemiluminescence assays.[47] A detailed description of these methods, however, is beyond the scope of this chapter. NADPH oxidases have been shown to play an important role in ROS-dependent signaling in SMC and are involved in TF expression in response to activated platelets. The flavin inhibitor diphenylene iodonium (DPI) in a concentration of 10 μM has been used frequently as an inhibitor of these enzymes. However, DPI is a relatively unspecific inhibitor of a variety of flavin-containing enzymes, including NO synthase, NADH oxidase, and the mitochondrial complex I.[53–55] Apocynin is another NADPH oxidase inhibitor that appears to interrupt assembly of the NADPH oxidase after stimulation.[56] Although apocynin has been reported to require conversion by peroxidases to exert its inhibitory effect, several investigators have seen an inhibitory effect of this reagent in nonphagocytic cells.

However, a more specific way to elucidate the role of the NADPH oxidase in TF expression in SMC is to use antisense oligonucleotides against the ubiquitously expressed subunit p22phox.[47]

Sequences of the phosphothiorate-modified oligonucleotides are as follows:

p22phox-scrambled: 5′-TAGCATAGCCCTCCGCTGGGGA-3′
p22phox-antisense: 5′-GATCTGCCCCATGGTGAGGACC-3′

These oligonucleotides are transfected into SMC with the help of liposomes using the Superfect reagent (Qiagen, Hildesheim, Germany). SMC are seeded in 10-cm dishes and grown until 80% confluency. Per dish, 3 μg of oligonucleotides is mixed with 300 μl medium without supplements (e.g., MEM) and 40 μl Superfect reagent. This mixture is vortexed gently and incubated for 10 min at room temperature. Three milliliters complete medium is added, and cells are incubated for 3 hr with this mixture. Subsequently, cells are washed with medium without supplements and then cultivated in complete growth medium for another 18 to 24 hr, followed by an incubation period with serum-free medium for 24 hr prior to stimulation.

Transactivation Assays. Using these assays, the contribution of one or more transcription factor binding to elements in the TF promoter to redox-sensitive transactivation can be monitored. Potential redox-sensitive transcription factors are located in a distal enhancer (−227 to −172) containing two AP-1 sites and an

[53] D. J. Stuehr, O. A. Fasehur, N. S. Kwon, S. S. Gross, J. A. Gonzalez, R. Levi, and C. F. Nathan, *FASEB J.* **5,** 98 (1991).

[54] K. M. Mohazzab-H, P. M. Kaminski, and M. S. Wolin, *Circulation* **96,** 614 (1997).

[55] Y. Li and M. A. Trush, *Biochem. Biophys. Res. Commun.* **253,** 295 (1998).

[56] J. Stolk, T. J. Hiltermann, J. H. Dijkman, and A. J. Verhoeven, *Am. J. Respir. Cell Mol. Biol.* **11,** 95 (1994).

NF-κB site, whereas proximal enhancers in human (-109 to -59) and rat (-103 to -80) TF promoters contain Egr-1 and Sp1 sites.[29]

To investigate redox sensitivity of these promoter elements, reporter genes containing either the distal and proximal enhancer or the proximal enhancer alone coupled to the luciferase gene are transfected in SMC using the Superfect reagent. Cells can be treated with antioxidants as outlined earlier, and luciferase activity can be determined after stimulation. Because expression of the reporter genes is relatively slow, care has to be taken when incubating cells with potentially toxic components over a longer timer period.

To study the involvement of NADPH oxidase in TF transactivation, p22phox sense or antisense vectors are cotransfected with the TF reporter genes using the Superfect reagent as outlined earlier, and luciferase activity is determined after stimulation of the transfected cells.[47]

Concluding Remarks

ROS have been identified as signaling molecules in a variety of cell types, including SMC and other vascular cells. TF has been added to the growing list of redox-sensitive genes in response to growth factors, cytokines, and proatherogenic factors such as oxLDL. Coagulation factors, growth factors, and stimuli or mediators of inflammation released from activated platelets and endothelial cells, as well as SMC, are found in response to vascular injury, e.g., after balloon catheterization, or in atherosclerotic plaques. Many of them, including PDGF, TGF-β, IL-1β, and TNF-α, have been shown to induce ROS production, which might be involved in the upregulation of TF, and increase procoagulant activity, as well as enhance migration of SMC at these sites of diseases. However, the sources of ROS production activated by the various agents are widely unknown.

The NADPH oxidase has been identified as the enzyme mainly responsible for the induction of TF in response to activated platelets in SMC. Because this enzyme can be activated by many of the same factors that also induce TF, it may be speculated that activation of the NADPH oxidase is a more general pathway leading to redox-sensitive TF expression in response to a variety of stimuli. In addition, knowledge about the signaling pathways linking NADPH oxidase-derived ROS production to TF expression is limited, but will be required to fully understand redox-sensitive signaling cascades involved in the regulation of TF.

Acknowledgments

The authors thank Professor R. Busse for continuous support and Isabel Winter for technical assistance.

[20] Redox Processes Regulate Intestinal Lamina Propria T Lymphocytes

By BERND SIDO, RAOUL BREITKREUTZ, CORNELIA SEEL, CHRISTIAN HERFARTH, and STEFAN MEUER

Introduction

Due to the special architecture of the gut, the mucosal surface is more than 200 times larger than the surface of the skin. The largest cellular immune system can be found on this "inner" surface of the body. At this mucosal barrier, the effector T-cell compartment (lamina propria) is permanently exposed to exogenous antigens of nutritional and microbial origin. This antigen challenge will not lead to pathology only if a systemic immune response with proliferation and cytokine secretion is prevented. The physiological hyporeactivity of lamina propria T lymphocytes (LP-T) in the normal gut *in vivo* is indicated by their low proliferative potential.[1] Analogously, LP-T are defective in their ability to proliferate *in vitro* in response to antigen receptor stimulation,[2,3] although these cells are predominantly of the memory phenotype ($CD45R0^+$) and express a full range of cell surface receptors necessary for immune activation.[4,5] Evidence is accumulating that the special mucosal environment regulates the functional state of effector cells. We have shown that coculture of peripheral blood T lymphocytes with the intestinal mucosa supernatant induces a similar functional behavior as found in freshly recovered LP-T.[6] It was suggested that small, nonprotein, nonpeptide molecules with oxidative capacities downregulate antigen receptor-induced T lymphocyte proliferation. In contrast, the antioxidant 2-mercaptoethanol (2-ME) could reverse the suppressive effect of mucosa supernatant and could restore the CD3 reactivity of LP-T. This finding suggests that regulation of the intracellular redox state in LP-T may represent a versatile physiological control mechanism to adjust the mucosal lymphocyte reactivity to particular local requirements.

Glutathione (GSH) is the most abundant intracellular low molecular weight thiol. Due to its strong antioxidative capacities, it controls several cellular immune

[1] F. Autschbach, G. Schürmann, L. Qiao, H. Merz, R. Wallich, and S. C. Meuer, *Virch. Arch.* **426,** 51 (1995).

[2] M. Zeitz, T. Quinn, A. S. Graeff, and S. P. James, *Gastroenterology* **94,** 353 (1988).

[3] L. Qiao, G. Schürmann, M. Betzler, and S. C. Meuer, *Gastroenterology* **101,** 1529 (1991).

[4] S. P. James, W. C. Kwan, and M. Sneller, *J. Immunol.* **144,** 1551 (1990).

[5] U. Pirzer, G. Schürmann, S. Post, M. Betzler, and S. C. Meuer, *Eur. J. Immunol.* **20,** 2339 (1990).

[6] L. Qiao, G. Schürmann, F. Autschbach, R. Wallich, and S. C. Meuer, *Gastroenterology* **105,** 814 (1993).

0076-6879/02 $35.00

functions, such as lymphocyte proliferation and cytotoxic activity, and modulates the activation of redox-regulated transcription factors such as NF-κB and AP-1. A 10–30% decrease of the GSH content in lymphocytes disrupts the proximal signal cascade after TCR stimulation with complete inhibition of the increase of intracellular free calcium.[7] Because lymphocytes are defective in cystine uptake,[8,9] the availability of the reduced derivative cysteine becomes limiting for the synthesis of GSH. However, cysteine circulates at extremely low concentrations in plasma and is not contained in standard tissue culture medium. Therefore, dynamic changes in the local supply of cysteine by other cells that can take up cystine (and subsequently release cysteine) may evolve as a physiological mechanism involved in the regulation of intestinal lymphocyte reactivity.

Preparation of Lamina Propria Cells from Human Gut

Large bowel specimens are obtained from patients undergoing resection for colon cancer. Normal mucosa (5 × 5 cm) is dissected from the submucosa near the resection margin. Lamina propria mononuclear cells are isolated according to a modification of the method of Bull and Bookman.[10] The fresh tissue is washed extensively in HBSS, without Ca^{2+} and Mg^{2+} (GIBCO, Paisley, Scotland), containing penicillin (100 U/ml; Sigma, Taufkirchen, Germany), streptomycin (100 μg/ml; Sigma, Taufkirchen, Germany), gentamycin (59 μg/ml; Sigma, Taufkirchen, Germany), and amphotericin B (2.5 μg/ml; GIBCO) at 4°. The mucus can be removed by incubation in HBSS, without Ca^{2+} and Mg^{2+}, containing 1 m*M* dithiothreitol (DTT) for 15 min at 37°. For the study of redox regulation, however, the use of the strong antioxidant DTT should be avoided and, instead, the mucus should be scraped off gently with a scalpel. The tissue is cut into 2- to 4-mm pieces and incubated in a shaking water bath in HBSS, without Ca^{2+} and Mg^{2+}, containing 0.7 m*M* EDTA (Sigma, Taufkirchen, Germany) and the aforementioned antibiotics at 37° for 45 min. This incubation is repeated twice with fresh medium until the supernatant is free of epithelial cells. The tissue is then washed four times for 10 min at 37° in HBSS, without Ca^{2+} and Mg^{2+}, until the supernatant becomes clear. Subsequently, the mucosal tissue is enzymatically digested in RPMI 1640 (GIBCO) containing 2% fetal calf serum (FCS; Sigma) 45 U/ml collagenase (type IV; Sigma), 27 U/ml deoxyribonuclease I (Sigma), 2% glutamine, antibiotics, and amphotericin B in a shaking water bath at 37° for 10 hr. The digest is passed through a 70-μm nylon mesh (Becton Dickinson, Heidelberg, Germany)

[7] F. J. Staal, M. T. Anderson, G. E. Staal, L. A. Herzenberg, C. Gitler, and L. A. Herzenberg, *Proc. Natl. Acad. Sci. U.S.A.* **91,** 3619 (1994).

[8] T. Ishii, Y. Sugita, and S. Bannai, *J. Cell. Physiol.* **133,** 330 (1987).

[9] H. Gmünder, H.-P. Eck, and W. Dröge, *Eur. J. Biochem.* **201,** 113 (1991).

[10] D. M. Bull and M. A. Bookman, *J. Clin. Invest.* **59,** 966 (1977).

and washed in RPMI 1640/2% FCS. After washing twice (4°), the cells are resuspended in 67.5% Percoll (Pharmacia Biotech, Uppsala, Sweden) and overlayed by 30% Percoll. After centrifugation at 3000 rpm and 4°, lamina propria mononuclear cells are recovered from the interphase and washed twice. Viable cells are separated by Ficoll–Hypaque (Sigma) density gradient centrifugation at room temperature (2500 rpm, 20 min) and washed twice. Cells are resuspended in a 1 : 1 mixture of RPMI 1640/2% FCS and autologous serum and are allowed to adhere to a plastic tissue culture petri dish (Greiner, Frickenhausen, Germany) for 3 hr at 37°. Adherent cells are harvested with a rubber policeman (Becton Dickinson) and used as lamina propria macrophages (LP-MO). Nonadherent cells are pelleted, mixed with 40 μl of a 5% suspension of sheep red blood cells (SRBC; ICN Biomedicals, Eschwege, Germany) per 10^6 cells, spun at 700 rpm for 5 min, and incubated for 50 min at room temperature to allow E rosette formation. Subsequently, the pellet is resuspended gently and centrifuged on Ficoll–Hypaque for 20 min at 1000 rpm and at 1800 rpm. Supernatant and E rosette-negative cells are discarded, and the pellet is treated with lysis buffer (155 m*M* ammonium chloride, 10 m*M* potassium bicarbonate, 0.13 m*M* EDTA, pH 7.27) for 5 min to lyse SRBC. E rosette-positive cells are washed three times and finally resuspended in RPMI 1640 supplemented with 10% FCS, penicillin/streptomycin, and 2% glutamine for use. This cell population is 90% positive for CD3 as shown by immunofluorescent staining and is used as lamina propria T cells (LP-T). Viability is consistently more than 95%.

Venous blood is collected from the same patient, and peripheral blood mononuclear cells are obtained by Ficoll–Hypaque density gradient centrifugation. Adherent cells are isolated according to the adherence step described earlier and are used as peripheral blood monocytes (PB-MO).

Stimulation of Intestinal Lamina Propria T Lymphocytes *in Vitro*

LP-T (5×10^4/well) are cultured in 96-well round-bottomed microtiter plates (Costar, Bodenheim, Germany) at 37° and 7% CO_2. PB-MO and LP-MO, respectively, are irradiated immediately before use (50 Gy) and added to LP-T at 30% of total cell number. LP-T cells are stimulated via CD3 according to standard procedures employing mitogenic monoclonal antibodies in the absence or presence of recombinant human IL-2 (10 U/ml; Biotest, Dreireich, Germany). For CD3 stimulation, we immobilize the mouse antibody OKT3 (IgG_{2a}) on beads (Irvine Scientific, Santa Ana, CA) or, alternatively, on 96-well flat-bottomed microtiter plates (Costar) that are precoated with goat antimouse immunoglobulin. After 4 days of culture, wells are pulsed with 1 μCi [^{3}H]thymidine (Amersham, Karlsruhe, Germany) for 16 hr and then harvested on glass fiber filters using an automatic cell harvester (FilterMate, Packard, Meriden, CT). [^{3}H]Thymidine incorporation is measured in a microplate scintillation counter (TopCount, Packard, Meriden, CT).

Experimental Procedures to Study the Influence of Redox Milieu on Proliferation of Intestinal Lamina Propria T Lymphocytes

Hydrogen peroxide is a very simple, stable, and naturally occurring potent oxidant. Due to its lipophilicity, it can permeate cell membranes easily and react slowly with organic substances. It is physiologically produced intracellularly and is supposed to function as a second messanger in signal transduction in cells. Moreover, hydrogen peroxide is produced *in vivo* in large amounts in areas of inflammation by polymorphonuclear phagocytes due to the catalytic dismutation of superoxide anion radicals and can be converted to highly reactive hydroxyl radicals in the presence of transition metals. Micromolar concentrations (0–100 μM) of hydrogen peroxide (Merck, Darmstadt, Germany) suppress the proliferation of LP-T in a dose-dependent manner after stimulation with OKT3 plus IL-2 (10 U/ml). The suppressive effect of 10–25 μM is only marginal, whereas 50 μM hydrogen peroxide decreases proliferation by 61%. In the presence of 75 μM, the proliferation of LP-T drops down to nearly background values (Fig. 1).

Even a partial depletion of the intracellular GSH pool in lymphocytes has dramatic effects on blast transformation and proliferation and suppresses the

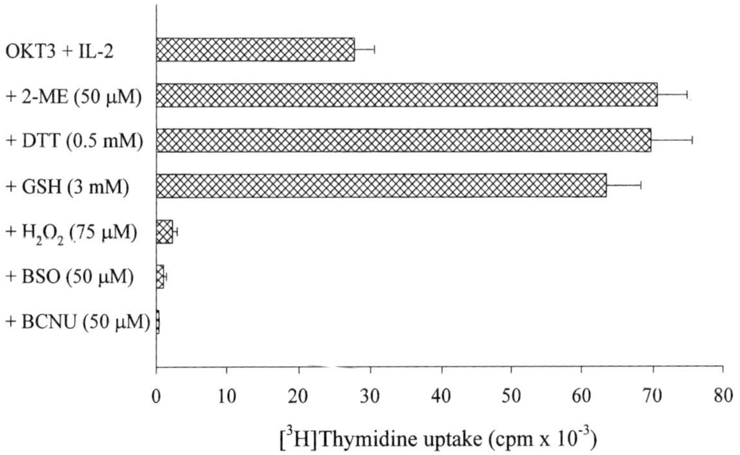

FIG. 1. Effect of antioxidant (2-ME, DTT, GSH) and prooxidant (H_2O_2, BSO, BCNU) culture conditions on the proliferative immune response of LP-T after stimulation via CD3. If added, the concentration of IL-2 was 10 U/ml. 2-ME, 2-mercaptoethanol (50 μM); DTT, dithiothreitol (0.5 mM); H_2O_2, hydrogen peroxide (75 μM); BSO, buthionine-[*S,R*]sulfoximine (50 μM); BCNU, 1,3-bis(2-chloroethyl)-1-nitrosourea (50 μM).

activation of cytotoxic T cells and LAK cells.[11–14] The water-soluble buthionine-[*S,R*] sulfoximine (BSO; Sigma, Taufkirchen, Germany) is a specific inhibitor of the rate-limiting enzyme of GSH synthesis (γ-glutamylcysteine synthetase)[15] and, therefore, can be used to study the specific consequences of GSH deficiency. BSO inhibits proliferation of LP-T by 48% at 10 μM when added at the beginning of culture and drops to background values at 50 μM (Fig. 1).

1,3-Bis(2-chloroethyl)-1-nitrosourea (BCNU; Bristol Laboratories, Evansville, IN) is an irreversible enzyme inhibitor of glutathione reductase and leads to long-lasting inhibition of the NADPH-dependent regeneration of GSH from glutathione disulfide (GSSG).[16] Inhibition of glutathione reductase in LP-T 120 min after the addition of 50 μM BCNU is more than 90% of the control. As a consequence, the concentration of GSSG increases and the redox balance of glutathione (ratio GSH/GSSG) is shifted toward a prooxidant state. In an attempt to restore a nearly physiological balance of GSH/GSSG, the cell exports GSSG, leading to a decrease in the total intracellular glutathione content. A stock solution of BCNU in 100% ethanol (40 m*M*) is stable at $-20°$ for at least 2 weeks.[16] BCNU inhibits proliferation of LP-T at concentrations as low as 10 μM and blocks proliferation completely at 50 μM (Fig. 1).

In contrast, an antioxidative environment increases the proliferation of LP-T. This can be demonstrated with the use of more than 0.1 m*M* DTT (Fig. 1). Concentrations above 1 m*M* DTT are toxic for the cells as proliferation decreases again. Similar stimulatory effects can be achieved with the addition of more than 1 m*M* GSH (Fig. 1). Note that GSH is acidic and the pH of the solution needs to be adjusted to 7.3 before use. The well-established potentiating effect of 2-ME on lymphocyte proliferation is due to an increased uptake of cysteine in lymphocytes.[17] 2-ME thus compensates for the low inherent membrane transport activity of lymphocytes for cystine (transport system X_c^-).[8,9] Although the membrane transport activity for cysteine is high in lymphocytes (transport system ASC),[8,9] the uptake of cysteine is insufficient because cysteine is not present in the culture medium. The oxidized derivative cystine is abundantly present in RPMI 1640 (204 $\mu M = 408$ cysteine equivalents). However, this cannot compensate for the cysteine deficiency because the activity of unstimulated and stimulated lymphocytes (both T and B cells) to take up cystine is more than 10 times lower than for cysteine.[8,9] Therefore, the

[11] J. P. Messina and D. A. Lawrence, *J. Immunol.* **143,** 1974 (1989).

[12] M. Suthanthiran, M. E. Anderson, V. K. Sharma, and A. Meister, *Proc. Natl. Acad. Sci. U.S.A.* **87,** 3343 (1990).

[13] A. Yamauchi and E. T. Bloom, *J. Immunol.* **151,** 5535 (1993).

[14] H. Gmünder and W. Dröge, *Cell. Immunol.* **138,** 229 (1991).

[15] O. W. Griffith and A. Meister, *J. Biol. Chem.* **254,** 7558 (1979).

[16] K. Becker and H. Schirmer, *Methods Enzymol.* **251,** 173 (1995).

[17] T. Ishii, S. Bannai, and Y. Sugita, *J. Biol. Chem.* **256,** 12387 (1983).

availability of cysteine becomes limiting for glutathione synthesis[8,18] and lymphocyte proliferation.[19,20] 2-ME forms a mixed disulfide with cysteine derived from cystine.[17] This mixed disulfide can be taken up easily by lymphocytes via the transport system shared by neutral amino acids such as leucine and phenylalanine.[17] The mixed disulfide 2-ME–cysteine is reduced intracellularly to liberate cysteine, whereas 2-ME recycles to the extracellular space as a carrier molecule. By promoting cysteine uptake, 2-ME preserves a critical intracellular GSH content,[8] which is a prerequisite for cell cycle progression from the G_1 to S phase.[11] At concentrations as low as 5 μM, 2-ME potentiates the proliferation of LP-T, which only slightly increases further at 50 μM 2-ME (Fig. 1).

Experimental Procedures to Analyze Differential Capacity of Peripheral Blood Monocytes versus Resident Intestinal Macrophages to Produce Cysteine

Because cysteine circulates at extremely low concentrations in human plasma (8–10 μM)[21] and because lymphocytes cannot produce cysteine by themselves, even moderate changes of extracellular cysteine concentrations influence lymphocyte immune functions.[11–14] LP-T, therefore, depend on an alternative cellular source of cysteine *in vivo* to establish a proliferative immune response like in inflammatory bowel disease. Lymphocytes deficient in cystine transport can be supplied with cysteine by other cells that can efficiently take up cystine; mouse peritoneal macrophages have a strong membrane transport activity for cystine[22] and, by release of substantial amounts of cysteine, these cells augment the GSH content and DNA synthesis in murine lymphocytes *in vitro*.[20] In inflammatory bowel disease, a sustained recruitment of PB-MO to severely inflamed areas has been well documented by immunohistochemical studies.[23,24] We, therefore, investigated the differential capacity of PB-MO versus LP-MO to release cysteine into the supernatant.

PB MO and LP-MO, respectively, are plated at 5×10^5 cells/well in a total volume of 1 ml/well in a 48-well culture plate (Costar, Bodenheim, Germany). Cells are either left untreated or stimulated with 1 μg/ml LPS from *Escherichia coli* (serotype 055:B5; Sigma). After 40 hr of culture (37°, 7% CO_2), the supernatant is harvested and centrifuged (4°) to remove cells.

[18] S. Bannai and N. Tateishi, *J. Membr. Biol.* **89,** 1 (1986).

[19] W. Dröge, R. Kinscherf, S. Mihm, D. Galter, S. Roth, H. Gmünder, T. Fischbach, and M. Bockstette, *Methods Enzymol.* **251,** 255 (1995).

[20] H. Gmünder, H. P. Eck, B. Benninghoff, S. Roth, and W. Dröge, *Cell. Immunol.* **129,** 32 (1990).

[21] M. A. Mansoor, A. M. Svardal, and P. M. Ueland, *Anal. Biochem.* **200,** 218 (1992).

[22] H. Watanabe and S. Bannai, *J. Exp. Med.* **165,** 628 (1987).

[23] J. Rugtveit, P. Brandtzaeg, T. S. Halstensen, O. Fausa, and H. Scott, *Gut* **35,** 669 (1994).

[24] V. L. Burgio, S. Fais, M. Boirivant, A. Perrone, and F. Pallone, *Gastroenterology* **109,** 1029 (1995).

Method 1: Determination of Cysteine in the Supernatant as Acid-Soluble Thiol

Principle. This spectrophotometric assay is simple to perform and very sensitive for the detection of reduced thiol compounds after protein sulfhydryls have been removed by acid. However, this assay is not specific for cysteine and does not discriminate between various acid-soluble thiols such as cysteine and GSH. 5,5′-Dithiobis(2-nitrobenzoic acid) (DTNB) reacts with thiols to give the yellow product 5′-thionitrobenzoic acid, which has an absorbance maximum at 412 nm.[18]

Reagents

NaOH, 1 *N*
EDTA, 80 m*M*
Trichloroacetic acid (TCA), 30%
Sodium phosphate buffer, 0.5 *M*, pH 7.0
DTNB (Serva, Heidelberg, Germany), 10 m*M* in buffer
L-Cysteine (Serva, Heidelberg, Germany), 10 m*M* in RPMI 1640 (for preparation of standards)

Procedure. The cell-free supernatant (600 μl) is mixed with 150 μl of EDTA and 150 μl of TCA to precipitate protein, followed by incubation on ice for 15 min and centrifugation. The deproteinized supernatant (267 μl) is mixed with 400 μl of buffer and 100 μl of NaOH. Finally, 33 μl of DTNB is added, and absorption is recorded spectrophotometrically at 412 nm. Standard solutions of cysteine are subjected to the same analytical procedure as described.

PB-MO constitutively produce micromolar amounts of cysteine that increase three to four times when the cells are stimulated with LPS. IFN-γ (200 U/ml), however, another potent activator of monocytes, does not enhance constitutive production of cysteine (Table I). We have demonstrated that receptor–ligand interactions

TABLE I
CYSTEINE (ACID-SOLUBLE THIOL) CONTENT IN SUPERNATANT OF PB-MO VERSUS LP-MO[a]

Treatment	PB-MO (μM)	LP-MO (μM)
Medium	6.06	0.08
+H_2O_2 (50 μM)	6.04	ND[b]
LPS (1 μg/ml)	20.05	0.16
+H_2O_2 (50 μM)	18.79	ND
IFN-γ (200 U/ml)	5.80	0.06
+H_2O_2 (50 μM)	4.85	ND

[a] After 40 hr of culture; PB-MO and LP-MO were isolated from the same patient and cultured at 5×10^5/ml.
[b] Not determined.

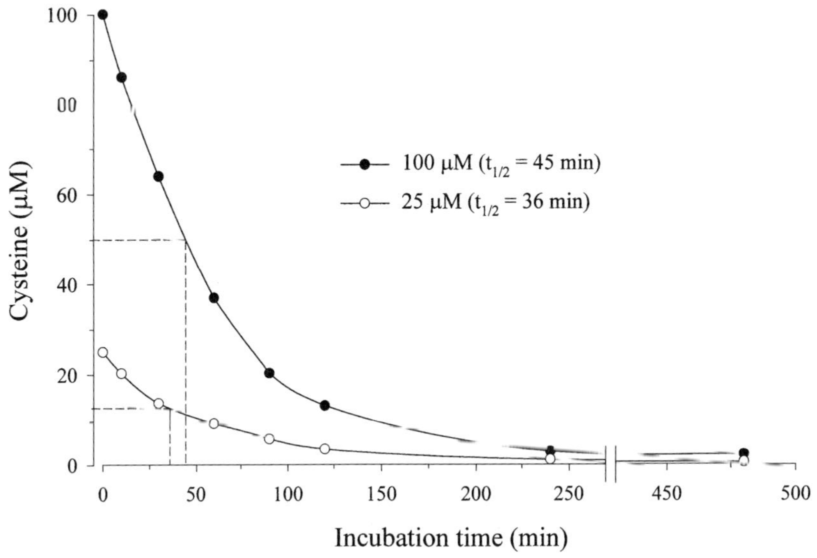

FIG. 2. Degradation of cysteine added to RPMI 1640 at 25 and 100 *μM*, respectively, under standard culture conditions (37°, 7% CO_2). Cysteine was determined as acid-soluble thiol (method 1). The dashed line indicates the half-life of cysteine.

between CD2 on LP-T and the ligand CD58 on PB-MO are involved in cysteine production comparable to LPS.[25] This indicates that CD2-mediated costimulation of T cells is not restricted to the CD2 signal transduced in T cells, but also induces metabolic changes in professional antigen-presenting cells, which by the release of cysteine enhance the proliferative immune response of activated lymphocytes. It is of note that LP-MO, in clear contrast to PB-MO, are defective in cysteine production both constitutively and after activation with LPS (Table I).

Cysteine oxidizes rapidly to cystine in culture. The half-life of cysteine depends on the concentration and is 36 and 45 min at 25 and 100 *μM* cysteine, respectively (Fig. 2). It can therefore be concluded that cysteine (acid-soluble thiol) does not accumulate in the supernatant during the 40-hr culture of PB-MO and that the cysteine level represents the actual and long-lasting production of cysteine.

Method 2: Determination of Cysteine by HPLC

Principle. This analytical procedure is based on the method described by Mansoor *et al.*[21] It is very sensitive and, in contrast to method 1, allows measurement of the reduced, oxidized, and protein-bound forms of cysteine,

[25] B. Sido, J. Braunstein, R. Breitkreutz, C. Herfarth, and S. C. Meuer, *J. Exp. Med.* **192,** 907 (2000).

cysteinylglycine, homocysteine, and glutathione. Thus, it is possible to determine the redox balance of various thiols as defined, e.g., by the ratio GSH/GSSG or cysteine/cystine. Here, we present the protocol for the specific detection of nonprotein-reduced thiols in the supernatant of PB-MO.

Reagents

NaOH, 0.05 *M*
Sulfosalicylic acid (SSA), 50%
Perchloric acid, 70%
N-Ethylmorpholine (Sigma), 1 *M* in water
Dithioerythritol (DTE; BioMol, Hamburg, Germany), 50 μM in 5% SSA
Monobromobimane (mBrB; Calbiochem, Bad Soden, Germany), 40 m*M* in acetonitrile
L-Cysteine (Serva, Heidelberg, Germany), 10 m*M* in 5% SSA containing 50 μM DTE
Glutathione (GSH; Serva), 10 m*M* in 5% SSA containing 50 μM DTE
Solution B: 65% DMSO and 35% water (v/v) containing 51 m*M* NaCl and 140 m*M* hydrobromic acid
Elution solvent A: 0.25% glacial acetic acid (pH adjusted to 3.4 with 2 *M* NaOH)
Elution solvent B: 80% acetonitrile (for chromatography; Merck, Darmstadt, Germany)

Procedure. The supernatant of PB-MO (930 μl) is deproteinized by the addition of 70 μl of SSA (50%), followed by incubation on ice (10 min) and centrifugation. To 60 μl of supernatant are added 10 μl of water, 30 μl of NaOH, 130 μl of solution B, 50 μl of *N*-ethylmorpholine, and 10 μl mBrB. The mixture is incubated for 30–40 min in the dark at room temperature. After centrifugation, the sample is stored at -80° and analyzed within 48 hr. Samples (40 μl) are injected into a 4.6×150-mm column packed with 3.5-μm particles (Symmetry C18; Waters, Milford, MA), equipped with a 3.9×20-mm guard column packed with 5-μm particles (Symmetry C18; Waters). The temperature is 20° and the flow rate is 1.0 ml/min. The elution profile is as follows: 0–13 min, 7.5% solvent B; 13–23 min, 8.75% solvent B; 23–30 min, 23.75% solvent B; 30–40 min, 25% solvent B; 40–50 min, 100% solvent B. Fluorescent material is detected by a Shimadzu RF-551 fluorometer detector at an excitation wavelength of 400 nm and an emission wavelength of 480 nm. The software GOLD Nouveau (Beckman, Coulter, Fullerton, CA) is used for plotting and integration of peaks. L-Cysteine and GSH are used as standards. The retention time for cysteine and GSH is 12 and 23 min, respectively. Linearity of the assay for both thiols is demonstrated in Fig. 3.

For the first time, acid-soluble thiol present in the supernatant of PB-MO has been identified as cysteine by HPLC. No reduced GSH can be detected in

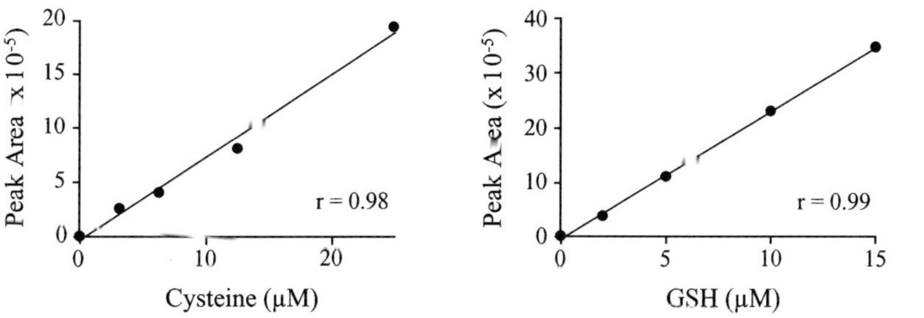

FIG. 3. Standard curves for cysteine and reduced glutathione (GSH). The standards were dissolved in 5% sulfosalicylic acid containing 50 μM dithioerythritol and were analyzed by HPLC (method 2).

the supernatant of PB-MO (Fig. 4). Indeed, under the experimental conditions described here, both analytical procedures yield identical results (Fig. 5). For routine analysis, method 1 is preferred. However, the specificity of the assay has to be evaluated by HPLC depending on the experimental setting and the questions to be answered.

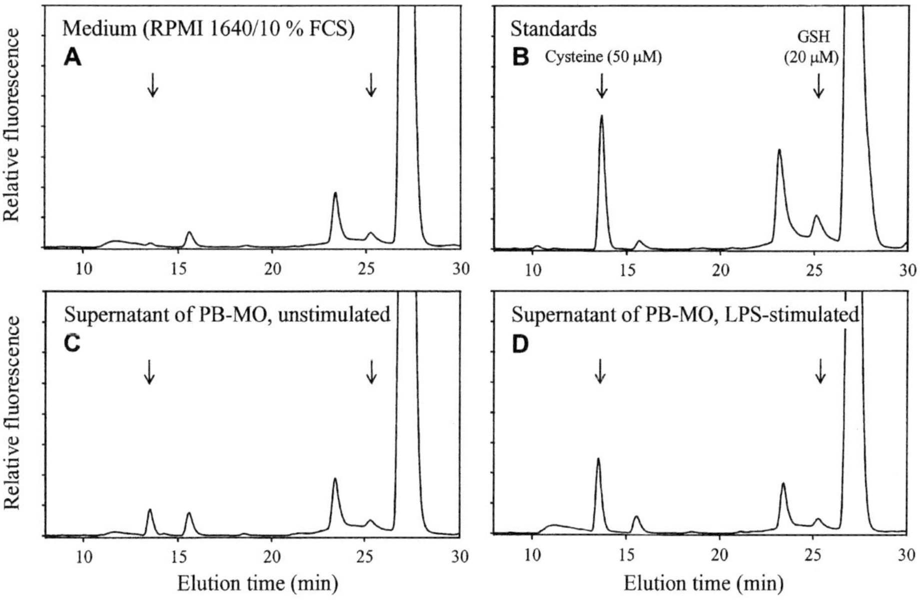

FIG. 4. Reversed-phase high-performance liquid chromatography of a deproteinized and monobromobimane-derivatized sample of (A) RPMI 1640/10% FCS (blank); (B) a standard solution of 50 μM cysteine and 20 μM reduced glutathione in 5% sulfosalicylic acid containing 50 μM dithioerythritol; (C) supernatant of unstimulated PB-MO in RPMI 1640/10% FCS after 40 hr of culture; and (D) supernatant of PB-MO simulated by LPS (1 μg/ml) in RPMI 1640/10% FCS for 40 hr.

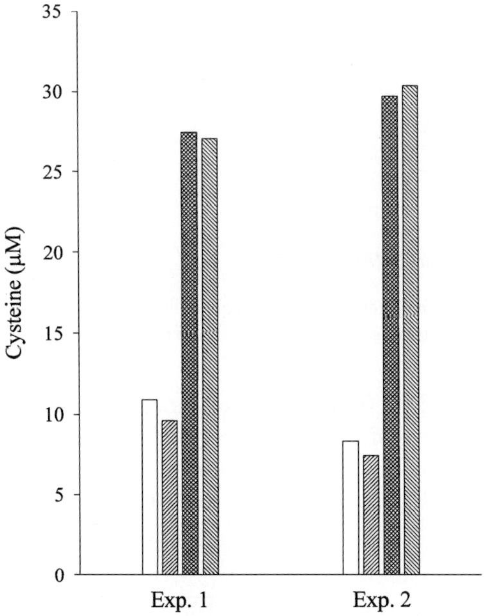

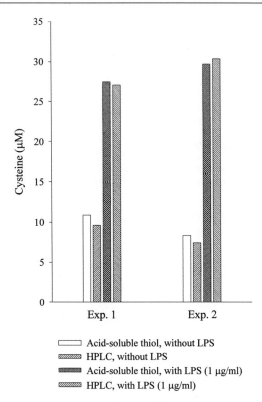

FIG. 5. Comparative analysis of cysteine as acid-soluble thiol (method 1) or by HPLC (method 2) in the supernatant of PB-MO after 40 hr of culture. Cells were either left untreated or stimulated with LPS (1 μg/ml). Results of two independent experiments are presented.

Experimental Procedures to Demonstrate Differential Capacity of Peripheral Blood Monocytes versus Intestinal Lamina Propria Macrophages to Restore CD3 Reactivity of Lamina Propria T Lymphocytes

PB-MO and LP-MO, respectively, are irradiated (50 Gy) immediately before use and are added to LP-T at 30% of total cell number. In accordance with their capacity to produce cysteine, PB-MO restore the CD3 reactivity of LP-T in the absence of IL-2, whereas LP-MO, in clear contrast, fail to do so (Fig. 6). PB-MO have to be added at more than 20% of total cell number to provide significant costimulation,[25] suggesting a metabolic or mediator-driven mechanism of costimulation. In conformance with the hypothesis of thiol (cysteine)-mediated redox regulation of LP-T by PB-MO, 2-ME (50 μM) is able to fully substitute for the

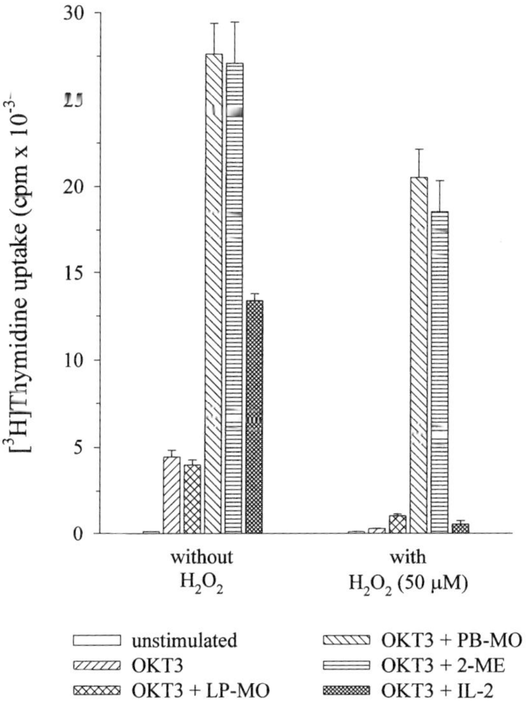

FIG. 6. Effect of 2-mercaptoethanol (2-ME; 50 μM) versus PB-MO versus LP-MO on the CD3 reactivity of LP-T (5×10^4/well) under standard and prooxidant culture conditions, respectively. PB-MO and LP-MO, respectively, were irradiated (50 Gy) and added to LP-T at 30% of total cell number in a total volume of 200 μl/well. If added, the concentration of IL-2 was 10 U/ml. Cells were cultured in either the absence or the presence of hydrogen peroxide.

costimulatory activity of PB-MO in the absence IL-2. Moreover, PB-MO, similar to the antioxidant 2-ME, allow proliferation of LP-T even under prooxidant conditions in the presence of 50 μM hydrogen peroxide (Fig. 6). LP-MO do not actively suppress the proliferation of LP-T, as LP-MO do not influence the restoration of CD3 reactivity of LP-T by 2-ME.

Experimental Procedures to Demonstrate Thiol-Mediated Regulation of DNA Synthesis of Lamina Propria T Lymphocytes

The use of cystine-deficient culture medium provides a simple and convenient experimental system to investigate the consequences of cysteine and/or GSH deficiency. To this end, experiments are set up in cystine-deficient RPMI 1640

(BioWhittaker, Verviers, Belgium) to which 10% FCS, 2% glutamine, and antibiotics are added. Cells are washed in cystine-free medium before use. For supplementation of cystine-free cultures with graded amounts of cystine, we use a 20 m*M* cystine stock solution (cystine 100 × for RPMI 1640; GIBCO, Paisley, Scotland). Contaminating cystine may result from the FCS. Therefore, it might be necessary to dialyze the FCS against cystine-free RPMI 1640 before use. Because such a procedure might change the biological activity of FCS in an unpredictable way, we tested batches of FCS from different companies to avoid dialysis. Best results were obtained in our experimental series using FCS from GIBCO (Paisley, Scotland).

PB-MO produce cysteine by uptake and intracellular reduction of cystine. Therefore, the costimulatory potential of PB-MO strictly depends on the availability of extracellular cystine (Fig. 7). In the absence of cystine, PB-MO completely fail to mediate any reactivity of LP-T to CD3 stimulation. According to

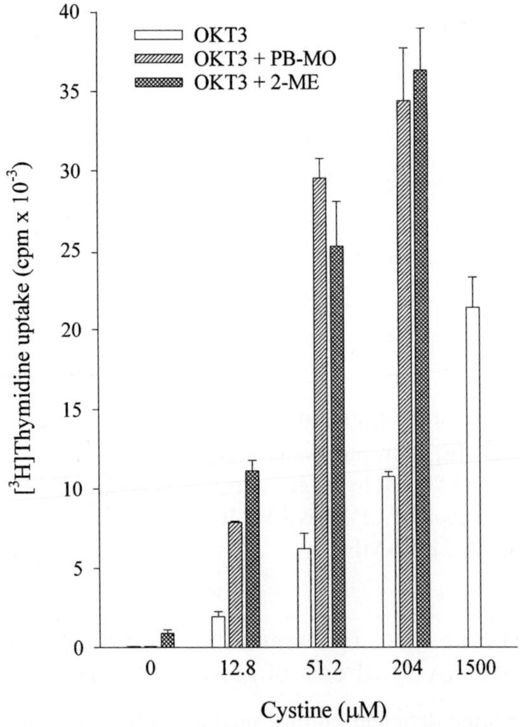

FIG. 7. Dependence of CD3 reactivity of LP-T on the availability of extracellular cystine in the absence and presence of 2-ME (10 μM) and PB-MO, respectively. Irradiated PB-MO were added at 30% of total cell number.

the molecular action, similar results are obtained with 2-ME (10 μM), although some minor proliferation of LP-T can be observed consistently in cystine-deficient medium. At physiological cystine concentrations (60–70 μM in human plasma), both PB-MO and 2-ME allow a significant proliferative immune response of LP-T after CD3 stimulation. In the absence of 2-ME or PB-MO, excessive high and unphysiological cystine concentrations would be needed (1.5 mM) to partly compensate for the low membrane transport activity for cystine in lymphocytes (Fig. 7).

Is it possible to substitute the thiol compound cysteine in cystine-deficient medium for the stimulatory function of PB-MO or 2-ME in cystine-containing medium? To address this question, LP-T (8×10^5/well) are cultured in 24-well plates (Costar, Bodenheim, Germany) in an initial volume of 1 ml/well. Cultures are primarily set up in cystine-free medium and are supplemented with cysteine in 15-μl volumes every 6 hr to reach a final concentration of 30 μM each time. Given the short half-life of cysteine of about 36 min in culture (see Fig. 2), it follows that the cysteine concentration is below 15 μM for nearly 90% of the culture period. Other cultures receive equimolar amounts of cystine (15 $\mu M = 30$ μM cysteine equivalents). After 84 hr of culture, wells are pulsed individually with [^{3}H]thymidine (5 μCi/ml). Cells are harvested at 18 hr postmetabolic labeling, at which time cumulative cystine concentrations have reached the unphysiologic level of 255 μM (510 cysteine equivalents). However, the proliferation of LP-T in the presence of very low but physiological relevant cysteine concentrations is nearly twofold higher as compared to supplementation with equimolar amounts of cystine (Table II). The specific requirement for cysteine for the synthesis of

TABLE II

INFLUENCE OF CYSTEINE VERSUS EQUIMOLAR AMOUNTS OF CYSTINE ON THE CD3 REACTIVITY OF LP-T[a]

Treatment	[^{3}H]Thymidine uptake (cpm $\times 10^{-4}$)
Cystine-deficient medium[a]	1.02 ± 0.11[d]
Cysteine (30 μM, every 6 hr)[b]	42.51 ± 1.44
+BSO (100 μM)[c]	0.98 ± 0.12
Cystine (15 μM, every 6 hr)[b]	22.43 ± 1.08

[a] Cultures were primarily set up in cystine-deficient RPMI 1640 in 24-well plates at 8×10^5/ml. FCS used in this experiment was from Sigma (Taufkirchen, Germany) and allowed some minor proliferation when added at 10% to cystine-deficient RPMI 1640.

[b] Cysteine was added every 6 hr at a final concentration of 30 μM each time during the entire culture period. Alternatively, equimolar amounts of cystine were added (15 $\mu M = 30$ μM cysteine equivalents).

[c] BSO was added at the beginning of culture.

[d] Results are presented as means ± SD of triplicate cultures.

GSH is demonstrated by the finding that addition of the glutathione synthesis inhibitor BSO (100 μM) abolishes the cysteine-driven restoration of CD3 reactivity of LP-T.

The increase in glutathione content in lymphocytes due to the increased availability of cysteine is well documented.[19,20] As an alternative to HPLC, intracellular GSH and GSSG can be differentially determined on a large scale by the spectrophotometric method of Griffith[26] using 96-well ELISA plates.

Principle. DTNB reacts with GSH to form the yellow product 5′-thionitrobenzoic acid, which can be detected at 412 nm. GSSG formed is converted to GSH in a cycling NADPH-dependent enzymatic reaction catalyzed by glutathione reductase. For determination of GSSG, GSH is derivatized by 2-vinylpyridine prior to reaction with DTNB.

Reagents

Sulfosalicylic acid, 2.5%
Sodium phosphate buffer, 150 mM containing 0.6 mM EDTA, pH 7.5
NADPH (Serva, Heidelberg, Germany), 0.6 mM in phosphate buffer
DTNB (Serva), 6 mM in phosphate buffer
Triethanolamine (Merck, Darmstadt, Germany)
2-Vinylpyridine (Sigma, Taufkirchen, Germany)
Glutathione reductase (Sigma), 14 U/ml in phosphate buffer
GSH (Serva), 20 mM in 2.5% SSA (for preparation of standards)
GSSG (Serva), 20 mM in 2.5% SSA (for preparation of standards)
Solution A: 9 ml NADPH and 2.17 ml DTNB are diluted with water to 20 ml

Procedure. At least 2×10^6 cells are lysed in 400 μl of SSA and incubated for 10 min on ice. After centrifugation, 10 μl of supernatant is mixed with 10 μl of sodium phosphate buffer and 165 μl of solution A in a 96-well ELISA plate. One minute later, absorbance is recorded at 412 nm using an ELISA reader (VICTOR^2, Wallac, Freiburg, Germany). The reaction is started by the addition of 40 μl of glutathione reductase 5 min later (total volume 225 μl/well). The total glutathione (tGSH) content is proportional to the increase in absorbance after 6 min at 25°. For determination of GSSG, 100 μl of each sample is preincubated with 4 μl of triethanolamine and 2 μl of vinylpyridine for 20 min at room temperature. Twenty microliters of this mixture is then used for the determination of glutathione according to the procedure described earlier. GSH and GSSG are used as standards. The amount of reduced GSH is calculated as follows: GSH = tGSH $-2 \times$ GSSG.

[26] O. W. Griffith, *Anal. Biochem.* **106,** 207 (1980).

Biological Implications

The availability of cysteine to lymhocytes is a critical parameter for the synthesis of glutathione, which, in turn, is essential for cell cycle progression.[11] The capacity of antigen-presenting cells to produce cysteine by uptake and intracellular reduction of cystine modulates the redox state in the microenvironment of intestinal T lymphocytes. This thiol-mediated redox regulation represents a novel mechanism that allows a dynamic modulation of the responsiveness of intestinal LP-T after antigen receptor-induced activation. The incapability of resident LP-MO to produce cysteine contributes to the hyporesponsiveness and low proliferative potential of LP-T in the normal gut, despite their continuous exposure to luminal antigens. In inflammatory bowel disease, macrophage populations with a different phenotype dominate in severely inflamed areas, especially near vessels, and result from a sustained recruitment of monocytes from peripheral blood.[23,24] As a consequence, the release of cysteine by recently recruited PB-MO in the microenvironment of LP-T might contribute to the increased lymphocyte reactivity in inflammatory bowel disease. In addition to bacterial wall products (LPS), receptor binding (CD2) to CD58 on PB-MO—as it occurs during cellular interaction with LP-T—enhances cysteine release by PB-MO considerably,[25] thereby initiating a bidirectional signal that provides sufficient constimulation to LP-T to increase intracellular glutathione synthesis and to allow antigen receptor-induced T-cell proliferation. Oxidative stress due to the infiltration of large numbers of granulocytes is regarded to be an important mediator of cytotoxic tissue damage in inflammatory bowel disease.[27] PB-MO continue to produce cysteine even in the presence of hydrogen peroxide and thus maintain a balanced redox state in their microenvironment, allowing mononuclear cells to escape oxidative damage and to carry out a proliferative immune response.

Acknowledgments

This work was supported by a grant from the Medical Research Council of the University of Heidelberg and from the Deutsche Forschungsgemeinschaft (SFB 405).

[27] V. Gross, H. Arndt, T. Andus, K. D. Palitzsch, and J. Schölmerich, *Hepato-Gastroenterol.* **41,** 320 (1994).

[21] Linker for Activation of T Cells: Sensing Redox Imbalance

By SONJA I. GRINGHUIS, FERDINAND C. BREEDVELD, and CORNELIS L. VERWEIJ

Introduction

The role of T lymphocytes in the pathogenesis of rheumatoid arthritis (RA) remains unclear, although the prolonged presence of hyporesponsive T lymphocytes within the inflamed joints of patients with RA is considered to contribute to and perpetuate the disease. T lymphocytes interact with other immune cells, such as B lymphocytes, monocytes, and neutrophils, that have invaded the synovium of the inflamed joints of RA patients and hence play a role in the production of mediators that instigate the destruction of the joint architecture. However, T lymphocytes are unable to respond to antigen-dependent stimulation and to undergo activation-induced cell death (AICD, apoptosis) and thus seem to accumulate in the inflamed joints where they persist in inducing damage of the joint cartilage.[1,2]

The hyporesponsiveness of synovial fluid (SF) T lymphocytes seems to correlate with markers of chronic oxidative stress, such as significantly decreased intracellular levels of the antioxidant glutathione (GSH) in the SF T lymphocytes and increased extracellular levels of another important redox regulator thioredoxin (TRX) within the synovial fluid, and also with the diminished tyrosine phosphorylation of substrates on T-cell receptor (TCR) stimulation.[3–5] The environment of oxidative stress in inflamed joints of RA patients prompted us to look at the effect of redox imbalances on proteins involved in the TCR signaling to determine the cause of the hyporesponsiveness of SF T lymphocytes. We focused our studies on the adaptor protein linker for activation of T cells (LAT), as LAT remains unphosphorylated in SF T lymphocytes on TCR stimulation.[6]

LAT plays a crucial role in the proximal signaling pathways induced by ligation of the TCR/CD3 complex. It is present in the so-called lipid rafts, glycolipid

[1] I. B. McInnes, B. P. Leung, R. D. Sturrock, M. Field, and F. Y. Liew, *Nature Med.* **3,** 189 (1997).

[2] G. S. Panayi, *Curr. Opin. Rheumatol.* **9,** 236 (1997).

[3] M. M. Maurice, H. Nakamura, E. A. van der Voort, A. I. van Vliet, F. J. Staal, P. P. Tak, F. C. Breedveld, and C. L. Verweij, *J. Immunol.* **158,** 1458 (1997).

[4] M. M. Maurice, A. J. Lankester, A. Z. Bezemer, M. F. Geertsma, P. P. Tak, F. C. Breedveld, and C. L. Verweij, *J. Immunol.* **159,** 2973 (1997).

[5] M. M. Maurice, H. Nakamura, S. I. Gringhuis, T. Okamoto, S. Yoshida, F. Kullmann, S. Lechner, E. A. van der Voort, A. Leow, J. Versendaal, U. Muller-Ladner, J. Yodoi, P. P. Tak, F. C. Breedveld, and C. L. Verweij, *Arthritis Rheum.* **42,** 2430 (1999).

[6] S. I. Gringhuis, A. Leow, E. A. M. Papendrecht-van der Voort, P. H. J. Remans, F. C. Breedveld, and C. L. Verweij, *J. Immunol.* **164,** 2170 (2000).

0076-6879/02 $35.00

(GPI)-enriched microdomains, in the plasma membrane of T lymphocytes and natural killer (NK) cells as an integral membrane protein. LAT consists of an α-helical transmembrane structure and a cytoplasmic tail, which functions as a docking site for other signaling proteins on T-cell activation. The presence of LAT within the lipid rafts is necessary for its function in the TCR signaling cascade and is regulated through the palmitoylation of a cysteine residue, Cys-26, just proximal of the α helix.[7,8] Upon TCR ligation, several tyrosine residues in the cytoplasmic tail of LAT become phosphorylated by the tyrosine kinase ZAP-70 (ζ-associated kinase of 70 kDa).[9] The crucial step in receptor signaling is the congregation of signaling proteins in multiprotein complexes at the cellular membrane to bring them into proximity of each other so that they are able to transduce, amplify, and enhance the initial stimulus through mechanisms such as phosphorylation and conformational changes.[10] The phosphorylated tyrosine residues of LAT serve as docking sites for SH2 domain containing signaling proteins such as phospholipase C-γ1 (PLC-γ1), growth factor receptor-bound protein 2 (Grb2), phosphatidylinositol (PI) 3-kinase, Grb2-related adaptor protein (Grap), the SH2 domain-containing leukocyte protein of 76 kDa (SLP-76), and inducible T-cell kinase (Itk), which will ultimately lead to the transcriptional activation of genes encoding proteins necessary for T lymphocytes to exert their function in the immune response.[11–14]

This chapter describes methods to determine the subcellular localization, conformation, and phosphorylation status of LAT under variable intracellular redox conditions.

Procedure

T Lymphocyte Isolation

T lymphocytes from heparin-collected peripheral blood or synovial fluid from RA patients are isolated through a negative selection procedure. Mononuclear cell suspensions are prepared by Ficoll–Hypaque density gradient centrifugation. Monocytes, B lymphocytes, and NK cells are depleted by incubating with MAbs against CD14, CD16, and CD19 [10 μg of each MAb per 20×10^6 of mononuclear

[7] W. Zhang, R. P. Trible, and L. E. Samelson, *Immunity* **9,** 239 (1998).

[8] J. Lin, A. Weiss, and T. S. Finco, *J. Biol. Chem.* **274,** 28861 (1999).

[9] W. Zhang, J. Sloan-Lancaster, J. Kitchen, R. P. Trible, and L. E. Samelson, *Cell* **92,** 83 (1998).

[10] M. G. Tomlinson, J. Lin, and A. Weiss, *Immunol. Today* **21,** 584 (2000).

[11] S. K. Liu, N. Fang, G. A. Koretzky, and C. J. McGlade, *Curr. Biol.* **9,** 67 (1999).

[12] N. J. Boerth, J. J. Sadler, D. E. Bauer, J. L. Clements, S. M. Gheith, and G. A. Koretzky, *J. Exp. Med.* **192,** 1047 (2000).

[13] K. A. Ching, J. A. Grasis, P. Tailor, Y. Kawakami, T. Kawakami, and C. D. Tsoukas, *J. Immunol.* **165,** 256 (2000).

[14] W. Zhang, R. P. Trible, M. Zhu, S. K. Liu, C. J. McGlade, and L. E. Samelson, *J. Biol. Chem.* **275,** 23355 (2000).

cells; Central Laboratory of the Netherlands Red Cross Blood Transfusion Service (CLB), Amsterdam, The Netherlands] and sheep anti-mouse IgG-coated dynabeads (Dynal, Oslo, Norway) for 1.5 hr, after which cells rosetted with immunomagnetic beads are removed with a Dynal magnetic particle concentrator. The remaining cell preparations contain >95% T lymphocytes as assessed by flow cytometric analysis after staining with a PerCP-conjugated anti-CD3 MAb (Becton Dickinson, San Jose, CA).

Control human peripheral blood T lymphocytes are obtained from healthy volunteer platelet donors. T lymphocytes are isolated from mononuclear cell suspensions after Ficoll–Hypaque density gradient centrifugation by 2-aminoethylisothiouronium bromide-treated sheep red blood cell (SRBC) rosetting. SRBC are lysed with 155 m*M* NH_4Cl, 10 m*M* $KHCO_3$, 0.1 m*M* EDTA, according to standard procedures. The remaining cell preparations contain >92% T lymphocytes as assessed by FACS analysis after staining with a PerCP-conjugated anti-CD3 MAb (Becton Dickinson) and are allowed to rest for 16 hr to ensure that they fully return to an unactivated state.

After isolation, T lymphocytes are kept at 37°, 5% CO_2 in Iscove's modified Dulbecco's medium (IMDM) (GIBCO BRL/Life Technologies, Gaithersburg, MD) containing 10% fetal calf serum (FCS; GIBCO BRL) supplemented with 100 U/ml penicillin and 100 μg/ml streptomycin (Roche, Mannheim, Germany). Stimulation of T lymphocytes (5×10^6/ml) is performed with 1 μg/ml anti-CD3 MAb (IXE; CLB).

Modulation of Intracellular GSH Levels by NAC and BSO

A key characteristic of cells in circumstances of chronic oxidative stress is the depletion of intracellular levels of glutathione. GSH is the major cellular antioxidant, and disturbances in intracellular GSH levels cause major redox imbalances.

N-Acetyl-L-cysteine (NAC; Sigma, St. Louis, MO) can serve as a precursor in the synthesis of GSH, and hence the treatment of cells with NAC results in the supplementation of intracellular GSH levels. NAC is used at a final concentration of 5 m*M*. (DL-Buthionine-(*S*,*R*)-sulfoximine (BSO; Sigma) is an inhibitor of γ-glutamylcysteine synthetase, an essential enzyme that catalyzes the rate-limiting step in the synthesis of GSH. Treatment of T lymphocytes with BSO thus leads to a significant depletion of their intracellular GSH levels. BSO is added at a final concentration of 200 μM.

Preparation of Membrane and Cytoplasmic Cell Fractions, Followed by Immunoprecipitation and Western Blotting

For subcellular fractionation of T lymphocytes and analysis of the subcellular localization of LAT, 10×10^6 purified T lymphocytes are disrupted by shearing through a 25-gauge needle in 500 μl ice-cold extraction buffer [50 m*M* Tris–HCl (pH 7.0), 10 m*M* KCl, 1 m*M* $CaCl_2$, 1 m*M* $MgCl_2$, 1 m*M* Na_3VO_4] supplemented with protease inhibitors [10 μg/ml leupeptin (Sigma), 10 μg/ml pepstatin A (Sigma), 0.4 m*M* phenylmethylsulfonyl fluoride (PMSF; Sigma)]. To maximize the disruption of cells, insoluble debris after the first round of shearing is spun down at 1600*g* for 10 min at 4°, and the extraction step is repeated on the pellet after adding another 500 μl ice-cold extraction buffer. Supernatants from both extraction rounds are pooled afterward. Separation of the membrane and cytoplasmic fractions is performed by ultracentrifugation of the pooled supernatants at 100,000*g* for 60 min at 4°. The supernatant after ultracentrifugation contains the cytoplasmic proteins. Triton X-100 is added to this fraction to an end concentration of 1% to solubilize the proteins. The pellet containing the membrane proteins is resuspended in 500 μl buffer [10 m*M* TEA (pH 7.8). 150 m*M* NaCl, 5 m*M* EDTA, 1 m*M* Na_3VO_4, 1% Nonidet P-40] supplemented with protease inhibitors (10 μg/ml leupeptin, 10 μg/ml pepstatin A, 0.4 m*M* PMSF).

To determine in which fraction LAT is present, LAT is immunoprecipitated from both the membrane and cytoplasmic extracts by adding 4 μg rabbit anti-LAT pAb (06-807, Upstate Biotechnology, Lake Placid, NY) and left for 16 hr at 4° while rotating and an additional 2 hr after the addition of 25 μl of protein A-agarose beads (50% slurry; Santa Cruz Biotechnology, Santa Cruz, CA). The immunoprecipitated LAT–agarose complexes are then spun down in a microcentrifuge for 30 sec at 4°, washed twice with lysis buffer, and resuspended in 30 μl of 1× reducing SDS–PAGE sample buffer. After boiling for 5 min, the samples are loaded and separated on 10% SDS–PAA gels using Rainbow-colored protein molecular weight markers (Amersham, Little Chalfont, UK) as a reference. The proteins are transferred onto a polyvinylidene difluoride (PVDF) membrane (Millipore, Bedford, MA) for Western immunodetection. The membrane is blocked in phosphate-buffered Saline (PBS) containing 5% skim milk and 0.01% Tween 20 for 1 hr to inhibit specific binding of the antibodies to the membrane. LAT detection is performed by incubating the membranes with a rabbit pAb against LAT (1 : 1000) for 16 hr, subsequently with a secondary antibody [HRP (horseradish peroxidase)]-conjugated swine anti-rabbit Ig-HRP (1 : 5000; DAKO, Glostrup, Denmark) for 3 hr, and then assayed using the ECL detection system (Amersham).

Immunofluorescence Staining and Microscopy

For visualization of the subcellular localization of LAT in intact cells, after incubation in medium containing BSO or NAC as indicated, 1×10^5 T lymphocytes are mounted onto adhesive microscope slides, air dried, and kept frozen until staining. Prior to staining, cells are fixed in 4% *p*-formaldehyde in PBS for 15 min at room temperature (RT). After three washes in PBS containing 5% bovine serum albumin (BSA), cells are permeabilized using 0.1% Triton X-100 in PBS for 4 min at RT, washed three times (PBS/BSA), and preblocked for 45 min at RT in PBS containing 10% FCS. Cells are then incubated with a rabbit pAb against LAT (1 : 250) in PBS/FCS for 45 min at RT, washed three times (PBS/BSA), and incubated with fluorescein isothiocyanate (FITC)-conjugated swine anti-rabbit Ig (1 : 100; DAKO) in PBS/FCS for 45 min at RT. A negative control is incubated with the secondary antibody only. After three final washes (PBS), cells are imbedded in 1 mg/m (*p*-phenyleendiamine (PPD), Sigma) in 90% glycerol/10% PBS to retain the fluorescence intensity of the FITC fluorochrome and covered with a coverslip. To ensure maintenance of the cell membrane integrity after BSO treatment, cells are stained immediately after fixation with a mouse MAb against CD3′Ω (1 : 25; Becton Dickinson) and detected using tetramethylrhodamine isothiocyanate (TRITC)-conjugated rabbit anti-mouse Ig (1 : 200; DAKO). Cells are viewed using a Leitz Aristoplan microscope (Leica, Wetzlar, Germany) equipped with a 100× objective and optics for FITC and TRITC. Images are taken with a Sony 3 CCD color video camera (Model DXC-9508).

Preparation of Whole Cell Lysates for Native Polyacrylamide Gel Electrophoresis, Followed by Western Blotting

For analysis of the effect of redox imbalances on the conformational structure of LAT, whole cell lysates are prepared from 20×10^6 T lymphocytes, after incubation in medium containing BSO as indicated, by lysis in 1 ml ice-cold TX lysis buffer [25 m*M* Tris–HCl (pH 7.6), 150 m*M* NaCl, 5 m*M* EDTA 1 m*M* Na_3VO_4, 50 m*M* β-glycerophosphate, 1% Triton X-100] supplemented with protease inhibitors (10 μg/ml leupeptin, 10 μg/ml pepstatin A, 0.4 m*M* PMSF) for 30 min on ice. Insoluble debris is subsequently spun down in a microcentrifuge for 10 min at 4°.

LAT is immunoprecipitated from whole cell lysates as described earlier, and immunoprecipitates are resuspended in 30 μl 1× nondenaturing sample buffer [110 m*M* Tris–HCl (pH 6.8), 40 m*M* EDTA, 8% glycerol, 0.01% bromphenol blue]. Samples are loaded and separated by native PAGE on a 12% nondenaturing PAA gel using electrophoresis buffer without SDS. LAT immunodetection is performed as described earlier after transfer of the proteins onto a PVDF membrane.

Preparation of Whole Cell Lysates for Detection of Phospho-LAT, Followed by Western Blotting

For detection of the phosphorylation status of LAT, whole cell lysates are prepared from 5×10^6 T lymphocytes, after incubation in medium containing NAC as indicated, either unstimulated or stimulated with anti-CD3 for 3 min to induce the TCR-mediated phosphorylation of LAT, by lysis in 300 μl ice-cold lysis buffer [10 m*M* triethanolamine (TEA) (pH 7.8), 150 m*M* NaCl, 5 m*M* EDTA, 1% Nonidet P-40] supplemented with phosphatase inhibitors to preserve the phosphorylation status of the proteins (10 m*M* Na_3VO_4, 50 m*M* β-glycerophosphate) and protease inhibitors (10 μg/ml leupeptin, 10 μg/ml pepstatin A, 0.4 m*M* PMSF) for 45 min on ice. Insoluble debris is spun down in a microcentrifuge for 15 min at 4°.

LAT is immunoprecipitated from whole cell lysates as described earlier. Immunoprecipitates are resuspended in 30 μl 1$\times$ reducing SDS–PAGE sample buffer and separated on 10% SDS–PAA gels as described previously. After transfer of the proteins onto a PVDF membrane and blocking of the membrane in PBS containing 5% skim milk and 0.01% Tween 20, immunodetection of tyrosine-phosphorylated proteins is performed by incubating the membranes with a mouse MAb against phospho-Tyr (PY99) (1 : 1000; sc-7020, Santa Cruz Biotechnology) for 16 hr, followed by incubation with a secondary antibody (HRP)-conjugated rabbit anti-mouse Ig (1 : 5000; DAKO) for 3 hr and then assayed using the ECL detection system (Amersham).

To confirm that equal amounts of LAT protein are precipitated from whole cell lysates, membranes are stripped of bound antibodies by incubating the membranes for 30 min at 50° in stripping buffer [100 m*M* 2-mercaptoethanol, 2% SDS, 62.5 m*M* Tris–HCl (pH 6.7)] and used for LAT immunodetection.

Results and Discussion

The subcellular localization of LAT in peripheral blood and synovial fluid T lymphocytes from RA patients as determined by Western blotting after fractionation of the membrane and cytoplasmic compartments is shown in Fig. 1.[6] In contrast to SF T lymphocytes, PB T lymphocytes from RA patients are not exposed to chronic oxidative stress and intracellular GSH levels are similar to those in T lymphocytes from healthy controls.[3] As expected, LAT was detected in the plasma membrane of these cells. However, in SF T lymphocytes that contain

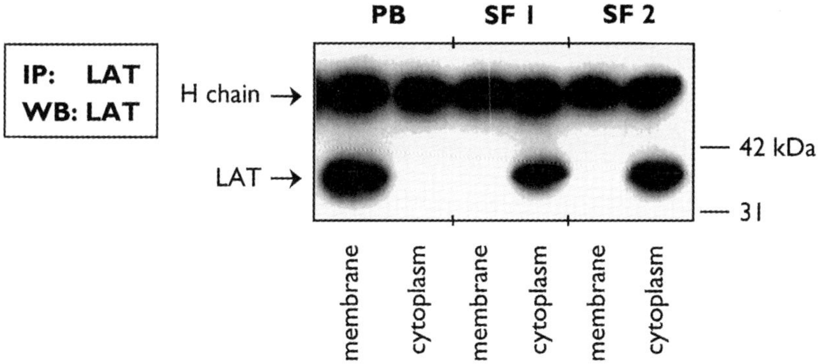

FIG. 1. Membrane and cytoplasmic fractions of peripheral blood (PB) and synovial fluid (SF) T lymphocytes were prepared. LAT was immunoprecipated and then detected by ECL Western blotting with anti-LAT antibodies and HRP-conjugated swine anti-rabbit antibodies. The detected heavy (H) chain comes from LAT antibodies used for immunoprecipitation. Reproduced with permission from S. I. Gringhuis, A. Leow, E. A. M. Papendrecht-van der Voort, P. H. J. Remans, F. C. Breedveld, and C. L. Verweij, *J. Immunol.* **164,** 2170 (2000); copyright © 2000, The American Association of Immunologists.

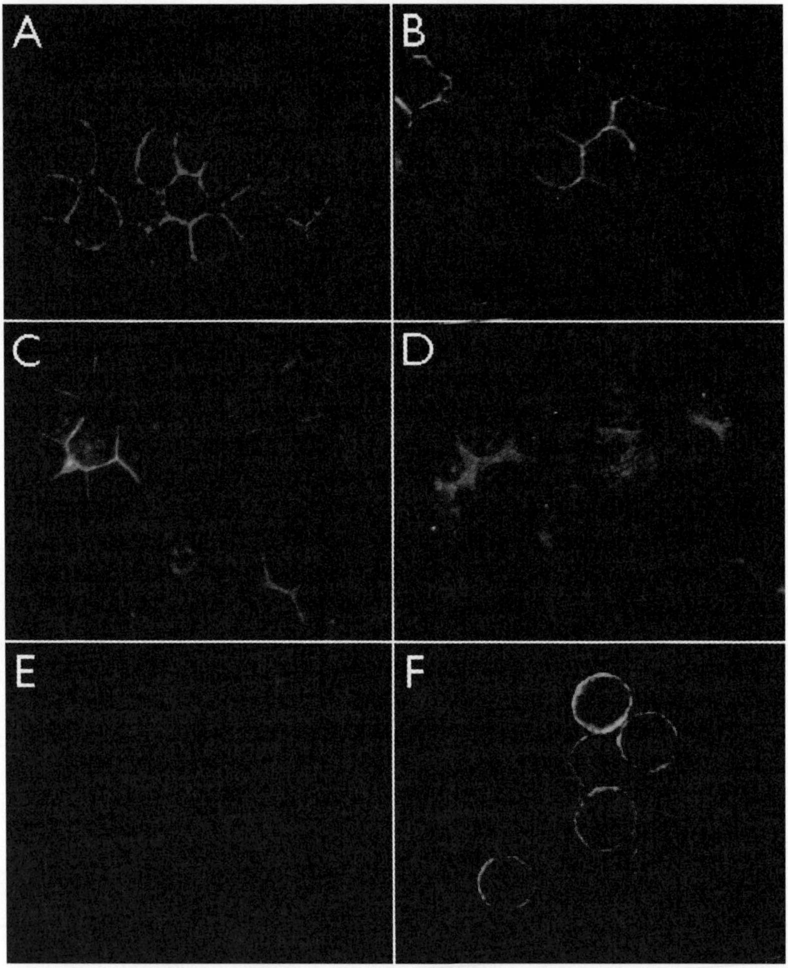

FIG. 2. LAT is gradually displaced from the plasma membrane in T lymphocytes from healthy controls due to treatment with BSO. T lymphocytes from healthy controls were cultured in medium for 72 hr (F) or treated with BSO and after (A) 0, (B) 16, (C) 48, or (D) 72 hr, fixed, permeabilized, labeled with specific antibodies for LAT, and then stained with FITC-conjugated swine anti-rabbit antibodies. (E) Fixed T lymphocytes from healthy controls after 72 hr of treatment with BSO were also labeled with monoclonal anti-CD3 antibodies followed by staining with TRITC-conjugated rabbit anti-mouse antibodies to check the integrity of the plasma membrane. Reproduced with permission from S. I. Gringhuis, A. Leow, E. A. M. Papendrecht-van der Voort, P. H. J. Remans, F. C. Breedveld, and C. L. Verweij, *J. Immunol.* **164,** 2170 (2000); copyright © 2000, The American Association of Immunologists.

significantly reduced intracellular levels of GSH, LAT is completely displaced from the membrane and sequestered in the cytoplasm of the cells. The distinction between membrane-anchored LAT in PB T lymphocytes from both RA patients and healthy controls and the cytoplasmic distribution observed in SF T lymphocytes was also clearly visible after immunofluorescence staining of LAT in intact cells on cytospins (Figs. 2A and 4A).[6]

To examine whether we could find an explanation for the membrane displacement of LAT in the decreased intracellular GSH levels of SF T lymphocytes, we performed experiments in which we modulated *in vitro* the redox balance of the cells and determined the effect of the variable intracellular GSH levels on the subcellular distribution of LAT by immunofluorescence staining. To see if the membrane displacement of LAT could be mimicked by depleting the GSH levels, T lymphocytes isolated from healthy controls were subjected to treatment with BSO over a period of 72 hr in which the intracellular GSH levels will become gradually reduced because the synthesis of GSH is abrogated. Figure 2B shows that after 16 hr of BSO treatment, LAT still resided in the plasma membrane of the cells. However, after 48 hr of BSO treatment, LAT was detected in both the membrane and the cytoplasm of the cells (Fig. 2C). After 72 hr of BSO treatment, LAT was completely absent from the plasma membrane and detectable exclusively in the cytoplasm (Fig. 2D). The displacement of LAT from the membrane was not induced by a breakdown of the membrane due to BSO treatment, as staining of T lymphocytes with an anti-CD3ε antibody showed that this component of the TCR/CD3 complex was still localized in the membrane (Fig. 2E). The membrane displacement of LAT could not be attributed to the lapsed time period of 72 hr, as in cells cultured in medium without BSO, LAT was still present in the cellular membrane (Fig. 2F).[6]

Because one function of GSH in the cell is to maintain the reduced state of sulfhydryl groups of proteins and thus enable them to gain their functional conformation, we tried to analyze whether the depletion of intracellular GSH levels by BSO might result in a conformational change of LAT, which could possibly interfere with the positioning of the α helix of LAT in the bilipid layer of the plasma membrane. T lymphocytes from healthy controls were cultured for 72 hr either in medium without BSO or in medium containing BSO, and LAT was immunoprecipitated from whole cell lysates and separated on a native PAA gel based on conformational and charge differences and detected after Western blotting. As shown in Fig. 3, LAT isolated from BSO-treated cells migrated with a considerably higher mobility compared with LAT present in cells with normal intracellular GSH levels, which is consistent with a conformational change of LAT. The results of these combined analyses suggest that newly translated LAT is affected by the reduced intracellular levels of GSH and, as a result, adapts an incorrect conformation, which hinders the transmembrane localization of LAT.[15]

[15] S. I. Gringhuis, E. A. M. Papendrecht-van der Voort, A. Leow, E. W. N. Levarht, P. H. J. Remans, F. C. Breedveld, and C. L. Verweij, submitted for publication.

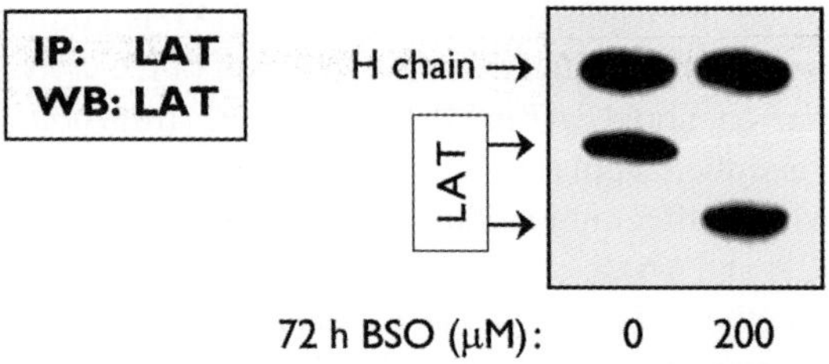

FIG. 3. Conformational changes of LAT are induced in T lymphocytes upon depletion of intracellular GSH levels. T lymphocytes from healthy donors were cultured for 72 hr in normal medium (0 μM BSO) or medium containing 200 μM BSO. LAT was immunoprecipitated from whole cell lysates and loaded on a native PAA gel. LAT was detected by ECL Western blotting with anti-LAT antibodies and HRP-conjugated swine anti-rabbit antibodies. The detected heavy (H) chain comes from LAT antibodies used for immunoprecipitation.

To determine whether supplementation of reduced intracellular GSH levels in SF T lymphocytes from RA patients would also be sufficient to restore the membrane localization of LAT, these cells were treated with NAC, a precursor in the synthesis of LAT. As shown in Fig. 4, immunofluorescence staining of LAT revealed that LAT was localized exclusively in the cellular membrane of SF T lymphocytes after culturing in the presence of NAC for 48 hr.[6]

The immunofluorescence staining of LAT on cytospins, however, does not distinguish between LAT present within lipid rafts, which has been shown previously

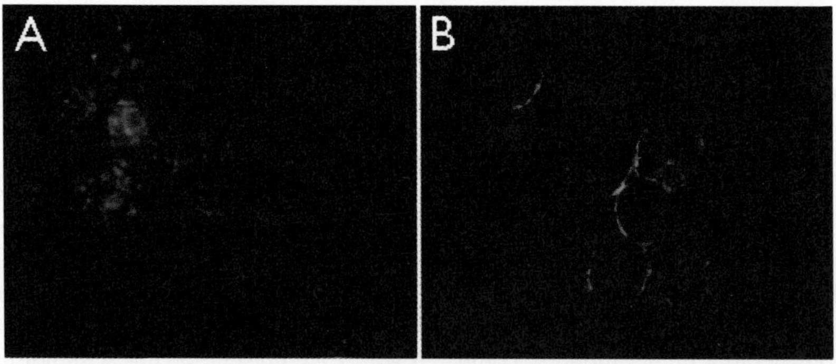

FIG. 4. Membrane localization of LAT is restored in synovial fluid T lymphocytes from RA patients after treatment with NAC. Synovial fluid T lymphocytes from RA patients were cultured in (A) medium or (B) treated with NAC for 48 hr and afterward fixed, permeabilized, labeled with specific antibodies for LAT, and then stained with FITC-conjugated swine anti-rabbit antibodies. Reproduced with permission from S. I. Gringhuis, A. Leow, E. A. M. Papendrecht-van der Voort, P. H. J. Remans, F. C. Breedveld, and C. L. Verweij, *J. Immunol.* **164,** 2170 (2000); copyright © 2000, The American Association of Immunologists.

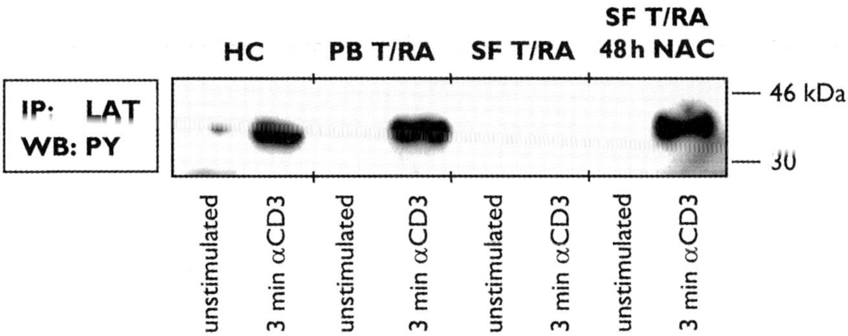

FIG. 5. Phosphorylation of LAT is restored in synovial fluid T lymphocytes from RA patients after treatment with NAC. T lymphocytes from healthy controls (HC), peripheral blood T lymphocytes from RA patients (PBT/RA), and synovial fluid T lymphocytes from RA patients (SF T/RA) either untreated or treated for 48 hr with NAC were left unstimulated or stimulated with anti-CD3 for 3 min. LAT was then immunoprecipitated from whole cell lysates. Phosphorylated LAT was detected by ECL Western blotting with antiphosphotyrosine (PY) antibodies and HRP-conjugated rabbit antimouse antibodies. Reproduced with permission from S. I. Gringhuis, A. Leow, E. A. M. Papendrecht-van der Voort, P. H. J. Remans, F. C. Breedveld, and C. L. Verweij, *J. Immunol.* **164,** 2170 (2000); copyright © 2000, The American Association of Immunologists.

to be a prerequisite for the functioning of LAT in TCR-mediated signaling pathways, or outside of these GPI-enriched membrane microdomains.[7,8] The phosphorylation of LAT is absent in SF T lymphocytes from RA patients on TCR stimulation.[4] To examine whether the treatment of SF T lymphocytes with NAC restored not just the membrane placement of LAT, but also its localization in lipid rafts, we immunoprecipitated LAT from whole cell lysates and determined the phosphorylation status of LAT by Western blotting and immunodetection with antiphosphotyrosine antibodies. Figure 5 shows that after 48 hr treatment of SF T lymphocytes with NAC, not only the membrane localization of LAT was restored, but also the phosphorylation of LAT on TCR stimulation.[6]

We conclude that the subcellular localization of LAT and its functioning in the TCR-mediated signaling pathways is very sensitive to redox imbalances, and hence leads to hyporesponsiveness of T lymphocytes on exposure to chronic oxidative stress due to significantly reduced intracellular levels of the antioxidant GSH.

Acknowledgments

We are grateful to the Department of Cell Biology and Genetics at the Erasmus University, Rotterdam, The Netherlands for the use of their immunofluorescence microscope/video equipment. This study was supported by a grant from the Dutch Arthritis Association.

[22] Generation of Prooxidant Conditions in Intact Cells to Induce Modifications of Cell Cycle Regulatory Proteins

By FRANCA ESPOSITO, TOMMASO RUSSO, and FILIBERTO CIMINO

Introduction

Reactive oxygen species (ROS) are free radical species containing an unpaired electron associated with oxygen (for a review, see Halliwell and Gutteridge[1]). Due to their unstable electronic configuration, ROS can capture an electron from other molecules. In cells, this often triggers a cascade of oxidations that can damage proteins, lipids, and nucleic acids. In proteins, ROS can cause changes in thiol groups, nitration of phenolic groups, conversion of some amino acid residue side chains to carbonyl derivatives, and aggregation and cross-linking reactions. In lipids, ROS can provoke the formation of lipid peroxides by oxidizing polyunsaturated fatty acids, whereas in nucleic acids, ROS can cause fragmentation, base modifications, and scission of the deoxyribose ring. The oxygen species typically responsible for these reactions are the hydroxyl radical, the superoxide anion, nitric oxide, and peroxynitrite. Although hydrogen peroxide is not a true radical, it can cause severe oxidative damage because it is transformed easily into the highly reactive hydroxyl radical through metal-catalyzed reactions. The effects exerted by ROS largely depend on their concentration and diffusion capability.

Very efficient antioxidant cellular defense systems scavenge and lower the damaging potential of ROS. The most noted are enzyme systems (superoxide dismutase and catalase) and enzymes involved in the glutathione (GSH) redox cycle, fat-soluble compounds (vitamins and bilirubin), and water-soluble compounds (vitamin C, uric acid, and GSH). Metal-chelating proteins (ferritin, transferrin, and ceruloplasmin) also belong to the antioxidant repertoire of the cell.

ROS found in living cells and organisms derive from exogenous and endogenous sources. The main exogenous sources of ROS are tobacco smoke, pesticides, pollutants, and ionizing reactions. An endogenous source of ROS is the mitochondrial electron transport chain in which a molecule of O_2 is reduced by four electrons to $2H_2O$, which, as a consequence of moderate electron leakage, produces ROS. Similarly, peroxisomes produce hydrogen peroxide as a by-product of fatty acid degradation, and the microsomal P450 cytocrome, monoamine oxidase, NO synthase, and other enzymes also yield oxidant by-products.

[1] B. Halliwell and J. M. C. Gutteridge, *Trends Neurosci.* **8,** 22 (1985).

0076-6879/02 $35.00

The same ROS perform specific "useful" functions. For example, stimulated polymorphonuclear leucocytes and macrophages produce superoxide ions to destroy bacteria. Moreover, an increasing body of evidence indicates that ROS are generated in most living cells by specific machineries and can function as subcellular messengers. To explore this possibility, we developed a procedure that causes redox perturbations in intact cells.

Redox Conditions Affect Protein Structure and Function

Redox conditions can be manipulated to alter protein structure and function *in vitro*. Exposure of purified proteins or protein extracts to hydrogen peroxide or to dithiothreitol (DTT) is widely used to study the role of disulfide bonds in protein function. For example, the key cysteine sulfydryl groups in DNA-binding domains of Zn finger transcription factors, such as the glucocorticoid receptor (GR), are very sensitive to redox modifications. Sulfydryl-modifying reagents inhibit receptor binding to DNA; this effect is reversed by DTT.[2] Similarly, binding to DNA of Sp1, another Zn finger transcription factor, is decreased remarkably in nuclear extracts of aged rats and is restored by DTT.[3] The oxidation of methionine residues in some proteins may affect their biological activity,[4] whereas it may have little or no effect on others.[5]

Various compounds and experimental systems have been used to alter redox conditions in intact cells. Hydrogen peroxide is the most common oxidizing agent used *in vivo*. The xanthine/xanthine oxidase system has also been used to generate ROS in intact cells.[6] This flavoprotein catalyzes the oxidation of hypoxanthine to xanthine and then to uric acid by a complex reaction in which molecular oxygen is the final acceptor of electrons.[7] Other oxidants (quinones, paraquat, and iron) generate ROS in the presence of O_2 and an electron donor, e.g., ascorbate or a thiol compound. Oxidative stress can also be induced in living systems by γ rays and high-energy electrons that produce hydroxyl radicals through the radiolysis of water.[8] Moreover, ROS scavenging systems can be manipulated *in vivo* to induce the accumulation of endogenous ROS. The most common of these methods exploits agents that inhibit enzymes involved in the synthesis of antioxidant

[2] W. Tienrungroj, S. Meshinchi, E. R. Sanchez, S. E. Pratt, J. F. Grippo, A. Holmgren, and W. B. Pratt, *J. Biol. Chem.* **262,** 6992 (1987).

[3] R. Ammendola, M. Mesuraca, T. Russo, and F. Cimino, *J. Biol. Chem.* **267,** 17944 (1992).

[4] W. Vogt, *Free Radic Biol. Med.* **18,** 93 (1995).

[5] R. L. Levine, L. Mosoni, B. S. Berlett, and E. R. Stadtman, *Proc. Natl. Acad. Sci. U.S.A.* **93,** 15036 (1996).

[6] B. Halliwell, *FEBS Lett.* **92,** 321 (1978).

[7] L. S. Terada, D. Piermattei, G. N. Shibao, J. L. McManaman, and R. M. Wright, *Arch. Biochem. Biophys.* **348,** 163 (1997).

[8] A. J. S. C. Vieira, J. P. Telo, and R. M. B. Dias, *Methods Enzymol.* **300,** 194 (1999).

molecules. A case in point is 1-chloro-2,4-dinitrobenzene, an irreversible inhibitor of thioredoxin reductase, which plays an important role in signal transduction and in the defense against oxidative stress.[9] However, most procedures devised to induce an intracellular oxidized environment are based on GSH. This molecule is essential for the maintenance of the thiols of many compounds (i.e., proteins and antioxidants), for reduction of ribonucleotides to deoxyribonucleotide precursors of DNA, and for protection against oxidative stress.[10] GSH synthesis can be selectively inhibited *in vivo* by various methods, e.g., by buthionine sulfoximine, which is a transition-state inactivator of γ-glutamylcysteine synthetase that catalyzes the first limiting step of GSH synthesis. Alternatively, it can be inhibited by nonspecific agents: diamide (a thiol-oxidizing agent), *N*-ethylmaleimide (a thiol-alkylating compound), and butylhydroperoxide (reviewed in Meister[10]). Diethylmaleate (DEM) also modifies intracellular redox conditions. This compound decreases the levels of intracellular GSH, thereby resulting in an increased GSSG/GSH ratio, which is normally 0.01. Diethylmaleate decreases the GSH concentration through a reaction catalyzed by the enzyme glutathione-*S*-transferase.[11]

$$\text{GSH} + \text{DEM} \xrightarrow{\text{GSH transferase}} \text{GSH/DEM conjugate}$$

This reaction entraps GSH, thereby resulting in the accumulation of ROS. Here we describe a procedure to induce *in vivo* modifications of the cellular redox state by DEM.

Materials

Diethylmaleate (DEM, Cat. No. M-5887) and *N*-acetylcysteine (NAC, Cat. No. A-9165) are from Sigma. The compound is added to the culture medium in the dish directly from the manufacturer's stock solution (6 *M*), and the solution is stored at 4°. The *N*-acetylcysteine (NAC) solution (30 m*M* final concentration) is made in phosphate-buffered saline (PBS) and is freshly prepared for each experiment. The pH of the NAC solution (approximately pH 2) must be increased to 7.4 by adding drops of sodium hydroxide. The intracellular GSH concentration is measured using the Bioxytech GSH 400 kit (Oxis International Inc., Portland, OR).

All the cell lines or primary cells used are from ATCC. Cells are cultured at 37° in a 5% CO_2 atmosphere in Dulbecco's modified Eagle's medium (DMEM, from GIBCO) or RPMI (GIBCO) containing 10% fetal calf serum (Hiclone), penicillin (50 IU/ml), and streptomycin (50 μg/ml). The p53-specific monoclonal antibody (Pab421) is from Oncogene Science. The anti-p21 polyclonal antibody is from Santa Cruz Biotechnology. [γ-^{32}P]ATP and [α-^{32}P]ATP (3000 Ci/mmol) are from

[9] A. Holmgren, *Annu. Rev. Biochem.* **54,** 237 (1985).

[10] A. Meister, *Pharmacol. Ther.* **51,** 155 (1991).

[11] E. Boyland and L. F. Chasseaud, *Biochem. J.* **104,** 95 (1967).

Amersham Pharmacia Biotech. The cDNA radioactive probes for Northern blot analysis are prepared with the random-priming method.

Generation of Prooxidant Conditions By Diethylmaleate

Recommended Experimental Procedures and Protocols

Prefixed amounts of DEM (from the manufacturer's 6 *M* stock solution) are added to the culture medium in the plastic dish. The dishes must be rocked gently several times to ensure that the reagent is dissolved completely and distributed uniformly. With this procedure, the effects of DEM on the molecular parameters examined are efficient and reproducible.

The amount of intracellular GSH is a measure of the direct effect of DEM. Several methods are available for this assay, but many are labor intensive and poorly sensitive, which means a large number of cells must be used. We measure the intracellular level of GSH after DEM treatment with an HPLC-based procedure[12] or a colorimetric assay (see Materials). The HPLC procedure reveals a 70% GSH depletion in HL60 human cell lines after a 1-hr exposure to 1 m*M* DEM.[13] We use the Bioxytech GSH-400 kit, which, unlike other colorimetric assays, is specific for GSH. In fact, while GSH is measured at 400 nm, total mercaptan (RSH) is measured at 356 nm. Because of its simplicity, the GSH-400 method can be used to measure GSH concentrations in large series of biological samples.

A typical result of GSH depletion in our experimental conditions is shown in Fig. 1. We examined the effect of 1 m*M* DEM on IMR90 human primary fibroblasts at various incubation times. Glutathione is decreased by about 50% after a 30-min exposure to DEM.

We evaluated the toxicity of various DEM concentrations in cultured cells by analyzing cell viability and cell growth curves. Concentrations of DEM in a range of 0.25–1 m*M* are well tolerated by mammalian cells even for many hours and produce biological effects. Tolerance to DEM varies among the cell lines examined: HeLa, Saos-2, Hep3B, and HL60 cells are by far the most resistant to DEM-associated toxicity that we have analyzed; 293 and Rat2 fibroblasts and IMR90 primary human fibroblasts are the least resistant. The effects of DEM are detectable after 5 min of treatment (see later), whereas toxic effects (decreased cell viability) become evident after 6 hr of exposure.

In yeast, DEM concentrations below 10 m*M* are ineffective; biological effects are seen at 10 m*M*.[14]

[12] D. J. Reed, J. R. Babson, P. W. Beatty, A. E. Brodie, W. W. Ellis, and D. Potter, *Anal. Biochem.* **106,** 55 (1980).

[13] F. Esposito, V. Agosti, G. Morrone, F. Morra, C. Cuomo, T. Russo, S. Venuta, and F. Cimino, *Biochem. J.* **301,** 649 (1994).

[14] V. Wanke, K. Accorsi, D. Porro, F. Esposito, T. Russo, and M. Vanoni, *Mol. Microbiol.* **32,** 753 (1999).

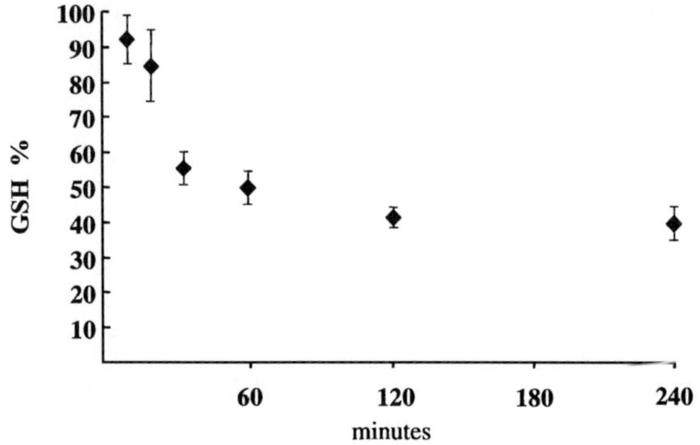

FIG. 1. Measurement of intracellular GSH during DEM treatment. Human primary fibroblasts (IMR 90) were cultured as described in the text and were exposed to 1 m*M* DEM for the indicated times. After DEM treatment, cell pellets were suspended in 5% metaphosphoric acid and homogenized, passing the cell suspension through a syringe; GSH levels of untreated cells were arbitrarily fixed to 100%. GSH depletion was calculated, at the indicated times, as a percentage of decrease respect to untreated cells. Results represent the means of three independent experiments.

Effects of Diethylmaleate on Cell Cycle Regulatory Proteins

We used the procedure described herein to study the effects of perturbation of intracellular redox conditions on proteins that regulate gene expression and cell cycle progression. Treatment of living cells with DEM modifies the DNA-binding activity of several transcription factors.[13,15] These proteins (Sp1, Egr1, and GR) interact with DNA through Zn finger DNA-binding domains, whose stability is probably affected by the oxidation of Cys residues interacting with the Zn ion. Zn–cysteine bonds are also present in the DNA-binding domain of p53. Here we demonstrate that DEM interferes with the binding to DNA of this transcription factor and, consequently, with its ability to regulate transcription *in vivo*.[16]

Nuclear extracts from COS7 cells, transfected with GR and p53 expression vectors and exposed to 1 m*M* DEM for different times, were analyzed for their binding to DNA by electrophoretic mobility shift assays. The intensity of the specific retarded bands clearly depends on the length of treatment (Fig. 2).[15,16]

DEM also induces quantitative and qualitative modifications in two important cell cycle regulatory proteins; p21^{waf1} and pRb. p21^{waf1} is an inhibitor of cyclin/cyclin-dependent kinase complexes. It is required for cell cycle progression and is the main target through which p53 negatively regulates cell growth.

[15] F. Esposito, F. Cuccovillo, F. Morra, T. Russo, and F. Cimino, *Biochim. Biophys. Acta* **1260,** 308 (1995).

[16] T. Russo, N. Zambrano, F. Esposito, R. Ammendola, F. Cimino, M. Fiscella, P. M. O'Connor, J. Jackman, C. W. Anderson, and E. Appella, *J. Biol. Chem.* **270,** 29386 (1995).

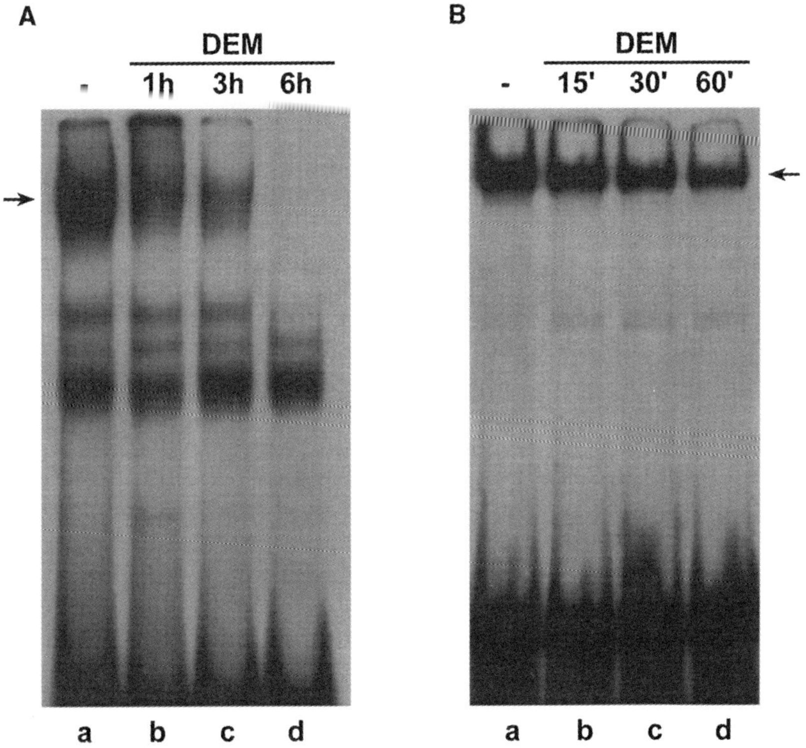

FIG. 2. Effects of DEM treatment on the DNA-binding efficiency of GR and p53. Cells were transfected with GR (A) and p53 (B) expression vectors; 36 hr after transfection, cells were treated with 1 m*M* DEM for various times (see later) and harvested for the preparation of cellular extracts. Electrophoretic mobility shift assays were done using [γ-^{32}P]oligonucleotide probes containing the glucocorticoid receptor (A) or p53 (B) responsive elements. (A) Lane a, labeled glucocorticoid receptor responsive element (GRE) oligonucleotide incubated with COS 7 extracts from untreated cells; lanes b–d, GRE oligonucleotide incubated with extracts from COS 7 cells treated with DEM for 1, 3, and 6 hr before harvesting, respectively. From F. Esposito, F. Cuccovillo, F. Morra, T. Russo, and F. Cimino, *Biochim. Biophys. Acta* **1260,** 308 (1995), with permission of Elsevier Science. (B) Lane a, p53 oligonucleotide incubated with extracts from Hep 3B cells transfected with a p53 expression vector; lanes b–d, as in a, but Hep 3B cells were exposed to DEM for 15, 30, and 60 min before harvesting, respectively. Arrows indicate the specific protein–DNA complexes. From T. Russo, N. Zambrano, F. Esposito, R. Ammendola, F. Cimino, M. Fiscella, P. M. O'Connor, J. Jackman, C. W. Anderson, and E. Appella, *J. Biol. Chem.* **270,** 29386 (1995), with permission of the American Society for Biochemistry and Molecular Biology.

Figure 3A shows a Northern blot experiment in which p21^{waf1} mRNA expression was studied in p53-null cells (T98G glioblastoma and SAOS-2 osteosarcoma cell lines, respectively).

In the presence of DEM, p21^{waf1} mRNA levels increased, whereas the concentration of GADD45, a DNA damage-induced gene, was unchanged. These

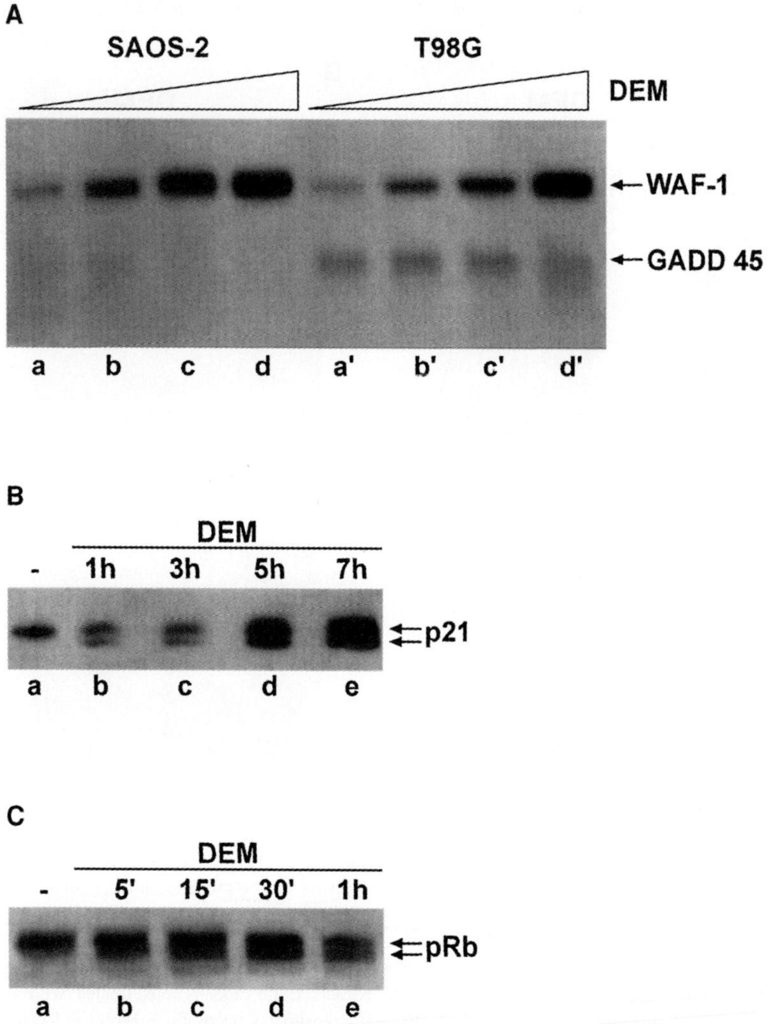

FIG. 3. Effects of DEM treatment on cell cycle regulatory proteins. (A) Northern analysis of p21^{waf1} and GADD45 mRNAs. SAOS-2 and T98G p53 null cells were exposed to different concentrations of DEM for 3 hr, and total RNA was analyzed by Northern blot using [α-^{32}P]p21^{waf1} and GADD45 cDNAs as probes. Lanes a and a′, untreated cells; lanes b–d and b′–d′, cells treated with 0.25, 0.5, and 1 m*M* DEM, respectively. From T. Russo, N. Zambrano, F. Esposito, R. Ammendola, F. Cimino, M. Fiscella, P. M. O'Connor, J. Jackman, C. W. Anderson, and E. Appella, *J. Biol. Chem.* **270,** 29386 (1995), with permission of the American Society for Biochemistry and Molecular Biology. (B) Western blot analysis of p21^{waf1} protein. HeLa cells were exposed to 1 m*M* DEM for the indicated times and harvested for protein extract preparation. Western blot analysis was performed using an anti-p21^{waf1} polyclonal antibody. (C) Western blot analysis of the retinoblastoma protein. HL60 cells were exposed to 1 m*M* DEM for the indicated times and harvested for protein extract preparation. Western blot analysis was performed using an antiretinoblastoma (pRb) monoclonal antibody. From F. Esposito, L. Russo, T. Russo, and F. Cimino, *FEBS Lett.* **470,** 211 (2000), with permission of Elsevier Science.

experiments also demonstrate that even though the DNA-binding efficiency of p53 is decreased greatly by DEM, the cell cycle is blocked through a p53-independent pathway.[16]

Because oxidative stress activates MAPK, a key enzyme in the regulation of mitogenic pathways, we evaluated the effect of DEM on p21^{waf1} mRNA expression in cells in which the MAPK pathway was blocked. The expression of *ras* and MEK-dominant negative mutants in cells exposed to DEM prevented the accumulation of p21^{waf1} mRNA.[17] The p21^{waf1} protein was also affected by DEM treatment. Figure 3B shows Western blot experiments, using anti-p21^{waf1} antibodies, with HeLa cells exposed to DEM for different times. Two distinct effects were observed: a late event (i.e., after DEM treatment for 3 hr) consisting of accumulation of p21^{waf1} in DEM-treated cells as a consequence of accumulation of p21^{waf1} mRNA (Fig. 3B) and an early event consisting in the appearance, very shortly after DEM exposure, of a faster migrating band of p21^{waf1} protein (FMp21), probably consequent to the activation of some serine–threonine phosphatases.[18] The hypothesis that this FM isoform corresponds to the hypophosphorylated protein was prompted by the finding that the Ser–Thr phosphatase inhibitor okadaic acid inhibits the appearance of the FM p21 isoform following DEM treatment.[18] Analogous dephosphorylation was demonstrated for the retinoblastoma protein (pRb), a well-known tumor suppressor, whose regulatory role in cell cycle progression depends on the degree of its phosphorylation.[19] Western blot analyses of protein extracts from HL60 cells, exposed for different times to DEM, showed the accumulation of faster migrating pRb isoforms that correspond to the hypophosphorylated protein (Fig. 3C).

Mechanisms of the Effects of Diethylmaleate in Cultured Cells

To evaluate whether DEM exerts a direct or an indirect oxidizing effect, we treated protein extracts *in vitro* with DEM or with hydrogen peroxide at the same concentrations used with intact cells.[15] As shown in Fig. 4A, the DNA-binding efficiency of GR is still present after incubation of cellular extracts on ice for 1 hr with DEM (lane b), but absent from extracts treated with hydrogen peroxide in the same conditions (lane c). This experiment demonstrates that DEM does not affect GR/DNA binding *in vitro,* suggesting that DEM is not a direct oxidant.

To test whether DEM increases intracellular ROS concentrations by depleting GSH levels, we pretreated cell cultures with NAC. Being a GSH precursor, NAC

[17] F. Esposito, F. Cuccovillo, M. Vanoni, F. Cimino, C. W. Anderson, E. Appella, and T. Russo, *Eur. J. Biochem.* **245,** 730 (1997).

[18] F. Esposito, F. Cuccovillo, L. Russo, F. Casella, T. Russo, and F. Cimino, *Cell Death Differ.* **5,** 940 (1998).

[19] F. Esposito, L. Russo, T. Russo, and F. Cimino, *FEBS Lett.* **470,** 211 (2000).

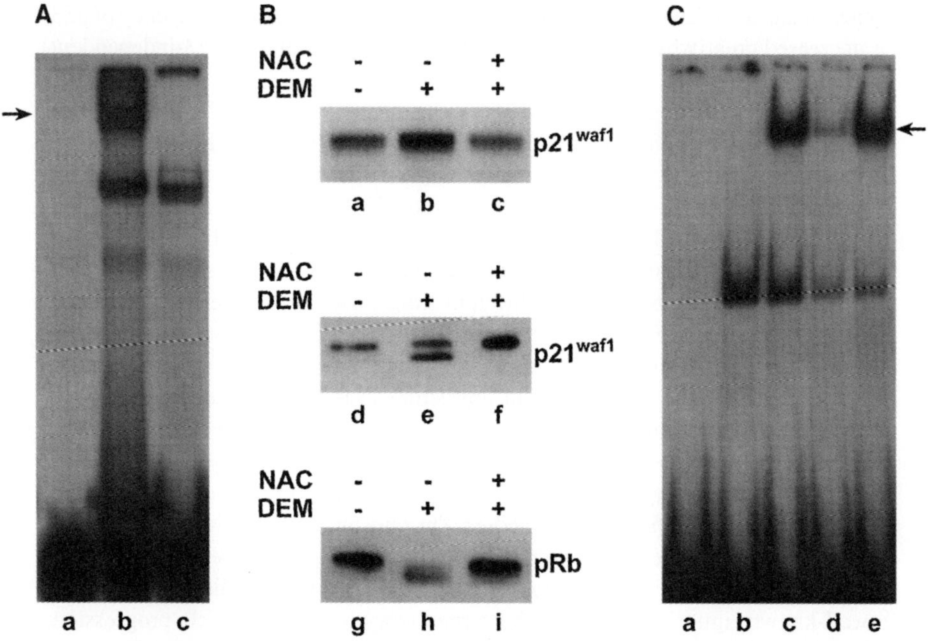

FIG. 4. Characterization of DEM effects. (A) Electrophoretic mobility shift assay of GR/DNA complexes. COS 7 cells were treated as described in the legend of Fig. 2. Lane a, GRE oligonucleotide alone; lane b, GRE oligonucleotide incubated with extracts from untreated cells incubated for 60 min on ice with 1 m*M* DEM; and lane c, GRE oligonucleotide incubated with extracts from untreated cells incubated for 60 min on ice with 20 m*M* hydrogen peroxide. From F. Esposito, F. Cuccovillo, F. Morra, T. Russo, and F. Cimino, *Biochim. Biophys. Acta* **1260,** 308 (1995), with permission of Elsevier Science. (B) Lanes a–c: Northern blot analysis of $p21^{waf1}$ mRNA. HeLa cells were pretreated with 30 m*M* NAC for 90 min and then with 1 m*M* DEM for other 90 min. Lane a, untreated cells; lane b, cells treated with DEM for 90 min; lane c, cells pretreated with NAC and then with DEM. From F. Esposito, F. Cuccovillo, M. Vanoni, F. Cimino, C. W. Anderson, E. Appella, and T. Russo, *Eur. J. Biochem.* **245,** 730 (1997), with permission of Blackwell Science Ltd. Lanes d–f: Western blot analysis of $p21^{waf1}$ protein. HeLa cells were pretreated with 30 m*M* NAC for 1 hr and then with 1 m*M* DEM for 30 min. Lane d, untreated cells; lane e, cells treated with DEM; lane f, cells pretreated with NAC and then with DEM. From F. Esposito, F. Cuccovillo, L. Russo, and F. Casella, T. Russo, and F. Cimino, *Cell Death Differ.* **5,** 940 (1998), with permission of Stockton Press. (Lanes g–i): Western blot analysis of the retinoblastoma protein. HL60 cells were pretreated with 30 m*M* NAC for 1 hr and then with 1 m*M* DEM for 30 min. Lane a, untreated cells; lane b, cells treated with DEM; lane c, cells pretreated with NAC and then with DEM. From F. Esposito, L. Russo, T. Russo, and F. Cimino, *FEBS Lett.* **470,** 211 (2000), with permission of Elsevier Science. (C) Electrophoretic mobility shift assay of p53/DNA complexes. Hep 3B cells, transfected with a p53 expression vector, were treated with DEM for 3 hr, washed with fresh medium and then harvested after 2 hr for the preparation of cellular extracts. Lane a, p53 labeled oligonucleotide alone; lanes b and c, p53 oligonucleotide incubated with extracts from Hep 3B cells transfected with an empty vector (lane b) or with a p53 expression vector (lane c); lane d, p53 oligonucleotide incubated with extracts from DEM treated cells; lane e, p53 oligonucleotide incubated with extracts from DEM treated cells and then washed extensively.

counteracts the effects of DEM by increasing intracellular GSH concentrations. NAC was added to the culture medium to obtain a final concentration of 30 m*M*. Treatment of cells with NAC solution must precede DEM exposure: because pretreatment longer than 3 hr with 30 m*M* NAC can be toxic for human fibroblasts and 293 cell lines, we pretreat for 1 hr. After exposure of HeLa cells to DEM for 3 hr, the level of $p21^{waf1}$ mRNA increased (Fig. 4B, lane b) in relation to DEM-induced GSH depletion. In fact, when cells were pretreated with NAC, the $p21^{waf1}$ mRNA level in DEM-treated cells decreased (Fig. 4B, lane c). This experiment shows that the accumulation of $p21^{waf1}$ mRNA on DEM treatment is counteracted by a GSH precursor and suggests that the effects of DEM depend on GSH depletion. Similarly, $p21^{waf1}$ and pRb dephosphorylation processes are prevented by NAC treatment (Fig. 4B, lanes f and i). The appearance of the FM p21 isoform after exposure to DEM for only 5 min and the abrogation of this effect by NAC suggest that even a minor decrease in GSH concentration, such as that observed after 5 min (about 10% decrease, see Fig. 1), can provoke significant changes in the intracellular environment and significantly affect cell metabolism.

Thus, with DEM-based procedures, one may analyze biological phenomena induced by small variations in intracellular ROS concentrations. Furthermore, most of the effects induced by DEM seem to be reversible. p53/DNA binding is completely restored within 2 hr of withdrawal of DEM from the culture medium (Fig. 4C).

Conclusions

This chapter described a method by which mild prooxidant conditions are generated in intact cells. We also described the effects of this manipulation on the regulation of the expression of some proteins involved in cell cycle progression. The method is based on the exposure of eukaryotic cell lines to DEM. Advantages of the DEM procedure are as follows: (i) low toxicity (good viability up to 6 hr after 1 m*M* DEM exposure); (ii) the effects are reversible (withdrawal of DEM from the culture medium completely restores the biological functions examined); and (iii) indirect action of DEM (it generates an oxidant environment only through GSH depletion).

It is generally acknowledged that the redox-based regulation of gene expression is an important regulatory mechanism and that ROS are subcellular messengers involved in the transduction of mitogenic signals. Given the modifications of cell cycle regulatory proteins caused by DEM (e.g., decreased DNA-binding activity of transcription factors, accumulation of $p21^{waf1}$ mRNA and protein, and activation of Ser–Thr protein phosphatases that induce rapid dephosphorylation of $p21^{waf1}$ and pRb), the procedure described here can be used to study the role of ROS as natural mediators of normal cellular functions.

Acknowledgments

We thank Dr. L. Russo (Ph.D. student) for the measurement of GSH. We are indebted to Jean Ann Gilder for editing. This work was supported by grants from Italian National Research Council (CNR) PF "Biotecnologie" and PSt/74 "Biologia dell'invecchiamento e sue conseguenze sul sistema assistenziale," from Associazione Italiana per la Ricerca sul Cancro (AIRC), and from Ministero dell'Università e della Ricerca Scientifica e Tecnologica (COFIN 98 and 2000; piano "Biomedicina" Prog. N. 1).

[23] Analysis of Transmembrane Redox Reactions: Interaction of Intra- and Extracellular Ascorbate Species

By MARTIJN M. VANDUIJN, JOLANDA VAN DER ZEE, and PETER J. A. VAN DEN BROEK

Introduction

The pivotal role of ascorbate in the defense against oxidants, as well as in other physiological processes, is generally recognized. To preserve ascorbate, it should remain in the reduced form. A number of systems inside the cell ensure the quick reduction of the oxidation products of ascorbate, which are ascorbate-free radical (AFR) and dehydroascorbic acid (DHA). These reduction reactions are very efficient, and AFR and DHA will therefore be virtually absent in a healthy cell. Extracellular DHA can be regenerated by transport into the cell followed by intracellular reduction. Alternatively, it has been shown that AFR and DHA can be reduced on the extracellular face of the cell.[1,2] This reaction involves a redox system in the plasma membrane, which uses intracellular NADH as an electron source. We found that intracellular ascorbate can also be an electron donor for this reaction.[3] However, the nature of the ascorbate-driven system in the plasma membrane remains uncertain. Most likely the electron transfer is mediated by a protein in the plasma membrane. However, it has also been suggested that small lipid-soluble molecules, such as α-tocopherol and coenzyme Q, can shuttle electrons from the intra- to the extracellular side of the membrane.[4–6] Irrespective

[1] J. M. Villalba, A. Canalejo, J. C. Rodriguez-Aguilera, M. I. Buron, D. J. Morre, and P. Navas, *J. Bioenerg. Biomembr.* **25,** 411 (1993).

[2] U. Himmelreich, K. N. Drew, A. S. Serianni, and P. W. Kuchel, *Biochemistry* **37,** 7578 (1998).

[3] M. M. VanDuijn, K. Tijssen, J. VanSteveninck, P. J. A. Van Den Broek, and J. Van Der Zee, *J. Biol. Chem.* **275,** 27720 (2000).

[4] J. M. May, Z. C. Qu, and J. D. Morrow, *J. Biol. Chem.* **271,** 10577 (1996).

[5] I. L. Sun, E. E. Sun, F. L. Crane, D. J. Morre, A. Lindgren, and H. Low, *Proc. Natl. Acad. Sci. U.S.A.* **89,** 11126 (1992).

0076-6879/02 $35.00

of the mechanism, this plasma membrane redox system efficiently helps maintain the concentration of extracellular ascorbate and is one of the main mediators in the interaction of intra- and extracellular ascorbate.

This chapter highlights techniques that can be used to study ascorbate-related redox reactions across cell membranes. First, the quantification of intra- and extracellular ascorbate species is discussed. Subsequently, we present methods to establish the proper intra- or extracellular concentrations of some ascorbate species, as well as methods for the detection of redox reactions between these intra- and extracellular molecules.

Materials

Ascorbate

Ascorbate is stable as a solid, and should only be dissolved on the day of use because of its susceptibility to oxidative degradation. When kept on ice, stock solutions have sufficient stability for many hours, especially in acidic buffers. Ascorbate concentrations can be checked by measuring the absorbance at 265 nm ($\varepsilon = 14{,}500\ \mathrm{cm}^{-1}M^{-1}$).

DHA

DHA is purchased as a solid. In solution it can degrade rapidly, especially at pH values above 5.[7] To avoid decomposition, solutions are made on ice and are used immediately. The purity of some commercial preparations has been questioned. It may therefore be required to test the DHA. An easy and convenient way to do this is by reducing DHA to ascorbate with, e.g., an excess amount of dithiothreitol (DTT; 5 m*M*), and measuring the resulting absorbance at 265 nm. More than 95% of the DHA should be recovered as ascorbate. DHA can also be prepared *in situ* by mixing ascorbate and a large amount of ascorbate oxidase (>0.5 U/ml). Alternatively, DHA can be prepared by oxidizing ascorbate with bromine. On ice, 5 μl bromine is added to 1 ml of 1 m*M* ascorbate. After mixing and a 30-sec reaction time, the solution is bubbled with nitrogen or argon to remove the bromine. After 10 min, when all the bromine is removed, the solution should have lost its typical brown color.

Ascorbate Oxidase

Ascorbate oxidase (EC 1.10.3.3) is a useful enzyme in the study of ascorbate and its free radical. Unfortunately, it does not retain activity in solution and must be

[6] A. Ilani and T. Krakover, *Biophys. J.* **51,** 161 (1987).
[7] A. M. Bode, L. Cunningham, and R. C. Rose, *Clin. Chem.* **36,** 1807 (1990).

prepared freshly before use. The enzyme can be purchased as a lyophilized solid. However, we prefer to use ascorbate oxidase adsorbed to small spatulas (Roche Diagnostics, Almere, The Netherlands). The spatulas contain 17 U of ascorbate oxidase, and the enzyme can be dissolved conveniently in a buffer of choice.

$Ni(en)_3^{2+}$

Tris-(ethylenediamine)-nickel(II) chloride 2-hydrate [$Ni(en)_3^{2+}$] can be prepared by dissolving 12 g $NiCl_2 \cdot 6H_2O$ in 60 ml H_2O and subsequently adding 14 ml of 70% ethylenediamine in water (v/v).[8] After reducing the volume to about 60% by evaporation in a boiling water bath, the solution is cooled, and purple crystals are formed. Crystallization can be promoted by the addition of ethanol. The crystals are obtained by filtration, two washes with ethanol, and air drying.

Measurement of Ascorbate Species

Ascorbate

Ascorbate has a strong absorption band at 265 nm, allowing simple and convenient spectrophotometric quantification. However, a prerequisite for such assays is the absence of interfering compounds that scatter or absorb light at that wavelength. When direct spectroscopic measurements are not possible, high-performance liquid chromatography (HPLC) can be used to separate ascorbate from other material absorbing at 265 nm. Many aspects of HPLC analysis of ascorbic acid were discussed in an earlier volume of this series.[9,10] HPLC analysis requires careful preparation of the samples to prevent oxidation of ascorbate. Precautions include cooling or freezing of samples, acidification, chelation of metal ions by EDTA or DTPA, flushing vials with inert gas, and deproteinizing the sample.

Both C-18 and SAX column packings have been employed for the separation of ascorbate species.[10,11] We have had good experience with Partisil-SAX material, which yields a single ascorbate peak from cell extracts. The system comprises an (auto)injector with a 100-μl loop, a Partisil SAX column (10 μm, 250 × 46 mm) with a 20-mm guard column, and a UV detector set at 265 nm. A fraction collector or a radiochemical detector can be added for the measurement of radio-labeled ascorbate or DHA. Ascorbate and DHA elute isocratically from the column with 7 m*M* potassium phosphate, 7 m*M* KCl, pH 4.0. However, after a run with cell

[8] H. M. State, *Inorgan. Syntheses* **6,** 200 (1960).

[9] M. Levine, Y. Wang, and S. C. Rumsey, *Methods Enzymol.* **299,** 65 (1999).

[10] A. M. Bode and R. C. Rose, *Methods Enzymol.* **299,** 77 (1999).

[11] L. F. Liebes, S. Kuo, R. Krigel, E. Pelle, and R. Silber, *Anal. Biochem.* **118,** 53 (1981).

extracts, the column should be flushed with high salt (0.25 *M* potassium phosphate, 0.5 *M* KCl, pH 5.0) for 5 min to remove other cellular anions. Thus, the use of a gradient controller is recommended. For more demanding applications, UV detection can be replaced by coulometric or amperometric electrochemical detection, which have superior sensitivity. The HPLC analysis of multiple samples can take several hours. If an autosampler is used, cooling of the sample vials during analysis is required to prevent sample degradation pending analysis. In addition, vials may be flushed with an inert gas to prevent oxidation. When many samples are analyzed, checking for degradation by adding standard samples at the beginning and end of a run is recommended.

HPLC analysis of ascorbate (and DHA) samples from cells requires techniques that preserve ascorbate and DHA. Due to the instability of these compounds, the extraction of ascorbate from cells should be performed under conditions where minimal degradation occurs. Controls should be performed to check this. The following procedure was used for a leukemic cell line, but it can be applied to most cultured cells. After washing the cells to remove extracellular ascorbate, about 10^6 cells are collected as a pellet in a microcentrifuge tube. Extraction and deproteinization are achieved by the addition of 600 μl methanol, dispersion of the pellet, and the subsequent addition of 400 μl water. To ensure the stability of ascorbate, EDTA and HCl are added to final concentrations of 50 μM and 50 m*M*, respectively. After centrifugation of the extract to remove precipitated proteins, the supernatant can be analyzed immediately by HPLC or frozen for later analysis.

Analysis and sample pretreatment of ascorbate from erythrocytes differ from other cell types because of the presence of hemoglobin. It has been reported that denatured hemoglobin can catalyze the oxidation of ascorbate.[12] Hemoglobin must therefore not be removed by precipitation with methanol. Instead, we lyse 200 μl packed erythrocytes in 3 volumes of 7 m*M* potassium phosphate, pH 4.0, remove membrane fragments by microcentrifugation, and subsequently remove the hemoglobin from the supernatant by ultrafiltration. The reusable Millipore micropartition system with 30-kDa cutoff membranes is suitable for this purpose. Ultrafiltration devices in a microcentrifuge format were found to clog during filtration. Apparently, the increased membrane surface of the Millipore unit prevents this problem and yields adequate amounts of a clear colorless filtrate for further analysis. After filtration, methanol is added to the ultrafiltrate up to 60% (v/v) to precipitate any remaining small proteins, and 50 μM EDTA and 50 m*M* HCl (final concentrations) are added to stabilize ascorbate. After centrifugation, the sample can be frozen or analyzed directly by HPLC. When desired, it is also possible to rupture cells by freeze-thawing instead of hypotonic lysis.[13]

[12] E. N. Iheanacho, N. H. Hunt, and R. Stocker, *Free Radic. Biol. Med.* **18,** 543 (1995).

[13] J. M. May, Z. C. Qu, and S. Mendiratta, *Arch. Biochem. Biophys.* **349,** 281 (1998).

Dehydroascorbic Acid

Several different approaches are possible for the analysis of DHA. Four of them are described here. The preparation of samples for DHA analysis needs even greater care than ascorbate samples. This is due to the swift decomposition of DHA, which has a half-life of about 10 min under physiological conditions. An acid pH and cold storage can slow down the hydrolysis reaction, but samples should be analyzed quickly.[7] The use of autosamplers should therefore be avoided when DHA is not derivatized before, e.g., HPLC analysis. Furthermore, it should be kept in mind that an assay for intracellular DHA may be of limited value. Although it is possible to extract DHA from cells, a part of it may be lost on lysis of the cells, either by hydrolysis or by a redox reaction with a cellular component.

Reduction of DHA. Many protocols for the analysis of DHA are derived from methods for the determination of ascorbate. They involve the conversion of DHA to ascorbate by the addition of 5 m*M* of a reducing agent such as DTT or 2-mercaptoethanol to the sample. The difference in ascorbate content in samples with and without a reductant corresponds to the amount of DHA that is present. After a quick reduction, reduced and control (nonreduced) samples may be stored under ascorbate-preserving conditions, such as an acid pH, low temperatures, and in the presence of EDTA.

Derivatization of DHA. DHA can be analyzed without a reduction step when a suitable detection technique follows HPLC separation. DHA does not absorb light at useful wavelengths, nor can DHA be detected by electrochemical detection. It has been reported that pre- or postcolumn derivatization using *o*-phenylenediamine yields a stable fluorescent compound that is readily detectable.[14] Other methods of derivatization allow separation and detection by gas chromatography-mass spectroscopy (GC-MS). However, it has been suggested that derivatization reactions are prone to produce artifacts, indicating that this technique must be handled with great care.[9]

Radioactive Labeling of DHA. A convenient method to quantify unmodified DHA is the use of ^{14}C-labeled DHA. In the HPLC system described earlier for ascorbate, DHA elutes from the column before ascorbate. An in-line radioactivity detector can be used for detection or, alternatively, fractions can be collected for liquid scintillation counting. In the latter case, sufficient fractions should be collected to allow peak separation.

NMR and ^{13}C-Labeled DHA. An alternative that can be used for intact cells is ^{13}C-labeled nuclear magnetic resonance (NMR). This technique was described by Himmelreich *et al.*[2] for erythrocyte suspensions containing [^{13}C]ascorbate and [^{13}C]DHA.[2] NMR spectra of these suspensions contain specific bands for ascorbate

[14] W. Lee, S. M. Roberts, and R. F. Labbe, *Clin. Chem.* **43,** 154 (1997).

and DHA. Moreover, it was found that these bands have a shift that correlates with the concentration of hemoglobin in the solution. Thus, in erythrocytes, distinct peaks can be observed for intra- and extracellular ascorbate or DHA. Drawbacks of this method are that it requires relatively high concentrations of both erythrocytes and ascorbate of DHA. Moreover, the erythrocytes must be pretreated with carbon monoxide to produce carbon monoxyhemoglobin, which has a more stable diamagnetic nature than the oxy and deoxy forms. It is unclear whether the band-shift phenomenon can also be used in other cell types.

Ascorbate Free Radical

AFR is an unstable molecule, with a lifetime of about 1 sec. This labile nature requires its measurement *in situ,* without any sample pretreatment. Thus, the use of chromatographic techniques is excluded. Only spectroscopic techniques can offer real-time measurement of the sample. In principle, AFR can be measured spectrophotometrically at 360 nm ($\varepsilon = 4900\ \mathrm{cm}^{-1}M^{-1}$).[15] However, in biological samples, AFR only reaches concentrations in the nanomolar range, which results in absorption values that will usually not exceed the detection limit. Moreover, spectrophotometric measurements cannot be performed easily in turbid cell suspensions.

Electron spin resonance (ESR) spectroscopy does not have these drawbacks and allows the identification and quantification of paramagnetic species (such as free radicals) in turbid suspensions. However, the measurement of aqueous samples at room temperature requires the use of a flat quartz sample cell and a special resonance cavity to accommodate this cell. The equipment must be calibrated after the cell is positioned in the cavity. A sampling device to load samples into the cell while it is still positioned in the cavity, e.g., by aspiration,[16] is recommended. In this way, successive samples can be measured without having to adjust the position of the cell or the settings of the spectrometer. This also allows scanning within seconds after mixing of the samples. The equipment used for our experiments consisted of a JEOL RE2X X-band spectrometer operating at 9.36 GHz with a 100-kHz modulation frequency. Samples were transferred to a quartz flat cell in a TM_{110} cavity with a rapid sampling device. The ESR spectrometer settings were as follows: microwave power, 40 mW; modulation amplitude, 1 G; time constant, 0.3 sec; scan time, 5 min; and scan width, 15 G.

A typical AFR signal consists of a doublet with a hyperfine splitting $\alpha^{H4} = 1.8$ G (Fig. 1A). The ESR signal is proportional to the amount of paramagnetic species in the sample, which can be used as a quantitative assay. The concentrations

[15] B. H. J. Bielski, D. A. Comstock, and R. A. Bowen, *J. Am. Chem. Soc.* **93,** 5624 (1971).

[16] R. P. Mason, *Methods Enzymol.* **105,** 416 (1984).

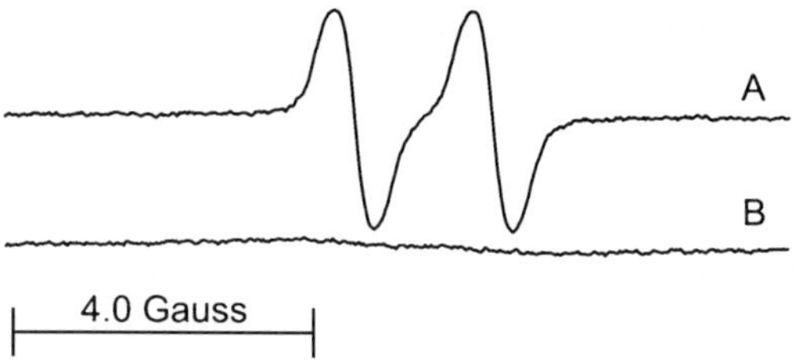

FIG. 1. ESR spectra of ascorbate-free radical: (A) 100 μM ascorbate and 4 mU/ml ascorbate oxidase in phosphate-buffered saline and (B) same as A, but with 5 mM $Ni(en)_3^{2+}$. Adapted from M. M. VanDuijn, K. Tijssen, J. VanSteveninck, P. J. A. Van Den Broek, and J. Van Der Zee, *J. Biol. Chem.* **275,** 27720 (2000) with permission.

of AFR can be determined by double integration of the ESR spectra, using the spectrum of a stable radical such as 2,2,6,6-tetramethylpiperidine-*N*-oxyl (TEMPO), with known concentration, as a standard. It is important to use identical spectrometer settings while scanning the AFR spectrum and the TEMPO spectrum.

Scanning of an ESR spectrum can take several minutes, depending on, e.g., signal intensity. When AFR concentrations need to be followed for a prolonged period of time or a short time frame, it is possible to lock the spectrometer to the Gauss value of the top of one of the peaks of the AFR doublet. Although this approach may result in a decrease in the signal-to-noise ratio, it allows the continuous tracking of AFR signal levels.

ESR spectroscopy does require expensive and specialized equipment, as well as a trained operator. However, it is the most powerful technique for studying free radical molecules, such as AFR, and their interactions with the living cell.

Interaction of Intracellular and Extracellular Ascorbate

Several requirements should be met to study the interaction of intra- and extracellular ascorbate, DHA, and AFR. First, cells have to be properly conditioned, resulting in the desired intracellular concentration of ascorbate. These cells can subsequently be used in an assay that can measure the interaction of the intra- and extracellular ascorbate forms.

Methods to Modify Intracellular Ascorbate Levels

The best way to modify ascorbate levels in the cell is by using the cells' transport systems in the membrane. Transport pathways exist for both ascorbate and DHA. However, many cells do not express the ascorbate transporter, but can quickly transport DHA through the GLUT-1 glucose transporter. Thus, loading with DHA is the preferred method. After transport, DHA is reduced to ascorbate, which only slowly leaks out of the cell. Good results have been obtained with an incubation period of 30 min at room temperature in a buffer containing DHA, although incubation at 37° might prove to be superior for some cell types. After 30 min, most of the DHA will have been degraded, and no additional ascorbate will accumulate in the cells. The cells can then be washed and used for an experiment. For erythrocytes, we typically incubate a 20% suspension of washed cells in phosphate-buffered saline (PBS) containing up to 500 μM DHA and 2.5 mM adenosine at room temperature. Adenosine (or glucose) improves the accumulation of ascorbate in erythrocytes by supplying energy needed for the reduction of DHA. After 30 min of incubation under gentle mixing conditions, the cells are washed three times with PBS and used for an experiment within an hour. The resulting intracellular ascorbate concentration depends on the concentration of DHA added. For example, erythrocytes require about 500 μM DHA to reach an intracellular concentration of 1 mM ascorbate. However, similar levels could be achieved in leukemic HL-60 cells after incubation with only 25 μM DHA. It is, therefore, important to determine the result of the incubation by HPLC analysis.

Intracellular ascorbate can be removed by treatment with 4-hydroxy-2,2,6,6-tetramethylpiperidine-*N*-oxyl (TEMPOL). TEMPOL can diffuse freely into the cell and oxidize ascorbate to DHA, which subsequently diffuses out of the cell through the GLUT-1 transporter. Three consecutive 5-min incubations of cells in buffer containing 1 mM TEMPOL remove over 90% of the ascorbate in the cells. It is important to subsequently wash the cells three times in regular buffer to remove all TEMPOL from the cells. It has been reported that intracellular reductants such as NADH and NADPH were not affected by this treatment, but we observed a decrease in NADH levels under some conditions.[3,17] One should therefore check this in each cell type used.

When studying ascorbate transport, it might be necessary to quickly separate cells from the incubation medium. For this purpose, cells can be spun down through a layer of dibutyl phtalate or another oil of appropriate density. When a suspension is layered on top of the oil, centrifugation quickly sediments the cells to the bottom of the tube, while the medium remains on top. Thus, the cells are immediately separated from the medium, but remain intact. When radio-labeled substrates are used, it is convenient to use small microcentrifuge tubes. The bottom of the tube can be cut off after centrifugation to collect the pellet for liquid scintillation counting.

Methods to Generate AFR

In several studies, AFR has been generated by mixing ascorbate and DHA. To calculate the resulting AFR concentration, the equilibrium constant of the

[17] J. M. May, Z. C. Qu, and R. R. Whitesell, *Biochemistry* **34,** 12721 (1995).

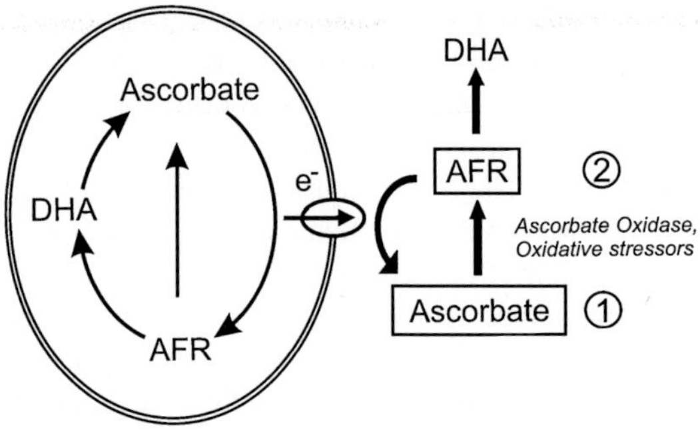

FIG. 2. Principle of AFR reduction assays. The reduction of extracellular AFR in a cell suspension can be measured by monitoring the decrease in the concentration of extracellular ascorbate (1) or by monitoring AFR using ESR spectroscopy (2).

disproportionation reaction [Reaction (1)] was used:

$$\text{Ascorbate} + \text{DHA} \rightleftharpoons 2\,\text{AFR} \tag{1}$$

$$\text{Ascorbate} + \text{O}_2 \xrightarrow[\substack{\text{Metal ions}\\ \text{Ascorbate oxidase}}]{\text{Spontaneous}} \text{AFR} + \text{O}_2^{\cdot -} \tag{2}$$

However, at micromolar or low millimolar concentrations of ascorbate and DHA, this method is seriously hampered by redox-active metals (Fe or Cu) present in the buffers used.[18] Under these conditions, the AFR concentration is determined by metal-catalyzed reactions [Reaction (2)] and not by the equilibrium reaction. The most reliable and reproducible method to produce AFR is the incubation of ascorbate with moderate amounts of ascorbate oxidase. The concentration of ascorbate oxidase can be varied to generate different concentrations of AFR, which may be quantified by ESR spectroscopy. In our laboratory, 100 μM ascorbate and 1–50 mU/ml ascorbate oxidase give a useful range of AFR concentrations for our experiments.

Methods to Measure the Ascorbate-Dependent Reduction of Extracellular AFR

Extracellular ascorbate can be regenerated from AFR by a one-electron reduction step. This can be measured by two different methods. The first method monitors the oxidation of ascorbate in the suspension [Fig. 2, (1)]. Ascorbate is lost by oxidation when AFR is generated. The reduction of AFR regenerates ascorbate and will thus appear to slow down the oxidation of ascorbate. The second method

[18] M. M. Van Duijn, J. Van Der Zee, and P. J. A. Van Den Broek, *Protoplasma* **205,** 122 (1998).

uses ESR spectroscopy to directly measure AFR in a cell suspension [Fig. 2, (2)]. Reduction of AFR should decrease its concentration in the suspension.

Measuring Extracellular Ascorbate Oxidation by UV Spectroscopy. This assay measures the extracellular ascorbate concentration in a cell suspension where ascorbate is oxidized by ascorbate oxidase, thus directly measuring the capacity of the cell to prevent the loss of ascorbate by oxidation. When AFR is reduced, the rate of ascorbate oxidation will appear to have decreased. The technique has therefore also been referred to as an ascorbate stabilization assay.

In each cell system, ascorbate and ascorbate oxidase concentrations, as well as cell densities, need to be optimized. The rate of AFR formation by ascorbate oxidase and its corresponding steady state concentration are critical when the ascorbate-dependent reduction of AFR has to be detected. When the rate is too high, the decrease in AFR concentration due to the ascorbate-dependent reductase will be relatively small and hard to detect. However, a low rate of AFR formation will be overwhelmed by the reduction rate of the plasma membrane redox system. In this case, it is not possible to show the capacity of the ascorbate-dependent reductase activity.

For experiments with erythrocytes, ascorbate (100 μM) and ascorbate oxidase (4 mU/ml) are added to 8 ml of a 10% ascorbate-loaded cell suspension in a 10-ml screw-cap tube, while rocking gently to prevent erythrocyte sedimentation. The oxidation of ascorbate by ascorbate oxidase proceeds in a linear fashion for more than 15 min (Fig. 3). Thus, the oxidation rate can be determined

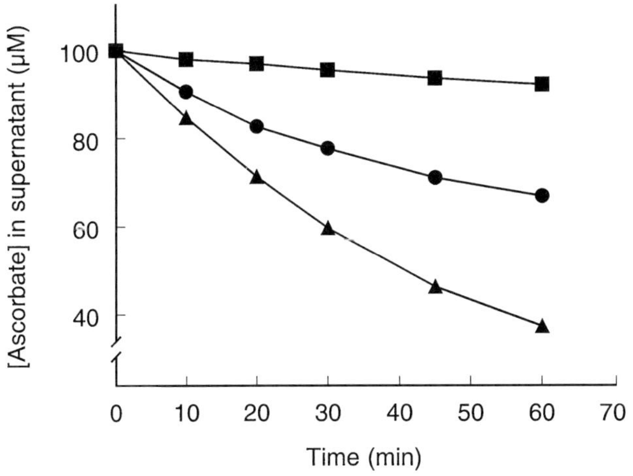

FIG. 3. Oxidation of extracellular ascorbate in erythrocyte suspensions. Ascorbate (100 μM) and ascorbate oxidase (4 mU/ml) were incubated in 10% suspensions of ascorbate-loaded erythrocytes (■), control erythrocytes (●), or in the absence of cells (▲). Samples were centrifuged, and the absorption at 265 nm was measured in the supernatant at different time points. All experiments were performed in phosphate-buffered saline. Adapted from M. M. VanDuijn, K. Tijssen, J. VanSteveninck, P. J. A. Van Den Broek, and J. Van Der Zee, *J. Biol. Chem.* **275,** 27720 (2000) with permission.

from samples taken after 0 and 15 min incubation. Duplicate samples (1.5 ml) are drawn from the tubes and centrifuged, and the supernatant is transferred to a quartz cuvette to measure the absorbance at 265 nm. An aliquot can also be used for HPLC analysis. Samples must be processed promptly, as the oxidation of ascorbate will continue in the supernatant. One should ensure that the activity of ascorbate oxidase is not affected by the cells, e.g., by verifying enzyme activity in the supernatant after an incubation with cells. In addition, control experiments with [^{14}C]ascorbate-loaded cells should show that no intracellular ascorbate is leaking from the cells. The involvement of intracellular ascorbate in the reduction of AFR can be inferred from the comparison of cells with different internal concentrations of the vitamin. Figure 3 illustrates the effect of control and ascorbate-loaded erythrocytes on the oxidation of ascorbate by ascorbate oxidase. Both affected the oxidation rate, but the effect was more pronounced in the presence of ascorbate-loaded cells. The protective effect of control erythrocytes can most likely be attributed to endogenous ascorbate and a NADH-dependent redox system in the cells.

Measuring AFR Reduction by ESR Spectroscopy. The capacity of cells to reduce extracellular AFR and thus to prevent oxidation of ascorbate can also be measured by determining the AFR concentration by ESR spectroscopy. Ascorbate-loaded erythrocytes are resuspended at a 10% hematocrit in PBS and exposed to an AFR-generating system, consisting of 100 μM ascorbate and 4 mU/ml ascorbate oxidase. Immediately after mixing, the suspension is aspirated into a flat cell in an ESR spectrometer, and the AFR signal intensity is determined. The AFR signal should be stable for the duration of the scan (5 min). By varying the intracellular ascorbate concentrations, the effect of various ascorbate concentrations on AFR signal intensity can be determined. It was found that, with, e.g., 1 m*M* intracellular ascorbate, the extracellular AFR signal intensity decreased by 45% relative to control erythrocytes.[3]

Discrimination between intracellular and extracellular AFR is possible with nonpermeant line-broadening agents. One potent and useful line-broadening agent is the nickel ion. Due to its toxicity, it must be chelated when used in biological samples, e.g., with ethylenediamine to $Ni(en)_3^{2+}$.[19] The chelate is not toxic and does not affect the redox properties of AFR, while it preserves the line-broadening properties. The addition of $Ni(en)_3^{2+}$ broadens a sharp ESR signal, resulting in a negligible amplitude compared to an unaffected signal (Fig. 1B). When added to a suspension, it will only broaden extracellular AFR signals and leave intracellular signals unaffected. Figure 4 illustrates how 5 m*M* $Ni(en)_3^{2+}$ allows the detection of a small intracellular AFR signal in the presence of a large amount of extracellular AFR. The signal of extracellular AFR, generated by mixing 1 m*M*

[19] L. M. Wakefield, A. E. Cass, and G. K. Radda, *J. Biol. Chem.* **261,** 9746 (1986).

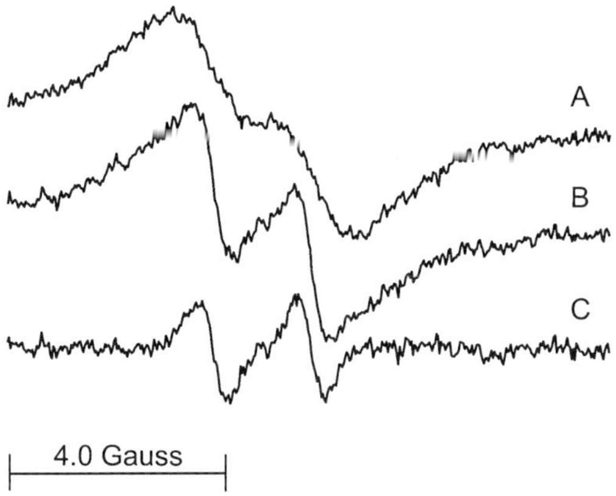

FIG. 4. Detection of intracellular AFR using ESR: (A) 1 m*M* ascorbate, 20 mU/ml ascorbate oxidase, and 5 m*M* $Ni(en)_3^{2+}$ in phosphate-buffered saline; (B) same as A, but with 20% ascorbate-loaded erythrocytes; and (C) A subtracted from B. Erythrocytes were loaded with ascorbate, using 500 μM DHA, as described in the text. Adapted from M. M. VanDuijn, K. Tijssen, J. VanSteveninck, P. J. A. Van Den Broek, and J. Van Der Zee, *J. Biol. Chem.* **275,** 27720 (2000) with permission.

ascorbate with 20 mU/ml ascorbate oxidase, is broadened by the presence of 5 m*M* $Ni(en)_3^{2+}$, but is still visible due to the high concentration of AFR (Fig. 4A). The same signal is observed in the presence of 20% control erythrocytes (not shown). However, the addition of ascorbate-loaded erythrocytes results in the signal given in Fig. 4B. Comparison of A and B shows that superimposed on the signal in Fig. 4A, a sharp signal can be observed that is unaffected by $Ni(en)_3^{2+}$. Subtraction of spectra 4A and 4B reveals that, indeed, a small AFR signal is present. Because this signal is unaffected by $Ni(en)_3^{2+}$, it must be of intracellular origin, which nicely illustrates the formation of intracellular AFR as an intermediate in the reduction of extracellular AFR. Thus, the methods described in this chapter enabled us to show that intracellular ascorbate can be an electron donor for the reduction of extracellular AFR.

[24] Regulation of Endothelial Cell Proliferation by Nitric Oxide

By CYNTHIA J. MEININGER and GUOYAO WU

Introduction

Nitric oxide (NO) plays an important role in the vascular system.[1] NO induces vasodilation, inhibits platelet aggregation and adherence to sites of vascular injury, inhibits neutrophil adherence to endothelial cells, inhibits vascular smooth muscle cell migration and proliferation, regulates apoptosis of cells, and maintains endothelial cell barrier function. Thus NO is critical for normal vascular homeostasis under a variety of physiological and pathological conditions.

NO stimulates angiogenesis,[2] the formation of new blood vessels from preexisting vessels, and has been shown to stimulate endothelial cell proliferation *in vitro*.[3] Pharmacologic agents that act as NO donors also stimulate endothelial cell proliferation.[2] Vascular endothelial cell growth factor, an angiogenic factor, stimulates endothelial cell proliferation via NO production[4,5] and by upregulating NO synthase (NOS), the enzyme that catalyzes the synthesis of NO from L-arginine.[6] Indeed, blocking the formation of NO with inhibitors of NOS prevents the proliferative response of endothelial cells to this mitogen.[4] Wound-healing studies performed in NOS-deficient mice have provided further evidence for the critical role of NO in angiogenesis.[7,8] Taken together, these data suggest that NO may act as an endogenous proliferative signal for endothelial cells.

We have found that endothelial cells from the spontaneously diabetic BB rat have an impaired ability to produce NO due to a deficiency in tetrahydrobiopterin, a necessary cofactor for NOS.[9] These cells also proliferate more slowly

[1] L. J. Ignarro, *Biosci. Rep.* **19,** 51 (1999).

[2] M. Ziche, L. Morbidelli, E. Masini, S. Amerini, H. J. Granger, C. A. Maggi, P. Geppetti, and F. Ledda, *J. Clin. Invest.* **94,** 2036 (1994).

[3] M. Ziche, L. Morbidelli, E. Masini, H. J. Granger, C. A. Maggi, P. Geppetti, and F. Ledda, *Biochem. Biophys. Res. Commun.* **192,** 1198 (1993).

[4] L. Morbidelli, C. H. Chang, J. G. Douglas, H. J. Granger, F. Ledda, and M. Ziche, *Am. J. Physiol.* **270,** H411 (1996).

[5] A. Papapetropoulos, G. Garcia-Cardena, J. A. Madri, and W. C. Sessa, *J. Clin. Invest.* **100,** 3131 (1997).

[6] J. D. Hood, C. J. Meininger, M. Ziche, and H. J. Granger, *Am. J. Physiol.* **274,** H1054 (1998).

[7] P. C. Lee, A. N. Salyapongse, G. A. Bragdon, L. L. Shears, S. C. Watkins, H. D. J. Edington, and T. R. Billiar, *Am. J. Physiol.* **277,** H1600 (1999).

[8] K. Yamasaki, H. D. J. Edington, C. McClosky, E. Tzeng, A. Lizonova, I. Kovesdi, D. L. Steed, and T. R. Billiar, *J. Clin. Invest.* **101,** 967 (1998).

[9] C. J. Meininger, R. S. Marinos, K. Hatakeyama, R. Martinez-Zaguilan, J. D. Rojas, K. A. Kelly, and G. Wu, *Biochem. J.* **349,** 353 (2000).

METHODS IN ENZYMOLOGY, VOL. 352
0076-6879/02 $35.00

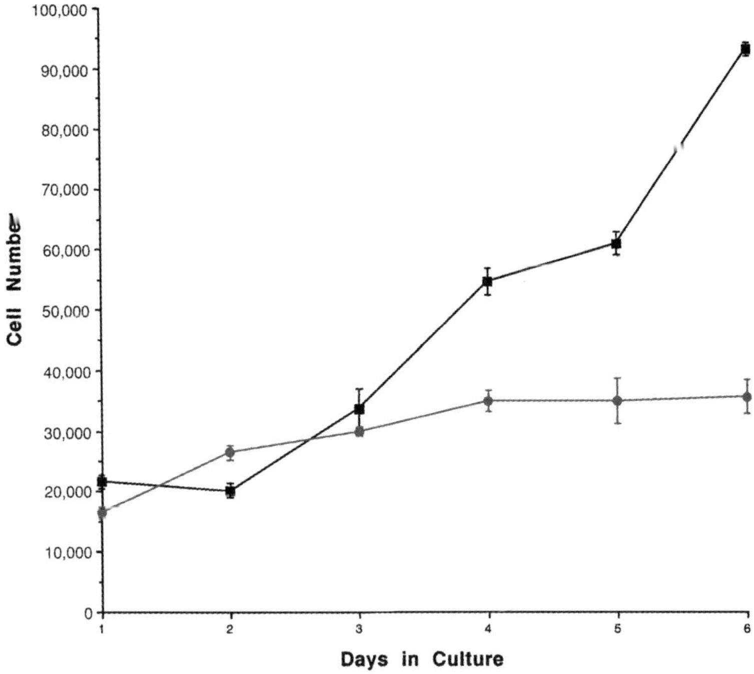

FIG. 1. Growth curves of endothelial cells (EC) from diabetic and nondiabetic BB rats. Endothelial cells from diabetic rats (●) and nondiabetic rats (■) were cultured in DMEM containing 10% FBS. Triplicate sets of wells were trypsinized each day, and cell numbers were determined using a hemacytometer. Data are means ± SEM from one representative experiment.

than endothelial cells taken from nondiabetic animals (Fig. 1). Replenishing tetrahydrobiopterin levels in the endothelial cells from diabetic animals restores their ability to make NO.[9] As a result of increased NO production, these endothelial cells proliferate more rapidly.[10] Conversely, blocking the production of tetrahydrobiopterin in endothelial cells from nondiabetic animals decreases their NO synthesis and concomitantly reduces their proliferation. These data further suggest that NO acts as a homeostatic regulator of endothelial cell proliferation.

Isolation of Endothelial Cells

Endothelial cell cultures have been employed extensively in the study of blood vessel growth and its modulation by various angiogenic and antiangiogenic factors. Endothelial cells were initially obtained by direct scraping of the luminal

[10] R. S. Marinos, W. Zhang, G. Wu, K. Kelly, and C. J. Meininger, *Am. J. Physiol.* **281,** 11482 (2001).

surface of vessels[11] or by the use of various enzymatic agents. Collagenase, for example, has been used for the isolation of cells from both large and small vessels.[12,13] Refinements in these techniques have allowed the isolation and culture of endothelial cells from virtually any tissue of interest and from specific vascular sites.

Rat Coronary Endothelial Cells

Principle. Endothelial cells of coronary vessels are released enzymatically by collagenase treatment of ventricular muscle. Endothelial cells are enriched by elimination of nonendothelial cells through differential separation steps.

Materials

Joklik's buffer: Joklik's modified minimal essential medium (GIBCO/BRL, Rockville, MD) containing 60 m*M* taurine, 20 m*M* creatine, and 5 m*M* HEPES. [Unless indicated, all chemicals are obtained from Sigma Chemical Co. (St. Louis, MO).]

Dialyzed bovine serum albumin (BSA): BSA, class V, prepared in Joklik's salts solution (111 m*M* NaCl, 4 m*M* KCl, 10 m*M* NaH_2PO_4, 60 m*M* taurine, 20 m*M* creatine, and 5 m*M* HEPES) to a final concentration of 0.1 g/ml. This solution is dialyzed overnight (12,000 dalton cut-off) at 4° against the same salt solution.

Collagenase, type II (Worthington, Freehold, NJ): different lots of enzyme should be tested for sufficient activity.

Dulbecco's modified Eagle's medium (DMEM) with 2 m*M* L-glutamine and 25 m*M* glucose (Sigma).

Acetylated low-density lipoprotein labeled with 1,1′-dioctadecyl-3,3,3′,3′-tetramethylindcarbocyanine perchlorate (Biomedical Technologies, Stoughton, MA).

Protocol

Rats are injected with heparin (130 U/100 g body weight, ip) 20 min before anesthesia with sodium pentobarbital (Nembutal, 140 mg/kg ip). Hearts are removed and placed in ice-cold Joklik's buffer. Aortas are cannulated with stainless steel tubing (2 mm I.D.), and hearts are lanced once with a 16-gauge needle to allow blood removal from the ventricles. Hearts are perfused from a static 40-mm Hg hydrostatic pressure head using oxygenated Joklik's buffer supplemented with

[11] U. S. Ryan, M. Mortara, and C. Whitaker, *Tissue Cell* **12,** 619 (1981).

[12] E. A. Jaffe, R. L. Nachman, C. G. Becker, and C. R. Minick, *J. Clin. Invest.* **52,** 2745 (1973).

[13] J. Folkman, C. C. Haudenschild, and B. R. Zetter, *Proc. Natl. Acad. Sci. U.S.A.* **76,** 5217 (1979).

0.1% dialyzed BSA and heparin (1 U/ml). After 10 min, new perfusate containing collagenase (0.7 mg/ml) is introduced and allowed to recirculate 30–40 min until aortic perfusion pressure decreases below 40 mm Hg. The ventricles are cut from the hearts, minced, and placed in fresh collagenase-containing perfusate. The tissue is placed in a shaking water bath (37°) for 20 min. $CaCl_2$ (50 μM) is added and digestion is continued for an additional 10 min. The cells are dispersed, filtered through a double layer of cheesecloth, diluted to 150 ml with Joklik's buffer containing 0.1% dialyzed BSA, and poured into three 50-ml conical centrifuge tubes. The resulting suspension is allowed to settle by gravity for 5–10 min, which separates myocytes from endothelial cells (myocytes are heavier and settle more quickly). Endothelial cells in the supernatant are further purified by sequential filtration through a series of 90-, 45-, 25-, and 15-μm nylon screens and pelleted by centrifugation (100*g* for 7 min). The cell pellets are resuspended and combined in 3–4 ml Joklik's buffer, layered over a 6-ml BSA solution (6% in Joklik's buffer), and centrifuged (100*g*, 7 min). This pellet is resuspended in 6 ml Joklik's buffer, layered over a 6-ml BSA solution (3% in Joklik's buffer), and centrifuged (100*g*, 7 min). Cells are washed three times with phosphate-buffered saline (PBS) followed by centrifugation. Finally, cells are resuspended in DMEM supplemented with 20% fetal bovine serum (FBS), 20 U/ml sodium heparin, 1 m*M* sodium pyruvate, 100 IU/ml penicillin, 100 μg/ml streptomycin, and 0.25 μg/ml amphotericin B and cultured on gelatin-coated (1.5% in PBS) dishes at 37° under 10% CO_2. Cells are passaged by trypsinization (0.25% trypsin and 0.02% EDTA). The endothelial identity of propagated cells is confirmed by positive staining for factor VIII-related antigen and/or uptake of modified low-density lipoprotein.[14]

Rat Skeletal Muscle Endothelial Cells

Principle. Endothelial cells from rats exhibit an affinity for Griffonia simplicifolia lectin (GSL). Biotinylated GSL, isolectin B4, will bind to endothelial cells released from enzymatically digested skeletal muscle. Endothelial cells can be retrieved from this mixture by allowing the interaction of streptavidin-coated polystyrene paramagnetic beads with the biotinylated lectin and utilizing a magnetic separation system.

Materials

Liberase Blendzyme 3 (Roche, Indianapolis, IN)
Cell strainers with 100 μm mesh (BD Falcon, Lincoln Park, NJ)
Biotinylated GSL, isolectin B4 (Vector, Burlingame, CA)
Dynabeads M-280, streptavidin coated, washed as directed by manufacturer (Dynal, Lake Success, NY)

[14] J. C. Voyta, D. P. Via, C. E. Butterfield, and B. R. Zetter, *J. Cell Biol.* **99,** 2034 (1984).

Protocol. A portion of skeletal muscle (<1 cm^3) is removed, minced for 1–2 min with fine surgical scissors, and placed in 1–3 ml DMEM containing 25 m*M* HEPES and Liberase Blendzyme 3 (0.02 mg/ml) in a sterile tube. The tissue is incubated for 30–40 min in a 37° shaking water bath. The resulting cell digest is passed through a 100-μm nylon mesh strainer suspended over a sterile 50-ml centrifuge tube to eliminate large tissue fragments. To remove medium enzymes, cells are pelleted (200*g*, 4 min), washed with HEPES-buffered DMEM, and pelleted again. The final pellet of cells is resuspended in 0.5 ml of HEPES-buffered DMEM containing 10 μg/ml biotinylated GSL isolectin B4 and incubated in an Eppendorf tube for 30 min in a 37° water bath. Cells are centrifuged (200*g*, 4 min), washed with HEPES-buffered DMEM, and pelleted again. Cells are resuspended in 0.5 ml HEPES-buffered DMEM, and Dynabeads M-280 are added to a concentration of 1–2 × 10^7/ml. Tubes are placed on a rotator and incubated at 4° for 15–20 min. Tubes are then placed in a magnetic stand for 1 min to collect endothelial cells bound to beads along the side of the tube. The medium is aspirated carefully, the tube is removed from the stand, and beads are resuspended in 1 ml HEPES-buffered DMEM. The tube is placed back into the magnetic stand, left for 1 min, and the medium is carefully aspirated again. The procedure is repeated once more. Finally, cells are resuspended in growth medium and cultured as described earlier for coronary endothelial cells.

Human Vascular Segment Endothelial Cells

Principle. Endothelial cells exhibit platelet–endothelial cell adhesion molecule-1 (PECAM-1, CD31) on their surface. Paramagnetic beads with antibodies recognizing CD31 attached to the surface can be utilized for rapid immunomagnetic isolation of endothelial cells from enzymatically digested vascular segments.

Materials

Gentamicin reagent solution (GIBCO/BRL)
Dynabeads CD31, washed as per manufacturer's instructions (Dynal)

Protocol. Human vascular segments are transported in PBS containing gentamicin (50 μg/ml). Segments are minced for 1–2 min with fine surgical scissors and placed in 1–3 ml DMEM containing 25 m*M* HEPES and Liberase Blendzyme 3 (0.02 mg/ml) in a sterile tube. The tissue is incubated for 30–40 min in a 37° shaking water bath. The resulting cell digest is passed through a 100-μm nylon mesh strainer suspended over a 50-ml centrifuge tube to eliminate large tissue fragments. To remove medium enzymes, cells are pelleted (200*g*, 4 min), washed with HEPES-buffered DMEM, and pelleted again. Cells are then resuspended in 0.5 ml HEPES-buffered DMEM, and Dynabeads CD31 are added to a concentration of 1–2 × 10^7/ml. Tubes are placed on a rotator, incubated at 4° for 15–20 min, and then placed in a magnetic stand for 1 min to collect endothelial cells bound to beads. The medium is aspirated carefully, the tube is removed from the stand, and beads are resuspended in 1 ml HEPES-buffered DMEM. The tube is placed back into the magnetic stand for 1 min, and the medium is carefully aspirated again. The procedure is repeated once more. Finally, cells are resuspended in growth medium and cultured as described earlier for coronary endothelial cells.

Techniques for Assessing Endothelial Cell Proliferation

Determination of the number of cells in a culture dish is crucial for assays demonstrating endothelial cell proliferation. Several methods are available to measure cell proliferation. Cell counts may be made manually using a hemacytometer. Alternatively, proliferation can be assessed by measuring DNA synthesis, monitoring cellular redox activity, or by detecting an antigen present in proliferating cells.

Cell Counting

Endothelial cells are plated at a density of 5000–10,000 cells/cm^2 in DMEM containing 5% FBS (and supplemented as described earlier). Cells are given overnight to attach. After attachment, cells in one triplicate set of wells are released by treatment with trypsin (0.25% trypsin and 0.02% EDTA in PBS), loaded into a hemacytometer, and counted to determine plating density. The remaining cells are rinsed with PBS and cultured for 72 hr in DMEM containing 0.1% FBS to induce quiescence. Agents to be tested are then added to the cultures. Positive control cultures receive FBS (final concentration 10%), whereas negative control cultures remain in DMEM with 0.1% FBS. Cell counting is generally done 48–96 hr after the addition of test agents. This method is simple but time-consuming. A more automated method is desirable when the quantitation of large numbers of samples is necessary.

Tritiated Thymidine Incorporation

Principle. DNA synthesis is active in proliferating cells. Thymidine occurs in DNA but not in RNA. When supplied to proliferating cells, [^{3}H]thymidine is taken up and incorporated into newly synthesized DNA.

Materials

[*methyl*-^{3}H]Thymidine, specific activity 6.7 Ci/mmol (ICN, Costa Mesa, CA)
Scintillation cocktail, Ultima Gold (Packard, Meriden, CT)

Protocol. Endothelial cells are plated at 10,000–20,000 cells/well in 96-well trays and treated with test agents as described earlier. [^{3}H]Thymidine is added (1 μCi/well) 23 hr later. After an additional hour, the medium is aspirated, cells are rinsed three times with PBS, twice with ice-cold 5% trichloroacetic acid (TCA) (5 min/wash) to precipitate labeled DNA, and once with distilled water. Precipitated DNA is solubilized with 0.3 *M* NaOH (0.2 ml/well). Samples are transferred to scintillation vials containing 5 ml of scintillation cocktail and ^{3}H radioactivity is measured by liquid scintillation spectroscopy. Because of its simplicity, incorporation of [^{3}H]thymidine has traditionally been used to measure DNA synthesis in cells.

Bromodeoxyuridine (BrdU) Incorporation

Principle. The incorporation of BrdU (a thymidine analog) into nacently synthesized DNA can be measured by the enzyme-linked immunosorbant assay of BrdU using an anti-BrdU monoclonal antibody.[15]

Materials

5-Bromo-2-deoxyuridine (BrdU, Sigma)
Mouse anti-BrdU antibody (Dako, Carpinteria, CA)
Normal goat serum (Jackson, West Grove, PA)
Peroxidase-conjugated goat anti-mouse IgG (Pierce, Rockford, IL)
3,3′,5,5′-Tetramethylbenzidine (Kirkegard Perry, Gaithersburg, MD)

Protocol. Endothelial cells are plated in 96-well trays as described earlier and treated with test agents. BrdU is added (final concentration 10 μM) 16–22 hr later. After an additional 2-hr incubation, the culture medium is aspirated, and cells are washed with PBS and fixed with 70% ethanol (200 μl/well) for 20 min at room temperature. Wells are washed once with water, and DNA is denatured by incubation with 2 *M* HCl (100 μl/well) for 10 min at 37°. The HCl is aspirated, and 0.1 *M* borate buffer (pH 9) is added (200 μl/well) to neutralize the residual acid. Wells are washed once with PBS. Cells are then treated with 50 μl blocking buffer (PBS containing 2% normal goat serum and 0.1% Triton X-100) for 15 min at 37° and with monoclonal anti-BrdU antibody (50 μl/well, 1 μg/ml in blocking buffer) for 60 min at 37°. Unbound antibody is removed by four washes with PBS containing 0.1% Triton. Bound antibody is detected using the peroxidase-conjugated anti-mouse IgG antibody (50 μl/well, 0.4 μg/ml in blocking buffer) incubated for 30 min at 37°. Cells are washed four times with the PBS/Triton solution followed by one wash with PBS. The peroxidase substrate, tetramethylbenzidine, is added (100 μl/well), and cells are incubated for 15–30 min at room temperature to allow color development. Absorbance of the samples is measured at 650 nm using a microplate reader to allow a large number of samples to be processed simultaneously. The BrdU assay is at least as sensitive as [^{3}H]thymidine incorporation,

[15] D. Muir, S. Varon, and M. Manthorpe, *Anal. Biochem.* **185,** 377 (1990).

but avoids the use of radioactive materials. Alternatively, the peroxidase reaction can be stopped by the addition of 50 μl of 1 *M* H_2SO and the resulting solution (yellow) is measured at 450 nm. Acidification can increase the sensitivity of the assay two- to fourfold.

MTT Assay

Principle. The MTT assay, originally developed by Mosmann,[16] is based on the cleavage of MTT by living cells to yield a dark blue insoluble formazan product. Isopropanol dissolves formazan crystals to give a homogeneous blue solution suitable for absorbance measurement. HCl converts phenol red in tissue culture medium to a yellow color that will not interfere with the measurement of formazan.

Materials

MTT, 3-(4,5-dimethylthiazol-2-yl)-2,5-diphenyltetrazolium bromide (Sigma), is dissolved in DMEM to a concentration of 5 mg/ml, filtered through a 0.2-μm filter, stored at 4° protected from light, and used within a month.

Protocol. Cells are plated in 96-well trays and treated with test agents as described earlier. MTT is added to each well (final concentration 0.5 mg/ml), and trays are incubated at 37° for 4 hr to allow time for cleavage of MTT. The medium is aspirated, and the cleavage reaction is stopped by the addition of 100 μl extraction buffer (0.1 *N* HCl in absolute isopropanol). Absorbance is measured at 570 nm (with background subtraction at 630 nm) using a microplate reader. MTT plus extraction buffer is utilized as a blank. This assay easily detects 1000 to 50,000 cells, with absorbance being directly proportional to the number of cells. It is rapid and reproducible. However, cleavage of MTT can occur via cellular superoxide.[17] Because NOS can synthesize superoxide from O_2 when tetrahydrobiopterin is deficient,[18] a change in cell metabolism (not necessarily linked with a change in cell proliferation) can affect this assay. Thus, the MTT assay is not the method of choice for studies involving NO production in endothelial cells.

Proliferating Cell Nuclear Antigen (PCNA)

Principle. PCNA is present in proliferating cells but absent in nonproliferating cells. The amount of PCNA, analyzed using immunochemical detection methods, is proportional to the number of proliferating cells.

[16] T. Mosmann, *J. Immunol. Methods* **65,** 55 (1983).

[17] R. H. Burden, V. Gill, and C. Rice-Evans, *Free Radic. Res. Commun.* **18,** 369 (1993).

[18] G. Wu and S. M. Morris, Jr., *Biochem. J.* **336,** 1 (1998).

Materials

Protease inhibitor cocktail, Complete, Mini (Roche)
Nitrocellulose (15 × 15-cm sheets, Bio-Rad, Hercules, CA)
Mouse monoclonal anti-PCNA antibody (Santa Cruz, Santa Cruz, CA)
Peroxidase-conjugated donkey anti-mouse IgG (Jackson)
SuperSignal West Dura Extended Duration Substrate (Pierce)
Kodak Biomax ML film (Kodak, Rochester, NY)

Protocol. Cells are plated in 100-mm dishes as described previously. The culture medium is aspirated, and cells are scraped with a rubber policeman/tissue scraper in 500 μl PBS. The cell suspension is collected in an Eppendorf tube and pelleted in a microcentrifuge (2000*g*, 2 min). The supernatant is aspirated, and lysis buffer (1% sodium deoxycholate, 1% NP-40, 0.1% sodium dodecyl sulfate, 10 m*M* Tris, pH 8.0, 140 m*M* NaCl, plus protease inhibitor cocktail) is added to the cell pellet. Protein concentration is determined using the bicinchoninic acid protocol (Pierce) with BSA as a standard. Protein (1–10 μg/lane) is loaded onto a 9.5–16% polyacrylamide gel. Separated proteins are transferred to nitrocellulose, and blots are blocked for 2 hr at room temperature with blocking buffer [5% nonfat dried milk in Tris-buffered saline (TBS, 25 m*M* Tris, 150 m*M* NaCl, pH 7.5) with 0.1% Tween 20 (TTBS)]. Primary antibody (mouse anti-PCNA) is used at 1 : 1000 in blocking buffer, and blots are incubated at 4° overnight. Blots are washed with TTBS and incubated with secondary antibody (peroxidase-conjugated donkey anti-mouse IgG) diluted 1 : 30,000 in blocking buffer at room temperature for 2 hr. Blots are washed with TTBS, and peroxidase activity is visualized using the SuperSignal substrate, according to the manufacturer's directions, and by exposing blots to film. The amounts of PCNA protein are quantified by scanning densitometry.

Endothelial Cell NO/Tetrahydrobiopterin Production

Measurement of Endothelial NO Production

NO, a free radical molecule with a short half-life, is released from cells in picomolar to nanomolar amounts, making direct measurement of its production difficult. Since the first report of electrochemical detection of NO production by cells in 1990,[19] a variety of NO-selective microelectrodes have been developed. In addition, an *in vivo* spin-trapping technique combined with electron paramagnetic resonance (EPR) spectroscopy was employed to measure NO production.[20] This method was based on the trapping of NO by a metal–chelator complex to form a

[19] K. Shibuki, *Neurosci. Res.* **9,** 69 (1990).

[20] A. Komarov, D. Mattson, M. M. Jones, P. K. Singh, and C. S. Lai, *Biochem. Biophys. Res. Commun.* **195,** 1191 (1993).

stable complex with characteristic EPR spectra that can be monitored continuously with a spectrometer. A chemiluminescence assay, initially developed to measure NO in air samples[21] on the basis of the reaction of NO with ozone to generate light, has also been employed to determine NO production in cells or tissues. All of these methods require sophisticated and expensive equipment. Alternatively, nitrite and nitrate (the stable products of NO oxidation) can be measured as an indicator of NO synthesis. In this method, nitrate is reduced to nitrite chemically or enzymatically[22] prior to analysis. We have developed a rapid, sensitive, and specific HPLC method for detecting picomole levels of nitrite and nitrate in biological samples.[22]

Principle. Nitrite reacts with 2,3-diaminonaphthalene (DAN) under acidic conditions to yield 2,3-naphthotriazole (NAT), a highly fluorescent product, which is stable in alkaline solution. Reversed-phase HPLC separates NAT from DAN (and other fluorescent compounds present in biological samples) before fluorescence detection of NAT.

Materials

Nitrate reductase (Roche)
2,3-Diaminonaphthalene (DAN, Sigma)
C_8 column (15 cm × 4.6 mm, 5 μm) and C_{18} column (5 cm × 4.6 mm, 40 μm) (both from Supelco, Bellefonte, PA)
HPLC grade methanol and HPLC grade water (both from Fisher Scientific, Houston, TX)
Waters HPLC apparatus including a Model 600E Powerline multisolvent delivery system with 100-μl heads, a Model 712 WISP autosampler, a Model 474 fluorescence detector, and a Millenium-32 Workstation (Waters, Milford, MA)

Protocol. All samples to be tested are filtered through 10-KDa cut-off ultrafilters to remove large molecular weight proteins. Filters are washed four times with deionized and double-distilled water (DD-H_2O) prior to use. Nitrate is converted to nitrite using nitrate reductase as follows: 200 μl of diluted sample or nitrate standard (0–2 μM), 10 μl of 1 U/ml nitrate reductase, and 10 μl of 120 μM NADPH are mixed and incubated at room temperature for 1 hr. This solution is then used directly for nitrite analysis. The conversion of nitrate to nitrite is 98% as determined with known amounts of both standards.

One hundred microliters of diluted sample (diluted with DD-H_2O), diluted blank medium, or sodium nitrite standard (0–2 μM) is mixed with 100 μl of DD-H_2O and 20 μl of 316 μM DAN (in 0.62 M HCl). These reaction mixtures are incubated at room temperature for 10 min, followed by the addition of 10 μl

[21] O. C. Zafiriou and M. McFarland, *Anal. Chem.* **52,** 1662 (1980).
[22] H. Li, C. J. Meininger, and G. Wu, *J. Chromatogr. B* **746,** 199 (2000).

of 2.8 *M* NaOH. After mixing, 15 μl of the derivatized nitrite–DAN solution is injected into a 5-μm C_8 column guarded by a 40-μm C_{18} column for chromatographic separation of NAT. The mobile phase (1.3 ml/min) is 15 m*M* sodium phosphate buffer (pH 7.5) containing 50% methanol (1 liter of 30 m*M* Na_2HPO_4 and 125 ml of 30 m*M* NaH_2PO_4 mixed with 1.125 liter of 100% methanol) (0.0–3.0 min), followed sequentially by 100% HPLC grade water (3.1–5.0 min), 100% methanol (5.1–8.0 min), 100% HPLC grade water (8.1–10.0 min), and the initial 15 m*M* sodium phosphate buffer (pH 7.5)–50% methanol solution (10.1–15.0 min). The use of 100% HPLC grade water before and after 100% methanol is necessary to prevent abrupt marked increases in column pressure and is sufficient to regenerate the columns for automatic analysis of multiple samples. All chromatographic procedures are carried out at room temperature. Fluorescence is monitored with excitation at 375 nm and emission at 415 nm. The retention time for NAT is 4.4 min.

NOS Activity Assay

Principle. NOS catalyzes stoichiometric formation of NO and citrulline from L-arginine. Thus, determination of NOS activity has been based on the detection of either product. Because of the simplicity and sensitivity of [^{3}H]- or [^{14}C]citrulline analysis, most of the published NOS assays have employed [^{3}H]- or [^{14}C]arginine. In tissues or cell lysates containing arginase activity, valine (an inhibitor of arginase) is used to improve assay sensitivity and specificity. Because endothelial NOS is present in both the plasma membrane and cytosol, whole cell lysates are used for enzyme activity assays.

Materials

- Protease inhibitor solution [50 m*M* HEPES buffer containing 5 μg/ml aprotinin, 5 μg/ml chymostatin, 5 μg/ml pepstatin A, 5 μg/ml phenymethylsulfonyl fluoride, 1 m*M* EDTA, and 1 m*M* dithiothreitol (DTT).]
- L-[U-^{14}C]Arginine, specific activity 240 mCi/mmol (American Radiolabeled Chemicals, St. Louis, MO)
- Reagent mixture: 2.5 ml of 2 m*M* DTT/4 m*M* $MgCl_2$ (in 400 m*M* HEPES, pH 7.4), 2.5 ml of 0.4 m*M* arginine/0.4 m*M* citrulline/20 m*M* valine (in water), 1.7 mg NADPH, 0.7 mg tetrahydrobiopterin, 1.7 mg FAD, 1.0 mg FMN, 20 μg calmodulin, and 50 μl [U-^{14}C]arginine
- N^G-Monomethyl-L-arginine (L-NMMA, Sigma)
- AG 50W-X8 ion-exchange resin (H^+ form, 200–400 mesh, Bio-Rad)

Protocol. Endothelial cells (2×10^7 cells/ml) are suspended in 0.3 ml of protease inhibitor solution. Cells are lysed by three cycles of freezing in liquid nitrogen and thawing in a 37° water bath. The resulting whole cell lysate is used for the NOS assay. Total NOS activity is assessed by mixing 100 μl of lysate with 50 μl of reagent mixture and 50 μl of 8 mM $CaCl_2$. iNOS activity is assessed by mixing 100 μl of the cell lysate with 50 μl of reagent mixture, 25 μl of 16 mM EGTA, and 25 μl H_2O. A blank is prepared by mixing 100 μl of cell lysate with 50 μl of reagent mixture, 25 μl of 16 mM EGTA, and 25 μl of 16 mM L-NMMA. All assay tubes are incubated at 37° for 30 min. Reactions are terminated by the addition of 50 μl of 1.5 M $HClO_4$. After 2 min, the acidified solution is neutralized by the addition of 25 μl of 2 M K_2CO_3 and 1 ml of 20 mM HEPES (pH 5.5). The supernatant is used for [^{14}C]citrulline separation using AG 50W-X8 resin (Na^+ form).

For converting AG 50W-X8 resin from the H^+ to the Na^+ form, 10 ml of the resin (suspended in 1 M NaOH) is placed in a plastic column. NaOH (1 M) is pumped through the resin for 15 min at 2 ml/min. The column is then washed with DD-H_2O for 15 min at 2 ml/min until the pH of the eluate is <9.0. This resin is used to pack Pasteur pipettes (tips are blocked with glass wool) (0.55 × 6 cm resin bed) and 0.5 ml of the samples are loaded into the resin. Water (4 ml) is pumped through the resin at 2 ml/min and the 4-ml eluate is collected in a 20-ml scintillation vial (recovery of citrulline is 96%). Fifteen milliliters of scintillation cocktail is added to the vial, and ^{14}C radioactivity is measured using liquid scintillation spectroscopy. NOS activity is calculated on the basis of [^{14}C]citrulline production and [^{14}C]arginine specific activity (SA) in the assay medium. Total NOS activity = (^{14}C dpm in the total NOS tube − ^{14}C dpm in the blank tube)/[^{14}C]arginine SA. iNOS activity = (^{14}C dpm in the iNOS tube − ^{14}C dpm in the blank tube)/[^{14}C]arginine SA. cNOS activity = total NOS activity − iNOS activity.

Western Blot Analysis of Endothelial Nitric Oxide Synthase

Principle. The amount of NOS in cell lysates can be determined by western blot analysis using antibodies specific for the different NOS isoforms.

Materials

Mouse monoclonal anti-human ecNOS primary antibody (Transduction, Lexington, KY)
Peroxidase-conjugated donkey anti-mouse IgG secondary antibody (Jackson)

Protocol. Endothelial cells are processed as described earlier for PCNA analysis. Primary and secondary antibody are diluted 1 : 2500 and 1 : 50,000–100,000, respectively, in blocking buffer. Blots are processed as described for PCNA analysis.

Measurement of Tetrahydrobiopterin in Endothelial Cells

Principle. Acidic oxidation converts both tetrahydrobiopterin and dihydrobiopterin to biopterin, whereas alkaline oxidation converts dihydrobiopterin to biopterin and tetrahydrobiopterin to pterin.[23] Thus, the amount of tetrahydrobiopterin in cells can be calculated by subtracting the biopterin measured under alkaline conditions from the biopterin measured under acidic conditions.

Materials

6*R*-5,6,7,8-Tetrahydro-L-biopterin (Sigma)
Biopterin (Sigma)
Phenosphere 5 ODS-1 reversed-phase column (4.6 mm × 25 cm, 5 μm) (Phenomenex, Torrance, CA)
Dithioerythritol (Sigma)

Note: It is essential that HPLC grade water be deoxygenated with helium or nitrogen for at least 30 min prior to use for preparing all reagent solutions. All standards and reagent solutions are freshly prepared before use.

Protocol. Endothelial cells (3×10^6 in a brown microcentrifuge tube) are washed three times with 1 ml PBS and resuspended in 0.3 ml of 0.1 *M* phosphoric acid containing 5 m*M* dithioerythritol (an antioxidant) to which 35 μl of 2 *M* TCA is added. The suspension is vortexed for 1 min, and cell debris is removed by centrifugation (10,000*g*, 1 min). One hundred microliters of tetrahydrobiopterin standard (50 pmol/ml), biopterin standard (50 pmol/ml), or cell extract is mixed with 15 μl of 0.2 *M* TCA and 15 μl of acidic oxidizer (1% I_2/2% KI in 0.2 *M* TCA) (acidic oxidation) or with 15 μl of 1 *M* NaOH and 15 μl of alkaline oxidizer (1% I_2/2% KI in 3 *M* NaOH) (alkaline oxidation). Samples are incubated at room temperature in the dark for 1 hr. Excess iodine is removed by adding 25 μl of 20 mg/ml ascorbic acid, followed by the addition of 30 μl of 2 *M* NaOH to the acidic oxidation solution or addition of 30 μl of 0.2 *M* TCA to the alkaline oxidation solution. All solutions are centrifuged (10,000*g*, 1 min). Fifty microliters of supernatant is analyzed on a Phenosphere 5 ODS-1 column (4.6 mm × 25 cm, 5 μm) using isocratic elution (flow rate of 1 ml/min) and fluorescence detection (excitation 350 nm and emission 440 nm). The mobile phase solvent is 5% HPLC grade methanol, 95% HPLC grade water, 7.5 m*M* sodium phosphate, pH 6.35 (running time is 15 min). The retention time for biopterin is 8.6 min. The conversion of tetrahydrobiopterin to biopterin is essentially 100% as determined with known amounts of both standards. The amount of tetrahydrobiopterin in the cell extract is determined by subtracting the amount of biopterin measured after alkaline oxidation from the amount of biopterin measured after acidic oxidation.

GTP Cyclohydrolase I Activity Assay

Principle. GTP cyclohydrolase I catalyzes the formation of dihydroneopterin triphosphate from GTP (Fig. 2). This is the first and rate-limiting step in the

[23] T. Fukushima and J. C. Nixon, *Anal. Biochem.* **102,** 176 (1980).

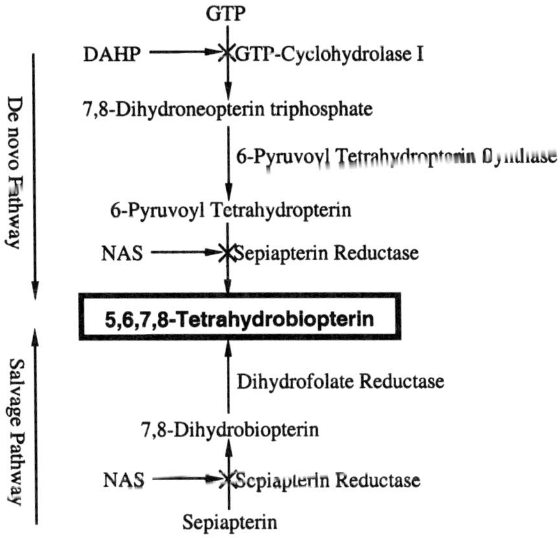

FIG. 2. *De novo* and salvage pathways of tetrahydrobiopterin synthesis. DAHP, 2,4-diamino-6-hydroxypyrimidine; NAS, *N*-acetylserotonin.

de novo synthesis of tetrahydrobiopterin. Dihydroneopterin triphosphate is oxidized by iodine to neopterin triphosphate, which is dephosphorylated by alkaline phosphatase to yield neopterin, a highly fluorescent molecule. When crude cell extracts are used for the determination of GTP cyclohydrolase I activity, desalting of enzyme extracts through a Sephadex G-25 column is necessary to remove tetrahydrobiopterin and phenylalanine (regulators of GTP cyclohydrolase I activity). In addition, EDTA is included in the assay mixture to inhibit the activities of 6-pyruvoyltetrahydropterin synthase (which converts dihydroneopterin triphosphate to 6-pyruvoyltetrahydropterin) and some GTP-degrading phosphatases.

Materials

Sephadex G-25 (Pharmacia, Piscataway, NJ)
Guanosine triphosphate (GTP, Sigma)
Alkaline phosphatase (Roche)
Neopterin (Sigma)

Protocol. Endothelial cells (1×10^7) are washed three times in PBS and then resuspended in 0.45 ml of 100 μM phenylmethylsulfonyl fluoride in 0.1 M Tris buffer (pH 7.8, 0.3 M KCl, 2.5 mM EDTA, 10% glycerol). Cells are lysed by three cycles of freezing in liquid nitrogen and thawing in a 37° water bath. The cell lysate is centrifuged (10,000g, 4°, 10 min), the supernatant is loaded into a

Sephadex G-25 column (5 × 60 mm) equilibrated with 0.1 *M* Tris buffer, and the eluate is discarded. The column is then washed with 0.45 ml of 0.1 *M* Tris buffer, and the resulting eluate is discarded. An additional 0.45 ml of 0.1 *M* Tris buffer is added to the column, and the resulting 0.45-ml eluate is collected for enzyme assay.

The desalted enzyme preparation (200 μl) is mixed with 100 μl of 6 m*M* GTP in a brown microcentrifuge tube, and the solution is incubated at 37° in the dark. After 90 min, 25 μl of 1% I_2/2% KI (in 1 *M* HCl) is added to the tube. A separate blank is prepared with 200 μl of the desalted enzyme preparation and 25 μl of 1% I_2/2% KI (in 1 *M* HCl); after 5 min, 100 μl of 6 m*M* GTP is added to the blank tube. All tubes are left standing for 30 min and then centrifuged (10,000*g*, 1 min). The supernatants are mixed with 25 μl of 1% ascorbic acid and then neutralized with 25 μl of 1 *M* NaOH (pH should be 7.5–8.5). Alkaline phosphatase is added to each tube (10 U/tube). After a 60-min incubation at 37° in the dark, samples are analyzed for neopterin by HPLC as described previously for tetrahydrobiopterin determination. The retention time for neopterin is 4.5 min. Neopterin in samples is quantitated on the basis of neopterin standards.

NADPH Analysis

Principle. NADPH is a highly fluorescent molecule at an excitation wavelength of 340 nm and an emission wavelength of 460 nm. Thus, NADPH can be measured by fluorescence detection after chromatographic separation. Because of its instability under acidic conditions and oxidation by iron, NADPH is extracted from cells using a KOH solution containing bathophenanthrolinedisulfonic acid (BPTD, a bivalent metal chelator).

Materials

Bathophenanthrolinedisulfonic acid (Sigma)

Protocol. Endothelial cells (1 × 10^6 cells in a brown microcentrifuge tube) are washed three times with 1 ml PBS containing 5 m*M* glucose and pelleted. Cells are resuspended in 100 μl of ice-cold 1 m*M* BPTD/250 m*M* KOH and vortexed for 1 min. Ice-cold 1 *M* KH_2PO_4 (25 μl) and 100 μl of 1 m*M* BPTD (in 150 m*M* potassium phosphate buffer, pH 7.5) are added sequentially to the tubes, followed by centrifugation (10,000*g*, 1 min). An aliquot of the supernatant (25 μl) is analyzed on a Phenosphere 5 ODS-1 column (4.6 × 25 cm, 5 μm) using isocratic elution (flow rate of 1 ml/min) and fluorescence detection (excitation 340 nm and emission 460 nm). The mobile phase solution is 150 m*M* potassium phosphate/5 m*M* tetrabutylammonium hydrogen sulfate/23% methanol (running time is 10 min). The retention time for NADPH is 3.8 min.

Manipulation of Endothelial Tetrahydrobiopterin Production

Principle. The *de novo* production of tetrahydrobiopterin is a multistep process (Fig. 2). Tetrahydrobiopterin may also be salvaged from dihydrobiopterins inside cells via the action of dihydrofolate reductase. Compounds that increase or inhibit the enzymes responsible for tetrahydrobiopterin synthesis can be utilized to manipulate the levels of this cofactor inside endothelial cells.

Materials

2,4-Diamino-6-hydropyrimidine (DAHP, Sigma)
N-Acetylserotonin (NAS, *N*-acetyl-5-hydroxytryptamine, Sigma)

Protocol. Endothelial cells are cultured with DAHP (final concentration 10 m*M*) or NAS (final concentration 2 m*M*) in medium containing 5% FBS for 48 hr. Proliferation assays with test compounds are then performed as described earlier.

Conclusion

Endothelial NOS catalyzes the formation of NO and L-citrulline from L-arginine in a reaction that requires seven cofactors (NADPH, FAD, FMN, tetrahydrobiopterin, heme, calcium, and calmodulin). Tetrahydrobiopterin and NADPH are known to modulate NO synthesis by endothelial cells[9,24] and thus have the potential to regulate endothelial cell proliferation. In addition, NADPH is required for the conversion of oxidized glutathione to reduced glutathione, and therefore plays an important role in the cellular defense against oxidants. Similarly, tetrahydrobiopterin can function as a reducing agent and antioxidant. As such, measurements of cellular NADPH and tetrahydrobiopterin concentrations, as well as the activity and expression of NOS, are crucial to understanding the mechanisms for endothelial cell function and proliferation. Because endothelial cell proliferation is modulated by cellular tetrahydrobiopterin availability through an increase in NO production, agents that can enhance tetrahydrobiopterin synthesis may be of clinical importance in a variety of physiological and pathological conditions.

Acknowledgments

The authors thank Dr. Harris J. Granger and Dr. Bryan H. Johnson for their support, as well as Katherine Kelly, Wene Yan, Tony Haynes, and Hui Li for dedicated technical assistance. This work was supported by grants from the American Heart Association (AHA) and the Juvenile Diabetes Research Foundation. G. Wu is an Established Investigator of the AHA.

[24] G. Wu, T. E. Haynes, H. Li, W. Yan, and C. J. Meininger, *Biochem. J.* **353,** 245 (2001).

[25] Fluorescent Imaging of Mitochondrial Nitric Oxide in Living Cells

By MANUEL O. LÓPEZ-FIGUEROA, CLAUDIO A. CAAMAÑO, M. INÉS MORANO, HUDA AKIL, and STANLEY J. WATSON

Introduction

Nitric oxide (NO) is an important modulator of mitochondrial respiration and membrane potential.[1,2] The presence of a mitochondrial NO synthase (mtNOS) has been described in the inner membrane of isolated rat liver mitochondria.[3–5] However, because NO is a highly diffusible, short-lived (0.5–5 sec), highly reactive radical, its determination poses considerable technical problems. In fact, until recently, the production of NO by mtNOS has only been indirectly characterized in broken mitochondria using techniques such as spectroscopy or the citrulline assay.[6,7] Two novel fluorometric detection systems have been developed allowing the demonstration of the presence of NO within mitochondria via microscopy.[8] 4,5-Diaminofluorescein diacetate (DAF-2/DA) permits the direct detection of NO production,[9,10] and the MitoTracker allows the detection of mitochondrial potential changes.[11,12] Thus, whereas DAF-2/DA can be used as a semiquantitative method of evaluation of NO production in living cells, the combination of these two systems provides an ideal tool for the spatial and direct visualization of NO within mitochondria.

[1] A. Boveris, L. E. Costa, J. J. Poderoso, M. C. Carreras, and E. Cadenas, *Ann. N.Y. Acad. Sci.* **899,** 121 (2000).

[2] G. C. Brown, *Biochim. Biophys. Acta* **1411,** 351 (1999).

[3] P. Ghafourifar and C. Richter, *FEBS Lett.* **418,** 291 (1997).

[4] C. Giulivi, J. J. Poderoso, and A. Boveris, *J. Biol. Chem.* **273,** 11038 (1998).

[5] A. Tatoyan and C. Giulivi, *J. Biol. Chem.* **273,** 11044 (1998).

[6] A. Boveris, L. E. Costa, E. Cadenas, and J. J. Poderoso, *Methods Enzymol.* **301,** 188 (1999).

[7] C. Richter, M. Schweizer, and P. Ghafourifar, *Methods Enzymol.* **301,** 381 (1999).

[8] M. O. López-Figueroa, C. Caamano, M. I. Morano, L. C. Ronn, H. Akil, and S. J. Watson, *Biochem. Biophys. Res. Commun.* **272,** 129 (2000).

[9] H. Kojima, N. Nakatsubo, K. Kikuchi, S. Kawahara, Y. Kirino, H. Nagoshi, Y. Hirata, and T. Nagano, *Anal. Chem.* **70,** 2446 (1998).

[10] N. Nakatsubo, H. Kojima, K. Kikuchi, H. Nagoshi, Y. Hirata, D. Maeda, Y. Imai, T. Irimura, and T. Nagano, *FEBS Lett.* **427,** 263 (1998).

[11] A. J. Krohn, T. Wahlbrink, and J. H. Prehn, *J. Neurosci.* **19,** 7394 (1999).

[12] M. Poot, Y. Z. Zhang, J. A. Kramer, K. S. Wells, L. J. Jones, D. K. Hanzel, A. G. Lugade, V. L. Singer, and R. P. Haugland, *J. Histochem. Cytochem.* **44,** 1363 (1996).

METHODS IN ENZYMOLOGY, VOL. 352
0076-6879/02 $35.00

Materials

Chemicals and drugs can be obtained from a variety of commercial suppliers: Dulbeco's modified Eagle's media (DMEM), neuronal growth factor (NGF), and horse serum (HS) from Life Technologies (Rockville, MD); fetal bovine serum (FBS) from Hyclone Labs, Inc. (Logan, UT); and sodium nitroprusside (SNP) from Fisher Scientific (Fair Lawn, NJ). Phosphate-buffered saline (PBS) is from Sigma (St. Louis, MO) and MitoTracker Red CM-H_2XRos from Molecular Probes (Eugene, OR). 2-Phenyl-4,4,5,5-tetramethylimidazoline-1-oxyl-3 oxide (PTIO), *N*-[3-(aminomethyl)benzyl]acetamidine, dihydrochloride (1400W), N^{W}-propyl-L-arginine (NPA), and DAF-2/DA are available at Calbiochem (San Diego, CA). All of the drug solutions are prepared on the day of the experiment.

Characteristics and Preparation of DAF-2/DA

DAF-2/DA (MW 497.08) can be obtained from commercial suppliers (Calbiochem, Sigma, and Alexis, San Diego, CA, or Daiichi in Japan) and is typically shipped as a 10 m*M* solution in dimethyl sulfoxide (DMSO). The stock solution is stored at 4° and brought to room temperature a few minutes prior to the experiment in a light-protected environment. Once liquid, the DAF-2/DA solution is diluted in the desired buffer or media to a recommended working concentration of 10 μM and is added to the cells for incubation. DAF-2/DA penetrates through the cell membrane, and acetate esters are hydrolyzed by cytosolic esterases, producing DAF-2 (Fig. 1), which is retained within the cells. In the presence of O_2, the diamine groups of DAF-2/DA or DAF-2 react with NO to produce a green fluorescent triazole, DAF-2 T. NO does not react directly with DAF-2, and oxygen oxidation is required to convert NO into its highly nitrosating form N_2O_3. However, DAF-2 is highly specific for NO and does not react with other reactive species, such as $O_2^{\cdot -}$, H_2O_2, $ONOO^-$, NO_2^-, and NO_3^-.[9] The detection limit of the DAF technique is about 5 n*M*. The excitation maximum of DAF-2 is at 495 nm and results in light emission at 515 nm.

Characteristics and Preparation of MitoTracker

A variety of products from Molecular Probes allow the detection of mitochondrial activity. We chose MitoTracker because it diffuses passively across the cell plasma membrane and then reacts with accessible thiol groups of peptides and proteins in mitochondrial membranes (Fig. 1). This reaction forms an aldehyde-fixable conjugate that restricts the compound within the mitochondria. It is recommended that MitoTracker be kept at −20° in its lyophilized and reduced form under inert gas. Just prior to the experiment, sealed vials are brought to room temperature in a light-protected environment and the compound is then dissolved in DMSO at

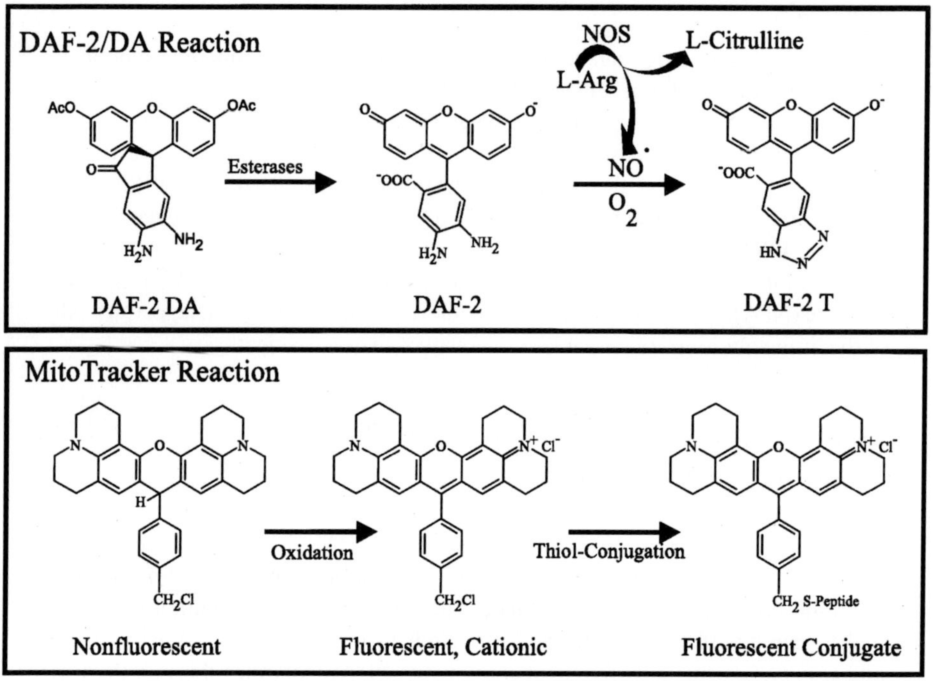

FIG. 1. Reactions of DAF-2/DA and MitoTracker. Adapted from the product data sheet with permission of Calbiochem and Molecular Probes.

a concentration of 1 m*M*. This stock is finally diluted with incubation medium to a recommended working concentration of 25–500 n*M*. Due to the variability introduced by the limited solubility and stability of the MitoTracker, its final concentration is difficult to assess. Therefore, both the optimal concentration and the incubation time should be established empirically for each experiment. Typically, we incubate for 10–15 min prior to washing the cells. Prolonged incubation times and high concentrations increase the background signal, and also the risk of labeling other subcellular structures. The MitoTracker is neutrally charged and nonfluorescent until its cell entrance and oxidation. Intracellular oxidation transforms it to a positively charged and fluorescent form that is sequestered readily by active mitochondria. The correspondent fluorescence excitation and emission maximal signals are 579 and 599 nm, respectively.

Cell Cultures

A variety of cell lines, including PC12, Cos-1, and Neuro2A, have been used in our experiments. The cells are grown under standard conditions (37° in a humidified incubator equilibrated with 5% CO_2); however, some cell lines require supplementation with specific growth factors. In general, cells seeded to reach a density of ~30,000/well are cultured in eight-well chamber slides (Nalge-Nunc Int. Corp., Naperville, IL) in phenol red-free DMEM (to avoid interference with the red MitoTracker signal). Media are supplemented with 50 U/ml penicillin, 25 μg/ml streptomycin, and 10% FCS (Cos 1 and Neuro2A) or 10% HS/5% FCS (PC12). In the case of PC12 cells, NGF is also added to media at a 50-ng/ml concentration the day before processing. Moreover, in order to minimize cell loss during subsequent washing steps, it is critical that the cells be firmly attached to the slide surface. Therefore, the slides are coated with collagen and poly-lysine prior to cell seeding, particularly in the case of PC12 cells.

It should be noted that mitochondrial activity and abundance depend on the metabolic stage of the cells. Therefore, the culture conditions should be adapted for each cell type in order to obtain optimal results with the NO detection method.

Incubation with DAF-2/DA and MitoTracker

The order in which the compounds are added is very important. First, cells grown as described earlier are incubated in 10 *μM* DAF-2/DA for 30–60 min under standard conditions.

Assuming that all of the substrates and cofactors needed for NOS optimal activity are present in the cells, further supplementation of the media is omitted. Next, MitoTracker is added to a final concentration of 25–500 n*M*, and the cells are further incubated for 10–15 min under the same conditions. In order to prevent an increase in background signal, longer incubation times should be avoided. The cells are rinsed gently with warm DMEM followed by two rinses in PBS. Finally, the slides are coverslipped using a water-soluble mounting medium such as Aqua Poly/mount (Polysciences, Inc., Warrington, PA).

Controls

The interpretation of results obtained with this method should be done with caution. A series of controls are performed to assess the specificity of the signal and, if required, to characterize the subtype of NOS producing NO. As a negative control, cells are incubated in medium lacking DAF-2/DA or MitoTracker. This should result in the absence of signal. In cases in which a signal is observed, it is likely due to autofluorescence. Other controls for DAF-2/DA include preincubation for 1 hr in DMEM medium containing NOS inhibitors (e.g., 1400W and L-NPA), a NO scavenger (e.g., PTIO), or a NO donor (e.g., SNP) followed by incubation for 30 min in DAF-2/DA, as described previously.[8] The intensity of the signal should decrease following the addition of increased concentrations of the NOS inhibitor or NO scavenger, whereas addition of a NO donor should result in a dose-dependent increase of the green fluorescent signal. We have found that inhibitor/donor concentrations from 0.1 to 10 m*M* produce a clear concentration response. Another adequate control is to use arginine-free media, as well as media supplemented with either D-arginine or L-arginine. Media lacking L-arginine or supplemented with D-arginine should exhibit negligible fluorescence.

Microscopy and Image Analysis

A Leica DHR epifluorescence microscope is utilized in our laboratory to assess the colocalization of the DAF-2 T green signal and the MitoTracker red signal. The microscope is equipped with an excitation (450–490 nm) and emission (515–560 nm) green filter for fluorescein and an excitation (515–560 nm) and emission (590 nm-LP) red filter. This filter combination ensures the individual detection of each chromophore. We used a color video camera (Sony DXC-970MD), but a black-and-white unit is also suitable. The video camera should be adjusted for gain, offset, and speed (integration). Once the camera is adjusted for optimal conditions, the same setup is used throughout each experiment. In our experience, images obtained with a high-power fluorescence objective should be digitized rapidly, starting by the MitoTracker. For assurance that both fluorescence signals are in the same plane, images can be captured with a confocal microscope. We use a Nikon Diaphot 200 microscope equipped with a Noran confocal laser-scanning imaging system; both fluorescence signals can be acquired in dual channel mode. It is important to note that in the DAF-2 reaction, the time course of the fluorescence signal is a reflection of the *de novo* synthesis of NO, and therefore of its accumulation over time. In addition, the signal is affected by other variables, including the duration and intensity of illumination. We typically observe that irradiation has two opposite effects: activation and photobleaching of fluorescence. Initially, the positive effect of activation predominates, causing a net increase in fluorescence. This increase is proportional to the intensity of epillumination, thus being more noticeable with higher magnification objectives. In contrast, photobleaching limits the life span of fluorescence. Therefore, for the comparative control experiments using DAF-2/DA, the images are taken at a fixed time of less than a minute of constant light exposure. For mitochondria colocalization studies, we chose the peak of green fluorescence signal, which is normally reached in approximately 1–2 min, depending of the objective used. In the case of the MitoTracker, longer exposure times are particularly undesirable due to its faster photobleaching rate, and also the possibility of artifactual labeling of structures other than mitochondria.

We have created a macro in our image analysis system MCID (Imaging Research Inc., Ontario, Canada) that allows a rapid acquisition of the images, which are saved in TIFF format so that they can be further analyzed or manipulated. The optical values of the images are expressed in relative optical densities (ROD), allowing quantitative analysis and comparison of the control experiments.

ROD is the mean optical density above the background, multiplied by the total target area. The background is established by multiplying the mean of all pixels with the lowest optical density outside a cell multiplied by 3.5 times the standard deviation. At least eight images, containing an average of 20 cells, are analyzed in each experimental condition. It is important to note that for each experiment, images are digitized under constant exposure time, gain, and offset. For demonstration of the colocalization of the two fluorophores, images are deconvoluted using a NoNeighbor deblourring algorithm from the MCID software to reduce the signal from the NO produced in the cytosol, and improve sharpening. Images are fused, and yellow represents areas of colocalization (Fig. 2).[8]

Advantages and Limitations of the Method

DAF-2 does not distinguish NO production among different cellular compartments. However, the combination of DAF-2/DA and MitoTracker is a method that allows the visualization and potential "semiquantitation" of mitochondrial NO production. Moreover, although the DAF-2/DA method does not reveal the type of NOS producing the NO, the information could be obtained indirectly by other techniques, such as immunocytochemistry or *in situ* hybridization, as demonstrated previously.[13,14]

The combined method is sensitive, direct, and simple. Both compounds, DAF-2/DA and MitoTracker, are cell permeable and highly specific for NO and active respiring mitochondria, respectively, and only become fluorescent upon activation. This is very advantageous because it avoids background signal and restricts the signal to the target. In addition, this system allows the visualization of NO production in an ample range of concentrations (5 n*M* to m*M*), and fluorescence excitation is performed with visible light, minimizing cell death.

Although we have described the method for real-time measurements in living cells, the DAF-2/DA method can be used in cells or tissue sections that are fresh or frozen.[14,15] However, whereas results by others differ,[16] our data demonstrate that the distribution and the quality of the fluorescence signal are compromised by cell fixation for 1 hr in 4% paraformaldehyde or harder fixatives. Consequently, we do not recommend this method in combination with other techniques that require prior fixation because the NOS activity and also the mitochondrial respiration may be significantly compromised. This mitochondrial impact of fixation is particularly undesirable, as MitoTracker labels actively respire mitochondria only, whereas inactive ones remain undetected.

[13] U. Frandsen, M. López-Figueroa, and Y. Hellsten, *Biochem. Biophys. Res. Commun.* **227,** 88 (1996).

[14] M. O. López-Figueroa, H. E. W. Day, S. Lee, C. Rivier, H. Akil, and S. J. Watson, *Brain Res.* **852,** 239 (2000).

[15] L. A. Brown, B. J. Key, and T. A. Lovick, *J. Neurosci. Methods* **92,** 101 (1999).

[16] K. Sugimoto, S. Fujii, T. Takemasa, and K. Yamashita, *Histochem. Cell. Biol.* **113,** 341 (2000).

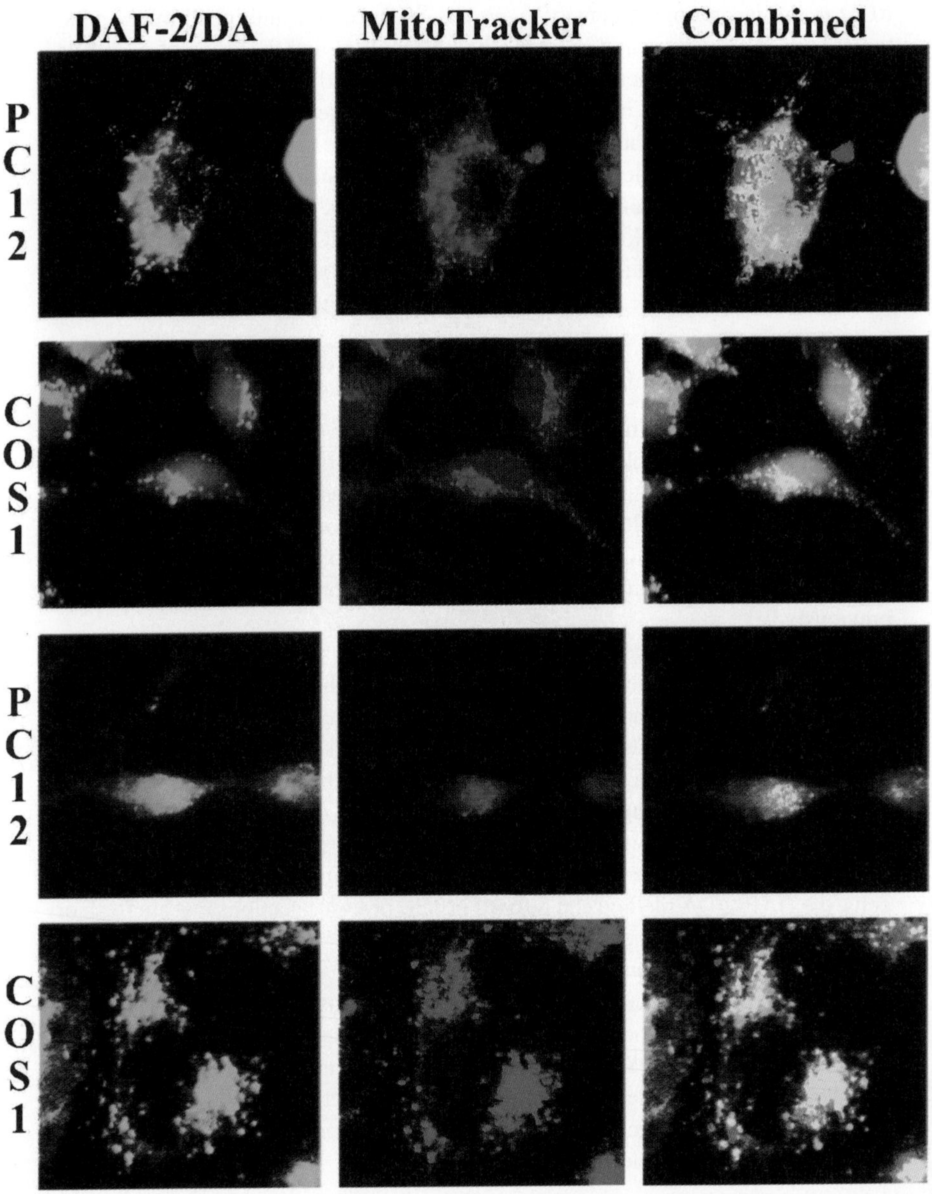

FIG. 2. Fluorescence micrographs of cells loaded with DAF-2/DA (left column) and MitoTracker (middle column) (×63). The right column represents a superimposed image in which yellow areas are indicative of colocalization. The top two rows represent images obtained from a conventional microscope and the lower two from a confocal microscope. Reprinted from M. O. López-Figueroa, C. Caamano, M. I. Morano, L. C. Ronn, H. Akil, and S. J. Watson, *Biochem. Biophys. Res. Commun.* **272,** 129 (2000).

The significant wavelength difference between fluorescence emission maxima of MitoTracker and the DAF-2 T(84 nm) makes the combination ideal for colocalization studies in mitochondria. However, the DAF-2/DA method alone provides sufficient resolution to allow the visualization of the NO sources to other subcellular locations by conventional and confocal microscopic analysis. In addition, this method permits the analysis of subcellular fractions using a fluorometer, although the cytological distribution is lost. Then, it is important to remark that although we used DAF-2/DA in combination with MitoTracker in this particular case, the former can be used in combination with other fluorescence markers, allowing the elucidation of the production of NO in other subcellular compartments.

Albeit the many advantages of the combined method, it also has some limitations. For instance, the dose of incident light and the power of the objective could affect the intensity of the fluorescence signal as mentioned earlier. Thus, the longer the exposure and the higher the objective aperture, the faster the increased signal is followed by a decrease. Moreover, MitoTracker photobleaches faster than DAF-2. Consequently, because the speed and the order of image acquisition are crucial, we strongly recommend acquiring the images corresponding to the MitoTracker first followed by the DAF-2 T.

A common concern with the DAF-2 method is whether this compound reacts with other NO derivatives or other reactive species. However, according to Kojima and colleagues,[9] DAF-2 is highly specific for N_2O_3. Another minor concern is the low solubility of Mitotracker, which makes it very difficult to assess its exact dilution, which could affect the reproducibility of the results.

In conclusion, the DAF-2/DA method combined with MitoTracker provides a novel and useful approach to demonstrate the specific production of NO within mitochondria. We have provided a protocol that can be used with a conventional or confocal fluorescence microscope.

Acknowledgments

We thank Dr. Lars C. Rønn for his contribution to this paper. NIMH Program Project (MH42251) and The Pritzker Depression Network supported the research from our laboratory presented in this chapter.

Section II

Tissues and Organs

[26] Detection of Reactive Oxygen and Nitrogen Species in Tissues Using Redox-Sensitive Fluorescent Probes

By LI ZUO and THOMAS L. CLANTON

Introduction

Many methods are currently available or are emerging for the detection of reactive oxygen and nitrogen species in tissue. Few methods have the appeal of fluorescent imaging techniques, which have the potential for semiquantitative detection within individual cells. However, despite their widespread use, there are significant limitations to their application that are very often overlooked and can lead to erroneous conclusions. This chapter outlines common methods of use for these probes with discussion of their individual strengths and pitfalls.

The most frequently used fluorescence probes for measuring reactive oxygen species (ROS) and reactive nitrogen species (RNS) are summarized in Table I, along with their relative sensitivities to a variety of intracellular redox signals. It is important to realize that none of the available fluorescent probes for ROS are particularly quantitative or specific when used in living cells and tissues. The specificity of the RNS probes may be somewhat better, as will be discussed later. The loss of quantitation of both RNS and ROS fluorescent probes arises from unpredictable loading, varying intracellular retention, heterogeneity of intracellular localization,[1] varying penetration of excitation and emission light, parallel chemical reactions that may arise from peroxidases or other redox-sensitive intracellular elements,[2] varying amounts of photobleaching and photooxidation,[3] additional ROS production from oxidized forms of the probes,[4] and common nonlinearities of many fluorescent detection systems. Nevertheless, fluorescent detection has proven invaluable in countless experiments for qualitative and rough quantitative estimates of ROS and RNS activity at the cellular level using estimates of "relative changes" in fluorescence. Critical to this approach, however, are careful standardizations of experimental conditions between experiments and controls, including handling of the tissue, loading procedures, and adjustment of the gain of the fluorescent detection systems to a common standard. Whenever possible, it is also advisable to (1) repeat experiments with multiple probes that have different spectral characteristics and specificities or preferably run parallel experiments with

[1] J.-B. LePecq and C. Paoletti, *J. Med. Biol.* **27,** 87 (1967).

[2] S. L. Hempel, G. R. Buettner, Y. Q. O'Malley, D. A. Wessels, and D. M. Flaherty, *Free Radic. Biol. Med.* **27,** 146 (1999).

[3] E. Marchesi, C. Rota, Y. C. Fann, C. F. Chignell, and R. P. Mason, *Free Radic. Biol. Med.* **26,** 148 (1999).

[4] C. Rota, C. F. Chignell, and R. P. Mason, *Free Radic. Biol. Med.* **27,** 873 (1999).

0076-6879/02 $35.00

TABLE I
COMMONLY USED FLUORESCENT PROBES FOR DETECTION OF REACTIVE OXYGEN AND NITROGEN SPECIES[a,b]

Probe[c]	Ex/Em (nm)	$O_2^{\cdot -}$	H_2O_2	$^{\cdot}OH$	$ONOO^-$	Cyt C	Fe^{2+}	$^{\cdot}NO$	Peroxidases[m]
HE/ET[d]	460–490/ LP 590	●●● (7, 15, 9, 16)	○ (7, 15, 9, 16)	? ●●●(7) ○ (9, 16)	? ●●●(7) ○ (9, 16)	? ○ (15)	○ (7, 15)	○ (9)	HRP: ○ (17) XO: ◖ (9)
DCFH/DCF[e]	480–500/ 510–560 or LP 515	○ (2, 30, 29)	●●●[k] (2)	◖ (2)	●●● (2, 30)	● (2, 61)	● (2)	●● (2, 10)	HRP: ● Cat: ● SOD: ● GPX: ○ XO: ● LO: ● (2)
HFLUOR/ FLUOR[f]	480–500/ 510–560 or LP 515	◖ (2)	●●●[k] (2)	○ (2)	● (2)	○ (2)	○ (2)	○ (2)	HRP: ● Cat: ● SOD: ◖ GPX: ○ XO: ● LO: ○(2)
5&6CDCFH/ 5&6CDCE[g]	480–500/ 510–560 or LP 515	◖ (2)	●●●[k] (2)	○ (2)	● (2)	○ (2)	○ (2)	○ (2)	HRP: ○ Cat: ● SOD: ◖ GPX: ○ XO: ● LO: ○(2)
DHR 123/ Rh 123[h]	480–500/ 510–560 or LP 515	○ (2, 30)	●●●●[k] (2)	●● (2)	●● (2, 30)	● (2)	●● (2)	○ (2, 30)	HRP: ●● Cat: ● SOD: ● GPX: ○ XO: ●● LO: ●●(2)
DAF-2/ DAF-2-T[i]	480–500/ 510–560 or LP 515	○ (44)	○ (44)	?	○ (44)	?	?	●●●	?

DAF-FM/ DAF-FM-T[j]	480–500/ 510–560 or LP 515	○?[l]	○?[l]	?	○?[l]	?	?	•••	?

[a] Ex/Em, excitation and emission of the oxidized form of the probes; ○,◐,●, respectively, low to high estimates of probe sensitivity to species, based on citation (in parentheses).

[b] Key to references: (2) S. L. Hempel, G. R. Buettner, Y. Q. O'Malley, D. A. Wessels, and D. M. Flaherty, *Free Radic. Biol. Med.* **27**, 146 (1999). (7) A. B. Al-Mehdi, H. Shuman, and A. B. Fisher, *Am. J. Physiol* **272,** L294 (1997). (9) V. P. Bindokas, J. Jordan, C. C. Lee, and R. J. Miller, *J. Neurosci.* **16,** 1324 (1996). (15) L. Benov, L. Sztejnberg, and I. Fridovich, *Free Radic. Biol. Med.* **25,** 826 (1998). (16) T. L. Vanden Hoek, C. Li, Z. Shao, P. T. Schumaker, and L. B. Becker, *J. Mol. Cell.Cardiol.* **29,** 2571 (1997). (29) H. Zhu, G. L. Bannenberg, P. Moldeus, and H. G. Shertzer, *Arch. Toxicol.* **68,** 582 (1994). (30) J. P. Crow, *Nitric Oxide* **1,** 145 (1997). (44) H. Kojima, K. Sakurai, K. Kikuchi, S. Kawahara, Y. Kirino, H. Nagoshi, Y. Hirata, and T. Nagano, *Chem. Pharm. Bull (Tokyo)* **46,** 373 (1998). (61) M. J. Burkitt and P. Wardman, *Biochem. Biophys. Res. Commun.* **282,** 329 (2001).

[c] Probes expressed in their deacetylated form, where appropriate. For cell loading the diacetate forms are used (e.g., DCFH-DA).

[d] Hydroethidine/ethidium.

[e] 2′,7′-dichlorodihydrofluorescein/2′,7′-dichlorofluorescein.

[f] Dihydrofluorescein/fluorescein.

[g] 5-(and 6)-carboxy-2′,7′-dichlorodihydrofluorescein/5-(and 6)-carboxy-2′,7′-dichlorofluorescein.

[h] (Dihydrorhodamine 123/Rhodamine 123).

[i] 4,5-Diaminofluorescein/Triazole form of 4,5-diaminofluorescein.

[j] (3-Amino-4-(*N*-methylamino)-2′,7′-difluorofluorescein/triazole form of 3-Amino-4-(*N*-methylamino)-2′,7′-difluorofluorescein).

[k] Reaction with H_2O_2 in the absence of peroxidases is essentially nonexistent.

[l] Reactivity assumed the same as other DAF probes.

[m] Peroxidase/oxidase abbreviations: HRP, horseradish peroxidase; XO, xanthine oxidase; Cat, catalase; SOD, superoxide dismutase; GPX, glutathione peroxidase; LO, lipoxygenase.

[?] Contradictory or uncertain findings in literature.

completely different ROS or RNS detection methods; (2) run positive controls with intracellular ROS-producing agents such as menadione[2] or nitrazepam[5]; (3) repeat exact experimental paradigms without the fluorescent probe in question to resolve erroneous data due to tissue autofluorescence, particularly if probes are used with excitation and emission wavelengths in the region of NADH fluorescence[6]; and (4) repeat control experiments in the presence of antioxidants, ROS scavengers, or nitric oxide synthase inhibitors, where appropriate.

Fluorescence data have most often been expressed as the fractional change or absolute output voltage of the fluorescence signal from "baseline" fluorescence, measured following loading. Therefore, common units are "percentage change" and "relative units." The absolute values of these measures have no particular meaning when compared across experimental models, paradigms, or detection devices, but when averaged together under identical experimental conditions, they can provide some approximation of grouped responses for comparison. At the present time, none of the probes provide the opportunity for a ratiometric determination, such as the methods that exist for pH- or Ca^{+2}-sensitive probes. Although far beyond the scope of this chapter, a variety of techniques have been used to monitor the fluorescence changes with these probes. Some examples include reflectance fluorometry,[7,8] epifluorescence fluorescent microscopy,[9,10] flow cytometry[11] and extraction of oxidized probes from the tissue followed by fluorometry,[7,12] HPLC with fluorescence detection,[13] and laser-scanning confocal microscopy.[14]

General Guide to Loading Probes into Cells and Tissues

Some important initial points can be applied to all the fluorescent probes discussed in this chapter. First, the final intracellular dye concentrations after loading are generally much greater than the dye concentrations used in the extracellular environment during loading. This occurs due to intracellular compartmentalization or to conversion of the loading form of the probe to a new species, resulting in a sustained concentration gradient across the cell membrane. The clearest example

[5] G. D. Ford, *FASEB J.* **15,** 1132A (2001).

[6] T. A. Fralix, F. W. Heineman, and R. S. Balaban, *FEBS Lett.* **262,** 287 (1990).

[7] A. B. Al-Mehdi, H. Shuman, and A. B. Fisher, *Am. J. Physiol.* **272,** L294 (1997).

[8] L. Zuo, L. J. Berliner, and T. L. Clanton, *Free Radic. Biol. Med.* **25,** S25 (1998).

[9] V. P. Bindokas, J. Jordan, C. C. Lee, and R. J. Miller, *J. Neurosci.* **16,** 1324 (1996).

[10] C. L. Murrant, F. H. Andrade, and M. B. Reid, *Acta Physiol. Scand.* **166,** 111 (1999).

[11] A. Macho, R. F. Castedo, P. Marchetti, J. J. Aguilar, D. Decaudin, N. Zamzami, P. M. Girard, J. Uriel, and G. Kroemer, *Blood* **86,** 2481 (1995).

[12] D. Nethery, D. Stofan, L. Callahan, A. DiMarco, and G. Supinsky, *J. Appl. Physiol.* **87,** 792 (1999).

[13] Y. Itoh, F. H. Ma, H. Hoshi, M. Oka, K. Noda, Y. Ukai, H. Kojima, T. Nagano, and N. Toda, *Anal. Biochem.* **287,** 203 (2000).

[14] L. Zuo, F. L. Christofi, V. P. Wright, C. Y. Liu, A. J. Merola, L. J. Berliner, and T. L. Clanton, *Am. J. Physiol Cell Physiol.* **279,** C1058 (2000).

of this applies to the acetoxymethyl ester (AM) or diacetate (DA) forms of certain dyes, which are used to improve penetration of the cell membrane. Once inside the cell, these attached groups are cut off by intracellular esterases, resulting in entrapment and accumulation of the dye inside the cell. The use of a minimal loading dose, resulting in a relatively strong fluorescent signal, is recommended. If too much fluorescent dye is loaded into the tissue, the probes can work as either antioxidants or prooxidants, disturbing the intracellular environment. The leakage of dyes may also be increased during the experiments due to overloading. Therefore, careful adjustment of loading concentrations in specific experimental conditions is imperative and must be determined for each model and each new probe.

Use of Specific Fluorescent Dyes for ROS and RNS

Hydroethidine/Ethidium (HE/ET)

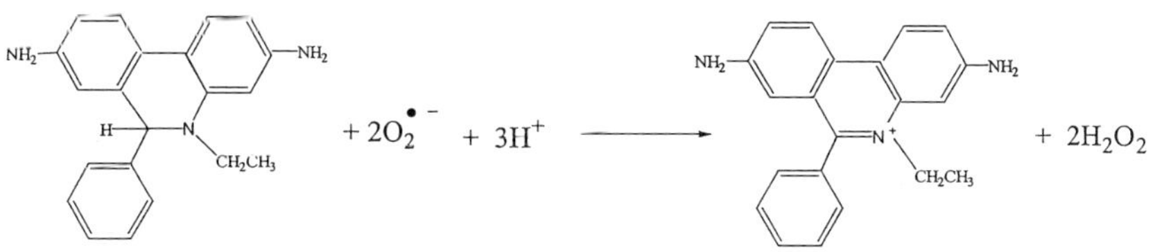

Hydroethidine

Ethidium

(1)

Description and General Sensitivity to ROS

Hydroethidine (HE), or dihydroethidium, is an uncharged hydrophobic molecule, taken up readily by living cells or tissues (Molecular Probes, Inc., Eugene, OR; Polysciences, Inc., Warrington, PA). It is very sensitive to superoxide $(O_2^{\cdot -})$,[7,9,15] as shown in Eq. (1).[15] It is very insensitive to H_2O_2.[7,9,15] In the process of oxidation, it forms one or more free radical intermediates (not shown) that may be important for some secondary reactions.[15] Although some *in vitro* experiments have demonstrated that it can be oxidized by $^{\cdot}OH$ to ET,[7,16] others have not been able to show this.[9] Regardless, this is probably not quantifiable *in vivo* because of the low reaction probability and the fact that $^{\cdot}OH$ simultaneously reduces the fluorescence of ET when intercalated with DNA.[16] Some studies have reported that the probe is insensitive to peroxynitrite $(ONOO^-)$ or HOCl.[9] However, others have found that HE is oxidized easily by $ONOO^-$ using the $ONOO^-$ generator,

[15] L. Benov, L. Sztejnberg, and I. Fridovich, *Free Radic. Biol. Med.* **25,** 826 (1998).

[16] T. L. Vanden Hoek, C. Li, Z. Shao, P. T. Schumaker, and L. B. Becker, *J. Mol. Cell. Cardiol.* **29,** 2571 (1997).

SIN-1.[7] One author has suggested that SIN-1 may oxidize HE by separate formation of $O_2^{\cdot -}$.[9] Unlike the fluorescein and rhodamine-based probes, discussed later, HE oxidation is not sensitive to and does not require peroxidases, such as horseradish peroxidase (HRP),[17] but it has a small sensitivity to xanthine oxidase.[9] ET apparently does not appreciably autooxidize, thus avoiding the additional production of $O_2^{\cdot -}$,[15] although this may require further evaluation in the presence of cellular reducing reactions.[4] As shown in Eq. (1), the oxidized product is positively charged, contributing to its tendency to accumulate in the nucleus and intercalate with the negatively charged DNA phosphate backbone, as well as in other negatively charged compartments, such as the mitochondrial matrix and inner cell membrane.

Procedure for Loading HE

With poor solubility in water, HE stock must be made in an organic solvent such as dimethyl sulfoxide (DMSO) or *N,N*-dimethylacetamide (DMAM). Based on our experience, DMAM is preferred for this probe, as the stock stays in aqueous phase at −10 to −20°, whereas HE-DMSO stock is frozen at 3–4°. In general, it is advisable to avoid repeated freeze–thaw cycles with these probes. There are other considerations when using DMSO as a solvent. It is a well-known $^{\cdot}OH$ scavenger that could possibly attenuate HE oxidation. However, for this reason, it might also aid in separating HE oxidation via $O_2^{\cdot -}$ vs oxidation from $^{\cdot}OH$.

The exact concentrations and incubation times for staining in whole tissue preparations should be optimized for each experimental model. However, one recommended procedure (Polysciences, Inc., Warrington, PA) appears to work well for a large range of experimental conditions from cell preparations[18] to whole tissues.[14]

STOCK AND LOADING SOLUTIONS. Use 7 mg HE in 1 ml DMAM. Store, sealed in −10 to −20°. Fresh loading solution should be made each day. Use 20 μl HE stock/10 ml buffer (44 μM HE solution). Buffer can be physiologic salt solution for tissues or phosphate-buffered saline (PBS) containing Ca^{2+} for cell preparations.[18] Successful loading has been reported for cell preparations as low as 2 μM HE.[11] The loading solution should be reasonably clear to light pink prior to loading. As the stock or loading solution ages, autooxidation of HE to ET is apparent by the change to a darker pink and ultimately to a red color. This increases the background fluorescence and can decrease sensitivity. In whole tissue experiments, loading concentrations of up to 88 μM HE may be necessary for adequate signal intensities.

LOADING PROCEDURE. Successful loading of the HE dye does not appear to be extremely temperature dependent. However, at times it is convenient for preservation and oxygenation of thick tissues to load in buffer, on ice. Cells or well-perfused

[17] W. O. Carter, P. K. Narayanan, and J. P. Robinson, *J. Leukocyte Biol.* **55,** 253 (1994).
[18] C. Bucana, I. Saiki, and R. Nayar, *J. Histochem. Cytochem.* **34,** 1109 (1986).

preparations can load sufficiently within 15–30 min at 37°.[7,11] Our laboratory loads isolated diaphragm preparations for periods of 1 hr, in the dark, in physiologic saline, on ice.[14] This is followed by 10–15 min of rinsing in oxygenated buffer. With these loading conditions, a variable level of baseline ET fluorescence remains, prior to initiation of the experiment, presumably due to autooxidation of HE or oxidation caused by basal ROS production.[14]

PRECAUTIONS. Although HE is relatively nontoxic to cells during the course of most experiments, it is very cytotoxic after extended periods and is absorbed easily by the skin and other tissues, combining with DNA as ET. Positively charged ET is also absorbed and is considered toxic. Care should be taken in handling and disposing of all loading solutions and tissues to completely avoid skin exposure, with appropriate disposal of solutions. Additional care should be taken with loading solutions that are being bubbled with O_2, as these can produce aerosols in the immediate environment.

Detection

Theoretically, HE/ET can be used as a dual fluorescent probe, i.e., the loss of HE can be monitored as a blue signal as it is dehydrogenated to ET, a red emission.[7] Although intracellular HE is relatively stable for hours in some isolated cell preparations,[19] in other tissues, such as isolated skeletal muscle at 37°,[14] significant leakage begins to occur after about 15 min. The HE excitation occurs at 360–380 nm and emission is at 430–460 nm, with the latter depending on the specific excitation wavelength used and the cell or tissue microenvironment. Furthermore, HE shares similar excitation and emission ranges as NADH,[6] a major source of autofluorescence. The short excitation (high energy) wavelengths can promote photooxidation of the probe. These factors somewhat limit the usefulness of HE detection as an indicator of ROS formation, but do provide a useful method for monitoring HE loading. In contrast, ET is trapped easily by negatively charged organelles or it can be dissolved well in aqueous cytosol, reducing its membrane penetration. Once formed, it is extremely stable in the cell. Importantly, tissue autofluorescence is weak at ET excitation and emission ranges, light penetration is better because of the longer wavelengths at lower energy, and there is good separation between excitation and emission wavelengths, allowing for a variety of filter options. Therefore, the measurement of relative increases in ET fluorescence has been preferred in most studies to decreases in HE or changes in the ET/HE ratio as an indicator of ROS.[7,9,12,14,20]

The excitation wavelengths for ET are actually very broad,[7] usually in the 460- to 490-nm range, but can even be stimulated with an argon/krypton laser line of

[19] I. Saiki, C. D. Bucana, J. Y. Tsao, and I. J. Fidler, *J. Natl. Cancer Inst.* **77,** 1235 (1986).
[20] S. L. Budd, R. F. Castilho, and D. G. Nicholls, *FEBS Lett.* **415,** 21 (1997).

568 nm.[14] It is worth mentioning that in practice, the excitation and emission wavelengths of most probes are not fixed and depend on numerous variables, including the buffer solutions, pH, the thickness of the tissue, or even the characteristics of the excitation light and the fluorometer. For example, in reflectance tissue fluorometry, using isolated diaphragm, we have found that both HE emission and ET emission shift to ~10- to 30-nm longer wavelengths than are measured *in vitro* in solution. This "shift" strategy can sometimes be used to advantage, avoiding the background noise that arises from the Raman scatter of H_2O molecules. Therefore, when possible, pretesting of optimum excitation/emission spectra for the best signal is highly recommended for each new experimental preparation. The long wavelength emission spectra for ET and the low autofluorescence in this range allow for the use of long-pass filters (LP 590).

Although HE/ET is one of the more specific and quantitative of the fluorescent probes for ROS, it still shares some of the limitations of other more commonly used probes. For example, it can be oxidized by ferricytochrome *c*, but this results in a new product, distinct from ET.[15] The fluorescence properties of this product are not well described. It also catalyzes the dismutation of $O_2^{\cdot-}$, thus diminishing the ET signal in the presence of high concentrations of $O_2^{\cdot-}$.[15] This may not be relevant under most intracellular conditions because HE does not compete well with superoxide dismutase (SOD) or nitric oxide $^{\cdot}$NO.[17]

The accumulation in negatively charged compartments and the binding to DNA/RNA have some additional effects on the ET signal. For example, with DNA binding, there is a 20-fold increase in quantum efficiency of the ET fluorescence signal measured *in vitro*.[1] However, it is possible that this may be countered somewhat *in vivo* by the influence of histones, preventing ET from accessing the DNA and thus possibly influencing the emission signal (unpublished results). The net effect of these influences is difficult to quantify but is certain to result in a very nonlinear output signal. The trade-off, however, is the stability of the accumulating signal in the cell, which can even be measured in preserved tissue.[7,9] Another complicating feature is the possibility that ET accumulation in mitochondria can act as an indicator for the mitochondrial membrane potential, i.e., ET fluorescence is influenced by the depolarization of the membrane.[20] For this reason, the lowest concentration of HE should be used that still provides an adequate image or signal.

Fluorescein- and Rhodamine-Based ROS Probes

General Description and Sensitivity to ROS

Fluoresceins and rhodamines, which share a similar structure, are highly fluorescent compounds, but when reduced to their dihydroxyl forms (Fig. 1), they exhibit little or no fluorescence. The reduced derivatives of fluorescein are normally attached with diacetate (DA) for better penetration into cells or tissues (Fig. 2). This is not necessary for dihydrorhodamine 123. Once the diacetate molecules

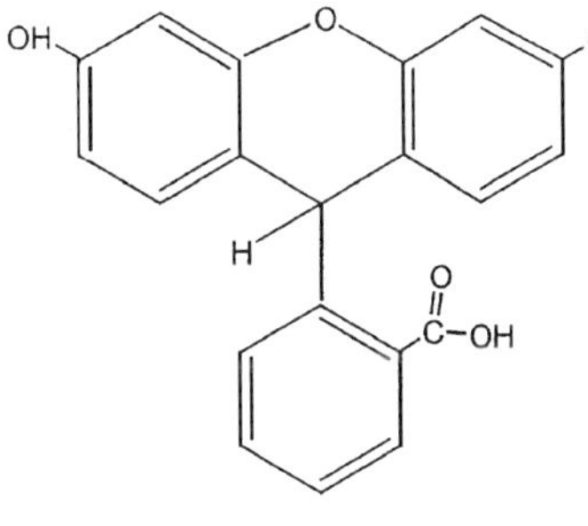

Dihydrofluorescein (HFLUOR)
Oxidized: Fluorescein (FLUOR)

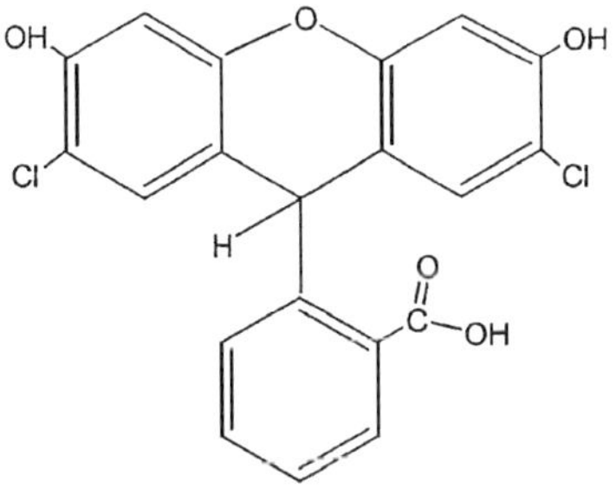

2',7'-dichlorodihydrofluorescein (DCFH)
Oxidized: 2',7'-dichlorofluorescein (DCF)

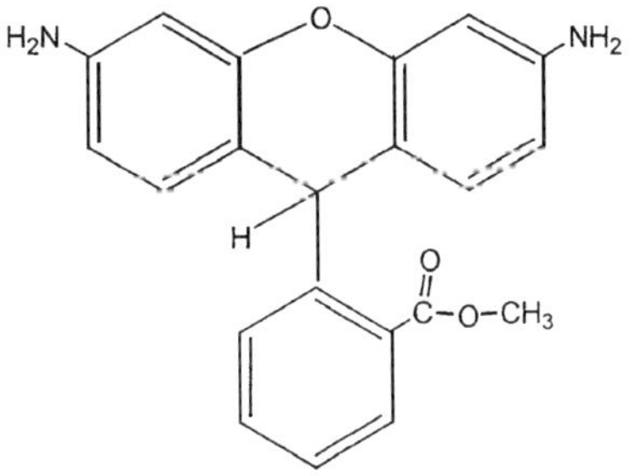

Dihydrorhodamine 123 (DHR123)
Oxidized: Rhodamine 123 (Rh123)

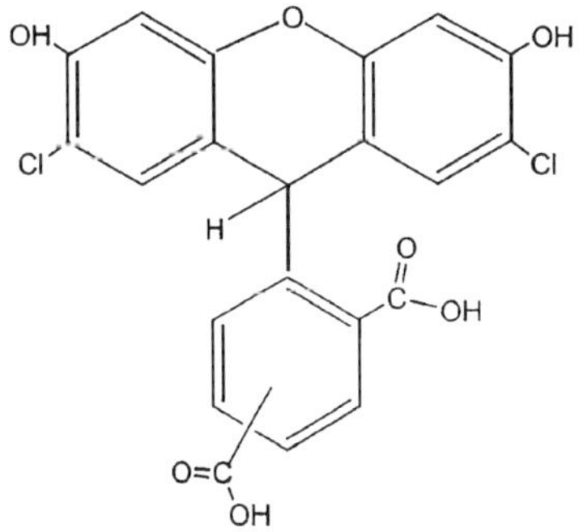

5-(and 6)-carboxy-2',7'-dichlorodihyrofluorescein (5&6CDCFH)
Oxidized: 5-(and 6)-carboxy-2',7'-dichlorofluorescein (5&6CDCF)

FIG. 1. Commonly used fluorescein- and rhodamine-based probes for detecting ROS. The acetylated forms, used for loading of fluorescein probes, are not shown. The simple reduced forms are shown in the schematic drawings and subtitles. Beneath each subtitle, in italics, are nomenclatures for the oxidized forms used in the text.

get into the cytosol, intracellular esterases cut off the acetate groups, resulting in better retention within the cytosol. Oxidation in the cell results in conversion to fluorescent species. Four commonly used probes are illustrated in Fig. 1.

In general, these probes have been considered indicators of the presence of H_2O_2.[21–27] However, their potential reaction with a number of other oxidative

[21] A. S. Keston and R. Brandt, *Anal. Biochem.* **11,** 1 (1965).

[22] S. Burow and G. Valet, *Eur. J Cell Biol.* **43,** 128 (1987).

[23] D. A. Bass, J. W. Parce, L. R. Dechalet, P. Szedja, M. C. Seeds, and M. Wallace, *J. Immunol.* **130,** 1910 (1983).

[24] M. B. Reid, K. E. Haack, K. M. Franchek, P. A. Valberg, L. Kobzik, and M. S. West, *J. Appl. Physiol.* **73,** 1797 (1992).

[25] A. Al-Mehdi, H. Shuman, and A. B. Fisher, *Lab. Invest.* **70,** 579 (1994).

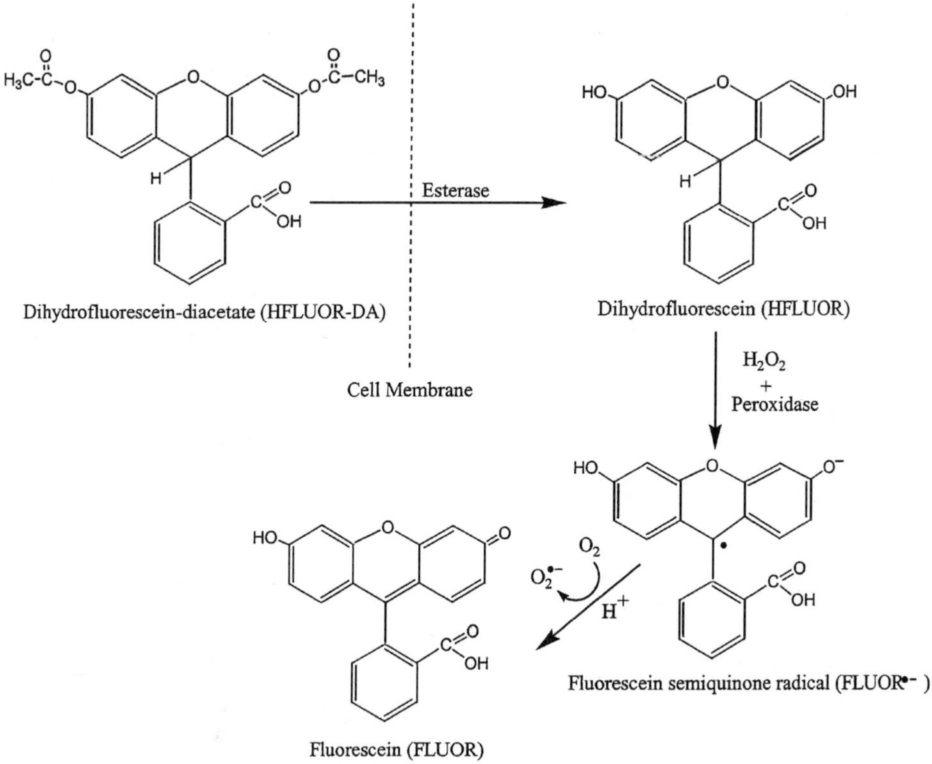

FIG. 2. Proposed scheme for the oxidation steps of dihydrofluorescein by H_2O_2 and peroxidase based on the studies of Rota and colleagues [C. Rota, C. F. Chignell, and R. P. Mason, *Free Radic. Biol. Med.* **27,** 873 (1999)], using dichlorodihydrofluorescein. Note that superoxide ($O_2^{\bullet -}$) is formed during this reaction.

enzymes and intermediates and their ability to self-propagate free radical reactions have resulted in some skepticism regarding their specificity for endogenous H_2O_2.[3,4,28] This is discussed in more detail later. Our laboratory shares the view of Hempel *et al.*[2] that these probes "act as detectors of a broad range of intracellular oxidizing reactions." The probe, 2′,7′-dichlorodihydrofluorescein (DCFH), is by far the most utilized in biological systems and the most studied with regard to its oxidation–reduction pathways. It is discussed later as a general model for fluorescein-like probes and a standard for comparison of the behavior of the other probes illustrated in Fig. 1.

[26] J. A. Royall and H. Ischiropoulos, *Arch. Biochem. Biophys.* **302,** 348 (1993).
[27] M. Tepel, M. Echelmeyer, N. N. Orie, and W. Zidek, *Kidney Int.* **58,** 867 (2000).
[28] C. Rota, Y. C. Fann, and R. P. Mason, *J. Biol. Chem.* **274,** 28161 (1999).

2′,7′-DICHLORODIHYDROFLUORESCEIN. The sensitivity of DCFH to oxidation by H_2O_2 was demonstrated in 1965 by Keston and Brandt.[21] The presence of some form of a peroxidase enzyme is required in the H_2O_2 reaction for all DCFH-like probes, as illustrated in Fig. 2. None are appreciably responsive to H_2O_2 alone.[2] Zhu *et al.*[29] showed that iron/H_2O_2-induced oxidation could be prevented by the H_2O_2 scavenger catalase or the $^{\bullet}OH$ scavenger dimethyl sulfoxide (DMSO), suggesting that DCFH is oxidized by both species. The probe is clearly not sensitive to $O_2^{\bullet -}$.[29] Importantly, it is capable of spontaneous oxidation with peroxidase enzymes, SOD, catalase, and lipoxygenase[2] and can serve as a substrate for xanthine oxidase (XO), competing with hypoxanthine.[2,29] This can present problems with interpretation, e.g., in conditions of ischemia where XO may be activated.

An important but generally unappreciated observation is that DCFH is sensitive to nitric oxide ($^{\bullet}NO$) in the presence of O_2,[2,10,30] a reaction that may be unique compared to other fluorescein-based ROS dyes[2] (Table I). The exact chemical reaction is not entirely clear, but it likely involves other redox forms of $^{\bullet}NO$, such as nitrogen dioxide ($^{\bullet}NO_2$).[30] Of note, not all investigations have come to this conclusion.[31] In cultured neuronal cells it has been estimated that ~50% of the DCFH fluorescence could be attributed to $^{\bullet}NO$.[32] Peroxynitrite strongly oxidizes DCFH and requires no other cofactors.[2,30] Iron (Fe^{+2}), in the presence of O_2, has a relatively slow oxidizing effect on DCFH, possibly because of its capacity to independently generate ROS species.[33] This may be true, particularly in the presence of ascorbate, thus keeping Fe^{+2} in the reduced form.[2]

DIHYDROFLUORESCEIN (HFLUOR). Dihydrofluorescein is one of the oldest fluorescent probes and has been reevaluated by Hempel *et al.*,[2] who found it to have some superior characteristics to DCFH, DHR123, and 5&6CDCFH (Fig. 1). Its oxidized product, fluorescein, has a considerably higher molar fluorescence, (i.e., highest molar extinction coefficient at 490 nm) and, probably more importantly, much better intracellular loading compared to any of the other probes. Our own experience in isolated skeletal muscle has confirmed that it results in very high fluorescence intensity in isolated diaphragm muscle (unpublished results). By comparison with DCFH, HFLUOR has a similar sensitivity to HRP, a reduced sensitivity to Fe^{2+}, even in the presence of ascorbate, a reduced sensitivity to SOD, catalase, lipoxygenase, and XO, and is essentially unresponsive to cytochrome *c* and $^{\bullet}NO$.[2] It is oxidized by peroxynitrite, but at a lower rate compared to DCFH or DHR123.[2]

[29] H. Zhu, G. L. Bannenberg, P. Moldeus, and H. G. Shertzer, *Arch. Toxicol.* **68,** 582 (1994).
[30] J. P. Crow, *Nitric Oxide* **1,** 145 (1997).
[31] N. W. Kooy, J. A. Royall, and H. Ischiropoulos, *Free Radic. Res.* **27,** 245 (1997).
[32] P. G. Gunasekar, A. G. Kanthasamy, J. L. Borowitz, and G. E. Isom, *J. Neurosci. Methods* **61,** 15 (1995).
[33] S. Y. Qian and G. R. Buettner, *Free Radic. Biol. Med.* **27,** 1447 (1999).

DIHYDRORHODAMINE 123 (DHR123). DHR123 also has similar reaction characteristics as DCFH. The diacetate form is not used because DHR123 already has good cell penetration. Like DCFH, it is thought of as a H_2O_2 probe.[26,34,35] However, it is sensitive to $ONOO^-$,[2,30] lipoxygenase activity, and xanthine oxidase activity, but not to ·NO.[2,36] DHR123 can also be used for staining mitochondria[34,37] because the oxidation product, rhodamine 123 (Rh123), is positively charged, making it preferably bind to the negatively charged mitochondrial matrix. This has certain advantages in some experiments where there is a need for improved cellular retention of the oxidized product.[26] However, it can also cause problems with interpretation, as discussed later.

5-(AND 6)-CARBOXY-2′,7′-DICHLORODIHYDROFLUORESCEIN (5&6CDCFH). 5&6CDCFH has been used less frequently in studies of oxidative stress compared to DCFH or DHR123. It is part of a series of DCFH derivatives designed to improve cellular retention. This group also includes a diacetoxymethyl ester version and a chloromethyl DCFH-DA version (Molecular Probes, Inc.). Unfortunately, little independent information is available regarding the reaction characteristics of these probes. However, Hempel *et al.*[2] compared 5&6CDCFH to the other probes listed in Fig. 1 and found that it had a low fluorescence intensity in cell culture, but also a lower spontaneous oxidation in the presence of HRP, XO, and other enzymes and the lowest sensitivity to iron or cytochrome *c*, but comparable responses to H_2O_2 in the presence of peroxidase. It was not sensitive to ·NO and similarly responsive to $ONOO^-$ compared to HFLUOR. It is possible that this probe may have some advantages by providing a low background signal while maintaining good cellular retention and good sensitivity to H_2O_2.[2] However, to our knowledge, this has not been tested thoroughly in variety of tissues or cell cultures.

Procedure for Loading

STOCK SOLUTIONS. DHR123 and the diacetate forms of fluorescein-based probes must be prepared in organic solvents prior to dissolving in buffers (Molecular Probes, Inc.). For fluorescein-based probes, many investigators have preferred using stock solutions of 10–33 m*M* in ethanol.[24,26,38] These are stored in the dark at $-20°$ and are often purged with N_2 to prevent autooxidation.[26] Of note, diluting 10 m*M* stock in pure ethanol into a buffer will result in an extracellular

[34] C. Garcia-Ruiz, A. Colell, M. Mari, A. Morales, and J. C. Fernandez-Checa, *J. Biol. Chem.* **272,** 11369 (1997).

[35] H. J. Chae, J. S. Kang, J. I. Han, B. G. Bang, S. W. Chae, K. W. Kim, H. M. Kim, and H. R. Kim, *Immunopharm. Immunotox.* **22,** 317 (2000).

[36] N. W. Kooy, J. A. Royall, H. Ischiropoulos, and J. S. Beckman, *Free Radic. Biol. Med* **16,** 149 (1994).

[37] C. Sobreira, M. Davidson, M. P. King, and A. F. Miranda, *J. Histochem. Cytochem.* **44,** 571 (1996).

[38] L. M. Swift and N. Sarvazyan, *Am. J Physiol. Heart Circ. Physiol.* **278,** H982 (2000).

environment of 0.1% ethanol if the tissue is incubated at 10 μM and 0.4% at 40 μM. As a comparison, in many states within the United States, the legal alcohol limit for driving is 0.08%. Therefore, it is important to realize that this level of alcohol could have a significant influence on the biology of the cells in question. This may be particularly relevant because alcohol can promote oxidant formation in some tissues.[39] Therefore, care should be taken to treat control tissues with the same vehicle concentration. Our laboratory prefers using DMSO to dissolve HFLUOR and related probes. For example, for HFLUOR we use 10 mg/100 μl DMSO stock. We then load the tissue with a 20 μM solution. This maintains the total DMSO in the bathing solution far below the concentration that has any measurable effects on skeletal muscle function.[40] Although the use of DMSO could underestimate ROS formation, it is unlikely to give false-positive results. If necessary, for some probes it is possible to use a low toxicity, dispersing agents to facilitate loading and possibly reduce the concentration of the DMSO or ethanol (Pluronic F-127, Molecular Probes, Inc.). For DHR123, most studies have used the methods of Royall and Ischiropoulos.[26] A stock solution of 28.9 mM DHR123 is made in dimethylformamide and stored at $-20°$ in the dark. Storage vessels are generally purged with He or N_2.

LOADING SOLUTIONS AND PROCEDURE. As with HE/ET, for each experimental preparation, it is best to experiment with different loading concentrations and times with the object to keep the loading concentration as low as practical. Most cell culture studies have loaded these probes at concentrations of 5–20 μM for 20–30 minutes in room temperature or at 37° in the dark.[2,16,17,26,38,41] However, whole perfused or unperfused tissues require higher concentrations for diffusion and as much as 50 μM for periods of 1 hr at room temperature have been used successfully.[10] These procedures are generally followed by 15–30 min of rinsing in fresh, oxygenated buffers.

Temperature is a critical variable for all diacetate-linked probes, as the activity of the esterases necessary for cleavage is highly temperature dependent. However, one recommended procedure for all diacetate-loaded fluorophores is to load first at low temperature (on ice), allowing accumulation of the probe into the cytosol. This is followed by incubation at room or body temperature to activate the esterases. This procedure has been effective in poorly vascularized tissues or organs, as reviewed in Ref. 42.

[39] D. Mantle and V. R. Preedy, *Adverse Drug React. Toxicol. Rev.* **18,** 235 (1999).

[40] M. B. Reid and M. R. Moody, *J. Appl. Physiol.* **76,** 2186 (1994).

[41] M. H. Ali, S. A. Schlidt, M. S. Chandel, K. L. Hynes, P. T. Schumaker, and B. L. Gewertz, *Am. J. Physiol.* **277,** L1057 (1999).

[42] D. N. Bowser, S. H. Cody, P. N. Dubbin, and D. A. Williams, *in* "Fluorescent and Luminescent Probes for Biological Activity" (W. T. Mason, ed.), 2nd Ed., p. 66. Academic Press, San Diego, 1999.

PRECAUTIONS. In organic solvents, because all of these probes can be absorbed by the skin, precautions should be followed when handling the loading solutions. However, any dangers of exposure to these compounds, if they exist, have not been well described.

Detection

Unlike the HE/ET probe, the reduced forms of the fluorescein-like probes do not fluoresce. Therefore, it is not possible to fully estimate the degree of loading of the reduced probe, except by extraction after the experiment.[26,41] For testing and calibration purposes, it is sometimes desirable to perform chemical hydrolysis to remove the diacetate. This can be done by treatment with 0.01 *N* NaOH at room temperature for 20 min and readjusting the pH back to physiologic values or by exposing the probe to purified esterases.[10]

All of the probes listed in Fig. 1 have similar excitation waveforms and can be excited between 480 and 500 nm. Emission waveforms can be detected at 510–560 nm or at LP 515 (Table I). Peak emission and excitation wavelengths for HFLUOR *in vitro* are shifted approximately 10 nm lower than the other probes,[2] but this may not greatly affect the use of the probe with more common filter sets. Because the excitation and emission wavelengths appreciably overlap for these probes, it is important to ensure that emission filter systems completely block out the excitation light to minimize background noise. One effective combination used for confocal measurements is excitation with the 488-nm line of an argon laser. Emitted light can then be collected with a >515-nm long-pass filter.[38]

These probes have a very strong tendency to photooxidize and photobleach, particularly when excited with laser light and this can, by itself, produce ROS.[3] Therefore, every attempt should be made to limit the time of light exposure and data collection. Procedures can be used to estimate the extent of photobleaching and photooxidation, as described by Murrant *et al.*[10]

The oxidized DHR123 (Rh123), having a positive charge, tends to have good cellular retention, possibly because of accumulation in the mitochondria.[26] Most investigators have utilized DCFH at or below room temperature and have demonstrated good retention of the reduced probe for periods of an hour or more.[10,38] At higher temperatures, 37° or above, our experience has been that 5&6CDCF is poorly retained in diaphragm cells but FLUOR is retained relatively well at these temperatures (unpublished results). There appear to be large differences in the leakage between various cell types, as endothelial cells leak DCF rapidly but cardiac myocytes do not.[38] In myocytes it appears that DCF may also concentrate in mitochondria, thus contributing to retention.[38]

One of the major difficulties in interpreting the results from experiments using any of the fluorescein- or rhodamine-based redox probes is the apparent lack of specificity of the reactions. Much of this has been described in recent years by

Mason and colleagues using electron spin resonance techniques to identify the intermediate reactions.[3,4,28] The pathway for the reaction of dihydrofluorescein (HFLUOR) with H_2O_2 and peroxidase is illustrated in Fig. 2. This is based on the work of Rota *et al.*[4] in studies on DCFH, assuming that the reaction is essentially the same for HFLUOR, as the reaction characteristics with HRP are similar. H_2O_2 reacts with peroxidase enzymes to form an enzyme intermediate and, in the process, oxidizes the DCFH to the semiquinone free radical, $DCF^{\cdot -}$. This is further oxidized by O_2, forming new $O_2^{\cdot -}$ and DCF. Thus, DCFH oxidation is inherently autocatalytic.[4] However, there are other sources of ROS in these reactions. (1) The oxidized form, DCF, can undergo new reactions with a peroxidase and H_2O_2 to form a new phenoxyl radical, which has the same valence but comprises a different molecular species from the semiquinone in Fig. 2.[28] When formed in the presence of reducing compounds such as glutathione (GSH) or NADH, $O_2^{\cdot -}$ is again generated.[28] (2) Photooxidation of DCF in the presence of NADH also produces additional $O_2^{\cdot -}$.[3] Thus, we have numerous sources of self propagation of the ROS when using this category of probe. This tendency to self-amplify the reactions could lead to erroneous conclusions in conditions in which cell injury or decompartmentalization of peroxidase enzymes or probes could generate very nonspecific increases in fluorescence.

One important question that remains to be addressed is exactly what peroxidase or oxidase-like enzymes are important in these reactions in the cell and are they ever rate limiting? For example, the fact that xanthine oxidase (XO) can alone oxidize DCFH is troubling because this enzyme is known to be increased in many tissues during conditions of hypoxia, ischemia, or ischemia/reperfusion, some of the most studied forms of oxidative stress. Catalase, SOD, and lipoxygenase all seem to contribute to spontaneous DCFH oxidation. Zhu *et al.*[29] have suggested that peroxisomal sugar and amino acid oxidases, as well as heme proteins, could provide the enzymatic machinery for these oxidations, and as cells are stressed, these enzymes could change their activity. Even some antioxidants could affect the oxidation of DCFH to DCF. For example, Trolox, a hydrophilic analog of vitamin E, can oxidize DCFH to DCF through the hydrogen abstraction by the phenoxyl radical of Trolox.[43] Therefore, the use of these probes should be interpreted with caution.

Diaminofluorescein (DAF) Probes for Measurement of ˙NO

General Description and Sensitivity

Kojima *et al.* have designed and synthesized a series of novel fluorescence indicators that are derivatives of DAF, including 4,5-diaminofluorescein (DAF-2)[44]

[43] J. F. Kalinich, N. Ramakrishnan, and D. E. McClain, *Free Radic. Res.* **26,** 37 (1997).

[44] H. Kojima, K. Sakurai, K. Kikuchi, S. Kawahara, Y. Kirino, H. Nagoshi, Y. Hirata, and T. Nagano, *Chem. Pharm. Bull* (*Tokyo*) **46,** 373 (1998).

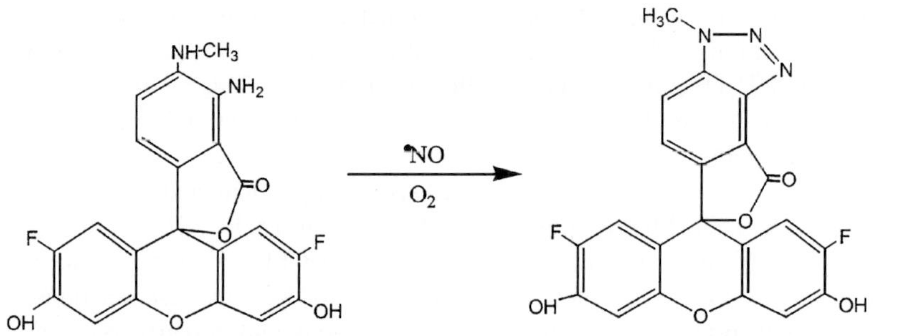

3-Amino-4-(*N*-methylamino)-2',7'-difluorofluorescein
DAF-FM

Triazole form of 3-Amino-4-(*N*-methylamino)-2',7'-difluorofluorescein
DAF-FM T

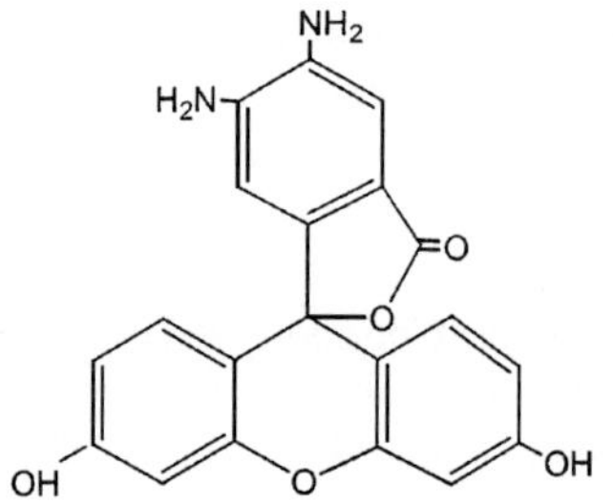

4,5- Diaminofluorescein
DAF-2

FIG. 3. Reaction pathway of DAF-FM with ˙NO and O_2 to form the triazole form of DAF-FM. From Y. Itoh, F. H. Ma, H. Hoshi, M. Oka, K. Noda, Y. Ukai, H. Kojima, T. Nagano, and N. Toda, *Anal. Biochem.* **287,** 203 (2000). (Bottom) A similar and commonly used probe, DAF-2.

and 3-amino,4-aminomethyl-2′,7′-difluorofluorescein (DAF-FM) illustrated in Fig. 3.[45–47] DAFs react with ˙NO in the presence of O_2 to form triazole derivatives, emitting a green fluorescence with a high sensitivity (5 n*M* for DAF-2 and 3 n*M* for DAF-FM). This has made it practical in the study of dynamic alterations in ˙NO of cells or tissues.[46,48,49] Both DAF-2 and DAF-FM come in diacetate

[45] N. Nakatsubo, H. Kojima, K. Kikuchi, H. Nagoshi, Y. Hirata, D. Maeda, Y. Imai, T. Irimura, and T. Nagano, *FEBS Lett.* **427,** 263 (1998).

[46] H. Kojima, N. Nakatsubo, K. Kikuchi, S. Kawahara, Y. Kirino, H. Nagoshi, Y. Hirata, and T. Nagano, *Anal. Chem.* **70,** 2446 (1998).

[47] H. Kojima, N. Nakatsubo, K. Kikuchi, Y. Urano, T. Higuchi, J. Tanaka, Y. Kudo, and T. Nagano, *Neuroreport* **9,** 3345 (1998).

[48] L. A. Brown, B. J. Key, and T. A. Lovick, *J. Neurosci. Methods* **92,** 101 (1999).

[49] M. O. Lopez-Figueroa, C. Caamano, M. I. Morano, L. C. Ronn, H. Akil, and S. J. Watson, *Biochem. Biophys. Res. Commun.* **272,** 129 (2000).

forms (DAF-DA and DAF-FM DA), which can be employed for better intracellular loading, as discussed earlier. The emission of DAFs was shown to be highly dependent on the intracellular ·NO concentration and was not sensitive to $O_2^{\cdot -}$, NO_2^-, NO_3^-, H_2O_2, or $ONOO^-$.[46] DAF-2 has been used in a number of biological studies, providing a fluorescent image with fine resolution.[44–51] Interestingly, even after aldehyde fixation, DAF-2-loaded cells still leave a record of ·NO production before fixation.[52] However, one of the problems with earlier DAF analogs, including DAF-2, is that they are pH sensitive, resulting in a rapid fall in fluorescence below pH 7.0.[13] DAF-FM shows little sensitivity to pH[53] and can detect ·NO linearly between 3 and 200 n*M*.[13] It has been suggested that DAF-FM may have some considerable advantages besides its lack of pH sensitivity, including better photostability of the triazole product and greater sensitivity to ·NO compared to DAF-2.[13,53]

Procedure for Loading

STOCK SOLUTIONS. DAF-FM and DAF-FM-DA are supplied in 7 and 5 m*M* solutions of DMSO, respectively; it is recommended that they be stored at −20° in the dark (Molecular Probes, Inc., Eugene, OR). DAF-2 or DAF-2 DA (Sigma, St. Louis, MO, or Calbiochem, La Jolla, CA) is supplied as a powder or in a 5 m*M* solution of DMSO and stored in the dark at −2 to −8°.

LOADING PROCEDURES. To our knowledge, DAF probes have not been used extensively in whole tissue preparations. However, the procedures for loading cells using DAF-FM-DA or DAF-2-DA generally employ concentrations of 10 μM in an appropriate buffer.[13,49] Cells are incubated for 30 min at 37° and are then rinsed thoroughly with several changes of buffer for 10 min. Waiting an additional 15–30 min for complete deesterification of the diacetates has also been recommended (Molecular Probes, Inc.).

Nonesterified versions of the probes can be used for measuring the extracellular release of ·NO.[13] In this application, 1 μM of DAF-FM or DAF-2, in the nonesterified form, is incubated with the cells and aliquots are removed and stored on ice for later HPLC measurement.

PRECAUTIONS. Handling requires similar precautions to that described previously for fluorescein- and rhodamine-based probes.

Fluorescent Detection of ·NO

DETECTION WITH DAF COMPOUNDS. For intracellular ·NO detection using DAF compounds, the peak excitation and emission wavelengths are similar to the

[50] I. Foissner, D. Wendehenne, C. Langebartels, and J. Durner, *Plant J.* **23,** 817 (2000).
[51] R. Berkels, C. Dachs, R. Roesen, and W. Klaus, *Cell Calcium* **27,** 281 (2000).
[52] K. Sugimoto, S. Fujii, T. Takemasa, and K. Yamashita, *Histochem. Cell Biol.* **113,** 341 (2000).
[53] H. Kojima, Y. Urano, K. Kikuchi, M. Higuchi, and K. Hirata, *Angew's Chem. Int. Ed.* **38,** 3209 (1999).

other fluorescein-based probes discussed earlier; approximately 490 nm excitation and 515 nm emission. Although the Ex/Em wavelengths listed in Table I are adequate for this probe, band-pass filtering of 450–490 excitation and 515–560 emission have also proved favorable.[49] Laser-scanning confocal detection can be used as described previously for fluorescein- and rhodamine-based probes or following procedures used for DAF compounds.[49] For extracellular detection, reversed-phase HPLC with fluorescence detection has been utilized.[13]

These probes are relatively new and have received less independent scrutiny regarding the biochemistry of their reactions than the probes for ROS. The reaction site on the molecule is quite different from that of DCFH or HE, as shown in Fig. 3, and therefore it is unlikely that the compound forms analogous radical intermediates, although it may form other radical species that may have important, as yet unidentified, side reactions. DAF probes also appear to be susceptible to reductants such as ascorbate, dithiothreitol, 2-mercaptoethanol, and glutathione, which can diminish the fluorescence induced by ˙NO.[54] Thus, the interpretation of results using DAF compounds may need to take into consideration changes in intracellular reducing conditions (e.g., during ischemia–reperfusion). The use of exogenous reducing agents should be avoided.

ALTERNATIVE APPROACHES FOR ˙NO DETECTION. DCFH was mentioned previously and has been used in a number of cell types for ˙NO detection, including neuronal cells,[32] rat cardiomyocytes,[55] rat skeletal muscle,[10] and rat macrophages.[56] However, it is very nonspecific and really needs to be coupled with applications of ˙NO scavengers (e.g., reduced hemoglobin) and ROS scavengers (e.g., superoxide dismutase or catalase).[32] Another probe, 4-[(3-amino-2-naphthyl)aminomethyl)] benzoic acid (DAN-1), has been developed by Kojima *et al.*[57] and used for the imaging of ˙NO in rat aortic smooth muscle cells.[57] The probe, 2,3-diaminonaphthalene (DAN), has been used to monitor the sum of nitrite and nitrate, indirect indices of ˙NO produced, resulting in the formation of highly fluorescent 2,3-naphthotriazole (NAT).[58] More recently, a lifetime-based ˙NO probe has been developed by Barker *et al.*[59] Cytochrome *c′*, which binds ˙NO with its heme component, is attached with a fluorescent reporter tail whose lifetime changes immediately once cytochrome *c′* "catches" the ˙NO molecule. They also

[54] N. Nagata, K. Momose, and Y. Ishida, *J. Biochem (Tokyo)* **125,** 658 (1999).

[55] M. N. Sharikabad, K. M. Ostbye, T. Lyberg, and O. Brors, *Am. J. Physiol Heart Circ. Physiol.* **280,** H344 (2001).

[56] A. Imrich and L. Kobzik, *Nitric Oxide* **1,** 359 (1997).

[57] H. Kojima, K. Sakurai, K. Kikuchi, S. Kawahara, Y. Kirino, H. Nagoshi, Y. Hirata, T. Akaike, H. Maeda, and T. Nagano, *Biol. Pharm. Bull.* **20,** 1229 (1997).

[58] T. E. Casey and R. H. Hilderman, *Nitric Oxide* **4,** 67 (2000).

[59] S. L. Barker, H. A. Clark, S. F. Swallen, R. Kopelman, A. W. Tsang, and J. A. Swanson, *Anal. Chem.* **71,** 1767 (1999).

attached cytochrome c' to the optical fiber with fluorescent reference micropheres through colloidal gold for the buildup of ratiometric sensors, which have a linear and fast response with high selectivity to $^{\cdot}$NO and excellent reversibility.[59] Finally, even the fluorescence quenching of pyrene derivatives by $^{\cdot}$NO has been used to measure the diffusion coefficient of $^{\cdot}$NO in the membrane.[60] Overall, fluorescent $^{\cdot}$NO probes are still in the development stage. DAF-FM appears to be superior among them.

Summary

The take-home message of this chapter is that the fluorescent probes for ROS and RNS have great potential in improving our understanding of redox behavior within cells and tissues. However, data obtained from studies using these probes must be expressed in the context of the limitations of the chemistry of the probes in the cellular microenvironment, which may change under different conditions, such as cell stress or injury. In most cases, as suggested,[2] results should be described in a general context of reflecting an increase in oxidizing reactions within the cell and not as a quantitative measure of the production of a specific oxidant species. It is highly recommended that results be verified, when possible, with alternative fluorescent probes or preferably using alternative methods, such as electron spin resonance or other newly emerging technology.

Acknowledgment

This work was supported by NIH NHLBI RO-1 53333.

[60] A. Denicola, J. M. Souza, R. Radi, and E. Lissi, *Arch. Biochem. Biophys.* **328,** 208 (1996).

[27] Simultaneous Detection of Tocopherols and Tocotrienols in Biological Samples Using HPLC-Coulometric Electrode Array

By SASHWATI ROY, MIKA VENOJARVI, SAVITA KHANNA, and CHANDAN K. SEN

Introduction

Vitamin E is a generic term for all tocopherols and tocotrienols derived from a chromanol structure having the biological activity of RRR-α-tocopherol.[1,2] In nature, eight substances have been found to have vitamin E activity: α-, β-, γ-, and δ-tocopherol and α-, β-, γ-, and δ-tocotrienol. These compounds are closely related homologues and isomers depending, respectively, on the number and position of methyl groups on the aromatic ring. Tocotrienols, formerly known as ζ-, ε-, or η-tocopherols, are similar to tocopherols except that they have an isoprenoid tail with three unsaturation points instead of a saturated phytyl tail. While tocopherols are found predominantly in corn, soybean, and olive oils, tocotrienols are particularly rich in palm, rice bran, and barley oils.[1,2] The structural complexity and the wide variation in biological activity of these compounds require reliable and sensitive analytical techniques for the isolation, separation, differentiation, and quantification of individual components in mixtures derived from various sample matrices.[3] Because of their low concentrations in biological samples, sensitivity is particularly a critical issue for the detection methods of tocotrienols.

Because of their low oxidative potential, the various forms of vitamin E can be analyzed by reversed-phase HPLC-electrochemical detection. Of importance, electrochemical properties of the various forms of vitamin E are not identical, requiring different detector potentials for the optimal detection of each form. Simultaneous detection of tocopherol and tocotrienol using single or dual channel channel electrochemical (EC) detection alone or in combination with UV detection has been reported.[4,5] Single or dual channel electrochemical detectors are typically used at settings that are suitable only for few analytes. In this way, sensitivity of detection of the other forms of vitamin E is compromised. Furthermore, gradient elution chromatography required for such multicomponent analyses has poor compatibility with the amperometric methods. The method reported herein

[1] M. G. Traber and L. Packer, *Am. J. Clin. Nutr.* **62,** 1501S (1995).

[2] M. G. Traber and H. Sies, *Annu. Rev. Nutr.* **16,** 321 (1996).

[3] S. L. Abidi, *J. Chromatogr. A* **881,** 197 (2000).

[4] M. Podda, C. Weber, M. G. Traber, R. Milbradt, and L. Packer, *Methods Enzymol.* **299,** 330 (1999).

[5] M. Podda, C. Weber, M. G. Traber, and L. Packer, *J. Lipid Res.* **37,** 893 (1996).

0076-6879/02 $35.00

was developed to simultaneously analyze various isoforms of tocopherol and tocotrienol from biological samples using gradient elution chromatography and a coulometric electrode array detector. This detector is based on the use of multiple coulometric electrochemical sensors in series, maintained at different potentials. This allows for a combination of detector potentials, each optimal for a specific analyte.

Principles of HPLC-EC Coulometric Detection

Electrochemistry involves heterogeneous chemical reactions between a compound and an electrode in which an electron is transferred from the solution to the electrode, or vice versa, and a measurable current is formed as a result. For such oxidation–reduction reactions to occur, energy in the form of an electric potential is required. In a traditional electrochemical detector, the potential is held constant (DC mode) and current is measured as a function of time. When an electroactive species flows through the electrode, current is formed. The magnitude of this current is proportional to the concentration of the compound in solution on the electrode. Most electrochemical detectors for HPLC operate in the amperometric mode. In such a mode, the solution of the compound only passes over a flat bed of working electrode. Under such conditions only a small flat surface area of the electrode is available for interaction with the analyte. As a result, only 5–15% of the electroactive species is oxidized or reduced by the electrode. In contrast, coulometric detectors use flow-through or porous graphite electrodes. The surface area of such electrodes is large, allowing almost 100% of the analyte to react with it. Thus, the efficiency in coulometric detection is approximately 100% compared to conventional amperometric detection, which has only 5–15% detection efficiency.

Coulometric Electrode Array Detector

HPLC-based spectrophotometric detection techniques were revolutionized with the development of the photodiode array (PDA) detector, which is capable of monitoring hundreds of wavelengths simultaneously. Matson *et al.*[6] conceptualized a PDA equivalent of an electrochemical detector and, in 1984, developed an HPLC-based detector containing a serial array of up to 16 coulometric electrodes. Like PDA, this detector provided the ability to evaluate peak purity, assign identity with higher confidence, and resolve coeluting solutes. Coulometric electrode series array sensors provide a route to multiply the resolving power of conventional LC by factors of 10 to 50. Femtogram level separations can now be performed for multiple components in both isocratic and gradient modes.

[6] W. R. Matson, P. Langlais, L. Volicer, P. H. Gamache, E. Bird, and K. A. Mark, *Clin. Chem.* **30,** 1477 (1984).

The coulometric electrode array offers several advantages over conventional single-channel detectors beyond its high resolution. Using a progressively increasing oxidative array of 16 electrodes, compounds can be made to react at three consecutive sensors. The upstream electrode oxidizes a small portion of the analyte, the second dominant electrode oxidizes the bulk of the analyte, and the downstream electrode oxidizes the remainder. A particular standard eluting at a given retention time will always provide a predictable response across these three electrodes. The ratio of the response across these three electrodes remains constant and is referred to as ratio accuracy. The comparison of ratio accuracy of standard versus sample is powerful and an immediate indicator of peak purity. Compared to the standard, a lower ratio accuracy in the sample will indicate presence of a coelution.

HPLC Apparatus and Analytical Cells

In this method for the detection of various forms of tocopherols and tocotrienols, the HPLC system (ESA Inc., Chelmsford, MA) consists of the following components: (1) coularray detector (Model 5600 with 12 channels), PEEK tubing for an inert connection to the HPLC system; (2) UV detector (Model 520 UV/Vis HPLC detector) set up in-line with the coularray detector, with the eluent passing first through the UV/Vis detector; (3) pumps (two ESA Model 582 dual-piston pumps with gradient option), HPLC dynamic gradient mixer, and PEEK pulse damper; and (4) an autosampler (Model 542) with sample cooling and an integrated column oven. CoulArray for Windows-32 software is used for data acquisition and processing.

Column and Mobile Phases

MDA-150 (C_{18} column, 150 mm long × 4.6 mm i.d., 5-μm pore size; ESA Inc.) is used for the separation of tocopherols and tocotrienols.

Mobile Phase

A gradient is used consisting of a mixture of A (methanol : 0.2 *M* CH_3COONH_3, pH 4.4; 90 : 10, v/v) and B (methanol : 1-propanol : 1.0 *M* CH_3COONH_3, pH 4.4; 78 : 2 : 20, v/v/v). Solvents used to prepare the mobile phase are of HPLC grade. Mobile phases are filtered through 0.22-μm pore size nylon filters. The flow rate is maintained at 0.5 ml/min throughout the analysis. The following gradient program is used: the initial condition is 100% A and 0% B. The mobile phase is changed linearly over 10 min to 20% A and 80%, after which the mobile phase is changed linearly over the next 10 min to 100% B. The system is reverted back linearly over the next 5 min to the initial conditions, i.e., 100% A, where it is continued for 5 min to equilibrate.

Standards and Standard Curve

Authentic and high purity (>98%) tocopherol and tocotrienol compounds are from Sigma-Aldrich (St. Louis, MO), BASF (Germany), and Carotech Inc. (Malaysia). The stock solution of standards is prepared in HPLC-grade ethanol, and their concentrations are determined spectrophotometrically.[5] These compounds are highly unstable and oxidize readily at room temperature. To avoid oxidation, stock standard solutions are stored at $-80°$ or in liquid nitrogen. The concentrations of tocopherol and tocotrienol versus the corresponding peak area response have been plotted in Fig. 1. Injection of large amounts of the compounds may overwhelm the redox capacity of the electrode and may cause a deviation from the linear relationship between peak area and sample quantity. Repeated regression analysis of standards over a several-day period results in high correlation ($R = 0.99$), indicating low variability of analysis in different days. The concentration of vitamin E components used to prepare the standard curve is kept in a range that matches the concentration of these compounds in biological cell samples.

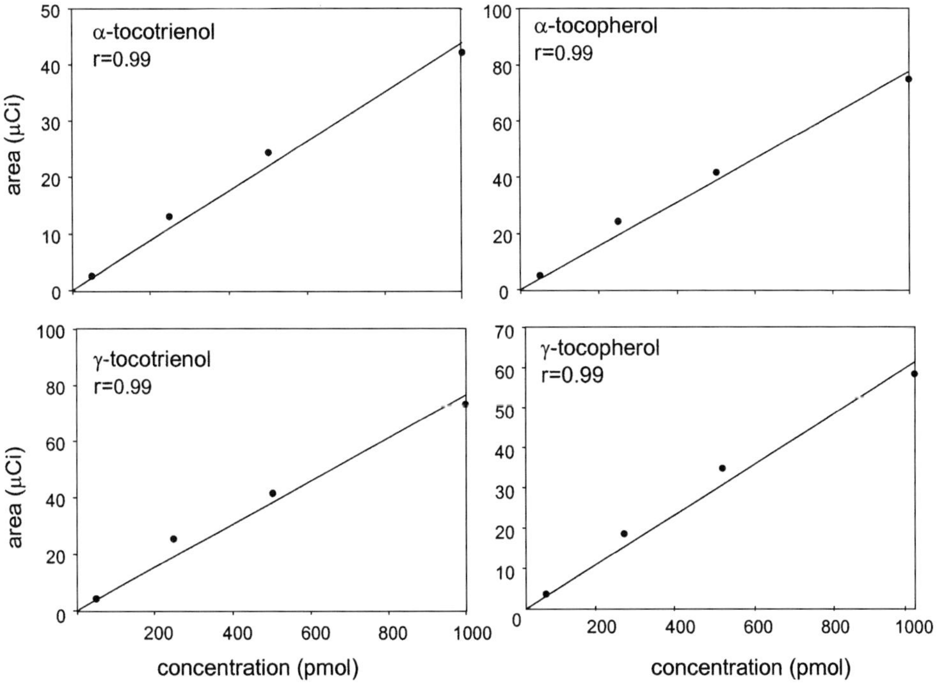

FIG. 1. Representative standard curves for α- and γ-tocopherols and tocotrienols. The conditions of chromatography are described in the text.

Current–Voltage Response Curve

The optimal use of an electrochemical detector for liquid chromatography requires knowledge of the appropriate potentials to drive the desired electrochemical reaction. This potential is dependent on a large number of factors, including the nature of the electrode surface, pH, composition of the mobile phase, and chemistry of the compound of interest. A plot of current generated (peak height) versus applied potential difference is commonly referred to as a hydrodynamic (HDV) voltammogram or a current–voltage (C-V) curve.

The optimum potential for the oxidation of α- and γ-tocopherol/tocotrienol is determined by injecting the compounds onto the column and by adjusting the potential difference across electrode 1–10 from 0 to 1000 mV. The optimum potential for the quantitative measurement of α-tocopherol/tocotrienol and γ-tocopherol/tocotrienol is 200 and 240 mV, respectively (Fig. 2). Using these settings of potential differences, the maximum peak area with minimal background in response to a given injection of vitamin E components is obtained. The C-V curve clearly indicates that the α forms (200 mV) of tocopherols/tocotrienols oxidize more readily than the corresponding γ forms (240 mV).

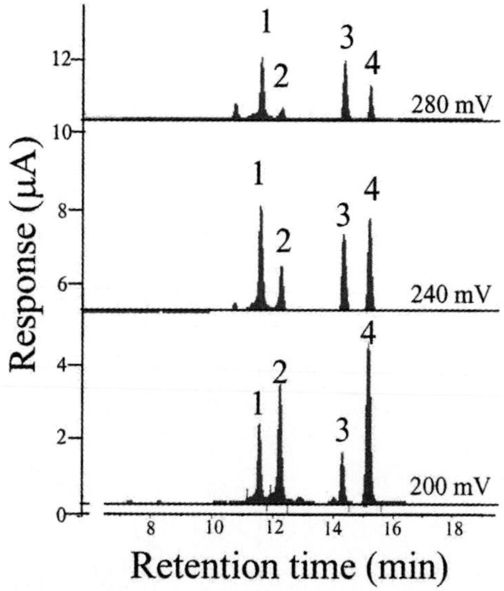

FIG. 2. Current–voltage response for (1) γ-tocotrienol, (2) α-tocotrienol, (3) γ-tocopherol, and (4) α-tocopherol. The responses of α-tocopherol and α-tocotrienols were maximal at 200 mV, whereas maximal responses of γ-tocopherol and γ-tocotrienols were observed at 240 mV.

Detection of Tocopherol and Tocotrienols in Biological Samples

To demonstrate that the method is applicable for biological samples, vitamin E components were analyzed from tissues of rats supplemented intragastrically with a tocotrienol-rich fraction (TRF) isolated from palm oil. TRF is provided in the form of Tocomin (78% tocotrienols and 22% tocopherols) by Carotech Inc. (Malaysia).

Supplementation and Tissue Collection

Eight-day pregnant rats are fed (intragastrically) daily with 1 g/kg body weight of TRF suspended in vitamin E-stripped corn oil for 9 days. The control group is fed intragastric daily vitamin E-stripped corn oil. Both groups of rats receive a standard laboratory diet (Harlan-Teklad, Indianapolis, IN). A third group of nonsupplemented rats is maintained on a vitamin E-deficient diet (Harlan-Teklad). On day 17 of pregnancy, rats are killed and tissues are collected from the mother. Tissues are rinsed in ice-cold phosphate buffered saline, pH 7.4 (PBS), and snap froze in liquid nitrogen. The samples are kept stored in liquid nitrogen until extraction and HPLC analysis. HPLC assays are done within 1 week of storage.

Extraction of Vitamin E from Biological Samples

Weigh approximately 150–200 mg tissue quickly, grind under liquid nitrogen using a mortal and pestle, and homogenize with a Teflon pestle on ice in a Potter–Elvenhjem tube containing PBS (1 ml/100 mg tissue) and butylated hydroxytoluene (BHT, 10 mg/ml stock, 50 μl BHT stock/100 mg tissue). Transfer the homogenate to a screw-cap glass tube, add 0.1 *M* sodium dodecyl sulfate (SDS, 1 ml/100 mg tissue), and vortex vigorously for 30 sec. Take 100 μl of homogenate for protein analysis (described later). To the rest of the homogenate, add ethanol (2 ml/100 mg tissue), mix briefly and sonicate on ice water, and extract homogenate with hexane (2 ml/mg tissue). Dry an appropriate aliquot under nitrogen and resuspend in mobile phase B. Filter the samples using microfilterfuge tubes (Rainin, Woburn, MA) fitted with a 0.22-μm nylon filter before injecting to HPLC.

A representative chromatogram form spleen of rats demonstrates that these vitamin E components are well separated using the present HPLC method (Fig. 3). A small amount of α- and γ-tocotrienols was observed in the spleen of a rat that was supplemented with vitamin E-stripped corn oil and maintained on a standard laboratory diet, suggesting the presence of these vitamin E forms in a standard rat diet. This is consistent with an earlier report in which the presence of these compounds was demonstrated in a standard laboratory mouse diet.[5] TRF feeding increased α- and γ-tocotrienol content in the rat tissue several folds (Fig. 3). A significant decrease in α-tocopherol, as well as other vitamin E forms, was observed in the spleen of nonsupplemented rats that were maintained on a vitamin E-deficient diet.

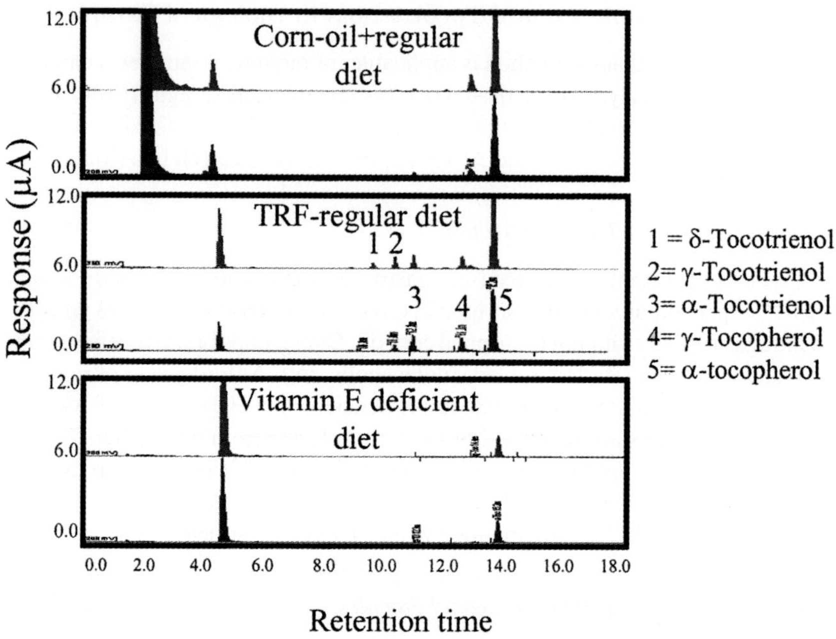

FIG. 3. Chromatograms of tocotrienols and tocopherols detected in spleen of rats supplemented (intragastrically) with (top) vitamin E-stripped corn oil and maintained on a standard rat diet; (middle) tocotrienol-rich fraction (TRF) derived from palm oil and maintained on a standard rat diet; and (bottom) nonsupplemented and maintained on a vitamin E-deficient diet. Individual peaks are labeled as 1–5.

In summary, the HPLC-EC coulometric electrode array detection method described here sensitively detects isoforms of tocopherols and tocotrienol from biological samples (Fig. 3). The detection limit for these compounds for this HPLC method is about 50 fmol. Using the extraction protocol, the recovery of tocopherol and tocotrienol from tissues was greater than 95%. Moreover, reproducibility of this HPLC method is excellent, with coefficient of variations ranging from 3 to 5%. Using this method, we showed that the optimum potentials for the detection of various isoforms of tocopherol/tocotrienols are different. Therefore, maintaining an electrochemical sensor in a series at various potentials, each optimal for a given vitamin E isoform, will provide enhanced sensitivity in the detection of these compounds in biological samples where the concentration of some of these vitamin E forms is very low.

[28] *In Vivo* Measurement of Oxidative Stress Status in Human Skin

By JÜRGEN FUCHS, NORBERT GROTH, and THOMAS HERRLING

Measuring Oxidative Stress Status in Humans

One of the greatest need in the field of free radical biology is the development of reliable methods for measuring oxidative stress status in humans.[1] Particularly noninvasive methods, which have the potential for clinical applications, need to be developed. Nitroxide-based electron paramagnetic resonance (EPR) spectroscopy and imaging are used extensively to measure the oxidative stress status in small laboratory animals.[2–5] In humans, the skin is a target organ for *in vivo* EPR applications mainly for two reasons. First, free radical generation and metabolism, which are key contributors to altered redox state, have been proposed to be of critical importance in the pathophysiology of cutaneous inflammation, photocarcinogenesis, photoaging, and a variety of skin diseases. Second, for technical reasons, body surface EPR spectroscopy and imaging can be performed at higher frequencies and thus with higher sensitivity and resolution. In addition, the superoxide anion radical is an important component of oxidative stress, and EPR is the most reliable technique for the measurement of superoxide production. Nitroxide reduction in tissues is a complex phenomenon involving enzymatic and nonenzymatic mechanisms, and nitroxides can be used for the study of redox metabolism.[6–8] Skin possesses "nitroxide radical reductase"-like activity.[9,10] The main reductants for nitroxides with piperidine structure in mouse and human skin were identified as ascorbic acid and thiol-dependent mechanisms.[9–11] Hydroxylamines, as well as secondary amines, have been detected as nitroxide metabolites in human keratinocytes,[12] indicating that reductants other than ascorbic acid are involved

[1] W. A. Pryor and S. S. Godber, *Free Radic. Biol. Med.* **10,** 177 (1991).

[2] V. Vallyathan, S. Leonhard, P. Kuppusamy, D. Pack, M. Chzhan, S. P. Sanders, and J. L. Zweier, *Mol. Cell. Biochem.* **168,** 125 (1997).

[3] S. Matsumoto, M. Mori, N. Tsuchihashi, T. Ogata, Y. Lin, H. Yohoyama, and I. Shin-Ichi, *Magn. Res. Med.* **40,** 330 (1998).

[4] N. Phumala, T. Ide, and H. Utsumi, *Free Radic. Biol. Med.* **26,** 1209 (1999).

[5] K. Takeshita, A. Hamada, and H. Utsumi, *Free Radic. Biol. Med.* **26,** 951 (1999).

[6] H. M. Swartz, *J. Chem. Soc. Farady. Trans. I* **83,** 191 (1987).

[7] A. Sotgiu, S. Colacicchi, G. Placidi, and M. Alecci, *Cell. Mol. Biol.* **43,** 813 (1997).

[8] L. H. Sutcliffe, *Phys. Med. Biol.* **43,** 1987 (1998).

[9] J. Fuchs, R. J. Mehlhorn, and L. Packer, *Methods Enzymol.* **186,** 670 (1990).

[10] J. Fuchs, R. J. Mehlhorn, and L. Packer, *J. Invest. Dermatol.* **93,** 633 (1989).

[11] J. Fuchs, N. Groth, T. Herrling, and G. Zimmer, *Free Radic. Biol. Med.* **22,** 967 (1997).

[12] C. Kroll, A. Langner, and H. H. Borchert, *Free Radic. Biol. Med.* **26,** 850 (1999).

0076-6879/02 $35.00

in the reaction. Therefore, the exact composition of the nitroxide reductants in human skin remains to be characterized. Nevertheless, measurement of nitroxide reduction in skin can be considered as an objective but not universal indicator characterizing oxidative stress status. A variety of oxidizing conditions have been shown to decrease antioxidant activity of the skin and thus the nitroxide-reducing activity. These conditions comprise, e.g., ultraviolet radiation, exposure to organic hydroperoxides, transition metal ions, and thiol-depleting agents. This chapter describes an EPR-based assay to evaluate the oxidative stress status in human skin before and after exposure to solar-simulated light.

EPR Technique for Human *in Vivo* Measurements

For EPR *in vivo* measurements in human skin, two different experimental approaches have been used. The first strategy employs an X-band microwave system. The X-band bridge system is based on the concept of Furusawa and Ikeya,[13] who used microwave cavities with a small hole from which the microwave field leaks out to a small cross-sectional area of the object. The magnet system is a conventional electromagnet with a 100-mm gap, which is wide enough to accommodate human limbs (e.g., the forearm) between the pole faces. The microwave penetration depth in the X-band region is about 0.5–1.0 mm in human skin and about 5 mm for S-band frequency. Thus the X band is restricted to the upper layer of the skin (the human epidermis), whereas the S band is suited for deeper layers (the human dermis and subcutis). At the S band, the probe head is a 90° bent surface coil with an electronically matched system. A quartz plate is mounted on one side of the surface coil, which defines a plane-parallel measuring area on the skin. Matching is accomplished by placing a piezoelectric element at a distance of 1/4 of the wavelength (microwave) outside the loop. Two 100-kHz modulation coils near the surface coil generate a modulation field B_m in the skin layer. The rapid scan coils are mounted on the surface of the magnet pole plates.[14] We have used this experimental setup at S-band frequency for the measurement of nitroxide biokinetics in skin of human limbs.

Experimental Conditions

The EPR spectrometer ERS 221 was built in the former Academy of Science, Berlin. A block diagram of the set up is shown in Fig. 1. The surface coil and modulation coils in the pole gap of the electromagnet are shown in detail in Fig. 2.

The experimental parameters are as follows. Magnetic field strength Bo (T): 0.11, field gradient ΔB(T/m): 0, microwave frequency υ (GHz): 3.08 (S band), microwave power P (mW): 60, modulation amplitude Bm (mT): 0.1 (100 kHz),

[13] M. Furusawa and M. Ikeya, *Jpn. J. Appl. Phys.* **29,** 270 (1990).

[14] T. Herrling, N. Groth, and J. Fuchs, *Appl. Magn. Res.* **11,** 47 (1996).

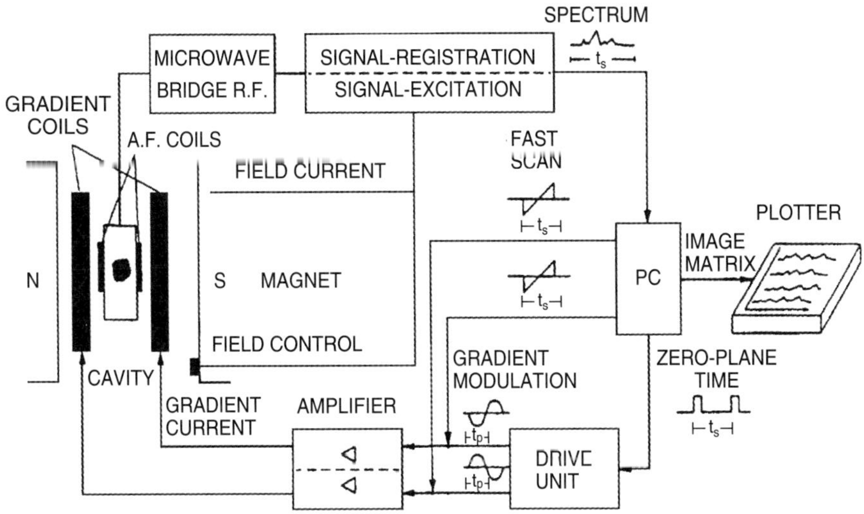

FIG. 1. Block diagram of the experimental setup.

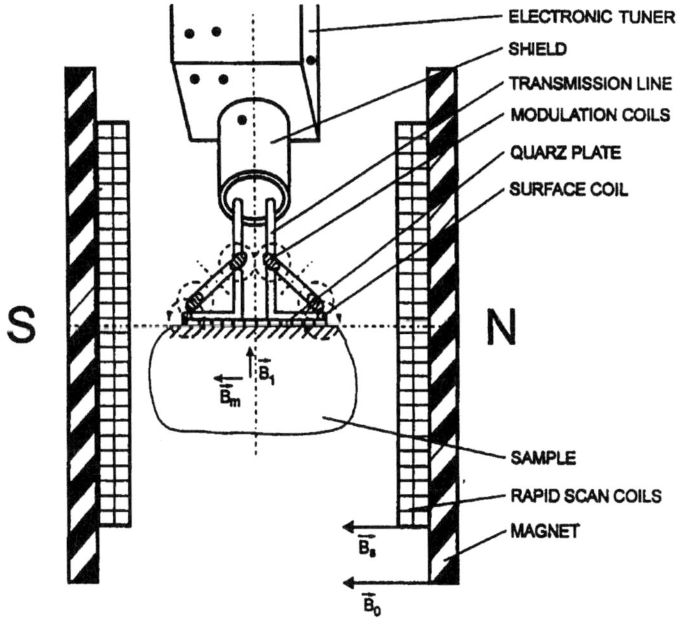

FIG. 2. Surface coil and modulation coils in the pole gap of the electromagnet.

field sweep ΔB (mT): 2.5, sweep time t_B (sec): 0.2, accumulations n_A: 25, time interval t_i (sec): 50, projections n_p: 16, image matrix: 16×128, total acquisition time t_a (min): 15.0.

S-Band *in Vivo* EPR Spectroscopy in Human Skin

Due to their redox properties, uncharged piperidine-type nitroxides are used preferentially to measure tissue antioxidant activity. They possess the right reduction potential, polarity, and permeability through the wall of membranes and the horny layer of the skin. Nitroxides such as 2,2,6,6-tetramethyl piperidine-*N*-oxyl (TEMPO) can be used at low concentrations (<10 m*M*) for EPR-based skin measurements in humans without the risk of serious cutaneous side effects.[15] For the most recent applications on human skin, we have utilized TEMPO for measurement of the oxidative stress status. Spectroscopy measurements are performed on the forearm of human subjects. The skin is cleaned with isotonic sodium chloride, and 10 μl of 1 m*M* nitroxide stock solution is then applied to the epidermal surface and incubated for 4 min at room temperature. The skin is washed with isotonic sodium chloride prior to EPR measurements. A typical spectrum is shown in Fig. 3.

Midfield (h_0) and low-field (h_{-1}) peaks are essentially unchanged in line shape from which a quantitative distribution of spin label may be estimated. Accurate concentration measurements are not possible for the high-field peak h_{+1} as it is too sensitive to mobility and polarity, which can even result in additional line splittings. A typical spin label reduction kinetic plot of the midfield peak h_0 is presented in Fig. 4 for TEMPO in the forearm skin of a human subject. The reduction curve was fit to a simple exponential, which is typical for these kinetic processes. Clearance of the applied nitroxide radicals can occur through metabolism, predominantly reduction, or vascular washout. Previously we have analyzed the time course of nitroxide distribution as a function of skin depth in mouse skin *in vitro* and found significant penetration of the nitroxide into the vascularized dermis after 20 min.[16] In human skin biopsies, nitroxide signals were obtained in the dermis after about 60 min, which is presumably due to the thicker epidermis. In these experiments, high concentrations (50–100 m*M*) of perdeuterated N^{15}-2,2,5,5-tetramethyl-3-pyrrolin-d_{13}, 1-^{15}N-1-oxyl-3-carboxamide were used, which is less subject to bioreduction than piperidine-type nitroxides and provides for enhanced sensitivity and narrower line width than nondeuterated nitroxides. At low concentrations (1–10 m*M*), piperidine-type nitroxide radicals were reduced rapidly at the epidermal surface or in the epidermis (within 10 min) and no signal

[15] J. Fuchs, N. Groth, T. Herrling, and G. Zimmer, *Free Radic. Biol. Med.* **24,** 643 (1998).

[16] J. Fuchs, R. Milbradt, N. Groth, T. Herrling, G. Zimmer, and L. Packer, *J. Invest. Dermatol.* **98,** 713 (1992).

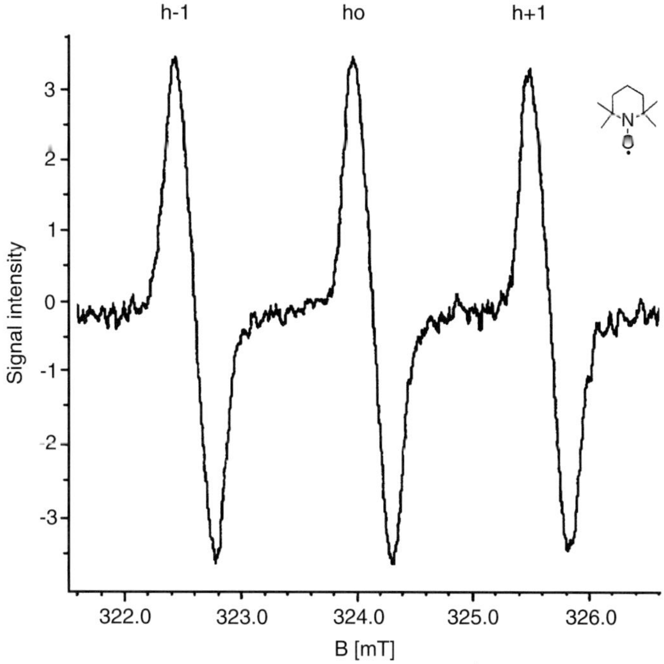

FIG. 3. Typical spectrum of the nitroxide TEMPO on the surface of human skin.

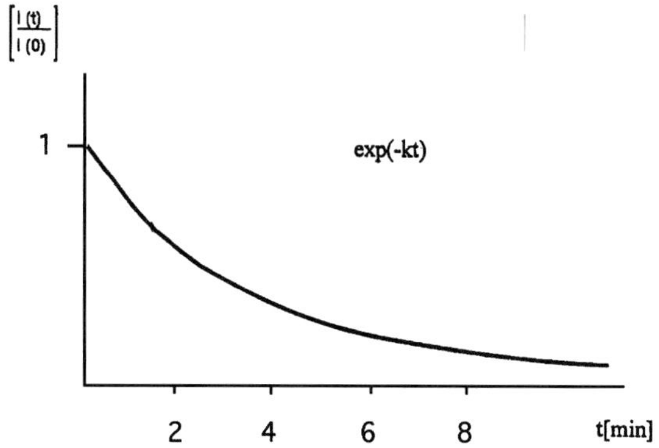

FIG. 4. Kinetic plot of a reduction curve expressed by and exponential function and characterized by the reduction factor k (min^{-1}).

was obtained in the dermis. Thus, washout does not contribute to the radical clearance under these conditions.

Both the rate constant and the half-life time $t_{1/2}$ give information about the antioxidative potential (AOP) of the skin against free radicals. The radical clearance process corresponds to an exponential function of the form: $I(t) = I(0)\ \exp(-kt)$. Observed intensity data, I, are fit by nonlinear exponential least squares. The AOP is based on the rate constant k, where the dimensions of k are min^{-1}. The nitroxide is reduced by the skin antioxidant system to the corresponding hydroxylamine and other products. The clearance rate constant k, which depends on the concentration of antioxidants and their clearance properties, describes the radical-scavenging activity of the skin. For practical applications, the antioxidative factor (AOF) is defined by $\text{AOF} = 100k$ per min. We have measured the AOP before and after solar-simulated irradiation. Figures 5 and 6 represent two-dimensional spectral-time images S (B, t) of the nitroxide TEMPO in human skin before (Fig. 5) and after (Fig. 6) exposure to two minimal erythema doses (MED) of solar-simulated light. An oriel xenon arc solar simulator (Oriel Corp., Stratford, CT) with a spectral output in the ultraviolet region region very similar to natural sunlight was used as the source of ultraviolet radiation.

Ultraviolet radiation is well known to cause oxidative stress, which decreases the antioxidant capacity and the radical-scavenging activity of skin. In order to calculate the rate constant k of the nitroxide reduction process in skin, the slope of the signal intensities is plotted versus the time as shown for TEMPO in Fig. 4. The slopes for the two reduction kinetics were fit to a simple exponential function for the determination of the rate constant k. Before ultraviolet exposure, the reduction curve of TEMPO in human skin reveals a half-life time $t_{1/2} = 2.6$ min

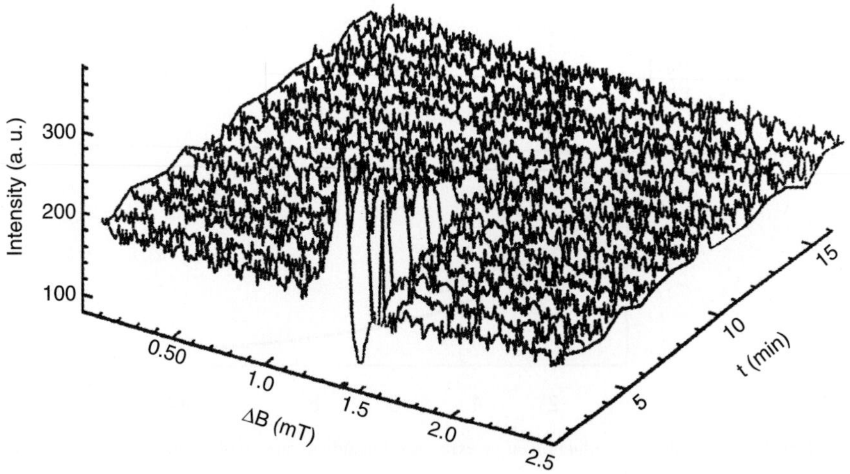

FIG. 5. TEMPO reduction in human skin before solar-simulated radiation.

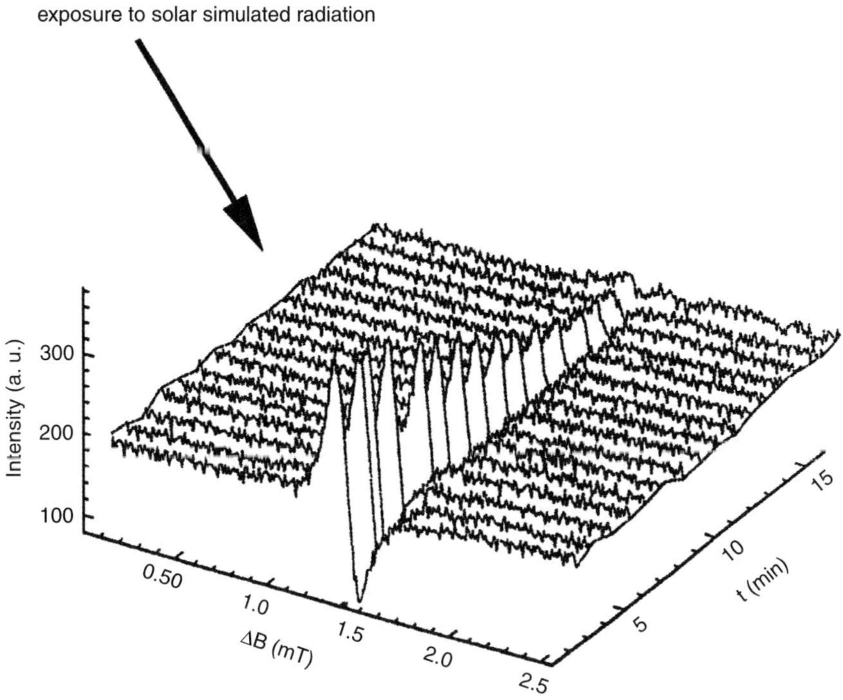

FIG. 6. TEMPO reduction in human skin after solar-simulated radiation.

and a reduction factor $k = 0.27\ \text{min}^{-1}$. The determined AOF is 27. After exposure to solar-simulated light, the TEMPO half-life time $t_{1/2} = 3.95$ min and the reduction factor $k = 0.18\ \text{min}^{-1}$. The determined AOF is 18. This clearly indicates ultraviolet-induced depletion of antioxidant equivalents and thus allows assessment of the oxidative stress status in human skin *in vivo*. This assay was used for the measurement of the radical protection factor of cosmetic and pharmaceutical products in human skin *in vivo*.[17,18] The group of Zweier and Kuppusamy has developed a topical S-band EPR instrumentation suitable for noninvasive localized EPR spectroscopy and imaging of human skin. In their excellent paper they reported the successful imaging of the distribution and metabolism of nitroxides in human skin *in vivo*.[19] It is believed that this technique and application hold great promise in the study of cutaneous redox state, skin diseases, and development of topical pharmaceutical and cosmetic products.

[17] T. Herrling, L. Zastrow, N. Groth, J. Fuchs, and K. Stanzl, *Seifen Öle Fette Wachse J.* **122,** 472 (1996).

[18] T. Herrling, L. Zastrow, and N. Groth, *Seifen Öle Fette Wachse J.* **124,** 282 (1998).

[19] G. He, A. Samouilov, P. Kuppusamy, and J. L. Zweier, *J. Magn. Res.* **148,** 155 (2001).

[29] Localization of Oxidation-Specific Epitopes in Tissue

By GREGORY D. SLOOP

Introduction

Oxidation of lysine results in molecules such as malondialdehyde-lysine (MDA-lysine) and 4-hydroxynonenal-lysine (4-HNE-lysine), which creates neoepitopes that can be recognized by antibodies and localized by immunohistochemistry. This chapter provides a rationale for determining the localization of oxidation-specific epitopes and offers an overview of the methods by which these epitopes can be localized with particular attention to immunohistochemistry. Additionally, it describes technical issues associated with immunohistochemistry for oxidation-specific epitopes and provides a protocol for performing immunohistochemistry. Finally, there is a brief overview of previous work regarding the immunolocalization of oxidation-specific epitopes in atherosclerosis, and the author's opinion of which issues remain to be settled.

Why Determine the Sites Where Oxidation-Specific Epitopes Accumulate?

The spectrum of illness in which oxidative stress is proposed to play a pathogenic role includes chronic diseases associated with aging, such as atherosclerosis, diabetes mellitus, Alzheimer disease, and aging itself, as well as acute diseases such as ischemia/reperfusion injury and bacterial meningitis. Of these, perhaps the most research regarding the role of oxidative stress has been in atherosclerosis, the leading cause of mortality in the industrialized world. Although demonstration of oxidation-specific epitopes can help define the role of oxidative stress in any of these diseases, atherosclerosis will be used to illustrate the insight that can be provided by localizing oxidation-specific epitopes.

MDA-lysine and 4-HNE-lysine have been identified in low-density lipoprotein (LDL) extracted from human atherosclerotic plaques.[1] Many putative proatherogenic properties of oxidatively modified LDL (oxLDL), such as induction of monocyte chemotaxis, inhibition of macrophage migration, and foam cell formation, are local, not systemic. Thus, the location of oxLDL in the vascular tree should correspond to the presence of vascular lesions. Atherosclerosis is a focal disease in which intimal lesions are next to uninvolved normal intima. Therefore, the *in vivo*

[1] S. Ylä-Herttuala, W. Palinski, M. E. Rosenfeld, S. Parthasarathy, T. E. Carew, S. Butler, J. L. Witztum, and D. Steinberg, *J. Clin. Invest.* **84,** 1086 (1989).

0076-6879/02 $35.00

relevance of *in vitro* data ascribing pathologic activity to oxLDL can be tested by determining the *in vivo* location of oxLDL. This has been the rationale for several studies examining the localization of oxLDL.

If oxidation-specific epitopes are widely distributed, particularly in veins that do not develop atherosclerosis, then the existence of discrete areas of "oxidative stress" and the role of oxLDL in the pathogenesis of atherosclerosis should be questioned. Data showing that oxLDL induces apoptosis *in vitro*[2] are not relevant *in vivo*.[3,4] Considering the enthusiastic and burgeoning literature about the role of "oxidative stress" and oxLDL in atherogenesis, proving that oxLDL has pathologic activity *in vivo* is important.[5]

Methods of Determining the *in Vivo* Distribution of Oxidization-Specific Epitopes

The distribution of oxidization-specific epitopes can be determined in two ways: biochemically or histologically using immunofluorescence or immunohistochemistry. Biochemical localization requires extraction of proteins from tissue and identification by gas chromatography/mass spectroscopy. This method has the advantage of allowing chemical identification and quantitation of the molecule in question.[6] Biochemical localization has been used to demonstrate the accumulation of oxidatively modified carbohydrate adducts on glycated lysine residues in skin collagen[7] with aging. The drawback of this method is that it does not allow localization of oxidization-specific epitopes to specific locations within the tissue under investigation. Therefore, this chapter focuses on histologic methods of determining the localization of oxidization-specific epitopes.

The histologic identification of oxidization-specific epitopes uses antibodies that bind to neoepitopes created by the oxidative modification of a protein. In immunofluorescence microscopy, tissue is incubated with a fluorescently labeled antibody, and bound antibody is visualized with a fluorescent microscope. Drawbacks include the higher cost and limited availability of fluorescent microscopy compared with conventional microscopy and the relatively rapid decay of fluorescent signals. In immunohistochemistry, bound antibody is detected by a colorimetric

[2] Y. Okura, M. Brink, H. Itabe, K. J. Scheidegger, A. Kalangos, and P. Delafontaine, *Circulation* **102,** 2680 (2000).

[3] M. R. Bennett and J. J. Boyle, *Atherosclerosis* **138,** 3 (1998).

[4] G. D. Sloop, J. C. Roa, A. G. Delgado, J. T. Balart, M. O. Hines III, and J. M. Hill, *Arch. Pathol. Lab. Med.* **123,** 529 (1999).

[5] G. M. Chisolm and D. Steinberg, *Free Radic. Biol. Med.* **28,** 1815 (2000).

[6] J. R. Requen, M. X. Fu, M. U. Ahmed, A. J. Jenkins, T. J. Lyons, J. W. Baynes, and S. R. Thorpe, *Biochem. J.* **322,** 317 (1997).

[7] J. A. Dunn, D. R. McCance, S. R. Thorpe, T. J. Lyons, and J. W. Baynes, *Biochemistry* **30,** 1205 (1991).

reaction and is visualized by conventional microscopy. This technique is routine in diagnostic pathology, and all pathologists receive training in interpreting immunohistochemical findings. The equipment for performing immunohistochemistry is available in most pathology laboratories in the United States. Immunohistochemistry also has the advantage over immunofluorescence microscopy, at least in the opinion of this author, of allowing better visualization of morphology.

A mouse monoclonal antibody to oxidatively modified LDL (OXL41.1) is available commercially (Lab Vision Corporation, Fremont, CA, or Research Diagnostics, Inc., Flanders, NJ). Data about the specific epitope that this antibody binds are not available, although the antibody does bind MDA-modified LDL and acetyl-LDL.[8] This antibody can be used for immunofluorescence or immunohistochemistry, depending on the secondary antibody used (vide infra). Antibodies not distributed commercially include MDA2, which recognizes MDA-lysine, and NA59, which recognizes 4-HNE-lysine.[9] These antibodies bind all oxidatively modified proteins, but are selective for oxLDL because of the close physical association of fatty acids to lysine and the high lysine content in LDL. The lack of specificity for a single protein is not a drawback in using these antibodies to identify areas of putative "oxidative stress" because they are specific for oxidative modification.

Performing Immunohistochemistry

Obtaining Tissue

Because of the important biological differences between the arteries of experimental animals and humans (vide infra), animal tissue is not suitable for determining the localization of oxidation-specific epitopes in atherosclerosis. However, lipid-rich intimal lesions from animals make excellent positive controls. This tissue should be handled in the same fashion as human autopsy tissue.

Human tissue can be obtained fresh at autopsy and should be placed immediately in 10% neutral-buffered zinc formalin. Reducing the postmortem interval is advisable, although postmortem oxidation has been shown not to affect immunohistochemistry for oxidation-specific epitopes.[10] Twenty-four to 48 hr should be the maximum postmortem interval. Alternatively, formalin-fixed, paraffin-embedded tissue can be obtained from the Pathobiological Determinants of

[8] N. Sugiyama, S. Marcovina, A. M. Gown, H. Seftel, B. Joffe, and A. Chait, *Am. J. Pathol.* **141,** 99 (1992).

[9] M. E. Rosenfeld, W. Palinski, S. Ylä-Herttuala, S. Butler, and J. L. Witztum, *Arteriosclerosis* **10,** 336 (1990).

[10] G. Jürgens, Q. Chen, H. Esterbauer, S. Mair, G. Ledinski, and H. P. Dinges, *Arterioscler. Thromb.* **13,** 1689 (1993).

Atherosclerosis in Youth (PDAY) archive.[11] This material was collected from individuals aged 15 to 34 who died of trauma and has been used in two papers describing the location of oxidized lysine by immunohistochemistry. There are several advantages to using this material. First, specimens were obtained from standard locations of coronary arteries and aortas using standardized protocols. Second, risk factor data, including age, sex, serum cholesterol, body mass, surrogate markers for blood pressure, glucose tolerance, and smoking status, are available on these subjects. Applications for use of specimens in the archive should be addressed to Dr. Jack Strong, Department of Pathology, Louisiana State University Health Sciences Center, 1901 Perdido Street, New Orleans, LA 70112 (jstron@lsuhsc.edu).

Histotechnology

Following fixation, tissue must be dehydrated through a series of alcohols, embedded in paraffin, cut into sections 5 μm thick, and then placed on glass slides. Polylysine slides are preferred, as tissue adheres better to these slides. Histotechnology services are available in any histology laboratory for a charge of usually less than $10 per slide. Because of the skill needed in cutting and mounting slides and the irreplaceable nature of specimens, it is highly recommended that investigators use a trained histotechnologist to produce slides.

Protocol for Immunohistochemistry

Once sections are placed on glass slides, the protocol for immunohistochemistry is as follows.

1. Deparafinize tissue by incubating in xylene for 10 min, in 100% ethanol for 4 min, in 95% ethanol for 1 min, and finally in water for at least 5 min. Slides can be held in water for several hours if necessary.
2. Incubate in 3% hydrogen peroxide for 3 to 5 min to neutralize endogenous peroxidase activity in tissue.
3. Wash in phosphate-buffered normal saline, pH 7.0 (PBS), for 5 min.
4. Incubate with primary antibody for 32 min.
5. Wash in PBS for 5 min.
6. Incubate with secondary antibody, diluted 1 : 50, for 32 min. The secondary antibody must be isotype specific and should be species specific. For example, because the commercially available anti-oxLDL antibody OXL41.1 is a mouse IgM, the secondary antibody should be an anti-mouse IgM. The secondary antibody should be labeled with biotin for immunohistochemistry. (A fluorescent molecule

[11] Pathobiological Determinants of Atherosclerosis in Youth (PDAY) Research Group, *Arterioscler. Thromb.* **13,** 1291 (1993).

could be used, but this is not described herein.) For the reasons stated earlier, a biotin-labeled secondary antibody is recommended.

7. Wash in PBS for 5 min.

8. Incubate with avidin–horseradish peroxidase conjugate, diluted 1 : 50, for 32 min.

9. Wash in PBS for 5 min.

10. Incubate in 3,3′-diaminobenzidine, which should be prepared fresh according to manufacturer's directions.

11. Rinse in water.

12. Counterstain with hematoxylin for 30 sec.

13. Incubate in 95% ethanol for 1 min, 100% ethanol for 4 min, and xylene for 10 min. Then apply mounting media and a coverslip.

It is recommended that steps 2 through 11 be performed using a coverplate (Shandon Lipshaw, Pittsburgh, PA). When attached properly to a standard glass slide, this plastic device creates a small gap over the histologic section, which can be filled with buffer or two drops (100 μl) of reagent. The coverplate ensures that reagents are distributed uniformly over the section with minimum use of reagent. Each subsequent reagent in the procedure displaces the previous reagent by gravity. Coverplates can be held in a Sequenza slide rack, available from the same source.

Except for the primary antibody, immunohistochemical reagents are available from multiple sources. Automated immunohistochemical stainers are available if it is envisioned that immunohistochemistry will become a major part of a research program. Because only one monoclonal antibody to an oxidation-specific epitope is available commercially, investigators may want to contact investigators who have used "home brew" antibodies. Therefore, this chapter includes a section on previous work in this area (vide infra).

Technical Issues

In diagnostic immunohistochemistry, the expected pattern of staining (i.e., cytoplasmic, nuclear, or extracellular) is always known and can be used to distinguish genuine staining from artifact. A major technical issue in interpreting immunohistochemical staining for oxidation-specific epitopes is that the expected pattern of staining has not been determined systematically. Most investigators would agree that staining in intimal lesions in the Watanabe heritable hyperlipidemic (WHHL) rabbit is real. However, the same dilution of antibody that labels intimal lesions in the WHHL rabbit also labels the vascular adventitia, which is not affected by the pathologic changes attributed to oxidative stress, such as monocyte chemotaxis, foam cell formation, and apoptosis. Nevertheless, the selective staining of intimal lesions in the WHHL rabbit makes this a good positive control. As a negative control, the protocol can be performed using, instead of the primary antibody, either an isotype and species-matched irrelevant antibody (available commercially) or a diluent.

A second issue that has not been resolved is the correct dilution of primary antibody. Thus, use of a positive control is mandatory. The same dilution of antibody that gives a strong reaction in rabbit intimal lesions should be used with human tissue. Comparison of results with previously published color photomicrographs of immunoreactivity in WHHL rabbit lesions[9,12,13] should allow determination of the proper dilution. A trial with several dilutions, such as 1 : 50, 1 : 250, 1 : 500, and 1 : 1000, is recommended because antibody concentration may vary between lots. Other variables that can be manipulated include the concentration of other reagents and incubation times.

Previous Work

Animal Studies

Vascular staining for oxidation-specific epitopes was initially reported in the WHHL rabbit in 1988 by Haberland *et al.*[13] using immunohistochemical methods and the monoclonal antibody MDAlys. Those investigators found negligible staining in normal arteries.[13] Boyd *et al.*[14] reported similar findings using OXL41.1 and immunofluorescence microscopy. Using immunohistochemistry with the antibodies MDA2 and NA59, Rosenfeld *et al.*[9] showed that immunoreactivity in the WHHL rabbit was present in intimal lesions and adventitia. Most recently, *in vivo* uptake of radiolabeled MDA2 in the WHHL rabbit was shown to be greater in intimal lesions than in uninvolved aorta.[15]

In addition to the major differences in morphology between human atherosclerotic lesions and intimal lesions in animal models, a second drawback is the absence of diffuse intimal thickening in animals.[12] Diffuse intimal thickening is an accumulation of smooth muscle, collagen, and matrix present to some degree in all normal adult human arteries. The significance of this change is that apoB can bind to collagen and glycosaminoglycans, allowing the accumulation of atherogenic lipoproteins such as LDL in normal adult human intima. This results in immunoreactivity for LDL and the accumulation of cholesterol, known as perifibrous lipid, in normal human intima. Because oxLDL circulates, it is possible

[12] G. D. Sloop, K. B. Fallon, G. Lipscomb, H. Takei, and A. Zieske, *Atherosclerosis* **148,** 255 (2000).

[13] M. Haberland, D. Fong, and L. Cheng, *Science* **241,** 215 (1988).

[14] H. C. Boyd, A. M. Gown, G. Wolfauer, and A. Chait, *Am. J. Pathol.* **135,** 815 (1989).

[15] S. Tsimikas, B. P. Shortal, J. L. Witztum, and W. Palinski, *Arterioscler. Thromb. Vasc. Biol.* **20,** 689 (2000).

that it may accumulate in nonatherosclerotic intima as the result of diffuse intimal thickening. Using the antibody FOH1a/DLH3, Itabe *et al.*[16] demonstrated immunoreactivity for oxidation-specific epitopes in "swollen" collagen fibers in lesions from human coronary arteries, supporting this possibility.

Human Studies

Jurgens *et al.*[10] showed that epitopes of 4-HNE-protein adducts were not generated postmortem, suggesting that autopsy specimens are useful for determining the localization of oxidation-specific epitopes. Using an antibody to 4-HNE-lysine, those investigators showed that immunoreactivity was localized to "thickened" intima of early, transitional and advanced atherosclerotic lesions but not "normal" intima. Examination of their published photomicrographs reveals immunoreactivity in diffuse intimal thickening, which those investigators apparently considered an "early" atherosclerotic lesion. Immunoreactivity in diffuse intimal thickening next to an atherosclerotic plaque is also seen in published photomicrographs of Palinski *et al.*[17] Using the antibody Ox5, O'Brien *et al.*[18] found that immunoreactivity was significantly more prevalent in approximately 60% of coronary atherosclerotic plaques and only 4% of control arteries. In contrast, Sloop *et al.*[12] found some degree of immunoreactivity using MDA2 and NA59 in all vessels examined, including thoracic aorta and coronary arteries and veins. Immunoreactivity was most intense in coronary veins, which do not develop atherosclerosis. Sloop *et al.*[12] used postmortem tissue from the PDAY archive and found no correlation between postmortem interval and immunoreactivity, confirming the usefulness of postmortem specimens. Using MDA2 and NA59 and specimens from the PDAY archive, Scanlon *et al.*[19] also found significant immunoreactivity in "lesion-resistant" areas, i.e., intima with a low prevalence of fatty streaks and atherosclerotic plaques, again raising questions about the *in vivo* relevance of *in vitro* data ascribing pathologic activity to oxLDL. Finally, Napoli *et al.*[20] demonstrated widespread immunoreactivity for oxidation-specific epitopes using MDA2 and NA59 in fetal aortas. Because atherosclerotic plaques are not prevalent until the third or fourth decades of life, their results also call into question the relevance of *in vitro* data in which pathologic effects of oxLDL are manifest within hours or days.

[16] H. Itabe, E. Takeshima, H. Iwasaki, J. Kimura, Y. Yoshida, T. Imanaka, and T. Takano, *J. Biol. Chem.* **269,** 15274 (1994).

[17] W. Palinski, S. Hörkkö, E. Miller, U. P. Steinbracher, H. C. Powell, L. K. Curtiss, and J. L. Witztum, *J. Clin. Invest.* **98,** 800 (1996).

[18] K. D. O'Brien, C. E. Alpers, J. E. Hokanson, S. Wang, and A. Chait, *Circulation* **94,** 1216 (1996).

[19] C. E. O. Scanlon, B. Berger, G. Malcom, R. W. Wissler, and PDAY Research Group, *Atherosclerosis* **121,** 23 (1996).

[20] C. Napoli, F. P. D'Armiento, F. P. Mancini, A. Postiglione, J. L. Witztum, G. Palubo, and W. Palinski, *J. Clin. Invest.* **100,** 2680 (1997).

Regarding the localization of oxidation-specific epitopes in vessels, two important questions need definitive answers. Are oxidation-specific epitopes present in normal intima with diffuse intimal thickening? Are oxidation-specific epitopes abundant in veins? The definitive answer to these questions will have a bearing on the validity of theories invoking "oxidative stress" in atherosclerosis. A related question is whether the distribution of oxLDL is similar to that of unmodified LDL. LDL is widely distributed, being present in both blood and lymph. Finally, what is the role of "oxidative stress" in other diseases that increase in prevalence with aging? Immunolocalization of oxidation-specific epitopes could help provide an answer to that question as well.

[30] Quantitation of *S*-Nitrosothiols in Cells and Biological Fluids

By VLADIMIR A. TYURIN, YULIA Y. TYURINA, SHANG-XI LIU, HÜLYA BAYIR, CARL A. HUBEL, and VALERIAN E. KAGAN

Introduction

S-Nitrosothiols (RSNO) are a class of chemical compounds that are widespread *in vivo,* thus providing a reliable source of nitric oxide (NO).[1–3] RSNO have been detected in plasma, erythrocytes, platelets, polymorphonuclear leukocytes, and brain cerebellum.[4–8] The abundance of glutathione (GSH; 1 to 5 mM) in the intracellular space renders the formation of *S*-nitrosoglutathione, structurally the simplest nitrosothiol, kinetically favorable.[9] Formation of *S*-nitrosoproteins occurs posttranslationally by the *S*-nitrosylation of protein cysteines. Steady-state concentrations of nitrosothiols are tissue specific[6–8] and depend on the

[1] J. S. Stamler, *Cell* **78,** 931 (1994).
[2] J. S. Stamler, L. Jia, J. P. Eu, T. J. McMahon, I. T. Demchenko, J. Bonoventura, K. Gernert, and C. A. Piantadosi, *Science* **276,** 2034 (1997).
[3] J. N. Smith and T. P. Dasgupta, *Nitric Oxide Biol. Chem.* **4,** 57 (2000).
[4] J. S. Stamler, O. Jaraki, J. Osborne, D. I. Simon, J. Keaney, J. Vita, D. Singel, C. R. Valeri, and J. Loscalzo, *Proc. Natl. Acad. Sci. U.S.A.* **89,** 7674 (1992).
[5] I. Kluge, U. Gutteck-Amsler, M. Zollinger, and K. Q. Do, *J. Neurochem.* **69,** 2599 (1997).
[6] B. Gaston, *Biochim. Biophys. Acta* **1411,** 323 (1999).
[7] M. Kelm, *Biochim. Biophys. Acta* **1411,** 273 (1999).
[8] M. T. Gladwin, J. H. Shelhamer, A. N. Schecheter, M. E. Pease-Fye, M. A. Waclawiw, J. A. Panza, F. P. Ognibene, and R. O. Cannon, *Proc. Natl. Acad. Sci. U.S.A.* **97,** 11482 (2000).
[9] R. J. Singh, N. Hogg, J. Joseph, and B. Kalyanaraman, *J. Biol. Chem.* **271,** 18596 (1996).

0076-6879/02 $35.00

physiological state.[6,10] For example, endogenous levels of *S*-nitrosoglutathione in human airways are known to be elevated during pneumonia and are depleted in asthmatics.[6,11] Major RSNO compounds in blood differ with respect to their circulating concentrations and half-life (*S*-nitrosoalbumin, 0.05–7 μM; *S*-nitrosoglutathione, 0.02–0.2 μM; *S*-nitrosocysteine, 0.2–0.3 μM; *S*-nitrosohemoglobin 0.17–0.3 μM).[7,8] Most studies report that plasma concentrations of *S*-nitrosoalbumin exceed that of low molecular weight nitrosothiols by at least an order of magnitude.[12] Overall, however, the assessment of nitrosothiol status has been clouded by differences in the sampling and method of analysis. Reported levels of *S*-nitrosothiols in biological fluids and tissues vary from low nanomolor to micromolar concentrations. With some notable exceptions (see later), the refinement of assay methodologies has resulted in a decrease in reported concentrations. This is quite similar to the lowering of accepted values for plasma and tissue lipid hydroperoxides with improvements in detection systems for these bioactive molecules.

Formation and Decomposition of S-Nitrosothiols

Several biochemical pathways may yield *S*-nitrosothiols. They can be formed by direct recombination of NO with thiyl radicals, but not with thiols.[13] Production of *S*-nitrosothiols may be facilitated through electron transfer in iron-nitrosyl groups. Additionally, hemoglobin may serve as a catalyst for RSNO formation.[14] Interaction of nitric oxide with ceruloplasmin in the presence of thiols (GSH) also stimulates the formation of *S*-nitrosothiols (GSNO).[15] Furthermore, activation of nitric oxide synthase itself may form *S*-nitrosothiols through intermediate generation of $ONOO^-$ and its interactions with thiols.[6] Finally, protein nitrosylation by nitrous acid (HNO_2) may take place in acidic organelles.[6]

Decomposition of *S*-nitrosothiols occurs via heterolytic (in the presence of transition metals) and homolytic (photolysis) mechanisms, forming NO^+, NO, and NO^- [16] (Fig. 1). Heterolytic pathways of RSNO decomposition appear to predominate over the homolytic release of NO.[7] Photolysis of *S*-nitrosothiols results

[10] R. J. Schlosser, W. D. Spotnitz, E. J. Peters, K. Fang, B. Gaston, and C. W. Gross, *Otolaryngol. Head Neck Surg.* **123,** 357 (2000).

[11] B. Gaston, J. Reilly, J. M. Drazen, J. Fackler, P. Ramdev, D. Arnelle, M. E. Mullins, D. J. Sugarbaker, C. Chee, D. J. Singel, J. Loscalzo, and J. S. Stamler, *Proc. Natl. Acad. Sci. U.S.A.* **90,** 10957 (1993).

[12] D. Jurd'heuil, L. Gray, and M. Grisham, *Biochem. Biophys. Res. Commun.* **273,** 22 (2000).

[13] M. P. Murphy, *Biochim. Biophys. Acta* **1411,** 401 (1999).

[14] A. F. Vanin, I. V. Malenkova, and V. A. Serezhenkov, *Nitric Oxide* **1,** 193 (1997).

[15] K. Inoue, T. Akaike, Y. Miyamoto, T. Okamoto, T. Sawa, M. Otagiri, S. Suzuki, T. Yoshimura, and H. Maeda, *J. Biol. Chem.* **274,** 27069 (1999).

[16] N. Hogg, *Free Radic. Biol. Med.* **28,** 1478 (2000).

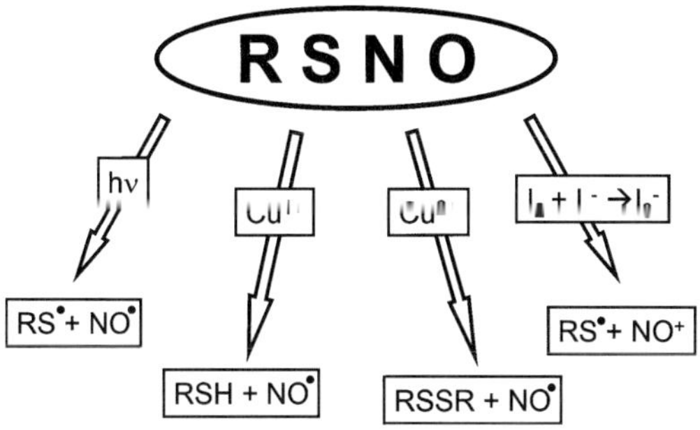

FIG. 1. Scheme illustrating pathways of decomposition of *S*-nitrosothiols important for their assays based on the analysis of released NO.

in the formation of NO and disulfide via the intermediacy of thiyl radicals. In the presence of transition metals, however, thiyl radicals are not likely intermediates. Reducing agents (thiols, ascorbate) can stimulate the decomposition of *S*-nitrosoglutathione (GSNO) by the reduction of contaminating transition metal ions.[17] In the case of ascorbic acid, accelerated GSNO decomposition is metal dependent, whereas GSH enhances GSNO decomposition predominantly via metal-independent pathways.[17] Decomposition of *S*-nitrosothiols by thiols also occurs via transnitrosation.[3] The major pathways of *S*-nitrosothiol decomposition, which are important in the quantification of these compounds based on NO release, are presented in Fig. 1.

Biologic Effects of *S*-Nitrosothiols

S-Nitrosylation of protein cysteines is a common mechanism of posttranslational regulation. *S*-Nitrosylation may control the activity of enzymes[18,19] and regulatory proteins (transcription factors, G proteins, ion channels, receptors, and kinases)[13] and modulate energy metabolism[20] and apoptosis.[21] *S*-Nitrosothiols, nitric oxide, and NO-derived species regulate palmitoylation of a variety of signaling proteins.[22] Neuroprotective and antioxidant effects of RSNO have been

[17] A. Xu, J. A. Vita, and J. F. Keaney, Jr., *Hypertension* **36,** 291 (2000).
[18] H. M. Abu-Sound and S. L. Hazen, *J. Biol. Chem.* **275,** 5425 (2000).
[19] L. J. Marnett, T. L. Wright, B. C. Crews, S. R. Tannenbaum, and J. D. Morrow, *J. Biol. Chem.* **275,** 13427 (2000).
[20] R. M. Clancy, D. Levartovsky, J. Leszezynska-Piziak, J. Yegudin, and S. B. Abramson, *Proc. Natl. Acad. Sci. U.S.A.* **91,** 3680 (1994).
[21] J. B. Mannick, A. Hausladen, L. Liu, D. T. Hess, M. Zeng, Q. X. Miao, L. S. Kane, A. J. Gow, and J. S. Stamler, *Science* **284,** 651 (1999).
[22] T. L. Baker, M. A. Booden, and J. E. Buss, *J. Biol. Chem.* **275,** 22037 (2000).

also described.[23] Formation of RSNO and oxidation of protein SH groups may act as switches in intracellular and intercellular signaling pathways, including gene transcription.[1,24] By providing a reservoir for the release of NO, nitrosothiols play a critical role in NO-mediated vasodilatory action and normal regulation of vascular tone.[2]

Techniques to Assess S-Nitrosothiols

Overview

A number of techniques have been developed to assay *S*-nitrosothiols. One of them is direct spectrophotometric measurement of the characteristic absorbance of the RSNO at 340 nm. This approach is excellent for simple chemical and biochemical model systems, but it has low sensitivity and cannot be used for measurements of nitrosothiols in biological samples.[6]

Most approaches to assay *S*-nitrosothiols in biological systems are based on their indirect detection involving their decomposition and subsequent analysis of the products formed. Homolytic decomposition of RSNO (by UV irradiation) or heterolytic cleavage of RSNO (using excess mercury or copper ions, often >100 μM) yields nitric oxide or its end product, nitrite. Detection of either of these products can be performed by a number of techniques, including chemiluminescence, electrochemical determination with a Clark-type NO electrode, gas chromatography–mass spectrometry (GS-MS), spectrophotometry, fluorometry, and HPLC.[4–6,25–28]

Photolysis-based methods are not specific for *S*-nitrosothiols and can also release NO from C-, N-, and O-NO derivatives and organometal complexes.[29,30] To exclude these nonspecific confounding factors, nitrosothiols (e.g., *S*-nitrosoalbumin, GSNO) may be separated before UV irradiation by chromatography (using Sephadex G-25) or by cut-off filters with centrifugation.[4,8] Decomposition of *S*-nitrosothiols can also be performed in the presence of iodine with subsequent chemiluminescent detection of NO.[8,30] The sensitivity of this procedure is in the

[23] P. Rauhala, K. P. Mohanakumar, I. Sziraki, A. M.-Y. Lin, and C. C. Chiueh, *Synapse* **23,** 58 (1996).
[24] R. Gopalakrishna, Z. H. Chen, and U. Gundimeda, *J. Biol. Chem.* **268,** 27180 (1993).
[25] J. A. Cook, S. Y. Kim, D. Teague, M. C. Krishna, R. Pacelli, J. B. Mitchell, Y. Vodovotz, R. W. Nims, D. Christodoulou, A. M. Miles, M. B. Grisham, and D. A. Wink, *Anal. Biochem.* **238,** 150 (1996).
[26] T. Akaike, K. Inoue, T. Okamoto, H. Nishino, M. Otagiri, S. Fujii, and H. Maeda, *J. Biochem.* **122,** 459 (1997).
[27] J. F. Ewing and D. R. Janero, *Free Radic. Biol. Med.* **25,** 621 (1998).
[28] D. V. Vukomanovic, A. Hussain, D. E. Zoutman, G. S. Marks, J. F. Brien, and K. Nakatsu, *J. Pharmacol. Toxicol. Methods* **39,** 235 (1998).
[29] L. Jia, C. Bonaventura, J. Bonaventura, and J. S. Stamler, *Nature* **380,** 221 (1996).
[30] A. Samouilov and J. L. Zweier, *Anal. Biochem.* **258,** 322 (1998).

range of 10–25 nM for GSNO. Using this protocol (chemiluminescent detection after pretreatment with iodine), the concentration of S-nitrosoalbumin in plasma was estimated to be $\approx$60 nM.[8] The amounts of nitrite in plasma samples estimated by this procedure were in the 0.2–0.9 μM range,[8] significantly lower than levels of 10–100 μM obtained by other methods.[12,30–32] Lower levels of RSNO obtained by workers using the iodine method may be explained by unwanted quenching of the iodine-catalyzed process by reducing agents (GSH, cysteine, protein thiols, etc.).[30] It should also be noted that iodine can be captured by double bonds of lipids in the samples.

Nanomolar levels of S-nitrosothiols in plasma were also described by Marley and co-authors.[33] Their assay is based on the release of NO from RSNO in reaction with Cu^+, iodine, and iodide with subsequent quantification by chemiluminescence. The low level of RSNO detected by these workers may be explained by insufficient amounts of copper available for the decomposition of RSNO, as high-affinity sites on albumin and ceruloplasmin are strong chelators of copper. These concentrations of RSNO are significantly lower than the values obtained for purified S-nitrosoalbumin after its UV irradiation and detection by chemiluminescence ($\approx$7 μM).[4] It has been confirmed that S-nitrosothiols in normal plasma are roughly 2.2 ± 1.3 μM.[34] This supports earlier data by Stamler *et al.*[4] A fluorometric method based on the use of 2,3-diaminonaphthalene as a fluorogenic NO scavenger has a sensitivity of 50 nM with GSNO, but the reaction may be inhibited in the presence of proteins and confounded by the presence of nitrite at low pH.[6]

Fluorescence Assay of S-Nitrosothiols with 4,5-Diaminofluorescein (DAF-2)

DAF-2 has been introduced as a highly sensitive and specific fluorogenic scavenger of NO in the presence of dioxygen, yielding the fluorescent product DAF-2 triazole (DAF-2T) with a sensitivity of 5 nM for NO.[35] In contrast to other NO scavengers, DAF-2 does not react with stable oxidized forms of NO, such as NO_2^- and NO_3^-, or with other reactive oxygen species, such as O_2^-, H_2O_2, or $ONOO^-$.[35] We utilized this procedure for NO detection and adapted it for measurements of S-nitrosothiols. Our assay is based on the decomposition of S-nitrosothiols by copper[25] or UV irradiation (>330 nm) in the presence of DAF-2 (Fig. 2). At physiological pH, DAF-2 is relatively nonfluorescent. Released NO reacts with DAF-2 to form a stable fluorescent product DAF-2T, whose fluorescence intensity can be

[31] S. Ferlito and M. Gallina, *Minerva Endocrinol.* **24,** 117 (1999).

[32] A. Balat, M. Cekmen, M. Yurekli, O. Kutlu, I. Islek, E. Sonmezgoz, M. Cakir, Y. Turkoz, and S. Yologlu, *Pediatr. Nephrol.* **15,** 266 (2000).

[33] R. Marley, M. Feelish, S. Holt, and K. A. Moore, *Free Radic. Res.* **32,** 1 (2000).

[34] O. A. Strand, A. Leone, K.-E. Giercksky, and K. A. Kirkeboen, *Crit. Care Med.* **28,** 2779 (2000).

[35] H. Kojima, N. Nakatsubo, K. Kikuchi, S. Kawahara, Y. Kirino, H. Nagoshi, Y. Hirata, and T. Nagano, *Anal. Chem.* **70,** 2446 (1998).

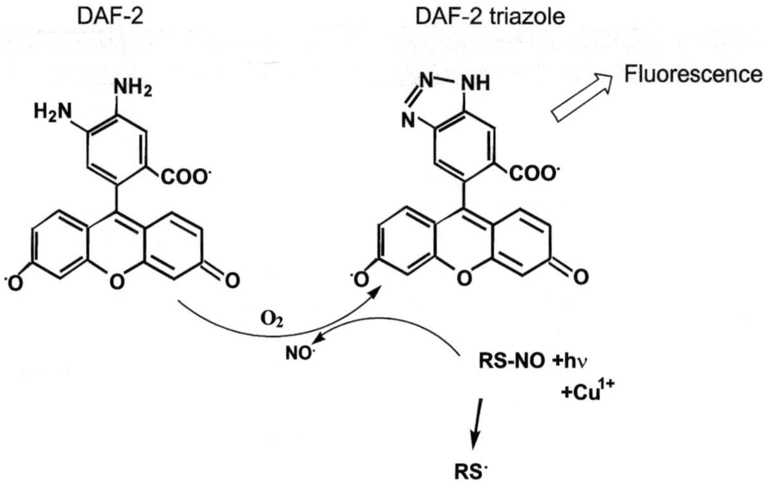

FIG. 2. Scheme illustrating the formation of fluorescent DAF-2 triazole (DAF-2T) complex with nitric oxide released by decomposition of *S*-nitrosothiols.

determined readily. For these assays, we used a Shimadzu RF-5301 PC spectrofluorophotometer (Kyoto, Japan) and the following conditions: excitation at 495 nm and emission at 515 nm (slits: excitation 1.5 nm, emission 5.0 nm). Data obtained were exported and treated using Shimadzu RF-5301 PC software.

To validate the technique, we performed several experiments in model systems. Initially, we tested whether the protocol can be used for the quantitative analysis of synthetic nitrosothiols in solutions. Synthesized GSNO (50 μl) is incubated in phosphate buffer (450 μl) (pH 7.4) in the presence of DAF-2 (5 μM) and $CuSO_4$ (300 μM) up to 1 hr at 37°. At the end of incubation, acetonitrile (500 μl) is added, and samples are centrifuged at 10,000*g* for 5 min. The supernatant is mixed with 1.5 ml PBS and used for fluorescence measurements. Typical fluorescence spectra of DAF-2T formed under these conditions are presented in Fig. 3A. The fluorescence response is dependent on the amount of $CuSO_4$ added (Fig. 3A, inset) and reaches a plateau at the copper concentration of 300 μM. The fluorescence intensity of DAF-2T increases rapidly during the first 20 min of incubation and subsequently plateauesd, indicating that NO is released from GSNO (Fig. 3B). In order to quantify the amounts of NO released from GSNO, we utilized an independent NO donor with known decay characteristics: (*Z*)-[*N*-(3-ammoniopropyl)-*N*-(*n*-propyl)-amino]diazen-1-ium-1,2-diolate, PAPANONOate (NOC-15). Different concentrations of NOC-15 were decomposed completely in the presence of DAF-2 within 5.5 hr, yielding 0.5–3 nmol of NO. Using NOC-15 for calibration of the amounts of NO, we found that our protocol with DAF-2 was linear in the range of GSNO concentration from 0.04 to 0.8 μM (Fig. 3C).

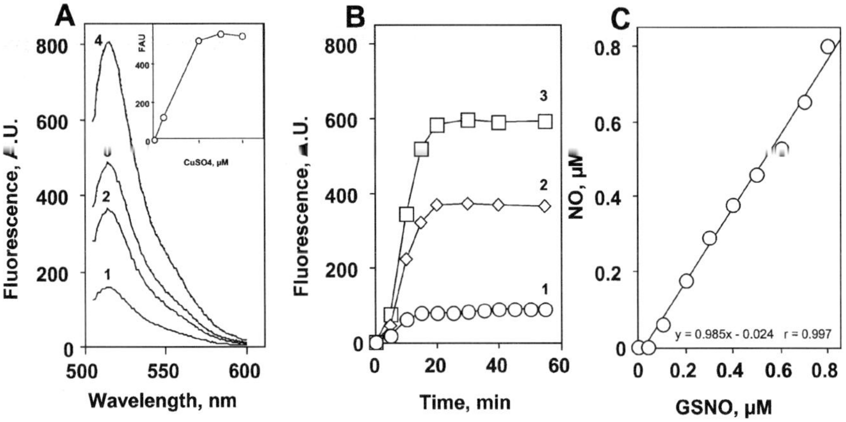

FIG. 3. Quantitative characterization of DAF-2 based assay of GSNO utilizing Cu-catalyzed decomposition of *S*-nitrosothiol. (A) Typical fluorescence emission spectra of *DAF-2T* obtained with different GSNO concentrations in the presence of $CuSO_4$ (300 μM) in PBS (pH 7.4): (1) 0.15 μM GSNO; (2) 0.3 μM GSNO; (3) 0.5 μM GSNO; (4) 0.75 μM GSNO. (Inset) Dependence of DAF-2T fluorescence response after a 20-min incubation with GSNO (0.5 μM) at various concentrations of $CuSO_4$. (B) Time course of DAF-2T formation with different concentrations of GSNO: (1) 0.09 μM GSNO; (2) 0.3 μM GSNO; and (3) 0.6 μM GSNO. (C) Comparison of the intensity of DAF-2T fluorescence produced by the reaction of DAF-2 with the No donor NOC-15 (completely decomposed over a 5.5-hr incubation) with that generated after a 20-min incubation of different concentrations of GSNO (as indicated) with $CuSO_4$ (300 μM) in PBS at 37°.

Similar results were obtained when UV light was used for the decomposition of GSNO instead of copper. Figure 4A shows typical fluorescence spectra of DAF-2T formed by UV irradiation of GSNO in phosphate buffer (pH 7.4) in the presence of DAF-2. A standard curve was established using GSNO as the standard. A plot of GSNO concentration vs fluorescence intensity was linear in the concentration range from 0.04 to 0.8 μM (Fig. 4B). We further attempted to use the protocol for assay of *S*-nitrosothiols in albumin and plasma samples. Samples (50 μl) containing DAF-2 (5 μM) in 2.5 ml PBS are exposed to UV irradiation (80 $\mu W/cm^2$, 10 min) using the Oriel UV light source (Model 66002) and a cut-off filter (>330 nm, Balzers, Pittsburgh, PA). Samples are heated at 80° for 4 min and centrifuged at 10,000*g* for 5 min, and the supernatant is used for measurements. The fluorescence response of DAF-2T developed during 2–5 min of irradiation, and NO was completely released from *S*-nitrosothiols (Fig. 4C).

Because proteins have been reported to interfere with fluorescence-based assays of *S*-nitrosothiols,[6] we next performed a series of experiments in which we studied the effect of proteins on the fluorescent response of DAF-2T formed from *S*-nitrosothiols (GSNO or human *S*-nitrosoalbumin, hSANO). Human *S*-nitroso-albumin was prepared as described by Stamler.[4] We found that, indeed, fluorescence

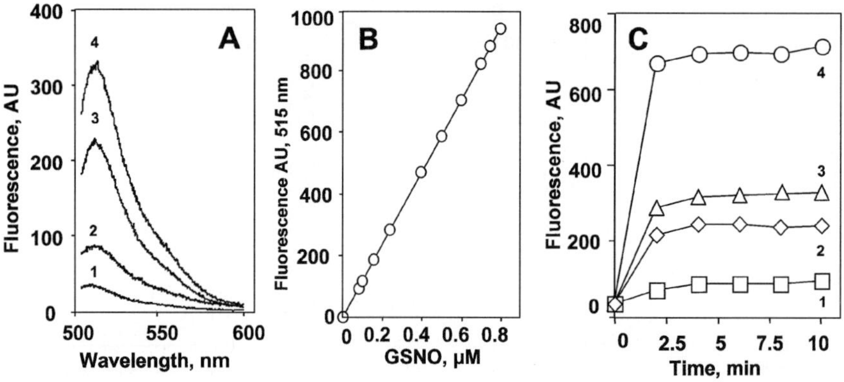

FIG. 4. Quantitative characterization of DAF-2-based assay of GSNO and *S*-nitrosothiols in plasma utilizing UV irradiation. (A) Typical fluorescence emission spectra of DAF-2T in PBS formed after UV irradiation (80 $\mu W/cm^2$, 10 min)-induced decomposition of *S*-nitrosothiols in the presence of DAF-2 (5 μM). (1) DAF-2 alone; (2) normal (nonpregnancy) plasma (50 μl) + DAF-2; (3) *S*-nitrosoalbumin (0.16 μM) + DAF-2; and (4) GSNO (0.24 μM) + DAF-2. (B) A plot of GSNO concentration vs fluorescence response of DAF-2T obtained after UV irradiation-induced decomposition of GSNO. (C) Time course of DAF-2T fluorescence response recorded at 515 nm after UV-induced decomposition of *S*-nitrosothiols. (1) normal (nonpregnancy) plasma (50 μl); (2) *S*-nitrosoalbumin (0.16 μM); (3) GSNO (0.24 μM); and (4) GSNO (0.7 μM).

responses from GSNO and hSANO were significantly quenched on addition of nonnitrosylated human albumin (Figs. 5A and 5B). This quenching, however, was completely abolished by heat treatment of the samples (80° for 4 min) after UV irradiation and subsequent centrifugation (10,000g for 5 min) to remove the proteins. As shown in Figs. 4A and 4B, complete recovery of DAF-2T fluorescence intensity was achieved by denaturing proteins followed by removal of the denatured proteins by centrifugation. The detection limit for *S*-nitrosothiols was estimated as 50 n*M* under these conditions.

Analysis of S-Nitrosothiols in Plasma Samples Using 4,5-Diaminofluorescein

Using the combined procedure, including UV-induced decomposition of *S*-nitrosothiols with the subsequent quantitation of released NO by the DAF-2-based fluorescence assay, we performed measurements of *S*-nitrosothiols in whole plasma as well as in plasma fractions. We used affinity column chromatography[36] to obtain albumin-enriched and albumin-free fractions of plasma. To this end, plasma (100 μl) is loaded on a HiTrap column (HiTrap Blue 1 ml, Pharmacia Biotech, Piscataway, NJ), and two fractions are collected after chromatography.

[36] J. Travis, J. Bowen, D. Tewsbury, D. Johnson, and R. Pannel, *Biochem J.* **157,** 301 (1976).

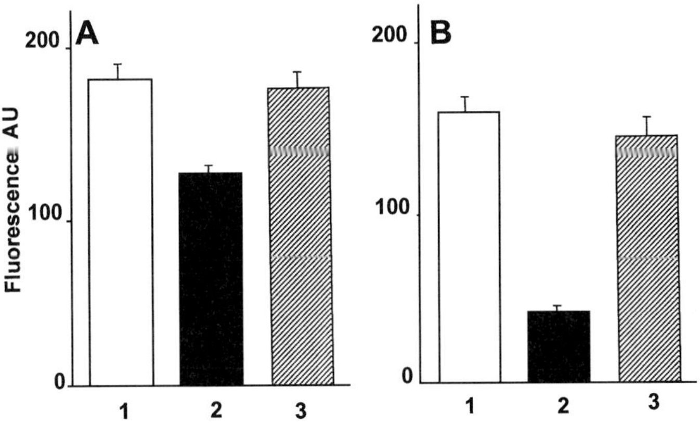

FIG. 5. Effect of albumin on the fluorescence response of DAF-2T after UV irradiation-induced (80 μW/cm^2, 10 min) decomposition of GSNO (A) and *S*-nitrosoalbumin (B) in PBS. (A) GSNO (0.15 μM) alone (1); GSNO (0.15 μM) + hSA (0.5 mg) (2); and GSNO (0.15 μM) + hSA (0.5 mg) (3) after UV irradiation were heated at 80° for 4 min and centrifuged at 10,000*g* for 5 min. (B) hSANO (0.098 μM) alone (1); hSANO (0.098 μM) + hSA (2.5 mg) (2); and hSANO (0.098 μM) + hSA (2.5 mg) (3) after UV irradiation were heated at 80° for 4 min and centrifuged at 10,000*g* for 5 min.

Fraction 1 (albumin free) is collected in 5 ml of buffer A (50 m*M* phosphate buffer, 0.1 *M* KCl, pH 7.0), and fraction 2 (albumin-enriched fraction) is collected in 5 ml of buffer B (50 m*M* phosphate buffer, 1.5 *M* KCl, pH 7.0). The elution rate is 0.5 ml/min. When the proteins in each plasma fraction are resolved by native polyacrylamide gel electrophoresis using an 8% acrylamide gel,[37] we found that fraction 1 was essentially albumin-free and that fraction 2 was enriched (>95%) with albumin (not shown). The protein concentration in the plasma samples and chromatographic fractions of plasma is determined with the Bio-Rad protein assay kit (Bio-Rad, Hercules, CA). A standard curve is generated using bovine serum albumin. In separate control experiments, we tested whether the chromatography procedure affected levels of *S*-nitrosothiols. To this end, a mixture of pure GSNO (125.0 n*M*) plus hSANO (70.5 n*M*) is applied to the HiTrap column. The level of nitrosylation in each fraction is determined using UV irradiation in the presence of DAF-2. The results presented in Fig. 6A show that the contents of GSNO in fraction 1 and hSANO in fraction 2 are 123.0 ± 3.0 and 64.2 ± 8.9 n*M*, respectively, indicating that 97% of *S*-nitrosothiols were recovered after affinity chromatography. Importantly, full recovery of *S*-nitrosothiols from fractions 1 and 2 was also achieved when whole plasma samples were subjected to affinity chromatography. Data on RSNO content in normal human plasma and in fractions 1 and 2 obtained

[37] J. Sambrook, E. F. Fritsch, and T. Maniatis, "Molecular Cloning: A Laboratory Manual," (2nd Ed.). Cold Spring Harbor Laboratory Press, Cold Spring Harbor, NY, 1989.

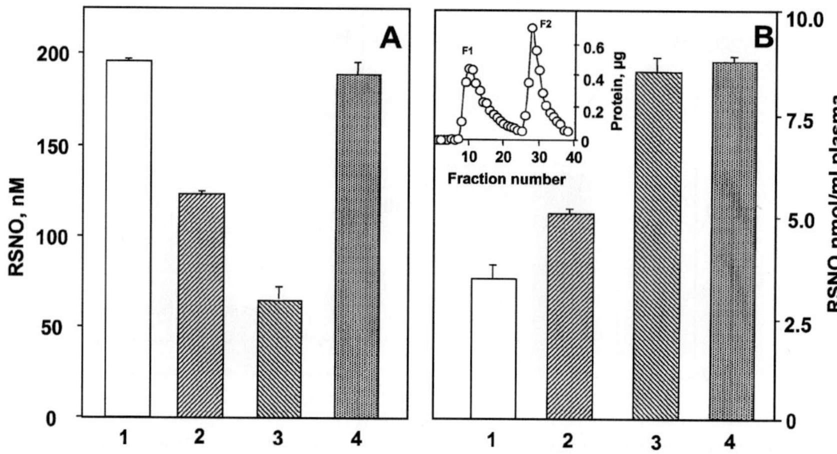

FIG. 6. Content of *S*-nitrosothiols in the model system (A) and in the plasma sample (B) before and after affinity chromatography. (A) Content of GSNO (125 n*M*) and hSANO (70.5 n*M*) before and after separation on a HiTrap column. (1) GSNO + hSANO before loading on the column; (2) fraction 1 containing GSNO; (3) fraction 2 containing hSANO; and (4) the sum of fraction 1 plus fraction 2. (B) Content of *S*-nitrosothiols in total plasma and in fractions obtained after affinity chromatography. (1) Fraction 1; (2) fraction 2 (albumin enriched); (3) the sum of fraction 1 plus fraction 2; and (4) total plasma before loading on the column. (Inset) Chromatographic profile of proteins in fractions 1 and 2.

after its chromatographic separation are presented in Fig. 6B. *S*-Nitrosoalbumin in fraction 2 was found to be the major *S*-nitrosothiol of plasma (>60%, 5.4 ± 1.0 nmol/ml plasma), whereas fraction 1 contained 3.5 ± 1.0 nmol/ml plasma, i.e., <40% of total plasma *S*-nitrosothiols. The recovery of *S*-nitrosothiols from plasma fractions after affinity chromatography was ≈99% as compared with the direct measurements of *S*-nitrosothiols in unfractionated plasma samples.

It is important to emphasize that decomposition of *S*-nitrosothiols in plasma should be accomplished by UV irradiation rather that by Cu or Hg, as use of these ions may lead to erroneous results. This is mainly due to the fact that plasma contains high concentrations of proteins (50–70 mg/ml) capable of binding Cu and Hg. In addition, when EDTA (4 m*M*) is used as an anticoagulant, even greater amounts of Cu and Hg may be bound by the chelator and thus are unavailable for the decomposition of *S*-nitrosothiols. As a result, lower concentrations of *S*-nitrosothiols are detectable in plasma samples under conditions when suboptimal Cu (Hg) may result in incomplete decomposition of nitrosthiols. This can be overcome, at least in part, by the addition of very high concentrations of Cu (or Hg), in large excess of the binding capacity of plasma proteins and chelators. In our experiments, addition of 6 m*M* Cu to normal human plasma resulted in the detection of *S*-nitrosothiols at the level of 3.7 ± 0.7 μM consistent with the data obtained by

TABLE I
CONTENT OF S-NITROSOTHIOLS IN TOTAL PLASMA SAMPLES AND FRACTIONS OBTAINED FROM PLASMA BY AFFINITY COLUMN CHROMATOGRAPHY[a]

	S-Nitrosothiols			
	nmol/ml plasma		pmol/mg protein	
Sample	Control	Normal pregnancy	Control	Normal pregnancy
Total plasma	8.9 ± 1.6	9.5 ± 1.5	133.0 ± 15.0	186.0 ± 25.0[d]
Fraction 1[b]	3.5 ± 1.0	4.4 ± 1.0	229.0 ± 43.0	290.0 ± 76.0[d]
Fraction 2[c]	5.4 ± 1.0	5.1 ± 0.7	94.0 ± 13.0	137.0 ± 15.0[d]

[a] Data are presented as means ± SD; $n = 11$ for control (nonpregnancy), $n = 21$ for normal pregnancy.
[b] Fraction 1 includes LMW and other proteins.
[c] Fraction 2 is the albumin-enriched fraction.
[d] $p < 0.001$ vs control (nonpregnancy).

other workers.[34] It has been shown that a divalent cation concentration in the medium, medium, as well as the incident light, affects the ability of DAF-2 to detect NO. A combination of UV light and DAF-2 would avoid these problems, as our assay is not dependent on chelators.[37]

Analysis of S-Nitrosothiols in Pregnancy Plasma Fractions

In plasma, *S*-nitrosoalbumin is the major reservoir of functionally important NO.[4,38,39] Therefore, monitoring nitrosylation/denitrosylation of albumin may be important for understanding its role in normal vascular regulation as well as its dysregulation in disease. We combined affinity chromatography to obtain albumin-enriched fractions (see earlier discussion) with a DAF-2-based fluorescence assay of NO released after UV irradiation of plasma fractions to determine the levels of *S*-nitrosylation of albumin. The results of our measurements of *S*-nitrosothiol content in whole plasma samples and in albumin-enriched and albumin-free plasma fractions are shown in Table I. Plasma samples were obtained from women with normal pregnancy and from women with the hypertensive pregnancy disorder preeclampsia. Samples from these two groups were matched by gestational age. Samples were also obtained from nonpregnant women of similar age during the follicular phase of the menstrual cycle. All were nonsmokers. We found that the level of *S*-nitrosothiols in control (nonpregnancy) plasma samples

[38] T. Minamiyama, S. Takemura, and M. Inoue, *Arch. Biochem. Biophys.* **341,** 186 (1997).
[39] R. K. Goldman, A. A. Vlessis, and D. Trunkey, *Anal. Biochem.* **259,** 98 (1998).

was $8.9 \pm 1.6\ \mu M$ (133 ± 15 pmol/mg protein). This is well within the range of concentrations reported previously for human plasma.[4,38,39] As indicated earlier *S*-nitrosoalbumin was responsible for more than 60% of total *S*-nitrosothiols in normal plasma. The content of *S*-nitrosothiols was increased significantly in preeclampsia plasma ($11.1 \pm 2.5\ \mu M$ or 317 ± 110 pmol/mg protein; $n = 21$) as compared with normal pregnancy and nonpregnancy control plasma samples (Table I). The level of *S*-nitrosoalbumin was significantly higher in preeclampsia samples ($6.3 \pm 1.4\ \mu M$ or 256 ± 105 pmol/mg protein) than in normal pregnancy plasma or nonpregnancy plasma (Table I). This increased concentration of *S*-nitrosothiols in preeclampsia plasma was almost completely explained by the increased levels of *S*-nitrosoalbumin in total plasma. Further studies are necessary to reveal whether elevated levels of *S*-nitrosoalbumin in preeclampsia plasma are due to increased rates of NO production and nitrosylation of albumin or caused by decreased rates of decomposition of *S*-nitrosoalbumin. It is tempting to speculate that the impaired release of physiologically relevant NO from *S*-nitrosoalbumin may be involved in vascular dysfunction, which is a central pathophysiologic feature of preeclampsia.

S-Nitrosylation of Metallothioneins

The low molecular weight (LMW) cysteine-rich proteins—metallothioneins (MTs)—readily release metals in a redox-dependent fashion (see [25] by Fabisiak *et al.,* Volume 353). MT binding/release of Cu can be regulated by oxidation and nitrosylation of MT cysteines *in vitro*.[40,41] Furthermore, redox-driven release and delivery of copper from MT to target proteins may be realized through nitrosative attack on their thiolate clusters, suggesting that *S*-nitrosylation of MT cysteines may be important for metal trafficking.[42] Therefore, analysis of MT *S*-nitrosylation may be instrumental in understanding the physiological and toxicological properties of MT.

We utilized our DAF-2-based fluorescence assay of *S*-nitrosothiols to evaluate the nitrosylation of MTs in HL-60 cells incubated in the presence of the NO donor, *S*-nitroso-*N*-acetyl penicillamine (SNAP). We were also interested in the ability of Cu-MTs (formed as a result of increased Cu-MT binding) to catalyze the decomposition of *S*-nitroso-MTs. To this end, HL-60 cells are first pretreated with 150 μM $ZnCl_2$ for 24 hr to induce MT expression as described previously.[41] After incubation of cells with 2 m*M* copper nitrilotriacetate (Cu-NTA) for 14 hr

[40] J. P. Fabisiak, L. L. Pearce, G. G. Borisenko, Y. Y. Tyurina, V. A. Tyurin, J. Razzack, J. S. Lazo, B. R. Pitt, and V. E. Kagan, *Antiox. Redox Signal.* **1,** 309 (1999).

[41] S. X. Liu, J. P. Fabisiak, V. A. Tyurin, G. G. Borisenko, B. R. Pitt, J. S. Lazo, and V. E. Kagan, *Chem. Res. Toxicol.* **13,** 922 (2000).

[42] S. X. Liu, K. Kawai, V. A. Tyurin, Y. Y. Tyurina, G. G. Borisenko, J. P. Fabisiak, P. J. Quinn, B. R. Pitt, and V. E. Kagan, *Biochem. J.* **354,** 397 (2001).

TABLE II

CONTENT OF *S*-NITROSO-MT IN $ZnCl_2$-PRETREATED AND $ZnCl_2$-PRETREATED/Cu-NTA-LOADED HL-60 CELLS AFTER EXPOSURE TO NO DONOR, SNAP[a]

	S-Nitrosylated MT, pmol/mg protein	
Sample	−SNAP	+SNAP
$ZnCl_2$-pretreated cells	18.4 ± 2.3	79.2 ± 28.0
$ZnCl_2$-pretreated/Cu-NTA-loaded cells	55.0 ± 13.0[b]	115.0 ± 32.0[c]

[a] Data are means ± SD, $n = 6$.
[b] $p < 0.05$ vs $ZnCl_2$-pretreated cells/(−SNAP).
[c] $p < 0.05$ vs $ZnCl_2$-pretreated/Cu-NTA-loaded cells/(−SNAP).

and removing excess Cu-NTA, the cells are incubated again with 100 μM SNAP at 37° for 4 hr.

Assessment of *S*-nitrosylation of MT cysteines in cells cannot be performed without their prior isolation. Therefore, we used size-exclusion chromatography on Sephadex G-75 to obtain an MT-enriched fraction from HL-60 cell lysates.[43] MT content in the fractions was estimated by immunodot blotting with modifications of the method as described by Liu *et al.*[42] To determine the content of *S*-nitrosylated cysteines in MTs, aliquots (500 μl) of MT fractions are mixed with DAF-2 (5 μM) in 2.5 ml PBS, and the fluorescence of DAF-2-T is recorded (background). Another aliquot (500 μl) is exposed to UV irradiation (10 min) to release nitric oxide from *S*-nitroso-MT to yield DAF-2-T fluorescence. The content of *S*-nitroso-MT is normalized to the amount of MT protein in the MT-enriched fraction obtained after the chromatography of HL-60 cell lysates. We found low but significant amounts of *S*-nitroso-MTs in lysates from Zn-pretreated HL-60 cells even prior to exposure to SNAP (Table II). Interestingly, exposure of Zn-pretreated HL-60 cells to Cu-NTA increased the level of MT nitrosylation almost threefold, suggesting that Cu was likely involved in the production of *S*-nitroso-MTs utilizing endogenous sources of NO. Exposure of HL-60 cells to SNAP significantly increased the amounts of *S*-nitrosylated MTs in both Zn-pretreated cells (fourfold) and Zn-pretreated/Cu-NTA-loaded cells (twofold) (Table II). The relative amounts of *S*-nitrosylated MTs, however, remained relatively low even after incubation of cells with SNAP. Based on our slot-blot estimates of MT content, only a very small fraction (<0.3%) of MT was retained as *S*-nitroso-MT after a 4-hr exposure to SNAP. As has been presented in [25], Volume 353,[43a] we found that 26–28% of

[43] K. Kawai, S. X. Liu, V. A. Tyurin, Y. Y. Tyurina, G. G. Borisenko, J. F. Jiang, C. M. St. Croix, J. P. Fabisiak, B. R. Pitt, and V. E. Kagan, *Chem. Res. Toxicol.* **13,** 1275 (2000).

[43a] J. P. Fabisiak, G. G. Borisenko, S.-X. Liu, V. A. Tyurin, B. R. Pitt, and V. E. Kagan, *Methods Enzymol.* **353,** [25], in preparation.

MT-cysteines in HL-60 cells were lost after exposure to SNAP. Comparison of these results suggests that only ≈2–3% of cysteines that underwent oxidative modification upon exposure to SNAP could be recovered as *S*-nitrosothiols. This indicates that SNAP-induced *S*-nitrosylation of MTs in HL-60 cells occurs transiently and that denitrosylation of MTs likely results in oxidative loss of their SH groups (likely to sulfinic and sulfonic acids[44]). One can speculate that being a redox-active metal, Cu can facilitate both nitrosylation and denitrosylation of protein cysteines and *S*-nitrosocysteine in MTs, respectively.[45,46]

In summary, there are several advantages of the combined procedure of UV-induced decomposition of *S*-nitrosothiols followed by quantitation of the released NO by DAF-2-based fluorescence. In contrast to other fluorogenic scavengers, DAF-2 does not react with NO_2^-, NO_3^-, or other reactive oxygen species. Released NO reacts with DAF-2 to form a stable fluorescent product (DAF-2-T)[47] whose fluorescence intensity can be determined readily. UV-induced decomposition of *S*-nitrosothiols is applicable for measurements in whole plasma or plasma fractions. Using this method, we have found potentially important elevations in plasma *S*-nitrosothiol concentrations in women with the pregnancy disorder preeclampsia. In contrast to UV, the use of Cu or Hg for the decomposition of *S*-nitrosothiols is problematic for measurements in plasma due to potential interference by high concentrations of transition metal-binding proteins (albumin, ceruloplasmin).

Acknowledgments

Supported by NIH 1RO1HL64145-01A1, 5PO1HD30367-06, and 1RO1HL56829-01, as well as by the Magee-Womens Research Institute Fellowship (V.A.T.).

[44] M. D. Percival, M. Ouellet, C. Campagnolo, D. Claveau, and C. Li, *Biochemistry* **38,** 13574 (1999).

[45] M. P. Gordge, D. J. Meyer, J. Hothersall., G. H. Neild, N. N. Payne, and A. Noronha-Dutra, *Br. J. Pharmacol.* **114,** 1083 (1995).

[46] G. Stubauer, A. Giuffre, and P. Sarti, *J. Biol. Chem.* **274,** 28128 (1999).

[47] M. Broillet, O. Randin, and J. Chatton, *FEBS Lett.* **491,** 227 (2001).

[31] Peroxisomal Fatty Acid Oxidation and Cellular Redox

By INDERJIT SINGH

Introduction

Peroxisomes are known to play an important role in the cellular catabolism of fatty acids by the α-oxidation of substituted fatty acids (e.g., phytanic and cerebronic acids) and by the β-oxidation of unsubstituted fatty acids, including the catabolism of H_2O_2 produced by various oxidases. The importance of peroxisomes in the metabolism of fatty acids is underscored by the identification of genetic disorders that result from defects in α-oxidation and/or β-oxidation enzyme systems present in peroxisomes. Deficiency in the catabolism of H_2O_2 in peroxisomes because of an abnormality in targeting of catalase to peroxisomes results in changes in peroxisomal cellular redox, leading to inactivation of peroxisomal function, including oxidation of fatty acids.

Fatty Acid β-Oxidation

Mitochondria were considered to be the only site of β-oxidation of fatty acids[1] until the discovery of a similar system in peroxisomes.[2] This demonstration of a fatty acid β-oxidation system in peroxisomes followed the earlier discovery of such a system by Cooper and Beever[3] in glyoxysomes. The peroxisomal β-oxidation system is functionally similar to the mitochondrial β-oxidation system (Fig. 1) in that acyl-CoA esters are β-oxidized by sequential steps of dehydrogenation, hydration, dehydrogenation, and thiolytic cleavage to acetyl-CoA, and acyl-CoA is shortened by two carbon atoms in each cycle. However, peroxisomal and mitochondrial enzymes are different proteins.[4,5] The role of the peroxisomal β-oxidation system was further enhanced by the demonstration of preferential or possibly exculsive oxidation of saturated[6] and unsaturated[7] very long chain (VLC) fatty acids (>C_{22}) in peroxisomes. In addition, the importance of peroxisomal

[1] F. Lynen, "Physiology or Medicine Nobel Lecturers 1963–1970," p. 103. American Elsevier, New York, 1973.

[2] P. B. Lazarow and C. de Duve. *Proc. Natl. Acad. Sci. U.S.A.* **73,** 2043 (1976).

[3] T. G. Cooper and H. Beever, *J. Biol. Chem.* **244,** 3514 (1969).

[4] T. Hashimoto, *in* "Peroxisomes in Biology and Medicine" (H. D. Fahimi and H. Sies, eds.), p. 97. Springer-Verlag, Berlin, 1987.

[5] I. Singh, *Mol. Cell. Biochem.* **167,** 1 (1997).

[6] I. Singh, A. E. Moser, S. Goldfishner, and H. W. Moser, *Proc. Natl. Acad. Sci. U.S.A.* **81,** 4203 (1984).

[7] R. Sandhir, M. Khan, A. S. Chahal, and I. Singh, *J. Lipid Res.* **39,** 2161 (1998).

0076-6879/02 $35.00

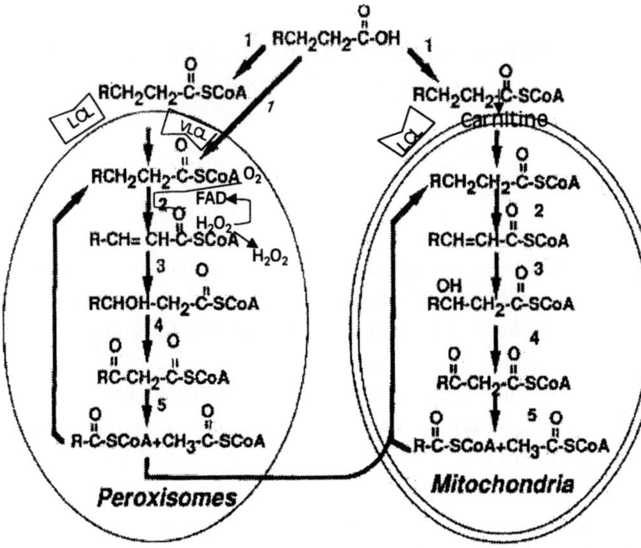

FIG. 1. Fatty acid β-oxidation pathway in peroxisomes and mitochondria. The various steps of peroxisomal β-oxidation are catalyzed by the following enzymes: (1) long chain acyl-CoA ligase (LCL) for long chain fatty acids (C_{14}-C_{20}) and very long chain acyl-CoA ligase (VLCL) for very long chain fatty acids ($>C_{22}$); (2) acyl-CoA oxidase; (3 and 4) multifunctional enzyme; and (5) 3-ketoacyl-CoA thiolase. Various steps of the mitochondrial β-oxidation of fatty acids are catalyzed by the following enzymes: (1) long chain acyl-CoA ligase; (2) acyl-CoA dehydrogenase; (3) enoyl-CoA hydratase; (4) 3-hydroxylacyl-CoA dehydrogenase; and (5) 3-ketoacyl-CoA thiolase.

β-oxidation is recognized in cleavage of the cholesterol side chain in the synthesis of bile acids, catabolism of dicarboxylic acids, arachidonic acid, and its metabolites, pristanic acid, glutaric acid, pipecolic acid, and xenobiotic compounds with a fatty acid acyl side chain.[5,8,9]

Peroxisomal disorders with abnormalities in these metabolic pathways result in excessive accumulation of these metabolites.[5,8–12]

Enzymatic Organization of Fatty Acid β-Oxidation System

Acyl-CoA ligases are associated with the peroxisomal membrane, and β-oxidation system enzymes are components of the peroxisomal matrix. Activation

[8] P. P. Van Veldhoven and G. P. Mannaerts, *in* "Current Views of Fatty Acid Oxidation and Ketogenesis: From Organelles to Point Mutation" (P. A. Quant and S. Eaton, eds.). Kluwer Academic/Plenum Publishers, New York, 1999.

[9] R. J. A. Wanders, E. G. Van Grunsven, and G. A. Jansen, *Biochem. Soc. Trans.* **28,** 141 (2000).

[10] P. B. Lazarow and H. W. Moser, *in* "The Metabolic and Molecular Basis of Inherited Disease" (C. R. Scriber, A. L. Beudet, W. S. Sly, and D. Valle, eds.), p. 2287. McGraw-Hill, New York, 1995.

[11] J. Vamecq, *in* "Peroxisomes in Biology and Medicine" (H. D. Fahimi and H. Sies, eds.). Springer-Verlag, Berlin, 1987.

[12] F. R. Brown, R. Voigt, A. K. Singh, and I. Singh, *Am. J. Dis. Child.* **147,** 617 (1993).

of fatty acids to acyl-CoA derivatives is a prerequisite for their degradation by β-oxidation (Fig. 1). So far, two acyl-CoA ligases are reported to be present in peroxisomes: long chain acyl-CoA ligase (palmitoyl-CoA ligase) for the activation of long chain fatty acids (C_{14}-C_{20}) and very long chain acyl-CoA ligase (lignoceroyl-CoA ligase) for the activation of VLC fatty acids ($>C_{22}$).[13–16] The palmitoyl-CoA ligase (LCL) is a component of the peroxisomal membrane with its active site on the cytoplasmic surface of the peroxisomal membrane.[15–17] Topology of the active site and intraperoxisomal localization of lignoceroyl-CoA ligase (VLCL) have been the subject of controversy.[16,17] Studies[18] have confirmed earlier findings[17] that the active site of lignoceroly-CoA ligase is on the luminal side of peroxisomal membrane and that, in fact, lignoceroyl-CoA ligase is a membrane-associated protein on the luminal side of the peroxisomal membrane.[18] The fatty acid transport mechanism and associated transporter proteins have not been well established. Unlike mitochondria, the fatty acid transport into peroxisomes does not require the carnitine system.[4,5] Although the abnormality (mutation/deletion) in the adrenoleukodystrophy (ALD) gene product (ALDP), a peroxisomal membrane component,[19,20] land deficient oxidation of VLC fatty acids as compared to normal oxidation of VLC acyl-CoA[21,22] suggest a role for ALDP in the translocation of VLC fatty acids across the peroxisomal membrane, similar transport of lignoceric acid was observed *in vitro* studies in peroxisomes isolated from fibroblasts from ALD patients and controls.[23] Moreover, consistent with luminal localization of lignoceroyl-CoA ligase,[16,18] a higher transport rate was observed for lignoceric acid as compared to lignoceroyl-CoA.[23]

The β-oxidation cycle in mitochondria consists of four different enzymes (acyl-CoA dehydrogenases, enoyl-CoA hydratase, 3-hydroxyacyl-CoA dehydrogenase, and 3-ketoacyl-CoA thiolase) (Fig. 1).[4,5] Similar reactions, in peroxisomes

[13] H. Singh and A. Poulos, *Arch. Biochem. Biophys.* **266,** 486 (1988).

[14] O. Lazo, M. Contreras, Y. Yoshida, A. K. Singh, W. Stanley, M. J. Weise, and I. Singh, *J. Lipid Res.* **31,** 583 (1990).

[15] G. P. Mannaerts, P. Van Veldhoven, A. Van Broekhoven, G. Vandebroek, and L. J. Debeer, *Biochem. J.* **204,** 17 (1982).

[16] O. Lazo, M. Contreras, and I. Singh, *Biochemistry* **29,** 3981 (1990).

[17] W. Lageweg, J. M. Tager, and R. J. A. Wanders, *Biochem. J.* **276,** 53 (1991).

[18] B. T. Smith, T. Sengupta, and I. Singh, *Exp. Cell Res.* **254,** 309 (2000).

[19] J. Mosser, A. M. Douar, C. O. Sarde, P. Kioschis, R. Feil, H. Moser, A. M. Poustka, J. L. Mandel, and P. Aubourg, *Nature* **361,** 726 (1993).

[20] M. Contreras, T. Sengupta, F. Sheikh, P. Aubourg, and I. Singh, *Arch. Biochem. Biophys.* **334,** 369 (1996).

[21] O. Lazo, M. Contreras, M. Hashmi, W. Stanley, C. Irazu, and I. Singh, *Proc. Natl. Acad. Sci. U.S.A.* **85,** 7647 (1988).

[22] R. J. A. Wanders, C. W. T. Van Roermund, M. J. A. Van Wijland, R. B. H. Schutgens, H. Van den Bosch, A. W. Schram, and J. M. Tager, *Biochem. Biophys. Res. Commun.* **153,** 618 (1988).

[23] I. Singh, O. Lazo, G. Dhaunsi, and M. Contreras, *J. Biol. Chem.* **267,** 13306 (1992).

are carried out by three different enzymes (acyl-CoA oxidases, multifunctional enzyme, and 3-ketoacyl-CoA thiolases), components of the peroxisomal matrix.[4,5] The first reaction, conversion of acyl-CoA to α, β-enoyl-CoA, in the peroxisomal β-oxidation system is catalyzed by FAD-containing acyl-CoA oxidases and is associated with the reduction of FAD bound to the enzyme followed by the direct transfer of electrons to molecular oxygen to yield H_2O_2.[4,5] Rat peroxisomes contain three different acyl-CoA oxidases as compared to two in humans for different substrates.[8,9] The next two reactions of the β-oxidation system in peroxisomes are catalyzed by multifunctional enzyme.[5,8,9] In addition to these two functions, the multifunctional enzyme also has Δ^{2-3}-enoyl-CoA isomerase activity required for the oxidation of unsaturated fatty acids. The last reaction of the peroxisomal β-oxidation system, the conversion of 3-ketoacyl-CoA (straight chain fatty acids) to acetyl-CoA plus acyl-CoA shortened by two carbon atoms is catalyzed by thiolases (A and B) and by SCPx-thiolase.[8,9] Several types of fatty acids (e.g., long chain and very long chain saturated, unsaturated, arachidonic acid metabolites, dicarboxylic acids, and branched chain fatty acids) and side chains of xenobiotic carboxylic acids are degraded to shorter chain fatty acids in peroxisomes and are further degraded by mitochondrial β-oxidation.

Assays for Peroxisomal Fatty Acid β-Oxidation

Isolation of Peroxisomes by Gradient Centrifugation

Because peroxisomes are relatively fragile as compared to other subcellular organelles, their isolation from tissue requires special attention.[7,14,23,24] Different organs/tissues contain different sizes and numbers of peroxisomes. In liver and kidney, peroxisomes are larger in size (0.3–1.0 μm in diameter) and account for 1–2% of the total cellular protein, whereas in other tissues (e.g., brain, muscle, and cultured skin fibroblasts), peroxisomes are fewer in number and smaller in size (0.1–0.25 μm in diameter), commonly referred to as microperoxisomes, and they represent a relatively smaller percentage of total protein. For isolation of peroxisomes, basically two different concentrations of Nycodenz gradients have been used: for peroxisomes from liver or kidney vs those for microperoxisomes from brain or cultured skin fibroblasts. Tissue or cells are homogenized in 10 volumes of homogenizing buffer containing 0.25 *M* sucrose, 1 m*M* EDTA, antipain (1 mg/ml), aprotinin (2 mg/ml), leupeptin (2 mg/ml), pepstatin A (0.7 mg/ml), 0.1 m*M* phosphoramidon, 0.2 m*M* phenylmethylsulfonyl fluoride, 0.1% ethanol, and 3 m*M* imidazole buffer, pH 7.4, essentially as described previously for cultured skin fibroblasts[24] and liver tissue.[23] The homogenate is first fractionated by differential centrifugation to prepare the λ fraction (light mitochondrial fraction enriched with

[24] O. Lazo, A. K. Singh, and I. Singh, *J. Neurochem.* **56,** 1343 (1991).

TABLE I

RELATIVE SPECIFIC ACTIVITIES OF ENZYME MARKERS AND PERCENTAGE CONTAMINATION IN SUBCELLULAR FRACTIONS ISOLATED FROM CULTURED SKIN FIBROBLASTS[a]

	Relative specific activity (mU/mg protein)			% Contamination		
	Peroxisomes	Mitochondria	Endoplasmic reticulum	Peroxisomes	Mitochondria	Endoplasmic reticulum
Catalase	19.6 ± 4.4	0.6 ± 0.3	0.2 ± 0.1		1.5 ± 0.6	0.3 ± 0.1
Cytochrome *c* oxidase	0.3 ± 0.1	4.8 ± 0.9	0.6 ± 0.2	5.6 ± 2.4		12.4 ± 4.2
NADPH cytochrome *c* reductase	0.11 ± 0.06	0.9 ± 0.4	5.9 ± 0.9	2.2 ± 1.2	18.6 ± 8.2	
N-Acetyl-β-glucosaminidase	0.4 ± 0.2	1.6 ± 0.5	2.3 ± 0.5	0.9 ± 0.4	3.3 ± 1.0	4.5 ± 1.1

[a] Values are expressed as mean ± standard deviation. Percentage contamination was calculated as described previously.

peroxisomes and lysosomes). The λ fraction is further fractionated by isopycnic equilibrium centrifugation in a continuous Nycodenz gradient (0–50% for λ fraction from liver or Nycodenz gradient of 0–30% for postnuclear fraction or λ fraction from cultured skin fibroblasts) overlying a cushion of 55% of Nycodenz for a 0–50% gradient and a cushion of 35% of Nycodenz for a 0–30% gradient. The tubes are sealed and then centrifuged at 33,700*g* for 60 min at 8° in a J2-21 centrifuge using a JV20 vertical rotor at low acceleration and deacceleration. The gradient fractions are collected from the bottom of the gradient, and the density of the gradient fractions is analyzed with a refractometer. The gradient fractions are analyzed for marker enzymes for various subcellular organelle; e.g., cytochrome *c* oxidase for mitochondria,[25] NADPH-cytochrome *c* reductase for endoplasmic reticulum,[26] *N*-acetyl-β-glucoraminidase for lysosomes,[27] phosphoglucomutase for cytoplasm,[28] and catalase for peroxisomes.[29] The specific activities of these marker enzymes and the relative contamination by other organelles are reported in Table I.[14,31]

[25] S. J. Cooperstein and P. Lazarow, *J. Biol. Chem.* **189,** 665 (1971).

[26] H. A. Beaufay, A. Amar-Costesec, E. Feytmans, M. Thines-Sempoux, M. Wibo, M. Robbi, and J. Berthet, *J. Cell Biol.* **61,** 188 (1974).

[27] O. Z. Sellinger, H. Beaufay, P. Jacques, A. Doyer, and C. de Duve, *Biochem. J.* **74,** 450 (1960).

[28] M. Bronfman, N. C. Inestrosa, F. O. Nervi, and F. Leighton, *Biochem. J.* **224,** 709 (1984).

[29] P. Baudhuin, Y. Beaufay, L. O. Rahman, Z. Sellinger, R. Wattiaux, P. Jacques, and C. de Duve, *Biochem. J.* **92,** 179 (1964).

[30] Y. Fujiki, A. L. Hubbard, S. Folwer, and P. B. Lazarow, *J. Cell Biol.* **93,** 97 (1982).

[31] G. H. Luers, R. Hartig, H. Mohr, M. Hausmann, H. D. Fahimi, C. Cremer, and A. Volk, *Electrophoresis* **19,** 1205 (1998).

Isolation of Peroxisomes by Immunomagnetic Sorting

Immunoisolation of peroxisomes using magnetic beads coated covalently with antibodies against the PMP-70 protein from rat liver and human blastoma cell line (Hep G2) has been performed.[31] In a typical isolation procedure, 1 ml of the light mitochondrial fraction corresponding to 1 g of rat liver or 1 ml of postnuclear supplement prepared from Hep G2 cells (1–2 × 10^7 cells) was incubated with anti-PMP70-labeled beads (1 × 10^7/ml) on ice for 30–60 min and was injected into the separation chamber (600 × 50 × 0.5 mm^3) of the CIMS apparatus with a peristaltic pump at a rate of 4.5 ml/hr. A thin film of separating medium (125 m*M* sucrose, 0.25 m*M* MOPS, 0.05 m*M* EDTA, 54 m*M* NaCl, 1.7 m*M* KCl, 5 m*M* sodium phosphate, and 0.5% BSA, pH 7.4) was directed upward, matching laminar flow conditions, and its velocity was maintained at 2.1 mm/sec by a second peristaltic pump. The sample, transported by the separation medium in two fine streams, passes on in a halogenous magnetic field perpendicularly oriented to the direction of the flow. The separation medium is split into multiple fractions containing the magnetic beads, which are easily identifiable due to their bright red color.

Fatty Acid β-Oxidation Assay for Isolated Peroxisomes

Enzyme activity for β-oxidation of [1-^{14}C]-labeled fatty acids is measured by incubation of α-cyclodextin solubilized fatty acid with peroxisomal preparation in media (0.5 ml) containing 20 m*M* MOPS–HCl buffer, pH 7.8, 30 m*M* KCl, 1 m*M* $MgCl_2$, 5 m*M* ATP, 0.171 m*M* FAD, and 0.081 m*M* CoASH. The reaction is started by the addition of substrate or peroxisomes and is stopped with 1.25 ml of 1 *N* potassium hydroxide in methanol, and the denatured protein is removed by centrifugation. The supernatant is incubated at 60° for 1 hr, neutralized with acid (0.25 ml of 6 *N* HCl), and partitioned by the addition of 2.5 ml of chloroform. The amount of activity in the upper phase is an index of the amount of [1-^{14}C]-labeled fatty acid oxidized to acetate. For solubilization of fatty acid with α-cyclodextrin, the fatty acid (20 × 10^6 dpm) is first dried in a tube under nitrogen and resuspended in 3.5 ml (20 mg/ml) of α-cyclodextrin in 20 m*M* MOPS–HCl buffer, pH 7.8, by sonication for 1 hr. The specific activities for β-oxidation of various substrates in isolated and in different subcellular organelles are reported in Table II.

Fatty Acid β-Oxidation in Suspended Cells and Cells in Monolayers

The β-oxidation of specific substrates can be studied by the addition of α-cyclodextrin solubilized [1-^{14}C]-labeled fatty acid to media of cultured monolayers or to cells suspended in PBS buffer. The reaction is terminated, and reaction products are processed as described for studies with isolated peroxisomes.

TABLE II
RATE OF β-OXIDATION OF FATTY ACIDS IN ISOLATED SUBCELLULAR ORGANELLES FROM CULTURED SKIN FIBROBLASTS[a]

	Rate of β-oxidation (nmol/hr/mg protein)			
	Homogenate	Peroxisomes	Mitochondria	Endoplasmic reticulum
Palmitic acid	1.97 ± 0.16	10.53 ± 1.63	12.6 ± 1.39	0.92 ± 0.33
Lignoceric acid	0.08 ± 0.02	0.19 ± 0.05	0.02 ± 0.01	0.01 ± 0.00
Nervonic acid	0.23 ± 0.07	0.69 ± 0.14	0.09 ± 0.02	0.03 ± 0.02

[a] Values are expressed as mean ± standard deviation.

Synthesis of [1-^{14}C]Lignoceric Acid

[1-^{14}C]Lignoceric acid is synthesized from tricosanyl bromide as described earlier[32] with some modifications.[7] 1-Tricosanol is converted to 1-tricosanyl bromide in the presence of Pbr_3 as reported.[33] The tricosanyl bromide (0.024 mmol) and $K^{14}CN$ (0.018 mmol, 1 mCi) are dissolved in 1.5 ml absolute ethanol, followed by the addition of 0.2 ml water, and the contents are heated in a Teflon-coated screw-capped glass test tube at 85° for 48 hr. The reaction mixture is worked up with ether, and the product is dried under nitrogen. The radiolabeled nitrite is hydrolyzed with 0.2 g KOH in 2 ml 95% ethanol at 100° for 72 hr. The contents are acidified with 10 m*M* HCl, and the radiolabeled product is extracted with ether and labeled lignoceric acid is purified on TLC plates. The identity of the lignoceric acid is established by TLC and Co-TLC with nonradioactive standard, infrared spectroscopy, high-performance TLC, capillary GC, and mass spectrometry.

Synthesis of [1-^{14}C]Nervonic Acid

[1-^{14}C]Nervonic acid is synthesized from erucic acid by two carbon chain elongation in two steps. The first step involves the synthesis of unlabeled *cis*-14-tricosenyl methanesulfonate from methyl erucate as described by Richter and Mangold.[34] In the second step, *cis*-14-tricosenyl methanesulfonate (0.070 mmol) dissolved in 2.5 ml of ethanol and 2 mCi of $K^{14}CN$ (0.037 mmol) dissolved in 40 μl of 50 m*M* KOH are mixed and refluxed in a Teflon-coated screw-capped test tube for 30 hr. The reaction mixture is worked up with ether, and the product is dried under nitrogen. The radiolabeled nitrile is dissolved in 2 ml of 95% ethanol

[32] P. E. Morell, E. Costantino-Ceccarini, and N. S. Radin, *Arch. Biochem. Biophys.* **141,** 738 (1970).
[33] M. Hoshi and Y. Kishimoto, *J. Biol. Chem.* **248,** 4123 (1973).
[34] I. Richter and H. K. Mangold, *Chem. Phys. Lipids* **11,** 210 (1973).

followed by the addition of 0.2 g KOH, and the reaction mixture is refluxed at 100° for 60 hr. The contents are acidified with 10 mM HCl, the radiolabeled product is extracted with ether, and labeled nervonic acid is purified on TLC. The structure of the product is established based on chromatographic and spectral data, including mass spectrometry.

Fatty Acid α-Oxidation

Fatty acids substituted at the α carbon (e.g., cerebronic acid; α-hydroxylignoceric acid) and fatty acids substituted at the β carbon (e.g., phytanic acid; 3,7,11,15-tetramethyl hexacosanoic acid) are constituents of mammalian tissues. Cerebronic acid is a major constituent of brain myelin lipids (e.g., cerebrosides and sulfatides) and is synthesized by α-hydroxylation of lignoceric acid. Phytanic acid originates mainly from dietary sources. Phytanic acid, cerebronic acid, and similar xenobiotic compounds cannot be oxidized by the β-oxidation enzyme system, but only by α-oxidation (Fig. 2).

Although excessive accumulation of phytanic acid in disorders of peroxisomal biogenesis indicates a role for peroxisomes in phytanic acid degradation, the

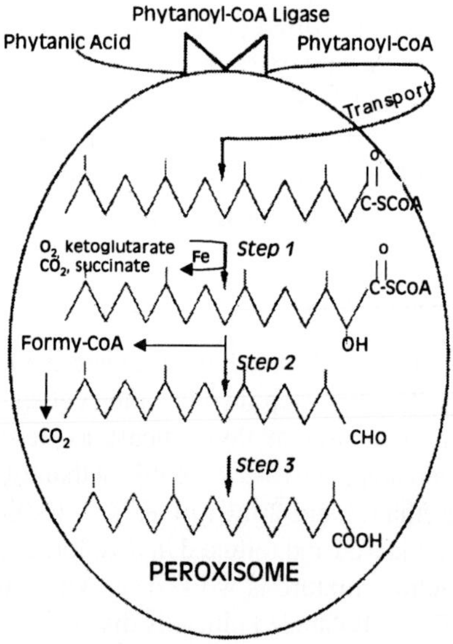

FIG. 2. Catabolic pathway for phytanic acid in peroxisomes. Step 1 is conversion of phytanoyl-CoA to α-hydroxyphytanoyl-CoA by phytanoyl-CoA hydroxylase. Step 2 is oxidative carboxylation of α-hydroxyphytanoyl-CoA to pristanal and formic acid. Step 3 is oxidation of pristanal to pristanic acid.

identification of the subcellular organelle and related enzyme system responsible has remained a subject of controversy.[35–39]

Studies from various laboratories[38–44] have demonstrated that phytanic acid is α-oxidized to pristanic acid in peroxisomes and that this enzyme system activity is deficient in peroxisomal disorders (e.g., disorders of peroxisomal biogenesis, rhizomelic chondrodysplasia punctata, and refsum disease).[40] Phytanic acid is converted to phytanoyl-CoA by acyl-CoA ligase with an active site on the cytoplasmic surface of peroxisomes.[40] The phytanoyl-CoA is transported into peroxisomes and is first converted to 2-hydroxyphytanoyl-CoA by phytanoyl-CoA hydroxylase, which in turn undergoes cleavage to produce formyl-CoA and pristanal followed by their oxidation to CO_2 and pristanic acid, respectively.[8,9,40–44] The pristanic acid is further catabolized by the peroxisomal β-oxidation enzyme system. The intraperoxisomal organization of an enzyme system for the α-oxidation of phytanic acid is not very well established. Phytanoyl-CoA ligase is a component of the peroxisomal membrane, whereas phytanoyl-CoA hydroxylase is a component of the peroxisomal matrix (Fig. 2).

The enzyme system for oxidative decarboxylation of α-hydroxyphytanic acid has not been characterized so far. Cerebronic acid synthesized by α-hydroxylation of lignoceric acid is a major fatty acid of myelin cerebrosides and sulfatides. Studies[45] have reported that the enzyme system for oxidative decarboxylation of cerebronic acid to tricosanoic acid and CO_2 is localized in the peroxisomal-limiting membrane but not in the matrix. The highest specific activity for α-oxidation was observed in peroxisomes as compared to other organelles (Table III) and this activity was deficient in Zellweger cells, which lack peroxisomes.[45] It is not known so far whether the same or a different enzyme system is responsible for oxidative decarboxylation of phytanic and cerebronic acids.

[35] P. A. Watkins and S. J. Mihalik, *Biochem. Biophys. Res. Commun.* **167,** 580 (1990).

[36] R. J. Wanders, C. W. T. Van Roermund, C. Jakobs, and H. J. ten Brink, *J. Inherit. Metab. Dis.* **14,** 349 (1991).

[37] S. Haung, P. P. Van Veldhoven, F. Vanhoutte, G. Parmentier, H. F. Eyssen, and G. P. Mannaerts, *Arch. Biochem. Biophys.* **296,** 214 (1992).

[38] I. Singh, O. Lazo, K. Pahan, and A. K. Singh, *Biochem. Biophys. Acta* **1180,** 221 (1992).

[39] I. Singh, K. Pahan, G. Dhaunsi, O. Lazo, and P. Ozand, *J. Biol. Chem.* **268,** 9972 (1993).

[40] K. Pahan, M. Khan, and I. Singh, *J. Lipid Res.* **37,** 1137 (1996).

[41] S. J. Mihalik, A. M. Rainville, and P. A. Watkins, *Eur. J. Biochem.* **232,** 545 (1995).

[42] K. Croes, M. Casteels, E. de Hoffman, G. P. Mannaerts, and P. P. Van Veldhoven, *Eur. J. Biochem.* **240,** 674 (1996).

[43] H. J. ten Brink, D. M. S. Schor, R. M. Kok, B. T. Poll-The, R. J. A. Wanders, and C. Jacobs, *J. Lipid Res.* **33,** 1449 (1992).

[44] A. Poulos, P. Sharp, H. Singh, D. W. Johnson, W. F. Carey, and C. Easton, *Biochem. J.* **292,** 457 (1993).

[45] R. Sandhir, M. Khan, and I. Singh, *Lipids* **35,** 1127 (2000).

TABLE III

RATE OF α-OXIDATION IN SUBCELLULAR ORGANELLES ISOLATED FROM CULTURED SKIN FIBROBLASTS FOR PHYTANIC ACID AND FROM RAT LIVER FOR CEREBRONIC ACID[a]

	Rate of α-oxidation			
	Homogenate	Peroxisomes	Mitochondria	Endoplasmic reticulum
Phytanic acid (pmol/hr/mg protein)	2.29 ± 0.88	37.82 ± 2.42	1.44 ± 0.13	0.29 ± 0.15
Cerebronic acid (nmol/hr/mg protein)	0.259 ± 0.02	2.42 ± 0.25	0.152 ± 0.05	0.142 ± 0.01

[a] Values are expressed as mean ± standard deviation.

Assay for α-Oxidation in Isolated Peroxisomes

The activity for α-oxidation of fatty acids is measured as release of tritium from [2,3-^{3}H]phytanic acid and release of $^{14}CO_2$ from [1-^{14}C]cerebronic acid or [1-^{14}C] phytanic acid.

α-Oxidation of [2,3-^{3}H] Phytanic Acid

The assay mixture (0.25 ml) for α-oxidation of phytanic acid contains 50 mM Tris–HCl, pH (7.4), 5 mM $MgCl_2$, 0.2 mM CoASH, 10 mM ATP, 0.2 mM DTT, 1 mM ketoglutarate, 1 mM Fe^{2+}, 1 mM ascorbate, 20 μm cold phytanic acid, and 250,000 dpm of [2,3-^{3}H]phytanic acid solubilized with β-cyclodextrin. After 1 hr of incubation at 37°, the reaction is stopped by the addition of 0.625 ml of 1 M methanolic KOH. The reaction mixture is warmed at 60° for 1 hr and then acidified with 0.125 ml of 6 N HCl. The aqueous phase is separated by the addition of 1.25 ml chloroform. The amount of tritium exchange between fatty acid and water in the aqueous phase is taken as an indication of α-oxidation.

α-Oxidation of [1-^{14}C]Cerebronic Acid or [1-^{14}C] Phytanic Acid

For α-oxidation of cerebronic acid, the reaction mixture (0.25 ml) contains 150,000 dpm of [1-^{14}C]cerebronic acid, 30 mM KCl, 5 mM $MgCl_2$, 8 mM ATP, 0.25 mM NAD, 0.08 mM CoASH, and 20 mM MOPS–HCl buffer, pH 7.8. The reaction is started by the addition of the peroxisomal fraction. Following different periods of incubation at 37°, the reaction is stopped by injecting 50 μl of 5 N H_2SO_4, and $^{14}CO_2$ is collected in KOH-wetted filter paper by shaking overnight at 37°. The radioactivity in the KOH-wetted filter is measured as an index of α-oxidation of cerebronic acid.

Assay for α-Oxidation in Suspended Cells or Cells in Cultured Monolayers

The α-oxidation of peroxisomal-specific substrates can be studied by the addition of cyclodextrin-solubilized substrate to media of cultured monolayers of cells or cells suspended in PBS. The reaction is terminated, and reaction products are processed as described for assays with isolated peroxisomes.

Synthesis of [1-^{14}C]Cerebronic Acid

[1-^{14}C]Cerebronic acid is not available commercially. [1-^{14}C]Lignoceric acid with a specific activity of 54.5 mCi/mmol is synthesized and characterized as described in an earlier publication.[7,24] [1-^{14}C]Cerebronic acid is synthesized from [1-^{14}C]lignoceric acid.[45] [1-^{14}C]Lignoceric acid (1 mCi) is allowed to melt in a screw-capped (16 × 100-mm) glass tube equipped with a stirring bar at 80°. PBr_3 (5 μl) is added slowly to the bottom of this tube. Bromine (8 μl) is added dropwise in the bottom of the tube with stirring, and the tube is closed immediately after addition of the final drop of bromine. The tube is then left at room temperature for 20 hr, after which water (1 ml) is added to the reaction mixture, and the tube is left open overnight in a fume hood. The reaction mixture is worked up with ether, and the product is dried under nitrogen. The radioactive product, α-bromolignoceric

TABLE IV

RATES OF α-OXIDATION AND β-OXIDATION OF FATTY ACIDS IN PEROXISOMES FROM CULTURED SKIN FIBROBLASTS FROM PATIENTS WITH PEROXISOMAL DISORDERS[a]

	α-Oxidation		β-Oxidation	
	Phytanic acid[b]	Cerebronic acid[c]	Lignoceric acid	Nervonic acid
	(pmol/hr/mg protein)	(nmol/hr/mg protein)		
Control	42 ± 4	0.363 ± 0.050	0.18 ± 0.07	0.69 ± 0.14
X-ALD	39 ± 4	0.321 ± 0.035	0.02 ± 0.01	0.22 ± 0.04
RD	1.5 ± 0.3	0.296 ± 0.030	0.17 ± 0.05	nd
ZS	nd	0.184 ± 0.021	nd	nd
RCDP	0.00	0.334 ± 0.01	nd	nd

[a] X-ALD, X-adrenoleukodystrophy; RD, Refsum disease; ZS, Zellweger syndrome; RCDP, rhizomelic chondrodysplasia; nd, not determined.

[b] The results are expressed as release of [CO_2] from [1-^{14}C]phytanic acid [I. Singh, K. Pahan, G. Dhaunsi, O. Lazo, and P. Ozand, *J. Biol. Chem.* **268,** 9972 (1993)].

[c] Data for cerebronic acid α-oxidation is expressed as a ratio of oxidation of cerebronic acid/palmitic acids in human cultured skin fibroblasts (instead of isolated peroxisomes) suspended in Hanks' balanced salt solution [R. Sandhir, M. Khan, and I. Singh, *Lipids* **35,** 1127 (2000)].

acid, is obtained in quantitative yield as examined by thin-layer chromatography (TLC) radioscanning. α-Bromolignoceric acid is converted to the corresponding α-hydroxy acid with some modification on a microscale. The crude product is treated with 1.2 ml 95% ethanol, having 2 mg KOH/ml, and the content is refluxed for 24 hr. Alcohol is removed under nitrogen, 2 ml water is added, and the content is acidified with 2 *N* HCl. The product is extracted with ether. The labeled compounds are identified by TLC and co-TLC with nonradioactive standard compounds. The compounds are characterized based on their chromatographic and spectral properties.

Peroxisomal Redox and Peroxisomal Cellular Functions

The respiratory chain in peroxisomes removes electrons from various metabolites, including fatty acids for the reduction of oxygen to hydrogen peroxide (H_2O_2) during the action of various oxidases.[46] One cycle of fatty acid β-oxidation produces one molecule of H_2O_2. Peroxisomes are estimated to consume between 10 and 30% of total cellular oxygen in liver where each molecule of oxygen is converted to H_2O_2.[46] The reactive oxygen species (ROS; $O_2^- H_2O_2$) produced in the peroxisomes are degraded by antioxidant enzymes (catalase, glutathione peroxidase, Cu-Zn superoxide dismutase, and Mn superoxide dismutase) present in peroxisomes.[5,47] Deficient detoxification of ROS can result in oxidative injury. Studies have reported human cultured skin fibroblasts with defects in targeting of catalase to peroxisomes, resulting in catalase-negative peroxisomes.[48] Excessive *in situ* levels of H_2O_2 produced in catalase-negative peroxisomes were found to inactivate peroxisomal functions because targeting of catalase expressed from exogenous cDNA in these cell lines led to the protection of peroxisomal function. Moreover, inhibition of catalase by aminotriazole in control cultured cells was also found to inhibit peroxisomal functions.[48] These observations indicate that peroxisomal redox is important for cellular functions.

[46] C. De Duve, *Sci. Am.* **248,** 74 (1983).
[47] I. Singh, *Ann. N.Y. Acad. Sci.* **804,** 612 (1996).
[48] F. Seikh, K. Pahan, E. Barbosa, and I. Singh, *Proc. Natl. Acad. Sci. U.S.A.* **95,** 2961 (1998).

[32] Ultrastructural Localization and Relative Quantification of 4-Hydroxynonenal-Modified Proteins in Tissues and Cell Compartments

By TERRY D. OBERLEY

Introduction

Reactive oxygen species (ROS) are produced as byproducts of aerobic metabolism.[1] At high levels, ROS are toxic, but studies have shown that, at lower concentrations, ROS regulate physiologic processes, including activities of signal transduction pathways and transcription factors.[2]

Measurement of ROS has been difficult, as these chemical species are so reactive that they persist in tissues for only short periods of time. One approach to this problem is to rely on biochemical assays of oxidative damage products. One major problem with such assays is that they rely on analysis of tissue homogenates, thereby precluding measurement of levels of oxidative damage products in specific cell types from complex tissues composed of many different types of cells. To circumvent this problem, techniques have been developed to localize and quantify oxidative damage products using specific antibodies and immunomorphologic procedures. These techniques have then been applied to the study of a number of pathologic processes, including cancer,[3,4] neurodegenerative diseases,[5,6] and aging.[7]

Interactions of ROS with polyunsaturated fatty acids of membrane lipids result in the production of a variety of aldehydes and alkenals, including malonaldehyde and 4-hydroxy-2-nonenal (4HNE).[8] Lipid peroxidation products can subsequently react with other cellular constituents, including proteins.[9] Antibodies to 4HNE bound to specific amino acid residues have been generated and can be used to localize 4HNE-modified proteins in both physiologic and pathologic processes

[1] B. Chance, H. Sies, and A. Boveris, *Physiol. Rev.* **59,** 527 (1979).
[2] T. Finkel, *Curr. Opin. Cell. Biol.* **10,** 248 (1998).
[3] T. D. Oberley, S. Toyokuni, and L. I. Szweda, *Free Radic. Biol. Med.* **27,** 695 (1999).
[4] T. D. Oberley, W. Zhong, L. I. Szweda, and L. W. Oberley, *Prostate* **44,** 144 (2000).
[5] A. Yoritaka, N. Hattori, K. Uchida, M. Tanaka, E. R. Stadtman, and Y. Mizuno, *Proc. Natl. Acad. Sci U.S.A.* **93,** 2969 (1996).
[6] K. S. Montine, S. J. Olson, V. Amarnath, W. Whetsell, Jr., D. J. Graham, and T. J. Montine, *Am. J. Pathol.* **150,** 437 (1997).
[7] T. A. Zainal, T. D. Oberley, D. B. Allison, L. I. Szweda, and R. Weindruch, *FASEB J.* **14,** 1825 (2000).
[8] A. Benedetti, M. Comporti, and H. Esterbauer, *Biochim. Biophys. Acta* **620,** 281 (1980).
[9] H. Esterbauer, R. J. Schaur, and H. Zollner, *Free Radic. Biol. Med.* **11,** 81 (1991).

0076-6879/02 $35.00

in vitro and *in vivo*. This chapter presents methods to localize 4HNE-modified proteins *in vivo*. Methods for relative quantification of 4HNE-modified proteins in tissues and subcellular compartments are also provided. These latter analytical methods allow one to compare levels of 4HNE-modified proteins between an experimental and a control group, but do not allow determination of absolute concentrations of 4HNE-modified proteins. Two previous studies have documented the validity of the immunomorphologic approaches presented herein by documenting good agreement between biochemical and immunomorphologic analyses in an experimental model of oxidative stress (iron nitrilotriacetate-induced injury of rat kidney[10]) and analysis of oxidative damage in skeletal muscle of aging rhesus monkey.[7] Biochemical assays usually measure free aldehydes plus alkenals, whereas immunomorphologic assays specifically measure 4HNE bound to protein. The rate and efficiency of 4HNE binding to proteins are not known and may vary between proteins and cell types. Further, degradation rates of oxidatively damaged proteins may vary between cell types. Finally, the cell turnover rate is unique for each cell type. Thus, while previous studies in our laboratory showed good correlation between 4HNE levels measured biochemically and levels of 4HNE-modified proteins measured with immunomorphologic techniques, the levels are not strictly comparable for the reasons mentioned earlier.

Procedures

Antibody Specificity

The use of antibodies with proved specificity is of utmost importance. Both rabbit polyclonal and mouse monoclonal antibodies have been used in previous studies. The rabbit polyclonal antibody used was raised to 4HNE-modified keyhole limpet hemocyanin. The specificity of this antibody has been characterized by immunochemical techniques, and it was concluded that the epitope recognized by the antibody appears to be the hemiacetal form of the 4HNE-derived portion of protein–4HNE adducts.[11,12] The monoclonal antibody used was raised by immunizing mice with a 4HNE–keyhole limpet hemocyanin conjugate and was also characterized thoroughly with immunochemical techniques. These latter immunochemical studies suggest that the monoclonal antibody is directed against the Michael addition-type 4HNE–histidine adduct.[13] Procedures described herein

[10] T. A. Zainal, R. Weindruch, L. I. Szweda, and T. D. Oberley, *Free Radic. Biol. Med.* **26,** 1181 (1999).

[11] K. Uchida, K. Itakura, S. Kawakishi, H. Hiai, S. Toyokuni, and E. R. Stadtman, *Arch. Biochem. Biophys.* **324,** 241 (1995).

[12] K. Uchida, L. I. Szweda, H.-Z. Chae, and E. R. Stadtman, *Proc. Natl. Acad. Sci. U.S.A.* **90,** 8742 (1993).

[13] S. Toyokuni, N. Miyake, H. Hiai, M. Hagiwara, S. Kawakishi, T. Osawa, and K. Uchida, *FEBS Lett.* **359,** 189 (1995).

are for the polyclonal antibody, but we have obtained similar results with the monoclonal antibody.

Light Microscopy Techniques

The use of either of these antibodies with standard light microscopy immunomorphologic techniques allows one to determine which cell type(s) within a tissue contains significant levels of 4HNE-modified proteins. Immunomorphologic procedures can involve either immunogold or immunoperoxidase techniques. Techniques for light microscopy are relatively standard and procedures utilized can be found elsewhere.[3,4,7] Such techniques allow identification of which cell types within a tissue are positive for 4HNE-modified proteins, but do not allow identification of subellular localization, and quantification of levels of 4HNE-modified proteins using light microscopy techniques is less sensitive and precise than immunogold ultrastructural techniques.

Fixation

Fixation is always a major issue when analyzing any immunomorpholgic problem. For optimal studies, tissues should be perfused with the fixative of choice at physiologic pressures *in situ,* the tissues then removed, and the tissues postfixed in the same fixative for 1 hr. Practically, many studies involve tissues already fixed, so perfusion is not possible. The choice of fixative is very important because some fixatives destroy antigenicity whereas others result in a high false positive background. For 4HNE-modified protein staining at the ultrastructural level, tissues are fixed for 1 hr in Carson–Millonig's fixative (4% formaldehyde in 0.16 *M* monobasic sodium phosphate buffer, pH 7.2). For relative quantification of 4HNE levels in an experimental versus a control group, it is crucial that tissues be treated in a similar manner: specifically, fixation must by performed in an identical fashion. Further, immunostaining must be performed with the same reagents with experimental and control slides being stained at the same time.

Immunogold Electron Microscopy

Tissues are cut into 1-mm^3 blocks and fixed in Carson–Millonig's fixative for 1 hr. Samples are embedded in LR White resin (Electron Microscopy Sciences, Fort Washington PA), as this resin allows preservation of antigenicity and therefore allows postembedding staining. After rinsing for 30 min in 0.1 *M* phosphate buffer, pH 7.4, samples are dehydrated by 30-min changes each of 70, 80, and 90% ethanol and then LR White resin/90% ethanol (2/1 : v/v). After immersion and infiltration in undiluted LR White resin overnight, the samples are washed with fresh undiluted LR White resin for 1 hr. Resin polymerization is induced thermally in sealed gelatin capsules at 45° for 48 hr in the absence of an accelerator. Ultrathin sections

(70–80 nm) are cut and transferred to nickel grids (G300-NI, Electron Microscopy Sciences) for postembedding immunogold procedures. These sections are rinsed with Tris-buffered saline (TBS: 0.05 *M* Tris, 0.9% NaCl, pH 7.6) for 10 min and incubated with 2% bovine serum albumin and 0.2% Tween 20 in TBS for 30 min to block nonspecific antibody-binding sites. Sections are then incubated with primary antibody (rabbit anti-4HNE, 1 : 80) overnight at 4°. Each new lot of primary and/or secondary antibody requires a study of optimal antibody titers to determine the amount of primary and/or secondary antibody needed for maximal immunogold labeling. After washing four times with TBS wash buffer (1 : 10 dilution of TBS blocking buffer) for 5 min each, the sections are washed in one change of alkaline TBS (pH 8.2) for 10 min. The grids are incubated with diluted (1:75) gold-conjugated anti-rabbit IgG (EM.GAR15, Goldmark Biologicals, Phillipsburg, NJ) for 90 min at room temperature. The sections are then washed in two changes of TBS wash buffer for 10 min each followed by two changes of distilled water for 5 min each. Following counterstaining with 4% aqueous uranyl acetate for 10 min, sections are examined with a Hitachi H-600 transmission electron microscope operated at 75 kV. For quantification using counting of gold beads, a magnification of 25,000× is used. To make a measurement, a region of interest (0.1 μm^2) is outlined using the picture field. Gold beads within the region are then counted. Using low magnifications so that beads are not discernible, 10 regions are randomly selected for each sample. Mean bead counts for each sample are then obtained by taking the mean value of the 10 measurements. Data are thus expressed as mean gold beads/μm^2 area.

Improved Methods for Quantitative Data Analysis

The method just outlined for data analysis is accurate but tedious and provides data for the entire cell, not for the individual subcellular compartments. Statistical analysis is also very complex. Image analysis programs are now available that allow calculation of area of each cellular compartment (i.e., mitochondria, nucleus, cytoplasm) and provide a statistical analysis package to analyze data. One example of such a software package is Scion Image Beta 4.02 (Scion Corporation, Frederick, MD). Electron microscopy micrographs (8 × 10″) are scanned with a scanner in grayscale (75 dpi resolution), and the image is saved for study with image analysis software. The software allows outline of cell compartments of interest (mitochondria, nucleus, cytoplasm), with subsequent calculation of areas of each cellular compartment. Gold beads must still be counted manually.

Positive Controls

Positive controls should be used in any morphologic analysis. 4HNE is found in large quantities in organs with large amounts of lipofuscin, including seminal vesicle and prostate. Aging brain and heart may also be used.

Negative Controls

Preincubation of antigen with antibody should abolish staining of tissue sections. Normal rabbit serum in place of primary antibody is also always used as a negative control.

Conclusions

Results to date have demonstrated that several pathologic processes studied have unique expression patterns of 4HNE-modified proteins. In ischemia–reperfusion injury, 4HNE-modified proteins were observed primarily in mitochondria (unpublished observations), in human renal cancer in both mitochondria and nucleus,[3] and in aging skeletal muscle primarily in cytoplasm.[7] We have quantified 4HNE-modified proteins in phorbol ester-treated mouse skin of nontransgenic and transgenic manganese superoxide dismutase (MnSOD)-overexpressing mouse skin and demonstrated that mouse skin from nontransgenic mice had a twofold increase in 4HNE-modified proteins as soon as 6 hr after 12-o-tetradecanoyl phorbol-13-acetate (TPA) treatment compared to MnSOD transgenic mice (unpublished observations). Further, quantitative image analysis indicated an increase in oxidative damage following TPA treatment in nuclei and mitochondria, but not in the cytoplasm. These results demonstrated that TPA caused oxidative damage and showed that overexpression of MnSOD was able to reverse this damage. These results were of special interest, as transgenic mice had twofold less papilloma formation (unpublished observations). These results illustrate the important use of immunomorphologic techniques in quantifying oxidative damage to allow study of important biological problems.

Future studies will require knowledge of the physiologic roles and location of free 4HNE before pathologic analyses can be completely interpreted in a reliable fashion. Further, other factors that can regulate levels of free 4HNE and 4HNE-modified proteins must be taken into consideration in analysis of data, including antioxidant defense mechanisms and protein and whole cell turnover rates. Thus, cells and protein with low turnover rates and low antioxidant defenses would be expected to have higher levels of 4HNE-modified proteins. Future studies will thus need knowledge of both 4HNE formation and protein and cell turnover rate in order to fully delineate the physiologic and pathologic roles of 4HNE.

[33] Ultrastructural Localization of Light-Induced Lipid Peroxides

By PETER KAYATZ, GABRIELE THUMANN, and ULRICH SCHRAERMEYER

Commonly employed biochemical or light microscopical procedures for detecting lipid peroxides (LP) do not allow for ultrastructural localization. However, for analysis of transport and processing of lipid peroxides, it is necessary to be able to localize lipid peroxides at the ultrastructural level. A method, which uses benzidine-reactive substances, has been developed and shown to be able to detect lipid peroxides at the ultrastructural level in rat and mouce tissues.

Introduction

Intense light exposure leads to retinal damage,[1,2] elevation of retinal hydroperoxide levels, and a decrease in the levels of rod outer segment (ROS) docosahexaenoic acid.[3–6] Docosahexaenoic acid ($22:6\omega3$), the major polyunsaturated fatty acid in the retina, is particularly susceptible to lipid peroxidation.[7] Lipid peroxidation causes retinal degeneration[8,9] and may be a major risk factor in the pathogenesis of age-related macular degeneration (ARMD),[10] the most common cause for blindness in Western countries. Lipofuscin accumulation, which accompanies the development of ARMD, may be caused, in large part, by lipid peroxidation, although a different genesis has also been discussed.[11–13] An accumulation of lipid peroxides in the ROS diminishes the susceptibility of the associated proteins to enzymatic degradation,[14] increasing the amount of indigestible residual material

[1] W. K. Noell, V. S. Walker, B. S. Kang, and S. Berman, *Invest. Ophthalmol.* **5,** 450 (1966).
[2] R. A. Gorn and T. Kuwabara, *Arch. Ophthalmol.* **77,** 115 (1967).
[3] R. D. Wiegand, N. M. Giusto, L. M. Rapp, and R. E. Anderson, *Invest. Ophthalmol. Vis. Sci.* **24,** 1433 (1983).
[4] V. E. Kagan, I. Y. Kuliev, V. B. Sprirchev, A. A. Shvedova, and Y. P. Kozlov, *Bull. Exp. Biol. Med.* **91,** 144 (1981).
[5] V. E. Kagan, A. A. Shvedova, K. N. Novikov, and Y. P. Kozlov, *Biochim. Biophys. Acta* **330,** 76 (1973).
[6] J. S. Penn and R. E. Anderson, *Exp. Eye Res.* **44,** 767 (1987).
[7] L. A. Witting, *J. Am. Oil Chem. Soc.* **42,** 908 (1965).
[8] R. E. Anderson, F. L. Kretzer, and L. M. Rapp, *Adv. Exp. Med. Biol.* **366,** 73 (1994).
[9] R. E. Anderson, L. M. Rapp, and R. D. Wiegand, *Curr. Eye Res.* **3,** 223 (1984).
[10] M. A. De La Paz and R. E. Anderson, *Invest. Ophthalmol. Vis. Sci.* **33,** 2091 (1992).
[11] K. S. Chio, U. Reiss, B. Fletscher, and A. L. Tappel, *Science* **166,** 1535 (1969).
[12] D. Yin D, *Free Radic. Biol. Med.* **21,** 871 (1996).
[13] G. E. Eldred and M. L. Katz, *Free Radic. Biol. Med.* **10,** 445 (1991).
[14] P. C. Burcham and Y. T. Kuhan, *Arch. Biochem. Biophys.* **340,** 331 (1997).

METHODS IN ENZYMOLOGY, VOL. 352
0076-6879/02 $35.00

in the retinal pigment epithelium (RPE), resulting in the formation of lipofuscin[12] and drusen,[15] which would promote the development of ARMD.[16]

The appearance of electron-dense structures found in ROS segments of hamsters after incubation of the retinas with tetramethylbenzidine (TMB) in the presence of H_2O_2[17] was thought to be mediated by a peroxidase. However, it was found that tetramethylbenzidine reacts with endogenous lipid peroxides to form an electron-dense reaction product[18] named "benzidine-reactive substances" (BRS). Based on these results, a method for the ultrastructural localization of lipid peroxides in the retina of rats and mice was developed.[17–21]

The availability of a method for the localization of lipid peroxides at ultrastructural levels, coupled with the effect on the retina of modifications in light exposure, offers the possibility of investigating the formation, transport, and processing of benzidine-reactive substances (BRS) during the digestion of peroxidative-damaged shed tips of ROS in the RPE.

Methods

Peroxidation of Rat Retinas by Light Exposure

Three-month-old Long Evans rats are kept in single cages under a 12-hr light/dark regimen. The light period starts at 7:00 AM and ends at 7:00 PM and is achieved with fluorescent light at 50 lux. Light intensity in lux is measured in the cages with a Gossen Colormaster 3F photometer, Erlangen, Germany.

Constant Light Exposure at 6000 lux for 12 and 24 hr. Animals are exposed to constant light of a high-pressure mercury lamp (HPL-N 125 W, Philips, Eindhoven, the Netherlands) for the following times:

12-hr protocol: Animals are exposed for 12 hr (10:00 PM–10:00 AM)
24-hr protocol: Animals are exposed for 24 hr (10:00 AM–10:00 AM)
Control group: Animals are kept under the 12-hr light/dark regimen as described previously

The spectrum of the high-pressure mercury lamp consists of six peaks at 365, 405, 435, 545, 580, and 625 nm (UV-B: 7.1 μW/lm, UV-A: 784.6 μW/lm,

[15] J. P. Sarks, S. H. Sarks, and M. C. Killingsworth, *Eye* **8,** 269 (1994).
[16] R. W. Young, *Surv. Ophthalmol.* **31,** 291 (1987).
[17] U. Schraermeyer, *Comp. Biochem. Physiol. B* **103,** 139 (1992).
[18] U. Schraermeyer, P. Kayatz, and K. Heimann, *Ophthalmologe* **95,** 291 (1998).
[19] P. Kayatz, K. Heimann, and U. Schraermeyer, *Invest. Ophthalmol. Vis. Sci.* **40,** 2314 (1999).
[20] P. Kayatz, K. Heimann, and U. Schraermeyer, *Graefe's Arch. Clin. Exp. Ophthalmol.* **237,** 763 (1999).
[21] P. Kayatz, K. Heimann, P. Esser, S. Peters, and U. Schraermeyer, *Graefe's Arch. Clin. Exp. Ophthalmol.* **237,** 685 (1999).

UV-tot: 791.7 μW/lm, visible: 3092.6 μW/lm, total radiation: 3884.4 μW/lm). The distance from the bottom of the cage is 35 cm, corresponding to a light intensity of 6200 lux at the center of the cage and 5400 lux at the periphery, with the photometer probe pointed toward the light source. The light is not diffused in any way. Rats do not seem to take any precautions to protect themselves from the light, such as closing eyes or hiding in a corner of the cage.

Cages used for constant-light exposure are constructed of translucent plastic (36 × 21 ×15) and are not sandblasted. The tops of the cages are covered by a grate made of steel (2-mm bars and 7-mm spacing). Humidity is 60%, and due to the absence of infrared light in the spectrum of the light source, the temperature inside the cages during illumination rises to a maximum of 24° and there is no need for fanning.

Short Light Exposure of 150,000 lux for 1 hr. Three-month-old Long Evans rats are anesthetized using an intramuscular injection of ketamine hydrochloride (200 mg/kg). Pupils are dilated using a single application of phenylephrine hydrochloride (2.5%)/tropicamide (0.5%) and are prevented from drying by the frequent application of a 0.9% sodium chloride solution. Eyelids are open without fixation. The eye is exposed for 1 hr (9:00–10:00 AM) to laboratory light (control, about 50 lux) or to a cold light source (Schott KL1500, halogen reflector lamp HLX 64634 EFR 15 V 150 W, 3200 K, position 5) pointed directly at the eye. The light intensity at the surface of the eye is approximately 150,000 lux. The spectrum of the light source is a bell-shaped curve with a maximum at 620 nm and negligible proportions of UV-A and infrared light. None of the rats are dark adapted before light exposure, and all animals are sacrificed immediately after exposure under common laboratory light conditions (about 500 lux) by cervical dislocation under CO_2 anesthesia.

Constant Light Exposure of 6000 lux for 20 hr Followed by Dark Period

20/4-hr protocol: Animals are exposed for 20 hr (11:00 AM to 7:00 AM the following day) followed by a 4-hr dark period (7:00–11:00 AM)

Control groups: Animals are kept under a 12-hr light/dark regimen and sacrificed 1 (8:00 AM) and 4 hr after the beginning of the light period (11:00 AM)

Processing and Fixation of Eyes

Preparation and fixation of the eyes are performed following a modification of the protocol of Fisher *et al.*[22] Eyes are enucleated, and a small incision is made at the limbus to allow the fixative access to the interior of the eye and fixed with 0.1 *M* cacodylate buffer, pH 7.4, containing 2% glutaraldehyde at 4°. After 10 min of fixation, the cornea, iris, and lens are removed, and the remaining tissue is fixed

[22] S. K. Fisher, D. H. Anderson, P. H. Erickson, C. J. Guerin, G. P. Lewis, and K. A. Linberg, *in* "Methods in Neurosciences" (P. A. Hargrave, ed.), Vol. 15, p. 3. Academic Press, San Diego, 1993.

for an additional 2 hr. After fixation, the vitreous body is removed and the eyecups are bisected through the optic nerve head along the posterior ciliary artery. Retinal buttons with a diameter of about 2 mm and the center approximately 2 mm superior of the optic nerve head are cut out from each animal, washed in 0.1 *M* cacodylate buffer, pH 7.4, five times and treated with a freshly prepared BRS reaction mixture at 4° overnight.

Localization of Benzidine-Reactive Substances in the Eye

For the localization of BRS, 0.5 mg/ml 3,3′,5,5′-tetramethylbenzidine dihydrochloride (Sigma Chemical Co., Deisenhofen, Germany) is suspended in 4 parts McIlvaine's citric acid buffer, sonicated, and adjusted to pH 3.0 by adding 1 part 0.2 *M* Na_2HPO_4.

Aldehyde-fixed and washed pieces of eyecups are incubated with a freshly prepared BRS reaction mixture at 4° overnight. Control tissue is incubated with McIlvaine's citric acid buffer, pH 7.3, without TMB or is fixed with glutaraldehyde overnight, without using the BRS reaction mixture. After incubation, the tissue is washed in cold cacodylate buffer five times, postfixed with 1.5% osmium tetroxide (Paesel + Lorei, Hanau, Germany) in 0.1 *M* cacodylate buffer, pH 7.4, for 2 hr, bloc stained with 1% uranyl acetate (Merck, Darmstadt, Germany) in 70% ethanol, dehydrated in a graded series of ethanol, and embedded in Spurr's resin. Semithin sections (approximately 700 nm) are stained with toluidine blue and examined with a Zeiss Axiophot light microscope. Ultrathin sections (light gold, approximately 100 nm) are poststained with uranyl acetate and lead citrate and are studied with a Zeiss EM 902 A.

Results

Rats

Constant Light Exposure of 6000 lux for 12 and 24 hr. In rats exposed to about 6000 lux for 24 hr (24-hr protocol) without dilation of the pupils, electron-dense structures appear throughout the rod outer segments (Figs. 1 and 2) and are absent in animals kept under physiological (50 lux, control group) light conditions (Fig. 3a) or exposed to 6000 lux for 12 hr (12-hr protocol). Electron-dense structures appear as electron-dense sites on the disc membranes, bubble-like structures, or spherical structures localized in the space between individual discs. The discs are interrupted at sites of the electron-dense structures. Several discs are seen to contribute to a single electron-dense bubble-like structure (Fig. 2). No electron-dense structures are found when tissue is incubated in the absence of tetramethylbenzidine.

Remarkably, retinas of rats kept under physiological light conditions and incubated with the BRS reaction mixture show swollen and irregular disc membranes, closely resembling light damage (Fig. 3a), but no electron-dense structures are a

b

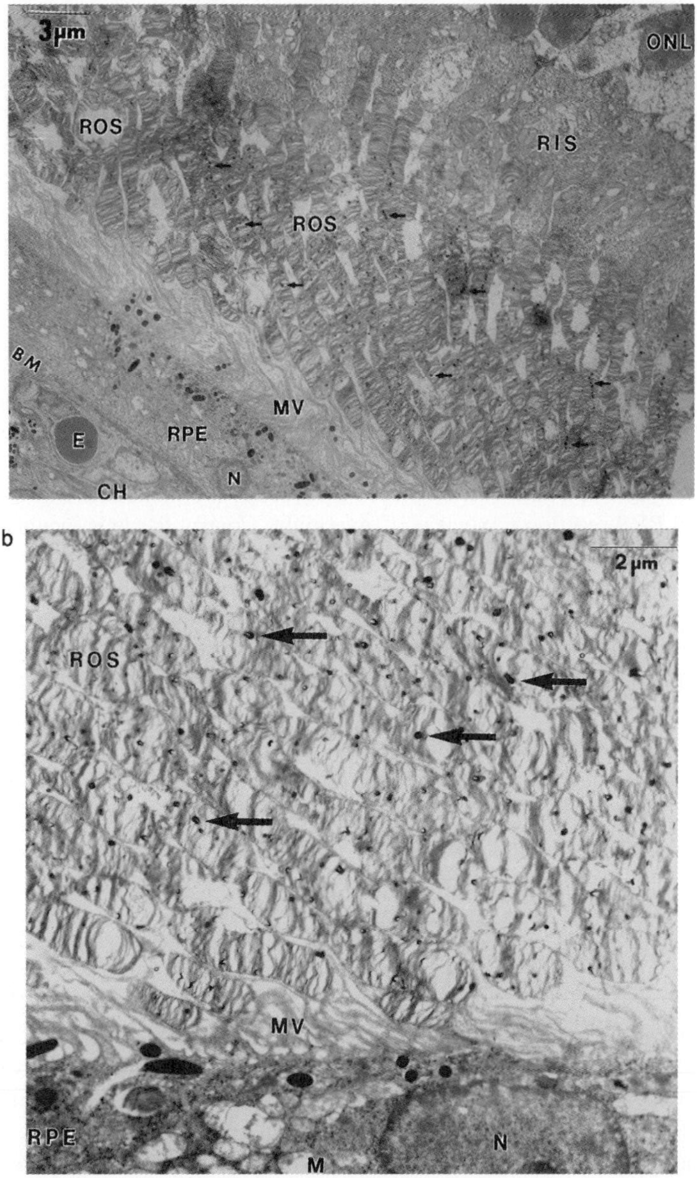

FIG. 1. (a) Overview of choroid (CH), RPE, and the photoreceptor cell layer. In rats with nondilated pupils exposed to intense light (about 6000 lux) for 24 hr, benzidine-reactive substances (BRS) (arrows) appeared after incubation with tetramethylbenzidine as electron-dense structures exclusively throughout the rod outer segments, but are missing in choroid, RPE, rod inner segments (RIS) and outer nuclear layer (ONL). (b) Higher magnification of RPE and ROS of the same eye. BRS (arrows) are seen exclusively in the ROS. No BRS appeared in the RPE. E, erythrocytes; N, nucleus; M, mitochondria. Reproduced with permission from P. Kayatz, K. Heimann, and U. Schraermeyer, *Invest. Ophthalmol. Vis. Sci.* **40,** 2314 (1999).

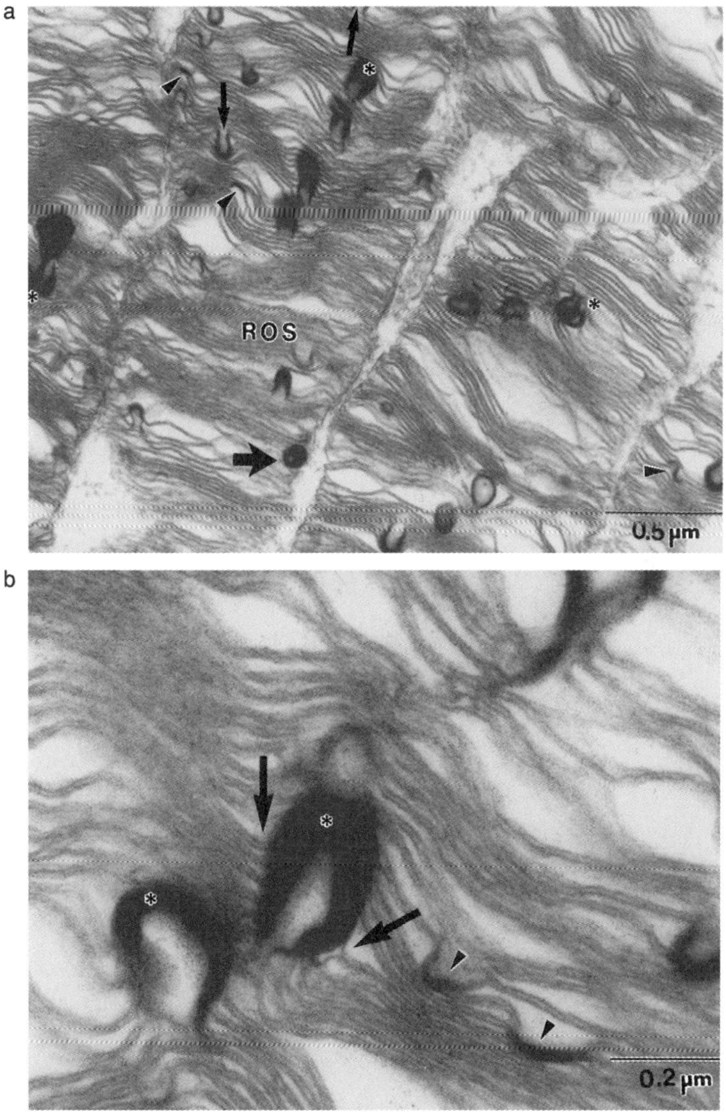

FIG. 2. (a) Benzidine-reactive substances (BRS) in the rod outer segments (ROS) appear as electron-dense sites of disc membranes (arrowheads) or bubble-like structures (asterisks). The electron-dense sites seem to extrude from the discs to form spherical structures that locate in spaces between individual discs (large arrows). It is possible that both appearances represent the same structures and differ only in the plain of section. The continuity of the discs is interrupted at sites were the electron-dense structures appear (small arrows). (b) A high magnification of the BRS reveals that several discs seem to contribute to one electron-dense bubble-like structure (arrows) and the interior of the bubbles appears as electron lucent. Benzidine-reactive substances appear as electron-dense sites of disc membranes (arrowheads) or bubble-like structures (asterisks). Reproduced with permission from P. Kayatz, K. Heimann, and U. Schraermeyer, *Invest. Ophthalmol. Vis. Sci.* **40,** 2314 (1999).

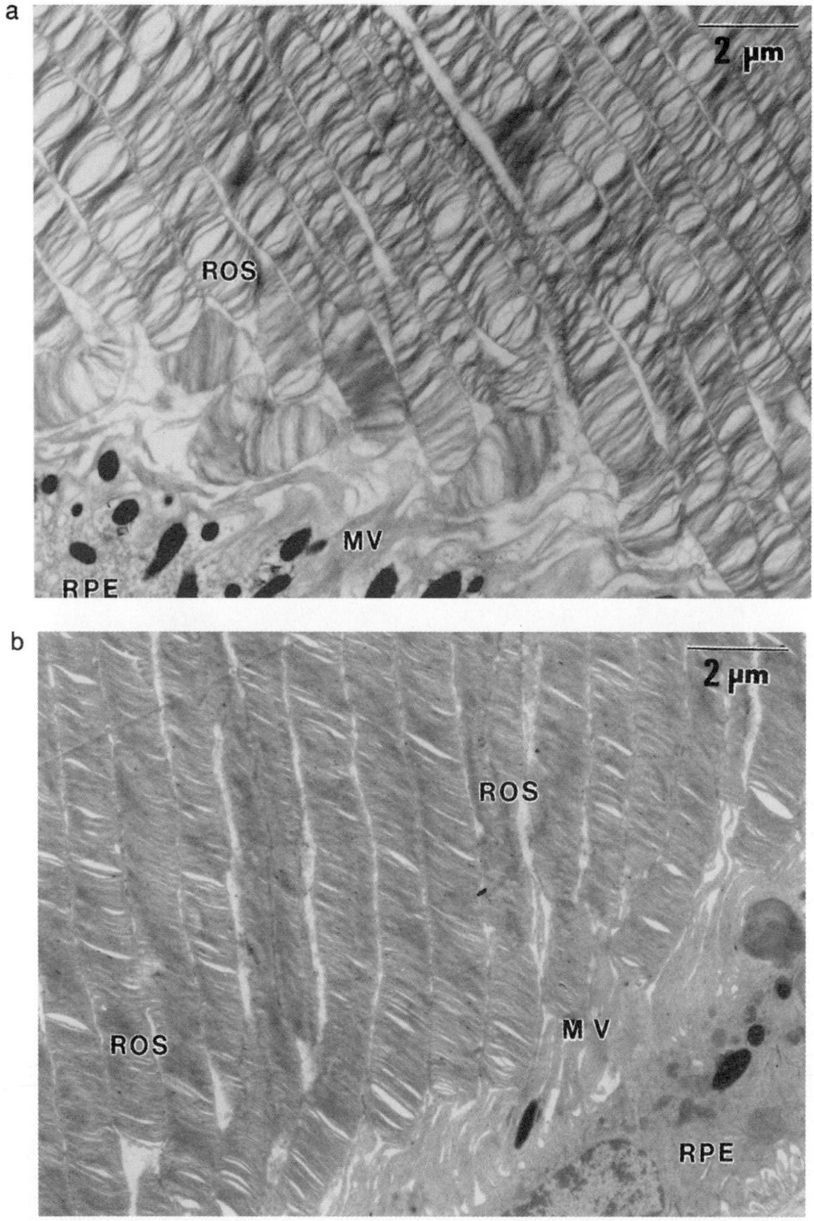

FIG. 3. (a) In rats kept at physiological light conditions (50 lux) with nondilated pupils, benzidine-reactive substances were absent in the rod outer segments after incubation with the BRS reaction mixture. The ROS exhibit swollen, irregular disc membranes, resembling light damage, but these aberrations are due to the incubation with the BRS reaction mixture and are not due to light damage. (b) If retinas of animals treated by the same light protocol are investigated without using the BRS reaction mixture, the irregularities are reduced to a minimum. Reproduced with permission from P. Kayatz, K. Heimann, and U. Schraermeyer, *Invest. Ophthalmol. Vis. Sci.* **40,** 2314 (1999).

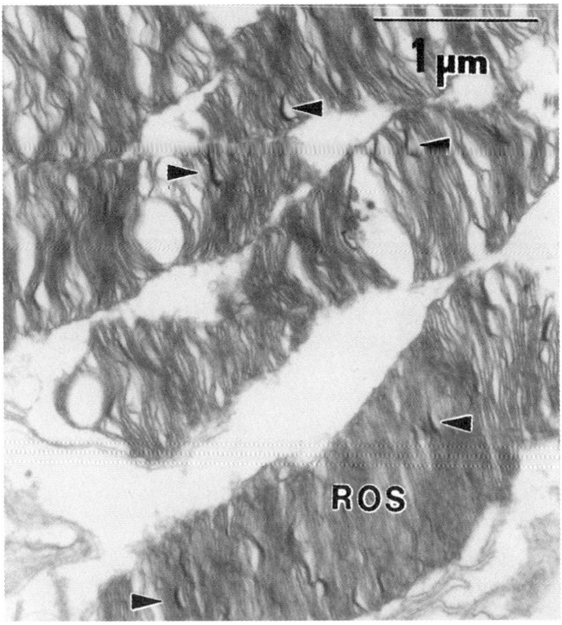

FIG. 4. In rats with dilated pupils, a light intensity of 50 lux for 1 hr results in the formation of electron-dense sites of disc membranes throughout the ROS (arrowheads). These electron-dense sites seem to represent early stages of bubble-like and spherical electron-dense structures that are not detectable using this soft light regimen. Reproduced with permission from P. Kayatz, K. Heimann, and U. Schraermeyer, *Invest. Ophthalmol. Vis. Sci.* **40,** 2314 (1999).

observed. However, retinas of the same animals treated with control buffer (without BRS reaction mixture) do not show these aberrations (Fig. 3b).

Short Light Exposure of 150,000 lux for 1 hr. Illumination of dilated pupils of rats with a light intensity of 50 lux for 1 hr leads to the appearance of small electron-dense sites on disc membranes throughout the ROS (Fig. 4). However, bubble-like structures or spherical structures are not detected.

Illumination with 150,000 lux led to a picture similar to that seen in eyes treated for 24 hr with 6000 lux (Fig. 5); the formation of bubble-like structures and spherical structures localized in the space between individual discs (Fig. 5). The appearance of bubble-like and spherical electron-dense structures following exposure of the eye to constant intense light is restricted to the ROS and is not seen in the RPE, in the inner segments, in the nuclear layer, or in the nerve fiber layer.

Constant Light Exposure of 6000 lux for 20 hr Followed by Dark Period. In animals exposed to light for 20 hr followed by a dark period of 4 hr, ellipsoid-shaped electron-dense structures appear mostly between the ROS in the extracellular space (Fig. 6), in the microvilli (MV) region of the RPE (Fig. 6a), and between the basal infoldings of the RPE (Fig. 7a). These structures are never seen in control tissues.

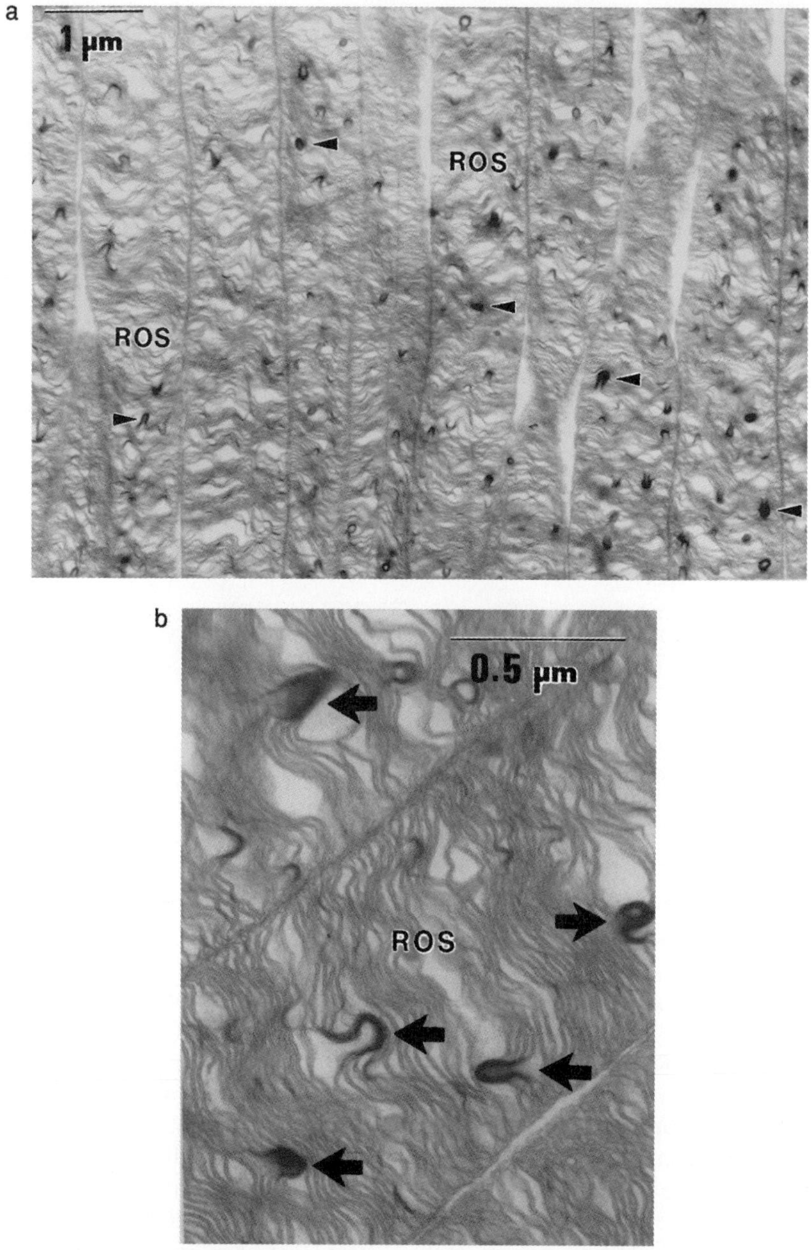

FIG. 5. (a) In rats with dilated pupils, an intense irradiation (150,000 lux) for 1 hr results in the formation of high numbers of bubble-like and spherical electron-dense structures (arrowheads) throughout the entire ROS. (b) A higher magnification reveals that these electron-dense structures resemble the structures that appear after an illumination of 6000 lux for 24 hr (see Fig. 2a). Reproduced with permission from P. Kayatz, K. Heimann, and U. Schraermeyer, *Invest. Ophthalmol. Vis. Sci.* **40,** 2314 (1999).

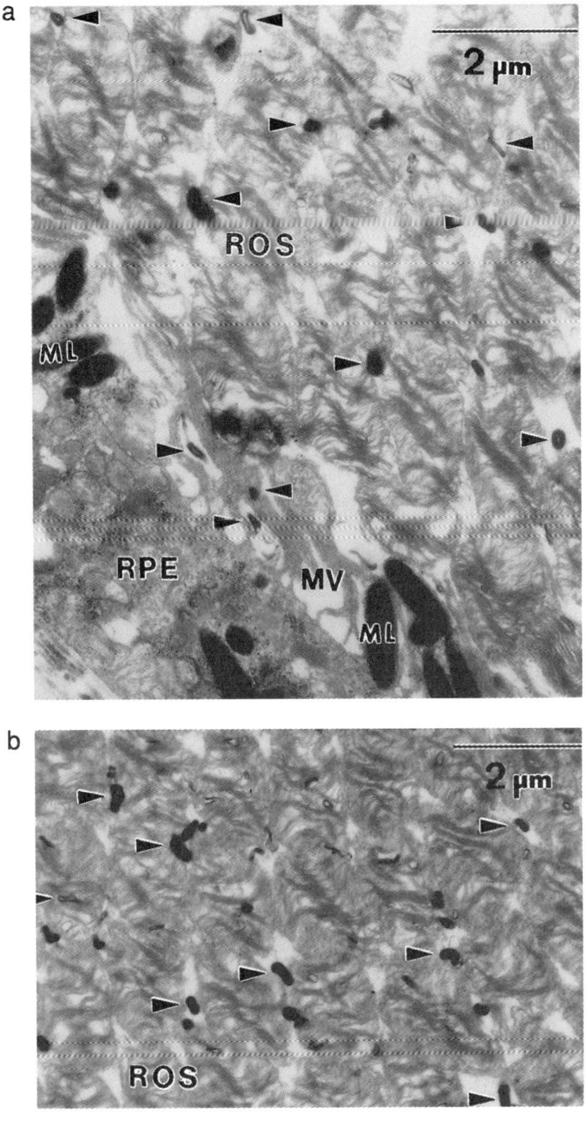

FIG. 6. Localization of BRS in the ROS and microvilli area after irradiation for 20 hr followed by a dark period of 4 hr (20/4-hr protocol). (a) Peroxidation of the rat retina by irradiation for 20 hr followed by a dark period of 4 hr leads to an accumulation of BRS (arrowheads) in the extracellular space of the ROS and extracellularly in the microvilli area (MV). These BRS appear as nonlamellar ellipsoid electron-dense structures without lamellae. ML, melanin granule (b). BRS (arrowheads) in the basal part of the ROS. Also, after this light protocol, BRS are distributed equally over the ROS, but locate in the extracellular space. Almost no BRS are seen intracellularly. The ellipsoid BRS in the extracellular space of the ROS emerge exclusively after intense irradiation and incubation with TMB. Reproduced with permission from P. Kayatz, K. Heimann, and U. Schraermeyer, *Graefe's Arch. Clin. Exp. Ophthalmol.* **237,** 763 (1999).

Ribbon-like electron-dense structures are found in electron-lucent vacuoles of the RPE (Fig. 7a), in the basal labyrinth of the RPE close to Bruch's membrane (BM) (Fig. 7b), and in the lumen (L) of the choroidal capillaries. These ribbon-like electron-dense structures appear as fine lamellas within a period of about 5 nm.

a

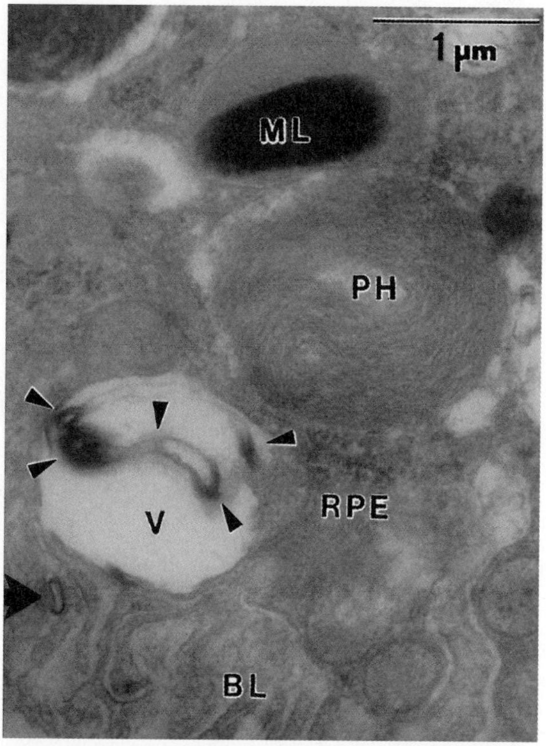

FIG. 7. Localization of BRS in the RPE, the basal labyrinth of the RPE and in the lumen of choroidal capillaries. (a) After the initiation of disc shedding by the 20/4-hr protocol, phagosomes (PH) and electron-lucent vacuoles (V) can be detected in the RPE. These vacuoles contain electron-dense ribbon-like lamellar structures (arrowheads) with a periodicity of about 5 nm. The electron-dense ribbons appear almost exclusively after the incubation with TMB. Vacuoles are similar in size and shape to phagosomes (PH), containing shed tips of ROS, that emerge due to the same light protocol. It is proposed that the fine lamellar electron-dense ribbons represent residual material from the degradation of shed tips of ROS. Nonlamellar ellipsoid BRS, similar to those in the extracellular space of the ROS and the microvilli area, are located in the extracellular space of the basal labyrinth (BL) of the RPE (arrow). This smooth, nonlamellar electron-dense appearance resembles the benzidine-reactive substances in the rod outer segments and in the extracellular space of the ROS. (b) Electron-dense ribbon containing vacuoles (V) located in the basal labyrinth of the RPE, close to Bruch's membrane. Reproduced with permission from P. Kayatz, K. Heimann, and U. Schraermeyer, *Graefe's Arch. Clin. Exp. Ophthalmol.* **237,** 763 (1999).

b

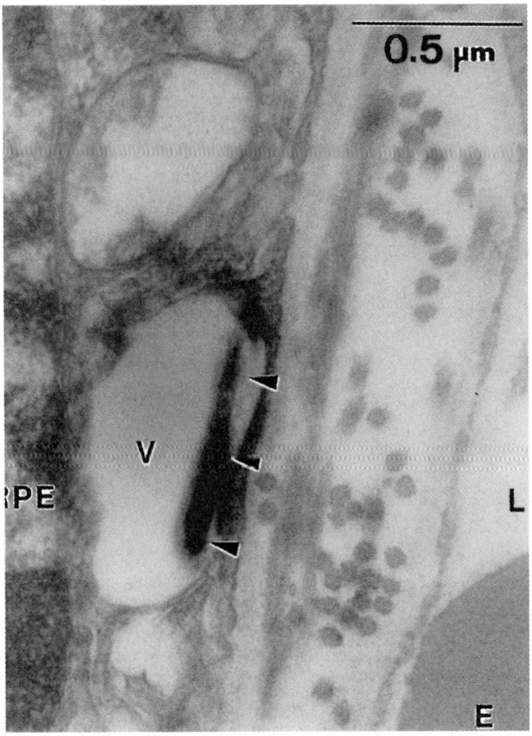

FIG. 7. (*Continued*)

Discussion

In 1927 Sehrt first discussed the possibility of detecting lipid peroxides by their ability to oxidize an indicator substance.[23] Lipid peroxides are capable of oxidizing tetramethylbenzidine (TMB) *in vitro*[24,25] as well as in fixed tissue.[18–21] This was demonstrated spectrophotometrically by Thomas and Poznansky[25] and in our laboratory[18–21] by the staining of synthesized lipid peroxides with TMB. Biochemically, TMB reacts selectively with lipid peroxides.[26] It has also been shown that lipoxigenase-generated lipid peroxides in porcine retinas, as well as light-induced lipid peroxides[10–12] in hamster retinas, can be detected as electron-dense structures, which are absent in the retinas of untreated animals.[18]

[23] E. Sehrt, *Münch. Med. W.* **74,** 139 (1927).

[24] S. S. Shibata, J. Terao, and S. Matsushita, *Lipids* **21,** 792 (1986).

[25] P. D. Thomas and M. J. Poznansky, *Anal. Biochem.* **188,** 228 (1990).

[26] M. R. Egmond, M. Brunori, and P. M. Fasella, *Eur. J. Biochem.* **61,** 93 (1976).

The various light regimens employed here produce retinal damage, which is detected following incubation of the tissue with TMB as electron-dense structures of various shapes and is localized in specific sites. The fact that these structures are not observed in tissues not incubated with TMB supports the hypothesis that treatment of tissue with TMB can demonstrate the formation and localize lipid peroxides at the electron microscopic level. The sensitivity of the method is evident from the results of anesthetized rats. When the eyes of these animals were dilated and exposed to laboratory light (50 lux) for 1 hr, lipid peroxide (benzidine reactive substances) was evident, whereas no lipid peroxides were detected in the eye of rats exposed to the same light for the same time, but without pupil dilation.

Because ROS are especially susceptible to lipid peroxidation, it is useful to work on the assumption that the majority of hydroperoxides generated under light-induced peroxidatic conditions consist of lipid peroxides. Taken together, it is likely that these electron-dense structures, detected in the ROS as BRS, represent lipid peroxides, although it cannot be ruled out that peroxidized proteins and carbohydrates also appear as BRS.

The irregularities of disc membranes observed in the retina of control animals kept under a 12-hr light/dark regimen of 50 lux without dilation of the pupils appear to be due to the overnight incubation with TMB, pH 3.0, prior to fixation with osmium tetroxide. In rats kept under the same light conditions, treated the same, but without overnight incubation with the BRS reaction mixture, these irregularities were rarely observed; and these might be abolished by a fixation by cardiac perfusion[22] instead of the 2-hr fixation used in this study. The typical signs of light damage are edema, disruption, and fragmentation of ROS, shrinking of inner segments, and nuclear pyknosis, leading to complete degeneration and photoreceptor cell death, evidenced by the disappearance of cell bodies; even the entire receptor layer may be missing.[27] However, despite the extreme light conditions used in this study, almost none of these indications of light damage were observed. This is consistent with both Li *et al.*,[28] who stated that cell death due to constant light takes some time to manifest itself after the initial light insult, and Shvedova *et al.*,[29] who stated that no ultrastructural changes were apparent after incubation of frog retinas with Fe^{II} and ascorbate for 20 min.

BRS appear as bubble-like electron-dense structures that seem to extrude from the disc membrane or as spherical structures that are located in the spaces between individual discs. The continuity of the discs is interrupted at the site of the electron-dense structures. These observations suggest that the integrity of the disc membranes is affected by the peroxidation and subsequent incubation with TMB at

[27] L. M. Rapp and T. P. Williams, *in* "The Effect of Constant Light on Visual Processes" (T. P. Williams and B. N. Baker, eds.), p. 135. Plenum Press, New York, 1980.

[28] Z. Y. Li, M. O. Tso, H. M. Wang, and D. T. Organisciak, *Invest. Ophthalmol. Vis. Sci.* **26,** 1589 (1985).

[29] A. A. Shvedova, A. S. Sidorov, K. N. Novikov, I. V. Galushchenko, and V. E. Kagan, *Vis. Res.* **19,** 49 (1979).

pH 3.0. Nevertheless, this effect is specific for irradiated retinas and does not occur in untreated animals. Furthermore, these structures appear only after incubation with the BRS reaction mixture and is not seen in irradiated eyes incubated with the BRS reaction mixture without TMB.

However, these considerations are not sufficient to explain the irregularities of the disc membranes in the retinas of control animals incubated with BRS reaction buffer, as no electron-dense structures were observed. One possibility for the irregularities in control retinas may be that the short fixation time (2 hr with 2% glutaraldehyde) leaves some enzymes functional, which could be responsible for the damage to the disc membranes. The procedures described here are a modification of methods developed for the localization of peroxides in pig and hamster retinas.[18] The buffer system used here to prepare the BRS reaction mixture is McIlvaine's Na_2HPO_4/ citric acid buffer, which allows TMB to be dissolved in citric acid and the pH is adjusted to 3.0 by adding Na_2HPO_4. A pH of 3.0 is optimal for the localization of lipid peroxides as BRS using TMB (unpublished). Modification of the buffer system makes the preparation of the BRS reaction mixture easier and results in a more sensitive method for the detection of BRS.

The method presented here offers a simple application for the ultrastructural localization of lipid peroxides, which can be used to investigate the formation, decomposition, or transport of lipid peroxides.

[34] A Survival Model for the Study of Myocardial Angiogenesis

By NILANJANA MAULIK, SHOJI FUKUDA, and HIROAKI SASAKI

Introduction

Angiogenesis is known to be the body's natural healing process in which new blood vessels grow in response to injury. Therefore, it is extremely important to develop the body's natural angiogenic process to create collateral circulation in areas in which blocked coronary arteries deprive the heart muscle of sufficient blood flow, e.g., in chronic myocardial ischemia. Thus, therapeutic induction of collateral vascularization in the ischemic heart in fast emerging as a highly attractive treatment modality in the realm of cardiovascular medicine.[1]

Various coronary interventions have made a remarkable contribution to the treatment of coronary artery disease. The ultimate goal of these interventions is

[1] A. Rivard and J. M. Isner, *Mol. Med.* **4,** 429 (1998).

0076-6879/02 $35.00

to improve arterial blood supply through the formation of coronary collateral vessels (angiogenesis) to potentially ischemic myocardium. Factors such as fibroblast growth factor (FGF) and vascular endothelial growth factor (VEGF), which stimulate collateral growth, are expected to exert a protective effect against myocardial infarction. Indeed, VEGF is a major regulator of angiogenesis and vasculogenesis.[2] A strong temporal and spatial correlation exists between VEGF expression and angiogenesis in both animals and humans.[3,4] The biological functions of VEGF, triggered by external stimuli, are initiated through the activation of intracellular signal transduction cascades involving specific kinases.[5] VEGF behaves as a classical stress-induced gene in this respect.

Among various triggers of angiogenesis, tissue hypoxia has been identified as being an important stimulus for the induction of new vessel growth, especially at the capillary level.[6] In the myocardium, hypoxia occurs during ischemic insult, usually followed by reoxygenation and revascularization. Preconditioning induced by cyclic episodes of brief periods of ischemia and reperfusion or hypoxia and reoxygenation has been found to possess profound cardioprotective ability in lowering myocardial infarction by reducing both cell necrosis and apoptosis and in improving postischemic contractile function.[7,8]

In the myocardium, hypoxia usually occurs during ischemic episodes. Such ischemic hypoxia is often followed by reperfusion. It is during this subsequent reperfusion period that reoxygenation occurs and reactive oxygen intermediates are formed.[9,10] In observations by Kuroki *et al.*,[11] reactive oxygen intermediates were found to increase VEGF expression *in vitro*. Reoxygenation of human retinal epithelial cells *in vitro* and ocular reperfusion *in vivo* increased retinal VEGF mRNA levels.[11,12] Administration of antioxidants prior to reoxygenation/reperfusion

[2] S. L. Starnes, B. W. Duncan, J. M. Kneebone, C. H. Fraga, S. States, G. L. Rosenthal, and F. M. Lupinetti, *J. Thorac. Cardiovasc. Surg.* **119,** 534 (2000).

[3] B. Millauer, L. K. Shawven, K. H. Plate, W. Risau, and A. Ullrich, *Nature* (*Lond.*) **367,** 576 (1994).

[4] A. P. Adamis, J. W. Miller, M. T. Bernal, D. J. D'Amico, J. Folkman, T. K. Yeo, and K. T. Yeo, *Am. J. Ophthalmol.* **118,** 445 (1994).

[5] K. Gupta, S. Kshirsagar, W. Li, L. Gui, S. Ramakrishnan, P. Gupta, P. Y. Law, and R. P. Hebbel, *Exp. Cell Res.* **247,** 495 (1999).

[6] D. T. Shima, A. P. Adamis, N. Ferrara, K. T. Yeo, T. K. Yeo, R. Allende, J. Folkman, and P. A. D'Amore, *Mol. Med.* **1,** 182 (1995).

[7] D. T. Engelman, M. Watanabe, R. M. Engelman, J. A. Rousou, E. Kisin, V. E. Kagan, N. Maulik, and D. K. Das, *Cardiovas. Res.* **29,** 133 (1995).

[8] N. Maulik, V. E. Kagan, V. A. Tyurin, and D. K. Das, *Am. J. Physiol.* **274,** H242 (1998).

[9] D. K. Das and N. Maulik, *Methods Enzymol.* **233,** 601 (1994).

[10] D. K. Das and R. M. Engelman, *in* "Patho-physiology of Reperfusion Injury" (D. K. Das, ed.). CRC Press, Boca Raton, FL, 1992.

[11] M. Kuroki, E. E. Voest, S. Amano, L. V. Beerepoot, S. Takashima, M. Tolentino, R. Y. Kim, R. M. Rohan, K. A. Colby, K. T. Yeo, and A. P. Adamis, *J. Clin. Invest.* **98,** 1667 (1996).

[12] J. Pe'er, D. Shweiki, and I. Itin, *Lab. Invest.* **72,** 638 (1995).

effectively inhibited such increases in VEGF mRNA. Lelkes *et al.*[13] have shown that hypoxia/reoxygenation, but not hypoxia alone, causes the formation of reactive oxygen species (ROS). In their study, the rate of tubular morphogenesis by human microvascular cells was increased threefold by hypoxia/reoxygenation. Tube formation in response to hypoxia/reoxygenation, as well as under normoxic conditions, was inhibited by various ROS antagonists in a dose-dependent manner, indicating the essential role of ROS in initiating angiogenesis.[14] A wide array of pathological conditions are known to generate ROS, including inflammation, atherosclerosis, ischemia, and reperfusion. The vascular endothelium is thus a target of oxidative stress under a variety of conditions. In fact, Lelkes *et al.*[13,14] have demonstrated that administration of ROS inhibitors such as pyrrolidine dithiocarbamate (PDTC) inhibits activation of NFκB and tube formation *in vitro*. It has been shown that following myocardial infarction various angiogenic growth factors are upregulated in the human heart as an intrinsic adaptive response.[15] Clinical gene trials with VEGF show promise as a potential means of augmenting such adaptive angiogenic responses. However, in order to achieve clinical relevance, the enhanced angiogenesis must be sufficient to effectively reduce the loss of cardiac function after ischemic injury.

Thus, to study the extent of angiogenesis in the infarcted myocardium, we have developed a rat survival surgery model of chronic myocardial infarction and have examined the left ventricular response during pharmacological stress, testing with dobutamine as a measure of cardiac reserve. The experimental protocol described here clearly demonstrates myocardial angiogenesis stimulated by nonlethal moderate hypoxic challenge by utilizing an animal model of chronic myocardial infarction progressing to heart failure. Such angiogenic effects of preconditioning were found to be due to its ability to augment VEGF protein expression, which presumably played a crucial role in reducing endothelial cell apoptosis, a proved determinant for congestive heart failure.

Chronic Myocardial Infarction Model

Male Sprague–Dawley rats (Harlan–Sprague–Dawley) weighing between 250 and 300 g, which have already been subjected either to prior hypoxic preconditioning (4 hr) or to time-matched normoxia under similar conditions, followed by 24 hr of reoxygenation, are anesthetized with ketamine · HCl (100 mg/kg ip) and

[13] P. I. Lelkes, K. A. Hahn, S. Karmiol, and D. H. Schmidt, *in* "Angiogenesis" (M. E. Maragoudakis, ed.), p. 321. Plenum Press, New York, 1998.

[14] P. I. Lelkes and C. R. Waters, *in* "Angiogenesis in Health and Disease: Basic Mechanisms and Clinical Applications" (G. M. Rubanyi, ed.). Dekker, New York, 2000.

[15] S. H. Lee, P. L. Wolf, R. Escudero, R. Deutsch, S. W. Jamieson, and P. A. Thistlethwaite, *N. Engl. J. Med.* **342,** 626 (2000).

xylazine (10 mg/kg ip). Cefazolin (25 mg/kg ip) is administered by ip injection as a preoperative antibiotic. The animal is placed in the supine position, and the body temperature is maintained at 37° by means of a water-circulating thermal heating pad. A midline neck incision is made, the trachea is identified, and a tracheostomy is then performed by making a vertical incision approximately 2 mm long in the trachea.

A polyethylene tube (2 mm o.d.)[16] is introduced through the opening after applying lidocaine jelly over the tip, and positive pressure ventilation with room air is begun with a stroke volume of 12 ml/kg[16–18,19] and a respiratory rate of 70/min using a small animal respirator (Model 683, Harvard Apparatus). The tube is then be secured in place by means of stay sutures.

An intercostal thoracotomy in the left fourth intercostal space is performed.[20–23] The heart is exposed and the pericardium incised. The left anterior descending (LAD) coronary artery is then identified. A 6-0 suture (Tevlek, Genzyme Fall River, MA) is passed beneath the vessel 1 to 2 mm from its origin.[20–23] Care is taken not to enter the ventricular cavity. In sham-operated controls, the suture is left in place up until chest closure but is not ligated. Permanent coronary occlusion is carried out by ligating LAD. Myocardial ischemia is confirmed by observing electrophysiological changes such as ST segment elevation and ventricular tachycardia on the continuously monitored electrocardiogram.

After completion of all protocols, the chest wall is closed in three layers using 4-0 silk sutures. A small red rubber catheter is used to evacuate air from the pleural space as the last stitch is being tightened. Bacitracin ointment is applied over the chest wound prior to bandaging the rat. Postoperative pain relief is achieved by the administration of buprenorphine (0.05–2.5 mg/kg sc twice a day for 3 days or more if needed). The tracheal tube is removed.

Postoperative Monitoring

The animals are allowed to recover in a small animal intensive care unit to assist monitoring for the first 24 hr. They are observed for signs of pain and

[16] Z. Xie, M. Gao, S. Batra, and T. Komyama, *Cardiovasc. Res.* **33,** 671 (1997).

[17] J. Leor, M. J. Quinones, M. Patterson, L. Kedes, and R. A. Kloner, *J. Mol. Cell Cardiol.* **28,** 2057 (1996).

[18] J. E. J. Schultz, Y. Z. Qian, G. J. Gross, and R. C. Kukreja, *J. Mol. Cell Cardiol.* **29,** 1055 (1999).

[19] R. J. Gumina, J. E. Schultz, Z. Yao, D. Kenny, D. C. Warltier, P. J. Newman, and G. J. Gross, *Circulation* **94,** 3327 (1996).

[20] R. L. Duerr, M. D. McKirnan, R. D. Gim, R. G. Clark, K. R. Chien, and J. Ross, *Circulation* **93,** 2188 (1996).

[21] R. L. Duerr, S. Huang, H. R. Miraliakbar, R. Clark, K. R. Chien, and J. Ross, *J. Clin. Invest.* **95,** 619 (1995).

[22] B.-H. Oh, S. Ono, H. A. Rockman, and J. Ross, *Circulation* **87,** 598 (1993).

[23] M. Assem, J. R. Teyssier, M. Benderitter, J. Terrand, A. Laubriet, A. Javouhey, M. David, and L. Rochette, *Am. J. Pathol.* **151,** 549 (1997).

distress, such as labored breathing, every half hour for the first 2 hr and thereafter every hour until being moved to their normal housing facility. After adequate recovery, they are moved to their normal housing facility, which is a climate-controlled environment subjected to 12-hr light/dark cycles. They are provided water and standard rat chow *ad libitum*. The animals are observed twice daily for signs of distress, such as decreased food and/or water intake, decreased response to stimuli, excessive lethargy, or labored breathing. If any such postoperative complications are detected in an animal, it is euthanized by subjecting it to CO_2 inhalation and is excluded from the study. To determine the degree of angiogenesis, animals are kept for different periods of time depending on the experimental protocol.

Determination of Myocardial Function, Tissue Injury, and Infarction

Approximately 1 hr prior to sacrifice the rats are anesthetized and ventilated as described earlier. A small incision is made to the right of the midline in the neck. The right internal jugular vein is identified. Two 3-0 silk sutures are passed beneath the vein. The cranial end of the vessel is ligated. The vein is then partially incised, and a PE 50 catheter is introduced into the vein. The proximal end of the catheter is connected to a low-pressure transducer. The inserted catheter tip is advanced until it reaches the right atrium and central venous pressure is recorded. The remaining ligature is used to secure the catheter in place. The same procedure is used to place a catheter in the right carotid artery where the catheter is used to measure left ventricular blood pressure. All cardiac functional parameters are monitored, analyzed, and recorded in real time using the Digimed data acquisition and analysis system (Micromed, Louisville, KY).

Measurement of Percentage Circumferential Left Ventricular Infarct Zone

After following a standard deparaffinization protocol, the extent of left ventricular myocardial infarct is delineated histochemically by utilizing the Masson's trichrome staining technique for collagen present in the infarct scar. Staining is performed on four sections, 12 μm apart for each level (above, surrounding, and below the occlusion site). After permanent mounting and coverslipping, the sections are scanned using a Hewlett Packard Scanjet 5p flatbed scanner and stored as digital images in tiff file format for later processing. In case of samples obtained earlier than 4 days postoperatively, frozen sections are stained with TTZ to delineate the extent of myocardial infarction and are similarly scanned. For each section, the circumferential arc of the left ventricular wall occupied by the infarct scar is determined using the NIH Image public domain software package and is expressed as a percentage of the left ventricular wall circumference. Values obtained from all four sections of the same level are averaged to obtain the percentage circumferential left ventricular infarct zone at that level. Similar steps are followed to yield

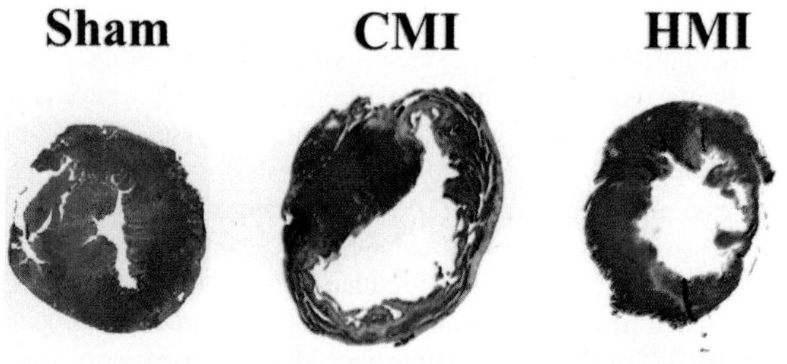

FIG. 1. Heart sections obtained 3 weeks postop after permanent coronary occlusion stained for collagen demonstrating circumferential infarct size. Collagen deposition, stained blue, is markedly pronounced in the control, CMI group compared to the hypoxic preconditioned group, HMI. Left ventricular wall thinning is also pronounced in the CMI group compared to HMI.

the percentage infarct zone measurements for each level to characterize the extent of myocardial infarction at three levels with respect to the level of occlusion. In our study, collagen deposition (blue) is markedly pronounced in the nonhypoxic, CMI group as opposed to the hypoxic preconditioned (HMI) myocardium. Left ventricular wall thinning is also apparent in the control (Fig. 1).

Nonsurvival Surgery and Baseline Cardiac Function

After the desired operative intervention, the rats are anesthetized and ventilated as described previously. A small incision is made to the right of the midline in the neck. The left internal jugular vein is identified, and a PE 50 catheter (Becton Dickinson, Franklin Lakes, NJ) is introduced into the vein. The proximal end of the catheter is connected to a low-pressure transducer. The inserted catheter tip is then advanced until it reaches the right atrium and the central venous pressure (CVP) signal is obtained. The same procedure is used to place a catheter in the right carotid artery where the catheter is used to measure left ventricular blood pressure. The inserted tip of this catheter is advanced down until it reaches the left ventricular lumen and the left ventricular pressure (LVP) signal is obtained. All pressure signals are monitored, analyzed, and recorded in real time using the Digimed data acquisition and analysis system (Micromed, Louisville, KY). Heart rate (HR), developed pressure, and $dP/dt_{\max}$ are calculated from the continuously obtained LVP signal.

Cardiac Stress Testing with Dobutamine

After baseline left ventricular functional parameters are recorded, rats are subjected to pharmacological cardiac stress testing with dobutamine infusion in progressively incremental doses to reveal the extent of the left ventricular contractile functional reserve. Dobutamine is infused iv through the venous catheter by means of a microinjection pump (Harvard Apparatus, Holliston, MA) to achieve doses of 1, 2, 3, and 5 μg/kg/min. Based on preliminary studies in our laboratory, we have found that administration to a dose of 5 μg/kg/min is necessary to observe the plateau of the dose–response curve. Administration is maintained at each dose for a duration of 2 min. dP/dt_{max} and other left ventricular functional parameters are recorded for each dose 90 sec after the start of each dose. Before measurement of all functional parameters, the animals are systemically heparinized (heparin sodium, 500 IU/kg body weight, iv injection, Elkins-Sinn Inc., Cherry Hill, NJ) to prevent intravascular coagulation. Blood samples are collected and centrifuged at 5000 rpm for 15 min.

Statistically significant differences were noticed in LVSP and LVEDP between the groups representative of the sham/baseline vs all the other permanently occluded groups (CMI and HMI) after 1, 2, and 3 weeks. No statistical (two-way ANOVA analysis) difference was found between the groups for heart rate even after 3 weeks of LAD occlusion (data not shown). Pharmacological cardiac stress testing with dobutamine infusion in incremental doses revealed differences in the extent of cardiac contractile reserve between groups (CMI and HMI). This was evident from differences in the extent of change in dP/dt_{max} values displayed by various groups during the course of such stress testing after 1 week of intervention. The differences remain statistically significant even after 3 weeks (Fig. 2). The hypoxic-preconditioned group displayed significantly elevated contractile reserve

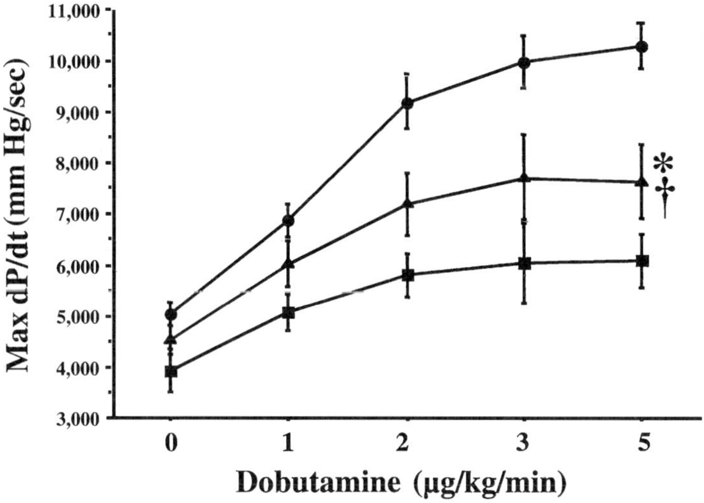

FIG. 2. Dose–response curves of dp/dt_{max} during the dobutamine stress test. Change in dp/dt_{max} in millimeters of mercury per second (mm Hg/sec) from baseline with increasing dobutamine boluses (0–5 μg/kg/min) for 3 weeks with MI hearts. The dose–response curve for dobutamine for the various doses was tested for significant differences by repeated measure of ANOVA. Sham [●], CMI [■], and HMI [▲]. Differences were considered significant at $p < 0.05$. *$p < 0.01$ compared to sham; $^{\dagger}p < 0.01$ compared to CMI.

at each dose point of evaluation compared to the CMI group, but the extent of such reserve was lower than that displayed by sham-operated groups or by control rats (all of which exhibited similar and statistically indifferent values of dP/dt_{max} throughout the course of stress testing). This enhanced preservation of contractile reserve in the hypoxic-preconditioned group was apparent at the 1-μg/kg/min dose (6419 ± 488 mm Hg/sec in the HMI group vs 5594 ± 197 mm Hg/sec in the CMI group after 3 weeks) and this trend persisted at all subsequent doses to 5 μg/kg/min.

Measurement of Blood Flow by Neutron Microsphere Technique

The microsphere technique has been used extensively for quantifying systemic and regional organ blood flows in the rat. Stable-labeled microspheres of BioPAL are used. The number of microspheres injected is calculated in such a way as to assure a sufficient number to accurately determine blood flow in the myocardium. The following equation is used to determine the minimum total number of microspheres needed per injection:

$$Y = 1.2 \times 10^6 + 1.9 \times 105X$$

where Y is the minimum number of microspheres needed for injection and X is the mass of the subject (i.e., approximately 1–1.5 million for a rat, approximately 2–2.5 million for a rabbit, approximately 5 million for a small canine, approximately 6–7 million for a large canine, and 8–9 million for a swine). The physiological state of the animal is not altered even with the large numbers of microspheres. Several grams of microspheres are needed to induce physiological effects. The nonradioactive microspheres of BioPAL provide a superior alternative to the use of radioactive and optical microspheres for measuring regional organ blood flow. They are neutron activated after sample collection and become temporarily radioactive. This is the most sensitive nonradioactive microsphere technique available. Approximately 15×10^5, 15-μm stable isotope-labeled microspheres (nonradioactive) (BioPAL, Inc., MA) are injected into an *in vivo* rat model of myocardial infarction through the left ventricle under anesthesia. Before, during, and after injection of the microspheres, blood is collected at a constant rate, starting 5 sec before and continuing 115 sec following injection of the spheres. Blood samples are withdrawn from the descending aorta at a rate of 1.0 ml/min. At the end of the experiments the animals are euthanized, and left ventricles (area at risk) are collected and weighed from the various groups. Tissue and blood samples are dried in a 70° oven overnight. The following equation is used to measure the blood flow.

$$\text{Organ (myocardium) flow (ml/min)} = \frac{\text{Known organ flow (ml/min)} \times \text{No. of microspheres in the myocardium}}{\text{No. of microspheres in the blood sample}}$$

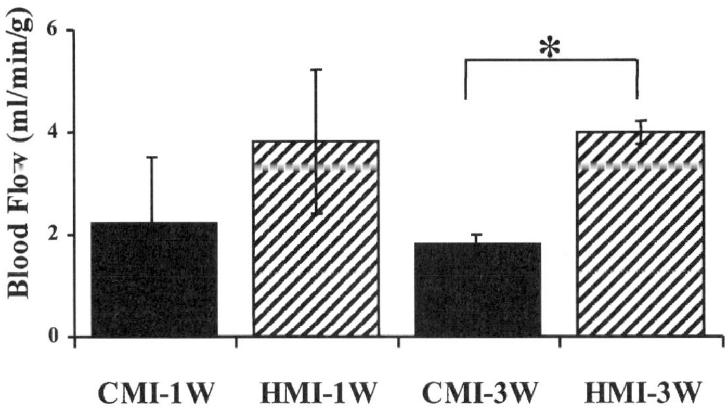

FIG. 3. Regional blood flow with neutron microsphere injection. Regional blood flow was estimated after 1 and 3 weeks of LAD occlusion. Differences were considered significant at $p < 0.05$. $^{*}p < 0.01$ compared to CMI. ■, CMI; ▨, HMI.

Left ventricular blood flow measured 1 and 3 weeks postop rat myocardium by the neutron microsphere technique revealed significantly higher blood flow in the experimental or HMI group compared to the CMI group. Therefore, in the CMI group the flow rate was 2.22 ± 1.31, whereas in the HMI group the blood flow rate was 3.83 ± 1.41 after 1 week. After 3 weeks, the blood flow rate in HMI group was increased further to 4.0 ± 0.22 vs 1.8 ± 0.20 in the CMI group (Fig. 3).

Tissue Retrieval and Processing

After measurement of all functional parameters, the animals are systemically heparinized to prevent intravascular coagulation. The animals are euthanized with 1 ml of 30 m*M* KCl injected iv. Hearts are excised, the ventricular lumen is flushed clean with Ringer's lactate solution, and the ventricles are either frozen in liquid nitrogen for later use in biochemical studies or sectioned approximately 5–6 mm from the apex into approximately 2-mm-thick transverse sections. The sections are then fixed in 6% buffered formalin and embedded in paraffin using standard procedures.

Measurement of Capillary and Arteriolar Density

An indirect immunohistochemistry technique is employed to visualize capillaries and arterioles. After following a standard deparaffinization protocol and enzyme pretreatment with pepsin at 37° to aid antigen unmasking, endothelial cells are labeled using mouse monoclonal anti-rat CD-31 (PECAM-1) (1 : 100, Innogenex, San Ramon, CA), and smooth muscle cells are labeled using mouse monoclonal antismooth muscle action (1 : 50, Biogenex, San Ramon, CA). Biotinylated goat anti-rabbit IgG (1 : 50, Vectorlabs, Burlingame, CA) and biotinylated horse anti-mouse IgG (1 : 50, Vectorlabs), secondary antibodies are used to visualize endothelial cells as brown and smooth muscle cells as purple when viewed under a light microscope (Olympus BH-2) after permanent mounting and coverslipping with the Prolong antifade kit (Molecular Probes, Eugene, OR). Capillaries are recognized by their size, a thin layer of endothelial cells, and a lack of smooth muscle cells. Similarly, arterioles are recognized by their size and a thin layer of smooth muscle cells surrounding endothelial cells. At a total magnification of 400×, 5 nonoverlapping random fields each are selected from both epicardial and endocardial regions of infarct, noninfarct, and borderline zones of the left ventricle of each section (30 fields). Images of tissue cross-sections are captured and stored in the digital tiff file format for later image analysis. Such images are captured for four sections, 12 μm apart, for each of the three levels (above occlusion site, surrounding occlusion site, and below occlusion site; total numbers of images: 5 per region, 10 per zone, 30 per section, 120 per level, 360 per heart). Counts of capillary density and arteriolar density per square millimeter are obtained after superimposing a calibrated morphometric grid on each image using Adobe Photoshop software. Blind counting is done by two separate investigators. Counts of a particular type (e.g., capillary) for each field in the same region (e.g., epicardial) of the same zone (e.g., infarct) from each of the 4 sections at the same level (e.g., above the occlusion site) (20 fields) are averaged to yield the numerical value of the particular density measurement. Similar steps are followed to yield density measurements for each region of each zone at each level (6 counts of capillary density and 6 counts of arteriolar density to characterize findings at three levels with respect to the level of occlusion). Arteriolar length density is estimated by obtaining the axial ratios of the vessels and multiplying the mean value by the numerical density. For this, systemic scanning is done very carefully. The arteriole is classified according to luminal diameter, and wall-to-wall lumen ratios are calculated.

The HMI group displayed a statistically significant increase in capillary density after 1 week of operation when compared to sham-operated groups and/or the control group [1862 ± 67 in HMI vs 1600 ± 16 counts/mm^2 in control baseline (BL)]. The increased capillary density was maintained even after 3 weeks postoperative when compared to sham-operated groups and/or the control group (Figs. 4 and 5). For the measurements of arteriolar density, five nonoverlapping random fields were selected from the endocardial regions of noninfarcted zones of the left ventricles and examined at 2000× magnification. Arteriolar counts (purple color) for each field in the same region were averaged to yield the value of the particular density measurement. The arteriolar density was significantly elevated in the HMI group after 1 week postop (2.23 ± 0.12 HMI vs 1.53 ± 0.02 counts/mm^2 in CMI). Again, the increased arteriolar density was maintained even after 3 weeks postop (Fig. 6).

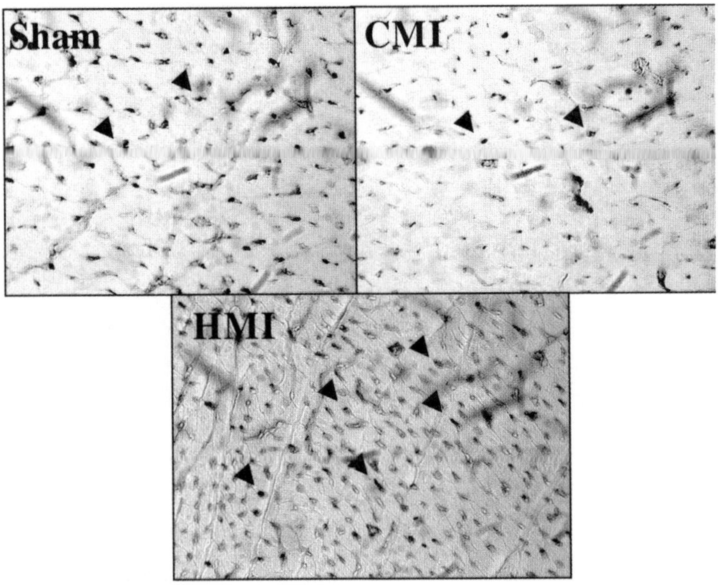

FIG. 4. CD-31 immunohistochemistry. Representative digital photomicrographs of noninfarcted endocardium treated immunohistochemically with anti-CD-31 to visualize capillary profiles (brown). Original magnification 400X.

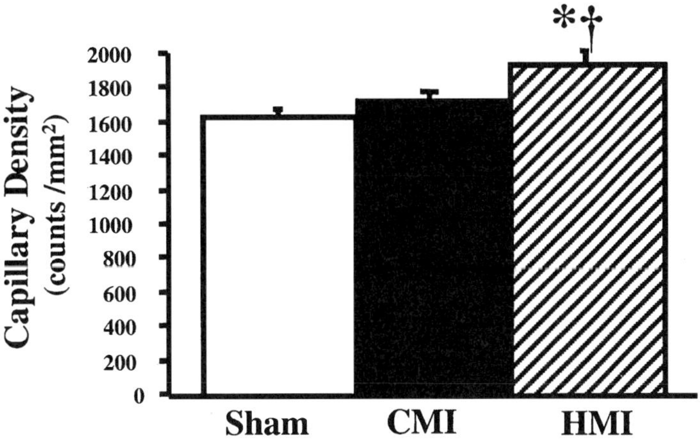

FIG. 5. Left ventricular endocardial capillary density. Tissue sections were processed for CD-31 staining, and 8 nonoverlapping random fields were selected from endocardial regions of noninfarcted region of the left ventricle. Two sections from each heart were used for the analysis (16 fields per region per heart, 64 fields per region per group; magnification 400X; $n = 6$). Images were captured and stored in a digital tiff file format for image analysis. $^{*}p < 0.001$ compared to sham control, $^{\dagger}p < 0.001$ compared to the CMI group. □, sham; ■, CMI; ▨, HMI.

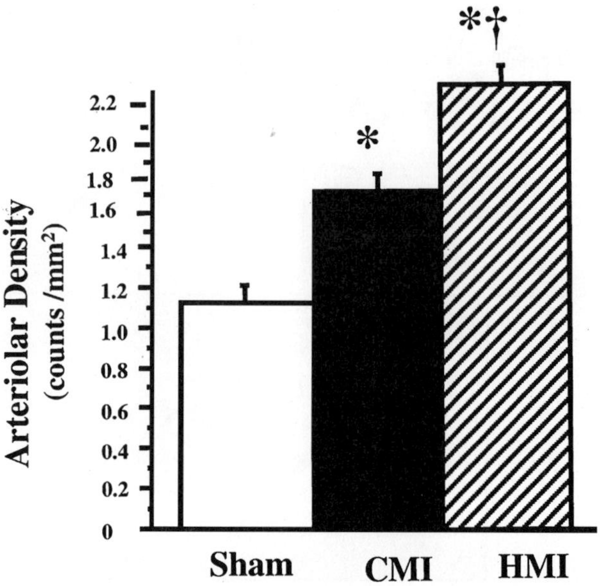

FIG. 6. Left ventricular endocardial arteriolar density. Left ventricular tissue sections were labeled using monoclonal antismooth muscle actin, and eight nonoverlapping random fields were selected from the endocardial region of the left ventricle. $*p < 0.001$ compared to sham control; $^{\dagger}p < 0.001$ compared to HMI. □, sham; ■, CMI; ▨, HMI.

Measurement of BrdU-Positive Endothelial Cells

While capillary and arteriolar density measurements help characterize the extent of vascularization, counting of BrdU-positive endothelial cells is considered to be a more accurate index of new blood vessel growth (neovascularization or angiogenesis). To this end, a double indirect immunofluorescence technique is employed to visualize endothelial cells whose nuclei stain positive for BrdU as a result of intraperitoneal administration of BrdU and FDU 2 hr prior to sacrifice. FDU (5′-fluoro-2′-deoxyuridine) is a known inhibitor of thymidylate synthetase whose concurrent administration helps enhance BrdU (bromodeoxy uridine) incorporation into DNA faster, allowing for quicker *in vivo* assays. After following a standard deparaffinization protocol and enzyme pretreatment with pepsin at 37°, as well as a 5-min incubation with 0.1 *N* HCl to aid antigen unmasking, endothelial cells are labeled using mouse monoclonal anti-rat CD-31 (1 : 100, Innogenex, San Ramon, CA), and BrdU is labeled using mouse monoclonal anti-BrdU (1 : 1000, Sigma Chemical Co., St. Louis, MO). Biotinylated goat anti-rabbit (1 : 50, Vectorlabs), and Texas red-conjugated horse anti-mouse (1 : 50, Vectorlabs) secondary antibodies are used to visualize endothelial cells as yellowish green and BrdU as red when viewed under a confocal laser microscope (Zeiss). At a total magnification of 400×, 5 nonoverlapping random fields each are selected from both epicardial and endocardial regions of infarct, noninfarct, and borderline zones of the left ventricle of each section (30 fields). Images are captured and stored in a digital tiff file format for later image analysis. Such images are captured for 4 sections, 12 μm apart, for each of three levels (above occlusion site, surrounding occlusion site, and below occlusion site; total number of images: 5 per region, 10 per zone, 30 per section, 120 per level, and 360 per heart). Counts of BrdU-positive endothelial cells per square millimeter are obtained after superimposing a calibrated morphometric grid on each image using Adobe Photoshop software. Blind counting is done by two separate investigators. Counts for each field in the same region (e.g., epicardial) of the same zone (e.g., infarct) from each of the 4 sections of the same level (e.g., above occlusion site) (20 fields) are averaged to yield the numerical value of the particular density measurement. Similar steps are followed to yield density measurements for each region of each zone at each level (6 counts of BrdU-positive endothelial cells to characterize findings at three levels with respect to the level of occlusion). A significant number of BrdU-positive cells is observed in the hypoxic-preconditioned group compared to the sham and/or CMI group (results not shown).

TUNEL Assay for Endothelial Cells

At the end of the experiments, the heart is perfusion fixed with 10 ml of 10% formaldehyde in 0.1 mol/liter phosphate buffer (pH 7.2) at room temperature. Transverse ventricular slices are embedded in paraffin, cut into 4-μm sections, and deparaffinized with a graded series of xylene and ethanol solutions. The sections are then assayed for *in situ* terminal transferase labeling (TUNEL) using Apop Tag Plus (Oncor Inc., Gaithersburg, MD). Negative control slides are processed with the TdT enzyme excluded. The slides are first stained with peroxidase-conjugated sheep polyclonal antidigoxigenin antibodies and diaminobenzidine and with methyl green as a counterstain. For the detection of endothelial cell apoptosis, the serial sections are first stained with TUNEL with fluorescein isothiocyanate (FITC) and then incubated with polyclonal rabbit anti-von Willebrand factor antibodies (Dako Japan, Tokyo, Japan), followed by incubation with tetrarhodamine isothiocyanate (TRITC)-conjugated goat anti-rabbit IgG (Dako Japan). FITC and TRITC fluorescences are viewed with a confocal laser microscopy (Fluo View, Olympus Co., Tokyo, Japan). The number of TUNEL-positive endothelial cells is counted on 60 high power fields (magnification ×600) from the endocardium through the epicardium of the left ventricular free wall. The degree of endothelial cell apoptosis was found to be inversely proportional to the extent of VEGF expression in our model. A higher proportion of endothelial cells was observed to be apoptotic in the CMI group of myocardium when compared to the HMI group after 2 days, 1 week, 2 weeks, and 3 weeks (9 ± 0.4 in CMI vs $1.2 \pm 0.4\%$ in HMI) after LAD occlusion. Thus, in both CMI and HMI groups the extent of apoptosis was found to be extremely significant in the early stage after LAD occlusion (2 days, 4 days, and 1 week). However, the number of apoptotic cells was reduced in the CMI group after 1, 2, and 3 weeks, but remained significantly higher when compared to the HMI group.

Western Blot Analysis for Angiogenic Factors

To quantify the abundance of the angiogenic factors VEGF, angiopoietins 1 and 2 (Ang-1, Ang-2), and their receptors (VEGFR1 or Flt-1, VEGFR2 or Flk-1/KDR, and Tie-1 and Tie-2), Western blot analysis is performed using various specific primary antibodies. Heart tissues from each treatment group are homogenized and suspended (5 mg/ml) in sample buffer [10 m*M* HEPES, pH 7.3, 11.5% sucrose, 1 m*M* EDTA, 1 m*M* EGTA, diisopropyl fluorophosphate (DFP), 0.7 mg/ml pepstatin A, 10 mg/ml leupeptin, 2 mg/ml aprotinin]. Homogenates are centrifuged at 3500 rpm, and the cytosolic fractions are used for protein analysis. Total protein concentrations are determined using the bicinchoninic acid (BCA) protein assay kit (Pierce, Rockville, IL). Cytosolic proteins (10 μg) are run on polyacrylamide electrophoretic gels (SDS–PAGE), typically using 10% (acrylamide to bis ratios) for VEGF, Ang-1, and Ang-2 and 8% for Tie-1, Tie-2, Flk-1, and Flt-1. Separated proteins are transferred electrophoretically to Immobilon-P membranes (Millipore Corp., Bedford, MA) using a semidry transfer system (Bio-Rad, Hercules, CA). Protein standards (Bio-Rad) are run for each gel. The blots are blocked in Tris-buffered saline/Tween 20 (TBS-T containing 20 m*M* Tris base, pH 7.6, 137 m*M* NaCl, 0.1% Tween 20) supplemented with 5% BSA for 1 hr. Blots are incubated for 2 hr with the specific primary rabbit antibodies (Santa Cruz Biotech, Santa Cruz, CA) against VEGF (1 : 200), Tie-1 (1 : 500), Tie-2 (1 : 500), Flk-1 (1 : 1000), and Flt-1 (1 : 1000). Blots being analyzed for the assessment of Ang-1 and Ang-2 are incubated with the appropriate primary goat antibodies (Santa Cruz) diluted 1 : 200. Blots are then incubated for 1 hr at room temperature with 1 : 10,000 diluted horseradish peroxidase (HRP)-conjugated secondary antibodies (Boehringer Mannheim Corp., Inc.), which are goat anti-rabbit IgG for all except Ang-1 and Ang-2 where rabbit anti-goat IgG is used. Direct reprobing with the anti-β-actin antibody as an internal control for Western blot is also performed. The blot is directly reprobed after washing with phosphate-buffered saline containing Tween (PBST) for 10 min at room temperature and then reprobed with a mouse monoclonal anti-β-actin as an internal control antibody (Clone AC-150; Sigma) with a dilution of 1 : 5000 in blocking solution after detection of the primary target. The secondary antibody used for β-actin is HRP-conjugated goat polyclonal anti-mouse IgG (Transduction Laboratories, Lexington, KY). After three washes of 5 min each, blots are treated with enhanced chemiluminescence (ECL from Amersham) reagent, and the required proteins are detected by autoradiography for

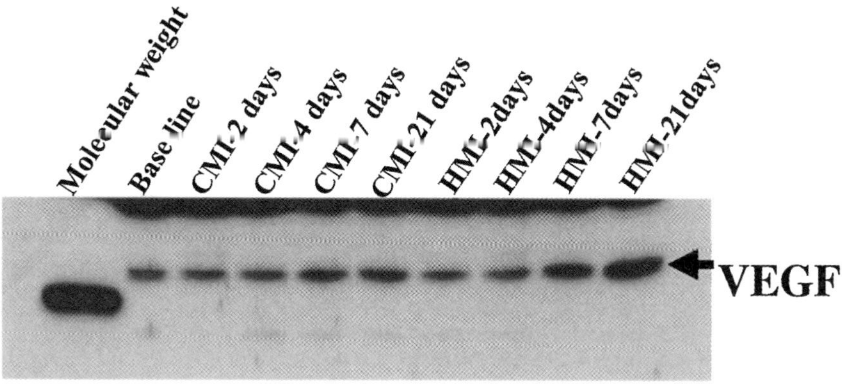

FIG. 7. Representative Western blots showing the effects of systemic hypoxia and LAD occlusion on the expression of VEGF in rat myocardium *in vivo* after 2, 4, 7, and 21 days. VEGF proteins were expressed as 40 kDa. Similar results were obtained in six independent experiments performed in triplicate.

variable lengths of time with Kodak X-Omat film. All the samples are tested for nonspecific labeling. Negative and positive controls are run to validate the results. The Protein expression profile of VEGF was found to be significantly elevated after 2 (33%) and 3 (63.3%) weeks of LAD occlusion in the HMI group compared to CMI and/or baseline control or sham group. However, in the CMI group, after 3 weeks of LAD occlusion, the VEGF expression level was increased moderately (20%) compared to the baseline control (Fig. 7). The protein level was measured by densitometry scanning and normalized with β-actin. The other angiogenic factors/ receptors were also found to be modulated in the HMI group compared to the CMI group (data not shown).

Immunohistochemistry

Serial sections of fresh frozen heart tissue obtained on poly-L-lysine-coated slides are allowed to air dry for 30 min before fixing in 100% acetone for 15 min. Sections are washed for 5 min 3× in phosphate-buffered solution (PBS) in between each step. Sections are immersed in 2% H_2O_2 in methanol for 25 min for endogenous peroxidase inactivation and are blocked in 5% goat serum in PBS for all sections except those being analyzed for Ang-1 and Ang-2 activities, which are incubated in 1% bovine serum albumin (BSA) instead. An additional avidin/biotin blocking step is performed by using a commercial kit (Vector Labs) according to the manufacturer's instructions. All primary antibodies are obtained from Santa Cruz Biotech. Secondary antibodies, streptavidin HRP, the peroxidase substrate kits Vector Nova Red and Vector VIP, and Nuclear Fast Red for counterstaining are all from Vector Labs. Harris' hematoxylin and permount mounting medium are obtained from Fisher Scientific Co. (Pittsburgh, PA). Sections are incubated overnight with the appropriate primary antibody diluted 1 : 50 in 5% goat serum in PBS except those being analyzed for Ang-1 and Ang-2 immunoreactivity, in which case the specific antibody is diluted 1 : 50 in 1% BSA in PBS. For assessing VEGF activity, the primary antibody dilution used is 1 : 100. Slides are then incubated in biotinylated secondary antibody solution (goat anti-rabbit IgG diluted 1 : 200 in 1% BSA; rabbit anti-goat IgG diluted 1 : 200 in 1% BSA for Ang-1 and Ang-2 assays) for 1 hr followed by incubation in streptavidin HRP (diluted 1 :100 in PBS). The biotinylated secondary antibody used in Ang-1 and Ang-2 assays is in PBS. Negative controls for both secondary antibodies are performed, and no HRP reaction product is observed. For visualizing VEGF, Ang-1, and Ang-2, Vector Nova Red is used as the HRP substrate with Harris' hematoxylin for counterstaining. Flk-1, Flt-1, Tie-1, and Tie-2 are visualized using VECTOR VIP as the HRP substrate with Nuclear Fast Red for counterstaining. Sections are viewed under an Olympus BH-2 microscope, and images are captured with an Olympus DP-10 digital camera.

Immunohistochemical analysis of VEGF revealed a diffuse pattern of distribution throughout the ventricular myocardium with strong localization around the coronary arterial wall where coronary endothelium as well as vascular smooth muscle appeared to stain positive for VEGF (data not shown). Hearts obtained from rats that had been subjected to whole body hypoxia followed by a 24-hr period of reoxygenation displayed a progressive increase in intensity of staining for VEGF with increasing durations of hypoxia. Although higher in intensity as compared to the control, the distribution pattern remained diffuse and there were no observable areas of localization around capillaries. However, VEGF remained strongly localized around the coronary arteries. The tissue distribution patterns of the VEGF receptors Flk-1 and Flt-1 were different from that of VEGF in that in addition to displaying strong localization around the coronary arteries, they both displayed intense staining along capillaries. Around the coronary arteries, Flk-1 was sharply localized almost exclusively to the coronary endothelium, whereas staining for Flt-1 was not as sharp in comparison. The intensity of staining for both receptors increased with increasing durations of hypoxia, but that of Flt-1 tended to persist while that of Flk-1 tended to decrease slightly after 2 hr of hypoxia (data not shown). Immunostaining for Ang-1 and Ang-2 was diffusely present throughout the ventricular myocardium and resembled the pattern of VEGF distribution except around the coronary arteries where they were less intense by comparison. Immunoreactivity to Ang-1 was increased after hypoxic exposure noticeably around the coronary vessels in contrast to its baseline pattern. The intensity of Ang-1 staining did appear to be decreased slightly after 1 hr of hypoxia, coinciding with the strong presence of Ang-2 at the same time point (data not shown). Thereafter, Ang-1 staining displayed a progressive increase in intensity in contrast to the steady decrease in Ang-2 immunoreactivity with an increasing duration of hypoxia.

Basal Tie-1 and Tie-2 distribution patterns were similar in that both were strongly localized around coronary arteries and both displayed reticular staining along capillaries. A sharp increase in Tie-1 immunoreactivity was observed after 1 hr of hypoxia, which decreased gradually to resemble its basal appearance after 3 hr of hypoxia. Tie-2 staining, however, displayed a modest increase by 1 hr of hypoxia, becoming very intense after 2 and 3 hr of hypoxia.

Acknowledgment

This study was supported by NIH-HL 56803.

[35] Determination of Angiogenesis-Regulating Properties of NO

By MARINA ZICHE and LUCIA MORBIDELLI

Introduction

Angiogenesis, the process by which new blood vessels are formed from pre-existing ones, plays a crucial role both in physiological (wound healing, embryonic development) and in pathological conditions (diabetic retinopathy, arthritis, tumor growth, and metastasis).[1] Angiogenesis is a tightly controlled process. In recent years, different families of regulators of angiogenesis, both stimulators and inhibitors, have been discovered and characterized by a molecular point of view.[2]

The steps required for new vessel growth are biologically complex. Sprout formation during the initial steps of the angiogenic process is commonly preceded by strong and persistent vasodilation and increased vascular permeability. Later events include endothelial cell proliferation, migration, and protease release. Following microscopic assessment of the modifications to the microvasculature elicited *in vivo* by a tumor and/or by a purified angiogenic factor, we concluded that the hemodynamic modifications of preexisting vessels were indispensable for triggering the angiogenic cascade.

Nitric oxide (NO) is the most important endothelial-derived relaxing factor and plays a major role in regulating vascular tone and vessel permeability. NO is produced by the enzymatic conversion of L-arginine to L-citrulline through NO

[1] J. Folkman, *Nature Med.* **1,** 27 (1995).

[2] E. Keshet and A. B. Ben-Sasson, *J. Clin. Invest.* **104,** 1497 (1999).

0076-6879/02 $35.00

synthase (NOS). Molecular cloning and sequence analyses revealed the existence of at least three main types of NOS isoforms. Both neuronal (nNOS or NOS I) and endothelial NOS (ecNOS or NOS III) are expressed constitutively, whereas inducible NOS (iNOS or NOS II) is expressed in response to lipopolysaccharide (LPS) and a variety of proinflammatory cytokines. High levels of NO have been described in pathophysiological processes, including various forms of circulatory shock, inflammation, and carcinogenesis, whereas conditions associated with a reduction of NO release include hypertension, congestive heart failure, disturbed vascular remodeling, atherosclerosis, and ischemia.

Using *in vitro* and *in vivo* methodologies of angiogenesis, we have demonstrated that NO is a trigger of angiogenesis[3,4] and mediates the angiogenic activity of vasoactive peptides and angiogenic factors.[4–8] The finding that ecNOS knock-out mice revealed an impaired angiogenesis in response to ischemia confirmed that NO is necessary for tissue neovascularization.[9] Moreover, it has been demonstrated that the metastatic behavior of human and experimental tumors is associated with high NOS activity and angiogenesis.[10–12]

Techniques for the Study of Angiogenesis *in Vivo*

The most used *in vivo* bioassay in our laboratory is the rabbit cornea assay (Fig. 1). Because the cornea is an avascular tissue, this assay avoids the problems of interpretation inherent in the other bioassays, such as the chick chorionallantoic membrane (CAM). The outgrowth of new vessels into the cornea stroma can be readily identified and their morphology documented. The corneal assay is performed in New Zealand white rabbits.

[3] M. Ziche, L. Morbidelli, E. Masini, H. J. Granger, P. Geppetti, and F. Ledda, *Biochem. Biophys. Res. Commun.* **192,** 1198 (1993).

[4] M. Ziche, L. Morbidelli, E. Masini, S. Amerini, H. J. Granger, C. A. Maggi, P. Geppetti, and F. Ledda, *J. Clin. Invest.* **94,** 2036 (1994).

[5] M. Ziche, L. Morbidelli, R. Choudhuri, H. T. Zhang, S. Donnini, H. J. Granger, and R. Bicknell, *J. Clin. Invest.* **99,** 2625 (1997).

[6] L. Morbidelli, C.-H. Chang, J. G. Douglas, H. J. Granger, F. Ledda, and M. Ziche, *Am. J. Physiol.* **270,** H411 (1996).

[7] A. Parenti, L. Morbidelli, X. L. Cui, J. G. Douglas, J. Hood, H. J. Granger, F. Ledda, and M. Ziche, *J. Biol. Chem.* **273,** 4220 (1998).

[8] A. Parenti, L. Morbidelli, F. Ledda, H. J. Granger, and M. Ziche, *FASEB J.* **15,** 1487 (2001).

[9] T. Murohara, T. Asahara, M. Silver, C. Bauters, H. Masuda, C. Kalka, M. Kearny, D. Chen, D. Chen, J. F. Symes, M. C. Fishman, P. L. Huang, and J. M. Isner, *J. Clin. Invest.* **101,** 2567 (1998).

[10] O. Gallo, E. Masini, L. Morbidelli, A. Franchi, I. Fini-Storchi, W. A. Vergari, and M. Ziche, *J. Natl. Cancer Inst.* **90,** 587 (1998).

[11] L. L. Thomsen and D. W. Miles, *Cancer Metast. Rev.* **17,** 107 (1998).

[12] L. C. Jadeski and P. K. Lala, *Am. J. Pathol.* **155,** 1381 (1999).

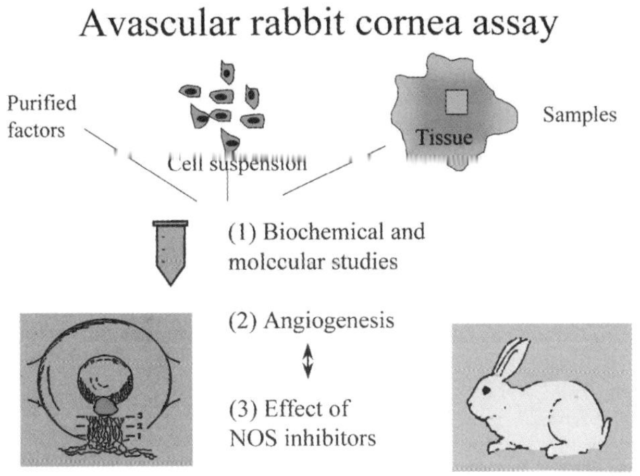

FIG. 1. Schematic representation of experimental procedures used to study the relevance of the NOS pathway in angiogenesis. Angiogenesis triggers, in the form of purified factors, cell suspensions, and tissue samples, are surgically implanted in the corneal stroma, and angiogenesis is monitored. The relevance of endogenous NO in mediating angiogenesis is evaluated by systemically treating the animals with NOS inhibitors. In parallel, the NOS pathway in cultured cells and tissues is studied *in vitro* by means of biochemical and molecular studies.

Surgical Procedure

After being anesthetized with sodium pentothal (30 mg/kg, iv), a micro pocket (1.5 × 3 mm) is surgically produced under aseptic conditions using a pliable iris spatula 1.5 mm wide in the lower half of the cornea. The implant is positioned 2.5–3 mm from the limbus to avoid false positives due to the mechanical procedure and to allow the diffusion of test substances into the tissue, with the formation of a gradient for the endothelial cells of the limbal vessels. Implants sequestering the test materials and the controls are coded and implanted in a double-masked manner.

Sample Preparation

The material being tested can be in the form of slow-release pellets incorporating recombinant growth factors, cell suspensions, or tissue samples.[10,13,14]

Slow Release Preparations

Recombinant growth factors/vasoactive molecules are prepared as slow-release pellets by incorporating the test substance into an ethylene vinylacetate

[13] M. Ziche, J. Jones, and P. M. Gullino, *J. Natl. Cancer Inst.* **69,** 475 (1982).
[14] M. Ziche, G. Alessandri, and P. M. Gullino, *Lab. Invest.* **61,** 629 (1989).

copolymer (Elvax-40) (DuPont de Nemours, Wilmington, DE). In order to avoid nonspecific reactions, Elvax-40 has to be carefully prepared as follows: after extensive washings of the Elvax-40 beads in absolute alcohol, 100-fold, at 37°, a 10% casting stock solution is prepared in methylene chloride and tested for its biocompatibility.[15] The casting solution is eligible for use if none of the implants derived from this preparation induces the slightest histological reaction in rabbit cornea. For testing, a predetermined volume of Elvax-40 casting solution is mixed with a given amount of the compound to be tested on a flat surface, and the polymer is allowed to dry under a laminar flow hood. After drying, the film sequestering the compound is cut into $1 \times 1 \times 0.5$-mm pieces. Empty pellets of Elvax-40 are used as controls.

Cell and Tissue Implants

Cell suspensions are obtained by the trypsinization of confluent cell monolayers. Five microliter-containing 2×10^5 cells in medium supplemented with 10% serum are introduced in the corneal micropocket. When the overexpression of growth factors by the stable transfection of specific cDNA is studied, one eye is implanted with transfected cells and the other with the wild-type cell line. When cells are pharmacologically treated, one eye is implanted with treated cells, while the controlateral one with untreated cells. When tissue samples are tested, samples of 2–3 mg are obtained by cutting the original fragments under sterile conditions. The angiogenic activity of tumor samples is compared with macroscopically healthy tissue.

Quantification

Subsequent daily observation of the implants is made with a slit lamp stereomicroscope without anesthesia. Angiogenesis, edema, and cellular infiltrate are recorded daily with the aid of an ocular grid by an independent operator who does not perform the surgery. An angiogenic response is scored positive when budding of vessels from the limbal plexus occurs after 3–4 days and capillaries progress to reach the implanted pellet in 7–10 days. Implants that fail to produce a neovascular growth within 10 days are considered negative, whereas implants showing an inflammatory reaction are discarded. The number of positive implants of the total implants performed is scored during each observation. The potency of angiogenic activity is evaluated on the basis of the number and growth rate of newly formed capillaries, and an angiogenic score is calculated by the formula vessel density $\times$ distance from limbus.[4] A density value of 1 corresponds to 0 to 25 vessels per cornea, 2 from 25 to 50, 3 from 50 to 75, 4 from 75 to 100, and 5

[15] R. Langer and J. Folkman, *Nature* **363,** 797 (1976).

for more than 100 vessels. The distance from the limbus (in mm) is graded with the aid of an ocular grid.

Histological Examination

Corneas are removed at the end of the experiment as well as at defined intervals after surgery and/or treatment and fixed in formalin for histological examination. Newly formed vessels and the presence of inflammatory cells are detected by hematoxylin/eosin staining or specific immunohistochemical procedures [i.e., anti-rabbit macrophages (RAM11), anti-CD-31 for endothelium].[5]

Considerations

Crucial points for a successful setup and outcome of the rabbit cornea assay are listed.

1. The body weight of animals is in the range of 1.8–2.5 kg for easy handling and prompt recovery from anesthesia.
2. Immobilization during anesthetic procedure and observation is important to avoid self-induced injury.
3. Sterility of materials and procedures is crucial to avoid nonspecific responses.
4. Elvax-40 beads should be washed carefully in absolute alcohol as indicated to avoid inflammatory reactions.
5. Make the surgical cut in the cornea stroma at pupil level and orient the micropocket toward the lower eyelid.
6. When two factors are tested, make two independent micropockets.
7. Drain a small amount of the aqueous humor when implanting cells or tissue fragments to reduce corneal tension.

Experimental Strategies to Study the Involvement of NOS on *in Vivo* Angiogenesis

Different experimental designs can be applied to study the role of NO in angiogenesis with the use of pharmacological tools (Table I).

1. To test the potentiation of the angiogenic response by exogenous NO, two adjacent pockets can be surgically produced in the same cornea: one bearing the angiogenic trigger and the other an NO donor drug (i.e., sodium nitroprusside, NaNP) or control. The angiogenic trigger is tested at doses that produce a weak angiogenic response. Pellets containing 1 μg NaNP have been used to potentiate angiogenesis by substance P (SP) and prostaglandin E1 (PGE1).[4]

TABLE I
PHARMACOLOGICAL TOOLS IN NO RESEARCH

NO system agonists	NO system antagonists
Nitro derivatives	NOS inhibitors
Nitroglycerin (TNG)	L$^{\omega}$-Nitro-L-arginine (L-NA)
Sodium nitroprusside (NaNP)	L$^{\omega}$-Nitromonomethyl-L-arginine (L-NMMA)
Isosorbide dinitrate	L$^{\omega}$-Nitro-L-arginine methyl ester (L-NAME)
Amyl nitrate	L-Canavaline
Linsidomine	*N*-Iminoethyl-L-ornithine (L-NIO)
SIN-1	Guanidine
S-Nitroso-*N*-acetylpenicellamine (SNAP)	Guanidine derivatives
NONOates or diazeniumdiolates	7-Nitroimidazole
NOC-5, -7, -9, -12, -18	Aminoguanidine
NOR-1, -3, -4	L-*N*(6)-(1-Iminoethyl)lysine hydrocloride (L-NIL)
Molecules that stimulate the release of NO	*N*-[3-(Aminomethyl)benzyl]acetamidine (W1800)
Acetylcholine	NO antagonists
ATP	Hemoglobin
Adenosine	Soluble guanylate cyclase inhibitors
Serotonin	Methylene blue
Substance P	LY83583
Bradykinin	ODQ
Endothelin	cGMP-dependent protein kinase G inhibitor (KT5823)
Histamine	
Vasculotropins (VEGF)	
Cytokines	
E. coli lipopolysaccharide (LPS)	
NO potentiators	
Antioxidants	
Inhibitors of phosphodiesterase V	

2. To evaluate the effect of NOS inhibition on the response to angiogenic effectors, N^{ω}-nitro-L-arginine methyl ester (L-NAME) or the inactive enantiomer N^{ω}-nitro-D-arginine methyl ester (D-NAME) are given in the drinking water *ad libitum*. Drug solutions (0.5–1 g/liter) are prepared freshly every day in tap water. Water intake is approximately 200 ml/day in treated animals and is not different from the control group. (*Note:* With higher doses of L-NAME, some sugar should be added to facilitate water intake by the animals.) Animals are kept under treatment 1 week before surgery and 10 days following corneal implant. The effect of L-NAME is compared to the treatment with both D-NAME and vehicle alone. Corneas are implanted with optimal doses of the angiogenic stimuli in all three groups. To evaluate the reversibility of systemic NOS inhibition, a group of rabbits receiving L-NAME treatment for 10 days can be returned to normal diet for 2 weeks and then tested for angiogenesis.

To ascertain the efficacy of NOS inhibition in treated animals, platelet cGMP levels and aggregation and the vasorelaxant response of aortic ring preparations isolated from treated and control rabbits are evaluated at the end of the experiments as well as at defined intervals during treatment.[4,5]

Assays for Angiogenic Activity *in Vitro*

Microvascular endothelial cells are the true players of the angiogenesis process. A great number of endothelial cell functions that mimic one or more steps contributing to capillary vessel formation *in vivo* can be studied *in vitro*. In our laboratory, biological functions activated during angiogenesis, i.e., endothelial cell migration, production of degradative enzymes, and cell proliferation, are routinely evaluated. Endothelial cells from both large and small vessels can be obtained from animal and human sources, and many are available commercially (American Type Tissue Collection, Rockville, MD, Clonetics Corporation, San Diego, CA). For endothelial cell proangiogenic activity, we routinely assay coronary venular endothelial cells (CVEC) obtained by a bead-perfusion technique through the coronary sinus of bovine heart,[16] although large vessel endothelium such as human umbilical vein endothelial cells can be used.

Cell Chemotaxis and Chemoinvasion

The Boyden chamber procedure is used to evaluate cell chemotaxis and chemoinvasion.[4,5] The method is based on the passage of endothelial cells across porous filters against a concentration gradient of the migration effector. While in chemotaxis the filter is coated with ECM proteins; during chemoinvasion, cells have to degrade a layer of ECM stratified on the filter. The Neuro Probe 48-well microchemotaxis chamber (Nuclepore) is used. The two wells are separated by a polyvinyl pyrrolidone (PVP)-free polycarbonate filter, 8 μm pore size, coated with type I collagen and fibronectin for chemotaxis, and with Matrigel for chemoinvasion. [The coating procedures of the filters are as follows: (1) For chemotaxis, incubate the filter with type I collagen (100 μg/ml) for 1 min at room temperature (RT); dry under a hood; incubate with bovine serum fibronectin (10 μg/ml) for 2–3 min at RT; wash with 0.1% bovine serum albumin (BSA) medium to remove aspecific binding. (2) For chemoinvasion, incubate the filter in 1 ml Matrigel, an extract of murine basement membrane proteins consisting predominantly of laminin, collagen IV, heparin sulfate, proteoglycan, and entactin (500 μg/ml) for 1 hr at 37°; dry under a hood overnight; reconstitute Matrigel with 1 ml serum-free medium.] Test solutions are dissolved in Dulbecco's modified Eagle's medium

[16] M. E. Schelling, C. J. Meininger, J. R. Hawker, and H. J. Granger, *Am. J. Physiol.* **254,** H1211 (1988).

DMEM + 1% fetal calf serum (FCS) and placed in the lower wells (30 μl/well). Then 50 μl of cell suspension (2.5×10^4 cells) is added to each upper well. The chamber is incubated at 37° for 4 hr, and the filter is then removed and fixed in methanol overnight. Nonmigrating cells on the upper surface of the filter are removed with a cotton swab. Cells migrated in the lower surface of the filter are stained with Diff-Quik (following manufacturer's instructions). The filter or part of it is mounted on a histological slide. The number of cells moving across the filter is counted using a light microscope (40×) in 10 random fields per each well.

SDS–PAGE Zymography for Gelatinases

The production of degradative enzymes is evaluated in subconfluent cell monolayers grown in 96-multiwell plates. Cells seeded at the density of 5×10^3/100 μl/well are serum starved overnight and then stimulated in medium containing 0.1% BSA for 1–24 hr in a working volume of 50 μl/well. Supernatants are collected and stored at −20°. Cell monolayers are fixed in methanol and stained in Diff-Quik. The total number of cells/well is counted as described in the proliferation assay. Protease activity in cell-conditioned media is assayed as described.[17] Briefly, 50 μl of medium is mixed with 12.5 μl of 4× Laemmli loading sample buffer. Zymography is carried out by electrophoresis in 9% polyacrylamide gel containing 0.1% gelatin, without heating the samples. After electrophoresis, gels are washed twice for 15 min with 2.5% Triton X-100, incubated overnight at 37° in 50 m*M* Tris–HCl (pH 7.4) containing 0.2 *M* NaCl and 5 m*M* $CaCl_2$, stained for 30 min with 30% methanol/10% acetic acid containing 0.5% Coomassie Brilliant Blue R-250, and destained in the same solution without dye. Bands of gelatinase activity appear as transparent areas against a blue background. Gelatinase activity is then evaluated by quantitative densitometry. HT1080 cells can be used as standard for the metalloproteinases MMP-2 and MMP-9. For gelatinase activity, optical density values are normalized to the number of cells counted/well.

Cell Proliferation

Cell proliferation is evaluated as cell replication and as DNA synthesis, measured as total cell number and 5-bromo-2′-deoxyuridine (BrdU) uptake, respectively.[4–6] For proliferation studies, CVEC are seeded onto 96-multiwell plates (1×10^3 cells/100 μl/well) in DMEM supplemented with 5% FCS and left to adhere for 4 hr. Media are removed and cells are incubated with increasing concentrations of the test substances for 4, 24, and 48 hr and 5 days. At the end of each incubation time, the supernatants are removed from the multiwell plates, and the

[17] H. Sato, T. Takino, Y. Okada, J. Cao, A. Shinagawa, E. Yamamoto, and M. A. Seiki, *Nature* **370,** 61 (1994).

cells are fixed by adding 100 μl of ice-cold methanol and kept at 4° overnight. Cells are then stained with Diff-Quik. Cell numbers are obtained by counting through microscopic examination at 10× magnification with the aid of an ocular grid (21 mm^2). Each well is divided into 10 fields, and cells are counted by a double-blind procedure in 7 randomly selected fields.

DNA synthesis is measured as the percentage of labeled nuclei counted over at least 300 cells/well after 24 hr of BrdU uptake and immunocytochemical processing, following the manufacturer's instructions (cell proliferation kit, Amersham).

Biochemical and Molecular Methodologies to Study Involvement of NOS in Angiogenesis

Commercially available NO donor drugs and NO-mediated drugs can be used as stimuli of endothelial cell functions in angiogenesis, i.e., cell chemotaxis, cell chemoinvasion, and cell proliferation (see Table I). Particular attention should be paid to solubilization media (water solution or organic solvent), pH, temperature, and light sensitivity. Appropriate controls should be run with the same amount of organic solvents to avoid misinterpretation of the results and aspecific toxic effects. Some drugs are not stable and should be prepared fresh each time, whereas others can be stored at −20° in small aliquots to avoid repetitive freezing/thawing. Proangiogenic activity has been shown with different NO donor drugs (NaNP, SNAP, TNG, isosorbide dinitrate), whereas SIN-1, which mostly releases peroxynitrites, shows a toxic effect. The dose range in which proproliferative and promigratory activity is obtained varies among the compounds in the nanomolar to micromolar range, whereas millimolar concentrations are toxic.[4,5] The involvement of the NO pathway in the responses observed can be corroborated by the use of synthetic cGMP analog 8-bromo-cyclic-GMP (8-br-cGMP). At concentrations of the order of 100 μM, the effects induced by NO donor or NO-mediated drugs can be reproduced.

When using NOS or soluble guanylate cyclase or PKG inhibitors (Fig. 2 and Table I), cells should be treated for 30 min to 1 hr with the drug before the addition of angiogenic factors. The effective concentrations devoid of any toxic effect are 200 μM–1 mM for L-NMMA or L-NAME, 10 μM for ODQ, and 2 μM for KT5823. For the proliferation assay, performed on adherent cells, cell cultures are preincubated with the drugs for 1 hr at 37°. Then angiogenic stimuli are added directly to the medium without removing the inhibitor. When chemotaxis and chemoinvasion assays are performed, cell suspensions are treated with the inhibitors for 30 min before challenging the cells toward the angiogenic factors. Cells in the upper compartment of the microchemotaxis apparatus remain in contact with the inhibitor for the duration of the entire experiment (2–4 hr). Parallel aliquots of cells should be checked for toxic effects by the inhibitors.

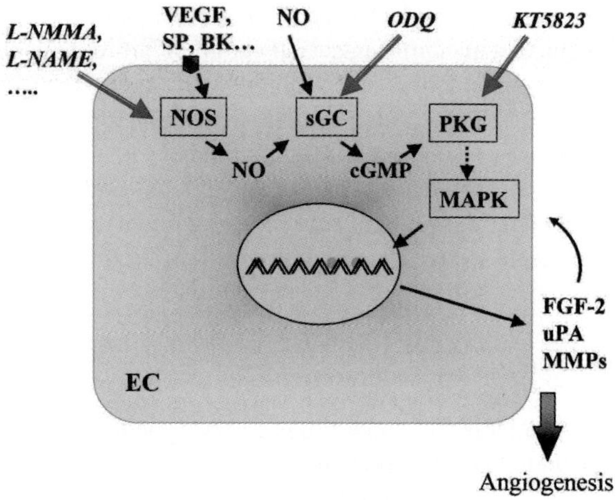

FIG. 2. Pharmacological tools for the study of the signaling cascade activated by endogenous and exogenous NO during angiogenesis occurrence. In endothelial cells (EC), the NOS pathway activated by angiogenic factors involves cGMP production, PKG activation, and gene transcription via the MAPK cascade. The production of degradative enzymes and survival growth factors (as FGF-2) is responsible for the switching on of an autocrine mechanism of cell survival and angiogenesis. L-NMMA and L-NAME are false substrates of NOS and inhibit NOS activity. ODQ is a selective inhibitor of soluble guanylate cyclase, inhibiting the pathway downstream from NO production. The contribution of PKG is assessed by the use of a selective inhibitor (KT5823). The switching on of the MAPK cascade is studied with the use of PD98059, an inhibitor of MEKK.

The use of medium depleted of L-arginine, to completely inhibit NOS synthase, is not recommended, as endothelial cells die in this condition.

Measurement of NOS Activity and Expression during Angiogenesis

Determination of Nitrate/Nitrite Levels

For a detailed description of the measurements of NO in biological systems and the problems associated with sample preparation, see Archer.[18]

Colorimetric Assay. The Griess reaction is useful for the rapid quantitative measurement of NO in aqueous solutions. It is based on the enzymatic conversion of nitrate to nitrite by nitrate reductase, followed by a spectrophotometric quantitation of nitrite levels using the Griess reagent.

Nitrite/nitrate production is evaluated in growing cells cultured in phenol red-free and with low nitrite/nitrate containing medium to which 1% FCS has been added. To improve sensitivity, supernatants can be lyophilized and reconstituted to

[18] S. Archer, *FASEB J.* **7,** 349 (1993).

a final 5× concentration. To evaluate total nitrite/nitrate content, purified nitrate reductase is added to the final concentration of 0.1–1 U/ml in the presence of 40 m*M* NADPH+ for 1 hr at RT. Nitrite production is measured by adding 0.5 ml of the sample to 0.25 ml of 1% sulfanilic acid in 5% phosphoric acid (or, alternatively, in 0.5 *M* HCl). After 3 min, 0.25 ml of 0.1% *N*-(1-naphthyl)ethylenediamine hydrocloride (NED) prepared in distilled water is added to each sample. After a 30-min incubation at room temperature in the dark, absorbance is measured at 540 nm in a spectrophotometer (UV/Vis, PerkinElmer, Norwalk, CT). Volumes can be adjusted to perform the test on an automated 96-well plate reader. A nitrite or nitrate calibration curve can be obtained in the culture medium treated as the samples by using sodium nitrite and sodium nitrate (range 0.1–10 μM). Alternatively, nitrite levels can be measured in cell lysates obtained by sonication of the cells detached from culture plates. Commercially available kits have been developed based on the just-described principle (Calbiochem).

Fluorimetric Assay. Sensitivity is increased compared to colorimetric determination. The assay is based on the enzymatic conversion of nitrate to nitrite by nitrate reductase, followed by the addition of 2,3-diaminonaphthalene (DAN), which converts nitrite to the fluorescent compound 1-(*H*)-naphthotriazole. Fluorescence measurements of this compound accurately determine the nitrite concentration (excitation maximum: 365 nm; emission maximum: 450 nm; minimum detectable quantitiy of $NO_2^-/NO_3^- = 10$ n*M*). Commercial kits are available (Cayman).

Measurement of NOS Activity

NOS activity is measured as the conversion of L-arginine in L-citrulline.[19]

Determination of NOS Activity in Tissue or Cell Lysates. By using this technique, activity from cytosolic and membrane-bound NOS can be measured separately.[4,5,10] The assay of tissue homogenates does not distinguish which particular cell type within a tissue is expressing NOS.

Fragments of tissues are frozen immediately in liquid nitrogen and stored at −80°. Fragments are homogenized in buffer containing 0.32 *M* sucrose, 20 m*M* HEPES (pH 7.2), 0.5 *M* EDTA acid, and 1 m*M* dithiotheitol (DTT).

Cells are seeded in 100-mm culture dishes, allowed to grow to 90% confluence, and serum starved overnight. Cells are treated with test substances at fixed times in DMEM plus 1% FCS. At the end of incubation, the plates are washed twice with cold phosphate-buffered saline (PBS), and cells are scraped in PBS without Ca^{2+} and Mg^{2+} (samples can be kept at −80°). Cells are centrifuged at 1500*g*, and the pellet is resuspended in homogenization buffer (containing 0.2 *M* sucrose, 20 m*M* HEPES, pH 7.2, 1 m*M* EDTA, 1 m*M* EGTA, 10 μg/ml leupeptin, 2 μg/ml aprotinin, and 1 m*M* DTT).

[19] D. S. Bredt and S. H. Snyder, *Proc. Natl. Acad. Sci. U.S.A.* **87,** 682, 1990.

Homogenized samples from either tissue or cell preparations are lysed by sonication (six bursts at 100 W of 10 sec each, working on ice) and then centrifuged at 100,000*g* for 60 min. Supernatants containing the cytosolic NOS (sample A) are collected, and the pellets, containing the membrane-bound NOS (sample B), are resuspended in 500 μl of homogenization buffer. Two hundred microliters from sample A and 200 μl from sample B are collected in new vials in duplicate to test both ecNOS and iNOS. Two hundred microliters of reaction buffer for total NOS [containing 2 m*M* $NADPH^+$, 0.5 m*M* $CaCl_2$, 10 μM calmodulin, 0.75 m*M* BH4, 200 m*M* arginine, [^{3}H]L-arginine (5 μCi/ml activity)] are added to each sample for total NOS activity measurement. In each sample for Ca^{2+}-independent NOS detection, 200 μl of reaction buffer for iNOS (containing 2 m*M* $NADPH^+$, 10 μM calmodulin, 0.75 m*M* BH4, 200 m*M* arginine, [^{3}H]L-arginine (5 μCi/ml activity), 1.5 m*M* EGTA, and 150 μM of the calmodulin inhibitor trifluperazine] is added. The reaction is carried out for 60 min at 37° and is stopped by adding 4 ml of stop buffer (50 m*M* HEPES, pH 5.5, and 5 m*M* EDTA) to each sample. Samples are applied to the ion-exchange resin previously activated and equilibrated (4 ml/column). The flow through from the columns is collected in 20-ml vials. Ten milliliters of scintillation fluid is added to each vial, and radioactivity is counted in a β counter. NOS activity is then expressed as dpm/mg protein measured in the homogenate. Blanks should be run by incorporating samples in media containing 1–2 m*M* L-NMMA. Ca^{2+}-dependent NOS activity can be calculated as total NOS activity minus Ca^{2+}-independent NOS.

Determination of NOS Activity in Cell Monolayers. Total NOS activity is measured on adherent cells, a more physiological condition.[7] Cells are grown in 6-cm dishes and serum starved overnight. Cells are washed with buffer A (containing 145 m*M* NaCl, 5 m*M* KCl, 1 m*M* $MgSO_4$, 10 m*M* glucose, 1 m*M* $CaCl_2$, 10 m*M* HEPES, pH 7.4, and 2 mg/ml L-arginine) and incubated at 37° for 20 min with 2 ml buffer A. Supernatant is removed, and cells are incubated for 30 min with 1 ml buffer A to which 2 μCi of [^{3}H]L-arginine has been added. Test substances are added for 5–30 min or longer times at 37°. When required, an NOS inhibitor (effective dose 2 m*M*) is added in the 20-min incubation, as well as during the following 30 min and then in the presence of the stimuli. The reaction is stopped by adding 1 ml cold stop buffer (containing 10 m*M* HEPES and 4 m*M* EDTA, pH 5.5). The supernatant is removed, and 0.5 ml ethanol/dish is added and allowed to evaporate to permeabilize the cell membrane. Two milliliters of stop buffer is added for 20 min under strong agitation at RT to extract the cell cytosol. The supernatant is collected, and 1 ml is applied to 1 ml of resin (twice) prepared previously. Vials are vortexed vigorously for 45 min. Then 0.8 ml of supernatant is collected and added to 4 ml of scintillation liquid, and radioactivity is counted in a β counter. NOS activity is expressed as cpm/mg protein measured in the cell monolayers remaining on the dish after extraction.

Preparation of Ion-Exchange Resin. The resin (Dowex 50Wx8 200–400 mesh, Aldrich) should be prepared before hand. Weigh 1 g resin/column and dissolve in distilled water. Remove water by decantation. Wash the resin with 2 *M* NaOH (100 ml for 30 g of resin) for 15 min under mild agitation. Leave to precipitate the resin and decant the supernatant. Wash with 1 *M* HCl (100 ml) and then with 1 *M* NaOH. Wash with water to pH 7 (use about 2 liters of distilled water) and then with 1× stop buffer. The resin can be kept in this buffer, but the pH (5.5) should be checked before use. At this pH value, arginine is linked to the resin, whereas citrulline remains in the flow through.

Measurement of cGMP Levels

cGMP is the intracellular trigger of NO signaling in target cells and can be used to measure the release of both NO and NOS activity. cGMP levels are measured on extracts from cell monolayers by radioimmunoassay.[3,6,7] Serum-starved and subconfluent cell monolayers (in 10 cm-plates) are treated with 1 m*M* 3-isobutyl 5-methylxanthine (IBMX) for 15 min before stimulation to block phosphodiesterase. After stimulation (5–30 min), cells are rinsed with PBS and removed by scraping in ice-cold 10% TCA (500 μl). An NO donor drug should be used as a positive control of cGMP production. After centrifugation at 13,000 rpm for 10 min, TCA from the supernatant is extracted with 0.5 *M* tri-N-octylamine dissolved in 1,1,2-trichlorotrifluoroethane (1 ml/sample) in glass vials. Following centrifugation, cGMP levels in the aqueous phase are measured by radioimmunoassay in duplicate with prior acetylation of samples with acetic anhydride following the manufacturer's instructions (Amersham). Protein content is measured in the pellet using Bradford's procedure. Data are expressed as fmol cGMP/mg protein.

Determination of Gene Expression of NOS Isoforms. NOS isoform mRNA expression in cell cultures and tissue samples can be monitored by semiquantitative RT-PCR.[7]

Total RNA Extraction and Reverse Transcription. After overnight starvation, cells (3.5×10^6/10 cm plate) are incubated for 6 hr with angiogenic stimuli in the presence of 0.1% FCS. At the end of incubation, total RNA is isolated by the standard guanidine thiocyanate-phenol-chloroform extraction. Total RNA from tissue samples, frozen immediately in liquid nitrogen after surgery, is obtained by homogenization and commercial kits. Quantity and purity of RNA are checked by spectrophotometry. cDNA is synthesized in 20 μl of reaction volumes containing 1 μg of total RNA, 2.5 μM oligo-dT16, 0.5 m*M* dNTPs, 50 m*M* Tris–HCl (pH 8.3), 75 m*M* KCl, 3 m*M* $MgCl_2$, and 200 U of M-MLV reverse transcriptase. After a 60-min incubation at 38°, samples are heated at 95° for 5 min and then chilled rapidly on ice.

Differential RT-PCR Analysis. Amplification for NOS isoforms is carried out by using 5 μl of cDNA and specific primers for NOS isoforms with sequences as

TABLE II
PRIMERS FOR NOS ISOFORM EXPRESSION BY RT-PCR

Species	NOS isoform	Sense/antisense	Sequence (5′-3′)	bp
Bovine	ecNOS	Sense	GCT TGA GAC CCT CAG TCA GG	296
		Antisense	GGT CTC CAG TCT TGA GCT GG	
Bovine	iNOS	Sense	TAG AGG AAC ATC TGG CCA GG	372
		Antisense	TGG CAG GGT CCC CTC TGA TG	
Human	ecNOS	Sense	GTG ATG GCG AAG CGA GTG AAG	422
		Antisense	CCG AGC CCG AAC ACA CAG AAC	
Human	iNOS	Sense	TCC GAG GCA AAC AGC ACA TTC A	462
		Antisense	GGG TTG GGG GTG TGG TGA TGT	

reported in Table II. Calibration is performed by coamplification of the same cDNA samples with primers for glyceraldehyde-3-phosphate dehydrogenase (GAPDH), with sequences as follows: sense (GAPDH-L), 5′-CCATGGAGAAGGCTGG-GG-3′; and antisense (GAPDH-R), 5′-CAAAGTTGTCATGGATGACC-3′ (194 amplification product). For PCR amplification, a PerkinElmer GeneAmp PCR System 2400 is used. The reaction mixture contains 10 n*M* Tris–HC1 (pH 8.3), 1.5 m*M* $MgCl_2$, 50 m*M* KCl, 0.25 m*M* of each dNTP, 1 μM of each primer, and 1.25 U Amplitaq DNA polymerase in a 80-μl final volume. The PCR cycles are 30 sec at 94°, 30 sec at 55° (for bovine genes) or 60° (for human genes), and 30 sec at 72°. After 30 (bovine) and 40 (human) cycles of amplification, aliquots of each sample product (20 μl) are electrophoresed on a 3% agarose gel and stained with ethidium bromide. The intensity of the two bands corresponding to the NOS isoform and GAPDH amplification products is measured by densitometry.

Quantitative RT-PCR (i.e., Taqman or real-time technologies) is now proposed as an absolute method to measure gene expression. However, when considering the economic outlay for instrument acquisition and the number of samples that have to be run for each cycle, there are no advantages in using this method in place of the routine clinical analyses.

Summary and Conclusions

Nitric oxide directs endothelial cells in each step of angiogenesis and its role is twofold. *In vitro* administration of NO donor drugs to sparse coronary endothelium increases their mitotic index and favors cell migration and matrix degradation.[3] As a result, *in vivo* NO donors speed up and potentiate neovascular growth.[4] Thus NO acts as the true effector of angiogenesis. However, mediators of angiogenesis signal the angiogenic switch in endothelium by releasing free NO and elevating cGMP levels.[6,7] Competent proangiogenic activity by the vascular endothelial growth factor (VEGF) is mediated by NO-cGMP signaling of MAPK activation in endothelium.[8,20] NO upregulates the expression of endogenous angiogenic factors such as FGF-2. As a consequence of the inability of FGF-2 to be secreted, its upregulation results in endogenous accumulation, which, by a paracrine/intracrine function, contributes to improving endothelial cell survival and possibly to inhibiting apoptotic events. As a result, impairment of endogenous NO signaling in endothelium is coupled to an inability to mount an angiogenic response to VEGF and may result in increased resistance to neovascularization.[5] In this respect, the role of the NOS pathway within endothelial cells can be regarded as an endogenous limiting step for the angiogenic switch, which might not be simply overcome by the redundance of angiogenic growth factors when the promotion of angiogenesis is foreseen or inhibited by the targeting of individual angiogenic effect when a block of angiogenesis is the goal.

Development of NO-targeting drugs with higher selectivity for the different isoforms, combined with targeted tissue delivery to minimize side effects, as well as qualitative assessment of NO function from activity to genomic expression, is a promising strategy for the exploitation of the findings on the role of NO on angiogenesis.

Acknowledgments

We thank Dr. Sandra Donnini for the helpful discussions and her skillful technical assistance. This work was supported by funds from the Italian Ministry of University and Scientific and Technological Research (Cofinanziamento Programmi di Ricerca Scientifica di Rilevante Interesse Nazionale, Grant 9906217877 and MM06037341), Italian Association for Cancer Research (AIRC), and the National Research Council (Target Projects "Biotechnology" and "Biomaterials") and PAR Research Project of the University of Siena to MZ.

[20] M. Ziche, A. Parenti, F. Ledda, P. Dell'Era, H. J. Granger, C. A. Maggi, and M. Presta, *Circ. Res.* **80,** 845 (1997).

[36] Hemangioma Model for *in Vivo* Angiogenesis: Inducible Oxidative Stress and MCP-1 Expression in EOMA Cells

By GAYLE M. GORDILLO, MUSTAFA ATALAY, SASHWATI ROY, and CHANDAN K. SEN

Hemangioma: An *in Vivo* Angiogenesis Model

Hemangiomas are the most frequently occurring tumors of infancy, occurring in approximately 0.54/1000 live births.[1] These tumors of endothelial cell origin can occasionally threaten vision, airway, or life of affected patients. We felt this was an important model of angiogenesis to study for two reasons: the angiogenic processes are extremely potent with this tumor and it is commonly encountered in humans, providing important clinical relevance. Two known murine models of hemangioma are based on the subcutaneous inoculation of transformed endothelial cells. One cell line (EOMA) was derived from a spontaneously arising hemangioma in the 129/J strain[2] and the other from newborn mice directly infected with a retroviral vector expressing the oncogenic papovavirus polyoma middle T antigen.[3] Hemangiomas arise at the site of local virus injection, e.g., skin, thymus, or brain, and cells isolated from these tumors (e.g., skin, s.End cells; brain, b.End cells) can be maintained *in vitro* and inoculated *in vivo* to generate tumors. Both of these cell lines are characterized with regard to the expression of endothelial cell markers[4,5] and demonstrate a branching pattern on Matrigel *in vitro* similar to nontransformed endothelial cells. We have chosen to work with EOMA cells over End cells because EOMA have a clearly identified strain of origin that allows work on syngeneic animals available from commercial vendors. Experiments can be done with mice with an intact immune system rather than having to resort to nude or SCID mice models or tolerate MHC class II differences with the End cell model.

Rapidly proliferating hemangiomas are associated with macrophage infiltration,[1] and we have observed this with EOMA model as well (not shown). Clinical findings correlate the presence of tumor-associated macrophages with increased mortality and vascularity, implying that these cells facilitate the angiogenesis

[1] J. B. Mulliken and A. E. Young, "Vascular Birthmarks: Hemangiomas and Malformations." Saunders, Philadelphia, 1988.

[2] J. C. Hoak, E. D. Warner, H. F. Cheng, G. L. Fry, and R. R. Hankenson, *J. Lab. Clin. Med.* **77,** 941 (1991).

[3] V. L. Bautch, S. Toda, J. A. Hassell, and D. Hanahan, *Cell* **51,** 529 (1987).

[4] J. Obeso, J. Weber, and R. Auerbach, *Lab. Invest.* **63,** 259 (1990).

[5] R. L. Williams, S. A. Courtneidge, and E. F. Wagner, *Cell* **52,** 121 (1988).

0076-6879/02 $35.00

process.[6–8] We are particularly interested in the role of oxidants in regulating angiogenesis. The EOMA model is well suited to address this subject for the following major advantages: (1) injection of EOMA cells results in tumor production in 4 days with 100% efficiency, resulting in quick reliable data generation, (2) EOMA cells can be pharmacologically or genetically treated *in vitro* to study the effects of variable manipulation on endothelial cell behavior *in vitro,* and (3) manipulating these cells *in vitro* followed by injection *in vivo* provides a unique model for studying the influence of tumor cell-derived signals that regulate angiogenesis. Our focus on monocyte chemoattractant protein-1 (MCP-1) is based on the observation that MCP-1 is the primary stimulus for macrophage chemotaxis in numerous conditions.[9] Additionally, signal transduction mediators, NF-κB and AP-1, responsible for MCP-1 expression are redox responsive.[10,11]

Endothelial cells are capable of generating reactive oxygen species (ROS) via NADPH oxidase. A functional NADPH oxidase system is present in human and bovine endothelial cells.[12,13] Furthermore, in endothelial cells, NADPH oxidase appears to be the primary source of ROS with little input from mitochondrial or other sources.[13,14] The described methods associate changes in the redox state of endothelial cells with changes in inducible MCP-1 expression.

Generation of Hemangiomas

1. EOMA cells[2] are grown at 37° and in a 95% relative humidified atmosphere containing 5% CO_2 in 175-cm^2 tissue culture flasks in Dulbecco's modified Eagle's medium (DMEM) supplemented with 10% fetal bovine serum (FBS), penicillin (100 units/ml), and streptomycin (100 μg/ml, Life Technologies, Inc., Carlsbad, CA). Cells are trypsinized when they reach ~85% confluency. The yield is approximately 5–6 million cells/flask.

2. Cells are washed three times in large volumes (e.g., 50 ml) of phosphate-buffered saline, pH 7.4 (PBS), counted, and resuspended in PBS at 5×10^7 cells/ml and kept on ice.

[6] R. D. Leek, C. E. Lewis, R. Whitehouse, M. Greenall, J. Clarke, and A. L. Harris, *Cancer Res.* **56,** 4625 (1996).

[7] R. P. M. Salcedo, H. A. Young, K. Wasserman, J. M. Ward, H. K. Kleinman, J. J. Oppenheim, and W. J. Murphy, *Blood* **96,** 34 (2000).

[8] T. Ueno, M. Toi, H. Saji, M. Muta, H. Bando, K. Kuroi, M. Koike, H. Inadera, and K. Matsushima, *Clin. Cancer Res.* **6,** 3282 (2000).

[9] B. Lu, B. J. Rutledge, L. Gu, J. Fiorillo, N. W. Lukacs, S. L. Kunkel, R. J. North, C. Gerard, and B. J. Rollins, *J. Exp. Med.* **187,** 601 (1998).

[10] C. K. Sen and L. Packer, *FASEB J.* **10,** 709 (1996).

[11] C. K. Sen, *Curr. Top. Cell. Regul.* **36,** 1 (2000).

[12] S. A. Jones, V. B. O'Donnell, J. D. Wood, J. P. Broughton, E. J. Hughes, and O. T. Jones, *Am. J. Physiol.* **271,** H1626 (1996).

[13] H. K. Mohazzab, P. M. Kaminski, R. P. Fayngersh, and M. S. Wolin, *Am. J. Physiol.* **270,** H1044 (1996).

[14] M. J. Somers, K. Mavromatis, Z. S. Galis, and D. G. Harrison, *Circulation* **101,** 1722 (2000).

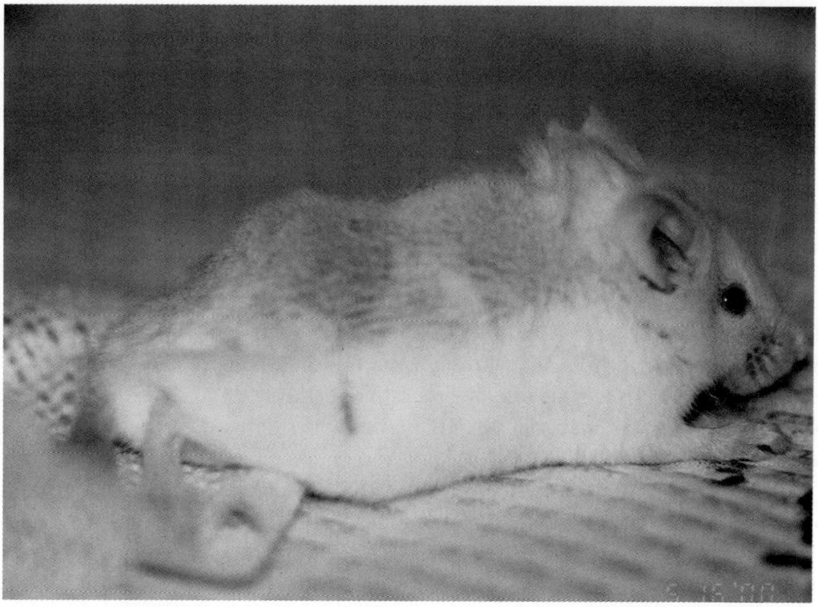

FIG. 1. Tumor-bearing mouse 7 days after injection of EOMA cells.

3. The cell suspension is loaded into 1-cc insulin syringes with a 28-gauge needle (Becton Dickinson, Franklin Lakes, NJ) to a final volume of 100 μl per syringe.

4. 129 P3 mice from Jackson Laboratories (Bar Harbor, ME) are syngeneic with EOMA cells. Six-week-old mice receive inhalation anesthesia and are injected with 100 μl of cell suspension for a total dose of 5×10^6 cells. The cells are injected subcutaneously and a large wheal is noted.

5. Visibly obvious tumor development is apparent 4 days after tumor injection (Figs. 1 and 2).

Note: A paper published in 1990,[4] which characterized the phenotype of EOMA cells, incorrectly identified these as MHC H-2^d, when in fact these cells are MHC H-2^b. The 129/J strain, which is the source of the EOMA, is originally from Jackson Laboratories and has been reclassified by them as 129 P3. They note an MHC background as H-2^b, which has been confirmed by flow cytometry A. VanBuskirk, personal communication. These tumors become very large in size, but do not metastasize. However, mice typically die 21–30 days after injection secondary to Kassabach–Merritt syndrome, which has been reported by others with this model as well.[15,16]

[15] E. D. Warner, J. C. Hoak, and G. L. Fry, *Arch. Pathol. Lab. Med.* **91,** 523 (1971).

[16] M. S. O'Reilly, T. Boehm, Y. Shing, N. Fukai, G. Vasios, W. S. Lane, E. Flynn, J. R. Birkhead, R. R. Olsen, and J. Folkman, *Cell* **88,** 277 (1997).

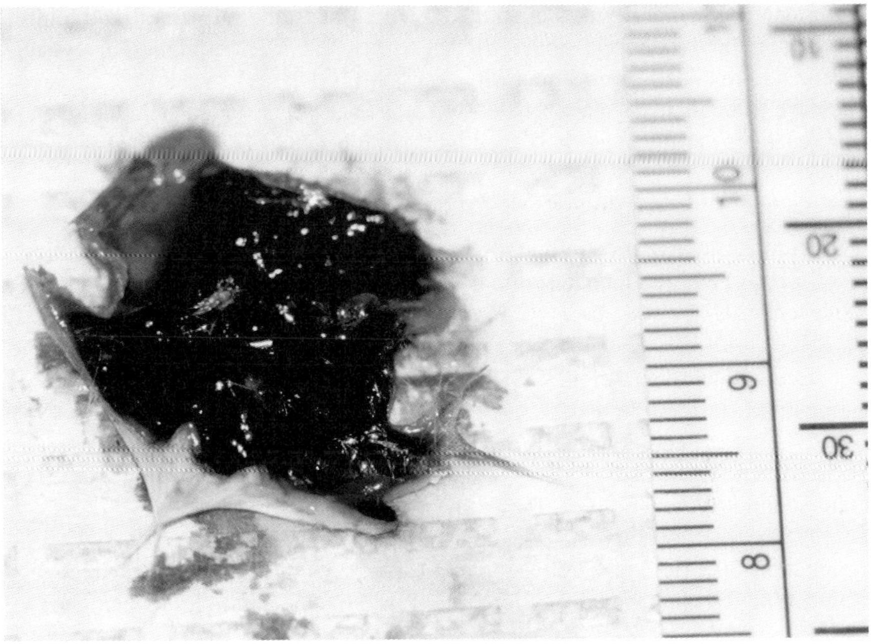

FIG. 2. Appearance of tumor at necropsy 7 days after injection of EOMA cells.

Western Blot Detection of Oxidized Proteins

Proteins are an important target for oxidative challenge. ROS modify amino acid side chains of proteins such as arginine, lysine, threonine, and proline residues to form protein carbonyls. Protein carbonyls can be detected in immunoassays by the reaction with dinitrophenyl hydrazine (DNPH) using monoclonal antidinitrophenyl (DNP) antibodies. This method has proven to be superior to the conventional spectrophotometric method because of its high sensitivity and reproducibility. Furthermore, in Western blot assays, relative oxidation of individual proteins can be determined.

Cell Culture

To perform Western blot analysis on EOMA cells, they are seeded at 1.5×10^5 cells/ml density in 100-mm plates. Cells are incubated for 24 hr in DMEM supplemented with 10% FBS, penicillin (100 units/ml), and streptomycin (100 μg/ml) at 37° in a humidified atmosphere containing 5% CO_2. Following 24 hr of seeding, the culture medium is replaced with fresh medium and the cells are treated with redox-modulating agents for a particular time period. After the treatment period, cells are washed three times with ice-cold PBS, harvested by scraping, resuspended, and lysed in 50 mM Tris–HCl buffer containing 5 μM phenylmethylsulfonyl fluoride (PMSF), 4 μg/ml leupeptin, 10 μg/ml pepstatin, 10 μg/ml aprotonin, and 2.5% sodium dodecyl sulfate (SDS). All reagents were of highest analytical grade and obtained from Sigma (St. Louis, MO). DNA is sheared by repeated passage through a 22-gauge needle. Cell lysates are centrifuged at 14,000*g*, and supernatants are stored at −80° overnight. All procedures are performed at 4°.

Extraction and Derivatization

Derivatization of protein-bound carbonyl groups with DNPH (Acros Organics, New Jersey) and immunodetection are performed according to the procedure of Shacter *et al.*[17]

1. Five microliters of 12% SDS (w/v) is added to a 5-μl aliquot of sample containing 20 μg of protein.
2. After the addition of 10 μl of 10 m*M* DNPH in 10% (v/v) trifluoroacetic acid (Acros Organics, New Jersey), the mixture is incubated for 15 min at room temperature.
3. The reaction mixture is neutralized by adding 7.5 μl of 2 *M* Tris base containing 30% (v/v) glycerol. Samples are treated with Laemmli loading buffer[18] containing 3% mercaptoethanol for 5 min at room temperature for reducing gels.

Immunodetection of Protein-Bound Carbonyl Groups

Proteins are separated by 10% SDS–PAGE under reducing conditions and are transferred onto a nitrocellulose membrane (Amersham Pharmacia Biotech, San Francisco, CA) using a tank transfer system (Hoeffer mini VE, Amersham Pharmacia Biotech). Additionally, equal transfer is checked and quantified by reversible protein staining of the nitrocellulose membrane with Ponceau S (Sigma). After blocking with a 5% (w/v) fat-free milk solution at 37° for 1 hr, membranes are treated with rat monoclonal antibody to DNP (Zymed Laboratories, San Francisco, CA) at 1 : 1000 dilution overnight at 4° and washed 3× for 10 min with Tris-buffered saline containing 0.05% Tween 20 (TTBS). The secondary antibody horseradish peroxidase-conjugated mouse monoclonal antirat antibody (Zymed Laboratories) is used at 1 : 10,000 dilution for 1 hr at room temperature. After 6× for 10 min of washing with TTBS, immunoblots are visualized using Renaissance Western blot chemiluminescence reagent (NEN Life Science Products, Boston, MA) with exposure times between 2 and 5 min (Fig. 3).

Note: Precipitated proteins should be dissolved directly in 6% SDS and reacted with an equal amount of DNPH. If the samples already contain SDS, the

[17] E. Shacter, J. A. Williams, M. Lim, and R. L. Levine, *Free Radic. Biol. Medi.* **17,** 429 (1994).
[18] U. K. Laemmli, *Nature* **227,** 680 (1970).

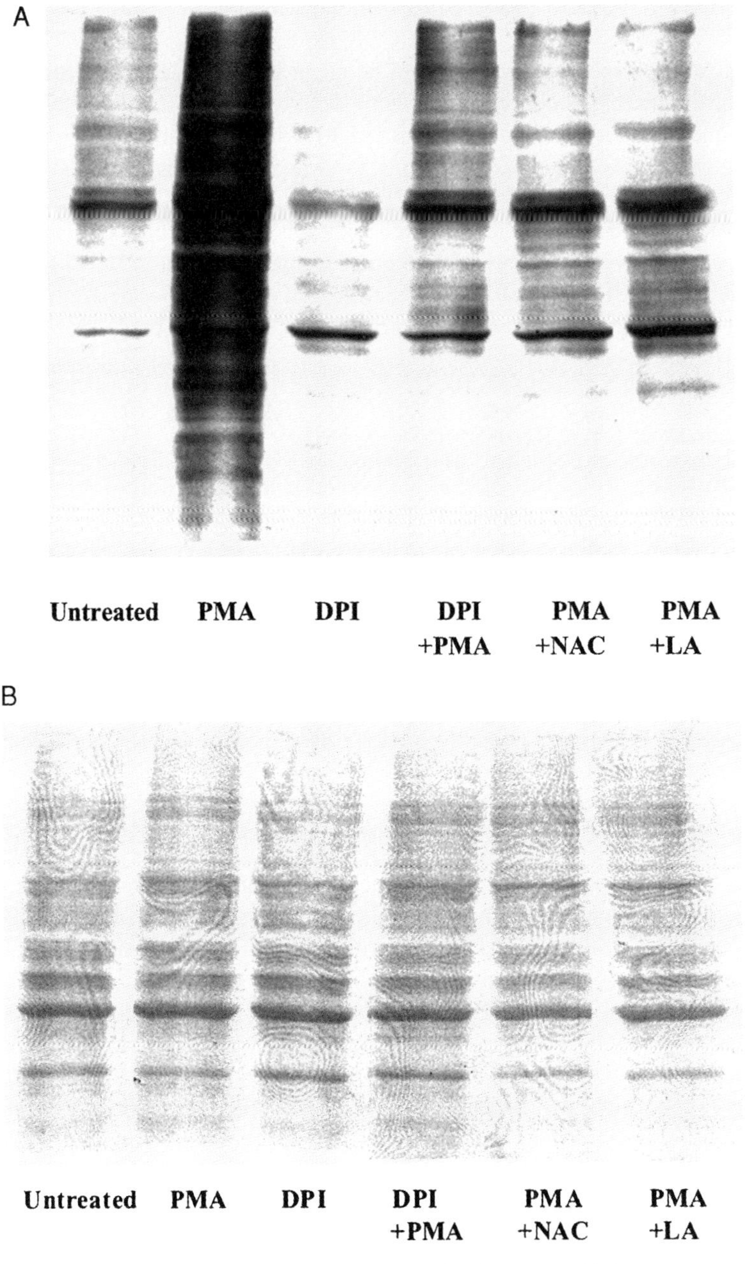

FIG. 3. (A) Protein oxidation in EOMA cells under varying degrees of oxidative stress documented by Western blot using an anti-DNP antibody to detect protein carbonyls. Cells are treated with diphenylene iodonium (DPI, 10 μM) for 1 hr prior to PMA stimulation. *N*-Acetylcysteine (NAC, 20 m*M*) and (*R*,*S*)-racemic lipoic acid (LA, 0.5 m*M*) are added at the same time as PMA. Cells are treated with PMA (100 n*M*) for 24 hr prior to harvesting. (B) Imaging of nitrocellulose membrane used in Western blot demonstrates equal transfer of proteins for all lanes of gel in (A).

amount of SDS in the samples should be taken into the account for the adjustment of the final SDS concentration. DNPH derivatization should be carried out at room temperature and should not exceed 30 min to avoid the formation of nonspecific reactions of unoxidized proteins. Derivatized samples are not boiled prior to SDS–PAGE. Samples without DNPH treatment are used as negative controls. For positive controls, cells are incubated with 200 μM hypochlorous acid (HOCl) for 30 min at 37° in a serum-free medium. The concentration of HOCl present in the diluted NaOCl solution (Fisher Scientific, Pittsburgh, PA) is determined spectrophotometrically using an extinction coefficient (350 M^{-1} cm^{-1}) at 290 nm as described previously.[19]

Footprints of Lipid Peroxidation

In this case, lipid peroxidation is detected using antiserum to 4-hydroxy-2-nonenal (HNE) protein adducts. A number of reactive aldehydes generated during lipid peroxidation process are potently cytotoxic and also have signaling functions[20] because of their stable nature and ability to diffuse within or from the cell. HNE, a major aldehyde produced during lipid peroxidation, has remarkable cytotoxic and mutagenic activities. One of the major mechanisms of HNE reactivity is the modification of sulfhydryl, histidine, and lysine residue of proteins. Although the method described is not suitable for the quantitative analysis, it provides direct comparisons of oxidative lipid damage in cells under varying conditions.

Immunodetection of 4-Hydroxy-2-nonenal Protein Adducts

For the detection of HNE protein adducts, cell lysates are treated with Laemmli sample buffer containing 3% mercaptoethanol for 5 min at 100°. Protein (20 μg) is run on 10% SDS–PAGE and is transferred to a nitrocellulose membrane. After blocking as described earlier for carbonyl detection, the membrane is incubated overnight with antiserum to HNE–protein adducts (Alexis, San Diego, CA) at a dilution of 1 : 500. After washing 3× for 10 min with TTBS containing 0.1% Tween 20, the membrane is incubated for 1 hr at room temperature with a 1 : 10,000 dilution of anti-rabbit horseradish peroxidase-conjugated secondary antibody (Upstate Biotechnology, Lake Placid, NY). The membrane is then washed 6× for 10 min with TTBS containing 0.1% Tween 20, and visualization is performed as described earlier. Equal protein loading is tested and quantified by probing the membrane for mouse β-actin using an antibody against mouse β-actin (Sigma) at

[19] L. J. Yan, M. G. Traber, H. Kobuchi, S. Matsugo, H. J. Tritschler, and L. Packer, *Arch. Biochem. Biophy.* **327,** 330 (1996).

[20] M. Parola, G. Robino, and M. U. Dianzani, *Int. J. Mol. Med.* **4,** 425 (1999).

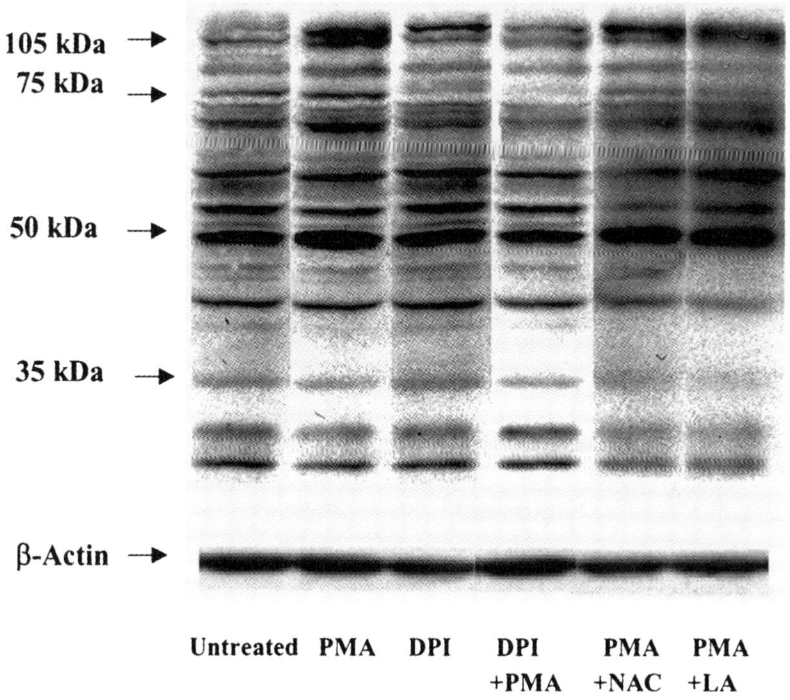

FIG. 4. Lipid peroxidation in EOMA cells under varying degrees of oxidative stress documented by Western blot using antibody to HNE–protein adducts. β-Actin imaging documents equal transfer of cellular proteins for all lanes.

a 1 : 5000 dilution and a 1 : 10,000 dilution of goat anti-mouse secondary antibody (NEN Life Science Products, Boston, MA) (Fig. 4).

Notes: Various protein bands show immunoreactivity with HNE–protein adducts. Specificity and sensitivity of the reactivity of the antiserum are tested using positive controls (50 μM 4-HNE-treated cells for 2 hr at 37°). Cells treated with phorbol 12-myristate 13-acetate (PMA) exhibited a strong HNE positive band at 48 kDa, which is consistent with previous reports.[21]

PMA-Induced Oxidation in EOMA Cells

Phorbol 12-myristate 13-acetate is a potent but nonspecific stimulus of oxidant production that is used to induce oxidant generation in endothelial cells[22] (Fig. 5).

[21] K. Uchida, L. I. Szweda, H. Z. Chae, and E. R. Stadtman, *Proc. Nat. Acad. Sci. U.S.A.* **90,** 8742 (1993).

[22] S. Roy, C. K. Sen, H. Kobuchi, and L. Packer, *Free Radic. Biol. Med.* **25,** 229 (1998).

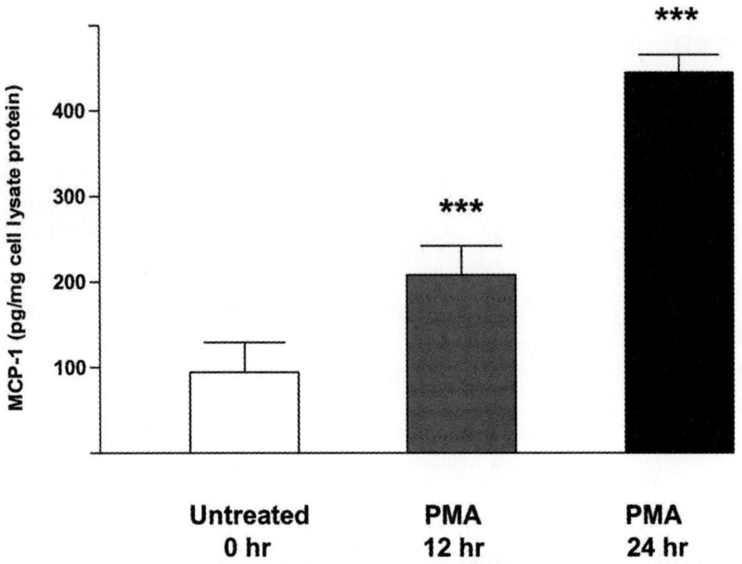

FIG. 5. Kinetics of PMA-induced MCP-1 expression. Selection of PMA concentration based on dose–response curve (data not shown). One-way analysis of variance (ANOVA) was used to compare the effects of PMA and the effects of antioxidants on MCP-1 expression versus untreated control cells. Results are log transformed before statistical analyses to correct for skewing and are presented as mean ± standard deviation (SD). Differences are compared to PMA-untreated cells: $^{***}p < 0.001$.

Western blots document the presence of oxidant footprints in untreated EOMA cells. PMA treatment significantly increased protein oxidation and lipid peroxidation (Figs. 3 and 4). Thus, EOMA cells can generate ROS in response to PMA treatment. PMA-induced oxidation in EOMA cells is sensitive to diphenylene iodonium (DPI) (Calbiochem, San Diego, California), a nonspecific inhibitor of NADPH oxidase.[23] PMA-induced oxidative events in EOMA cells are also inhibited by thiol-based antioxidants *N*-acetylcysteine (NAC) (Sigma) and (*R*,*S*)-racemic lipoic acid (LA, ASTA Medica, Frankfurt, Germany).[24]

Association of PMA-Induced Oxidation with Changes in MCP-1 Protein Expression

MCP-1 protein is detected using ELISA (Fig. 6). DPI and NAC, as well as LA treatments, suppressed PMA-induced MCP-1 expression, suggesting the involvment of redox-sensitive steps in the pathway of PMA-induced MCP-1 expression.

[23] Z. Wei, K. Costa, A. B. Al-Mehdi, C. Dodia, V. Muzykantov, and A. B. Fisher, *Circ. Res.* **85,** 682 (1999).

[24] C. K. Sen, *J. Nutr. Biochem.* **8,** 660 (1997).

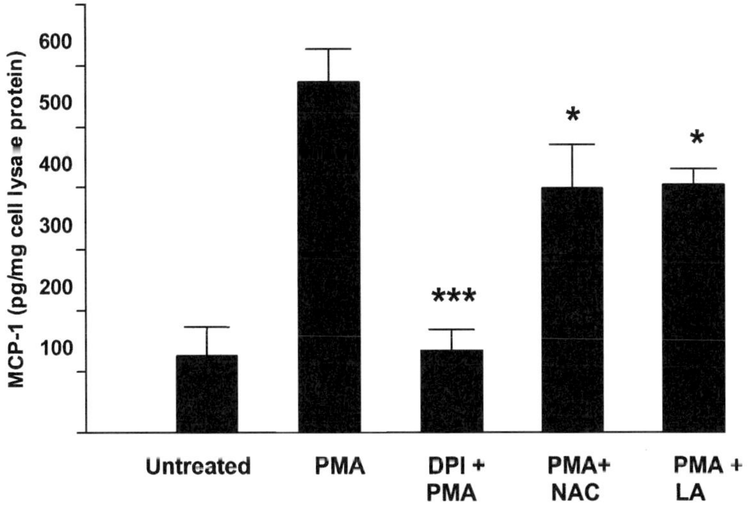

FIG. 6. Downregulation of PMA-induced MCP-1 expression by modulators of the cellular redox state. EOMA cells are treated with DPI (10 μM) for 1 hr prior to PMA stimulation. NAC (20 mM) and LA (0.5 mM) are added at the same time as PMA. Cells are treated with PMA (100 nM) for 24 hr prior to harvesting. When inhibitors are added to PMA, DPI results in a marked inhibition of MCP-1 expression (***$p < 0.001$), whereas NAC and LA treatments cause a significant but smaller inhibition of MCP-1 induction (*$p < 0.05$).

ELISA

To perform ELISA, EOMA cells are cultured in 100-mm plates as described earlier. At the end of the experiment, plates are washed three times with ice-cold PBS and are harvested by scraping. Harvested cells are centrifuged at 1000*g* for 10 min at 4° and are resuspended in 250 μl of extraction buffer [PBS, pH 7.4, containing 5 μM PMSF, 4 μg/ml leupeptin, 10 μg/ml pepstatin, and 10 μg/ml aprotonin]. Cells are disrupted by sonication (Sonic Dismembrator, Fisher Scientific, Pittsburgh, PA) on ice with three 15-sec bursts. Sonicates are centrifuged, and supernatants are stored at −80° overnight. All procedures are performed at 4°. ELISA plates are coated with 100 μl/well of affinity-purified anti-mouse JE/MCP-1 capture antibody (R&D Systems, Minneapolis, MN) at 0.4 μg/ml in PBS. Sealed plates are incubated overnight at room temperature. After incubation, wells are washed four times with 0.05% Tween 20 in PBS and blocked with 300 μl PBS containing 1% BSA, 5% sucrose, and 0.05% NaN_3. Plates are sealed and incubated at room temperature for a minimum of 1 hr. After four washes, 100 μl of standards (recombinant mouse JE/CCL2, R&D Systems), controls, or samples is diluted in dilution buffer (1% BSA in PBS) and added to each well. Standards are used in the range of 20–250 pg/ml. Following an overnight incubation at 4°, wells are washed four times, and 100 μl of the biotinylated anti-mouse JE antibody (R&D Systems) at a dilution of 0.1 μg/ml in dilution buffer is added to each well.

Sealed plates are incubated for 2 hr at room temperature. After washing the plate four times, 100 μl of streptavidin–HRP conjugate (Chemicon, Temecula, CA) at a dilution of 1 μg/ml in dilution buffer is added to each well. Covered plates are incubated for 30 min at room temperature. After six washes, 100 μl of HRP substrate solution (R&D Systems) is added, and plates are incubated for 20 min, protected from light, at room temperature. Finally, 50 μl of stop solution (1 *M* H_2SO_4) is added to the wells and mixed gently. Optical density is determined using a multiple well plate reader at 450 nm. Wavelength correction is done by a reference reading at 560 nm. ELISA results are normalized to total protein concentration using a BCA protein assay kit (Pierce, Rockford, IL).

Acknowledgments

G.G. and M.A. equally contributed to this work and are joint first authors. M.A. is on leave from University of Kuopio, Finland. Supported in part by NIH GM27345 and by U.S. Surgical Corporation grants to C.K.S.

[37] Redox Aspects of Vascular Response to Injury

By FRANCISCO R. M. LAURINDO, HERALDO P. DE SOUZA, MARCELO DE A. PEDRO, and MARIANO JANISZEWSKI

Introduction

The concerted alterations in vascular structure and function that occur in response to a mechanical injury represent more than a model to explore interactions among the endothelium, platelets, and smooth muscle cells.[1] With the development of invasive endovascular strategies for atherosclerosis therapy, the vascular repair reaction has taken an important clinical dimension as a major component of restenosis postangioplasty.[2] In addition, vascular response to injury displays all biological processes typical of vascular pathophysiology (Table I) and thus constitutes a relevant model for many vascular diseases.

The involvement of redox processes as mediators of the vascular repair reaction has been postulated on the basis of their increasingly evident role as signaling mediators of cell proliferation, differentiation, senescence, and apoptosis.[3–5] In

[1] R. Ross and J. Glomset, *N. Engl. J. Med.* **295,** 369 (1976).
[2] P. Libby and H. Tanaka, *Progr. Cardiovasc. Dis.* **40,** 97 (1997).
[3] J. I. Abe and B. C. Berk, *Trends Cardiovasc. Med.* **8,** 59 (1998).
[4] K. Irani, *Circ. Res.* **87,** 179 (2000).
[5] C. K. Sen and L. Packer, *FASEB J.* **10,** 709 (1996).

0076-6879/02 $35.00

TABLE I
TYPICAL GENERAL EVOLUTION OF VASCULAR RESPONSE TO INJURY

Time after injury	Major biological processes involved
Immediate	Massive early vascular cell apoptosis
24–48 hr	Platelet and leukocyte infiltration
	Edema
	Early wave of medial layer proliferation
4–7 days	Cell migration to incipient neointimal layer
	Second wave of proliferation, mainly in neointimal layer
	Early caliber loss and vasoconstriction
7–14 days	Cell migration to enlarged neointimal layer
	Maximal rate of caliber loss
	Extracellular matrix secretion
	Residual neointimal proliferation
14–28 days	Second wave of neointimal apoptosis
	Maximal extracellular matrix secretion
	Neointimal smooth muscle cell differentiation
	Final residual caliber loss
	Near-maximal neointimal layer size

addition, extra/intercellular redox processes may provide a regulatory integrative pathway for the overall coordination of such events at the tissue level. The occurrence of redox processes in a complex and basically homeostatic reaction such as tissue repair raises questions about the extent to which such processes can be defined as physiological vs pathological[2] or can be considered as oxidant signaling vs toxicity.[4] In fact, in parallel with a well-known less specific chemical toxicity, the biological toxicity of reactive oxygen species (ROS) may be viewed as disordered signaling resulting from increased, uncompensated, or decompartmentalized ROS production. While redox signaling involves oxidizing or reducing electron transfer reactions mediated by independent intermediates such as free radicals, reducing equivalents, or metals, production of ROS as second messengers is the hallmark of redox signaling.[3–5]

The purpose of this article is to critically analyze some specific aspects of vascular injury models and procedures for the assessment of ROS, particularly superoxide, and NAD(P)H oxidase activity at different time points after vascular injury.

Vascular Injury Model

The model of vascular response to injury should be chosen in line with the basic purpose of the study. Simple deendothelialization is useful to assess interactions between platelets and the vessel wall or as a means of creating a localized atherosclerotic plaque developed through hyperlipidemic diet. Models described as balloon deendothelialization in small animals (e.g., rat) are uniformly accompanied by significant medial smooth muscle injury and resemble overdistention models.[2,6] A number of vascular injury models in mice have been described,[7] which are useful with the caveat that they may involve peculiar mechanisms; e.g., a guide wire injury model involves neutrophil infiltration.[8]

Our laboratory has used and validated extensively an overdistention balloon injury in intact (i.e., nonatherosclerotic) rabbit iliac arteries, the basic aspects of which were described previously.[9–11] The time course of the vascular repair reaction in this model is representative of the overall pattern displayed in analogous models (Table I) in the same or different species (e.g., pig and rat), although there are significant differences as well.[12] A major question with all such models is the appropriate end point variable to be assessed. Frequently, results are reported as neointimal size in perspective with the assessment of smooth muscle cell proliferation. This approach is limited by two major drawbacks. First, neointimal size depends not only on proliferation, but also on cell migration and extracellular matrix accumulation. Also, the effect of neointimal proliferation is counteracted by apoptosis[13] (Table I). In addition, it is now accepted that a major determinant of vascular lumen caliber after injury, as well as in native atherosclerotic lesions, is the extent of strictu sensu remodeling, which may be either constrictive or compensatory.[14] Although the pathophysiology of remodeling may share common mechanisms with neointimal formation, usually there is little correlation between the extent of these two phenomena and, consequently, no correlation between extent of neointima and vascular lumen narrowing.[15,16] In summary, analysis of vascular caliber or remodeling is desirable in order to assess a more complete

[6] H. Perlman, L. Maillard, K. Krasinski, and K. Walsh, *Circulation* **95,** 981 (1997).

[7] P. Carmeliet, L. Moons, J. M. Stassen, M. D. Mol, A. Bouché, J. J. van den Oord, M. Kockx, and D. Collen, *Am. J. Pathol.* **150,** 761 (1997).

[8] M. Roque, J. T. Fallon, J. J. Badimon, W. X. Zhyang, M. B. Taubman, and E. D. Reis, *Arterioscler. Thromb. Vasc. Biol.* **20,** 335 (2000).

[9] M. Janiszewski, F. R. M. Laurindo, C. A. Pasqualucci, P. L. da Luz, and F. Pileggi, *Angiology* **47,** 349 (1996).

[10] L. C. P. Azevedo, M. A. Pedro, L. C. Souza, H. P. Souza, M. Janiszewski, P. L. da Luz, and F. R. M. Laurindo, *Cardiovasc. Res.* **47,** 436 (2000).

[11] H. P. Souza, L. C. Souza, V. M. Anastacio, A. C. Pereira, M. L. Junqueira, J. E. Krieger, P. L. da Luz, O. Augusto, and F. R. M. Laurindo, *Free Radic. Biol. Med.* **28,** 1232 (2000).

[12] J. Lefkovits and E. J. Topol, *Progr. Cardiovasc. Dis.* **40,** 141 (1997).

[13] K. Walsh, R. C. Smith, and S. Kim, *Circ. Res.* **87,** 184 (2000).

[14] M. R. Ward, G. Pasterkamp, A. C. Yeung, and C. Borst, *Circulation* **102,** 1186 (2000).

[15] F. R. M. Laurindo, P. C. Tufolo, M. Liberman, M. Janiszewski, and P. L. da Luz, *Eur. Heart J.* **17,** 115 (1996). [Abstract]

[16] A. M. Lafont, L. A. Guzman, P. L. Whitlow, M. Goormastic, J. F. Cornhill, and G. M. Chisolm, *Circ. Res.* **76,** 996 (1995).

picture of the vascular response. Such analysis must be performed either *in vivo* through arteriography or intravascular ultrasound or postmortem in some vascular beds under strict protocols for tissue processing.

From a vascular biology standpoint, these simple and reproducible injury models represent a connection between reductionistic studies in cell culture and *in vivo* pathophysiology. However, clinical restenosis is a complex phenomenon, likely a variable combination of (a) lumen obstruction by the original atheroma, (b) a *de novo* proliferative and remodeling process with neoplastic features, and (c) scar formation, i.e., homeostatic vessel wall repair, which is the major component of intact artery injury models.[10] Therefore, the assumption that intact artery injury is a model of restenosis is clearly an oversimplification. As a tool to test therapeutic procedures, our model is further limited by a lack of baseline atheroma and the innumerable reported discrepances between small animal and human responses to antirestenosis interventions.[2,12]

The homeostatic character of the vascular repair reaction must be kept in mind. This leads to redundancy of mediators, with consequent resistance to therapeutic inhibition of specific factors, e.g., a growth peptide or a kinase. Such character also leads to a balance among the components of vascular repair so that inhibition of proliferation or early sustained vasoconstriction may be counteracted by later increased remodeling.[15–17]

The extent of neointimal thickening in our and other's analogous models can be significantly nonuniform.[18] Also, similar to human atheromas and restenosis, the extent and pattern of remodeling may be quite focal,[14] e.g., adjacent segments in the same artery may sometimes exhibit constrictive and compensatory remodeling. Therefore, standardized and representative sampling of the whole artery is necessary. We usually exclude from the study rabbits that develop large mural thrombi, aneurysms, or evidences of dissection. The balloon-to-artery ratio at arteriography is 1.2 to 1.4. Increased uniformity can be obtained with more severe injury, but this may lead to nonphysiological or clinically unrealistic responses.

Methods for Characterization of Oxidant Stress after Injury

Prior observations suggested early oxidative stress and ROS production immediately after injury.[19,20] For example, oxidative stress induced by a single infusion of oxidized glutathione or cystine (4 mg/kg intraarterially) immediately after balloon injury significantly amplified the vascular response assessed 14 days later.[20]

[17] M. P. Bendeck, C. Irvin, and M. A. Reidy, *Circ. Res.* **78,** 38 (1996).

[18] C. Bauters and J. M. Isner, *Progr. Cardiovasc. Dis.* **40,** 107 (1997).

[19] F. R. M. Laurindo, P. L. da Luz, L. Uint, T. R. Rocha, R. G. Jaeger, and E. A. Lopes, *Circulation* **83,** 1705 (1991).

[20] M. Janiszewski, C. A. Pasqualucci, L. C. Souza, F. Pileggi, P. L. da Luz, and F. R. M. Laurindo, *Cardiovasc. Res.* **39,** 327 (1998).

Using a variety of techniques, our laboratory characterized the time course and pattern of oxidant stress early after *ex vivo* injury.[11] Further studies with similar techniques assessed redox status at different stages of the vascular repair.[10]

Choice of Control Arteries

As control arteries for acute injury experiments, uninjured arteries of the same animal, preferentially the matched contralateral artery segments, are used to allow paired design. For chronic injury studies, potential restrictions to such controls include (a) systemic reactions to manipulation and inflammation, leading to induction of stress proteins or nitric oxide synthase and (b) flow imbalances between injured vs control arteries due to occlusion of arterial branches by injury. For arteriographic caliber, we have found analogous results by comparing the injured artery caliber either with the uninjured contralateral artery or with its respective preinjury angiogram. Basal lucigenin reductase activity is similar in such controls vs intact arteries. However, we have observed some differences in NADPH oxidase activity and glutathione levels (unpublished data).

Procedures

Lucigenin Chemiluminescence

1. Buffer A: Krebs–HEPES bicarbonate buffer of the following composition (in m*M*): NaCl, 118.3; KCl, 4.69; $CaCl_2$, 1.87; $MgSO_4$, 1.20; KH_2PO_4, 1.03; $NaHCO_3$, 25.0; glucose, 11.1; and HEPES, 20.0. pH is adjusted with NaOH to 7.4.
2. After sacrifice and exsanguination, the abdominal aorta is canulated and gently perfused with buffer A at 4–8° until no gross blood is visible through an inferior vena cava cut down. The iliac arteries are dissected and freed from visible perivascular tissue *in situ*. Analogous procedures are performed for the carotids.
3. The arteries are removed and cut into 1.0- to 1.5-cm segments, which are incubated in buffer A, bubbled with 95% O_2/5% CO_2 at 37° for 30 min.
4. Segments are then transferred to a petri dish, and a deflated coronary-type balloon angioplasty catheter is introduced into the lumen. The artery segment (plus the balloon) is quickly transferred to the chemiluminescence vial and immersed in buffer A containing lucigenin. Light exposure is minimized during this procedure.
5. Keeping the artery totally immersed in the solution, the balloon (2.5 or 3.0 mm in diameter) is inflated at 8 atm during 1 min. After deflation, the artery is quickly removed from the balloon back to the same buffer and the counts are immediately performed.
6. Lucigenin concentration should be the smallest possible to yield consistent discrimination of signal vs background, typically 5 μM. We use a six-channel Berthold 9505 luminometer (EG&G Instruments GmbH, Munich, Germany) with the temperature set at 37°. Scintillation counters in an out-of-coincidence mode

are also quite sensitive, although temperature control is usually not possible and this option has become less available in new models. Luminometers designed for molecular biology applications are inadequate.

7. Strict control of pH is essential for accuracy.[21] Krebs–HEPES buffer, which provides more physiological milieu for arterial segments and whole cells, is generally adequate, but occasionally the pH may drift after 5–10 min reading. If a stronger acid or base must be added, the use of 50 m*M* phosphate-buffered saline (PBS) provides better control.

8. SOD inhibits luminescence signals by 40–50%. More complete blockade can be achieved with the cell-permeable SOD mimetic Mn(III) tetrakis (4-benzoic acid) porphyrin chloride (MnTBAP, 10–50 μM).

9. For studies of the late stages of vascular repair, similar techniques are used, except that after the early equilibration period, the injured and contralateral control segments are transferred to luminescence vials, followed by lucigenin.

Comments

A great deal of knowledge in redox vascular biology has been shaped by findings obtained with lucigenin chemiluminescence, formerly assumed to measure superoxide production.[22] Recent reinterpretation of this method demonstrated its major drawbacks for superoxide detection, particularly the typical low-output production by vessels.[23–26] A major limitation of lucigenin is artifactual luminescence generation due to redox cycling; i.e., the autooxidation of lucigenin cation-radical leading to superoxide production.[23,24] Because the extent of redox cycling is dependent on the relative balance between superoxide and lucigenin levels, it is possible to reduce such artifacts to a workable level by using small lucigenin concentrations, e.g., 1–10 μM.[27,28] It should not be assumed, however, that redox cycling is totally eliminated at such lower concentrations. After injury, while it is conceivable that the early luminescence increase[11] (Table II) is in part due to direct reduction by peak high levels of superoxide, a more conservative inference is solely that massive electron transfer, largely enzyme mediated, occurred during

[21] R. P. Brandes, M. Barton, K. M. H. Philippens, G. Schweitzer, and A. Mügge, *J. Physiol.* **500,** 331 (1997).

[22] K. Faulkner and I. Fridovich, *Free Radic. Biol. Med.* **15,** 447 (1993).

[23] J. Vásquez-Vivar, N. Hogg, K. A. Pritchard, P. Martasek, and B. Kalyanaraman, *FEBS Lett.* **403,** 127 (1997).

[24] S. I. Liochev and I. Fridovich, *Arch. Biochem. Biophys.* **337,** 115 (1997).

[25] M. M. Tarpey, C. R. White, E. Suarez, G. Richardson, R. Radi, and B. A. Freeman, *Circ. Res.* **84,** 1203 (1999).

[26] S. Liochev and I. Fridovich, *Free Radic. Biol. Med.* **25,** 926 (1998).

[27] Y. Li, H. Zhu, P. Kuppusamy, V. Roubaud, J. L. Zweier, and M. A. Trush, *J. Biol. Chem.* **273,** 2015 (1998).

[28] M. P. Skatchkov, D. Sperling, U. Hink, A. Mulsch, D. G. Harrison, I. Sindermann, T. Meinertz, and T. Munzel, *Biochem. Biophys. Res. Commun.* **254,** 319 (1999).

TABLE II
OXIDATIVE STRESS INDEXES AT DIFFERENT STAGES OF THE VASCULAR REPAIR REACTION: SUMMARY OF RESULTS[a,b]

Time after injury	Lucigenin reductase activity (250 μM)	Lucigenin reductase activity (5 μM)	Coelenterazine luminescence (5 μM)	Superoxide dismutase activity	GSH/GSSG ratio
Control					
Basal	19 ± 4	9 ± 2	608 ± 126	58.2 ± 1.9	2.18 ± 0.16
After DETC	72 ± 19				
Immediate	2070 ± 180[c]	225 ± 46[c]	1750 ± 250[c]		0.81 ± 0.16[c]
7 days					
Basal	101 ± 29[c]	7 ± 1	671 ± 190	34.1 ± 9.1[c]	1.71 ± 0.26
After DETC	95 ± 15				
14 days	27 ± 6				
28 days	9 ± 4				2.27 ± 0.13

[a] Chemiluminescence data are expressed as cpm $\times$ 10^2/mg/min and SOD activity data as units/mg protein. Data with lucigenin were obtained at lucigenin concentrations of 250 or 5 μM.
[b] Data are the compilation of works reviewed in L. C. P. Azevedo, M. A. Pedro, L. C. Souza, H. P. Souza, M. Janiszewski, P. L. da Luz, and F. R. M. Laurindo, *Cardiovasc. Res.* **47,** 436 (2000).
[c] $p < 0.05$ vs control. Data are means $\pm$ SEM.

or shortly after vascular injury. While such an electron flow in the presence of oxygen implicates a *high probability* of superoxide production, it cannot be ascertained *which fraction* of the electron flow was channeled toward superoxide. The term lucigenin reductase activity is therefore employed to convey this uncertainty regarding the detection of superoxide. The *in vivo* clarification of this problem is yet difficult, but an overall conclusion is that more than one technique should always be used to characterize ROS generation in biological systems.

Coelenterazine Chemiluminescence

1. The coelenterazine stock solution is prepared in ethanol at 1 or 2 mM final concentration (MW is 423.5) and divided into 0.5-ml aliquots, purged with nitrogen, and stored in small vials at $-20°$ protected from light, under nitrogen. It should be protected from direct light during use and purged again with nitrogen before storage. These precautions are important, as coelenterazine oxidizes and loses activity very quickly in contact with oxygen.

2. Final coelenterazine concentrations are typically 5–10 μM, although we have not tested lower concentrations. Assays are performed in PBS or potassium phosphate buffer.

3. The background is typically higher, and the signal-to-noise ratio is not as good as with lucigenin. Most of the background is given by coelenterazine

oxidation by molecular oxygen,[29] also possibly by metals. The signal-to-noise ratio improves if the buffer is treated with Chelex or DTPA 0.1 mM, but DTPA biological effects should be controlled.

4. Because superoxide levels are affected by nitric oxide output, the addition of *N*-methyl-L-arginine (L-NMMA, 0.1 m*M*) to the buffer enhances accuracy and, furthermore, minimizes coelenterazine oxidation by peroxynitrite, also a strong agonist of this probe. We found this step to be important to measure the NAD(P)H-driven increase in superoxide, i.e., NAD(P)H oxidase activity (see later). The increase in signals is again less than with lucigenin. Coelenterazine signals are antagonized by the SOD-mimetic MnTBAP.

5. A conveniently priced source of coelenterazine is PROLUME, Ltd. (www.prolume.com); 1085 William Pitt Way, Pittsburgh, PA 15238 [Tel: (412) 826-5055; FAX: (412) 826-5056].

Comments

Coelenterazine is a novel option for superoxide quantification in biological systems and undergoes substantially less redox cycling than lucigenin,[25] although it still seems far from being the ideal probe. Contrary to lucigenin, coelenterazine emits light when oxidized[29] and thus should focus on the oxidant chemistry of superoxide. Other probes have been developed with similar profile, e.g., CLA[30] and compound 5 [31,32] (soon to become available commercially). In general, these probes are less specific for superoxide than lucigenin, being potentially affected by molecular oxygen as well as a number of oxidants (e.g., metals, peroxynitrite, hydrogen peroxide).

Electron Paramagnetic Resonance Spectroscopy

1. Segments of thoracic aorta (4–5 cm long) are removed from rabbits killed as described earlier and briefly incubated in oxygenated Krebs–HEPES buffer (buffer A described earlier).

2. A balloon catheter, 6.0 mm in diameter, is introduced into the vessel lumen and both are quickly transferred to a tube containing 5 ml oxygenated Krebs–HEPES buffer, the spin trap DEPMPO (5-diethoxyphosphoryl-5-methyl-1-pyrroline-*N*-oxide, from Oxis, Oregon, 100 m*M*), NADH and NADPH (0.1 m*M* each) at 25° in the dark. After 3 min, the balloon is inflated at 8.0 atm for 60 sec

[29] K. Teranishi and O. Shimomura, *Anal. Biochem.* **249,** 37 (1997).

[30] M. P. Skatchkov, D. Sperling, U. Hink, E. Anggard, and T. Munzel, *Biochem. Biophys. Res. Commun.* **248,** 382 (1998).

[31] O. Shimomura, C. Wu, A. Murai, and H. Nakamura, *Anal. Biochem.* **258,** 230 (1998).

[32] I. Fleming, U. R. Michaelis, D. Bredenkötter, B. Fisslthaler, F. Dehghani, R. P. Brandes, and R. Busse, *Circ. Res.* **88,** 44 (2001).

and then is removed. The artery fragment is further incubated with the spin trap for 40 min.

3. Samples (500 μl) of the incubation buffer are collected at desired time points and immediately frozen in liquid N_2. We collected samples immediately and up to 40 min after injury. Experiments are repeated with added SOD or other antagonists.

4. Samples are kept at -80° until analysis. EPR spectra are recorded at room temperature in a Bruker EMX EPR spectrophotometer at 9.76 GHz employing 100 KHz field modulation (Fig. 1). Computer simulation of spectra can be performed with the software PEST (Public EPR Simulation Tools) WinSim version 0.96, by David Duling, obtained through the Internet from the National Institute of Environmental Health Sciences.

Comments

The novel spin trap DEPMPO is particularly useful to detect superoxide, as the DEPMPO–OOH adduct has a longer half-life in chemical systems than the DMPO–OOH adduct, from the parent spin trap DMPO.[33] However, in cells and tissues, the DEPMPO–OOH adduct is preferentially reduced to DEPMPO–OH, similarly to the known DMPO–OOH adduct.[34] An assay of endothelial oxidase with DEPMPO has been reported.[35] The complete inhibition of the adduct spectra observed after injury by native Cu,Zn superoxide dismutase (SOD, Fig. 1) suggests that superoxide or related species, at least in part extracellular, were involved in the genesis of the EPR signal. Analogous blockade of EPR adduct signals by SOD was found after shear stress stimulus in the rabbit aorta.[36]

Glutathione Levels

In our experiments of acute injury, balloon overdistention was performed *in vivo* and the rabbits were killed 10 min later. Injured and control iliac arteries were removed and homogenized with a mortar and pestle in liquid nitrogen. On the day of analysis, 5% metaphosphoric acid was added to the frozen homogenate, and the assay was performed as described in detail in this series,[37,38] consisting of spectrophotometric analysis of NADPH consumption following reaction with 5,5′-dithiobis(2-nitrobenzoic) acid (DTNB or Ellman's reagent). GSSG levels were analyzed after treatment of samples with NEM.

[33] C. Frejaville, H. Karoui, B. Tuccio, F. Le Moigne, M. Culcasi, S. Pietri, R. Lauricella, and P. Tordo, *J. Med. Chem.* **38,** 258 (1995).

[34] Y. Xia and J. L. Zweier, *Proc. Natl. Acad. Sci. U.S.A.* **94,** 6954 (1997).

[35] M. J. Somers, J. S. Burchfield, and D. G. Harrison, *Antiox. Redox Signal.* **2,** 779 (2000).

[36] F. R. M. Laurindo, M. A. Pedro, H. V. Barbeiro, M. H. C. Carvalho, O. Augusto, and P. L. da Luz, *Circ. Res.* **70,** 700 (1994).

[37] T. P. M. Akerboom and H. Sies, *Methods Enzymol.* **77,** 373 (1981).

[38] M. E. Anderson, *Methods Enzymol.* **113,** 548 (1985).

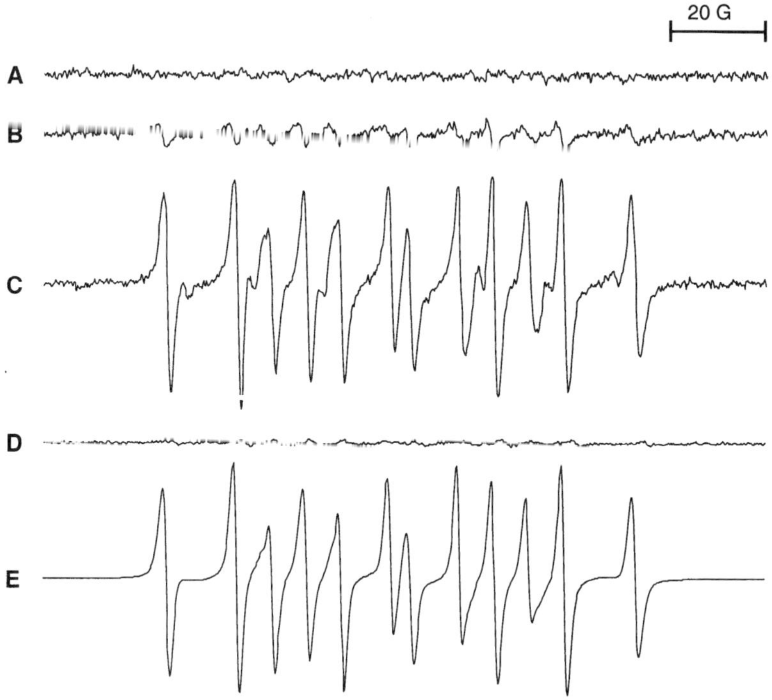

FIG. 1. EPR spin adduct spectra obtained with DEPMPO demonstrating vascular free radical generation after balloon injury. Little EPR signals were detected in the conditions depicted in A [buffer with DEPMPO (100 m*M*)] and in B (buffer plus intact vessel). After balloon overdistention injury, a robust EPR spectrum was detected (C) and shown by computer simulation (E) to be a composite of DEPMPO–OH radical adduct (45% contribution), $a_N = 14.0$ G; $a^P = 47.3$ G; $a_H = 13.2$ G, $a_H = 0.27$ (^{3}H), and two DEPMPO–carbon-centered radical adducts (34 and 21% contribution), $a_N = 14.6$ G; $a^P = 45.9$ G; $a_H = 21.6$ G and $a_N = 14.8$ G; $a^P = 47.8$ G; $a_H = 20.7$ G, respectively. These carbon-centered radical adducts might derive from oxidation of spin trap contaminants [H. Karoui, N. Hogg, C. Fréjaville, P. Tordo, and B. Kalyanaraman, *J. Biol. Chem.* **271,** 6000 (1996)]. Injury performed in the presence of SOD (500 units/ml) significantly antagonized adduct generation (D), confirming superoxide radical as the primary species underlying the observed spectra. Instrumental conditions: microwave power, 20 mW; modulation amplitude, 1 G; time constant 81×10^{-3}; scan rate, 0.466 G/sec; gain, 2×10^6. Reprinted from H. P. Souza, L. C. Souza, V. M. Anastacio, A. C. Pereira, M. L. Junqueira, J. E. Krieger, P. L. da Luz, O. Augusto, and F. R. M. Laurindo, *Free Radic. Biol. Med.* **28,** 1232 (2000) with permission from Elsevier Science.

Comments

Careful standardization and simultaneous analysis of controls are mandatory, as minor variations in sample manipulation, survival time after injury, and type of artery selected as control appear to affect the extent of glutathione oxidation. Despite precautions, the lower GSH/GSSG ratio found in arteries (Table II) vs other tissues[37] suggests that GSH oxidation can still occur during the more elaborate dissection required in vessels, although to some degree this may be a peculiarity of such cells. Depletion of total glutathione pool, which should be quantified in GSH equivalents, accompanies the fall in GSH/GSSG ratio and may be due to leaking, protein glutathiolation, or increased efflux, which is a feature of apoptotic cells.[39]

Vascular Superoxide Dismutase Activity

The cytochrome *c* (Cyt c) assay of SOD activity in vessel homogenates is similar to other tissues.[40] For the control curve, 5 or 10 μl xanthine oxidase diluted in either 995 or 990 μl buffer, respectively, should reduce Cyt c at 0.025–0.030 abs U/min. The homogenate concentration is adjusted so that the assay curve with 5 μl homogenate should inhibit Cyt c reduction rate by 20–40%. One unit of SOD activity is defined as 50% inhibition of Cyt c reduction rate vs control.

Comments

Decreased SOD activity/expression after vascular injury (Table II) may not only directly affect superoxide output, but also its detection with lucigenin. Decreased SOD expression/activity is uniformely observed in mice knockout for NAD(P)H oxidase components,[41] whereas other data show increased SOD levels with overexpression of the oxidase subunit rac.[42] The mechanisms of such SOD changes are unclear.

Vascular Redox Status at Different Stages of Vascular Repair: Summary of Results

A summary of results[10,11] is shown in Table II. *Ex vivo* acute injury in arterial rings yielded a massive oxygen-dependent luminescence peak, which decayed exponentially and was proportional to the degree of injury (with different balloon sizes). The early peak was not seen, although the signals were still increased, if the injury was not performed within lucigenin-containing buffer. Careful removal of endothelium before overdistention had little effect on luminescent signals.[11] With coelenterazine and lower lucigenin concentrations, analogous signal increases occurred, which were abolished by MnTBAP. EPR spin trapping provided more conclusive evidence for superoxide production (Fig. 1). Acute *in vivo* overdistension led to a fall in reduced and to an increase in oxidized glutathione, together

[39] D. J. Van den Dobbelsteen, C. S. I. Nobel, J. Schlegel, I. A. Cotgreave, S. Orrenius, and A. F. G. Slater, *J. Biol. Chem.* **271,** 15420 (1996).

[40] N. Inoue, S. Ramasamy, T. Fukai, R. M. Nerem, and D. G. Harrison, *Circ. Res.* **79,** 32 (1996).

[41] E. Hsich, B. H. Segal, P. J. Pagano, F. E. Rey, B. Paigen, J. Deleonardis, R. F. Hoyt, S. M. Holland, and T. Finkel, *Circulation* **101,** 1234 (2000).

[42] B. Selvakumar, H. H. Hassanaim, G. Nuovo, K. D. Sarkar, and P. J. Golschmidt-Clermond, *Circulation* **102,** II-134 (2000). [Abstract]

with 29% loss of the glutathione pool. A second injury performed *ex vivo* after a first *in vivo* or *ex vivo* injury yielded 75–95% attenuation in the luminescence counts in comparison with the first injury in a naive artery segment.[43]

After superoxide generation and oxidative stress immediately after balloon dilation, the injured segments examined 24 or 48 hr later showed variable but significant luminescence increase vs uninjured controls.[10] At days 4 and 7 after injury, the peak neointimal proliferative activity in this model, lucigenin reductase activity (250 μM) was significantly increased vs normal, although glutathione depletion was not significant any longer (Table II). However, there was no difference in lucigenin reductase activity assessed with 5 μM lucigenin concentration or with coelenterazine chemiluminescence. Overall, this suggests increased electron transfer activity detected with higher lucigenin concentration, but the fraction to which this is proportionally reflected in superoxide production is unclear. Some of this increased luminescence could also be favored by a deficit in SOD activity (Table II), allowing the redox cycling of lucigenin at higher concentrations to become more conspicuous. In fact, incubation with the SOD inhibitor diethyl dithiocarbamate (DETC) increased signals only in the control, and to the same level of the injured artery (Table II). In later stages after injury, despite substantial neointimal thickening, minor residual proliferation, and accelerated caliber loss, injured segments show little evidence of enhanced superoxide production and oxidative stress at days 14 and 28 after injury. Thus, the neointima per se is not a sufficient condition for increased superoxide production.[10]

Therefore, the major component of oxidative stress after injury appears to occur immediately after injury and is sustained in a way that roughly parallels the early inflammatory and proliferative phase of vascular repair. It should be considered, however, that given the limitations in methods for superoxide quantification, subtle, transient, or localized free radical production cannot be excluded in late vascular repair. In fact, data with hydroethydine fluorescent technique[44,45] depicted increased superoxide levels in connection with NAD(P)H oxidase activity (see later), although this technique, by involving tissue freezing and cutting, may not accurately reflect *in vivo* production.

Redox-Dependent Signaling Processes after Injury

The pathophysiological processes triggered by injury (Table I) are associated with a number of redox-dependent signaling events (Table III), which at first may

[43] M. A. Pedro, M. Janiszewski, H. P. Souza, and F. R. M. Laurindo, *Rev. Farm. Bioquím. Univ. S. Paulo* **34**(Suppl. 1), 95 (1998). [Abstract]

[44] F. J. Miller, A. E. Shriver. W. G. Li, T. He, and N. L. Weintraub, *Circulation* **102,** II-126 (2000). [Abstract]

[45] K. Szocs, B. Lassegue, D. Sorescu, L. L. Hilensky, L. Valppu, T. L. Couse, J. N. Wilcox, M. T. Quinn, J. D. Lambeth, and K. K. Griendling, *Arterioscler. Thromb. Vasc. Biol.* **22,** 21 (2002).

TABLE III

REDOX-DEPENDENT VASCULAR CELL SIGNALING PROCESSES TRIGGERED BY INJURY[a]

Model	Signaling pathway	Comments	Reference
Balloon overdistension, rabbit aorta	NF-κB	Increased activation early after injury, inhibited by diphenylene iodonium (see text)	H. P. Souza, L. C. Souza, V. M. Anastacio, A. C. Pereira, M. L. Junqueira, J. E. Krieger, P. L. da Luz, O. Augusto, and F. R. Laurindo, *Free Radic. Biol. Med.* **28,** 1232 (2000)
Balloon injury, rabbit aorta	Proinflammatory cytokines	Proliferating medial and neointimal cells expressed TNF-α, but not bFGF or IL-1	H. Tanaka, G. Sukhova, D. Schwartz, and P. Libby, *Arterioscler. Thromb. Vasc. Biol.* **16,** 12 (1996)
Balloon injury, rat carotid artery	MAPK (ERK and SAPK/JNK)	ERK2 and JNK1 activities ↑ early after injury. Increased ERK2 activity sustained 7–14 days in the media and neointima. MAPK activation is followed by ↑ c-fos and c-jun gene expression and ↑ AP-1 DNA binding	Y. Hu, L. Cheng, B. W. Hochleitner, and Q. Xu, *Arterioscler. Thromb. Vasc. Biol.* **17,** 2808 (1997)
Balloon injury, porcine arteries	MAPK (ERK)	MAPK activity ↑ 7.7-fold in injured carotid arteries after 5 min, but the ↑ in coronary arteries was variable	J. M. Pyles, K. L. March, M. Franklin, K. Mehdi, R. L. Wilensky, and L. P. Adam., *Circ. Res.* **81,** 904 (1997)
Balloon injury, rat carotid artery	ERK1/2 JNK	↑ ERK1/2 activity within 30 min of injury, persisting 12 hr in arterial wall cells. Anti-FGF2 monoclonal Ab ↓ ERK1/2 activation. PD98059 (MEK-1 inhibitor) did not block intimal cell replication. MKP-1 expressed within hours and after 7 and 14 days, associated with JNK activation	H. Koyama, N. E. Olson, F. F. Dastvan, and M. A. Reidy, *Circ. Res.* **82,** 713 (1998)
Balloon overdistension, rabbit carotid	Stress-activated protein kinases	↑ SAPK activity 10 min after injury followed by apoptosis of medial cells. Neointimal cells relatively resistant to apoptosis associated with ↑ *bcl-xL* protein levels	M. J. Pollman, J. L. Hall, and G. H. Gibbons, *Circ. Res.* **84,** 113 (1999)
Cultured endothelium	ERKs, JNK/SAPK and p38 MAPK	↑ ERK-1/2 activation 5 min after mechanical injury, abolished by anti-FGF-2 monoclonal Ab	G. Pintucci, B. M. Steinberg, G. Seghezzi, J. Yun, A. Apazidis, F. G. Baumann, E. A. Grossi, S. B. Colvin, P. Mignatti, and A. C. Galloway, *Surgery* **126,** 422 (1999)
Balloon injury, rat carotid artery	p38 MAPK	↑ p38 in majority of medial cells after injury. FR167653 (inhibitor of p38) 10 mg/kg/day ↓ proliferating medial cells at 48 hr and ↓ intima/media ratio 14 days after injury	N. Ohashi, A. Matsumori, Y. Furukawa, K. Ono, M. Okada, A. Iwasaki, T. Miyamoto, A. Nakano, and S. Sasayama, *Arterioscler. Thromb. Vasc. Biol.* **20,** 2521 (2000)

[a] bFGF, basic fibroblast growth factor; TNFα, tumor necrosis factor α; Il-1, interleukin 1; MAPK, mitogen-activated protein kinases; ERK, extracellular regulated protein kinase; SAPK, stress-activated protein kinase; JNK, c-jun N-terminal kinase; AP-1, activator protein 1; FGF-2, fibroblast growth factor 2.

result from the mechanical stimulus itself,[46] secondary cell injury, or from peptide growth factors released from vascular cells.[47] However, it is surprising that only a few such events have been specifically demonstrated in actual vascular injury models. We showed evidence for NF-κB activation inhibitable by the flavoprotein antagonist dyphenyleneiodonium (DPI—see next section).[11] Also, massive redox-dependent apoptosis associated with activation of SAPK was reported.[48] None of the signaling events described in Table III has yet been reported to correlate with the final extent of the vascular reaction, remodeling, or restenosis so they cannot be strictly taken as surrogate end points to the later phenomena. The acute luminescence peak (Table II) is proportional to the degree of injury, a known determinant of the later extent of response.

Activation of NF-κB after Injury: Procedures

1. Accurate separation of nuclear proteins is the key step in this assay. Five types of buffer (named A to E) should be prepared. Segments of rabbit thoracic aorta (wet weight ~100 mg) submitted to balloon injury as described earlier (for the EPR assay) are frozen in liquid nitrogen after incubation in Krebs–HEPES buffer for 30 min. Analogous uninjured control fragments from other rabbits are always processed in parallel. Additional vessels can be included, e.g., incubated with antagonists such as DPI (20 μM).

2. For nuclear protein extraction, vessel fragments are minced and homogeneized with a mortar and pestle for 2 min with 1 ml buffer A (10 mM Tris–HCl, pH 7.60, 5 mM $MgCl_2$) plus 0.25 M sucrose.

3. The homogenate is centrifuged at 2500g for 15 min; the supernatant is discarded, and 1 ml buffer B (similar to A, plus 0.5% Triton X-100) with 0.25 M sucrose is gently mixed. After another identical centrifugation, the supernatant is discarded and the pellet is gently homogeneized with 5 volumes of 2.4 M sucrose in buffer A.

4. After centrifugation at 16000g for 1 hr, the supernatant is discarded and the pellet is resuspended in 1 ml buffer A. Presence of disrupted cells and nuclei is assessed with microscopy with Trypan blue. The suspension is then diluted 1 : 0.5 (v/v) in buffer C 20 mM HEPES, 25% glycerol, 1.5 mM $MgCl_2$, 0.02 mM KCl, 0.2 mM EDTA, 0.2 mM phenylmethylsulfonylfluoride (PMSF), dithiothreitol (DTT) 0.5 mM and centrifuged at 3000g.

5. The pellet is then resuspended in half of its volume in buffer D (similar to C, except for 1.0 mM KCl), and the whole suspension is stirred for 30 min at 4°, followed by centrifugation at 12,500g for 30 min.

[46] G. De Keulenaer, D. C. Chappell, N. Ishizaka, R. M. Nerem, R. W. Alexander, and K. K. Griendling, *Circ. Res.* **82,** 1094 (1998).

[47] S. T. Crowley, C. J. Ray, D. Nawaz, R. A. Majack, and L. D. Horwitz, *Am. J. Physiol.* **269,** H1641 (1995).

[48] M. J. Pollman, J. L. Hall, and G. H. Gibbons, *Circ. Res.* **84,** 113 (1999).

6. The pellet is discarded and the supernatant is dialyzed against buffer E (20 m*M* HEPES, 20% glycerol, 100 m*M* KCl, and 0.2 m*M* EDTA) for 1 hr. After centrifugation at 12500*g* for 15 min, the supernatant is collected and the protein concentration determined.

7. Binding of nuclear extracts to DNA is assayed by gel retardation analysis by standard techniques with the oligonucleotide from Promega (Madison, WI). Supershift assays should be performed with the antibodies from Santa Cruz Biotechnology, Inc. (Santa Cruz, CA).

Comments

Some activation of NF-κB is found in control vessels, possibly reflecting the constitutive activity reported in smooth muscle cells,[49] as well as some activation during manipulation. We found an increase in NF-κB activation as early as 30 min after injury, which was completely antagonized by DPI,[11] although we have not found a consistent pattern of antagonism by SOD and catalase, possibly reflecting the more complex secondary redox pathways involved in NF-κB activation.[50]

Studies with Antioxidant Compounds

An approach to test the role of redox processes in vascular repair *in vivo* is through the use of antioxidant compounds. In experimental restenosis, some antioxidants such as vitamins C and E were shown to afford protection; in humans, relatively consistent results have been achieved with probucol, but not multivitamins, due to a favorable effect in vessel remodeling (reviewed in Ref. 10). Signaling pathways associated with probucol effects may include decreases in MAP kinase and protein kinase C activation.[51]

Overall, studies with antioxidants are the ultimate tool to test antirestenosis therapies at the clinical level. From the mechanistic standpoint, however, not much information can be derived from this approach, as (a) many redox processes transduce physiologically protective signals; (b) the antioxidants available at present are quite nonspecific; (c) redox processes depend at the same time on oxidative as well as reductive events and the effect of an antioxidant may provide contradictory results; (d) many antioxidants exhibit a concentration-dependent prooxidant effect; and (e) the antagonism of a specific signaling pathway by an antioxidant may be due to its nonspecific effect in another cell function rather than a direct cause–effect correlation.

[49] R. Lawrence, L. J. Chang, U. Siebenlist, P. Bressler, and G. E. Sonenshein, *J. Biol. Chem.* **269,** 28913 (1994).

[50] R. G. Allen and M. Tresini, *Free Radic. Biol. Med.* **28,** 463 (2000).

[51] K. Tanaka, K. Hayashi, T. Shingu, Y. Kuga, K. Nomura, and G. Kajiyama, *Cardiovasc. Drugs Ther.* **12,** 19 (1998).

Activation of Vascular NAD(P)H Oxidase after Injury

Several enzymes connected with redox-dependent processes are likely to be induced at different stages of the vascular response (Table IV). Among these, several potential sources of ROS are included. In general, the most consistent evidence points to the major importance of NAD(P)H oxidase(s) as a ROS source in vascular tissue.[52] In our model of vascular injury,[11] a major role for NAD(P)H oxidase activation immediately after injury is suggested by several types of evidence: (a) the antagonism of superoxide production and NF-κB activation by DPI; (b) marked amplification of injured-triggered luminescence signals by exogenous NADH or NADPH; and (c) negative effect of several antagonists designed to disclose other potential superoxide sources (including xanthine oxidase, mitochondria, lipoxygenases, and nitric oxide synthase). After the initial peak, oxidase activity decreases thereafter; a second injury induces much smaller activation in parallel with selective loss of NADPH but not NADH-driven signals.

The role of a phagocyte-simile NAD(P)H oxidase after injury is further suggested by the report of increased selective neointimal expression of some of its subunits *in vivo* or in a novel model of cultured neointimal cells (Table IV). Whether this increased enzyme expression is associated with increased *in vivo* ROS production is much less clear. In our model, in contrast to its early activation after injury, NAD(P)H-driven oxidase measurements in vessel rings 7 days after injury showed increased activity in injured vessels only with 250 μM lucigenin, but not with more superoxide-specific methods.[10]

Procedures

Particulate Fraction of Vascular Homogenates

1. Iliac and carotid arteries from two rabbits are collected as described earlier and are incubated in buffer A (50 m*M* Tris–HCl, 0.1% mercaptoethanol, 1 m*M* PMSF, 0.01 m*M* DTPA, at 4°, pH 7.4.
2. Vessels are homogeneized thoroughly with a mortar and pestle under liquid nitrogen, and the obtained powder is collected in 3 ml buffer A and centrifuged at 1000*g* for 9 min at 4° to remove large debris. The supernatant is centrifuged at 18,000*g* for 10 min at 4° to remove intact cells, nuclei, and mitochondria.
3. The supernatant is centrifuged at 100,000*g* for 45 min at 4°. The pellet (particulate fraction containing microsomes and plasma membrane vesicles) is resuspended in 1 ml ice-cold buffer A.
4. Protein concentration is assessed by the Bradford method and calculations are performed to yield a final concentration in the assay vial of 20–30 μg/ml.

[52] K. K. Griendling, D. Sorescu, and M. Ushio-Fukai, *Circ. Res.* **86,** 494 (2000).

TABLE IV

SOME REDOX-RELATED ENZYMES ACTIVATED BY MECHANICAL VASCULAR INJURY

Enzyme	Comment	Selected references
NAD(P)H oxidase	Major activation early after injury, decreasing thereafter (see text). Increased selective expression of p47^{phox} in neointima after injury. New data show increased *Nox1* expression	H. P. Souza, L. C. Souza, V. M. Anastacio, A. C. Pereira, M. L. Junqueira, J. E. Krieger, P. L. da Luz, O. Augusto, and F. R. Laurindo, *Free Radic. Biol. Med.* **28,** 1232 (2000); C. Patterson, J. Ruef, N. R. Madamanchi, P. Barry-Lane, Z. Hu, C. Horaist, C. A. Ballinger, A. R. Brasier, C. Bode, and M. S. Runge, *J. Biol. Chem.* **274,** 19814 (1999); K. Szocs, B. Lassegue, D. Sorescu, L. L. Hilenski, L. Valppu, T. L. Couse, J. N. Wilcox, M. T. Quinn, J. D. Lambeth, and K. K. Griendling, *Arterioscler. Thromb. Vasc. Biol.* **22,** 21 (2002).
Superoxide dismutase	Decreased SOD activity after injury	L. C. Azevedo, M. A. Pedro, L. C. Souza, H. P. Souza, M. Janiszewski, P. L. da Luz, and F. R. Laurindo, *Cardiovasc. Res.* **47,** 436 (2000); E. Y. Sozmen, Z. Kerry, F. Uysal, G. Yetik, M. Yasa, L. Ustunes, and T. Onat., *Clin. Chem. Lab. Med.* **38,** 21 (2000)
NO synthase isoforms	Loss of eNOS message after vascular injury with gradual but incomplete recovery thereafter, up to day 30. Sustained iNOS activity in the forming neointima and luminal vascular surface even 21 days after injury. In rats, pigs, and rabbits, augmented NO production diminishes neointima formation	L. Chen, G. Daum, R. Forough, M. Clowes, U. Walter, and A. W. Clowes, *Circ. Res.* **82,** 862 (1998); S. Janssens, D. Flaherty, Z. Nong, O. Varenne, N. van Pelt, C. Haustermans, P. Zoldhelyi, R. Gerard, and D. Collen, *Circulation* **97,** 1274 (1998)

Matrix metalloproteinases (MMP)	Increased expression and activation of many MMPs after vascular injury. MMP inhibition may decrease neointima and reduce constrictive lumen loss	B. J. de Smet, D. de Kleijn, R. Hanemaaijer, J. H. Verheijen, L. Robertus, Y. J. van Der Helm, C. Borst, and M. J. Post, *Circulation* **101,** 2962 (2000); M. Janiszewski, C. A. Pasqualucci, L. C. Souza, F. Pileggi, P. L. da Luz, and F. R. Laurindo, *Cardiovasc. Res.* **39,** 327 (1998)
Heme oxygenase (HO)	Enhanced expression of active inducible HO after vascular injury in rats. Inhibition of HO decreases neointima	D. A. Tulis, W. Durante, K. J. Peyton, G. B. Chapman, A. J. Evans, and A. Schafer, *Biochem. Biophys. Res. Commun.* **279,** 646 (2000)
Cyclooxygenase (Cox)	Increased Cox2 gene transcription early after vascular injury, related to increased proliferation and thrombus inhibition	J. A. Rimarachin, J. A. Jacobson, P. Szabo, J. Maclouf, C. Creminon, and B. B. Weksler, *Arterioscler. Thromb.* **14,** 1021 (1994)
Lipoxygenase	LO activation after injury; inhibition decreases neointima	R. Natarajan, H. Pei, J. L. Gu, J. M. Sarma, and J. Nadler, *Cardiovasc. Res.* **41,** 489 (1999)
Glutathione peroxidase	Aorta injury in rats induces sustained loss in activity	K. W. Gong, G. Y. Zhu, L. H. Wang, and C. S. Tang, *J. Vasc. Res.* **33,** 42 (1996)
Glutathione reductase	Prolonged loss in activity after injury	M. F. Chen, H. C. Hsu, C. S. Liau, and Y. T. Lee, *Prostagl. Lip. Mediat.* **56,** 219 (1998)
Aldose reductase (AR)	Increased neointimal expression in rat injured carotids. AR inhibition decreases neointima	J. Ruef, S. Q. Liu, C. Bode, M. Tocchi, S. Srivastava, M. S. Runge, and A. Bhatnagar, *Arterioscler. Thromb. Vasc. Biol.* **20,** 1745 (2000)
Cyclophylin A (CyPA)	CyPA secreted by injured smooth muscle cells mediates proliferation and ERK1/2 activation. Neointimal expression increased after vascular injury	Z-G. Jin, G. Matthew, D. F. Liao, Y. Chen, J. Haendeler, Y. A. Suh, D. J. Lamneth, and B. C. Berk, *Circulation* **102**(Suppl), II (2000)
Prostacyclin synthase	Gene transfection or PGI_2 analog administration reduces neointima formation after balloon injury in rats	M. Harada, Y. Toki, Y. Numaguchi, H. Osanai, T. Ito, K. Okumura, and T. Hayakawa, *Cardiovasc. Res.* **43,** 481 (1999)

5. It is essential to keep the temperature constantly low at all times, or there is a quick loss of oxidase activity. We have frozen the homogenate at −80° before pellet resuspension in step (3) for up to 15 days without a measurable loss in oxidase activity.

Chemiluminescence Assays for NAD(P)H Oxidase

1. Procedures are described both for arterial rings and homogenates, although there is concern over interpretation of the ring assays, as discussed later. Artery rings are transferred to luminescence vials containing Krebs–HEPES buffer. For the particulate fraction, aliquots are added to 1 ml buffer B (50 m*M* PBS, 0.01 m*M* EDTA, pH 7.40) and 50 μM rotenone.

2. Experiments are performed with either 5 or 250 μM final lucigenin concentration or 5 μM coelenterazine. NADH or NADPH (0.1–0.3 m*M*) is added and the luminescence counts measured continuously for 5–15 min at 37°; results are compared with homogenates in the absence of NAD(P)H, which are typically close to background with lucigenin alone. Some experiments are performed with 20 μM DPI, 10 μM MnTBAP, 500 units/ml SOD, or 500 units/ml catalase. These compounds are added to the luminescence vials and, for the arterial rings, also to the prior incubation.

3. Signals from homogenate/artery alone are subtracted from NAD(P)H-driven signals. Results are normalized for the dry weight of arteries or the protein concentration of homogenates and are reported as cpm/mg/min.

EPR Assay for NAD(P)H Oxidase

1. EPR experiments are performed with the particulate fraction. Aliquots (20–30 μg protein) are incubated for 10 min at 37° with the spin trap 5,5′-dimethyl-1-pyrroline-*N*-oxide (DMPO, 50 m*M*) in 1 ml PBS alone or followed by the addition of 0.3 m*M* NADPH or NADH.

2. The incubation buffer is transferred to a quartz flat cell, and EPR spectra are recorded at room temperature in a Brucker ESR 200D SRC spectrometer at 9.7 GHz frequency, employing field modulation of 100 KHz, modulation amplitude 1 G, microwave power 39 mW, time constant 200 msec, and scan rate 0.1 G/s (Fig. 2).

3. DMPO–OH spin adducts are quantified by double integration of their signals, using TEMPOL (4-hydroxy-2,2,6,6-tetramethylpiperidine-*N*-oxyl) as a standard.[36]

Some Problems Related to Evaluation of NAD(P)H Oxidase after Injury

Further elucidation of the role of NADPH oxidase after injury depends on a number of unanswered general questions about the oxidase, some of which will be addressed here.

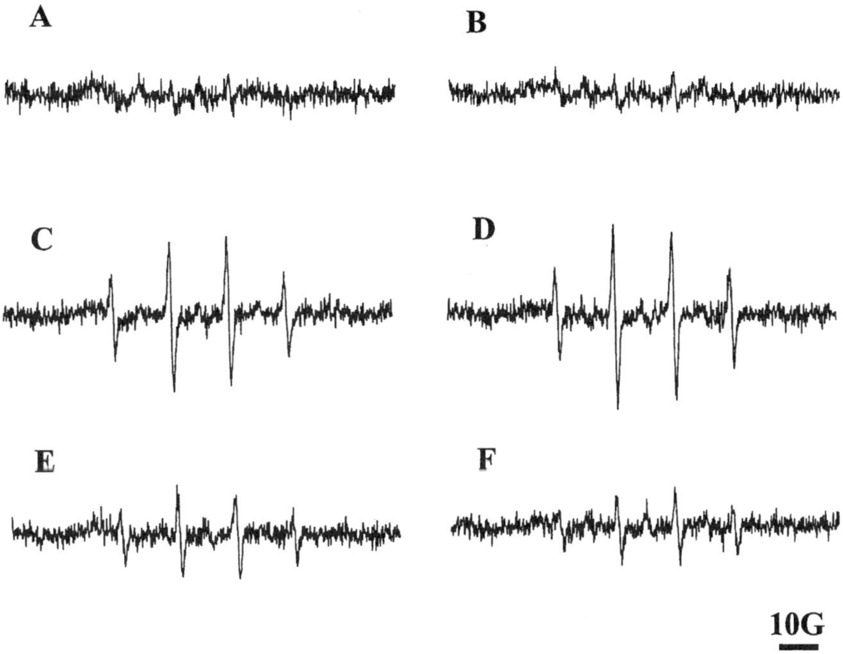

FIG. 2. Detection of NAD(P)H oxidase activity in homogenates of vascular segments through EPR spin trapping with DMPO, with the procedures and instrumental conditions described in the text. EPR adduct spectra are as follows: (A) DMPO plus PBS. (B) DMPO, PBS, and homogenate. There is no significant radical adduct generation. (C and D) The same as in B with 0.3 m*M* NADH and NADPH, respectively. There is a robust characteristic DMPO–OH adduct signal depicting four lines with a 1 : 2 : 2 : 1 ratio and hyperfine splitting constants $a_N = a_H = 14.9$ G. (E and F) The same as in C and D with SOD 600 units/ml. There is a significant inhibition of DMPO–OH adduct signals, indicating that superoxide was the primary radical species generated.

What Is the Real Meaning of NADPH Oxidase Activity?

NAD(P)H oxidase activity, as assessed by NAD(P)H consumption or dinucleotide-driven ROS generation, is a relatively nonspecific biochemical activity that can be displayed by many electron transfer enzymes, including, e.g., some dehydrogenases and oxygenases. Because many such proteins are flavoenzymes, a blocking effect of DPI in the target variable under study is by no means specific proof that a phagocyte-like enzymatic complex is involved, even if DPI is kept at the recommended low concentrations of 10–20 μM. Some enzymes reportedly inhibited by DPI include nitric oxide synthase,[53,54] xanthine oxidase,[54] mitochondrial

[53] D. J. Stuehr, O. A. Fasehun, N. S. Kwon, S. S. Gross, J. A. Gonzalez, R. Levi, and C. F. Nathan, *FASEB J.* **5,** 98 (1991).

[54] V. B. O'Donnell, G. C. M. Smith, and O. T. G. Jones, *Mol. Pharmacol.* **46,** 778 (1994).

electron transport chain dehydrogenases,[54,55] and cytochrome P450.[54] Interestingly, some flavoenzymes are resistant to DPI, e.g., glutathione reductase and some diaphorases, i.e., enzymes capable of a two-electron transfer activity.[54] A particularly puzzling problem is the interpretation of NAD(P)H oxidase activity detected in whole cells and vascular rings exposed to NAD(P)H.[10,11,52,56,57] Overall, there is similarity in the characteristics of the oxidase detected in such segments vs particulate homogenate fraction. With growth factor agonists, discrepancies may occur between observed increase in whole cell oxidase activity vs no increase in homogenates of the same cells (unpublished observations). However, the fact that NADH and NADPH are cell impermeable and almost exclusively intracellular makes it difficult to interpret activity in whole tissues in connection with intracellular ROS generation known to occur in vascular cells.[52] Some oxidases have been reported to exhibit dinucleotide-binding sites oriented to the extracellular milieu.[58] However, while this makes up for a convenient assay of these membrane surface enzymes, it cannot be assumed that their physiological function involves NAD(P)H oxidase activity and ROS generation. This should not be taken to assume the absence of extracellular ROS generation in vascular cells, which can be due to a transmembrane oxidase architecture (e.g., similar to the phagocyte enzyme). Interestingly, superoxide-dependent vasoconstriction in response to exogenous NADPH (but not NADH) has been consistently shown.[59]

What Is the Best Method to Assess NAD(P)H Oxidase Activity?

Contrary to the phagocyte, vascular oxidase does not seem to require major protein assembling for its activity, which is driven by exposure to NAD(P)H. Thus, many oxidase assays have been designed to assess ROS production before and after exposure to NAD(P)H, not infrequently using lucigenin chemiluminescence. Data from our laboratory suggest that lucigenin at higher concentrations undergoes significantly more redox cycling and other artifacts with NADH than with NADPH, so that the previously reported oxidase preference for NADH is artifactual.[60] EPR assays of the vascular oxidase may overcome some of these problems.[11,35,59] Spectrophotometric measurements of NAD(P)H consumption provide a useful, but less

[55] Y. Li and M. A. Trush, *Biochem. Biophys. Res. Commun.* **253,** 295 (1998).

[56] P. Pagano, Y. Ito, K. Tronheim, P. M. Gallop, A. I. Tauber, and R. A. Cohen, *Am. J. Physiol.* **268,** H2274 (1995).

[57] M. Janiszewski, M. A. Pedro, R. C. H. Scheffer, J. T. van Asseldonk, L. C. Souza, P. L. da Luz, and F. R. M. Laurindo, *Free Radic. Biol. Med.* **29,** 889 (2000).

[58] D. J. Morré and A. O. Brightman, *J. Bioenerg. Biomembr.* **23,** 469 (1991).

[59] H. P. Souza, F. R. M. Laurindo, R. C. Ziegelstein, C. O. Berlowitz, and J. L. Zweier, *Am J. Physiol.* **280,** H658 (2001).

[60] M. Janiszewski, H. P. Souza, X. Liu, M. A. Pedro, J. L. Zweier, and F. R. M. Laurindo, *Free Radic. Biol. Med.* **32,** in press (2002).

sensitive method.[61] The interpretation of such assays is clearer when performed with the particulate membrane fraction of specific cultured cell types, as opposed to vascular ring homogenates or whole vascular cells or segments. However, the composition of homogenate fraction is important, particularly concerning contamination by mitochondria. In these assays, it would be desirable to stress the oxidase with a specific and uniformly reproducible stimulus (analogous to PMA for the phagocyte). The mechanical stimulus discussed earlier and some growth factors, e.g., angiotensin II, platelet-derived growth factor, and thrombin, are useful in specific systems, but in general, such a universal stimulus is yet unavailable. *In vivo* angiotensin II infusion is also an interesting technique,[62] potentially applicable to study mice models.

Is the Vascular Oxidase Similar to That of the Phagocyte?

The presence and activity of phagocyte-simile oxidase has been characterized in endothelial cells[52,63] and possibly adventitial fibroblasts,[52,64] whereas smooth muscle cells appear to have an analogous but structurally different subunit(s).[52] This suggests that even the phagocyte-simile "vascular oxidase" is actually an unequal mixture of enzymes present in each cell type. Accordingly, vascular oxidase assessed in vessel segments or homogenates from mice with homozigous deficiency of the phagocyte subunit gp91^{phox} show normal superoxide production.[59] Molecular strategies toward components of the phagocyte-simile oxidase system have been developed, e.g., p22^{phox} antisense,[52] neutralizing antibodies against p47^{phox},[65] or p67^{phox},[64] development of models of rac subunit overexpression.[66] There is an expected trend to rely on such techniques to demonstrate a cellular signaling role for the vascular oxidase. The oxidase electron transport unit is likely a member of the gp91^{phox} family, shown to be involved in other activities such as proton[67] and iron[68] transport. Thus, those molecular antienzyme

[61] C. Patterson, J. Ruef, N. R. Madamanchi, P. Barry-Lane, Z. Hu, C. Horaist, C. A. Ballinger, A. R. Brasier, C. Bode, and M. S. Runge, *J. Biol. Chem.* **274,** 19814 (1999).

[62] S. Rajagopalan, S. Kurz, T. Munzel, M. Tarpey, B. A. Freeman, K. K. Griendling, and D. G. Harrison, *J. Clin. Invest.* **97,** 1916 (1996).

[63] A. Görlach, R. P. Brandes, K. Nguyen, M. Amidi, F. Dehghani, and R. Busse, *Circ. Res.* **87,** 26 (2000).

[64] P. J. Pagano, J. K. Clark, M. E. Cifuentes-Pagano, S. M. Clark, G. M. Callis, and M. T. Quinn, *Proc. Natl. Acad. Sci. U.S.A.* **94,** 14483 (1997).

[65] B. Schieffer, M. Luchtefeld, S. Braun, A. Hilfiker, D. Hilfiker-Kleiner, and H. Drexler, *Circ. Res.* **87,** 1195 (2000).

[66] H. H. Hassanain, Y. K. Sharma, L. Moldovan, V. Khramtsov, L. J. Berliner, J. P. Duvick, and P. J. Goldschmidt-Clermont, *Biochem. Biophys. Res. Commun.* **272,** 783 (2000).

[67] B. Banfi, A. Maturana, S. Jaconi, S. Arnaudeau, T. Laforge, B. Sinha, E. Ligeti, N. Demaurex, and K. Krause, *Science* **287,** 138 (2000).

[68] K. P. Shatwell, A. Dancis, A. R. Cross, R. D. Klausner, and A. W. Segal, *J. Biol. Chem.* **271,** 14240 (1996).

interventions should always be accompanied by the measurement of ROS production and NAD(P)H oxidase activity to document the mechanism of their effects. In other words, ROS production may not be the only activity of such an oxidase(s).

Is There More Than One Vascular Oxidase?

Considering the large number of ubiquitous membrane oxidases,[58] it is surprising that vascular oxidase is still generally thought of as a single enzyme system. However, evidence for vascular oxidases other than the phagocyte-simile enzyme is scarse and indirect, related solely to different effects of NADH vs NADPH-triggered effects. For example, NADPH-based superoxide production is more sensitive to DPI,[11,59,60] is selectively attenuated after a repeated balloon injury,[43] and is selectively inhibited by the thiol oxidant DTNB.[57] Distinct mechanisms for hydrogen peroxide formation driven by NADPH or NADH have been suggested due, respectively, to superoxide dismutation vs two-electron oxygen reduction.[69]

Concluding Remarks

Studies performed with the methods discussed allow the conclusion that oxidant stress with superoxide production likely due to the activation of NAD(P)H oxidase(s) can be documented after vascular injury. These alterations are prominent immediately after injury and are sustained to some undetermined extent throughout the vascular repair. The later stages of neointima formation appear to be associated with an absence or low levels of oxidative stress. Conclusive statements, however, are as yet substantially limited by constraints in the methods for the assessment of vascular ROS and NAD(P)H oxidase(s). As usual, further biological extrapolations in the redox/free radical field must be well grounded on chemical basis and tested by multiple analytical techniques. Therefore, conceptual progress in redox vascular biology will likely be possible only with such analytical advances, which will pave the way for molecular tools able to characterize the underlying actors of physiological and pathological signaling pathways.

Acknowledgments

The authors are indebted to Professors Ohara Augusto, Etelvino Bechara (Chemistry Institute, University of São Paulo, Brazil), and Jay L. Zweier (Johns Hopkins University, Baltimore, MD) for helpful advice and discussion. Supported by Fundação de Amparo à Pesquisa do Estado de São Paulo, Financiadora de Estudos e Projetos, Programa de Apoio a Núcleos de Excelência, Fundação E. J. Zerbini, and Fundação Faculdade de Medicina.

[69] H. P. Souza, X. Liu, A. Samouilov, F. R. M. Laurindo, and J. L. Zweier, *Free Radic. Biol. Med.* **29,** S22 (2000). [Abstract]

[38] Involvement of Superoxide in Pathogenic Action of Mutations That Cause Alzheimer's Disease

By MARK P. MATTSON

Introduction

Alzheimer's disease results from the degeneration of neurons in brain regions that function in learning and memory processes.[1] Considerable evidence suggests that increased levels of oxidative stress in neurons play a key role in their demise.[2] Mutations in the genes encoding the β-amyloid precursor protein (APP) and presenilin-1 (PS1) can cause early-onset inherited Alzheimer's disease, which is characterized by a rapid deterioration of cognitive function and premature death.[3] APP is a type I membrane protein that is the source of the 40–42 amino acid amyloid β-peptide (Aβ) that forms insoluble plaques in the brain and is also the source of a secreted form of APP (sAPP) that regulates synaptic transmission and neuron survival (Fig. 1A). APP mutations are located on either side of or within the Aβ sequence and, in all cases, increase the production of Aβ, which may promote dysfunction and degeneration of neurons by a mechanism involving oxidative stress and disruption of the ability of the cell to properly regulate intracellular calcium levels.[4] APP mutations also decrease the production of sAPP and, because sAPP activates signaling pathways that stabilize cellular calcium homeostasis and suppress oxidative stress, this effect of the APP mutations also promotes neuronal degeneration. Studies of experimental cell culture and animal models have provided evidence that the pathogenic action of APP mutations involves increased oxidative damage to proteins, lipids, and DNA.[5-9] Analyses of postmortem brain tissue from Alzheimer's disease patients and age-matched control patients have documented extensive oxidative modifications of DNA and proteins, and greatly increased membrane lipid peroxidation, in neurons in vulnerable brain regions. Oxidative damage is particularly prominent in neurons associated with Aβ deposits. Among

[1] J. L. Cummings, H. V. Vinters, G. M. Cole, and Z. S. Khachaturian, *Neurology* **51,** S2 (1998).

[2] W. R. Markesbery, *Free Radic. Biol. Med.* **23,** 134 (1997).

[3] J. Hardy, *Trends Neurosci.* **20,** 154 (1997).

[4] M. P. Mattson, *Physiol. Rev.* **77,** 1081 (1997).

[5] C. D. Smith, J. M. Carney, P. E. Starke-Reed, C. N. Oliver, E. R. Stadtman, R. A. Floyd, and W. R. Markesbery, *Proc. Natl. Acad. Sci. U.S.A.* **88,** 10540 (1991).

[6] M. A. Lovell, W. D. Ehmann, M. P. Mattson, and W. R. Markesbery, *Neurobiol. Aging* **18,** 457 (1997).

[7] M. A. Smith, P. L. R. Harris, L. M. Sayre, J. S. Beckman, and G. Perry, *J. Neurosci.* **17,** 2653 (1997).

[8] J. H. Su, G. Deng, and C. W. Cotman, *Brain Res.* **774,** 193 (1997).

[9] I. I. Kruman, C. Culmsee, S. L. Chan, Y. Kruman, Z. H. Guo, L. Penix, and M. P. Mattson, *J. Neurosci.* **20,** 6920 (2000).

0076-6879/02 $35.00

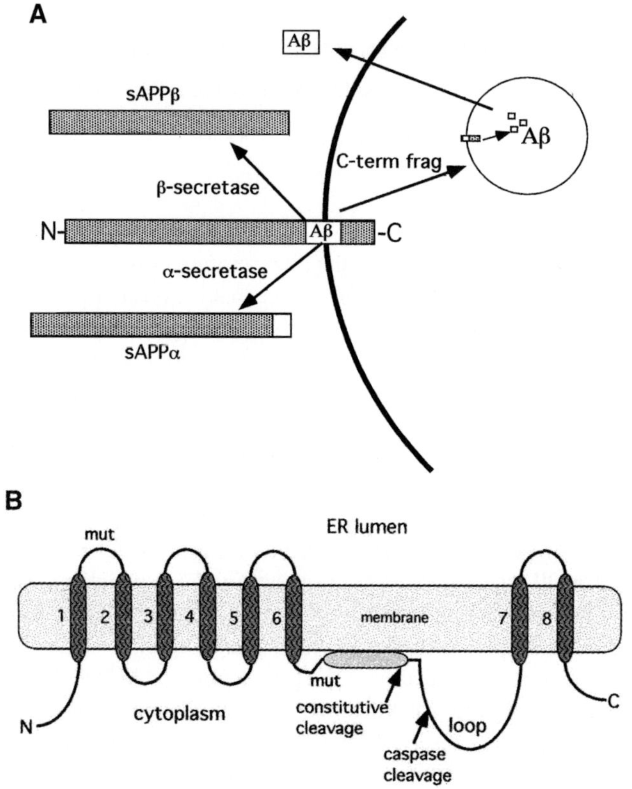

FIG. 1. Structure of the β-amyloid precursor protein (APP) and presenilin-1 (PS1). (A) APP is a transmembrane protein that contains the 40–42 amino acid amyloid β-peptide (Aβ), which, in Alzheimer's disease, forms insoluble diffuse and fibrillar aggregates (plaques) in the brain. Aβ is produced by neurons and other cells as the result of a cleavage at its N terminus by β-secretase, which is followed by internalization of the Aβ-containing C-terminal fragment of APP and a subsequent cleavage at the C terminus of Aβ by an enzyme called γ-secretase. Alternatively, APP is cleaved in the middle of the Aβ sequence by an enzyme called α-secretase, resulting in the release of a large N-terminal portion of APP (sAPPα), which serves important roles in modulating synaptic plasticity and in promoting the survival and growth of neurons. Mutations in PS1 that cause Alzheimer's disease occur immediately adjacent to the N- and C-terminal ends of the Aβ sequence, as well as within the Aβ sequence; each mutation alters proteolytic processing APP in a manner that increases Aβ production and decreases sAPPα production. (B) PS1 is an integral membrane protein with eight transmembrane domains; both N and C termini of PS1, as well as a hydrophilic loop, are localized on the cytoplasmic side of the membrane. Mutations in PS1 that cause Alzheimer's disease occur in several places in the protein with "hot spots" located adjacent to transmembrane domain 2 and in the initial portion of the loop domain. PS1 is localized primarily in the endoplasmic reticulum.

the reactive oxygen species that elicit cellular damage, the superoxide anion radical may play a particularly important role by promoting mitochondrial dysfunction and serving as a source for hydrogen peroxide and hydroxyl radical production.[10,11]

PS1 is an integral membrane protein that contains eight transmembrane domains with both the amino and the carboxy terminus of the protein, as well as a large "loop" domain, located on the cytosolic side of the membrane (Fig. 1B). PS1 mutations involve single base substitutions resulting in a change in one amino acid in the protein; more than 70 such mutations have been identified and are located throughout the protein with "hot spots" present near transmembrane domain 2 and within the cytosolic loop domain.[3] PS1 is localized primarily in the endoplasmic reticulum. Studies of cultured cells and transegenic mice expressing various PS1 mutations have provided evidence for two major abnormalities. First, mutant PS1 perturbs cellular calcium homeostasis such that more calcium is released from the endoplasmic reticulum upon stimulation of cells with various agonists or when cells are exposed to oxidative, metabolic, and excitotoxic insults.[12–15] By impairing their ability to properly regulate intracellular calcium levels, PS1 mutations may render neurons vulnerable to age-related changes that occur in the brain, including increased oxidative stress. The specific mechanism whereby PS1 mutations alter calcium homeostasis is not well understood, but appears to involve an overfilling of endoplasmic reticulum stores, which may, in turn, alter capacitive calcium entry through plasma membrane channels.[16,17] Second, mutant PS1 alters proteolytic processing of APP such that cells increase their production of Aβ[3]; accelerated deposition of Aβ would then lead to the neurodegenerative cascade described earlier and later (Fig. 2).

Experimental studies have shown that Aβ can induce oxidative stress in neurons, including production of superoxide and hydrogen peroxide, and membrane lipid peroxidation.[18,19] Evidence for mitochondrial dysfunction, associated with

[10] M. Lafon-Cazal and S. Pietri, *Nature* **364,** 535 (1993).

[11] M. P. Mattson, *Trends Neurosci.* **21,** 53 (1998).

[12] Q. Guo, B. L. Sopher, D. G. Pham, N. Robinson, G. M Martin, and M. P. Mattson, *NeuroReport* **8,** 379 (1996).

[13] Q. Guo, B. L. Sopher, K. Furukawa, D. G. Pham, N. Robinson, G. M. Martin, and M. P. Mattson, *J. Neurosci.* **17,** 4212 (1997).

[14] Q. Guo, S. Christakos, N. Robinson, and M. P. Mattson, *Proc. Natl. Acad. Sci. U.S.A.* **95,** 3227 (1998).

[15] Q. Guo, W. Fu, B. L. Sopher, M. W. Miller, C. B. Ware, G. M. Martin, and M. P. Mattson, *Nature Med.* **5,** 101 (1999).

[16] S. L. Chan, M. Mayne, C. P. Holden, J. D. Geiger, and M. P. Mattson, *J. Biol. Chem.* **275,** 18195 (2000).

[17] M. A. Leissring, Y. Akbari, C. M. Fanger, M. D. Cahalan, M. P. Mattson, and F. M. LaFerla, *J. Cell Biol.* **149,** 793 (2000).

[18] Y. Goodman and M. P. Mattson, *Exp. Neurol.* **128,** 1 (1994).

[19] R. J. Mark, M. A. Lovell, W. R. Markesbery, K. Uchida, and M. P. Mattson, *J. Neurochem.* **68,** 255 (1997).

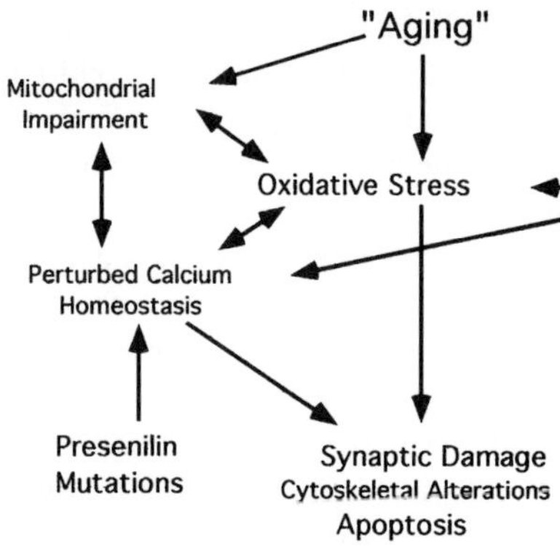

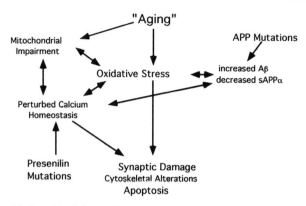

FIG. 2. Simplified model of the mechanisms whereby APP and PS1 mutations cause oxidative stress and neuronal degeneration. APP mutations cause an increase in the deposition of neurotoxic forms of $A\beta$ in the brain, which causes membrane lipid peroxidation and disrupts cellular calcium homeostasis, which in turn perturbs mitochondrial function, resulting in the excessive production of superoxide, hydrogen peroxide, and hydroxyl radical. APP mutations decrease levels of sAPPα, an important signaling molecule that normally enhances cellular calcium homeostasis and antioxidant defense mechanisms; e.g., sAPPα induces activation of the transcription factor NF-κB, which then stimulates the production of MnSOD. PS1 mutations perturb endoplasmic reticulum calcium regulation in a manner that promotes superoxide production. The overall effect of APP and PS1 mutations is to render neurons vulnerable to age-related changes in the brain such as increased oxidative stress and metabolic impairment. The impact of APP and PS1 mutations in neurons is particularly pronounced in synapses, where the mutations promote the activation of excitotoxic and apoptotic signaling cascades.

cellular energy deficits and increased production of mitochondrial reactive oxygen species, has also been obtained from studies of patients and experimental models of Alzheimer's disease.[20,21] Evidence that superoxide plays an important role in the death of neurons in Alzheimer's disease comes from studies showing that an increased expression of manganese superoxide dismutase (either induced by activation of cytokine signaling pathways or by ectopic gene expression) in cultured neural cells and transgenic mice can greatly increase resistance of the cells to death induced by $A\beta$ and other insults relevant to the pathogenesis of Alzheimer's disease.[22,23] Moreover, PS1 mutations enhance superoxide production following exposure of $A\beta$, and both superoxide and peroxynitrite play central roles in the endangering action of the PS1 mutations.[24]

[20] E. Bonilla, K. Tanji, M. Hirano, T. H. Vu, S. DiMauro, and E. A. Schon, *Biochim. Biophys. Acta* **1410,** 171 (1999).

[21] R. J. Mark, Z. Pang, J. W. Geddes, K. Uchida, and M. P. Mattson, *J. Neurosci.* **17,** 1046 (1997).

[22] M. P. Mattson, Y. Goodman, H. Luo, W. Fu, and K. Furukawa, *J. Neurosci. Res.* **49,** 681 (1997).

[23] J. N. Keller, M. S. Kindy, F. W. Holtsberg, D. K. St. Clair, H. C. Yen, A. Germeyer, S. M. Steiner, A. J. Bruce-Keller, J. B. Hutchins, and M. P. Mattson, *J. Neurosci.* **18,** 687 (1998).

[24] Q. Guo, W. Fu, F. W. Holtsberg, S. M. Steiner, and M. P. Mattson, *J. Neurosci. Res.* **56,** 457 (1999).

This chapter presents methods that lead to the establishment of the involvement of superoxide, and reactive oxygen species derived therefrom, in the pathogenesis of neuronal degeneration in Alzheimer's disease. Methods include the application of findings in the area of the molecular genetics of inherited forms of Alzheimer's disease in humans to the development of cell culture and mouse models that allow direct observation and manipulation of living neurons. Similar approaches may be applicable to many different neurodegenerative disorders, as well as inherited diseases that primarily affect other organ systems.

Cell Culture Models for Investigation of the Pathogenic Actions of APP and PS1 Mutations

Primary Cultures of Rat and Mouse Embryonic Hippocampal Neurons

One day prior to establishing dissociated cell cultures, plastic 35-mm culture dishes (Costar, Corning, Inc., Corning, NY) or 22-mm^2 glass coverslips (Gold Seal, Portsmouth, NH; one coverslip is placed in a 35-mm dish) are coated with polyethyleneimine [Sigma, St. Louis, MO; 1 : 10,000 dilution (v/v) in borate buffer, pH 8.5] by incubation for 1–2 hr. All procedures for the preparation of culture dishes are done under a laminar flow hood, and all solutions are filtered through 22-μm filters. Dishes and coverslips are rinsed four times (3 ml/wash) with ultrapure water and are allowed to dry. The sterility of the dishes is ensured by exposing them to ultraviolet light for 10 min with the light source provided by the manufacturer of the laminar flow hood (Baker Company, Sanford, ME). Culture medium (Neurobasal medium from Life Technologies, Rockville, MD) containing 10% fetal bovine serum (v/v) is then added to the dishes (1.5 ml/dish), and the dishes are placed in a humidified incubator in an atmosphere of 6% CO_2 and 94% room air at 37°. The next morning, pregnant female Sprague–Dawley rats (18 days pregnant) or mice (wild-type and transgenic; 16 days pregnant) are euthanatized by an overdose of isofluorane inhalation anesthesia, and the uterus is removed from the abdomen using asceptic technique and placed in a 100-mm-diameter petri dish containing 15 ml of calcium- and magnesium-free Hanks' balanced saline solution (HBSS; Life Technologies). Embryos are removed, decapitated, and brains removed. Hippocampi are removed from each brain with the aid of a dissecting microscope and collected in a 35-mm dish containing 2 ml HBSS. Hippocampi are then incubated for 15 min in 3 ml of a solution of 0.2% trypsin (Sigma) in HBSS, followed by one wash with 5 ml HBSS, a 1-min incubation in a solution of 0.2% (w/v) soybean trypsin inhibitor (Sigma), and another wash in 5 ml HBSS. HBSS is then added in a volume that results in 10 hippocampi/ml, and cells are dissociated by trituration through the narrowed bore of a fire-polished Pasteur pipette. Dissociated cells are seeded into cultured dishes (35 μl/dish), which are placed in the incubator for 3–4 hr to allow for cells to attach to the polyethyleneimine substrate. Cells are then washed once gently with serum-free medium, and then 1 ml of culture maintenance medium (Neurobasal with B27 supplements; Life Technologies) is added to each dish and the dishes are returned to the incubator. Experiments are performed after cultures are maintained for 7–10 days; approximately 90% of the cells are neurons and the remaining cells are astrocytes in such cultures.

PC12 Cells Stably Overexpressing Mutant PS1 and MnSOD

Two oligonucleotides (5-ACCTCAAGAGGCTTTGTTTTCTGTG-3′ and 5′-ACCTCCTGAAAAGTTCAGAATTCAG-3′) are used to polymerase chain reaction (PCR) amplify the PS1 gene from a human lymphoblastoid cDNA library using 30 cycles of Expand high fidelity PCR (Boehringer Mannheim, Indianapolis, IN) and cloned into the PCR cloning vector PCR3 (Invitrogen, Carlsbad, CA) yielding a vector designated PCRPS1. For *in vitro* mutagenesis, the *Kpn*I/*Xho*I fragment from PCRPS1 is subcloned into pAlter-1 (Promega, Madison, WI). Single base mutations are introduced into wild-type PS1 using the appropriate mutagenic oligonucleotide (e.g., 5′-CTTTTTCCAGCTGTCATTTACTCCTCA-3′ for the L286V mutation) according to the manufacturer's instructions (Promega). DNA sequences of wild-type and mutated PS1 genes are verified using AmpliTaq to dye terminator label the DNA prior to detection on an ABI373 DNA sequencer (Perkin-Elmer, Norwalk, CT). Wild-type and mutant PS1 cDNAs are further subcloned into an expression plasmid. As an example, the full-length human PS1 cDNA containing the L286V mutation is cloned into the *Bst*XI and *Sba*I cut pRcCMV expression vector to generate pCMV-PS1L286V. The full-length wild-type human PS1 cDNA is cloned into the *Hin*dIII and *Xba*I cut pRcCMV expression vector to generate pCMV-PS1. PC12 cells (American Type Culture Collection, Rockville, MD) are transfected using lipofectamine, and stable clonal transfectants from each group are selected with G418 and screened for the presence of the transfected mRNA using RT-PCR. Cultures are maintained at 37° (5% CO_2 atmosphere) in RPMI medium (Life Technologies) supplemented with 10% heat-inactivated horse serum and 5% heat-inactivated fetal bovine serum (Sigma). For RT-PCR analysis, mRNA from the cultured cells is isolated and reverse transcribed using a reverse transcription system (Promega) and the 3′ primer: 5′ GCTTCCCATTCCTCACTGAA 3′. The cDNA (2.5 μl) is used as a template in a 50-μl PCR using 15–40 cycles of 94° (1 min), 60° (2 min), and 72° (2 min), with a final extension time of 10 min at 72°. Reaction mixtures are as recommended for *Taq* polymerase (PerkinElmer), except that *Taq* is added after the mixtures are heated to 95° for 7 min. The 3′ primer used is the oligonucleotide used to prime the cDNA synthesis, the 5′ primer is as follows: 5′ GTGGCTGTTTTGTGTCCGAA 3′. PCR products are resolved and visualized by electrophoresis in a 3% agarose gel stained with ethidium bromide. Because the Leu-to-Val mutation at codon 286 creates a *Pvu*II site, the wild-type RT-PCR product cannot be cut by *Pvu*II and generates a single 251-bp fragment, while the mutation results in *Pvu*II cleavage of the product into 79- and 172-bp fragments. Lines of PC12 cells stably overexpressing human MnSOD are generated using a procedure similar to that used to generate PS1 mutant clones. Cells are transfected with an empty vector or a vector containing the human MnSOD cDNA.[25] The MnSOD cDNA is ligated into the mammalian expression vector pCB6 + neo, which is under the control of the cytomegalovirus promoter. The MnSOD expression vector or empty vector is transfected into subconfluent PC12 cells cultures using DOTAP, and G418-resistant clones are isolated. Analysis of hMnSOD mRNA expression is accomplished by Northern blot analysis. Total RNA is extracted from cells and size fractioned by electrophoresis on a 1% formaldehyde-agarose gel and transferred to nitrocellulose. The RNA is UV cross-linked to a filter and hybridized overnight with a ^{32}P-labeled hMnSOD cDNA in 50% formamide at 45°. The filter is washed three times in 1× SSC containing 0.1% SDS at room temperature for 15 min, followed by one wash in 0.25% SSC containing 0.1% SDS at 55° for 30 min. Cells are maintained in RPMI containing 10% horse serum and 5% fetal bovine serum. Relative levels of PS1 or MnSOD protein overexpression can be estimated by Western blot analysis using specific PS1 or MnSOD antibodies.[13,23] Figure 3 shows an example of Western blot analysis of clones of PC12 cells stably overexpressing wild-type or mutant PS1 alone or in combination with MnSOD.

Exposure of Cells to Aβ and sAPP

Synthetic Aβ1-42 is purchased from Bachem (Torrance, CA) and is stored lyophilized. A 1 m*M* stock solution of Aβ1-42 is prepared in sterile deionized water approximately 16 hr prior to use. Secreted APPα (sAPPα695) is purified from the culture supernatant of human embryonic kidney 293 cells transfected with the corresponding cDNA constructs. Purity is confirmed by Western blot analysis and silver staining after SDS polyacrylamide gel electrophoresis. The purified sAPP is diluted to a concentration of 5–10 μ*M* and is stored in aliquots of 10–50 μl at −80°. Immediately prior to treatment of primary neurons and PC12 cells with Aβ1-42 and/or sAPP, the culture maintenance medium is replaced with Locke's buffer, which contains (m*M*): NaCl, 154; KCl, 5.6; $CaCl_2$, 2.3; $MgCl_2$, 1.0; $NaHCO_3$, 3.6; glucose, 10; and HEPES buffer, 5 (pH 7.2). Examples of data obtained using Aβ1-42 and sAPP have been reported.[26–28]

[25] D. K. St. Clair, T. D. Oberley, and Y. S. Ho, *FEBS Lett.* **293,** 199 (1991).

[26] M. P. Mattson, B. Cheng, D. Davis, K. Bryant, I. Lieberburg, and R. E. Rydel, *J. Neurosci.* **12,** 376 (1992).

[27] M. P. Mattson, B. Cheng, A. R. Culwell, F. S. Esch, I. Lieberburg, and R. E. Rydel, *Neuron* **10,** 243 (1993).

[28] K. Furukawa, S. W. Barger, E. M. Blalock, and M. P. Mattson, *Nature* **379,** 74 (1996).

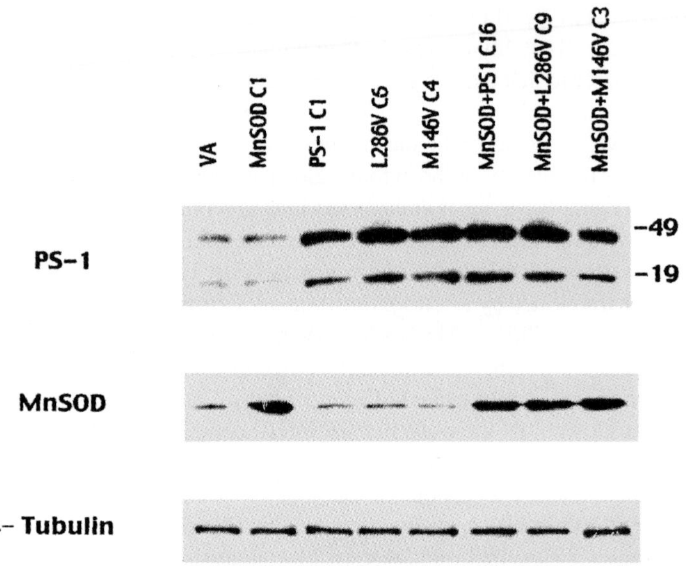

FIG. 3. Western blot analysis of PC6 cell clones (PC6 is a variant of the PC12 cell line) stably overexpressing wild-type or mutant (L286V and M146V mutations) alone or in combination with MnSOD. Solubilized proteins from the indicated cell clones (50 μg/lane) were separated by SDS–PAGE, transferred to a nitrocellulose sheet, and immunoreacted with antibodies to PS-1, MnSOD, and α-tubulin. VA, clone transfected with vector alone.

Quantification of Cell Survival and Apoptosis

In primary hippocampal cell cultures, neuron survival is assessed by counting the number of undamaged (surviving) neurons in premarked microscope fields prior to, and at indicated time points following exposure to, experimental treatments. Neurons with intact neurites and a cell body that is round to oval in shape and with a smooth surface are considered viable, whereas neurons with fragmented neurites and vacuolated or crenated cell body are considered nonviable (Fig. 4). To quantify apoptosis, primary neurons and PC12 cells are fixed in a solution of 4% paraformaldehyde in PBS, membranes are permeabilized with 0.2% Triton X-100, and cells are stained with the fluorescent DNA-binding dye Hoescht 33342. Hoescht-stained cells are visualized and photographed under epifluorescence illumination (340-nm excitation and 510-nm barrier filter) using a 40X oil immersion objective. Cells with condensed and fragmented nuclear DNA are considered apoptotic. Typically, 200 cells are scored in each culture, and the percentage of cells that are apoptotic is calculated. Analyses are performed without knowledge of the treatment history of the cultures to avoid possible experimenter bias. Examples of data obtained using these methods have been reported.[26,29]

[29] I. Kruman, A. J. Bruce, D. E. Bredesen, G. Waeg, and M. P. Mattson, *J. Neurosci.* **17,** 5089 (1997).

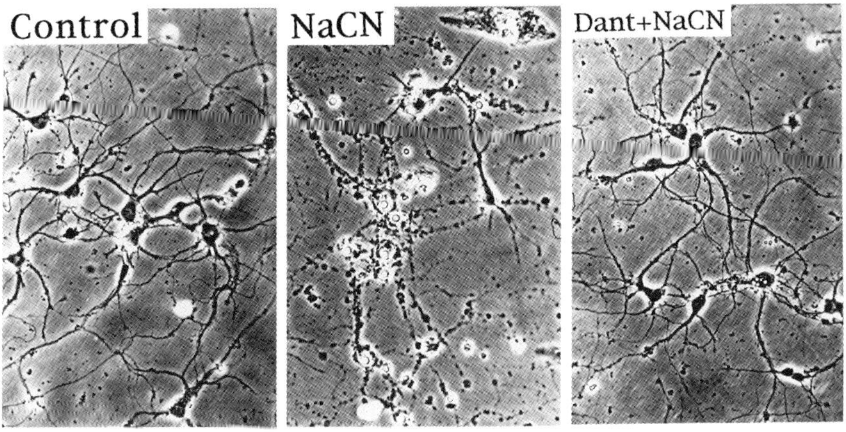

FIG. 4. Death of hippocampal neurons induced by chemical hypoxia requires calcium release from endoplasmic reticulum stores. Phase-contrast micrographs showing hippocampal neurons in a control culture not exposed to cyanide, a culture exposed to cyanide (NaCN; a mitochondrial toxin that causes hypoxia), and a culture pretreated with dantrolene (a blocker of calcium release from ryanodine-sensitive stores) and then exposed to cyanide. Note that cyanide induced the degeneration of neurons, as indicated by neurite fragmentation and cell body condensation, and dantrolene prevented such hypoxic damage.

Mouse Models for Investigation of Pathogenic Actions of APP and PS1 Mutations

The development of technologies for expressing a mutant gene that causes an inherited human disease in transgenic mice has been a major boon to biomedical research, facilitating the identification of the specific molecular and cellular alterations that lead to the disease phenotype and providing an animal model for testing of possible preventative and therapeutic treatments. In the case of Alzheimer's disease, the linkage of mutations in APP and PS1 to early-onset familial forms of the disease provided the opportunity to solve the mystery of the cellular and molecular events that result in neuronal degeneration and cognitive impairment.

APP Mutant Transgenic Mice

Transgenic mice that express mutant human APP primarily in neurons can be generated using established methods. The method involves microinjection of single-cell mouse embryos with and expression plasmid containing the gene intended to be expressed. One example in which transgenic mice were generated using a plasmid in which the human APP gene containing the "Swedish" FAD mutation (Lys670 $\rightarrow$ Asn, Met671 $\rightarrow$ Leu) is under the control of a prion promoter[30]

[30] K. Hsiao, P. Chapman, S. Nilsen, C. Eckman, Y. Harigaya, S. Younkin, F. Yang, and G. Cole, *Science* **274,** 99 (1996).

is described. All procedures use standard cloning methods and PCR-based site-directed mutagenesis methods (Clontech, Palo Alto, CA, www.clontech.com; or Invitrogen, Carlsbad, CA, www.invitrogen.com). A *Sal*I-flanked human APP cDNA harboring the Swedish mutation is inserted into a hamster PrP cosmid vector (the 5′ end of the APP sequence is preceded by a *Sal*I site and a translation initiation sequence 5′-GTCGACACCATGCTGCCC . . . , and the 3′ end of the APP sequence is followed by a stop codon and a *Sal*I site . . . AACTAGCAGCTG-3′). The prion protein–APP fusion gene is excised from the cosmid vector and isolated by size fractionation on an agarose gel followed by electroelution. The fusion gene is further purified using organic solvents and precipitation using ammonium acetate and ethanol. For embryo injection, the prion protein–APP fusion gene is dissolved in 10 m*M* Tris–Cl (pH 8.0) at a concentration of 2–4 μg/ml. Transgenic mice are generated and characterized using established methods (see Jackson Laboratories, Bar Habor, ME, www.jax.org). Other examples of APP mutant transgenic mice generated in a similar manner have been published.[31]

PS1 Mutant Knockin Mice

PCR primers designed to amplify a 180-bp product from the first coding exon of the murine PS1 gene are used to isolate a genomic DNA clone from a mouse 129/Svj P1 library (Genome System Inc., St. Louis, MO). A 1.2-kb *Sac*I–*Hin*dIII fragment containing exon 5 is subcloned into pAlter-1 (Promega Inc.) and is mutagenized with a 39-bp mutagenic oligonucleotide designed to introduce the I45V/M146V double mutation and a *Bst*EII restriction site.

Wild type: 5′-ATGATCAGTGTCATTGTCATTATGACCATCCTCCTGGTG-3′
Ile Met
*Bst*EII
I145V/M146V mutant: 5′-ATGATCAGTGTCATTGTCGTGGTGACCATCCTCCTGGTG-3′
Val Val

Introduction of the I145V substitution humanizes the only polymorphism (between mouse and man) in exon 5 of the murine PS1 gene and also allows the introduction of a unique restriction enzyme site (*Bst*EII). The mutagenized DNA and remaining 5′ and 3′ targeting arms are assembled in pZErO-2.1 (Invitrogen)-derived vectors into which additional cloning sites and loxP sites are inserted in the appropriate positions. These 5′ and 3′ targeting arms are then subcloned into the pNTK2 targeting vector[28] yielding the vector pNTKI. The assembled vector is linealized with *Pvu*I and electroporated into 129/Sv-derived R1 ES cells.[32]

[31] D. R. Borchelt, G. Thinakaran, C. B. Eckman, M. K. Lee, F. Davenport, T. Ratovitsky, C. M. Prada, G. Kim, S. Seekins, D. Yager, H. H. Slunt, R. Wang, M. Seeger, A. I. Levey, S. E. Gandy, N. G. Copeland, N. A. Jenkins, D. L. Price, S. G. Younkin, and S. S. Sisodia, *Neuron* **17,** 1005 (1996).

[32] A. Nagy, J. Rossant, R. Nagy, W. Abramow-Newerly, and J. C. Roder, *Proc. Natl. Acad. Sci. U.S.A.* **90,** 8424 (1993).

Genomic DNA is isolated from clones surviving double selection (250 mg/ml G418 and 2 mM FIAU), digested with *Hin*dIII and *Bgl*I, and analyzed with Southern blot using a 600-bp *Hin*dIII–*Hpa*I 5′ probe. Clones that produce the expected *Hin*dIII and *Bgl*I polymorphisms are injected into recipient blastocysts and are transferred to foster mothers to produce male chimeras, which are then mated with C57BL/6 females to produce heterozygous PS1mv(+/−) mice (129/SvXC57BL/6 F1s). Genotyping is performed as follows. PCR primers (5′-AGGCAGGAAGATCACGTGTT CAAGTAC-3′ and 5′ CACACGCACACTCTGACATGCACAGGC-3′) are used to amplify genomic DNA sequences flanking exon 5 prior to digestion of the amplified DNA with the restriction enzyme *Bst*EII. After an initial hot start at 94° for 2 min, 35 cycles (94°/40 sec, 62°/40 sec, 72°/1 min) are performed. The expected PCR product is 530 bp, and expected product sizes for the wild-type and targeted allele following *Bst*EII enzyme digestion are 530 and 350/180 bp, respectively (Fig. 5). Relative expression levels of mRNA from the wild-type and PS1mv allele are determined by RT-PCR. Total brain RNA is reverse transcribed, and the cDNA is subjected to PCR amplification using PCR primers (5′-TGCTCCAATGACAGAGATACCTGCACC-3′ and 5′-GATAAGAGCTGGAAA GAGAGTCTC-3′) designed to amplify the cDNA sequences encoding amino acids 1–287 of PS1 (sequences encoded by exon 3 to 8). The expected PCR product is 868 bp. Expected product sizes for the wild-type and targeted allele following *Bst*EII enzyme digestion are 868 bp and 426/442 bp, respectively. Analyses of the PS1 mutant knockin mice we have generated using these methods indicate that the expression of mRNA and protein from the targeted allele is normal (Fig. 5).[15] Examples of generation of transgenic mice overexpressing mutant PS1 are available in the literature.[33,34]

MnSOD Overexpressing Transgenic Mice

The plasmid used to generate MnSOD-overexpressing mice consists of human MnSOD cDNA and the human actin 5 flanking sequence and promoter. It is constructed by inserting the human MnSOD cDNA flanked by *Eco*RI restriction sites into the human actin expression vector pHAPr-1 as described previously.[35] The fragment of the construct without the plasmid is further purified for producing transgenic mice. The transgene is introduced into the pronuclei of mouse fertilized eggs by microinjection. Mice used for producing transgenic mice are

[33] K. Duff, C. Eckman, C. Zehr, X. Yu, C. M. Prada, J. Perez-Tur, M. Hutton, L. Buee, Y. Harigaya, D. Yager, D. Morgan, M. N. Gordon, L. Holcomb, L. Refolo, B. Zenk, J. Hardy, and S. Younkin, *Nature* **383,** 710 (1996).

[34] S. Qian, P. Jiang, X. M. Guan, G. Singh, M. E. Trumbauer, H. Yu, H. Y. Chen, L. H. Van de Ploeg, and H. Zhang, *Neuron* **20,** 611 (1998).

[35] H. C. Yen, T. D. Oberley, S. Vichitbandha, Y. S. Ho, and D. K. St. Clair, *J. Clin. Invest.* **98,** 1253 (1996).

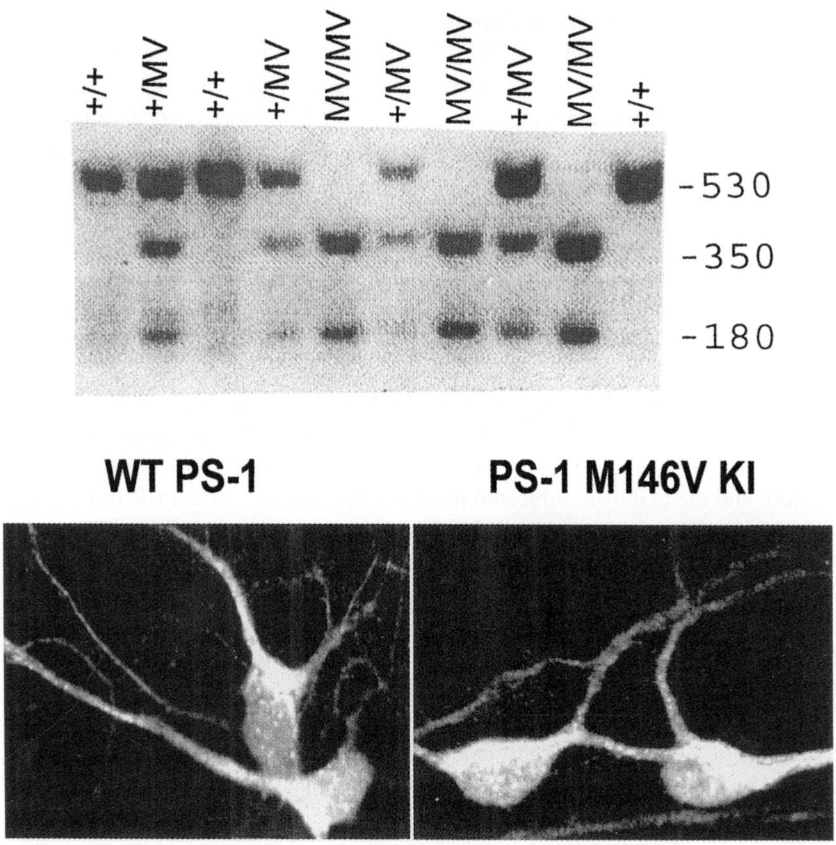

FIG. 5. Characterization of PS1 mutant knockin mice. (Top) Example of genotyping results. A 530-bp fragment (containing exon 5) was amplified by PCR, and the *Bst*EII polymorphism was revealed by cleavage of the PCR product into two diagnostic bands of 350 and 180 bp. (Bottom) Confocal laser-scanning microscope images showing PS1 immunoreactivity in cultured hippocampal neurons from wild-type mice and PS1 mutant knockin mice. Note that localization and intensity of the immunoreactivity are similar in the wild-type and PS1 mutant neurons.

the F1 progeny of C57BL6/C3H hybrids (B6C3), which are purchased from Harlan Sprague Dawley (Indianapolis, IN). Founders with stably integrated human MnSOD transgenes are identified by Southern analysis from mouse tail DNA. All transgenic mice are propagated as heterozygous transgenic mice. Relative levels of expression of the transgene in a given tissue can be determined by RT-PCR analysis of total RNA, MnSOD protein levels can be determined by Western blot analysis, and MnSOD enzyme activity can also be determined as described.[23]

Imaging of Superoxide and Other Reactive Oxygen Species in Cultured Primary Neurons and PC12 Cells Using Fluorescent Probes

A very valuable approach to establish the involvement of oxidative stress in physiological or pathological processes is to monitor relative levels of one or more reactive oxygen species in living cells under precisely controlled experimental conditions. To this end, we have employed confocal laser-scanning microscope imaging-based methods to visualize, in cultured primary neurons and neural cell lines, molecular probes that fluoresce when oxidized by specific reactive oxygen species. The approach allows for quantitative comparisons of relative levels of reactive oxygen species between cells with different genotypes (e.g., neurons from wild-type mice versus neurons from APP mutant or PS1 mutant mice) or cells exposed to different environmental conditions (e.g., amyloid β-peptide, glutamate, or iron). Because confocal microscope provides outstanding resolution of subcellular compartments, one can often arrive at conclusions as to where in the cell the reactive oxygen species are localized (e.g., mitochondria, cytoplasm, or nucleus). The latter information can then be correlated with measures of oxidative damage to particular types of molecules within the cell. For example, the presence of hydroxyl radical in mitochondria might be correlated with damage to mitochondrial DNA. We have focused our work on probes for four major reactive oxygen species: superoxide anion radical, hydrogen peroxide, hydroxyl radical, and peroxynitrite.

While imaging of fluorescent probes that report particular reactive oxygen species are of great value, they are not without caveats. One potential pitfall is due to the fact that exposing cells to ultraviolet light, which is required to excite the fluorochromes, induces photooxidation. In our experience, this is a particularly robust problem when imaging hydrogen peroxide levels using the probe 2,7-dichlorofluorescin diacetate, such that fluorescence becomes brighter and brighter with repetitive imaging of the cells. It is therefore important, when making quantitative comparisons between cells of different genotypes or with different experimental treatment histories, to acquire an image of an individual microscope field only once and then move to a different field. Another potential concern is differential dye loading among cultures, although we have found that careful attention to keeping medium volumes and probe incubation times constant when loading cells with the probes results in quite consistent levels of basal cellular fluorescence. Finally, it should be noted that it is always advisable to verify any results obtained using fluorescent probe-based analyses by applying another method for measuring the same reactive oxygen species. For example, electron paramagnetic resonance spectroscopy can be used to measure relative levels of hydroxyl and superoxide anion radicals.[10,36]

[36] K. Hensley, J. M. Carney, M. P. Mattson, M. Aksenova, M. Harris, J. F. Wu, R. A. Floyd, and D. A. Butterfield, *Proc. Natl. Acad. Sci. U.S.A.* **91,** 3270 (1994).

Superoxide Anion Radical

Relative levels of intracellular superoxide anion radical can be quantified by confocal laser-scanning microscope-based image analysis in cultured cells loaded with the fluorescent probe hydroethidine (Molecular Probes, Eugene, OR). When hydroethidine is oxidized by superoxide, it forms a fluorescent ethidium cation, which then binds to nuclear DNA. Neurons or other cell types are cultured on glass coverslips. Hydroethidine is prepared as a 5 mM stock in dimethyl sulfoxide and aliquots are stored at $-80°$; repeated freeze–thaw cycles should be avoided. The cells are incubated for 30 min in the presence of 5 μM hydroethidium (Molecular Probes), washed twice with Locke's buffer, and confocal images of cell-associated ethidium fluorescence are acquired using a Zeiss CSLM510 system with 488 nm excitation and 510 nm emission. The average pixel intensity in individual cell bodies is determined using Zeiss software. To avoid potential investigator bias, all cultures are coded and analyzed without knowledge of experimental treatment history of the cultures. An example of imaging of superoxide in cultured hippocampal neurons from wild-type and PS1 mutant knockin mice is shown in Fig. 6.

Mitochondrial Hydroxyl and Peroxynitrite Radicals

The dye dihydrorhodamine (Molecular Probes) is taken up selectively by mitochondria and fluoresces when oxidized to the positively charged rhodamine 123 derivative (Fig. 7); hydroxyl radical and peroxynitrite are believed to be the oxyradicals primarily responsible for the oxidation of dihydrorhodamine in mitochondria.[23,37] Dihydrorhodamine is prepared as a 5 mM stock in dimethyl sulfoxide, and aliquots are stored at $-80°$; repeated freeze–thaw cycles should be avoided. Cells are incubated for 30 min in the presence of 5 μM dihydrorhodamine, washed three times with Locke's buffer, and confocal images of cellular fluorescence are acquired (488 nm excitation and 510 nm emission) and analyzed as described for ethidium fluorescence. One problem with dihydrorhodamine is that because it can be oxidized by both hydroxyl radical and peroxynitrite, it cannot be used to accurately quantify either of these oxyradicals. However, by applying additional analytical methods, it is often possible to establish the relative contributions of hydroxyl radical and peroxynitrite to the physiological or pathological process in question. For example, one approach is to employ inhibitors of nitric oxide synthase to block peroxynitrite formation. Another approach is to evaluate damage to cellular molecules using antibodies that selectively recognize proteins

[37] N. W. Kooy, J. A. Royall, H. Ischoropoulos, and J. S. Beckman, *Free Radic. Biol. Med.* **16,** 149 (1994).

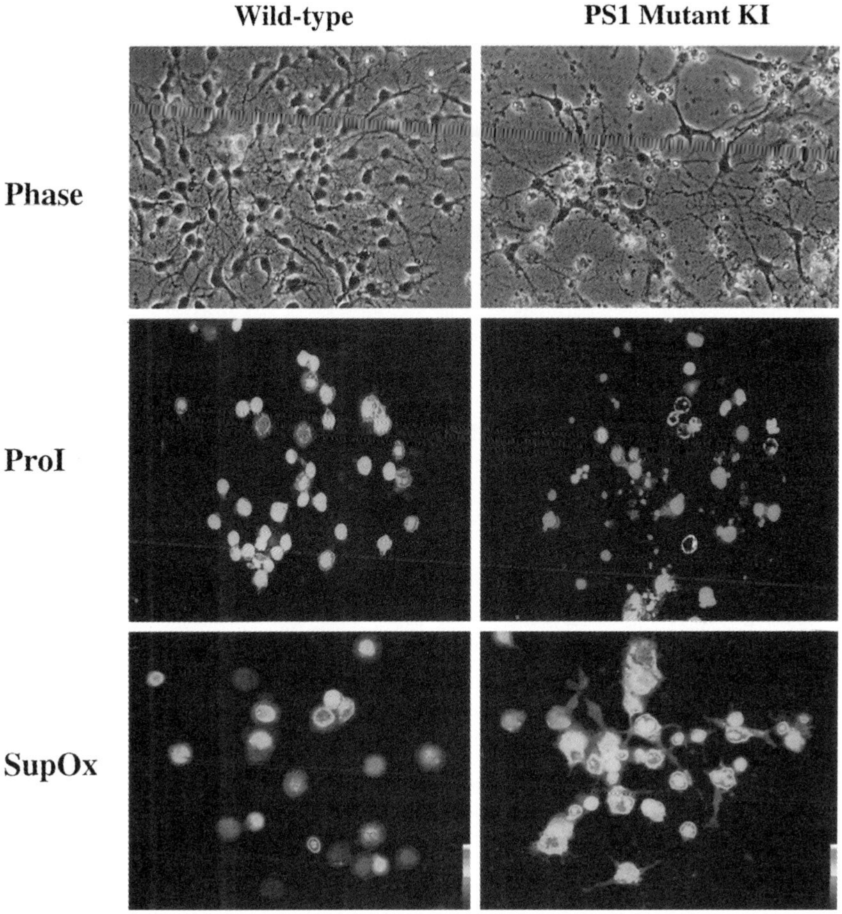

FIG. 6. Imaging of superoxide in cultured hippocampal neurons from wild-type and PS1 mutant knockin mice. Cultures from wild-type and PS1 mutant knockin mice were exposed to Aβ1-42. (Top) Phase-contrast images and (middle) propidium iodide fluorescence (propidium iodide is a fluorescent DNA-binding dye) 48 hr after exposure to Aβ1-42. (Bottom) Hydroethidine fluorescence (a probe for superoxide) 12 hr after exposure to Aβ1-42. Note that neurons from PS1 mutant mice (1) exhibited increased damage and nuclear DNA fragmentation and (2) increased accumulation of superoxide compared to neurons from wild-type mice.

modified as the result of actions of hydroxyl radical or peroxynitrite (Fig. 8). For example, antibodies against 4-hydroxynonenal (an aldehydic product of hydroxyl radical-induced membrane lipid peroxidation) or nitrotyrosine (a protein modification induced by peroxynitrite) can be used for immunostaining and Western blot analyses.[19,22,23,29]

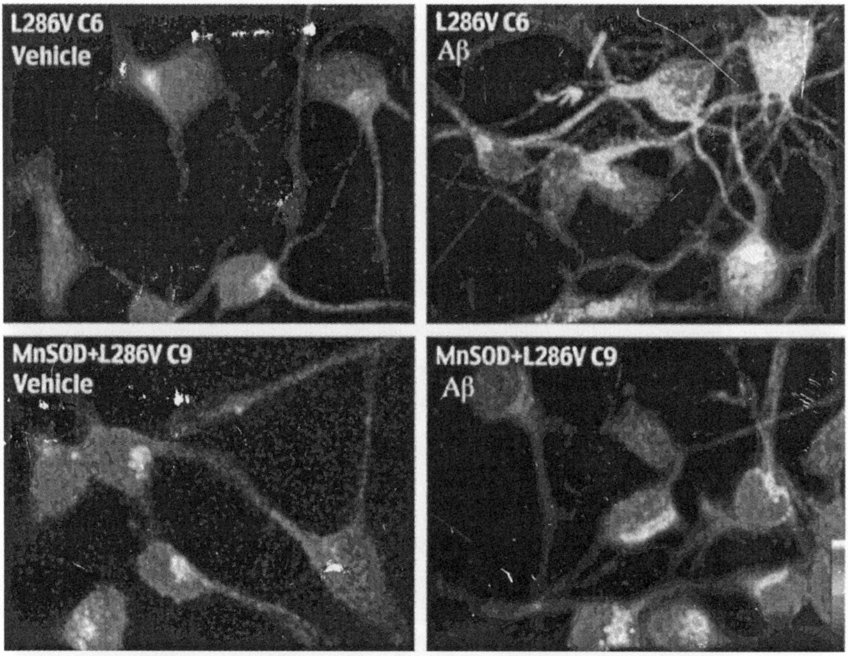

FIG. 7. Manganese superoxide dismutase overexpression suppresses intramitochondrial production of hydroxyl radical and peroxynitrite. Images of dihydrorhodamine fluorescence (a probe for mitochondrial hydroxy radical and peroxynitrite) in differentiated PC12 cells overexpressing mutant PS1 (L286V mutation) alone (top) or in combination with human MnSOD (bottom). Cultures had been exposed for 12 hr to either saline (vehicle) or Aβ. Note that Aβ increases the levels of mitochondrial oxyradicals and that the increase is suppressed greatly in cells overexpressing MnSOD.

Hydrogen Peroxide Levels

The probe 2,7-dichlorofluorescin diacetate (Molecular Probes) is converted to the fluorescent 2,7-dichlorofluorescein molecule on interaction with hydrogen peroxide. 2,7-Dichlorofluorescin diacetate is prepared as a 50 m*M* stock in dimethyl sulfoxide, and aliquots are stored at $-80°$; repeated freeze–thaw cycles should be avoided. Cells are loaded with by incubating for 30–50 minutes at 37° in the presence of 100 μM 2,7-dichlorofluorescin diacetate. Cultures are washed twice with Locke's buffer, and confocal images are acquired (488 nm excitation and 510 nm emission) and analyzed as described for ethidium fluorescence.

Membrane Lipid Peroxidation

Two methods for localization and quantification of membrane lipid peroxidation in cultured cells have been established in this laboratory. One method involves

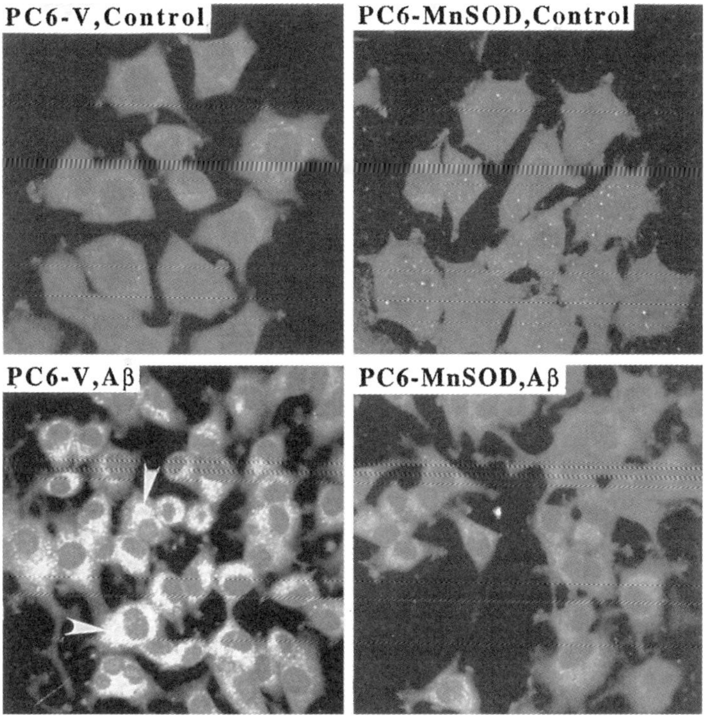

FIG. 8. Evidence for the involvement of superoxide in Aβ-induced accumulation of nitrated proteins. Vector-transfected control PC6 cells (PC6-V) and PC6 cells overexpressing MnSOD were exposed to Aβ for 12 hr. Cells were then immunostained with an antinitrotyrosine antibody, and confocal microscope images of immunofluorescence were acquired. Note that levels of nitrotyrosine immunoreactivity are increased greatly after exposure to Aβ in control cells (arrows point to regions of intense immunofluorescence), but not in cells overexpressing MnSOD.

a 2-thiobarbituric acid (TBA) fixation and fluorescence microscopy protocol. After experimental treatment, cells are fixed in an aldehyde free fixative containing 50% methanol, 10% glacial acetic acid, 2 m*M* EDTA, and 38 m*M* TBA. Cells are then heated to 85° for 45 min, the fixative is removed, antifade solution (100 μM propylgallate in PBS) is added, and cells are imaged using a Zeiss confocal laser-scanning microscope (excitation at 488 nm and emission at 510 nm). Values for average staining intensity/cell are obtained. The second method employs a mouse monoclonal antibody that binds selectively to proteins that have been modified covalently by 4-hydroxynonenal, an aldehydic product of membrane lipid peroxidation. After experimental treatment, cells are fixed for 30 min in a solution of 4% paraformaldehyde in PBS, and membranes are permeabilized by incubation for 5 min in a solution of 0.2% Triton X-100 in PBS.

The cells are then immunostained using a 1 : 500 dilution of mouse antibody against 4-hydroxynonenal-modified proteins[19] using a method essentially identical to that described in the section on immunocytochemistry in this chapter.

Measurements of Mitochondrial Transmembrane Potential, Intramitochondrial Calcium Levels, and Cytoplasmic Calcium Levels

Mitochondrial Membrane Potential

Mitochondrial transmembrane potential can be estimated using the fluorescent probe tetramethylrhodamine ethyl ester (Molecular Probes). Cells are incubated for 30 min in the presence of 100 n*M* tetramethylrhodamine ethyl ester (Molecular Probes). After washing in PBS, the cells are fixed for 20 min in 4% paraformaldehyde in PBS. Confocal images of cellular fluorescence are acquired (488 nm excitation and 510 nm emission), and the average pixel intensity/cell is quantified using the software supplied by the manufacturer (Zeiss). Measurements are typically made in 30–50 neurons/culture. Analyses are performed by an investigator unaware of the treatment history of the cultures. Examples of data obtained using this method have been published.[38,39]

Intramitochondrial-Free Calcium Levels

The concentration of free calcium within mitochondria can be estimated using the fluorescent probe rhod-2-AM (Molecular Probes). Cells are incubated for 60 min in the presence of 5 μM rhod-2, washed twice in Locke's solution, and confocal images of cellular fluorescence are acquired and analyzed as described for ethidium fluorescence. Examples of data obtained using this method have been published.[40]

Cytoplasmic-Free Calcium Levels

Cytoplasmic-free calcium levels are quantified by fluorescence ratio imaging of the calcium indicator dye fura-2 (Molecular Probes). Cells are loaded with the acetoxymethylester form of fura-2 (a 30-min incubation in the presence of 10 μM fura-2) and are imaged using a Zeiss AttoFluor system with a 40X oil objective. The average $[Ca^{2+}]_i$ in individual neuronal cell bodies is determined from the ratio of the fluorescence emissions obtained using two different excitation wavelengths

[38] A. J. Krohn, T. Wahlbrink, and J. H. Prehn, *J. Neurosci.* **19,** 7394 (1999).

[39] D. Liu, C. Lu, R. Wan, W. Auyeung, and M. P. Mattson, *J. Cereb. Blood Flow Metab.* **22,** in press (2002).

[40] I. I. Kruman and M. P. Mattson, *J. Neurochem.* **72,** 529 (1999).

(334 and 380 nm). The system is calibrated using solutions containing either no Ca^{2+} or a saturating level of Ca^{2+} (1 m*M*) using the following formula: $[Ca^{2+}]_i = K_d\,[(R\text{-}R_{min})/(R_{max}\text{-}R)](F_o/F_s)$. Examples of data obtained using this method have been published.[14,26,27]

Quantification and Subcellular Localization of APP, PS1, and MnSOD

Relative levels of APP, PS1, and MnSOD in PC12 cells and primary neurons are determined by Western blot analysis. Fifty to 100 μg of solubilized proteins is separated by electrophoresis in an SDS/polyacrylamide gel (7% acrylamide for APP, a 70- to 76-kDa protein; 10% acrylamide for PS1, a 46-kDa protein; and 12% acrylamide for MnSOD, a 23-kDa protein) and then transferred electrophoretically to a nitrocellulose sheet. After blocking with 5% milk and a 3-hr incubation in the presence of primary antibody, the nitrocellulose sheet is further processed using horseradish peroxidase-conjugated secondary antibody and a chemiluminescence system (Amersham). The primary antibodies used are a mouse monoclonal antibody against APP (clone IG5),[41] an affinity-purified polyclonal rabbit anti-human PS1 antibody (1 : 100 dilution),[13] and a rabbit anti-human MnSOD monoclonal antibody (1 : 1000 dilution).[23]

Immunocytochemistry

The cellular localization and relative levels of APP, PS1, and MnSOD in PC12 cells and primary neurons are determined by immunocytochemical analysis. Cells are fixed in 4% paraformaldehyde, and membranes are permeabilized by incubation for 5 min in a solution of 0.2% Triton X-100 in PBS. Cells are then incubated sequentially (with intervening washes in PBS) as follows: 1 hr in the presence of 1% blocking serum (normal goat or horse serum) in PBS; overnight in primary antibody in PBS containing blocking serum (4°); 1 hr in biotinylated secondary antibody (Vector Labs; 1 : 200 in PBS); and 30 min in FITC–avidin complex (Vector Labs; 4 μl/ml PBS). The primary antibodies are the same as those used for Western blot analysis. Images of immunofluorescence are acquired using a confocal laser-scanning microscope (excitation at 488 nm and emission at 510 nm) with a 60 X oil immersion objective. Levels of laser excitation and photodetector gain are held constant to allow direct comparisons of neuronal staining intensities among treatment groups. Relative fluorescence intensity/cell can be quantified using software supplied by the manufacturer of the confocal microscope (Zeiss).

[41] K. Furukawa, B. L. Sopher, R. E. Rydel, J. G. Begley, D. G. Pham, G. M. Martin, M. Fox, and M. P. Mattson, *J. Neurochem.* **67,** 1882 (1996).

Analyses are done by an investigator unaware of culture treatment history or the genotype of the cells being examined.

Conclusions

Oxidative stress is thought to contribute to the pathogenesis of many, if not all, age-related degenerative diseases, including diabetes, cardiovascular disease, cancers, and neurodegenerative disorders. We have shown how the identification of genetic mutations that cause rare inherited forms of Alzheimer's disease can lead to the development of cell culture and animal models that can be used to establish the molecular and cellular alterations that lead to the disease in humans. These same approaches and specific methods can be, and in several cases have been, applied to other diseases. Examples include the development of cell culture and animal models to elucidate roles for superoxide and/or other reactive oxygen species in the pathogenesis of amyotrophic lateral sclerosis,[42–45] Huntington's disease,[46,47] and Parkinson's disease.[48–50]

[42] M. Wiedau-Pazos, J. J. Goto, S. Rabizadeh, E. B. Gralla, J. A. Roe, M. K. Lee, J. S. Valentine, and D. E. Bredesen, *Science* **271,** 515 (1996).

[43] M. B. Bogdanov, L. E. Ramos, Z. Xu, and M. F. Beal, *J. Neurochem.* **71,** 1321 (1998).

[44] I. I. Kruman, W. A. Pedersen, J. E. Springer, and M. P. Mattson, *Exp. Neurol.* **160,** 28 (1999).

[45] R. Liu, R. K. Narla, I. Kurinov, B. Li, and F. M. Uckun, *Radiat. Res.* **151,** 133 (1999).

[46] A. J. Bruce-Keller, J. W. Geddes, P. E. Knapp, R. W. McFall, J. N. Keller, F. W. Holtsberg, S. Parthasarathy, S. M. Steiner, and M. P. Mattson, *J. Neuroimmunol.* **93,** 53 (1999).

[47] S. J. Tabrizi, J. Workman, P. E. Hart, L. Mangiarini, A. Mahal, G. Bates, J. M. Cooper, and A. H. Schapira, *Ann. Neurol.* **47,** 80 (2000).

[48] W. Duan, Z. Zhang, D. M. Gash, and M. P. Mattson, *Ann. Neurol.* **46,** 587 (1999).

[49] L. J. Hsu, Y. Sagara, A. Arroyo, E. Rochenstein, A. Sisk, M. Mallory, J. Wong, T. Takenouchi, M. Hashimoto, and E. Masliah, *Am. J. Pathol.* **157,** 401 (2000).

[50] G. I. Giasson, J. E. Duda, I. V. Murrary, Q. Chen, J. M. Souza, H. I. Hurtig, H. Ischiropoulos, J. Q. Trojanowski, and V. M. Lee, *Science* **290,** 985 (2000).

[39] Three-Dimensional Redox Imaging of Frozen-Quenched Brain and Other Organs

By AKIHIKO SHIINO, MASAYUKI MATSUDA, and BRITTON CHANCE

Introduction

NADH fluorometry has emerged as an available experimental technique for investigating the oxidative metabolism in many organs. The validity of the technique is dependent on the fact that the redox level of the mitochondrial NAD^+/NADH couple is related to the availability of oxygen to cytochrome oxidase, which, in turn, depends on the intracellular oxygen tension. However, if only the NADH signal is measured, the regional mitochondrial concentration would influence the results. Furthermore, it is known that the presence of hemoglobin interferes by absorbing the fluorescence signal. In order to provide a means for cancellation of these effects, we adapted two components from the respiratory chain—fluorescence of the oxidized flavoproteins and NADH fluorescence—for measuring the mitochondrial redox ratio.

The redox ratio is a sensitive parameter of the status of energy metabolism of the cell. For example, the shortage of oxygen, which is the terminal electron acceptor, shifts the redox ratio (NAD^+/NADH) toward reduction (metabolic state 5), whereas an accelerated consumption of ATP shifts the ratio toward oxidation (metabolic state 3). Furthermore, molecular biology has focused on the downregulation of mitochondria-mediated apoptosis. The redox imaging technique can also help elucidate research into redox signaling mechanisms.[1] This chapter describes a technique of a low temperature redox scanning developed in our laboratory.

Methods

Optimal Freezing of the Sample

An optimal quenching of the tissue metabolism must be introduced before measuring the redox state to provide a "snapshot" at the moment of freezing and to allow three-dimensional analysis by sectioning the frozen tissue block. To prevent the destruction of tissue morphology and hypoxic artifact due to clamping or premature circulatory arrest in the sample, an optimal freeze-quenching procedure needs to be designed for the organs being studied. In a small tissue sample or thin wafers of tissue, simply freeze clamping can be applied. However, for a large sample, such as rat brain, *in situ* funnel freezing must be prudently performed, as

[1] A. Shiino, M. Matuda, J. Handa, and B. Chance, *Stroke* **29,** 2421 (1998).

0076-6879/02 $35.00

the freezing time in the tissue increases exponentially with depth from the cooled surface. The procedure for funnel freezing is modified from the methods of Pontén *et al.*[2] Endotracheal intubation is performed, and a skin incision is made in the midline from the level of the eyes to the occiput, the length of which should be determined by the size of the opening of the funnel. The scalp is exfoliated with the periosteum and a funnel is placed over the skull. To prevent liquid nitrogen from leaking, the whole circumference of the bottom of the funnel needs to be covered by pulling up the skin. Needles are hooked at the ends of the skin, penetrating crosswise through the opening of the bottle, and they are sealed with glue (mixture of alcohol, glycerin, and water in 10 : 30 : 60 at volume %). Mechanical ventilation is begun just before the start of freezing. The respirator is adjusted so that the approximate end-tidal CO_2 concentrations are within the range of 35 to 40 mm Hg. If the freezing point extends to the neck or the respiratory airways early on, a premature obstruction of the cerebral blood flow occurs, resulting in an ischemic change in the base of the brain. Therefore, the neck and thorax must be kept warm during the procedure. After completion of the freezing procedure, the entire body is immersed in liquid nitrogen, and the head is dislocated, trimmed with an electric saw, mounted on a sample holder, and fixed with glue.

Tissue Cutting and Grinding

Any heating of the tissue above a temperature of -50°, even for a short period of time, will change the redox status toward a more reduced level, therefore samples are milled under cooling with liquid nitrogen. At low temperatures, tissue becomes hard enough to apply standard steel cutting or milling tools. A 16-bit angular cutter with a diameter of 40 mm (F&D Tool Co. Inc., Three Rivers, MA) is sufficiently cooled by immersion into liquid nitrogen before grinding a sample. When the targeted slice position is almost reached following rough trimming, the surface of the sample is milled slowly and thinned several times to make the surface of the sample as smooth as possible. A large change in the distance between the tip of the light guide and the sample surface affects fluorescent gain, so the apparent fluorescent ratio will change as a result.

Fluorescence Measurements

The simple ratio of the appropriately normalized signals of flavoprotein and pyridine nucleotide (PN) affords an oxidized-reduced measurement because of flavoprotein fluorescence in the oxidized form and pyridine nucleotide in the reduced form. When the flavoprotein/PN ratio is initially set to unity, it is relatively insensitive to variations in the concentration of mitochondria and to the screening of excitation and emission light by absorbing pigments, such as hemoglobin. A general design of the instrument for measuring redox has been reported

[2] U. Pontén, R. A. Ratcheson, L. G. Salford, and B. K. Siesjö, *J. Neurochem.* **21,** 1127 (1973).

previously.[3] The redox ratio is calculated by fluorescent signals obtained from intrinsic fluorochromes. For PN measurement, the peak excitation wavelength is 366 nm and the peak emission wavelength 450 nm. The excitation of flavoprotein is performed in a rather narrow region, around 463 ± 10 nm, to avoid interference due to changes in cytochrome absorption. The peak flavin emission is 520 nm. It is necessary to check whether cross talk between the excitation and emission filters is sufficiently excluded or not. The simplest way to check this is to put a mirror under the light guide. However, a very small amount of cross talk cannot be avoided in obtaining reasonable excitation-emission efficiency. It is better to measure reflectance signals before passing through the excitation filter.[3] The light source is provided by a 100-W mercury arc (UVP Inc., San Gabriel, CA) cooled by circulating water. The brightness of the light source needs to be constant; thus, attention must be paid to monitoring the current in applying a mercury lamp.

The light guide is composed of several quartz fibers whose caliber is 50 to 80 μm and allows UV transmission (for excitation of PN). The optimal distance between the light guide and the sample surface can be adjusted so that a maximum gain of emitted light can be obtained. The optimal distance is usually 100–200 μm. At this distance, the spatial resolution is about 100–160 μm. The excitation fiber guides the light onto the surface of the sample and emitted light is directed to the photomultiplier tube. The step size and number are selected in proportion to the spatial resolution and sample size. While studying the brain of a gerbil's, for example, we set the step size to 40 μm, with a total of 512 × 2048 steps to cover a whole coronal slice of the brain.

The signal intensity of each point is converted into digital data and is sent to a computer. Each point is displayed with the appropriate 256 gray or color scale, corresponding to the signal intensity of the point. Usually, black represents a low signal intensity or a low flavoprotein/PN ratio (reduced) and white represents a high signal or a high flavoprotein/PN (oxidized). In tissue samples, the flavoprotein signal is generally smaller than the PN signal, and the dynamic range of its fluorescent signal is smaller than that of PN. Therefore, the gain sensitivities of the two fluorescent sources need to be adjusted. A three-dimensional model of redox data from a set of continuous images is obtained by repeatedly scanning and shaving off the sample surface.

Calibration of Redox Ratio

The targeted organ is removed and placed into an ice-cold buffer (0.25 *M* sucrose, 1 m*M* EDTA, and 10 m*M* Tris–HCl, pH 7.4), chopped finely, and washed three times with the same solution. The minced tissue fragments are homogenized manually with a Dounce tissue grinder (Wheaton, Millville, NJ) and centrifuged, using Percoll based on a modification of the procedure by Sims.[4] An isolated

[3] B. Quistorff, J. C. Haselgrove, and B. Chance, *Anal. Biochem.* **148,** 389 (1985).
[4] N. R. Sims, *J. Neurochem.* **55,** 698 (1990).

mitochondrial pellet is made to a concentration of 30 mg of protein/ml in the isolation medium. Protein concentration is estimated by the biuret method, using bovine serum albumin as a standard.

Isolated mitochondria are moved to another ice-cold medium [0.2 *M* sucrose, 20 m*M* Tris–HCl, 10 m*M* Tris–phosphate, 15 m*M* KCl, 0.3 m*M* EDTA, 3% (w/v) bovine serum albumin, pH 7.2, 15 mg protein/ml], incubated in a small beaker, and mixed with a magnetic stirrer. Two samples of 0.15 ml each are withdrawn from this beaker and frozen by an aluminum block precooled with liquid nitrogen at different oxidation–reduction states in the following order: (1) incubated with 7 m*M* glutamate and 7 m*M* malate or 7 m*M* pyruvate and 7 m*M* malate and perfused with 95% oxygen (state 4); (2) plus 0.5 m*M* ADP (state 3); (3) plus 1 μM *m*-chlorocarbonylcyanide phenylhydrazone (fully oxidized); and (4) perfused with 100% nitrogen (fully reduced). Redox ratios of each frozen mitochondrial bead are measured by the same procedure as the gerbil brain except for using a smaller matrix size (64×64).

Sample Images of Redox Ratio

We reported the results of three-dimensional redox images of the normal brain previously.[5] There are regional redox differences, coinciding with the anatomical structures, suggesting a relationship between energy consumption and redox ratio in the resting state. Some interesting examples are presented as follows.

Redox Changes during and after Forebrain Ischemia

Arterial occlusion causes stasis of the electron transfer in the respiratory chain due to a decrease in the oxygen supply to the mitochondria. This can be seen as a reduction in the flavoprotein/PN ratio. Figure 1 shows redox images of a gerbil brain during ischemia induced by occlusion of the bilateral common carotid arteries. The flavoprotein/(flavoprotein + PN) ratio is decreased in the area fed by carotid arteries, in contrast to the territory of the vertebrobasilar artery, which feeds the medial side of the thalamus. Figure 2 shows a unique redox change after 12 hr of transient ischemia. This quite high flavoprotein/(flavoprotein + PN) ratio is seen in vulnerable areas, such as the striatum and layers 3 and 5–6 in the cortex, suggesting that a superoxidation mechanism occurs in these areas. The relation between redox change and metabolic condition is summarized in Table I.

Redox Changes during Seizures

Depolarization of neurons promotes a Ca^{2+} influx through voltage-activated and *N*-methyl-D-aspartate (NMDA)-gated ion channels, increasing intracellular and intramitochondrial Ca^{2+} concentrations. This causes a depolarization of

[5] A. Shiino, M. Haida, B. Beauvoit, and B. Chance, *Neuroscience* **91,** 1581 (1999).

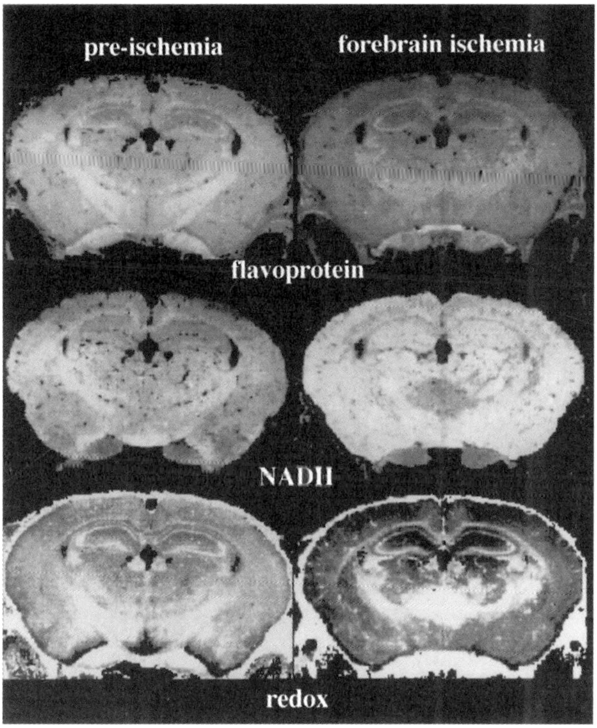

FIG. 1. Redox images of a gerbil brain before and during forebrain ischemia. Ischemia decreased the fluorescence signal of flavoprotein, increased that of pyridine nucleotide (PN), and decreased the ratio of flavoprotein/(flavoprotein + PN), suggesting inhibition of electron flow in the respiratory chain. Note that the area supplied by the basilar artery was not reduced.

mitochondrial membranes and an activation of Ca^{2+}-dependent enzymes, such as pyruvate dehydrogenase, NAD^+-isocitrate dehydrogenase, and α-ketoglutarate, resulting in a decrease in the NAD^+/NADH ratio. Figure 3 shows a reduction in the redox ratio in kainic acid-sensitive excitatory pathways. The flavoprotein/(flavoprotein + PN) ratio is first decreased in the dentate gyrus and the CA1, and is later decreased in the piriform cortex and amygdaloid complex.

Discussion

Theoretically, the effects of hemoglobin and protein concentrations on the flavoprotein/PN ratio will become negligible if the two fluorescent signals are both affected to the same extent. In fact, Chance *et al.*[6] reported that the flavoprotein/PN

[6] B. Chance, B. Schoener, R. Oshino, F. Itshak, and Y. Nalase, *J. Biol. Chem.* **254,** 4764 (1979).

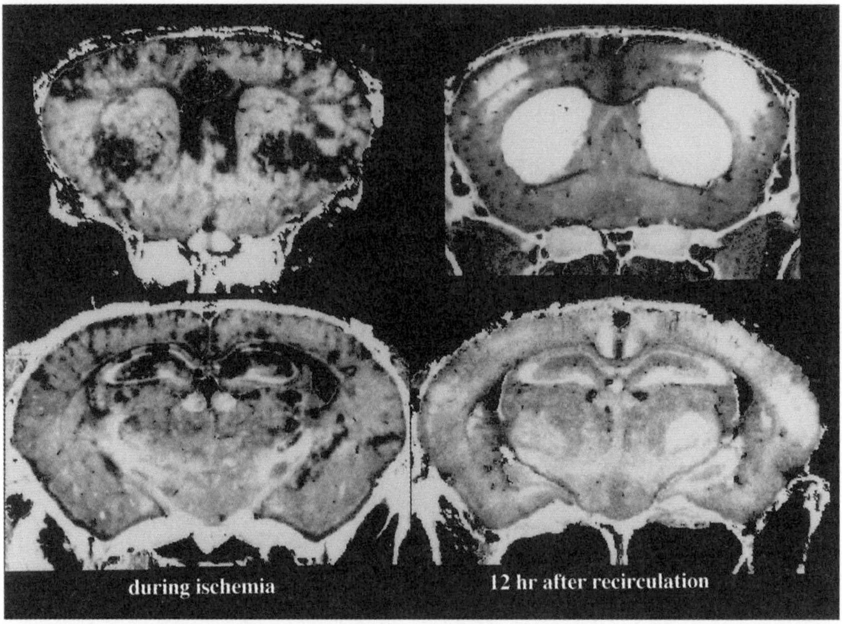

FIG. 2. Redox images of a gerbil brain during and 12 hr after forebrain ischemia. Note the hyperoxidized areas, which correspond to the "vulnerable" region, suggesting a relation between reactive oxygen intermediates and cell death.

ratio is only very slightly affected by hemoglobin and protein concentrations. Another benefit of determining the ratio of flavoprotein and PN fluorescence is an increase in sensitivity, as the two components of oxidation/reduction ratio vary inversely. Interpretation of the fluorescence of flavoproteins is a little complicated, as their fluorescence potential varies individually. Most of the enzyme-bound flavin is weakly fluorescent or nonfluorescent. The major flavoprotein fluorescence of mitochondria is the FAD of α-lipoamide dehydrogenase and Co-Q-linked flavin of

TABLE I
FACTORS ASSOCIATED WITH OXIDATION–REDUCTION LEVELS OF MITOCHONDRIA

$NAD^+/NADH$ ↑	$NAD^+/NADH$ ↓
Electron transport ↑	Electron transport ↓
ADP ↑ (state 3)	ADP ↓ (state 4)
Uncoupling	Dysfunction
Substrate ↓	Oxygen ↓

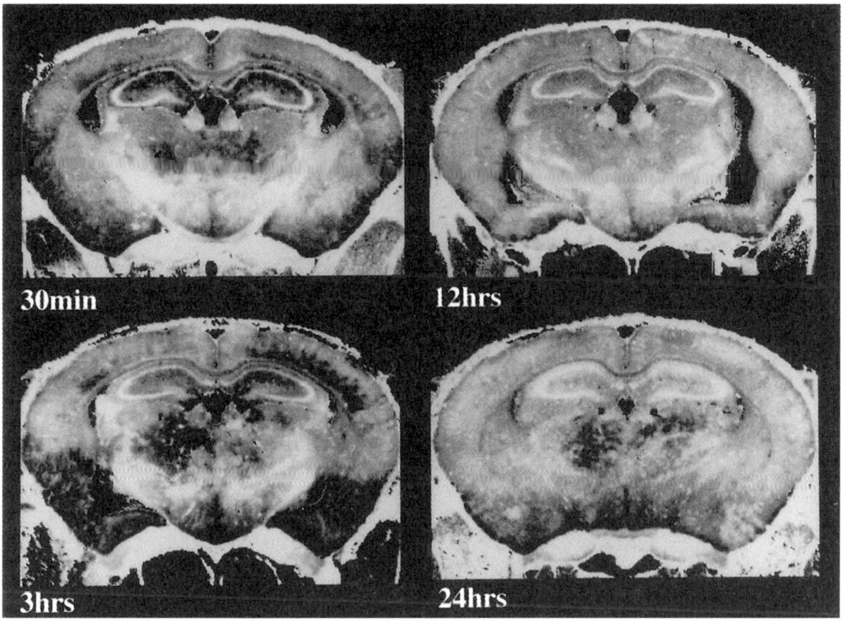

FIG. 3. Consecutive redox images of a gerbil brain after kainic acid injection. The redox was reduced in the dentate gyrus and the CA1 of the hippocampus 30 min after injection. The reduced area extended later to the piriform and amygdaloid complex.

the fatty acid-oxidizing system. Only the α-lipoamide dehydrogenase is thought to be a main source of flavin fluorescence and is in equilibrium with the mitochondrial NAD^+/NADH couple. The content rate of the NAD-linked α-lipoamide dehydrogenase and Co-Q-linked flavin differs according to the tissue.[7] For example, the Co-Q-linked flavin is rich in the liver, but almost absent in the brain.

The PN signal may originate from two different redox couples, i.e., cytosolic and mitochondrial components. In brain, heart, and skeletal muscle, data suggest that the PN signal predominantly originates from mitochondrial NADH. However, in the liver the PN signal suggests an almost equal contribution from the cytosol and mitochondria.

In many cases, the time resolution, the localization, and the selectivity of the analysis were insufficient, and the researcher resorted to freeze trapping. The merits of this low-temperature technique are as follows: (a) the nonradiative transfers from the excited state are often diminished at low temperatures and the quantum yield of several fluorochromes is enhanced 5- to 10-fold at liquid nitrogen temperatures

[7] W. S. Kunz and F. N. Gellerich, *Biochem. Med. Metab. Biol.* **50,** 103 (1993).

and (b) the integration time of the scanning significantly increases because of the complete arrest of metabolic processes, which allows a high signal-to-noise ratio by signal averaging. NADH and flavoprotein components can be effectively trapped at liquid nitrogen temperatures, and thus the redox state of this portion of the respiratory chain can be preserved for long periods and assayed at any time, without alteration.

Acknowledgment

This work was supported by NIH Grant NS27346.

[40] *In Vivo* Fluorescent Imaging of NADH Redox State in Brain

By ROBERT E. ANDERSON and FREDRIC B. MEYER

Introduction

The introduction by Chance *et al.*[1] of *in vivo* recording of reduced nicotinamide adenine dinucleotide (NADH) concentration has yielded a number of techniques to accurately ascertain the redox state of brain before, during, and after cerebral ischemia. NADH, which is an intrinsic fluorophore, is a primary intermediary of energy transfer from the tricarboxylic acid cycle to the electron transport chain of the mitochondria. When blood flow to the brain is compromised, i.e., due to an occlusion, the respiratory chain becomes inhibited when oxygen concentration is reduced. NADH fluoresces a blue color (440–470 nm) when tissue is excited by ultraviolet light. The oxidized cofactor NAD^+ does not fluoresce while the unbound or cytoplasmic form of NADH fluoresces minimally. Therefore, microfluorometry using NADH as a fluorescent marker can be used to ascertain mitochondrial function in brain.

Over the past several decades, a variety of optical techniques have been developed to measure NADH fluorescence in brain, muscle, liver, heart, kidney, and thyroid to assess mitochondrial function. Fluorescent microscopes using bright-field or dark-field illumination techniques, as well as fiber-optic devices, have been utilized. Fiber-optic microfluorometers have been developed by several research groups to measure NADH fluorescence in multiple tissue sites because microscopic techniques are limited by viewing only one site at a time. These microscopic techniques measure only spots or areas of tissue by the use of optical apertures, which

[1] B. Chance, P. Cohen, F. Jöbsis, and B. Schoener, *Science* **137,** 499 (1962).

0076-6879/02 $35.00

limit the field of view. The importance of the hemoglobin artifact is such that when organs or tissues have an intact blood supply, NADH fluorescent intensity is dependent on the concentration of hemoglobin in the field. It is essential that this artifact be minimized in the measurement of NADH redox state. This will be discussed in more detail later.

The development of optical fluorescent-imaging techniques has been facilitated over the past decade by the availability of computer image analysis systems and image pickup devices. In 1968, Engle and Freed[2] developed a flying spot microspectrophometer to image cultured cells based on their absorbance in the ultraviolet. This scheme made use of a microscope, photomultiplier, and scanning mirror. While the resultant images of the cultured cells were somewhat crude compared to the imaging techniques of the present day, it was nevertheless the first attempt to image living cells. Almost a decade later, Schuette *et al.*[3] developed a television fluorometer, which incorporated a vidicon television tube coupled to an image intensifier with long focal-length optics, which was used to view brain *in vivo*. An electronic window was used to select the area of interest to integrate the video signal for presentation as an analog signal for strip chart recording. This device was later used for identifying interictal epileptiform discharges in human epilepsy.[4] The distribution of NADH fluorescence has been also ascertained using *in vivo* photographic techniques.[5] This method was more useful when viewing slices of tissue rather than tissue *in vivo* because any tissue movement could be picked up by the required long exposure time of photographic film.

A decade ago, second generation image intensifier tubes coupled to CCD chip video cameras were developed. Using real-time video imaging, Anderson *et al.*[6] measured the NADH redox state along with simultaneous measurements of intracellular brain pH and regional cortical blood flow. Along similar lines, Ince *et al.*[7] developed a system for the measurement of NADH fluorescence in intact blood-perfused myocardium. The following is a description of the system that we currently use in our laboratory.

In Vivo Video Fluorescent Instrumentation

The system, illustrated in Fig. 1, consists of a honeycomb optical bench on which is mounted a light source, intravital type microscope, animal stage, and

[2] J. L. Engle and J. J. Freed, *Rev. Sci. Instrum.* **39,** 307 (1968).

[3] W. H. Schuett, W. C. Whitehouse, D. V. Lewis, M. O'Conner, and J. M. Van Buren, *Med. Instrum.* **8,** 331 (1974).

[4] J. M. Van Buren, D. V. Lewis, W. H. Schuette, W. C. Whitehouse, and C. Ajmone Marsan, *Neurosurgery* **2,** 114 (1978).

[5] S. Ji, B. Chance, B. H. Stuart, and R. Nathan, *Brain Res.* **119,** 357 (1977).

[6] R. E. Anderson, F. B. Meyer, and F. H. Tomlinson, *Am. J. Physiol.* **263,** H565 (1992).

[7] C. Ince, J. M. C. C. Coremans, and H. A. Bruining, *Adv. Exp. Med. Biol.* **317,** 227 (1992).

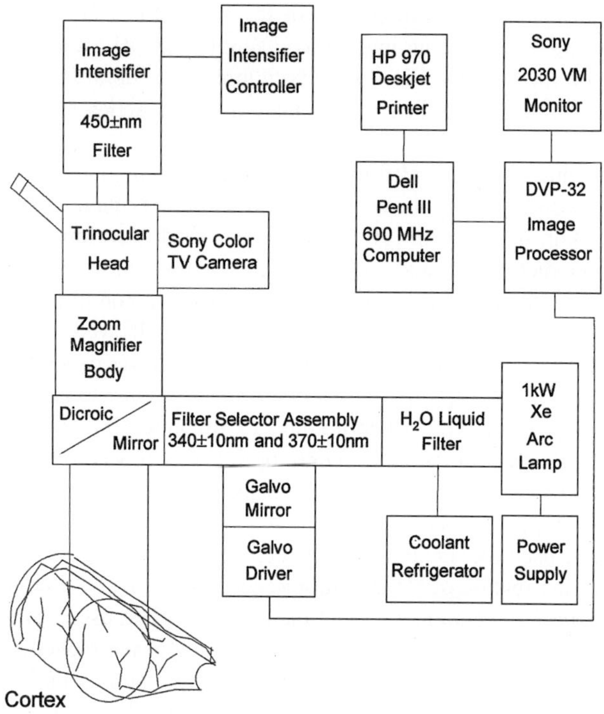

FIG. 1. Block diagram of a real-time computer video serial fluorescent system. Refer to the text for a detailed description of the various components.

camera assembly.[6] Illumination for excitation of umbelliferone is provided by a modular system consisting of a 1000-W xenon arc light source and exciting filter changer. The xenon arc lamp (LX1000UV, ILC Technology, Sunnyvale, CA) has a prealigned internal parabola reflector and a single-crystal sapphire window for maximizing the efficiency of the output in the ultraviolet. The lamp is powered by a constant current DC switching power supply (PSC-1000; ILC Technology) that provides tight current regulation (1.0% lamp regulation per 10% change in input voltage) and low current ripple (1.0% root mean square) for lamp stability. The lamp provides 20 W in the ultraviolet focused to 3-in. diameter, 750-mm focal length secondary lens. The excitation energy is passed through a 3-in. diameter, 100-mm liquid optical filter (Oriel, Stratford, CT) that is filled with distilled water and cooled by an external water chiller (RTE-100; Neslab, Portsmouth, NH) to reduce infrared wavelength energy.

A custom-built filter changer was constructed for selecting the appropriate excitation wavelength under computer control (XPS T600; Dell, Round Rock, TX). The exciting light coming from the liquid filter was split 50 : 50, using a metallic-fused silica beam splitter (Corion, Holiston, MA). Each beam was then directed separately to an interference filter, one 340 ± 5 nm and the other 370 ± 5 nm (Corion, Holiston, MA). These two beams were deflected at a 90° angle, intersecting a UV mirror (Corion) mounted on a UV mirror (Corion) galvanometer (General Scanning, Watertown, MA), which either deflected or passed the appropriate wavelength for excitation. This mirror was driven by a scanner controller (CX-660; General Scanning), which is controlled by the computer.

A Nikon SMZ-10 stereoscopic zoom microscope body is attached to a vertical stand mounted on an optical bench. A Ploem-type illuminator is attached to this stand below the objective lens. The filter changer is mounted behind the stand at the same optical axis as the illuminator. The beam-splitting mirror (Acton Research, Acton, MA) within the illuminator is coated so that at wavelengths below 400 nm, >90% of light would be reflected for the excitation, and at wavelengths above 400 nm, >90% would be passed for the fluorescent emission. It is image compatible so that secondary images cannot be formed. The image field size can be varied from 2.3 to 14 mm diameter, while the illumination field size is always 18 mm in diameter. The spatial resolution is 4.5 and 27 μm at maximum and minimum magnification, respectively. The working distance of the system is 50 mm, allowing room for the use of recording electrodes. The depth of focus is 100 and 600 μm at maximum and minimum magnification, respectively. The excitation energy at 340 and 370 nm as measured by a digital power meter (835 Optical Power Meter; Newport, Irvine, CA) is 200 and 790 $\mu W/cm^2$, respectively. The animal platform and head micromanipulator (Oriel, Stratford, CT) is mounted on the optical bench (Newport, Irvine, CA) below and lateral to the microscope unit.

A trinocular body is attached to the top of the microscope body (SMZ-10; Nikon, Tokyo, Japan). This body has a beam-splitting arrangement whereby 100% of the image is forwarded either to the observer for visual inspection or to an image-intensified camera. The secondary port of this trinocular body is interfaced to a Sony color TV camera (DXC102, Sony, Tokyo, Japan) for photo documentation of the cortical surface.

The image-intensified camera (ICCD-2525F: Videoscope International, Dulles, Washington, DC) utilizes a single stage microchannel plate intensifier with optical fiber optics coupling to a monochrome CCD chip camera. The characteristics of this camera are as follows: (a) sensitivity: 1×10^{-5} lux full video with a signal-to-noise ratio of 10; (b) resolution: 500 lines, 36 line pairs/mm; (c) spectral characteristics: S-25 photocathode, 300- to 800-nm range, peak at 450 nm, quantum efficiency of 11% typical; (d) lag: <1.0 msec for 67% response; (e) gain homogeneity: uniform gain within <1% of video field; (f) intensifier gain: 422,000 max; and (g) dynamic range, 256 levels one gain range. A 450 ± 5-nm interference filter (Corion; Holiston, MA) is inserted between the CCTV adapter attached to the trinocular body tube (SMZ-10; Nikon, Tokyo, Japan) and the camera assembly. This is the selected wavelength for recording NADH fluorescent emission and for the measurement of brain intracellular pH (pH_i), and regional cerebral blood flow (CBF) using a fluorescent dye indicator.

The fluorescent video image is processed by an image analyzer (DVP-32; Instrutech, Port Washington, NY) and a Dell XPS-T600 computer (Dell, Round Rock, TX). The acquired images are stored on an internal 25 GByte/sec SCSI disc system. The processed image is displayed on a Sony 20″ color video monitor (2030VM; Sony, Tokyo, Japan). Processed images are printed along with the calibration bars for brain pH_i and rCBF utilizing a color printer (HP-960Cxi; HP, Palo Alto, CA).

Software (MetaMorph; Universal Imaging Corporation, West Chester, PA) is customized to acquire images, perform image processing, display processed images, and store them on disc. This software is also used to determine NADH fluorescent pixel intensity values, values for brain pH_i and rCBF using region of interest functions. Paired images, one for each excited wavelength, 340 and 370 nm, are acquired with 16-frame averaging at 5-sec intervals for 180 sec and stored on disc. Sixteen frame averaging increases the signal-to-noise ratio by a factor of 4. Therefore, the overall signal-to-noise ratio given the just-described image intensifier is 40 at 1×10^{-5} lux. NADH fluorescence images excited at the 370-nm wavelength are used as the indicator of intramitochondrial redox state. The scale factor for the percentage change in NADH fluorescence from baseline is set so that 100% represents the level of NADH fluorescence in normal brain, whereas an increase to 300% represents brain death.

After obtaining NADH fluorescent images, umbelliferone, an intracellular pH indicator, is injected retrograde into the external carotid artery so it flows up the internal carotid artery into the area of brain destined for ischemia.[6,8,9,10] Because this has a very short half-life ($T_{1/2} < 2$ sec) when attached to cellular membranes, it can be used as a tissue perfusion indicator. Acquired images are then corrected for background NADH fluorescence prior to processing the rCBF and brain pH_i images. Images from the 340-nm excitation are processed to compute rCBF using the 1-min initial slope index.[6] rCBF images are then displayed and stored on disc for final analysis. Paired images from the 340- and 370-nm excitation are ratioed, and the resultant brain pH_i image is then displayed and stored on disc for final analysis.

Factors to Consider When Making NADH Fluorescence Measurements

Photodecomposition

A gradual decrease in the NADH fluorescence over time with constant exposure to the UV-exciting light can be avoided by keeping the intensity of the excitation

[8] T. M. Sundt, Jr., R. E. Anderson, and R. A. Van Dyke, *J. Neurochem.* **31,** 627 (1978).
[9] R. E. Anderson, J. D. Michenfelder, and T. M. Sundt, Jr., *Anesthesiology* **52,** 201 (1980).
[10] T. M. Sundt, Jr. and R. E. Anderson, *J. Neurophysiol.* **44,** 60 (1980).

beam to a level whereby a useable video signal can be obtained. However, at this level, intensified CCD cameras in general do show random noise in the video. Therefore, the use of real-time frame averaging improves picture quality greatly. The rationale for this is that intensified CCD cameras do not have the advantage of higher gain and signal-to-noise ratio than that of cooled photomultiplier tubes, which are used in nonimaging NADH fluorescence measurement applications. It is also of benefit to use a shutter in the excitation beam path when not illuminating the tissue between measurements for extended periods of time. By application of these techniques using the present optical configuration to whereby photodecomposition does not occur, we have found that the video images obtained are stable throughout long periods of observations in normal brain.

Hemoglobin Interference

One of the most important aspects to consider in measuring the NADH redox state *in vivo* in brain is the interference by hemoglobin in the measurement field. The spectral characteristics of hemoglobin are such that changes in blood volume, blood oxygenation, and/or dried red blood cells in the field of the measurement can significantly alter the absorption of the UV excitation, as well as the emitted fluorescence. Numerous investigators[5,11–17] in the field have devised a number of different compensation techniques to correct for hemoglobin artifact. These techniques range from subtracting the measured fluorescence from that of the recorded reflectance of the excitation, recording the difference between the measured NADH fluorescence and that of some isobestic wavelength from either the tissue itself or from the hemoglobin spectrum, or internal fluorescence standards such as fluorescein and rhodamine B. At best, compensation for hemoglobin interference can be quite difficult and the results are variable. Therefore, measurement of NADH fluorescence can only be considered as qualitative. Absolute measurements of NADH concentration *in vivo* cannot be made, as there are no appropriate methods of calibration.

In our original work,[18] we evaluated two different optical methods for excitation: bright-field illumination in which the excitation beam is on the same axis as the emitted fluorescence and dark-field illumination where the excitation is off axis to the emitted fluorescence. When using bright-field illumination, we observed[19]

[11] K. Harbig, B. Chance, A. G. B. Kovách, and M. Reivich, *J. Appl. Physiol.* **41,** 480 (1976).

[12] A. Mayevsky and B. Chance, *Brain Res.* **65,** 529 (1974).

[13] A. Mayevsky and B. Chance, *Science* **217,** 537 (1982).

[14] E. Dóra, L. Gyulai, and A. G. B. Kovách, *Brain Res.* **299,** 61 (1984).

[15] F. F. Jöbsis, M. O'Conner, A. Vitale, and H. Vreman, *J. Gen. Physiol.* **50,** 1009 (1971).

[16] R. S. Kramer and R. D. Pearlstein, *Science* **205,** 693 (1979).

[17] B. Vern, W. C. Whitehouse, and W. H. Schuette, *Brain Res.* **98,** 405 (1975).

[18] R. E. Anderson and T. M. Sundt, Jr., *Anal. Biochem.* **91,** 496 (1978).

[19] T. M. Sundt, Jr., R. E. Anderson, and F. W. Sharbrough, *J. Neurochem.* **27,** 1125 (1976).

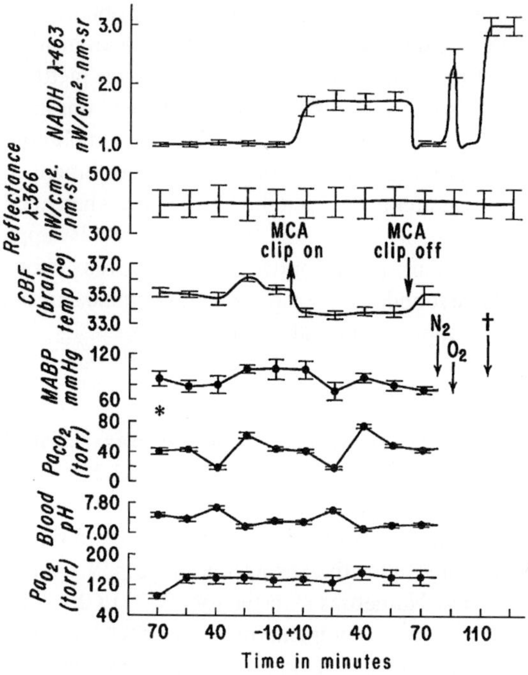

FIG. 2. NADH fluorescent measurements in five animals using bright-field illumination in which P_aCO_2 was altered before and during focal cerebral ischemia. In normal brain, NADH fluorescence remains unchanged through a wide range of changes in P_aCO_2, MABP, pH_a, and rCBF. NADH fluorescence in the beginning of this period did not change even though the P_aO_2 increased after the animal was placed on a ventilator and spontaneous respiration was abolished. NADH fluorescence is more variable during ischemia as it reflects the level of ischemic damage in individual animals. The reflectance signal remains unchanged throughout the experimental period with variations due to the subtle differences in the background coloration of brain. The asterisk indicates spontaneous respirations. Reproduced from T. M. Sundt, Jr., R. E. Anderson, and F. W. Sharbrough, *J. Neurochem.* **27,** 1125 (1976) with permission of Blackwell Science, Inc.

that there were no changes in NADH fluorescence or reflectance to changes in cerebral blood flow (Fig. 2) or mean arterial blood pressure (Fig. 2). Furthermore, the reflectance did not change during anoxia or at death (Figs. 2 and 3). The only changes that were observed were that of NADH fluorescence during focal cerebral ischemia, anoxia, and at death. We also observed using this scheme that there was very little variation in the measured values of NADH fluorescence or reflectance among groups of animals. Alternatively, when using dark-field illumination, there were changes in both NADH fluorescence and reflectance during changes in cerebral blood flow and at death (Fig. 4). The measured values using this scheme demonstrated much wider variation from one animal to another.

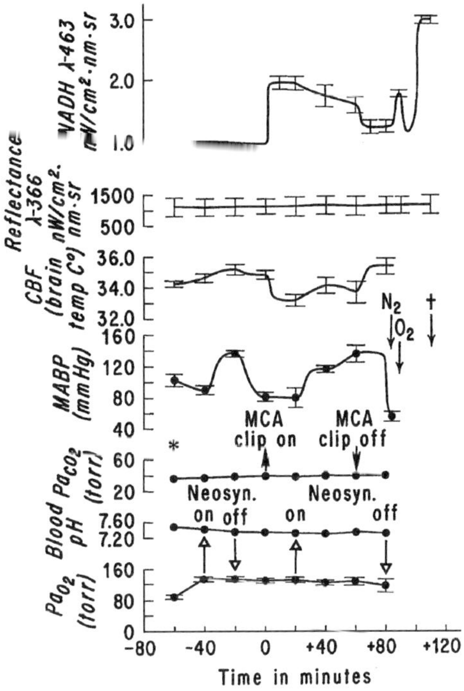

FIG. 3. NADH fluorescent measurements in five animals using bright-field illumination in which MABP was altered before and during focal cerebral ischemia. NADH fluorescence in normal brain remained unchanged with variations in MABP. NADH fluorescence in the beginning of this period did not change in which the P_aO_2 increased after the animal was placed on a ventilator and spontaneous respiration was abolished. During focal cerebral ischemia, CBF becomes pressure dependent when autoregulation is impaired, resulting in a decrease in NADH fluorescence when MABP is increased. Note that reflectance remains stable throughout the experimental period with variations due to subtle differences in the background coloration of brain. The asterisk indicates spontaneous respirations. Reproduced from T. M. Sundt, Jr., R. E. Anderson, and F. W. Sharbrough, *J. Neurochem.* **27,** 1125 (1976) with permission of Blackwell Science, Inc.

There are several possible reasons why we observed the changes as noted earlier under bright-field illumination. First, bright-field illumination minimizes shadows cast across surface-conducting vessels. Surface-conducting vessels change in caliber in response to hemodynamic alterations, resulting in changes in the reflectance signal with concomitant changes in the NADH fluorescence. Second, the use of narrow band excitation and emission filters minimizes the effects from other possible interfering substances and any leakage from specularly reflected tissue. Third, by using region-of-interest techniques in imaging software, areas of tissue can be analyzed independent of surface-conducting vessels. Based on this information, it was not necessary in our experience to incorporate compensation techniques in brain.

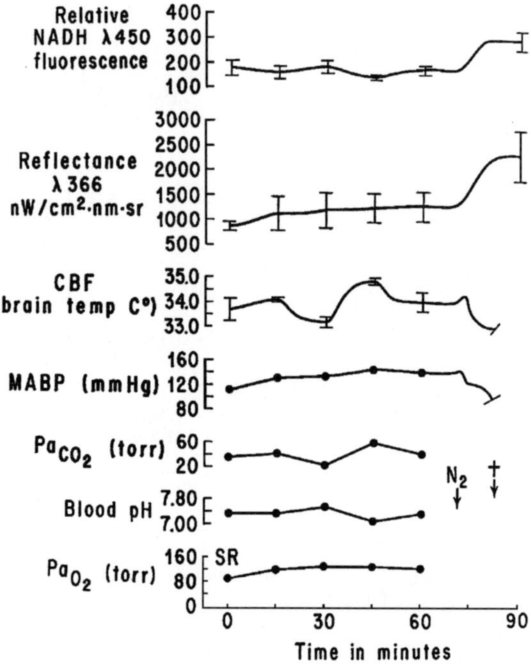

FIG. 4. NADH fluorescent measurements in five animals using dark-field illumination during alterations in P_aCO_2 and at death due to anoxia. NADH fluorescence in the beginning of this period decreased while reflectance increased when P_aO_2 increased after the animal was placed on a ventilator and spontaneous respiration was abolished. There are wide variations in NADH fluorescence and reflectance to alterations in P_aCO_2. A compensated NADH fluorescence curve could not be obtained, as there was not a consistent pattern between the raw NADH fluorescence and reflectance signals. SR, spontaneous respirations. Reproduced with permission from R. E. Anderson and T. M. Sundt, Jr., *Anal. Biochem.* **91,** 496 (1978).

Examples of *in Vivo* Measurements

Figures 5–7 are three groups of typical picture sets depicting animals that are moderate hypoglycemic, normoglycemic, and hyperglycemic before, during, and after focal cerebral ischemia. Because of anatomical variation of the microvasculature from animal to animal, images cannot be averaged frame by frame between different animals at the same physiological time frame according to the experimental protocol. The region of interest tool is used to delineate areas of the penumbra, core area of ischemia, and normal brain. Criteria for selecting these areas are, for example, (a) in the core area of ischemia NADH fluorescent intensity values would be near maximum, rCBF $<10\%$ of normal, and brain $pH_i < 6.5$; and (b) in the area of the penumbra, NADH fluorescent intensity, rCBF and brain pH_i varies between

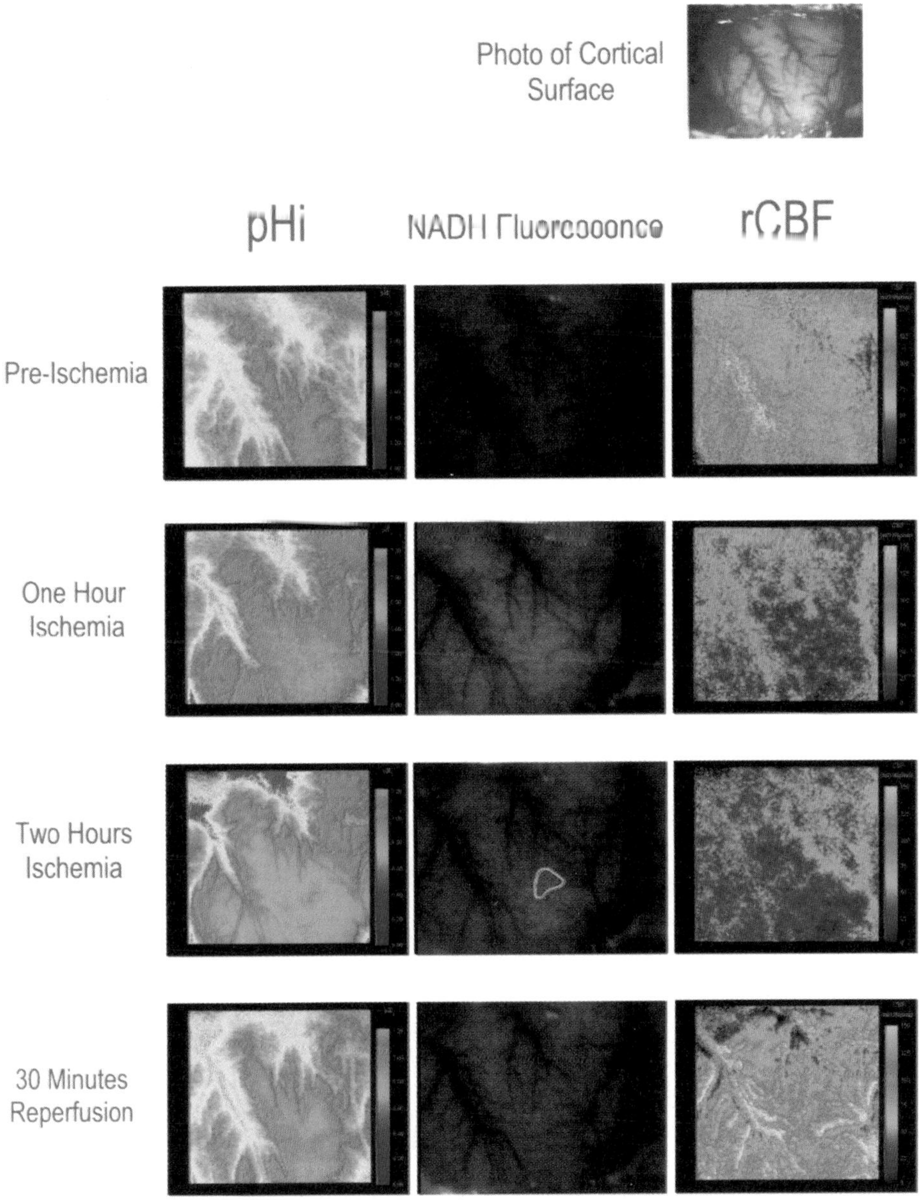

FIG. 5. Measurements of NADH fluorescence, brain pH_i, and rCBF in an animal that is moderate hypoglycemic (glucose, 95 mg/dl; P_aCO_2, 43 torr; pH_a, 7.397; MABP, 86). Prior to occlusion, this animal had a brain pH_i of 6.99, NADH redox state at 100%, and rCBF of 78 ml/100/min. Two hours into ischemia, brain pH_i decreased to 6.8, NADH redox state increased by 14%, and rCBF decreased to 35% of baseline. Thirty minutes after restoration of flow, brain pH_i went to 6.89, NADH redox state decreased by 17%, and rCBF returned to near baseline levels. The region of interest for measurements of brain pH_i, NADH redox state, and rCBF is outlined in yellow.

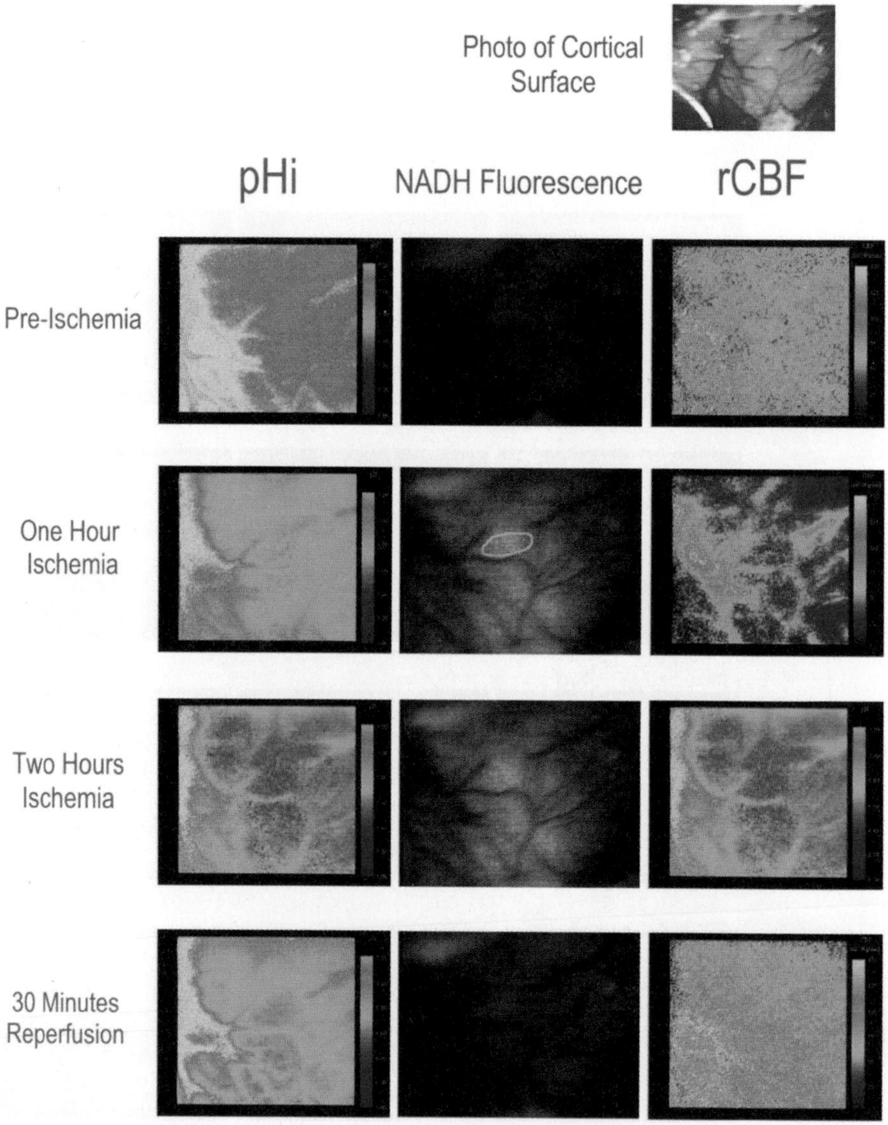

FIG. 6. Measurements of NADH fluorescence, brain pH_i, and rCBF in an animal that is normoglycemic (glucose, 219 mg/dl; P_aCO_2, 45.1 torr; PH_a, 7.329; MABP, 86). Prior to occlusion, this animal had a brain pH_i of 7.01, NADH redox state at 100%, and rCBF of 66 ml/100/min. Two hours into ischemia, brain pH_i decreased to 6.58, NADH redox state increased by 44%, and rCBF decreased to 36% of baseline. Thirty minutes after restoration of flow, brain pH_i went to 6.69, NADH redox state decreased to near baseline levels, and rCBF returned to near baseline levels. The region of interest for measurements of brain pH_i, NADH redox state, and rCBF is outlined in yellow.

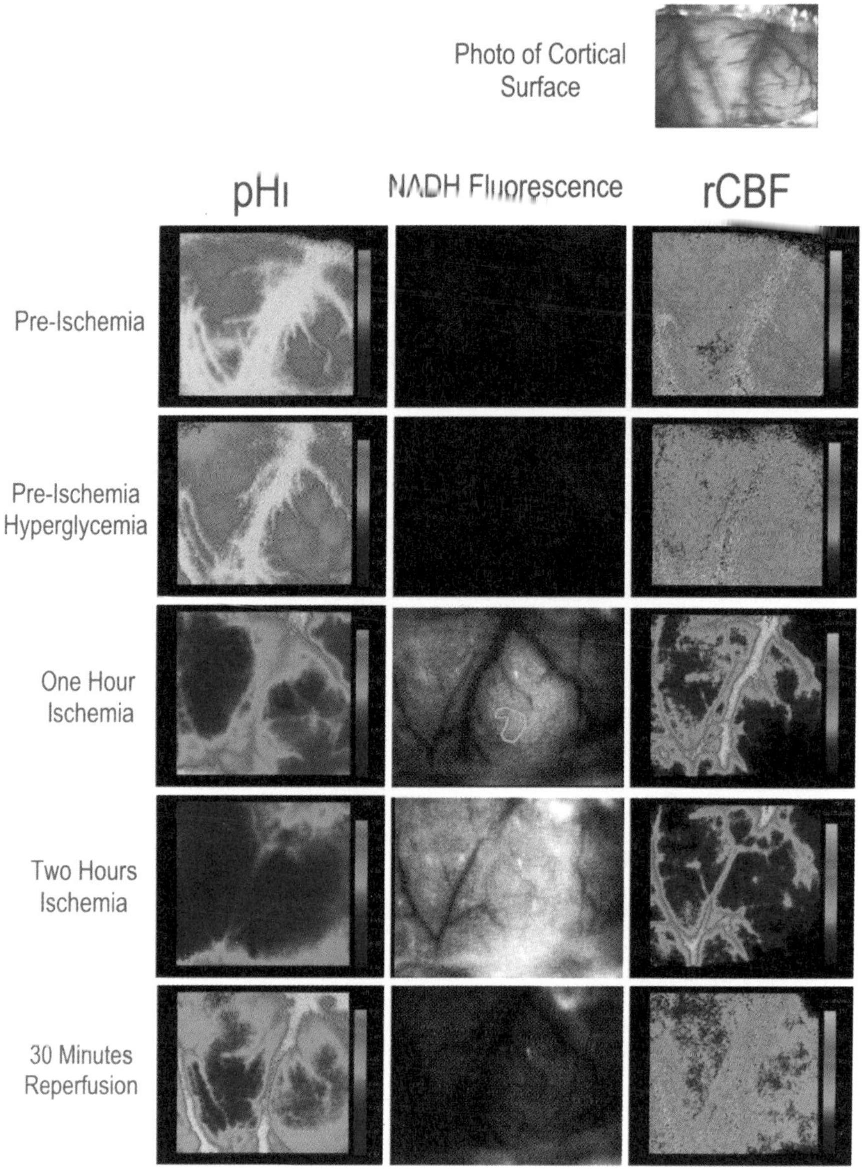

FIG. 7. Measurements of NADH fluorescence, brain pH_i, and rCBF in an animal that is hyperglycemic (glucose, 600 mg/dl; P_aCO_2, 48 torr; pH_a, 7.436; MABP, 89). Prior to occlusion, this animal had a brain pH_i of 7.00, NADH redox state at 100%, and rCBF of 82 ml/100/min. Two hours into ischemia, brain pH_i decreased to 6.16, NADH redox state increased by 75%, and rCBF decreased to 24% of baseline. Thirty minutes after restoration of flow, brain pH_i went to 6.58, NADH redox state decreased by 65%, and rCBF increased to 52 ml/100g/min. The region of interest for measurements of brain pH_i, NADH redox state, and rCBF is outlined in yellow.

normal and core ischemic values. The values of NADH fluorescent intensities, rCBF, and brain pH_i obtained from these areas are then averaged over each of the image frames from each animal at the same time intervals during the experimental protocol. The mean and standard deviation and/or error are then tabulated on a spreadsheet (Excel; Microsoft, Redmond, WA) and analyzed for statistical significance (CSS; Statsoft, Tulsa, OK).

Preischemic NADH fluorescence levels are similar in all three animals, whereas intracellular brain pH and rCBF are at their normal values. During ischemia, there is a close relationship between the levels of NADH fluorescence and the levels of serum glucose demonstrating increased NADH fluorescence to increasing levels of serum glucose. There is also a close relationship between the levels of NADH fluorescence and intracellular brain pH depicting increased acidosis and increasing levels of NADH fluorescence. Thirty minutes after restoration of flow, NADH fluorescence and intracellular brain pH return toward preischemic values. The degree of the return of these variables is a function of the severity and duration of ischemia.

Conclusion

The use of *in vivo* fluorescence to measure continuous changes in mitochondrial and cellular functions has facilitated an increased understanding of the coupling of cerebral blood flow and energy productions during various pathophysiological states.

[41] Nitroxyl Probes for Brain Research and Their Application to Brain Imaging

By Hideo Utsumi, Hiroaki Sano, Masaichi Naruse, Ken-ichiro Matsumoto, Kazuhiro Ichikawa, and Tetsuo Oi

Introduction

Free radical reactions in the brain are one of most interesting subjects for brain research because they are reportedly involved in various brain diseases, including ischemia reperfusion injury,[1] brain tumor,[2] aging,[3] familial amyotrophic lateral

[1] P. H. Chan, *Stroke* **27,** 1124 (1996).

[2] C. S. Cobbs, D. S. Levi, K. Aldape, and M. A. Israel, *Cancer Res.* **56,** 3192 (1996).

[3] E. R. Stadtman, *Science* **257,** 1220 (1992).

0076-6879/02 $35.00

sclerosis,[4] Alzheimer disease,[5] and other neurodegenerative diseases.[6] There have been, however, few reports on the direct determination of *in vivo* free radical reactions in the brain because of the lack of an adequate method to do so.

Nitroxyl radicals are susceptible to oxygen concentration,[7,8] reactive oxygen species (ROS),[9,10] and biological redox systems[11,12] and are widely used as probes for *in vivo* ESR/EPR (electron spin/paramagnetic resonance) spectroscopy. The *in vivo* ESR signal of exogenous nitroxyl radicals decreases gradually in living mice,[13] and its decay rate is affected by physiological conditions such as aging.[14] Signal decay is enhanced in ROS-related disease models such as ischemia reperfusion injury,[15] streptozotocin-induced diabetes,[16] cancer,[17] iron overload-induced liver injury,[18] and diesel exhaust particle-induced lung injury.[19] Enhanced signal decay is suppressed to the control level by pre- or simultaneous treatment with antioxidants and radical scavengers. These facts indicate that the generation of ROS may be determined as the enhanced signals decay with an *in vivo* ESR/nitroxyl probe method by utilizing specific radical scavenger.

The combination of *in vivo* ESR spectroscopy with an imaging technique enables two-dimensional (2D) or 3D imaging of radicals in living organisms.[20] Most conventional ESR-CT techniques utilize Fourier transformation (FT) deconvolution of a spectrum under a field gradient with that under zero field gradients

[4] D. R. Rose, T. Siddique, D. Patterson, D. Z. Figlewicz, P. Sapp, A. Hentati, D. Donaldson, J. Goto, J. P. O'Regan, H.-Q. Deng, Z. Rahmani, A. Krizus, D. McKenna-Yasek, A. Cayabyab, S. M. Gaston, R. Berger, R. E. Tanzi, J. J. Halperin, B. Herzfeld, R. van den Berg, W.-Y. Hung, T. Bird, G. Deng, D. W. Mulder, C. Smyth, N. G. Lang, E. Soriano, M. A. Perocak-Vance, J. Haines, G. A. Rouleau, J. S. Gusella, H. R. Horritz, and R. H. Brown, Jr., *Nature* **362,** 59 (1993).

[5] M. A. Smith, P. L. R. Harris, L. M. Sayre, and G. Perry, *Proc. Natl. Acad. Sci. U.S.A.* **94,** 9866 (1997).

[6] P. H. Evans, *Br. Med. Bull.* **49,** 577 (1993).

[7] M. Inoue, H. Utsumi, and Y. Kirino, *Chem. Pharm. Bull.* **42,** 2346 (1994).

[8] J. F. Glockner, H.-C. Chang, and H. M. Swartz, *Magn. Reson. Med.* **20,** 123 (1991).

[9] E. Finkelstein, G. M. Rosen, and E. J. Rauckman, *Biochim. Biophys. Acta* **802,** 90 (1984).

[10] M. C. Krishna, A. Russo, J. B. Mitchell, S. Goldstein, H. Dafni, and A. Samuni, *J. Biol. Chem.* **271,** 26026 (1996).

[11] H. Utsumi, A. Shimakura, M. Kashiwagi, and A. Hamada, *J. Biochem.* **105,** 239 (1989).

[12] A. Iannone, A. Tomasi, V. Vannini, and H. M. Swartz, *Biochim. Biophys. Acta* **1034,** 285 (1990).

[13] H. Utsumi, E. Muto, S. Masuda, and A. Hamada, *Biochem. Biophys. Res. Commun.* **172,** 1342 (1990).

[14] F. Gomi. H. Utsumi, A. Hamada, and M. Matsuo, *Life Sci.* **52,** 2027 (1993).

[15] H. Utsumi, K. Takeshita, Y. Miura, S. Masuda, and A. Hamada, *Free Radic. Res. Commun.* **19,** S219 (1993).

[16] T. Sano, F. Umeda, T. Hashimoto, H. Nawata, and H. Utsumi, *Diabetologia* **41,** 1355 (1998).

[17] P. Kuppusamy, M. Afeworki, R. L. Shankar, D. Coffin, M. C. Krishna, S. M. Hahn, J. B. Michell, and J. L. Zweier, *Cancer Res.* **58,** 1562 (1998).

[18] N. Phumala, T. Ide, and H. Utsumi, *Free Radic. Biol. Med.* **26,** 1209 (1999).

[19] J.-Y. Han, K. Keizo Takeshita, and H. Utsumi, *Free Rad. Biol. Med.* **30,** 516 (2001).

[20] K. Takeshita, H. Utsumi, and A. Hamada, *Biochem. Biophys. Res. Commun.* **177,** 874 (1991).

to obtain spatial projection. This deconvolution method is applicable to samples composed of a single radical species. The time-resolved ESR-CT techniques may provide imaging of ROS generation in living organisms. ESR-CT imaging of more than two radical species is possible with separable ESR-CT techniques.[21]

Most commercially available nitroxyl probes have no capability of passing through the blood–brain barrier. It would be desirable if nitroxyl probes were capable of passing through the blood–brain barrier, of accumulating in the brain and of being reactive toward ROS or the redox state.

In this chapter, various types of lipophilic nitroxyl spin probes, shown in Table I, are synthesized for the investigation of free radical reactions in the brain.[22,23] Alkyl ester compounds are utilized for positron emission tomography (PET) and single photon emission computed tomography (SPECT) because of their capability of passing through the blood–brain barrier.[24,25] Acetoxymethyl ester derivatives are also widely used as fluorescence probes in cultured cells. These compounds are characterized to be temporarily membrane permeable by masking their carboxyl group with ester, and then to be hydrolyzed inside the cells.[26] These facts indicate that ester compounds are ideal probes for brain research.

The lipophilic and hydrolytic properties of these nitroxyl spin probes are characterized in relation to the tissue distribution of the probes in living animals after intravenous injection. *In vivo* ESR imaging of the probes in the head of living mice has been reported.[22,23]

Methods

Synthesis of Blood–Brain Barrier-Permeable Spin Probes

Table I demonstrates the structures and acronyms of the nitroxyl spin probes used for imaging in the heads of living mice.

3-Methoxycarbonyl (CxP-M)- and 3-ethoxycarbonyl-2,2,5,5-tetramethylpyrrolidine-1-oxyl (CxP-E) are synthesized by esterification of 3-carboxy-2,2,5,5-tetramethylpyrrolidine-1-oxyl (CxP) as follows.[22] CxP (5.0 g, Aldrich Chemical Co. Inc., Milwaukee, WI) is dissolved in methanol or ethanol (200 ml), and concentrated hydrochloric acid (2.0 ml) is added to the solution, with stirring, in an ice bath. The mixture is kept at room temperature for 2 days. After reoxidizing the reduced products from nitroxyl radicals during reaction with bubbling oxygen gas,

[21] K.-I. Matsumoto and H. Utsumi, *Biophys. J.* **79,** 3341 (2000).

[22] H. Sano, K. Matsumoto, and H. Utsumi, *Biochem. Mol. Biol. Int.* **42,** 641 (1997).

[23] H. Sano, M. Naruse, K. Matsumoto, T. Oi, and H. Utsumi, *Free Radic. Biol. Med.* **28,** 959 (2000).

[24] T. Irie, K. Fukushi, Y. Akimoto, H. Tamagami, and T. Nozaki, *Nucl. Med. Biol.* **21,** 801 (1994).

[25] L. Friberg, A. R. Andersen, N. A. Lassen, S. Holm, and M. Dam, *J. Cereb. Blood Flow Metab.* **14,** S19 (1994).

[26] R. Y. Tsien, *Nature* **290,** 527 (1981).

TABLE I
STRUCTURES, ACRONYMS, AND PARTITION COEFFICIENTS (Po/w) OF NITROXYL PROBES[a]

	R	Po/w	Basic structure
CxP	—H	0.02	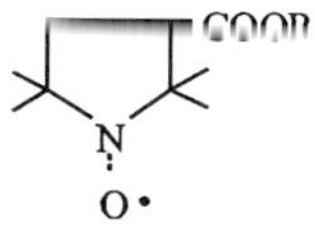
CxP-M	$-CH_3$	8.7	
CxP-E	$-CH_2CH_3$	4.1	
CxP-AM	$-CH_2OCOCH_3$	4.1	
Doxyl-butane		11.0	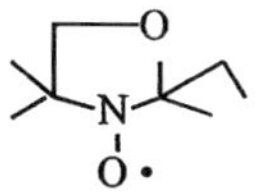

[a] Partition coefficients between *n*-octanol and PBS (Po/w) were determined from the amount of nitroxyl probe in *n*-octanol and PBS. The nitroxyl probe (1 m*M*) was dissolved in PBS. One milliliter of the nitroxyl probe solution was mixed vigorously with 1.0 ml of *n*-octanol, and the mixture was centrifuged at 3000 rpm for 5 min. The amounts of nitroxyl probe in both *n*-octanol and PBS were determined from double-integrated ESR signal intensity measured with an X-band ESR spectrometer at 5.0 mW of 9.45 GHz of microwave, 0.063 mT of 100-kHz field modulation, and a scan rate of 5 mT/min. Mn^{2+} was used as an external standard for correction of sensitivity between *n*-octanol and PBS.

100 ml of ice cold water is added, and then the products are extracted with ethyl ether. The amounts of CxP, its ester, and their reduced forms are determined by TLC and ESR experiments. The ether layer is washed once with saturated sodium bicarbonate and twice with saturated sodium chloride solution. After drying with magnesium sulfate, the ether is evaporated and applied to silica gel column chromatography (Wakogel C-200, Wako Pure Chemicals, LTD., Osaka; elution with hexane-ether). Purity is estimated with TLC, ESR, and MS measurements.

2-Ethyl-2,4,4-trimethyloxazolidine-3-oxyl (doxyl-butane) is synthesized by the method of Keana *et al.*[27] with slight modification because of the high volatility of butanone, and purified with column chromatography as described earlier.[22]

Acetoxymethyl-2,2,5,5-tetramethylpyrrolidine-1-oxyl-3-carboxylate (CxP-AM) is synthesized by the esterification of CxP with acetoxymethylbromide as follows.[23] Acetoxymethylbromide is prepared by stirring paraformaldehyde (2.44 g) in acetylbromide (6.0 ml) for 30 min at 80°. After removing paraformaldehyde, the mixture is distilled at 130–138°, and the product is obtained as a yellow oil (4.54 g, yield 37%). The resulting acetoxymethylbromide (1.31 g) is added at 0° to the mixture of CxP (1.0 g) with triethylamine (0.74 ml) in dimethylformamide (8 ml), and then

[27] J. F. W. Keana, S. B. Keana, and D. Beetham, *J. Amer. Chem. Soc.* **89,** 3055 (1967).

the solution is stirred for 17 hr at room temperature. After dilution with 40 ml of dichloromethane, washing once with water, and drying with manganese sulfate, the dichloromethane is evaporated. The residue is preliminary purified with silica gel column chromatography (Wakogel C-200, Wako Pure Chemicals LTD., Osaka) and then recrystallized from an ether-hexane [yellow rectangle crystal, yield 0.9 g (65%)]. The purity is estimated with ESR, IR (KBr; 3000, 2950, 2880, 1760, 1460, 1390, 1240, 1150, 1040, 840 cm^{-1}) and FAB-MASS [$M^+ = 258$, $(M\text{-}CH_2)^+ = 244$].

Partition Coefficients of Spin Probes between n-Octanol and PBS

Partition coefficients of nitroxyl spin probes between *n*-octanol and buffer are determined as indices of lipophilicity of the probes (Table I). The probe solution (1 m*M*) is prepared in phosphate-buffered saline (PBS, 20 m*M*, pH 7.4). One milliliter of the probe solution is mixed vigorously with 1.0 ml of *n*-octanol, and then the mixture is centrifuged at 3000 rpm for 5 min. ESR spectra of both *n*-octanol and PBS are measured with an X-band ESR spectrometer at 5.0 mW of 9.45 GHz of microwave, 0.063 mT of 100-kHz field modulation, and a scan rate of 5 mT/min. Amounts of the nitroxyl probe are calculated from double-integrated ESR signal intensity, using Mn^{2+} as an external standard for the correction of sensitivity between *n*-octanol and PBS.

Doxyl-butane has the highest partition coefficient and CxP the lowest. The introduction of ester to CxP increases lipophilicity. The coefficients of CxP-AM and CxP-E are about 200 times higher than that of CxP, but less than that of CxP-M. The high partition coefficients indicate the possibility that 4 lipophilic nitroxyl probes pass through the blood–brain barrier.

Determination of Hydrolytic Capability of Nitroxyl Spin Probes by Esterase

The hydrolytic capability of the probes by esterase facilitates its accumulation in the parenchymal cells of the brain. Moreover, the hydrolytic property of the nitroxyl probe, depending on the type of esterase, is required for the probe to pass through the blood–brain barrier.

Spin probes (2 m*M*) are mixed with either esterase or lipase (5 units) in PBS (pH 7.4) and are incubated at 37°. The amounts of intact nitroxyl probe and its hydrolyzed derivative are determined with high-performance liquid chromatography (HPLC, dual-pump CCPS, TOSOH) having a 35 × 4.6-mm precolumn packed with C_{18} material, 20-μm particle size, a 250 × 4.6-mm stainless-steel column packed with C_{18} material, 5-μm particle size, and a variable wavelength UV detector and integrator. The mobile phase is 40% methanol in 20 m*M* phosphate buffer, pH 2.2; the flow rate is 1.0 ml/min; and the column effluent is monitored at 245 nm.

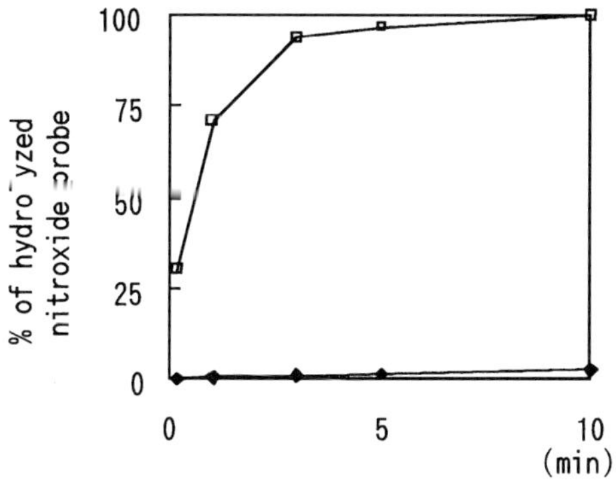

FIG. 1. Hydrolytic capability of CxP-M (◆) and CxP-AM (□) with carboxyl esterase. CxP-AM or CxP-M (2 mM) was mixed with esterase (5 units) in PBS (pH 7.4), and then incubated at 37° for various periods. The amount of intact nitroxyl probe and its hydrolyzed form was determined with HPLC. The mobile phase was 40% methanol in 20 mM phosphate buffer, pH 2.2, flow rate was 1.0 ml/min, and the column effluent was monitored at 245 nm. Data are means ± SE of three experiments.

Figure 1 shows the hydrolytic capability of CxP-M and CxP-AM with carboxyl esterase at 37°. CxP-AM was hydrolyzed very rapidly by carboxyl esterase: more than 90% was hydrolyzed within 3 min and 100% in 10 min. However, CxP-M was hardly hydrolyzed.

Table II demonstrates the hydrolytic property of CxP-AM with various esterase at 37°. CxP-AM was hydrolyzed very rapidly by carboxyl esterase: more than 90%

TABLE II
HALF-LIFE IN HYDROLYSIS OF CxP-AM WITH ESTERASE OR LIPASE[a]

	Half-life (min)
Carboxyl esterase	0.6
Butyrylcholinesterase	>200
Acetylcholinesterase	400
Lipase	850

[a] The half-life of CxP-AM was determined from the HPLC method as described in the legend of Fig. 1. CxP-AM (2 mM) was incubated at 37° with either esterase or lipase (5 units) in PBS, and the amounts of intact nitroxyl probe and its hydrolyzed derivative were determined with HPLC.

was hydrolyzed within 3 min and 100% at 10 min. CxP-AM was, however, hardly hydrolyzed with lipase, acetylcholinesterase, or butyrylcholinesterase. The half-life of CxP-AM was 0.6 min for carboxyl esterase and more than 200 min for the other esterases. These results suggest that CxP-AM is not likely to be hydrolyzed by lipase in blood after intravenous administration, pass through the blood–brain barrier, or hydrolyzed in the cytosolic phase of the brain cells and then accumulated in the brain. However, CxP-M and CxP-E were hardly hydrolyzed with lipase or any esterase, indicating that these two probes should keep their lipophilic property even after passing through the blood–brain barrier.

Using CxP-M, Miura *et al.*[28] examined the effect of X-ray irradiation on signal decay in the head of living mice, and Yokoyama *et al.*[29] reported the difference of the decay rate in each region of the brain. If CxP-M was hydrolyzed and retained in parenchymal cells of the brain, their results might be interpreted as free radical reactions in parenchymal cells of the brain. CxP-M is, however, not hydrolyzed with any esterase.

In Vivo ESR Measurement

The capability of nitroxyl probes to pass through the blood–brain barrier and to be hydrolyzed in the brain can be confirmed with *in vivo* ESR measurement after intravenous administration of the probes into mice.

Mice are anesthetized by intramuscular injection of Nembutal (75 mg/kg). The solution of the nitroxyl probe [100 μl of CxP-AM (25 m*M*) in PBS with 10% (v/v) of ethanol, 50 μl of CxP, CxP-M, CxP-E, or doxyl-butane (150 m*M*) in PBS] is administered into the tail vein of the mice. Then, ESR spectra of the head domain of mice are obtained with an L-band ESR spectrometer (JEOL, JES-RE-3L, Akishima, Japan) equipped with a loop-gap resonator under the following conditions: 1.0 mW of 1.1 GHz microwave, 0.1 mT of 100-kHz field modulation, 39 ± 5 mT of external magnetic field, and a scan rate of 5 mT/min. The sensitivity of the experiment is calibrated with the signal intensity of DPPH powder as an external standard. Figure 2 demonstrates typical ESR spectra of nitroxyl spin probes in the head of the mice 1 min after intravenous injection. The *in vivo* spectrum of CxP-AM in the head was the same as that of CxP, which was triplet lines having 1.60 mT of hfs and no shoulder, indicating that CxP-AM is retained in the head in free form, CxP. The intravenous injection of CxP-M, CxP-E, and doxyl-butane yielded two triplet components, whose hfs were the same as those in *n*-octanol and PBS.[23] These facts confirm that intravenously injected CxP-AM is hydrolyzed rapidly in the head of mice, resulting in long-term retention in the brain.

[28] Y. Miura, K. Anzai, S. Urano, and T. Ozawa, *Free Radic. Biol. Med.* **23,** 533 (1997).

[29] H. Yokoyama, O. Itoh, T. Ogata, H. Obara, H. Ohya-Nishiguchi, and H. Kamada, *Magn. Reson. Imag.* **15,** 1079 (1997).

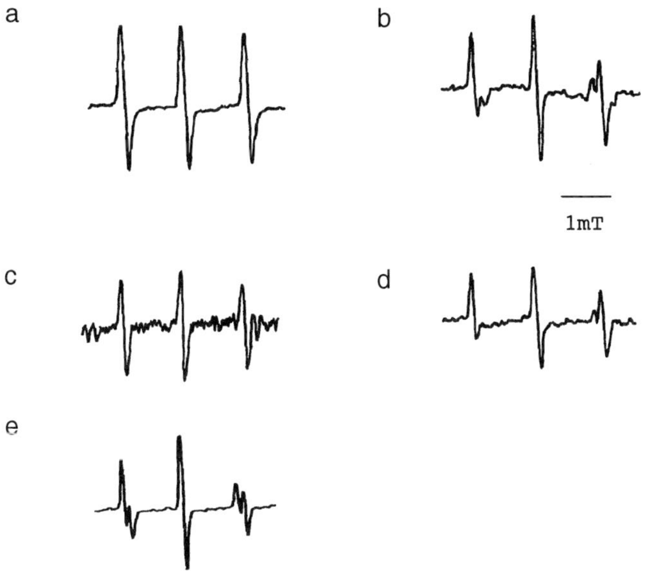

FIG. 2. *In vivo* ESR spectra of nitroxyl probes on the head of mice after intravenous injection. One minute after intravenous injection of nitroxyl probe [100 μl of CxP-AM (c) (25 mM) in PBS with 10% (v/v) of ethanol, 50 μl of CxP (a), CxP-M (b), CxP-E (d), or doxyl-butane (e) (150 mM) in PBS], L-band ESR measurement was carried out on the head of mice with a loop-gap resonator under the following conditions: 1.0 mW of 1.1 GHz microwave, 0.1 mT of 100-kHz field modulation, 39 $\pm$ 5 mT of external magnetic field, and a scan rate of 5 mT/min.

Tissue Distribution of Nitroxyl Spin Probes after Intravenous Injection

Tissue distribution of nitroxyl spin probes is determined after intravenous injection as follows. Blood is obtained by cutting the carotids of the mice after intravenous injection of the same nitroxyl probe solution as that used in the *in vivo* ESR measurement. Brain, liver, kidney, and heart are rapidly dissected and homogenized in ninefold (r/v) ice-cold PBS. The samples are transferred into capillary tubes and then observed with an X-band spectrometer (JEOL RE-1X) at 9.45 GHz of microwave frequency, 5.0 mW of the power, and 0.063 mT of 100-kHz field modulation. The amount of nitroxyl probe was determined from double-integrated ESR signal intensity by calibration of signal intensity using Mn^{2+} as a standard. Nitroxyl radicals are reduced easily to the corresponding hydroxylamine in living organism[13,30] or tissue.[31] The reduced forms, hydroxylamines, are

[30] H. Utsumi and K. Takeshita, *in* "Bioradicals Detected by ESR Spectroscopy" (H. Ohya-Nishiguchi and L. Packer, eds.), p. 321. Birkhauser Verlag, Basel, Switzerland, 1995.

[31] K. Takeshita, A. Hamada, and H. Utsumi, *Free Radic. Biol. Med.* **26,** 951 (1999).

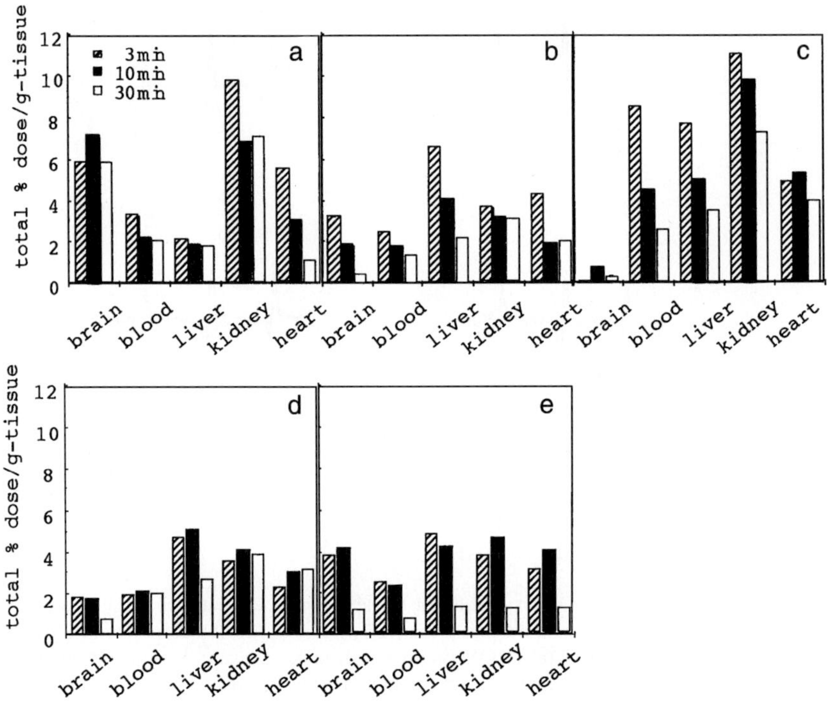

FIG. 3. Tissue distribution of nitroxyl probes in mice after intravenous injection of (a) CxP-AM, (b) CxP-M, (c) CxP, (d) CxP-E, and (e) doxyl-butane. Nitroxyl probes prepared as described in the legend of Fig. 1 were injected into tail vein of mice, and then blood, brain, liver, kidney, and heart were obtained 3, 10, and 30 min after the injection. Brain, liver, kidney, and heart were homogenized in ninefold ice-cold PBS. Blood and homogenates were treated with potassium ferricyanide (1 m*M*) and then observed with an X-band spectrometer (JEOL RE-1X) at 9.45 GHz of microwave frequency, 5.0 mW of the power, and 0.063 mT of 100-kHz field modulation. Total amounts of the intact and reduced nitroxyl probe were determined from double-integrated signal intensity after treatment with potassium ferricyanide, as described in the legend of Fig. 2. Data are means ± SE of three experiments.

reoxidized to the corresponding nitroxyl radicals by the addition of ferricyanide,[14] and the total amount of nitroxyl probe, intact and its reducing derivative, is calculated from the signal intensity after reoxidizing.

Figure 3 shows the total amount of nitroxyl probe in tissues and blood of mice 3, 10, and 30 min after intravenous injection. The percentage dose/g tissue of CxP-AM in the brain remained at a high level until 30 min after injection, although CxP-AM was mostly distributed in the kidney among the five tissues studied (Fig. 3a). The amount of CxP-AM was relatively low in both blood and liver, suggesting that CxP-AM may be excreted, chiefly in urine.

CxP-M was recovered chiefly from the liver (Fig. 3b). The percentage dose/g tissue of CxP-M in the brain was less than that of the other organs and decreased quickly as a function of time, indicating that CxP-M is removed rapidly from the brain. CxP-E and doxyl-butane showed tissue distribution similar to that of CxP-M (Figs. 3d and 3c). CxP, the free carboxy derivative, was distributed in organs except for the brain and showed no ability to be retained in a specific organ (Fig. 3c). The total amount in the brain was 44-fold larger in CxP-AM than that of CxP 3 min after injection and 1.8 times than that in CxP-M. Almost the same amount of CxP-AM remained in the brain after 30 min of injection, and the amount of CxP-AM in the brain after 30 min of injection was above 10-fold larger than that of CxP-M. These data suggest that CxP-AM passes through the blood–brain barrier, is metabolized specifically by carboxyl esterase into hydrophilic metabolites, and results in long-term retention in the brain.

Free radicals and ROS are generated in the brain in a region-dependent manner. For example, ischemia reperfusion injury occurs mainly in the hippocampus and cerebral cortex,[32,33] whereas in Parkinson's disease the substantial nigra is primarily affected.[34,35] If CxP-AM is localized in the hippocampus, it should be a suitable probe for investigating ischemia reperfusion injury. CxP-AM was retained in all regions of the brain.[23] The amount in the cerebral cortex was highest, being about 8% dose/g tissue following 30 min of intravenous injection. The amounts of CxP-AM in both the striatum and the hippocampus tended to increase with time.

ESR-CT Imaging

ESR-CT imaging was performed to determine the distribution of nitroxyl probes in the head of mice after intravenous injection. A 2D image of nitroxyl probes was obtained using an ESR-CT system as reported previously.[20] One hundred microliters of nitroxyl probes (25 m*M*) dissolved in PBS with 10% (v/v) of ethanol was injected into the tail vein of mice anesthetized with Nembutal. The head of the mouse is fixed in a loop-gap resonator (35 mm in diameter, 5 mm in axial length) of an L-band ESR spectrometer having field gradient coils (x and z axis, 0.45 mT/cm). Spectra with and without field gradient are alternately obtained, changing the direction in 10° steps, providing 18 projections. The lipophilic probes gave the two-signal components as shown in Fig. 1. Therefore, the central peak [h(0)] of triplet lines was used for the imaging to avoid the disturbance due to the occurrence of two-signal components in the deconvolution process. Spectral

[32] H. Kinouchi, C. J. Epstein, T. Mizui, E. Carlson, S. F. Chen, and P. H. Chan, *Proc. Natl. Acad. Sci. U.S.A.* **88,** 11158 (1991).

[33] X.-H. Liu, H. Kato, N. Nakata, K. Kogure, and K. Kato, *Brain Res.* **625,** 29 (1993).

[34] J. L. Cadet, S. F. Ali, R. B. Rothman, and C. J. Epstein, *Mol. Neurobiol.* **11,** 155 (1995).

[35] S. Przedborski, V. Jackson-Lewis, R. Yokoyama, T. Shibata, V. Dawson, and T. M. Dawson, *Proc. Natl. Acad. Sci. U.S.A.* **93,** 4565 (1996).

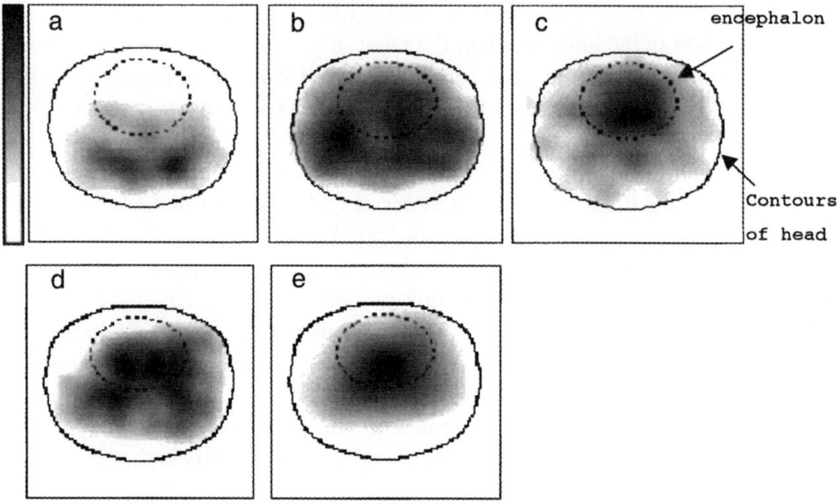

FIG. 4. Two-dimensional ESR-CT imaging at the heads of mice after intravenous injection of (a) CxP, (b) CxP-M, (c) CxP-AM, (d) CxP-E, and (e) doxyl-butane. Two-dimensional images of nitroxyl probes of mice heads were obtained with a homemade ESR-CT system as reported previously [H. Sano, K. Matsumoto, and H. Utsumi, *Biochem. Mol. Biol. Int.* **42,** 641 (1997)]. After intravenous injection of nitroxyl probes (25 mM, 100 μl) in PBS with 10% (v/v) of ethanol, the heads of the mice were fixed in a loop-gap resonator (35 mm in diameter, 5 mm in axial length) of an L-band ESR spectrometer having field gradient coils (x and z axis, 0.45 mT/cm). Then, spectra with and without field gradient were alternately obtained under the following conditions: 1.0 mW of 1.1. GHz microwave, 0.1 mT of 100-kHz field modulation, 39 ± 5 mT of external magnetic field, and a scan rate of 5 mT/min. 2D images were reconstructed from 18 projections of the deconvoluted data by a filtered back projection.

data under field gradient were deconvoluted with those under zero-field gradient by the Fourier transformation method. An image was reconstructed from 18 projections of the deconvoluted data by a filtered back projection. Figure 4 demonstrates an *in vivo* 2D image of CxP, CxP-M, CxP-AM, CxP-E, and doxyl-butane in the head of mice after intravenous injection, respectively. Prominent contrasts were obtained not only in the extracranial domain but also in the encephalon region by CxP-M, CxP-E, and doxyl-butane, whereas CxP was distributed only in the extracranial domain. These images clearly indicate that CxP-M, CxP-E, and doxyl-butane pass the blood–brain barrier and reach the encephalon region.

In vivo 2D imaging also provided direct evidence that CxP-AM is able to pass through the blood–brain barrier and accumulate in the brain after intravenous injection. CxP-AM provides clear contrasts only in the encephalon region. These images clearly demonstrate that intravenously injected CxP-AM reaches the encephalon and accumulates in that region.

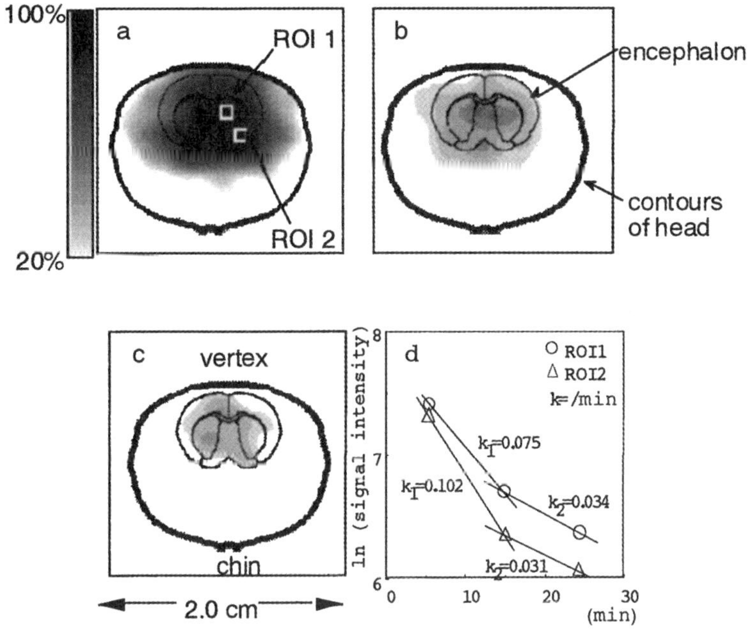

FIG. 5. Time-resolved 2D ESR-CT images (a–c) of mice heads, and semilogarithmic plot of the imaging intensity (d) of two different regions in the brain after intravenous injection of CxP-AM. One hundred microliters of CxP-AM (25 m*M*) in PBS with 10% (v/v) of ethanol was injected into the tail veins of mice, and then data for images (a), (b), and (c) were obtained between 0.9 and 9.9, 10.4 and 19.3, and 19.8 and 28.7 min, respectively. 2D images were reconstructed as described in the legend of Fig. 4. The logarithmic signal intensities at ROI 1 (○) and ROI 2 (△) of the 2D images were plotted against the intermediate time between data acquisition of (a), (b), and (c). Each ROI has 5 × 5 pixels, and symbols (○, △) indicate the mean and difference of two experiments, respectively.

Time-Resolved ESR-CT Imaging

Figures 5a–c show 2D images of CxP-AM signal intensity in the head of mice at different times after intravenous injection of CxP-AM. The encephalon region had distinguished intensity in the first image (Fig. 5a), but the contrast between the encephalon region and others gradually faded away (Figs. 5b and 5c). The total intensities of 5 × 5 pixels at ROI (region of interest) 1 and ROI 2 in the images were calculated from Fig. 5a–5c, and the logarithmic values of the total intensities were plotted against the intermediate time of data acquisition (Fig. 5d). The decline between 5 and 15 min was larger at ROI 2 than that at ROI 1, but there was no difference in the decline between 15 and 25 min. The decrease of the total intensity shown in Fig. 5d seems to reflect the reducing capability for nitroxyl radicals, implying that the reducing capability for nitroxyl radicals at ROI 2 is larger than that at ROI 1.

Conclusion

In order to estimate free radical reactions and its imaging in the brain of living animals, various types of lipophilic nitroxyl spin probes were synthesized. CxP-AM was designed to be hydrolyzed by esterase, but not by lipase, to pass through the blood–brain barrier and to be retained in the cytosolic phase of parenchymal cells in the brain after intravenous injection. Pharmacokinetics of CxP-AM were compared with CxP and CxP-M. CxP-AM was almost completely hydrolyzed by carboxyl esterase within 3 min. The retention of intravenously injected total CxP-AM in the brain was 1.8 times higher than that of CxP-M and lasted more than 30 min. ESR-CT imaging of CxP-AM in the head of mice generated clear contrasts in the encephalon region, whereas CxP was distributed only in the extracranial region. CxP-M was distributed in both regions, confirming the results of the pharmacokinetics of CxP-AM. The decay rate of CxP-AM determined with time-resolved ESR-CT imaging was different in the two brain regions, suggesting that the total reducing capability differs between brain regions. CxP-AM has the potential to serve as a powerful probe for the investigation and diagnosis of free radical reactions and its imaging in the brain.

[42] Analytical Implications of Iron Dithiocarbamates for Measurement of Nitric Oxide

By ALEXANDRE SAMOUILOV and JAY L. ZWEIER

Introduction

Nitric oxide (NO) is a gaseous paramagnetic substance with limited solubility in water. About 2 m*M* is in equilibrium at 100% atmospheric pressure and 24°. NO in biological objects can be produced by various pathways. The enzymatic pathways of NO formation have been of major interest in biological systems. There are three isoforms of nitric oxide synthase (NOS). This family of enzymes synthesizes NO using L-arginine, NADPH, and oxygen as substrates and requires calmodulin, Ca^{2+}, and tetrahydrobiopterin as cofactors.[1] The formation of NO also occurs by nonenzymatic pathways (without NOS) in biological tissues under pathological conditions such as ischemia. It was shown that in these cases the main

[1] Y. Xia, A. J. Cardounel, A. F. Vanin, and J. L. Zweier, *Free Radic. Biol. Med.* **29,** 793 (2000).

0076-6879/02 $35.00

source of NO is nitrite, which can be reduced rapidly to NO by various reducing equivalents. With these conditions, the formation of NO can reach the levels of enzymatic synthesis. Enzymes of types other than NOS can also be involved in the reduction of nitrite.[2–4] It has also been reported that nitrate can be transformed to NO.[3]

Once produced, NO can undergo different conversions, and nitrate and nitrite are the usual final products of these NO transformations. Several pathways can lead to nitrate formation. The observed lifetime of NO in biological samples is very short with more rapid decay than that which occurs only due to the reaction with dissolved oxygen. It has been shown that NO reacts rapidly with oxyhemoglobin, resulting in the formation of methemoglobin and nitrate. Reaction of NO with the superoxide anion leads to the formation of very potent oxidant peroxynitrite, and this reaction is known to occur at a near diffusion-controlled rate. Peroxynitrite transforms rapidly to nitrate at neutral pH.

There are many methods available to quantitate NO production. These include chemiluminescence[5,5a]; Griess reaction, which is based on the measurement of nitrite and nitrate concentrations, the main metabolites of NO; photometric monitoring of the formation of methemoglobin from the reaction of NO and oxyhemoglobin[6]; and EPR spectroscopy. EPR spectroscopy is a unique technique that enables direct quantitative detection and measurement of free radicals. It allows real-time measurements of paramagnetic substances both *in vitro* and *in vivo* and can even enable imaging of the presence of free radicals in biological tissues.[7,8] Because NO, like other free radicals produced in biological systems, is usually present in very small quantities and is able to interact rapidly with many substances, it needs to be stabilized, i.e., trapped, to facilitate its detection and measurement. Spin-trapping techniques have been developed to stabilize NO as the adduct of endogenous and exogenous spin traps, which enables an integrative measurement

[2] Z. Zhang, D. Naughton, P. D. Winyard, and N. Benjamin, *Biochem. Biophys. Res. Commun.* **249,** 767 (1998).

[3] T. M. Millar, C. R. Stevens, N. Benjamin, R. Eisenthal, R. Harrison, and D. R. Blake, *FEBS Lett.* **427,** 225 (1998).

[4] L. J. B. Godber, J. J. Doel, P. G. Sapkota, R. D. Blake, R. C. Stevens, R. Eisenthal, and R. Harrison, *J. Biol. Chem.* **275,** 7757 (2000).

[5] R. M. J. Palmer, A. G. Ferrige, and S. Moneada, *Nature* **327,** 524 (1987).

[5a] J. Collier and P. Vallance, *Trends Pharmacol. Sci.* **10,** 427 (1989).

[6] L. J. Ignarro, G. M. Buga, K. S. Wood, R. E. Byrns, and G. Chaudhuri, *Proc. Natl. Acad. Sci. U.S.A.* **84,** 9265 (1987).

[7] P. Kuppusamy, S. T. Ohnishi, Y. Numagami, T. Ohnishi, and J. L. Zweier, *Cereb. Blood Flow Metab.* **15,** 899 (1995).

[8] P. Kuppusamy, P. Wang, A. Samouilov, and J. L. Zweier, *Magn. Res. Med.* **36,** 212 (1996).

of NO formation over time.[9–14] To reach an efficient trapping reaction of NO the spin trap should react rapidly enough to effectively compete with all reactions in which NO is involved.

By the time of the discovery of NO as an important molecule in biology, EPR spectroscopists already had extensive experience in the studying of NO complexes of heme proteins. NO was used to probe the structure and function of these proteins based on its binding to these iron complexes and its property of rendering ferrous iron complexes "EPR visible." In these cases, NO added to the system in excess was trapped by the metal center and provided critical information regarding the properties of these centers.[15] After finding that NO has a vital role in many biological and biochemical processes, a new wave of interest in NO and EPR methods of its registration has developed over the last decade. Approaches have been developed for measurement of NO via intrinsic trapping by heme proteins or extrinsic trapping by the addition of exogenous metal complexes, including iron dithiocarbamates.

Complexes of NO with intrinsic heme proteins such as hemoglobin or myoglobin typically give a broad EPR signal best detected at liquid nitrogen or lower temperatures. They can be used for the detection of NO for *in vivo* and *ex vivo* experiments in some pathological conditions, such as ischemia, when the amount of oxyhemoglobin is negligible or greatly decreased so that deoxy complexes can effectively compete for NO. Because EPR measurements of these complexes at typical spectrometer frequencies of 9 to 10 GHz should be performed in the frozen state at liquid nitrogen temperatures or below, real-time measurement of the production of NO is generally not possible. Because the EPR signal of interest is relatively broad and EPR spectra of biological samples can be quite complex at cryogenic temperatures with a number of other overlying signals, the sensitivity of this approach is relatively low, of the order of micromolar to 0.1 μM at best. The biggest and only advantage of registering of *in vivo* NO production with heme proteins as NO traps is the absence of all interferences and alterations usually caused by the addition of spin traps.

Exogenous traps, such as nitrone- and nitroso-based spin traps, that are commonly used to trap oxygen-derived radicals are not capable of trapping NO as stable

[9] H.-G. Korth, R. Sustmann, P. Lommes, T. Paul, A. Ernst, H. de Groot, L. Hughes, and K. U. Ingold, *J. Am. Chem. Soc.* **116,** 2767 (1994).

[10] C. M. Arroyo and M. Kohno, *Free Radic. Res. Commun.* **14,** 145 (1991).

[11] K. Miyazaki, S. Ueda, and H. Maeda, *Biochemistry* **32,** 827 (1993).

[12] Y. Y. Woldman, V. V. Khramtsov, I. A. Grigorev, I. A. Kiriljuk, and D. Utepbergenov, *Biochem. Biophys. Res. Commun.* **202,** 195 (1994).

[13] P. Mordvintcev, A. Mulsch, R. Busse, and A. Vanin, *Anal. Biochem.* **199,** 142 (1991).

[14] M. Y. Obolenskaya, A. F. Vanin, P. I. Mordvintcev, A. Miflach, and K. Decker, *Biochem. Biophys. Res. Commun.* **202,** 571 (1994).

[15] H. Kon, *J. Biol. Chem.* **243,** 4350 (1968).

adducts.[16] Complexes of Fe^{2+} with various anion derivatives of dithiocarbamate ($^{-}S_2CNR_2$) have been the most widely used traps to detect and quantitate NO in enzymatic systems, cells, and tissues *in vivo* and *ex vivo*.[17–33] These traps enable sensitive detection of low levels of NO formation in biological tissue, with a detection limit down to 10 n*M* levels. They also readily enable isotope-labeling studies of the substrate source of NO with the use of ^{15}N-labeled substrates. The most commonly used derivatives are dithiocarbamate (DETC), *N*-methyl-D-glucamine dithiocarbamate (MGD), and the relatively new *N*-(dithiocarboxy) sarcosine (DTCS). Their structures are shown in Fig. 1. NO adducts of these ferrous dithiocarbamates give rise to characteristic EPR triplet signals in the case of natural abundance ^{14}N, with $g = 2.035$ and hyperfine splitting of $\approx$12.7 G (Fig. 2I spectrum A). Analysis of EPR spectra shows that the unpaired electron is localized mainly on the $d_Z{}^3$ orbital of the d^7 configuration iron (Fe^+).[23,30] Many authors, while interpreting results of EPR NO measurements of different systems, generally assume that the traps and their NO adducts are stable and that their trapping efficiency is not altered by the systems studied. Also, in many cases, potential system perturbation, which is introduced by the trap, is underestimated.

This chapter summarizes factors known to affect sensitivity and reliability of the dithiocarbamate EPR-trapping technique and points out approaches to address these problems, which enable reliable use of these traps for the quantitative detection of NO in biological systems.

[16] K. J. Reszka, P. Bilski, and C. F. Chignell, *J. Am. Chem. Soc.* **118,** 8719 (1996).
[17] A. F. Vanin, P. I. Mordvintcev, and A. L. Kleschyov, *Studia Biophys.* **102,** 135 (1984).
[18] L. N. Kuhrina, W. S. Caldwell, P. I. Mordvincev, I. N. Malenkova, and A. F. Vanin, *Biochim. Biophys. Acta* **1099,** 233 (1992).
[19] T. Tominaga, S. Sato, T. Ohnishi, and S. T. Ohnishi, *Brain Res.* **614,** 342 (1993).
[20] A. Mulsch, R. Busse, P. I. Mordvintcev, A. F. Vanin, E. O. Nielsen, J. Scheel-Kruger, and S.-P. Olesen, *NeuroReports* **5,** 2325 (1994).
[21] D. L. Laskin, M. R. DelValle, D. E. Heck, S. M. Hwang, S. K. Durham, S. T. Ohnishi, N. L. Goller, and J. D. Laskin, *Hepatology* **22,** 223 (1995).
[22] A. J. Bune, J. K. Shergil, R. Cammack, and H. T. Cook, *FEBS Lett.* **366,** 127 (1995).
[23] A. F. Vanin, *Methods Enzymol.* **301,** 269 (1999).
[24] C.-S. Lai and A. M. Komarov, *FEBS Lett.* **345,** 120 (1994).
[25] A. M. Komarov and C.-S. Lai, *Biochim. Biophys. Acta* **1272,** 29 (1995).
[26] J. L. Zweier, P. Wang, A. Samouilov, and P. Kuppusamy, *Nature Med.* **1,** 804 (1995).
[27] J. L. Zweier, P. Wang, and P. Kuppusamy, *J. Biol. Chem.* **270,** 304 (1995).
[28] P. Kuppusamy, P. Wang, A. Samouilov, and J. L. Zweier, *Magn. Res. Med.* **36,** 212 (1996).
[29] T. Yoshimura, H. Yokoyama, S. Fujii, F. Takayama, K. Oikawa, and H. Kamada, *Nature Biotechnol.* **14,** 992 (1996).
[30] H. Fujii, J. Koscielniak, and L. J. Berliner, *Magn. Res. Med.* **38,** 565 (1997).
[31] A. Goodman, J. B. Raynor, and M. Symons, *J. Chem. Soc.* (A), 2572 (1969).
[32] V. D. Mikoyan, L. N. Kubrina, V. A. Serezhenkov, R. A. Stukan, and A. F. Vanin, *Biochim. Biophys. Acta* **1336,** 225 (1997).

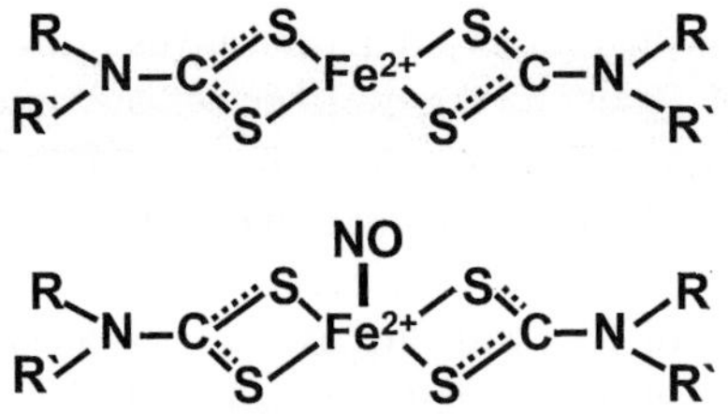

DETC $R,R' = -C_2H_5$

MGD $R = -CH_2\text{-}CH(OH)\text{-}CH(OH)\text{-}CH(OH)\text{-}CH(OH)\text{-}CH_2OH$
$R' = -CH_3$

DTCS $R = -CH_2\text{-}CO_2^-$
$R' = -CH_3$

FIG. 1. Simplified structure of iron dithiocarbamate complexes and their adducts with NO.

Methodological Aspects

The following reagents are used: sodium DETC, L-ascorbic acid, and sodium dithionite, all from Sigma; ammonium iron(II) sulfate hexahydrate, hydroquinone, ferric chloride, and dimethyl sulfoxide (DMSO), all from Aldrich. Sodium MGD is synthesized using carbon disulfide and *N*-methyl-D-glucamine, as described.[34] Gaseous NO and its mixtures with argon are purchased from Matheson.

For synthesis and purification of NO, the common method of preparing NO-saturated water solutions or synthesis of NO iron complexes is based on bubbling of commercially available compressed NO through water or through solutions of Fe^{2+} dithiocarbamate complexes. We have found this direct approach to be completely unacceptable. Compressed NO is subjected to disproportionation to NO_2 and N_2O. Even initially high purity gas in time will be severely contaminated with NO_2 and with its product of recombination with NO, N_2O_3. Purging of this kind of gas through water causes an accumulation of highly soluble nitrogen oxides in solution, which hydrolyze with the formation of nitrous acid. Purging of contaminated NO through a Fe^{2+} dithiocarbamate trap solution causes rapid oxidation of iron and "loss" of EPR signal. The best way to monitor NO_2 presence in gas is by measurement of the pH of its equilibrated solution (pH indicator strips are good enough for this purpose). Pure NO gas does not affect the pH of the solution. For removal of higher oxides, NO gas was passed through the gas-washing bottle filled with 1–2 *M* sodium or potassium hydroxide. Our experience shows

[33] Y. Kotake, D. R. Moore, H. Sang, and L. A. Reinke, *Nitric Oxide Biol. Chem.* **3,** 114 (1999).
[34] L. A. Shinobu, S. G. Jones, and M. M. Jones, *Acta Pharmacol. Toxicol.* **54,** 189 (1984).

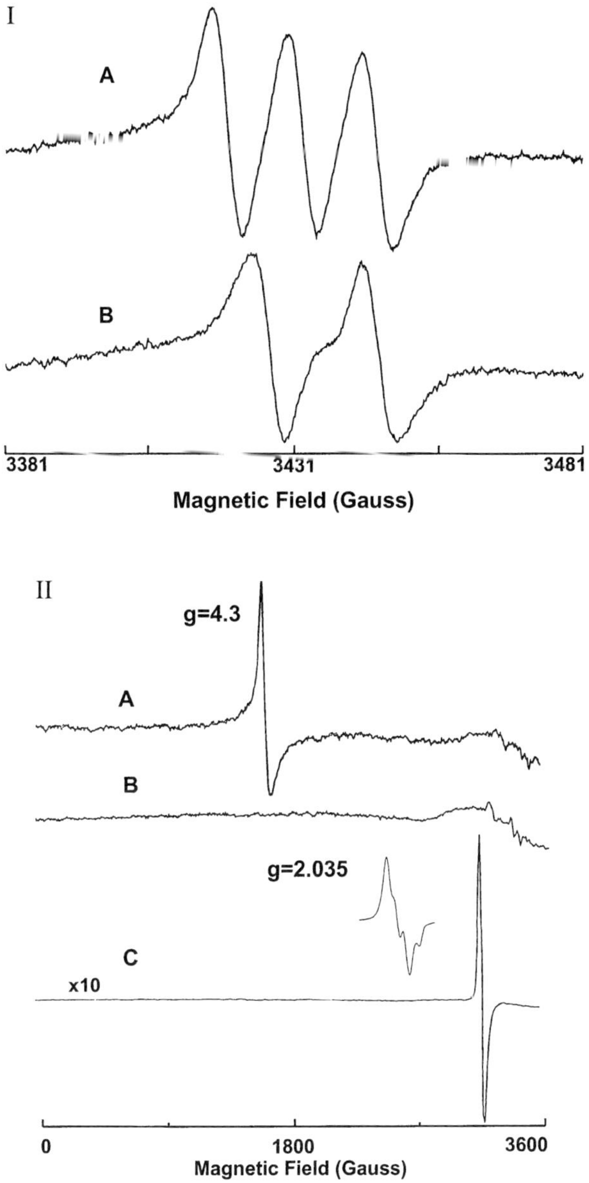

FIG. 2. (I). Characteristic EPR spectra of (A) NO^{14}–Fe^{2+} MGD and (B) NO^{15}–Fe^{2+} MGD recorded at room temperature. (II) Characteristic EPR spectra of (A) Fe^{3+}–MGD complex, (B) NO–Fe^{3+}–MGD complex obtained by treating of (A) with NO for 3 min, and (C) NO–Fe^{2+}–MGD, obtained by treatment of (B) with 10 m*M* ascorbate. (Insert) Expanded view of (C). Spectra were recorded at 77 K, with a microwave power of 20 mW, and modulation amplitude of 5 G.

that one-step purification is not enough for 100% compressed NO; gas should be passed through at least two gas-washing bottles. For mixtures of NO with less than 10%, one-step purification is usually sufficient. As a precaution against accumulation of nitrite, the addition of ion exchange resin to the water solution of NO can be done.[35] Another way of preparing high-purity NO gas is low-temperature sublimation of solidified (at liquid nitrogen temperature) NO in an evacuated system. The high efficiency of this method also allows synthesis and purification of NO gas in the laboratory by the reaction of $FeSO_4$ with sodium nitrite in 0.1 *M* HCl.[36]

Iron MGD(DETC) complexes are synthesized using a Thunberg tube apparatus with the addition of solid ferrous ammonium sulfate into degassed vacuumized distilled water at the top chamber. After revacuumization, the 1 m*M* Fe^{2+} solution is mixed with the anaerobic degassed 2–200 m*M* MGD solution in 100 m*M* phosphate buffer or DMSO placed in the main bottom chamber of the tube. Fe^{3+} complexes are synthesized in a similar way, but instead of ferrous ammonium sulfate, ferric chloride is used. Synthesis of NO complexes is performed in a similar way, but in the presence of gaseous NO in the headspace volume and 1–5 ml of ferrous sulfate or ferric chloride solution in oxygen-free distilled water or is placed in the upper part of the Thunberg tube apparatus and mixed with the dithiocarbamate solution in the presence of NO.[36]

Ferrous DETC complexes in soybean oil emulsion are synthesized by the addition of a degassed ferrous ammonium sulfate water solution (2–20 m*M*) in the emulsion (Liposyn III Abbott Laboratories) during purging by argon. This degassed solution of DETC (4–100 m*M*) is slowly added to the mixture. Ferric DETC complexes are obtained by a 10-min exposure of ferrous DETC complexes in soybean oil emulsion to the air. Prepared mixtures are used within 2 hr. The stoichiometric ratio between Fe^{2+} and dithiocarbamates in these complexes is 1 : 2, but because dithiocarbamates are weak ligands, complexes should be prepared in the presence of excess dithiocarbamates. In general, the minimum ratio should be 1 : 5. Lessening of this ratio leads to precipitation due to the hydrolysis of iron and also decreases the apparent affinity of NO to the complexes. Most probably, the limited life of DETC complexes synthesized in oil emulsion is caused by the unavailability of water-soluble DETC ligand to water-insoluble iron DETC complexes.

Experimental Concerns

Effect of Oxygen on Traps and Adducts

While strictly anaerobically prepared aqueous solutions of Fe^{2+} MGD or DMSO solutions of Fe^{2+} DETC are clear and colorless, similar solutions of Fe^{3+}

[35] W. A. Blumberg, *Methods Enzymol.* **76,** 312 (1981).

[36] A. F. Vanin, X. Liu, A. Samouilov, R. A. Stukan, and J. L. Zweier, *Biochim. Biophys. Acta* **1474,** 365 (2000).

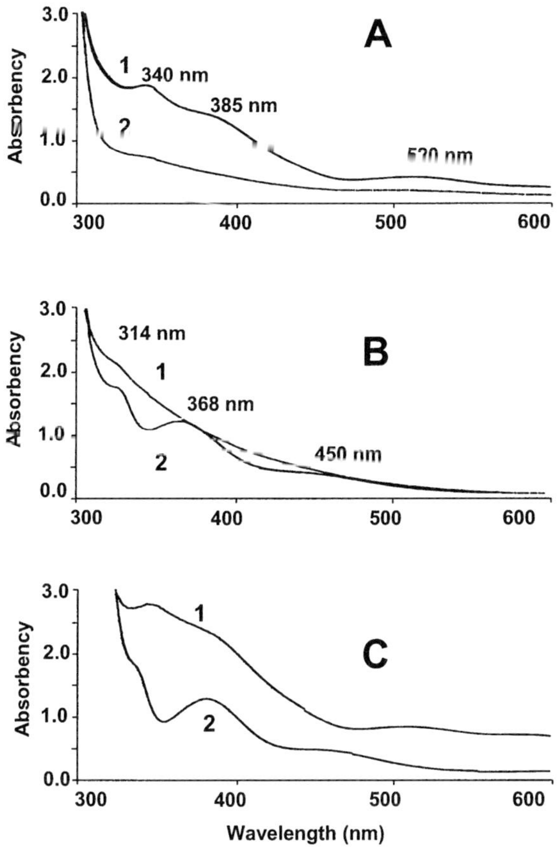

FIG. 3. Optical absorption spectra of ferrous and ferric MGD and DETC complexes and their adducts with NO. (A1) Spectra of Fe^{3+}–MGD water solution and (2) spectra of Fe^{2+}–MGD, obtained by reduction of A with ascorbate. (B1) spectra of NO–Fe^{3+}–MGD obtained by treatment of A(1) with NO, (2) spectra of NO–Fe^{2+}–MGD obtained by treatment of B(1) with ascorbate. 0.1 m*M* iron 1 m*M* MGD, pH 7.4. (C1) Spectra of Fe^{3+}–DETC in DMSO solution, (2) spectra of NO–Fe^{2+}–DETC in DMSO solution obtained by treatment of C(1) with NO, which instantly formed NO–Fe^{2+}–DETC 0.15 m*M* iron.

are clear, but their color varies from orange to dark brown depending on the concentration. Fe^{3+} complexes exhibit three absorption bands at 340, 385, and 520 nm (Fig. 3). Exposure of Fe^{2+} complexes to air or oxygen causes rapid change in color to orange-brown and to the appearance of characteristic absorption bands of Fe^{3+}. Also, Fe^{2+} complexes are diamagnetic and thus EPR silent, but frozen solutions of Fe^{3+} dithiocarbamates at cryogenic temperatures give rise to a characteristic peak at g = 4.3. After exposure to air, solutions of Fe^{2+} develop the characteristic peak at g = 4.3 (Fig. 2I). Thus, in the presence of oxygen, Fe^{2+} complexes

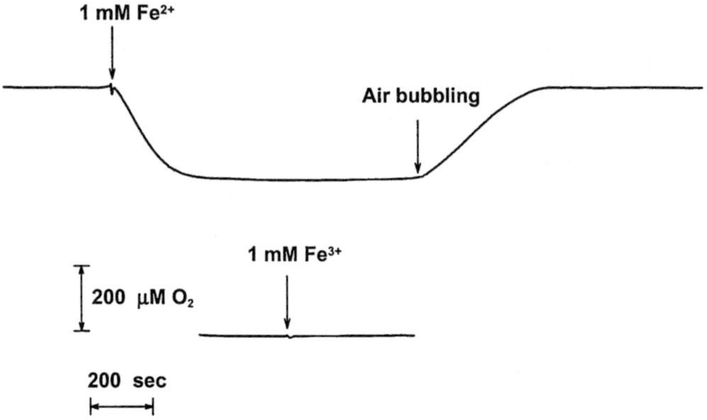

FIG. 4. Oxygen consumption in an aqueous solution of 1 m*M* Fe^{2+}–MGD and Fe^{3+}–MGD in 100 m*M* phosphate buffer, pH 7.4.

oxidize to Fe^{3+}. Oxygen consumption by Fe^{2+} complexes was detected in water solution by electrochemical measurements and by EPR measurement of oxygen-sensitive charcoal. It was observed that 1 m*M* Fe^{2+} completely consumes the ambient oxygen in a closed system without headspace within 3 min. Consumption of oxygen by Fe^{3+} was not seen (Fig. 4). The time, which is necessary for the complete oxidation of Fe^{2+} by oxygen, is variable and depends on the availability of oxygen. Treatment of Fe^{3+} with an excess of reducing equivalents, such as ascorbate and dithionate, removes the color from solution and eliminates the $g = 4.3$ EPR signal, thus reducing Fe^{3+} to Fe^{2+}.

Synthesized NO adducts of Fe^{2+} MGD and DETC have green color and show three absorbance bands at 314, 368, and 450 nm similar in water and DMSO (Fig. 3). Exposure of aqueous solutions of the NO–Fe^{2+}MGD complex to air turn its color to yellow with loss of optical absorption bands at 368 and 450 nm. Simultaneously, loss of the characteristic EPR triplet is observed. In the absence of oxygen, this complex is stable; only a 10% decrease in the intensity of the EPR signal is seen after 24 hr in a flat cell, most probably this loss is due to the leak of air into the flat cell. In contrast to aqueous solutions of Fe^{2+}MGD–NO, DMSO solutions of iron DETC and MGD NO complexes were not affected by exposure to oxygen. However, exposure of complexes in DMSO solutions to NO and oxygen simultaneously resulted in a loss of the EPR triplet signal and conversion of solutions to yellow.

Interaction of NO with Ferrous and Ferric Complexes

It was commonly accepted that trapping of NO occurs only by direct binding with Fe^{2+} dithiocarbamates, which occurs at high rates. It has been observed that exposure of Fe^{3+} dithiocarbamates to NO also leads to formation of the characteristic triplet EPR signal at g = 2.035.[36] However, contrary to the instant appearance of the NO signal with ferrous dithionate complexes, formation of this signal with ferric complexes occurred more slowly. Anaerobic exposure of water solutions of ferric complexes to NO resulted in a two-step process. First, there was a fast disappearance of the characteristic EPR spectrum of Fe^{3+} S = 5/2 at g = 4.3 with simultaneous rapid changing of color from brown to yellow. Second, after additional exposure, a gradual rise of the paramagnetic NO–Fe^{2+}–MGD triplet signal was observed. However, maximum intensity of the EPR signal did not exceed 50% of intensity obtained with ferrous complexes under similar conditions. Concurrently with the appearance of the triplet signal, the yellow complexes gradually (over an hour) change color from yellow to green with the appearance of the characteristic optical absorption of the ferrous NO–MGD complex. The addition of excess reducing reagents such as ascorbate or dithionite (10–100 m*M*) made this second transformation instant with an increased yield of NO–Fe^{2+}–MGD of up to 80%. In DMSO solutions, complexes with DETC or MGD in DMSO solutions transform rapidly to a green color when exposed to NO. This was accompanied by the appearance of the characteristic optical absorption spectra of the NO–Fe^{2+} complexes of DETC or MGD. This second process was completed within 2–3 min. The high rate of this transformation in DMSO prevented detection of the formation of the intermediate complexes with DETC or MGD that was readily detected in aqueous solutions for the Fe^{3+}–MGD complexes.

Interference with Nitrite

One of the main concerns, recently raised,[37] is the interaction of NO–Fe^{2+} complexes of MGD with nitrite. It is known that nitrite is able to disproportionate to NO and nitrate. The rate of this process is pH dependant and at physiological values is very slow.[37a] However, it was observed that the Fe^{2+}–MGD complex forms paramagnetic NO adducts significantly faster than can be produced from nitrite disproportionation.[37] The proposed mechanism included an overall fifth order reaction, where the rate of NO adduct formation is proportional to nitrite, square of Fe^{2+}, MGD, and hydrogen ions concentrations. The calculated rate constant allowed prediction of the rate of adduct formation from nitrite.

While relatively rapid NO complex formation occurs with Fe^{2+}–MGD, with Fe^{3+} complexes, the rate of the reaction is much slower, on the order of the magnitude of nitrite disproportionation. So with Fe^{3+}–MGD complexes, nitrite-related perturbation is negligible (Fig. 5). For Fe^{2+}–DETC in lipid emulsions, the rate of NO complex formation is also not directly affected by nitrite. The Fe^{3+}–DETC complex in lipid emulsions did not produce significant EPR signal from nitrite but quantitatively traps NO from an NO donor.

[37] K. Tsuchiya, M. Yoshizumi, H. Houchi, and R. Mason, *J. Biol. Chem.* **275,** 1551 (2000).
[37a] A. Samouilov, P. Kuppusamy, and J. L. Zweier, *Arch. Biochem. Biophys.* **357,** (1998).

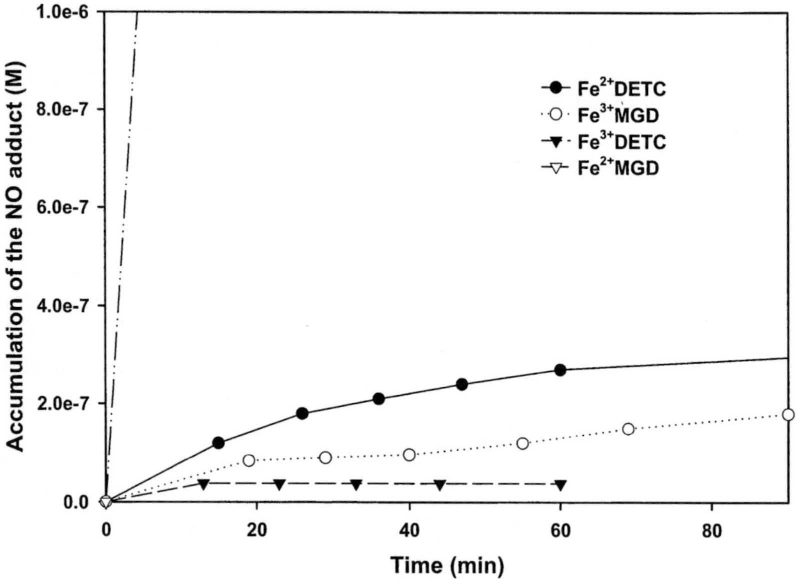

FIG. 5. Formation of nitric oxide iron dithiocarbamate complexes from nitrite with different iron dithiocarbamates. Measurements were performed anaerobically in sealed 100-μl capillary tubes in an X-band Bruker ESP 300 spectrometer. MGD complex solutions were prepared in HEPES buffer, pH 7.4. DETC complex experiments were performed in lipid emulsion as described elsewhere. Acquisition parameters: microwave power, 80 mW; modulation amplitude, 4 G; and sweep width, 80 G. Concentration was calculated from the intensity of the EPR signal compared to that of a known concentration standard.

Discussion

A major problem encountered with EPR spin trapping of NO by Fe^{2+} complexes is the presence of oxygen. Only solutions prepared strictly anaerobically truly contain the Fe^{2+} complex. Even brief exposure to air causes the transparent Fe^{2+} complex solutions to turn dark due to the oxidative formation of the Fe^{3+} complex. It has been shown that one of the intermediates of this process is superoxide,[37b,37c] the product of the one electron reduction of oxygen. Superoxide then transforms to hydrogen peroxide (scheme in Fig. 6). Due to the presence of iron, hydrogen peroxide is not stable under these conditions and decomposes to hydroxyl radicals. In fact, it is very difficult to observe the formation of superoxide or hydroxyl radicals using the conventional approach of EPR trapping, as

[37b] K. Tsuchiya, J. J. Jiang, M. Yoshizumi, T. Tamaki, H. Houchi, K. Minakuchi, K. Fukuzawa, and R. P. Mason, *Free Radic. Biol. Med.* **27,** 347 (1999).

[37c] S. Pou, P. Tsai, S. Porasuphatana, H. J. Halpern, G. V. R. Chandramouli, E. D. Barth, and G. M. Rosen, *Biochem. Biophys. Acta* **1427,** 216 (1999).

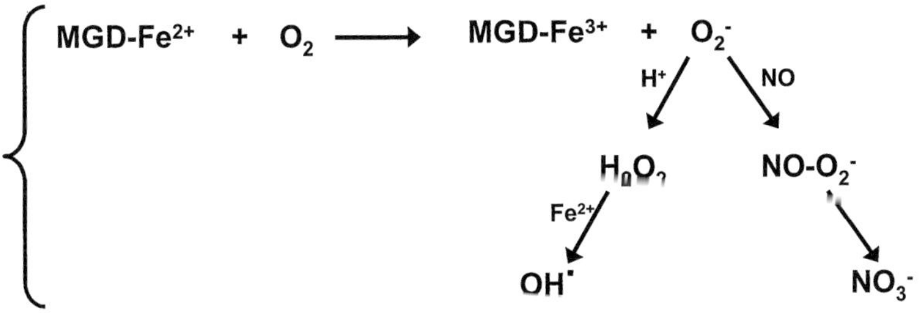

FIG. 6. Simplified reactions of oxidation ferrous MGD(DETC) complexes.

the presence of reduced iron dithiocarbamates makes spin adducts of DEMPO very unstable due to reduction by Fe^{2+} (unpublished data). That is why traces of hydroxyl and carbon centered adducts reported previously[37b,37c] can be considered as evidence of prominent superoxide production under these experimental conditions. The presence of such extremely reactive substances, including superoxide and hydroxyl radicals, can significantly alter any biological system, not only by interacting with released NO (Fig. 6), but also by interacting with enzymes and other biologically relevant molecules.

Beyond the effect of oxygen on the trap, it should also be remembered that introduction of the Fe^{2+} form of the trap (MGD or DETC complex) into any system causes a rapid consumption of surrounding oxygen. If an experimental system with an ambient concentration of oxygen is sealed or oxygen diffusion is very limited, complete depletion of all oxygen will occur over as little as a minute (Fig. 4) if the concentration of Fe^{2+} exceeds the concentration of available oxygen. Thus this form of the trap can cause significant alterations in experimental conditions. Performing experiments in excess of available oxygen will cause the transformation of all Fe^{2+}–MDG complexes to Fe^{3+} with the rate dependent on oxygen availability. It is possible to maintain the reduced state of iron (Fe^{2+}) even if the experimental setup is opened to the atmosphere. Addition of a significant excess of reducing equivalents (e.g., 10–100 m*M* of sodium ascorbate or dithionate) will maintain the complex in the reduced Fe^{2+} state for a time, which depends on oxygen flux into the experimental setup. However, with these conditions, oxygen concentration will be close to zero and there will be a significant flux of superoxide production.

The ability of oxygen to rapidly and efficiently oxidize Fe^{2+} complexes can be used for the convenient preparation of Fe^{3+}–dithiocabamate complexes. Because solutions of Fe^{2+} salts are much less subject to hydrolysis, it is more convenient to synthesize the Fe^{3+} complex by combining solutions of Fe^{2+} with the ligands followed by exposure to the air for about 10 min, with intense stirring to allow complete oxidation.

As shown previously,[36] ferric dithiocarbamate complexes are similar to ferrous complexes in that they also rapidly bind NO and result in the formation of NO–Fe^{3+}

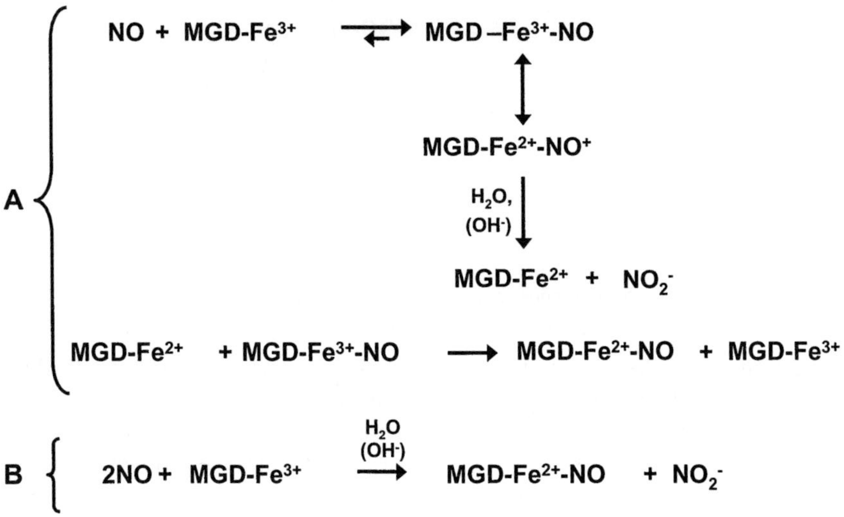

FIG. 7. Reactions of spontaneous formation of NO–Fe^{2+} dithiocarbamate complexes from NO and Fe^{3+} dithiocarbamate. (A) Formation occurs through formation of EPR-silent NO-Fe^{3+} dithiocarbamate complexes with consecutive reduction of Fe^{3+}MGD(DETC) to Fe^{2+}MGD(DETC). The latter accept NO from NO-Fe^{3+} dithiocarbamate, resulting in the formation of a paramagnetic complex. (B) Net reaction.

or, more precisely, (NO^+-Fe^{2+})–MGD(DETC). These complexes are EPR silent both at $g = 4.3$ and $g = 2.035$. Thus formed, they transform into a nitrosonium iron complex (NO^+–Fe^{2+}–MGD) that is hydrolyzed. This results in the liberation of NO^+ from this complex in the form of nitrite ion and in the transition of Fe^{3+}–MGD complexes to Fe^{2+}–MGD complexes. The latter accept NO from NO–Fe^{3+}–MGD, resulting in the formation of paramagnetic NO–Fe^{2+}–MGD (Fig. 7). Thus, these complexes can spontaneously transform to ferrous complexes with a predicted yield of 50%. This mechanism occurs in both aqueous and organic solutions of iron dithiocarbamates; however, differences occur with regard to the rate of this spontaneous transformation. In water solutions, it takes an hour to reach maximum conversion. In an organic solution, this process is completed within a minute. Addition of reducing agents such as ascorbate, cysteine, or hydroquinone to the samples can be performed to transform the diamagnetic complex to the EPR detectable paramagnetic state with the efficiency of "recovery" of the EPR signal from NO trapped by Fe–MGD of up to 100% (Fig. 8). This efficiency can be achieved if the reducing agent is added immediately after NO treatment.

Applications of organic solutions of iron dithiocarbamates have additional advantages over aqueous solutions. Adducts of NO in aqueous solutions are subject to oxidation by oxygen and, in general, the observed formation of the EPR signal of the water-dissolved adduct is a process determined by three reactions: production of

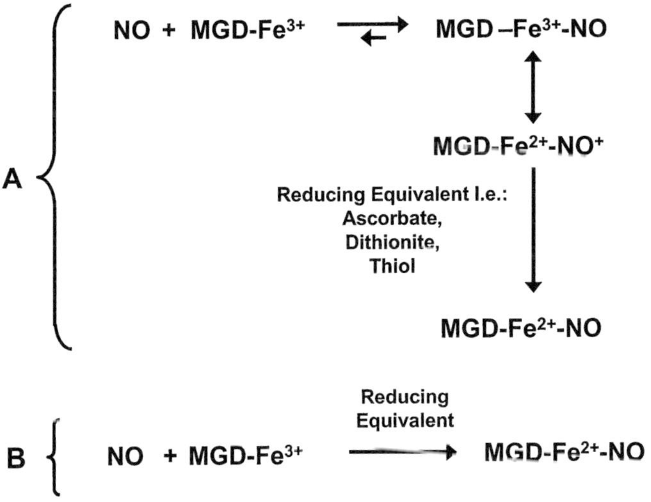

FIG. 8. Reactions of formation of NO–Fe^{2+} dithiocarbamate complexes from NO and Fe^{3+} dithiocarbamate in the presence of reducing equivalents. (A) Formation occurs through formation of EPR-silent NO-Fe^{3+} dithiocarbamate complexes that transform to nitrosonium complexes and then transform to paramagnetic NO–Fe^{2+}–MGD complexes. (B) Net reaction.

NO, transformation of ferric adduct to the ferrous state (spontaneous or by reaction with reducing equivalent), and oxidation by oxygen. However, in organic solutions, self-transformation of ferric adduct to the ferrous state occurs very rapidly and, furthermore, organic solutions of the adduct are not oxidized by oxygen (scheme in Fig. 9). Thus, the formation of water-insoluble NO adducts is determined entirely by the production of NO in the system.

Additional complications in the interpretations of the results of spin trapping of NO by Fe^{2+} dithiocarbamates can arise due to the direct interaction of nitrite with the trap leading to elevated levels of NO adduct signal. The estimated rate[37] of this process for a system containing 0.5 mM Fe^{2+}–MGD and 10 μM nitrite is 1.2–3.7 pM/sec. In comparison, in rat heart homogenates, maximally activated nitric oxide synthase produces NO on the level of 1.5 nmol/sec/kg of heart tissue.[37d] This is three orders of magnitude higher than the potential production of NO due to the interaction of 0.5 mM Fe^{2+}–MGD with 10 μM of nitrite, the concentration of nitrite that was found in heart tissue. In some circumstances, when the pH is lower (ischemic conditions[26] or the nitrite concentration is extremely high (septic shock,[37e] part of the observed signal could arise from Fe^{2+}–MGD-induced nitrite

[37d] R. R. Giraldez, A. Panda, Y. Xia, S. P. Sanders, and J. L. Zweier, *J. Biol. Chem.* **272,** 21420 (1997).

[37e] Y. Kamisaki, K. Wada, M. Ataka, Y. Yamada, K. Nakamoto, K. Ashida, and Y. Kishimoto, *Biochem. Biophys. Acta* **1362,** 24 (1999).

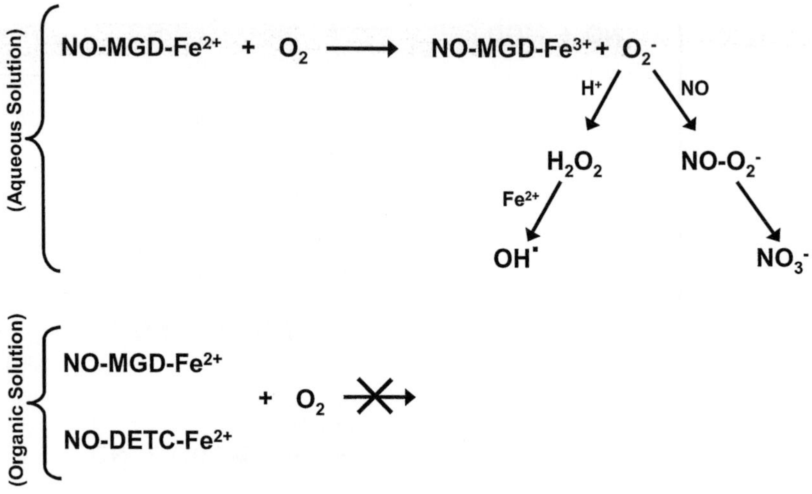

FIG. 9. Simplified reactions of oxidation ferrous MGD(DETC) complexes. Note difference in the behavior of adducts in aqueous and organic solutions. In organic solutions, oxidation does not occur.

reduction. In this case, nitrite could compete with NO and form an EPR silent NO–Fe^{3+}MGD complex, which then rapidly donates NO to Fe^{2+}–MGD, forming EPR visible NO–Fe^{2+}–MGD (Fig. 10). In general, nitrite-mediated NO formation can be estimated easily by measurement of the nitrite concentration in the system by any available method, and then the rate of NO production can be calculated.[37] Also, one can significantly reduce the nitrite interference by using the ferric complex–Fe^{3+}–MGD instead of the ferrous complex. As shown, the oxidized complex is not capable of reducing nitrite, but it traps NO and self-transforms to the EPR visible adduct with a 50% yield.

The Fe^{2+}–DETC complex has also been used commonly for trapping NO.[32,38–45] Because Fe^{2+}–DETC is not soluble in water, it requires special preparation. For *in vivo* experiments, one can inject iron and DETC water solutions

[38] G. Wei, V. L. Dawson, and J. L. Zweier, *Biochim. Biophys. Acta* **1455,** 23 (1999).

[39] S. Stao, T. Tominago, T. Ohnishi, and S. T. Ohnishi, *Biochim. Biophys. Acta* **1181,** 195 (1993).

[40] T. Tommaga, S. Stao, T. Ohnishi, and S. T. J. Ohnishi, *Cereb. Blood Flow Metab.* **14,** 715 (1994).

[41] L. N. Kubrina, W. S. Caldwell, P. I. Mordvintcev, I. V. Malenkova, and A. F. Vanin, *Biochim. Biophys. Acta* **1099,** 233 (1992).

[42] N. V. Voevodskaya and A. F. Vanin, *Biochem. Biophys. Res. Commun.* **86,** 1423 (1992).

[43] V. D. Mikoyan, N. V. Voevodskaya, L. N. Kubrina, I. V. Malenkova, and A. F. Vanin, *Biochim. Biophys. Acta* **1269,** 19 (1995).

[44] L. N. Kubrina, V. D. Mikoyan, P. I. Mordvintcev, and A. F. Vanin, *Biochim. Biophys. Acta* **1176,** 40 (1993).

[45] V. D. Mikoyan, L. N. Kubrina, and A. F. Vanin, *Biophysics* **39,** 953 (1995).

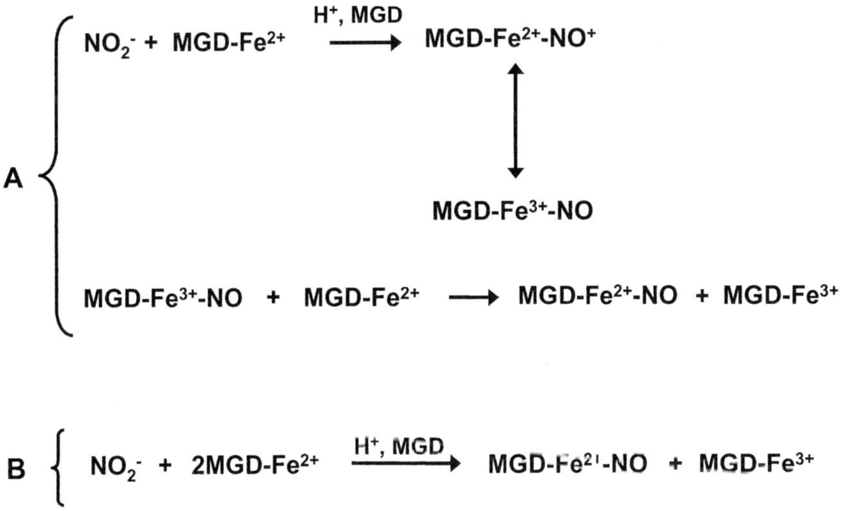

FIG. 10. Reactions of spontaneous formation of paramagnetic NO–Fe^{2+}–MGD complexes from NO–Fe^{2+}–MGD and nitrite [K. Tsuchiya, M. Yoshizumi, H. Houchi, and R. Mason, *J. Biol. Chem.* **275,** 1551 (2000)]. The overall rate of this reaction has been reported to depend on the concentration of nitrite, iron (squared), MGD, and hydrogen ion.

separately (intravenously and intraperitoneally), thus allowing synthesis inside the tissue. The same process can be simulated in water/lipid emulsion in a manner analogous to that described earlier. The Fe^{2+}–DETC complex, dissolved in the oil phase of the emulsion or localized in the lipophylic compartment of the tissue, is unavailable for direct interaction with lipid-insoluble nitrite. The slowly growing signal, which was observed after combining nitrite and Fe^{2+}–DETC emulsion or with Fe^{3+}–MGD solution, can be attributed to slow disproportionation of nitrite with formation of NO.[37a] The oxidized iron complex of DETC (Fe^{3+}–DETC) does not stimulate production of NO from nitrite. This reaction may not occur for two reasons. First, the lipid-dissolved complex is not available for direct interaction with nitrite. Second, the oxidized complex is not capable of nitrite reduction. At this time it remains unclear why the development of an NO signal from nitrite occurs even more slowly with Fe^{3+}–DETC than with Fe^{3+}–MGD (Fig. 5). However, trapping of NO from a donor such as SNAP occurs quantitatively, proving that Fe^{3+}–DETC is an efficient NO trap.

Thus, using the ferric form of the water-soluble trap can significantly minimize interference with nitrite, whereas the water-insoluble iron dithiocarbamates can completely exclude this unwanted effect. Advantages and disadvantages of applications of different redox states of water-soluble and water-insoluble traps are presented in Table I.

TABLE I
RELATIVE PROPERTIES OF FERROUS AND FERRIC DITHIOCARBAMATES

Iron dithiocarbamate	Solubility	Interaction with O_2	Interaction with nitrite	Stability of NO adduct in O_2
Fe^{2+}–MGD	Water ≫ lipid	+	+	−
Fe^{2+}–DETC	Lipid	+	−	+
Fe^{3+}–MGD	Water ≫ lipid	−	−	N/A
Fe^{3+}–DETC	Lipid	−	−	N/A

Furthermore, to exclude the effect of high concentrations of nitrite, isotope labeling of the substrate source can be used because EPR trapping techniques distinguish the nitrogen isotopes of NO. For enzymatic studies and some cellular applications, one can replace natural abundance ^{14}N L-arginine with the ^{15}N-substituted isotope and thus easily distinguish ^{15}NO produced from the enzymatic NOS reaction from ^{14}NO derived from nitrite (Fig. 2I).

Thus, from extensive studies over the last decade, it has been shown that dithiocarbamate complexes of iron can be used successfully for the measurement of NO production in a variety of chemical and biological systems. With careful consideration of the chemical and physical properties of these traps and their NO adducts, this EPR method of NO detection provides a unique and powerful tool for the measurement of NO in a broad range of applications, ranging from enzymatic to cellular systems. The technique also enables *in vivo* or *ex vivo* measurement and imaging of NO formation and the metabolic pathways by which it is formed.

Acknowledgment

This work was supported by NIH Grants HL38324, HL63744, and HL65608.

Author Index

Numbers in parentheses are footnote reference numbers and indicate that an author's work is referred to although the name is not cited in the text.

A

Abe, J., 55, 432, 433(3)
Abendschein, D. R., 223
Abidi, S. L., 326
Abramow-Newerly, W., 464
Abramson, S. B., 349
Abu-Sound, H. M., 349
Accorsi, K., 261
Adam, L. P., 444
Adam, S. A., 124, 130(18; 19)
Adamis, A. P., 392
Adams, E., 102(8), 103
Adams, G. E., 3, 14
Adams, R. N., 112(1), 113
Adams, S. R., 102(12), 103
Adan, J., 73
Ades, E. W., 201
Adler, V., 53, 54
Aebersold, R., 192
Agani, F., 32
Agosti, V., 261, 262(13)
Aguilar, J. J., 310, 312(11), 313(11)
Aguirre, L., 72
Ahmad, M., 208
Ahmed, M. U., 341
Ajmone Marsan, C., 483
Akaike, N., 71
Akaike, T., 324, 348, 350
Akbari, Y., 457
Akerboom, T. P. M., 440
Akil, H., 296, 299(8), 301, 302, 322, 323(49), 324(49)
Akimoto, Y., 496
Aksenova, M., 467
Alam, L., 217
Alameda, F., 114
Ala-Rämi, A., 135
Albertine, H., 100
Albig, W., 123
Aldape, K., 494
Alecci, M., 333
Alemany, S., 213
Alessandri, G., 410
Alexander, R. W., 445
Alexandre, A., 38
Ali, M. H., 319
Ali, S. F., 503
Allen, J. G., 24
Allen, R. G., 224, 446
Allende, R., 392
Allison, D. B., 373, 374(7), 375(7), 377(7)
Al-Mehdi, A. B., 309, 310, 311(7), 312(7), 313(7), 314(7), 315, 430
Alpers, C. E., 346
Alvai, A., 29, 30(86)
Amano, S., 392
Amar, M., 225(39), 226
Amar-Coretesec, E., 365
Amarnath, V., 373
Ambrose, J. A., 221
Ambrosio, G., 224, 225(37)
Amemiya, S., 116
Amerini, S., 280, 408, 411(4), 412(4), 413(4), 415(4), 416(4)
Amidi, M., 63, 453
Ammendola, R., 259, 262, 263, 264, 265(16)
Anastacio, V. M., 434, 436(11), 437(11), 441, 442(11), 444, 445(11), 446(11), 447(11), 448, 452(11), 454(11)
Andersen, A. R., 496
Anderson, C. W., 262, 263, 264, 265, 265(16), 266
Anderson, D. H., 380, 390(22)
Anderson, M. E., 236, 237(12), 440
Anderson, M. T., 233
Anderson, R. E., 378, 389(10), 482, 483, 484(6), 486, 486(6), 487, 488, 489, 490
Anderson, W. B., 114
Andrade, F. H., 310, 319(10), 320(10), 323(10)

Andus, T., 247
Anggard, E., 439
Angiolieri, M. R., 183
Ann, D., 167
Annex, B. H., 221
Anzai, K., 500
Anzevino, R., 91, 100
Apazidis, A., 444
Appella, E., 262, 263, 264, 265, 265(16), 266
Archer, S., 31, 416
Areworki, M., 495
Argyros, G., 163
Arnaudeau, S., 453
Arndt, H., 247
Arnelle, D., 348
Arrardi, G., 36
Arroyo, A., 474
Arroyo, C. M., 508
Asahara, T., 409
Ashida, K., 519
Assem, M., 394
Ataka, M., 519
Atalay, M., 422
Atanasiu, R., 134
Atkinson, M. A., 112(6), 113
Aubourg, P., 363
Aubry, J. M., 54
Auerbach, R., 422, 424(4)
Augé, N., 62
Augusto, O., 434, 436(11), 437(11), 440, 441, 442(11), 444, 445(11), 446(11), 447(11), 448, 450(36), 452(11), 454(11)
Austria, J. A., 124, 130(29), 131
Autschbach, F., 232
Avraham, H., 192
Avraham, S., 192
Avvedimento, V. E., 224, 225(37)
Azevedo, L. C. P., 434, 435(10), 436(10), 438, 442(10), 443(10), 446(10), 447(10), 448, 452(10)
Azzi, A., 226

B

Babson, J. R., 261
Bacher, S., 53
Badimon, J. J., 434
Bae, Y. S., 92
Baer, K. A., 14, 15(48)
Baeuerle, P. A., 208
Baggiolini, M., 92
Baines, I. C., 92
Baker, M. A., 207
Baker, T. L., 349
Balaban, R. S., 135, 141(3), 310, 313(6)
Balart, J. T., 341
Balat, A., 351
Balazs, R., 73
Baldwin, A. S., Jr., 211
Ballester, A., 213
Ballinger, C. A., 448, 453
Bando, H., 423
Banfi, B., 453
Bang, B. G., 318
Banker, G., 185
Bannai, S., 111, 190, 233, 236, 236(8), 237, 237(8), 238(18)
Bannenberg, G. L., 309, 317, 321(29)
Banni, S., 164
Baraban, J. M., 183, 187, 188(3; 9), 189
Barbeiro, H. V., 440, 450(36)
Barbosa, E., 372
Bard, A. J., 113, 114(10; 11), 115, 115(10), 116, 116(12; 13)
Barger, S. W., 203, 461
Barker, A. L., 113, 116(14)
Barker, S. L., 324, 325(59)
Barlow, C. H., 135
Barrett, J. N., 191
Barry-Lane, P., 448, 453
Barsky, S. H., 114
Barsukov, L. I., 160
Barth, E. D., 516, 517(37c)
Barton, M., 437
Barve, S., 203, 205
Bascoul, J., 67
Bass, D. A., 35, 315
Bassus, S., 225(47), 226, 227(47), 230(47)
Bast, A., 167
Batabyal, S. K., 220
Bates, G., 474
Batlle, E., 114
Batra, S., 394
Baudhnin, P., 365
Bauer, D. E., 249
Baughman, R. W., 186
Baumann, F. G., 444
Baunoch, D., 32
Baussmann, E., 4

Bautch, V. L., 422
Bauters, C., 409, 435
Bayir, H., 347
Baynes, J. W., 341
Beal, M. F., 474
Bean, J. M., 4, 5(14)
Beatty, P. W., 261
Beaufay, H. A., 365
Beauvoit, B., 478
Bechet, J. M., 92
Becker, C. G., 282
Becker, K., 236
Becker, L. B., 309, 311, 319(16)
Becker-Andre, M., 213
Beckman, J. S., 318, 455, 468
Bedogni, B., 91, 100
Beerepoot, L. V., 392
Beeson, C., 113
Beetham, D. J., 497
Beever, H., 361, 364(3)
Beg, A. A., 211
Begley, J. G., 203, 473
Beguin, S., 228
Belliveau, M. J., 191
Bendeck, M. P., 435
Benderitter, M., 394
Benedetti, A., 373
Benezra, M., 54
Benjamin, I., 21
Benjamin, N., 507, 509(3)
Bennett, M. R., 341
Benninghoff, B., 237, 246(20)
Benoist, H., 62, 69
Benov, L., 66, 309, 311, 312(15), 314(15)
Ben-Sasson, A. B., 407
Berd, D., 100
Bergelson, L. D., 160
Berger, B., 346
Berger, R., 495
Bergeron, M., 16
Berk, B. C., 55, 432, 433(3), 449
Berk, P. D., 201
Berkels, R., 323
Berlett, B. S., 259
Berlin, W. K., 124
Berliner, J. A., 208
Berliner, L. J., 310, 312(14), 313(14), 314(14), 453, 509
Berlowitz, C. O., 452, 454(59)
Berman, S. A., 189, 378
Bernal, M. T., 392
Berthelsen, A., 3(6), 4
Bethea, J., 225(45), 226
Betzler, M., 232
Beufay, Y., 365
Bevers, E. M., 160
Bevilacqua, M. P., 223
Bezemer, A. Z., 248, 257(4)
Bhatnagar, A., 449
Bialik, S., 16, 28(71)
Bicknell, R., 408, 411(5), 413(5), 415(5), 416(5), 421(5)
Bidgood, M. J., 199
Bielski, B. H. J., 273
Bierhaus, A., 223
Billiar, T. R., 280
Bilski, P., 41, 43, 44, 47(9), 48, 49, 50(16; 17), 51(17; 18), 509
Binder, B. R., 222
Bindokas, V. P., 309, 310, 311(9), 312(9), 313(9), 314(9)
Bird, E., 327
Bird, R. P., 80
Bird, T., 495
Bischoff, F. R., 123, 130
Bishop, A., 24
Bjorge, J. D., 192
Bkaily, G., 69
Blake, D. R., 507, 509(3)
Blake, R. D., 507
Blalock, E. M., 461
Bleehen, N. M., 24
Blobel, G., 123, 124
Bloom, E. T., 236, 237(13)
Blumberg, W. A., 512
Boag, J. W., 7
Boccia, A. D., 192
Bockstette, M., 237, 246(19)
Bode, A. M., 269, 270, 272(7)
Bode, C., 448, 449, 453
Bode, W., 222
Boehm, T., 424
Boerth, N. J., 249
Bogdanov, M. B., 474
Bohnensack, R., 136
Bohrer, H., 223
Boight, R., 362
Boirivant, M., 237, 247(24)
Bokoch, G. M., 97(16), 98
Bonaventura, J., 347, 350, 350(2)

Boneu, B., 224, 226(36)
Bonilla, E., 458
Booden, M. A., 349
Bookman, M. A., 233
Borchelt, D. R., 464
Borchert, H. H., 333
Borisenko, G. G., 163, 164, 358, 359, 359(42)
Bornkamm, G. W., 92
Borowitz, J. L., 317, 323(32)
Borrello, S., 91, 92, 100
Borst, C., 434, 435(14), 449
Bortolon, R., 124
Bos, J. L., 97(17), 98
Bosse, D. C., 201
Botero, A., 208
Bottcher, C. J. F., 167
Bouché, A., 434
Boullier, A., 160
Boveris, A., 38, 296, 373
Bowen, J., 354
Bowen, R. A., 273
Bowser, D. N., 319
Boyd, H. C., 345
Boyden, S., 175
Boyland, E., 260
Boyle, J. J., 341
Boyum, A., 137
Bradfield, C. A., 32
Bradford, M. M., 55
Bragdon, G. A., 280
Braigon, J. H., 67
Branch, R. A., 163
Brand, K., 222
Brandes, R. P., 63, 225(47), 226, 227(47), 230(47), 437, 439, 453
Brandt, R., 315
Brandtzaeg, P., 237, 247(23)
Brasier, A. R., 448, 453
Bratton, D. L., 160
Braun, R. D., 14
Braun, S., 453
Braunstein, J., 239, 247(25)
Breccia, A., 14
Bredenkötter, D., 439
Bredesen, D. E., 462, 469(29), 474
Bredt, D. S., 417
Breedveld, F. C., 248, 253, 253(3), 254, 255, 255(6), 256, 256(6), 257, 257(4; 6)
Breitkreutz, R., 53, 54(4), 232, 239, 247(25)
Bressler, P., 446
Bridey, F., 225(39), 226
Brielmeier, M., 92
Brien, J. F., 350
Brightman, A. O., 452, 454(58)
Brink, M., 341
Briviba, K., 54
Brizel, D. M., 4, 5(14)
Broder, G. J., 4
Brodie, A. E., 261
Broich, P., 141
Broillet, M., 360
Brontman, M., 365
Brors, O., 324
Broughton, J. P., 423
Brown, F. R., 362
Brown, G. C., 296
Brown, J. M., 4, 14
Brown, L. A., 301, 322, 323(48)
Brown, R. H., Jr., 495
Bruce, A. J., 462, 469(29)
Bruce-Keller, A. J., 458, 461(23), 466(23), 468(23), 469(23), 472(23), 474
Brugge, J. S., 191
Bruining, H. A., 483
Brunori, M., 389
Bryant, K. J., 199, 461, 462(26)
Bucana, C., 312, 313
Bücher, T., 135
Büchner, M., 139, 141(13)
Budd, S. C., 66
Budd, S. L., 313, 314(20)
Buee, L., 465
Buerk, D. G., 13(43), 14
Buettner, G. R., 64(10), 66, 307, 309, 316(2), 317, 317(2), 318(2), 319(2), 325(2)
Buga, G. M., 507
Bull, D. M., 233
Bump, E. A., 81, 86(18), 88(18), 89(18)
Bune, A. J., 509
Bunn, H. F., 31, 32
Burcham, P. C., 378
Burchfield, J. S., 440, 452(35)
Burden, R. H., 287
Burdick, M. D., 212
Burgess, J. K., 101, 102(6)
Burgio, V. L., 237, 247(24)
Burke, N., 183
Burkitt, M. J., 309, 325
Burne, J. F., 93

Buron, M. I., 268
Burow, S., 315
Burrows, N., 213
Bursztajn, S., 189
Busch, T. M., 27
Buss, J. E., 349
Busse, R., 63, 225(47), 226, 227(47), 230(47), 439, 453, 508, 509
Butler, S., 340, 342, 345(9)
Butterfield, C. E., 283
Butterfield, D. A., 467
Buttke, T. M., 151, 161
Byrns, R. E., 507

C

Caamano, C., 296, 299(8), 302, 322, 323(49), 324(49)
Cadenas, E., 296
Cadet, J. L., 503
Cadroy, Y., 224, 226(36)
Cahalan, M. D., 457
Cai, J., 161
Cakir, M., 351
Caldwell, W. S., 509, 520
Calkins, D. P., 24
Callahan, L., 310, 313(12)
Callis, G. M., 453
Camarasa, J., 72, 73, 74
Camins, A., 71, 72, 73, 74
Cammack, R., 509
Campagnolo, C., 360
Campbell, P. A., 160
Canalejo, A., 268
Candal, F. J., 201
Cannon, R. O., 347, 348(8), 350(8), 351(8)
Cao, J., 415
Caracciolo, M. R., 124
Cardounel, A. J., 506
Carew, T. E., 340
Carey, W. F., 369
Carlson, E., 503
Carmeliet, P., 223, 434
Carney, J. M., 455, 467
Carreras, M. C., 296
Carta, G., 163, 172(28)
Carter, W. O., 64(9), 66, 69(9), 70(9), 312, 319(17)
Carvalho, M. H. C., 440, 450(36)
Casella, F., 265
Casey, T. E., 324
Caspar-Bauguil, S., 69
Cass, A. E., 278
Castedo, R. F., 310, 312(11), 313(11)
Casteels, M., 360
Castilho, R. F., 313, 314(20)
Castillo, R. F., 66
Castle, L., 80
Castranova, V., 163, 170(25)
Cater, D. B., 4
Cayabyab, A., 495
Ceccarini, C., 367
Cekmen, M., 351
Cerasoli, F., Jr., 100
Cerniglia, G. J., 207
Chae, H. J., 318
Chae, H.-Z., 374, 429
Chae, S. W., 318
Chahal, A. S., 361, 364(7), 367(7), 371(7)
Chahine, R., 134
Chait, A., 206, 342, 345, 346
Chakrabari, B., 41
Chakraborti, S., 220
Chakraborti, T., 220
Chalian, A. A., 21
Champagne, C., 191
Chamulitrat, W., 41
Chan, C. Y., 5, 16, 16(27), 24(68), 27(68), 28(27)
Chan, P. H., 494, 503
Chan, S. L., 455, 457, 472
Chance, B., 38, 135, 138, 141(5), 373, 475, 477, 478, 479, 482, 483, 487, 487(5)
Chandel, M. S., 319
Chandel, N. S., 31, 32
Chandra, J., 160
Chandramouli, G. V. R., 516, 517(37c)
Chang, A., 222
Chang, C.-H., 280, 408, 415(6), 419(6), 421(6)
Chang, H.-C., 495
Chang, L. J., 446
Channon, K. M., 221
Chapman, G. B., 449
Chapman, J. D., 3, 7, 13(1), 14, 15, 15(48), 29
Chapman, P., 463
Chapman, W., 4
Chappell, D. C., 445
Chasseaud, L. F., 260
Chatton, J., 360
Chaudhuri, G., 507

Chavardjian, A., 165
Chavez, J. C., 32
Chee, C., 348
Chemtob, S., 16, 28(70)
Chen, A., 133(8), 136
Chen, D., 409
Chen, E. Y., 5
Chen, H. Y., 465
Chen, J., 449
Chen, J. B., 66
Chen, L., 448
Chen, L. B., 133(8), 136
Chen, M. F., 449
Chen, Q., 342, 346(10), 474
Chen, S. F., 503
Chen, Y.-Q., 5, 16, 16(27), 28(27; 70)
Chen, Z. H., 350
Cheng, B., 461, 462(26)
Cheng, H. C., 192
Cheng, H. F., 422
Cheng, J., 222
Cheng, L., 345, 444
Cheng, R., 16, 28(71)
Cheng, W., 118(26), 119(26), 122
Chensue, S. W., 212
Chesterman, C. N., 101, 102(6)
Chiariello, M., 224, 225(37)
Chida, K., 123
Chien, K. R., 394
Chignell, C. F., 41, 43, 44, 47(9), 48, 49, 50(16; 17), 51(17; 18), 307, 312(4), 316, 316(3; 4), 320(3), 321(3; 4), 509
Chikahisa, L., 71
Chin, L. K., 24
Ching, K. A., 249
Chio, K. S., 378, 389(11)
Chisolm, G. M., 224, 225(38; 45; 46), 226, 341, 434, 435(16)
Chiu, D., 81
Chiueh, C. C., 350
Chock, P. B., 92
Choi, D. W., 72, 191
Chomcznski, P., 227
Chong, B. H., 101, 102(6)
Choudhuri, R., 408, 411(5), 413(5), 415(5), 416(5), 421(5)
Christakos, S., 457
Christodoulou, D., 350
Christofi, F. L., 310, 312(14), 313(14), 314(14)
Chzhan, M., 333
Cifuentes-Pagano, M. E., 453
Cimino, F., 258, 259, 261, 262, 262(13), 263, 264, 265, 265(15; 16), 266
Cirillo, P., 224, 225(37)
Clancy, R. M., 349
Clanton, T. L., 307, 310, 312(14), 313(14), 314(14)
Clapham, D. E., 124
Clark, H. A., 324, 325(59)
Clark, J. K., 453
Clark, L. C. J., 6
Clark, R. G., 394
Clark, S. M., 453
Clarke, J., 423
Clarkson, M. J., 123
Claveau, D., 360
Claycamp, H. G., 163, 164
Cleland, L. G., 199, 200(1)
Clement, M. V., 151, 152, 152(5), 159(5; 11)
Clements, J. L., 249
Clerk, A., 54
Clowes, A. W., 448
Clowes, M., 448
Clyman, R. I., 5, 16, 16(27), 28(27; 70)
Cobb, L. M., 14, 15(57), 27(57)
Cobb, M. H., 58
Cobbs, C. S., 494
Cody, S. H., 319
Coffin, D., 495
Cohen, G., 81
Cohen, J. J., 160
Cohen, P., 482
Cohen, R. A., 452
Coia, L. R., 4
Colacicchi, S., 333
Colavitti, R., 91, 92, 100
Colby, H. D., 80(11), 81
Colby, K. A., 392
Cole, G. M., 455, 463
Cole, P. A., 192
Colell, A., 318
Coleman, C. N., 3(9), 4, 24, 24(9)
Collen, D., 434, 448
Collier, J., 507
Collins, J. L., 92
Collins, M. K. L., 153
Colvin, S. B., 444
Comas, J., 74
Comfurius, E. M., 160

Comporti, M., 373
Comstock, D. A., 273
Condorelli, M., 224, 225(37)
Condron, R., 101, 102(7)
Contreras, M., 363, 364, 364(23), 365(23; 24)
Cook, H. T., 509
Cook, J. A., 350
Cook, J. L., 217
Cooper, J. A., 54, 192
Cooper, J. M., 474
Cooper, P. H., 92
Cooper, T. G., 361, 364(3)
Cooperstein, S. J., 365
Copeland, N. G., 464
Corbett, A. H., 123
Cordes, A. W., 102(8), 103
Cordis, G. A., 80, 81
Coremans, J. M. C. C., 483
Cornhill, J. F., 434, 435(16)
Corongiu, F. P., 164
Costa, K., 430
Costa, L. E., 296
Cotgreave, I. A., 442
Cotman, C. W., 455
Cotran, R. S., 223
Cotton, P. C., 191
Coughlin, S. R., 223
Courtneidge, S. A., 422
Covey, J. M., 24
Cox, M. M., 168
Cox, W., 5, 16(27), 28(27)
Coyle, J. T., 72, 183, 188(2), 190(2), 191
Crane, F. L., 268
Crapo, J. D., 38
Cremer, C., 364, 366(25)
Creminon, C., 449
Crews, B. C., 349
Croes, K., 369
Cronshagen, U., 132
Crookall, J. O., 7
Cross, A. R., 453
Crow, J. P., 62, 63(5), 69(5), 70(5), 71(5), 99, 309, 317
Crowley, S. T., 445
Cruickshank, G., 4
Cruz, T. F., 54, 92
Cuccovillo, F., 262, 264, 265, 265(15), 266
Cui, M. Z., 224, 225(38; 45; 46), 226
Cui, X. L., 408, 419(7), 421(7)
Culcasi, M., 440
Culmsee, C., 455
Culwell, A. R., 461
Cummings, J. L., 455
Cunningham, L., 269, 272(7)
Cuomo, C., 261, 262(13)
Currie, A. R., 151
Curry, H. A., 208
Curtiss, L. K., 346
Czubryt, M. P., 123, 124, 130(29), 131

D

Dabestani, R., 44
Dachs, C., 323
Dafni, H., 495
Daleke, D., 160
da Luz, P. L., 434, 435, 435(10, 15), 436(10; 11), 437(11), 438, 440, 441, 442(10; 11), 443(10), 444, 445(11), 446(10; 11), 447(10; 11), 448, 449, 450(36), 452, 452(10; 11), 454(11; 57)
Dam, M., 496
D'Amico, D. J., 392
D'Amore, P. A., 392
Dancis, A., 453
Darley-Usmar, V. M., 229
Darmanyan, A. P., 41
D'Armiento, F. P., 346
Das, D. K., 80, 81, 392
Dasgupta, T. P., 347, 349(3)
Dastvan, F. F., 444
Dauar, A. M., 363
Daub, M. E., 41, 42, 47, 47(10), 48, 49, 49(10), 50(16; 17), 51(17; 18)
Daum, G., 448
Davenport, F., 464
David, M., 394
Davidson, M., 318
Davidson, W. S., 167
Davie, E. W., 221
Davies, K. J., 62
Davis, D., 461, 462(26)
Davis, J. B., 191
Davis, R. A., 53
Davis, R. J., 54, 58
Dawson, T. M., 503
Dawson, V., 503, 520
Day, B. W., 163, 164
Day, H. E. W., 301

Dean, R. T., 80
De Beer, L. J., 363
Decaudin, D., 310, 312(11), 313(11)
Dechalet, L. R., 315
Dechatelet, L. R., 35
Decker, K., 508
de Duve, C., 361, 364(2), 365, 372
DeFranco, D. B., 183
de Groot, H., 508
Dehghani, F., 63, 439, 453
de Hiffman, E., 369
Deighton, F., 365
De Keulenaer, G., 445
Dekkers, D. W. C., 160
de Kleijn, D., 449
de Laat, S. W., 97(17), 98
Delafontaine, P., 341
De La Paz, M. A., 378, 389(10)
Deleonardis, J., 442
Delgado, A. G., 341
Dell'Era, P., 421
del Mont Llosas, M., 114
Delphin, C., 130
DelValle, M. R., 509
Demaurex, N., 453
Demchenko, I. T., 347, 350(2)
Demer, L. L., 208
Dendorger, U., 208
Deneke, S. M., 229
Deng, G., 455, 495
Deng, H.-Q., 495
Deng, Y., 223
Denicola, A., 325
Denning, S. M., 221
Dennis, E. A., 160
Dennis, M. F., 15
de Prost, D., 225(39), 226
de Smet, B. J., 449
de Souza, H. P., 432
DeSouza, R., 189
Desterro, J. M., 123, 124(16)
Deutsch, R., 393
de Vries-Smits, A. M. M., 97(17), 98
Dewhirst, M. W., 4, 5(14), 14
Dhaunsi, G., 363, 364(23), 365(23), 369
Dianzani, M. U., 428
Dias, R. M. B., 259
Diaz, V., 114
Dickens, M., 58
Dickinson, C. D., 222
DiCorleto, P. E., 224, 225(38; 45), 226
Dietzmann, K., 141, 143(14)
Dijkman, J. H., 230
Dikic, I., 192
Diliberto, E. M., Jr., 112(4), 113
DiMarco, A., 310, 313(12)
DiMauro, S., 458
Dimmeler, S., 161
Dinges, H. P., 342, 346(10)
Dische, S., 24
Do, K. Q., 347, 350(5)
Doak, G. O., 103(14), 111
Dobrowolski, D. C., 47
Dodia, C., 430
Doel, J. J., 507
Doenecke, D., 123
Doi, H., 225(44), 226
Dolbier, W. R. J., 29, 30(86)
Dominguez, D., 114
Donaldson, D., 495
Donnini, S., 408, 411(5), 413(5), 415(5), 416(5), 421(5)
Donoghue, N., 101, 102(5), 103(5), 112(5)
Dóra, E., 487
Douglas, J. G., 280, 408, 415(6), 419(6; 7), 421(6; 7)
Dousset, N., 67
Doyer, A., 365
Drake, T. A., 222
Draper, H. H., 80
Drazen, J. M., 348
Drew, K. N., 268
Drexler, H., 453
Dreyfuss, G., 123, 124(7)
Dröge, W., 53, 54(4), 233, 236, 236(9), 237, 237(14), 246(19; 20)
Drummen, G. P., 162(30), 163, 164(30)
Dryfe, R. A. W., 116
Duan, W., 474
Dubbin, P. N., 319
Dubey, R. K., 163
Duda, J. E., 474
Duddy, S. K., 120
Dudman, N. P. B., 101, 102(6)
Duerr, R. L., 394
Duff, K., 465
Dumoulin, M. J., 134
Duncan, B. W., 392
Duncan, R., 167
Dunn, J. A., 341

Dunphy, E. P., 4
Dupouy, D., 224, 226(36)
Durante, W., 449
Durham, S. K., 509
Duriez, P. J., 152
Durner, J., 323
Dush, P., 112(6), 113
Duval, C., 62
Duval, D., 191
Duvick, J. P., 453
Dzeja, P. P., 124

E

Easton, C., 369
Eaton, J. W., 80
Ebert, B. L., 31
Echelmeyer, M., 315(27), 316
Eck, H.-P., 233, 236(9), 237, 246(20)
Eckman, C., 463, 464, 465
Eder, H. A., 67
Edgell, C. J. S., 201
Edgington, H. D. J., 280
Edgington, T. S., 222, 223
Edwards, P. A., 208
Egmond, M. R., 389
Eguchi, Y., 151
Ehmann, W. D., 455
Ehrenshaft, M., 42, 47(10), 49, 49(10), 50(17), 51(17; 18)
Eichhorn, K., 136
Eisenberg, P. R., 223
Eisenthal, R., 507, 509(3)
Eldred, G. E., 378
Elger, C. E., 135, 139, 141, 141(13)
Eliceiri, B. P., 192
Ellis, W. W., 261
Elner, V., 212
Elsayed, N. M., 163
Emaus, R. K., 66
Engel, R., 92
Engelhardt, E. L., 29
Engelman, D. T., 392
Engelman, R. M., 392
Engle, J. L., 483
Enomoto, T., 123
Epstein, C. J., 503
Erickson, P. H., 380, 390(22)
Erlich, J. H., 223
Ernst, A., 508
Esch, F. S., 461
Escubedo, E., 72, 73, 74
Escudero, R., 393
Esmon, C. T., 222
Esposito, F., 258, 261, 262, 262(13), 264, 265, 265(15; 16), 266
Esposito, N., 224, 225(37)
Esser, P., 379, 389(21)
Esterbauer, H., 205, 206(20), 342, 346(10), 373
Estrabrook, R. W., 135, 141(1)
Eu, J. P., 347, 350(2)
Evanoff, H. L., 212
Evans, A. J., 449
Evans, P. H., 495
Evans, S. M., 5, 16, 16(27), 17(81), 20(65), 21, 21(65), 24, 24(68), 26, 27, 27(65; 68), 28, 28(27; 71), 29, 30(86)
Evistigneeva, R. P., 160
Ewing, A. G., 112(5), 113
Ewing, J. F., 350
Eyssen, J. F., 369

F

Fabisiak, J. P., 159, 160, 160(9), 161, 163, 163(8), 163(8; 10; 11; 25; 27), 170, 170(8; 11; 25), 171, 171(39), 172, 172(8), 358, 359, 359(42)
Fackler, J., 348
Fadeel, B., 160
Fadok, V., 160
Fahimi, H. D., 364, 366(25)
Fais, S., 237, 247(24)
Falk, M. H., 92
Fallon, J. T., 221, 222, 223, 434
Fallon, K. B., 345, 346(12)
Fan, F.-R. F., 113, 114(10; 11), 115, 115(10), 116
Fan, G., 150
Fanburg, B. L., 150
Fang, K., 348
Fang, N., 249
Fanger, C. M., 457
Fann, Y. C., 307, 316, 316(3), 320(3), 321(3; 28)
Fasehun, O. A., 451
Fasehur, O. A., 230

Fasella, P. M., 389
Faulkner, K., 437
Fausa, O., 237, 247(23)
Faustino, R. S., 123
Fayngersh, R. P., 423
Faytamns, D., 365
Feelish, M., 351
Feid, R., 363
Feistner, H., 139, 141, 141(13), 143(14)
Feliciello, A., 224, 225(37)
Fenton, J. W., 223
Ferlito, S., 351
Fermont, D., 24
Fernandez-Checa, J. C., 318
Ferrans, V. J., 92
Ferrara, N., 392
Ferriero, D. M., 16
Ferrige, A. G., 507
Fidler, I. J., 313
Field, M., 248
Fielden, E. M., 14
Figlewicz, D. Z., 495
Finco, T. S., 211, 249, 257(8)
Fini-Storchi, I., 409, 410(10)
Finkel, T., 91, 92, 97(2), 373, 442
Finkelstein, E., 495
Fiorillo, J., 423
Fiscella, M., 262, 263, 264, 265(16)
Fischbach, T., 237, 246(19)
Fisher, A. B., 309, 310, 311(7), 312(7), 313(7), 314(7), 315, 430
Fisher, S. K., 380, 390(22)
Fishman, M. C., 409
Fisslthaler, B., 439
Fitzgerald, L. A., 225(42), 226
Fitzgerald, M., 101, 102(4), 111(4)
Fitzsimmons, S. A., 4
Flaherty, D. M., 64(10), 66, 307, 309, 316(2), 317(2), 318(2), 319(2), 325(2), 448
Flavell, R. A., 53
Fleisher, T. A., 64(11), 66
Fleming, I., 439
Fletscher, B., 378, 389(11)
Floyd, R. A., 455, 467
Flynn, E., 101, 102(7)
Fogelman, A. M., 208
Foissner, I., 323
Folch, J., 164, 173(32)
Folkman, J., 282, 392, 407, 410
Folwer, S., 365
Fong, D., 345
Foot, C. S., 41
Foote, C. S., 47
Ford, G. D., 310
Forough, R., 448
Fox, M., 473
Fraga, C. H., 392
Fraker, D., 17(81), 26
Fralix, T. A., 310, 313(6)
Franchek, K. M., 315, 318(24)
Franchi, A., 409, 410(10)
Frandsen, U., 301
Frank, B., 141, 143(14)
Frank, J. S., 208
Franke, H., 135
Franklin, M., 444
Franko, A. J., 3, 7, 13(1), 14, 15
Frasch, S. C., 160
Freed, J. J., 483
Freedman, H. I., 7
Freedman, L. D., 103(14), 111
Freeman, B. A., 38, 437, 453
Fréjaville, C., 440, 441
Freyer, J. P., 13(45), 14
Friberg, L., 496
Fridovich, I., 153, 309, 311, 312(15), 314(15), 437
Fridovich, J., 66
Friedman, P., 160
Friedrich, V., 223
Fritsch, C., 54
Fritsch, E. F., 355, 357(37)
Froissard, P., 191
Fry, A. M., 124
Fry, G. L., 422, 424
Fu, M. X., 341
Fu, W., 457, 458, 469(22)
Fuchs, J., 333, 334, 336
Fuchs, S. Y., 54
Fuciarelli, A. F., 15
Fujihara, S. M., 124
Fujii, H., 509
Fujii, S., 301, 323, 350, 509
Fujiki, Y., 365
Fujinaga, M., 5
Fujita, D. J., 192
Fukai, N., 424
Fukai, T., 442
Fukikawa, K., 221
Fukuda, S., 391

Fukudome, K., 222
Fukushi, K., 496
Fukushima, T., 292
Fukuzawa, K., 516, 517(37b)
Fuller, S. J., 54
Furukawa, K., 71, 457, 458, 461, 461(13), 469(22), 473, 473(13)
Furusawa, M., 334
Fyfe, B. S., 221, 222, 223
Fyles, A. W., 4

G

Gabriel, C., 72, 73, 74
Gacy, A. M., 124
Gal, D., 41
Galeotti, T., 91, 92, 100
Galis, Z. S., 423
Galkina, S. I., 100
Gallant, P., 124
Gallina, M., 351
Gallo, O., 409, 410(10)
Gallo, V., 73
Gallop, P. M., 452
Galloway, A. C., 444
Galter, D., 53, 237, 246(19)
Galushchenko, I. V., 390
Gamache, P. H., 327
Gandley, R., 163
Gandy, S. E., 464
Gao, M., 394
Garcia-Cardena, G., 280
Garcia de Herreros, A., 114
Garcia-Ruiz, C., 318
Garner, A., 80
Garrecht, B. M., 14, 15
Garrido, R., 212
Gash, D. M., 474
Gassmann, M., 32
Gaston, B., 347, 348, 348(6), 350(6), 351(6), 353(6)
Gaston, S. M., 495
Gatenby, R. A., 4
Gearhart, J. D., 32
Geddes, J. W., 458, 474
Gee, M. H., 100
Geenen, D. L., 16, 28(71)
Geertsma, M. F., 248, 257(4)
Geiger, J. D., 457
Gellerich, F. N., 136, 138, 481
George, V. G., 201
Geppetti, P., 280, 408, 411(4), 412(4), 413(4), 415(4), 416(4), 419(3; 4), 420(3; 4)
Gerace, L., 124, 130, 130(18)
Gerard, R., 448
Germeyer, A., 458, 461(23), 466(23), 468(23), 469(23), 472(23)
Gernert, K., 347, 350(2)
Gewertz, B. L., 319
Ghafourifar, P., 296
Gheith, S. M., 249
Ghoda, L., 123
Ghosh, S. K., 220
Ghoshal, A. K., 80
Giaccia, A. J., 5
Giasson, G. I., 474
Gibbons, G. H., 444, 445
Gibson, R. A., 199, 200(1)
Giercksky, K.-E., 351, 357(34)
Giesen, P. L., 222, 223
Gilbert, H. F., 101
Gill, V., 287
Gille, H., 58
Gillespio, D., 163
Gillotte, K. L., 160
Gim, R. D., 394
Gimbrone, M. A., 223
Gimple, L. W., 223
Gingras, M. C., 191
Giraldez, R. R., 519
Girard, P. M., 310, 312(11), 313(11)
Gitler, C., 102(10; 11), 103, 233
Giuffre, A., 360
Giulivi, C., 296
Gius, D., 208
Giusto, N. M., 378
Gladwin, M. T., 347, 348(8), 350(8), 351(8)
Glazer, P. M., 5
Gleiss, B., 160
Glockner, J. F., 495
Glomset, J., 432
Gmünder, H., 233, 236, 236(9), 237, 237(14), 246(19; 20)
Godber, L. J. B., 507
Godber, S. S., 333
Goldfiscner, S., 361
Goldman, R., 229
Goldman, R. K., 357, 358(39)
Goldschmidt-Clermont, P. J., 453

Goldstein, S., 495
Goldwasser, E., 32
Golino, P., 224, 225(37)
Goller, N. L., 509
Golschmidt-Clermond, P. J., 442
Gomi, F., 495, 502(14)
Gong, K. W., 449
Gonzalez, J. A., 230, 451
Goodman, A., 509
Goodman, Y., 204, 457, 458, 469(22)
Goormastic, M., 434, 435(16)
Gopalakrishna, R., 114, 350
Gorbunov, N. V., 163
Gordge, M. P., 360
Gordillo, G. M., 422
Gordon, D., 221
Gordon, M. N., 465
Görlach, A., 63, 220, 224, 225(47), 226, 227(47), 230(47), 453
Görlich, D., 123, 124, 130
Gorman, A. A., 41
Gorn, R. A., 378
Goslin, K., 185
Goto, J. J., 474, 495
Gotoh, Y., 54
Gow, A. J., 62, 63(5), 69(5), 70(5), 71(5), 349
Gown, A. M., 342, 345
Graeber, T. G., 5
Graeff, A. S., 232
Graf, E., 80
Graham, D. J., 373
Graham, J. B., 201
Gralla, E. B., 474
Granger, H. J., 280, 408, 411(4; 5), 412(4), 413(4; 5), 414, 415(4–6), 416(4; 5), 419(3; 4; 7), 420(3; 4), 421, 421(5–8)
Grant, C. M., 101, 102(4), 111(4)
Grasis, J. A., 249
Grasshoff, H., 136
Gratzl, M., 112(8; 9), 113
Gray, L., 348, 351(12)
Green, S. R., 160
Greenall, M., 423
Greene, E. L., 63
Greene, W. C., 123
Greten, J., 223
Griendling, K. K., 221, 443, 445, 447, 448, 452(52), 453, 453(52)
Griffin, B. A., 102(12), 103
Griffioen, A. W., 175
Griffith, O. W., 57, 236, 246
Grigorev, I. A., 508
Gringhuis, S. I., 248, 253, 254, 255, 255(6), 256, 256(6), 257, 257(6)
Grippo, J. F., 259
Grisham, M. B., 348, 350, 351(12)
Großhans, H., 123
Gross, C. W., 348
Gross, G. J., 394
Gross, S. S., 230, 451
Gross, V., 247
Grossi, E. A., 444
Grotendorst, G. R., 176
Groth, N., 333, 334, 336, 339
Gu, J., 32, 449
Gu, L., 423
Guan, T., 130
Guan, X. M., 465
Guerin, C. J., 380, 390(22)
Guha, A., 221, 223
Gui, L., 392
Gullino, P. M., 410
Gulyas, S., 14
Gumina, R. J., 394
Gunasekar, P. G., 317, 323(32)
Gundimeda, U., 350
Guo, Q., 457, 458, 461(13), 473(13)
Guo, Z. H., 455
Gupta, K., 392
Gupta, P., 392
Gupta, S., 58
Gusella, J. S., 495
Gutteck-Amsler, U., 347, 350(5)
Gutteridge, J. M. C., 81, 153, 258
Guzman, L. A., 434, 435(16)
Gyulai, L., 487

H

Haack, K. E., 315, 318(24)
Haase, M., 223
Haberland, M., 345
Hack, C. E., 222
Hacker, T., 14, 15(57), 27(57)
Haendeler, J., 161, 449
Hagaman, K. A., 167
Hagiwara, M., 374
Hahn, K. A., 393

Hahn, S. M., 17(81), 21, 24, 26, 27, 495
Haida, M., 478
Haines, J., 495
Hakim, J., 225(39), 226
Hall, A., 97(17), 98
Hall, D., 189
Hall, J. L., 444, 445
Hall, R. D., 41
Hallaway, P. E., 80
Halliwell, B., 80, 153, 258, 259
Halperin, J. J., 495
Halpern, H. J., 516, 517(37c)
Halstensen, T. S., 237, 247(23)
Hamada, A., 333, 495, 501, 501(13), 502(14), 503(20)
Hamilton, T. A., 225(45), 226
Hammer, R., 3(6), 4
Hampton, M. B., 151, 161
Han, D., 191
Han, J., 58
Han, J. I., 318
Han, J.-Y., 495
Hanahan, D., 422
Handa, J., 475
Hanemaaijer, R., 449
hang, Z. G., 507
Hankenson, R. R., 422
Hanks, G. E., 29
Hanover, J. A., 124, 130
Hansen, H., 3(6), 4
Hanson, S. R., 221
Hanzel, D. K., 296
Harada, M., 449
Harbig, K., 487
Hardy, J., 455, 457(3), 465
Harigaya, Y., 463, 465
Harker, L. A., 221
Harrelson, J. M., 4, 5(14)
Harris, A. L., 423
Harris, J. R., 24
Harris, M., 467
Harris, P. L. R., 455, 495
Harrison, D. G., 423, 437, 440, 442, 452(35), 453
Harrison, R., 507, 509(3)
Hart, P. E., 474
Hartig, R., 364, 366(25)
Hartz, W. H., 4
Harwell, L. W., 16, 18(67)
Haselgrove, J. C., 477
Hashimoto, M., 474, 495
Hashimoto, T., 361, 363(4), 364(4)
Hashmi, M., 363
Haskard, D. O., 213
Hassanaim, H. H., 442
Hassanain, H. H., 453
Hassell, J. A., 422
Hassinen, I., 135, 138
Hatakeyama, K., 280, 281(9), 295(9)
Hatakeyama, Y., 71
Hattori, N., 373
Haudenschild, C. C., 282
Haughey, N., 472
Haughland, R. P., 67
Haugland, R. P., 296
Haung, S., 369
Hausladen, A., 349
Hausmann, M., 364, 366(25)
Haustermans, C., 448
Havel, R. I., 67
Hawker, J. R., 414
Hay, R. T., 123, 124(16)
Hayakawa, T., 449
Hayashi, A., 71
Hayashi, K., 446
Haynes, T. E., 295
Hazen, S. L., 349
Hazlehurst, J. L., 14
He, G., 339
He, T., 443
Hebbel, R. P., 392
Heck, D. E., 509
Heck, R., 92
Hedley, D. W., 156
Hehner, S. P., 53, 54(4)
Heimann, K., 379, 389(18–21)
Heineman, F. W., 310, 313(6)
Heinze, H.-J., 141
Heist, E. K., 123
Hellsten, Y., 301
Hemker, H. C., 228
Hempel, S. L., 64(10), 66, 307, 309, 316(2), 317(2), 318(2), 319(2), 325(2)
Henderson, C. J., 54
Hengartner, M. O., 150
Henk, J. M., 3
Henn, V., 225(43), 226
Hennig, B., 198, 199, 200, 201, 203, 204, 205, 206, 207, 209(18), 212
Hensley, K., 467

Henson, P. M., 160
Hentati, A., 495
Henter, J. I., 160
Herbst, U., 206
Herfarth, C., 232, 239, 247(25)
Herkert, O., 220
Herrling, T., 333, 334, 336, 339
Herzenberg, L. A., 233
Herzfeld, B., 495
Hess, D. T., 349
Hiai, H., 374
Hibi, M., 53, 58(8)
Higuchi, M., 323
Higuchi, T., 322, 323(47)
Hikichi, T., 41
Hilderman, R. H., 324
Hilf, R., 207
Hilfiker, A., 453
Hilfiker-Kleiner, D., 453
Hill, J. M., 341
Hill, R. P., 4, 5
Hill, S. A., 15
Hiltermann, T. J., 230
Himmelreich, U., 268
Hines, M. O. III, 341
Hink, U., 437, 439
Hirano, M., 458
Hirata, K., 323
Hirata, Y., 296, 297(9), 303(9), 309, 321, 322, 323(44–47), 324, 351
Hirpara, J., 151, 152, 159(11)
Hirst, D. G., 14
Hissin, P. J., 207
Hjelm-Hansen, M., 3(6), 4
Ho, Y. S., 461, 465
Hoak, J. C., 422, 424
Hochleitner, B. W., 444
Hockel, M., 4
Hockenberry, D. M., 161
Hodd, J., 408, 419(7), 421(7)
Hodgkiss, R. J., 15
Hofer, E., 222
Hofer-Warbinek, R., 222
Hofstrand, P. D., 24
Hogg, N., 347, 348, 437, 441
Hogg, P. J., 101, 102(4–7), 103(5), 111(4), 112(5)
Hogstrand, K., 160
Hokanson, J. E., 346
Holcomb, L., 465
Holden, C. P., 457
Holland, S. M., 442
Holm, S., 496
Holmgren, A., 101, 259, 260
Holt, S., 351
Holtsberg, F. W., 458, 461(23), 466(23), 468(23), 469(23), 472(23), 474
Holzmuller, H., 222
Hood, J. D., 280
Hope, T. J., 123
Hopper, A. K., 123
Horaist, C., 448, 453
Horkko, S., 160, 346
Horner, J., 16, 28(71)
Horritz, H. R., 495
Horwitz, L. D., 445
Hoshi, H., 310, 322, 323(13), 324(13)
Hoshi, M., 367
Hotchkiss, K. A., 101, 102(6)
Hothersall, J., 360
Houchi, H., 515, 516, 517(37b), 520(37)
Housman, D. E., 5
Houston, D. S., 225(40), 226
Howell, B. W., 192
Howes, A. E., 24
Howorko, J., 14
Hoyt, R. F., 442
Hsi, R. A., 17(81), 26
Hsiao, K., 463
Hsich, E., 442
Hsu, H. C., 449
Hsu, L. J., 474
Hu, K., 14
Hu, Y., 444
Hu, Z. Y., 221, 448, 453
Huang, L., 112(6), 113
Huang, L. E., 32
Huang, P. L., 409
Huang, Q., 14
Huang, S., 394
Hubbard, A. L., 365
Hubel, C. A., 347
Huber, J., 132
Hübner, S., 123
Huettner, J. E., 186
Hughes, E. J., 423
Hughes, L., 508
Hung, W.-Y., 495
Hunt, J. V., 205
Hunt, N. H., 271

Huppa, J. B., 101
Hurt, E., 123
Hurtig, H. I., 474
Hussain, A., 350
Hustinx, R., 29, 30(86)
Hutchins, J. B., 458, 461(23), 466(23), 468(23), 469(23), 472(23)
Hutton, M., 465
Hwang, S. M., 509
Hynes, K. L., 319
Hyslop, P. A., 64(8), 66

I

Iaccarino, G., 224, 225(37)
Iannone, A., 495
Ichijo, H., 54, 58
Ichikawa, K., 494
Ide, T., 333, 495
Ignarro, L. J., 280, 507
Iheanacho, E. N., 271
Ikeya, M., 334
Ilani, A., 268(6), 269
Imai, Y., 296, 322, 323(45)
Imamoto, N., 123
Imanaka, T., 346
Imrach, A., 324
Ince, C., 483
Inestrosa, N. C., 365
Ingman, P., 135
Ingold, K. U., 508
Inoue, K., 348, 350
Inoue, M., 357, 358(38), 495
Inoue, N., 442
Irani, K., 62, 92, 432, 433(4)
Irazu, C., 363
Irie, T., 496
Irimura, T., 296, 322, 323(45)
Irvin, C., 435
Ischiropoulos, H., 62, 63(4; 5), 64(4), 66(4), 67(4), 69(4; 5), 70(5), 71(5), 92, 316, 317, 318, 318(26), 319(26), 320(26), 468, 474
Ishida, Y., 324
Ishihara, Y., 100
Ishii, T., 233, 236, 236(8), 237(8)
Ishizaka, N., 445
Islek, I., 351
Isner, J. M., 391, 409, 435
Isom, G. E., 317, 323(32)
Israel, M. A., 494
Itabe, H., 341, 346
Itakura, K., 374
Itin, I., 392
Ito, T., 449
Ito, Y., 452
Itoh, O., 500
Itoh, Y., 310, 322, 323(13), 324(13)
Itshak, F., 479
Iwasaki, H., 346
Iyer, N. V., 32

J

Jackman, J., 262, 263, 264, 265(16)
Jacks, T., 5
Jackson, E. K., 163
Jackson-Lewis, V., 503
Jacobson, J. A., 449
Jacobson, M. D., 93, 151
Jaconi, M., 124
Jaconi, S., 453
Jacques, D., 69
Jacques, P., 365
Jadeski, L. C., 409
Jaeger, R. G., 435
Jaffa, A. A., 63
Jaffe, E. A., 282
Jakel, S., 123
Jakobs, C., 369
James, M. J., 199, 200(1)
James, S. P., 232
Jamieson, S. W., 393
Janero, D. R., 350
Janiszewski, M., 432, 434, 435, 435(10; 15), 436(10), 438, 442(10), 443, 443(10), 446(10), 447(10), 448, 449, 452, 452(10), 454(57; 60)
Jankowski, J. A., 112(4), 113
Jans, D. A., 123
Jansen, G. A., 362, 364(9)
Janssens, S., 448
Jaraki, O., 347, 350(4), 351(4), 357(4), 358(4)
Jason Niu, Q., 41
Javouhey, A., 394
Jeffries, B. A., 163
Jenkins, A. J., 341
Jenkins, N. A., 464

Jenkins, W. T., 16, 17(81), 21, 24(68), 26, 27(68), 29, 30(86)
Jenns, A. E., 49
Jeter, D., 102(8), 103
Ji, S., 483, 487(5)
Jia, L., 347, 350, 350(2)
Jiang, J. F., 359
Jiang, J. J., 516, 517(37b)
Jiang, P., 465
Jiang, Q., 192
Jiang, S., 192
Jiang, X.-M., 101, 102(4; 5; 7), 103(5), 111(4), 112(5)
Jiang, Z. Y., 205
Jin, Z.-G., 449
Jo, H., 229
Jöbsis, F. F., 482, 487
Joffe, B., 342
Johnson, D., 354
Johnson, D. E., 160, 163(8), 170(8), 172(8)
Johnson, D. W., 369
Johnson, J. P., 208
Johnson, J. W., 183
Johnson, L. V., 66
Johnson, R. J., 3(8), 4
Johnson-Saliba, M., 123
Joiner, B., 16
Jones, D. P., 161
Jones, G., 15
Jones, J., 410
Jones, K. A., 186
Jones, L. J., 296
Jones, M. M., 288, 508, 510
Jones, N., 53
Jones, O. T., 423, 451
Jones, S. A., 423
Jones, S. G., 510
Jordan, J., 309, 310, 311(9), 312(9), 313(9), 314(9)
Jorgensen, K., 3(6), 4
Jorgensen, O. S., 73
Joris, I., 151
Joseph, J., 347
Joshi-Barve, S., 205
Junqueira, M. L., 434, 436(11), 437(11), 441, 442(11), 444, 445(11), 446(11), 447(11), 448, 452(11), 454(11)
Jurd'heuil, D., 348, 351(12)
Jürgens, G., 342, 346(10)

K

Kachur, A. V., 29, 30(86)
Kadokura, M., 132
Kagan, V. E., 159, 160, 160(9), 161(8), 161(10; 11), 163, 163(8), 163(8–11; 25–28), 164, 170, 170(8; 11), 170(11; 25), 171, 171(39), 172, 172(8), 172(28), 173(8), 229, 347, 358, 359, 359(42), 378, 390, 392
Kaiser, S., 198, 206, 212, 217
Kajino, H., 16, 28(70)
Kajiyama, G., 446
Kalangos, A., 341
Kalbas, M., 225(43), 226
Kalef, E., 102(10; 11), 103
Kalinich, J. F., 321
Kalka, C., 409
Kallinowski, F., 4
Kalyanaraman, B., 347, 437, 441
Kamada, H., 500, 509
Kamiike, W., 151
Kaminski, P. M., 230, 423
Kamisaki, Y., 519
Kane, L. S., 349
Kang, B. S., 378
Kang, J. S., 318
Kang, S. W., 92
Kanse, S. M., 222, 223(10)
Kanthasamy, A. G., 317, 323(32)
Kao, J. W., 92
Karin, M., 53, 58(8)
Karmiol, S., 393
Karoui, H., 440, 441
Karp, J. S., 29, 30(86)
Kashiwagi, M., 495
Kass, G. E., 120
Kassis, A. I., 81, 86(18), 88(18), 89(18)
Kato, H., 503
Kato, K., 503
Kato, S., 191
Katz, L. A., 135, 141(3)
Katz, M. L., 378
Kavanagh, M. C., 4
Kawagoe, K. T., 112(2; 4), 113
Kawahara, S., 296, 297(9), 303(9), 309, 321, 322, 323(44; 46), 324, 351
Kawai, K., 163, 170, 170(25), 171(39), 358, 359, 359(42)
Kawakami, T., 249
Kawakami, Y., 249

Kawakishi, S., 374
Kayatz, P., 378, 379, 389(18–21)
Keana, J. F. W., 497
Keana, S. B., 497
Keane, R. W., 101
Keane, T. J., 4
Keaney, J. F., Jr., 347, 349, 350(4), 351(4), 357(4), 358(4)
Kearny, M., 409
Kedes, L., 394
Kehrer, J., 91
Keizo Takeshita, K., 495
Kelleher, D. K., 14
Keller, J. N., 458, 461(23), 466(23), 468(23), 469(23), 472(23), 474
Kelley, D. A., 15
Kelly, A. B., 221
Kelly, K. A., 280, 281(9), 295(9)
Kelm, M., 347, 348(7)
Kennedy, A. S., 24
Kennedy, K. A., 3
Kennedy, R. T., 112(2; 4; 6), 113
Kenny, D., 394
Kentor, R., 183
Kerr, J. F., 151
Kerry, Z., 448
Keshet, E., 407
Kessler, H. B., 4
Keston, A. S., 315
Key, B. J., 301, 322, 323(48)
Khachaturian, Z. S., 455
Khajuria, A., 225(40), 226
Khan, A. A., 186
Khan, M., 361, 364(7), 367(7), 369, 371(7; 45)
Khanna, S., 183, 191, 326
Khechai, F., 225(39), 226
Khramtsov, V. V., 453, 508
Kikuchi, K., 296, 297(9), 303(9), 309, 321, 322, 323, 323(44–47), 324, 351
Killingsworth, M. C., 379
Kim, G., 464
Kim, H. M., 318
Kim, H. R., 318
Kim, K. W., 318
Kim, R. Y., 392
Kim, S., 434
Kim, S. Y., 350
Kimura, J., 346
Kindy, M. S., 458, 461(23), 466(23), 468(23), 469(23), 472(23)
King, M. P., 36, 93, 318
Kingsbury, A., 73
Kinnon, C., 153
Kinouchi, H., 503
Kinscherf, S., 237, 246(19)
Kioschis, P., 363
Kirches, E., 139, 141, 141(13), 143(14)
Kirchmaier, C. M., 225(47), 226, 227(47), 230(47)
Kiriljuk, I. A., 508
Kirino, Y., 296, 297(9), 303(9), 309, 321, 322, 323(44; 46), 324, 351, 495
Kirkeboen, K. A., 351, 357(34)
Kishimoto, Y., 367, 519
Kisiel, W., 221
Kisin, E., 392
Kisker, O., 101, 102(7)
Kitamura, E., 190
Kitchen, J., 249
Kitsis, R. N., 16, 28(71)
Klann, E., 183
Klaus, W., 323
Klausner, R. D., 453
Kleinman, H. K., 423
Kleschyov, A. L., 509
Klockgether, T., 141
Kloner, R. A., 394
Klotz, L. O., 54
Kluge, I., 347, 350(5)
Knapp, P. E., 474
Knapstein, P. G., 4
Kneebone, J. M., 392
Knoop, C., 4
Kobayashi, M. S., 191
Kobuchi, H., 428, 429
Kobzik, L., 315, 318(24), 324
Koch, C. J., 3, 5, 6, 7, 7(30), 9(38–40), 10, 11(40), 13(1), 14, 14(38), 15, 16, 16(27), 17(81), 18(67), 20(65), 21, 21(65), 23(58), 24, 24(68), 26, 27, 27(65; 68), 28, 28(27; 70; 71), 29, 30(86)
Kockx, M., 434
Koepp, D. M., 123
Kogure, K., 503
Kohno, M., 508
Kohtz, D. S., 222
Koide, S. I., 225(44), 226
Kojima, H., 296, 297(9), 303(9), 309, 310, 321, 322, 323, 323(13; 14; 45–47), 324, 324(13), 351

Kölch, W., 53
Kolis, J. W., 102(8), 103
Komarov, A. M., 288, 508, 509
Kommineni, C., 163, 170(25), 171
Komyama, T., 394
Kooy, N. W., 62, 63(5), 69(5), 70(5), 71(5), 317, 318, 468
Kopelman, R., 324, 325(59)
Koretsky, A. P., 135, 141(3)
Koretzky, G. A., 249
Kornblum, C., 141
Korsmeyer, S. J., 161
Kortenjann, M., 58
Korth, H.-G., 508
Kosaka, H., 151
Koscielniak, J., 509
Kostka, S., 130
Kotake, Y., 510
Kotch, L. E., 32
Kouzarides, T., 124
Kovách, A. G. B., 487
Kovesdi, I., 280
Koyama, H., 444
Kozlov, Y. P., 378
Kraft, R., 130
Krakover, T., 268(6), 269
Kramer, J. A., 296
Kramer, R. A., 24
Kramer, R. S., 487
Krasinski, K., 434
Krause, K., 453
Krebs, E. G., 59
Kren, B. T., 150
Kress, G., 183
Kretzer, F. L., 378
Krieger, J. E., 434, 436(11), 437(11), 441, 442(11), 444, 445(11), 446(11), 447(11), 448, 452(11), 454(11)
Krigel, R., 270
Krishna, M. C., 350, 495
Krizus, A., 495
Kroczek, R. A., 225(43), 226
Kroemer, G., 310, 312(11), 313(11)
Krohn, A. J., 296, 472
Kroll, C., 333
Kronemann, N., 225(47), 226, 227(47), 230(47)
Kruman, I. I., 455, 462, 469(29), 472, 474
Kruman, Y., 455
Kruuv, J., 7
Kshirsagar, S., 392
Kubrina, L. N., 509, 520, 520(32)
Kuchel, P. W., 268
Kudo, Y., 322, 323(47)
Kuga, Y., 446
Kuge, S., 53
Kugiyama, K., 225(44), 226
Kuhan, Y. T., 378
Kühn, H., 201
Kuhr, W. G., 112(7), 113
Kuhrina, L. N., 509
Kukielczak, B. M., 41, 47(9)
Kukreja, R. C., 394
Kukuruga, M. A., 99
Kuliev, I. Y., 378
Kullmann, F., 248
Kunkel, S. L., 212
Kunsch, C., 221
Kunz, D., 133(7), 135, 136, 139, 141, 141(7; 13), 149
Kunz, W. S., 133(7; 13), 135, 136, 138, 139, 141, 141(7; 13), 149, 481
Kuo, C. H., 191
Kuo, S., 270
Kuppusamy, P., 153, 333, 339, 437, 495, 507, 509, 509(7), 515, 519(26), 521(37a)
Kurinov, I., 474
Kuroki, M., 392
Kuroki, T., 123
Kurosawa, S., 222
Kurz, S., 453
Kutay, U., 123, 130
Kutlu, O., 351
Kuwabara, T., 378
Kuypers, F. A., 81, 89(20), 90, 162(29), 163
Kuznetsov, A. V., 133(7), 136, 141, 141(7), 143(14)
Kwak, J., 115
Kwan, W. C., 232
Kwon, N. S., 230, 451
Kyriakis, J. M., 54

L

Labbe, R. F., 272
Laemmli, U. K., 55, 426
LaFerla, F. M., 457
Lafon-Cazal, M., 457, 467(10)
Lafont, A. M., 434, 435(16)
Laforge, T., 453

Lageweg, W., 363
Lai, C.-S., 288, 508, 509
Lala, P. K., 409
LaManna, J. C., 32
Lambeth, D. J., Jr., 448
Lambeth, J. D., 443, 449
Lander, H. M., 224
Lang, N. G., 495
Langebartels, C., 323
Langer, R., 410
Langlais, P., 327
Langner, A., 333
Lankester, A. J., 248, 257(4)
Lanzen, J., 14
Laskin, D. L., 509
Laskin, J. D., 509
Lassegue, B., 443, 448
Lassen, N. A., 496
Lau, Y. Y., 112(5), 113
Laubriet, A., 394
Laughlin, K. M., 16, 24(68), 27(68)
Laughner, E., 32
Lauricella, R., 440
Laurindo, F. R. M., 432, 434, 435, 435(10; 15), 436(10; 11), 437(11), 438, 440, 441, 442(10; 11), 443, 443(10), 444, 445(11), 446(10; 11), 447(10; 11), 448, 449, 450(36), 452, 452(10; 11), 454, 454(11; 57; 59; 60)
Lavoie, J. N., 191
Law, P. Y., 392
Lawler, A. M., 32
Lawley, T. J., 201
Lawrence, D. A., 236, 237(11), 247(11)
Lawrence, R., 446
Lay, A. J., 101, 102(7)
Layfield, L. J., 4, 5(14)
Lazarow, P. B., 361, 362, 364(2), 365
Lazo, J. S., 160, 160(9), 161, 161(8; 10; 11), 163(8; 10; 11), 163(8–11), 170(8; 11), 172, 172(8), 358
Lazo, O., 363, 364, 364(23), 365(23; 24), 369
Lechner, S., 248
Ledda, F., 280, 408, 411(4), 412(4), 413(4), 415(5; 6), 416(4), 419(3; 4; 6; 7), 420(3; 4), 421, 421(6–8)
Ledinski, G., 342, 346(10)
Lee, C. C., 309, 310, 311(9), 312(9), 313(9), 314(9)
Lee, J., 14, 15(48), 16
Lee, J. H., 183
Lee, M. K., 464, 474
Lee, M. S., 123
Lee, P. C., 280
Lee, S., 301
Lee, S. H., 393
Lee, V. M., 474
Lee, W., 272
Lee, W. T., 472
Lee, Y. T., 449
Lee, Y. W., 198, 201, 204, 209, 209(18), 212
Leek, R. D., 423
Lefkovits, J., 434, 435(12)
Lehninger, A. L., 38, 168
Leisman, G. B., 47
Leissring, M. A., 457
Lelkes, P. I., 393
Le Moigne, F., 440
Lenox-Smith, I., 24
Lenton, K. J., 207
Leone, A., 351, 357(34)
Leonhard, S., 333
Leor, J., 394
Leow, A., 248, 253, 254, 255, 255(6), 256, 256(6), 257, 257(6)
LePecq, J.-B., 307, 314(1)
Less, M., 164, 173(32)
Leszczyszyn, D. J., 112(2; 4), 113
Leszezynska-Piziak, J., 349
Leung, B. P., 248
Leung, S. W., 32
Lev, O., 115
Levarht, E. W. N., 255
Levartovsky, D., 349
Levey, A. I., 464
Levi, D. S., 494
Levi, R., 230, 451
Levin, V. A., 3(8), 4
Levin, W., 4
Levine, M., 270, 272(9)
Levine, R. L., 259, 426
Levinsky, R. J., 153
Levy, J. G., 54
Lewis, A., 4
Lewis, C. E., 423
Lewis, D. V., 483
Lewis, G. P., 380, 390(22)
Li, B., 474
Li, C., 309, 311, 319(16)
Li, C. Y., 14, 360

Li, H., 289, 295
Li, M. Y., 48, 49, 50(16; 17), 51(17; 18)
Li, W., 392
Li, W. G., 443
Li, Y., 153, 183, 188(4), 191, 230, 437, 452
Li, Z. Y., 390
Liao, D. F., 449
Liau, C. S., 449
Libby, P., 432, 433(2), 434(2), 435(2), 444
Liberman, M., 434, 435(15)
Libermann, T. A., 208
Lieberburg, I., 461, 462(26)
Lieberman, M., 81
Liebes, L. F., 270
Liew, F. Y., 248
Ligeti, E., 453
Lim, M., 426
Lin, A., 53, 58(8)
Lin, A. M.-Y., 350
Lin, J., 249, 257(8)
Lin, K. I., 189
Lin, M., 133(8), 136
Lin, P., 14
Lin, X., 123
Lin, Y., 333
Linberg, K. A., 380, 390(22)
Lindgren, A., 268
Lindhout, T., 228
Lins, H., 141, 143(14), 149
Liochev, S. I., 153, 437
Lipscomb, G., 345, 346(12)
Lissi, E., 325
Liu, B., 112, 116, 118(26), 119(26), 122
Liu, C. Y., 310, 312(14), 313(14), 314(14)
Liu, H., 54
Liu, L., 349
Liu, R., 474
Liu, S. K., 249
Liu, S. L., 163
Liu, S. Q., 449
Liu, S. X., 347, 358, 359, 359(42)
Liu, X., 452, 454, 454(60), 512, 515(36), 517(36)
Liu, X.-H., 503
Liu, Y., 188, 191
Lizonova, A., 280
Lo, L.-W., 9(40), 10, 11(40)
Lo, Y. Y. C., 54, 92
Loh, K. W., 152
Loikkanen, J., 72
Lommes, P., 508
Long, A., 15
Lopes, E. A., 435
López-Figueroa, M. O., 296, 299(8), 301, 302, 322, 323(49), 324(49)
Lord, E. M., 5, 16, 16(27), 18(67), 20(65), 21, 21(65), 24(68), 27(65; 68), 28, 28(27; 71), 29, 30(86)
Loscalzo, J., 347, 348, 350(4), 351(4), 357(4), 358(4)
Loukili, N., 114
Lovell, M. A., 455, 457, 469(19), 472(19)
Lovick, T. A., 301, 322, 323(48)
Low, H., 268
Lowe, S. W., 5
Lowenstain, C. J., 92
Lu, B., 423
Lu, C., 472
Lu, H., 112(8), 113
Lubin, B. H., 81, 89(20), 90
Luchtefeld, M., 453
Ludolph, A. C., 139, 141(13)
Luers, G. H., 364, 366(25)
Lugade, A. G., 296
Luhrmann, R., 132
Lukacs, N. W., 212, 423
Luley, C., 149
Luo, H., 458, 469(22)
Lupinetti, F. M., 392
Lusis, A. J., 208
Luther, T., 223
Lyberg, T., 324
Lyles, J. V., 160
Lynen, F., 361
Lyons, T. J., 341

M

Ma, F. H., 310, 322, 323(13), 324(13)
Ma, X., 150
Macho, A., 310, 312(11), 313(11)
Mackman, N., 222, 223, 224(29), 231(29)
Maclouf, J., 449
Macpherson, J. V., 113, 116(14)
Madamanchi, N. R., 448, 453
Madden, E. A., 171
Madri, J. A., 280
Maeda, D., 296, 322, 323(45)
Maeda, H., 324, 348, 350, 508

Maekawa, K., 71
Maggi, C. A., 280, 408, 411(4), 412(4), 413(4), 415(4), 416(4), 421
Mahajan, R., 130
Mahal, A., 474
Maher, P., 183, 188, 188(4), 191
Mahiouz, D., 213
Mahmood, A., 81, 86(18), 88(18), 89(18)
Maillard, L., 434
Mair, S., 342, 346(10)
Majack, R. A., 445
Majeau, G. R., 223
Majno, G., 151
Makrigiorgos, G. M., 80, 81, 86, 86(18), 88(18; 19), 89(18)
Malcolm, S. L., 24
Malcom, G., 346
Malech, H. L., 64(11), 66
Malenkova, I. N., 509, 520
Malenkova, I. V., 348, 520
Maliakal, J. C., 175
Mallory, M., 474
Maltepe, E., 32
Manchul, L., 4
Mancini, F. P., 346
Mandel, J. L., 363
Mangiarini, L., 474
Mangold, H. K., 367
Maniatis, T., 355, 357(37)
Mannaerts, G. P., 362, 363, 364(8), 369
Mannick, J. B., 349
Mansfield, K. D., 28
Mansoor, M. A., 237, 239(21)
Manthorpe, M., 286
Mantle, D., 319
Manz, R., 4
March, K. L., 444
Marchesi, E., 307, 316(3), 320(3), 321(3)
Marchetti, P., 310, 312(11), 313(11)
Marcovina, S., 342
Margolis, L. B., 100
Mari, M., 318
Marinos, R. S., 280, 281, 281(9), 295(9)
Mark, K. A., 327
Mark, R. J., 203, 457, 458, 469(19), 472(19)
Markesbery, W. R., 455, 457, 469(19), 472(19)
Marks, G. S., 350
Marley, R., 351
Marmur, J. D., 221, 223
Marnett, L. J., 349
Marquis, J. C., 189
Marshall, R. S., 5, 10
Marshallsay, C., 132
Martasek, P., 437
Martin, G. M., 457, 461(13), 473, 473(13)
Martínez, J. M., 73
Martinez-Zaguilan, R., 280, 281(9), 295(9)
Masini, E., 280, 408, 409, 410(10), 411(4), 412(4), 413(4), 415(4), 416(4), 419(3; 4), 420(3; 4)
Masliah, E., 474
Mason, R. P., 41, 273, 307, 312(4), 316, 316(3; 4), 320(3), 321(3; 4; 28), 515, 516, 517(37b), 520(37)
Masuda, H., 409
Masuda, S., 495, 501(13)
Mateescu, M. A., 134
Mather, T., 222
Mathieu, C. E., 32
Matson, W. R., 327
Matsuda, H., 151
Matsue, T., 116
Matsugo, S., 428
Matsumoto, K.-I., 494, 496, 497(22; 23), 503(23)
Matsumoto, S., 333
Matsuo, M., 495, 502(14)
Matsushita, S., 389
Matthew, G., 449
Mattison, C. P., 123, 124(11)
Mattson, D., 288, 508
Mattson, M. P., 203, 204, 205, 206, 455, 457, 458, 461, 461(13; 23), 462, 462(26), 466(23), 467, 468(23), 469(19; 22; 23; 29), 472, 472(19; 23), 473, 473(13), 474
Matuda, M., 475
Maturana, A., 453
Maulik, G., 80
Maulik, N., 80, 81, 391, 392
Mauray, F., 5, 16(27), 28(27)
Maurice, M. M., 248, 253(3), 257(4)
Mavromatis, K., 423
Mawatari, K., 191
May, J. M., 268, 271, 275
Mayboroda, O., 133(7), 136, 141(7)
Mayevsky, A., 135, 141(5), 487
Mayne, M., 457
McAndrew, J., 229
McCance, D. R., 341
McClain, C. J., 200, 203, 205

McClain, D. E., 321
McClosky, C., 280
McConnell, H. M., 113
McCormick, F., 97(17), 98
McDonald, C. C., 201
McFall, R. W., 474
McFarland, M., 289
McGarvey, D. J., 41
McGlade, C. J., 249
McInnes, I. B., 248
McKenna, W. G., 17(81), 21, 26
McKenna, W. K., 24
McKenna-Yasek, D., 495
McKhann, G. M., 71
McKirnan, M. D., 394
McLaughlin, M. K., 163
McLean, A. J., 41
McLean, L. R., 167
McLeod, L. L., 161
McMahon, T. J., 347, 350(2)
McManaman, J. L., 259
McNamara, C. A., 223
McPhie, D. L., 189
Meagher, L., 213
Mechtcheriakova, D., 222
Medford, R. M., 208, 221
Meerarani, P., 201
Mehdi, K., 444
Mehlhorn, R. J., 333
Meinertz, T., 437
Meininger, C. J., 280, 281(9), 289, 295, 295(9), 414
Meister, A., 57, 236, 237(12), 260
Melchior, F., 130
Mendenhall, G. D., 41
Mendiratta, S., 271
Mendlowitz, M., 223
Merola, A. J., 310, 312(14), 313(14), 314(14)
Merz, H., 232
Meshinchi, S., 259
Messer, G., 208
Messina, J. P., 236, 237(11), 247(11)
Mesuraca, M., 259
Meuer, S. C., 232, 239, 247(25)
Meyer, D. J., 360
Meyer, F. B., 482, 483, 484(6), 486(6)
Meyer, R. A., 71
Miao, Q. X., 349
Michael, A., 54
Michael, J. R., 220
Michael, W. M., 132
Michaelis, U. R., 439
Michelakis, E. D., 31
Michell, J. B., 495
Michenfelder, J. D., 486
Middlesworth, T., 123
Miflach, A., 508
Mignatti, P., 444
Mihalik, S. J., 369
Mihm, S., 53, 237, 246(19)
Mikoyan, V. D., 509, 520, 520(32)
Milbradt, R., 326, 336
Miles, A. M., 350
Miles, D. W., 409
Miles, P. R., 80(11), 81
Millar, T. M., 507, 509(3)
Millauer, B., 392
Miller, E., 346
Miller, F. J., 443, 448
Miller, G. G., 15
Miller, J. W., 392
Miller, M. P., 189
Miller, M. W., 124, 130, 457
Miller, R. J., 309, 310, 311(9), 312(9), 313(9), 314(9)
Milliman, C. L., 161
Milosevic, M., 4
Minakuchi, K., 516, 517(37b)
Minamiyama, T., 357, 358(38)
Minden, A., 53, 58(8)
Minick, C. R., 282
Miraliakbar, H. R., 394
Miranda, A. F., 318
Mirkin, M. V., 112, 113, 114(10; 11), 115(10), 116, 116(12; 13), 118(26), 119(26), 122
Mirza, N., 21
Mishra, K., 123, 124(14)
Mitchell, J. B., 350, 495
Mitze, M., 4
Miura, Y., 495, 500
Miyake, N., 374
Miyamoto, Y., 123, 348
Miyashita, T., 93
Miyazaki, K., 508
Mizui, T., 503
Mizuno, Y., 373
Model, M. A., 99
Moellering, D., 229
Mogyoros, M., 102(10), 103
Mohanakumar, K. P., 350

Mohazzab, H. K., 423
Mohazzab, K. M., 230
Mohindra, J. K., 14
Mohr, H., 364, 366(25)
Mol, M. D., 434
Moldeus, P., 309, 317, 321(29)
Moldofsky, P. J., 4
Moldovan, L., 453
Molema, G., 175
Moll, T., 222
Momose, K., 324
Moneada, S., 507
Monrocq, H., 191
Montal, M., 72
Monteiro, H. P., 92
Montine, K. S., 373
Montine, T. J., 373
Moody, M. R., 319
Moomaw, C., 58
Moons, L., 434
Moore, D. R., 510
Moore, K., 205, 351
Moore, M. S., 123
Morales, A., 318
Morano, M. I., 296, 299(8), 302, 322, 323(49), 324(49)
Morbidelli, L., 280, 407, 408, 409, 410(10), 411(4; 5), 412(4), 413(4; 5), 415(4–6), 416(4; 5), 419(3; 4; 7), 420(3; 4), 421(5–8)
Mordvintcev, P. I., 508, 509, 520
Morgan, D., 465
Morgan, M., 21
Mori, M., 333
Morra, F., 261, 262, 262(13), 265(15)
Morre, D. J., 268, 452, 454(58)
Morrell, P. E., 367
Morris, S. M., Jr., 287
Morrissey, J. H., 221, 222
Morrone, G., 261, 262(13)
Morrow, J. D., 268, 349
Mortara, M., 282
Moser, A. E., 361
Moser, H. W., 361, 362, 363
Mosmann, T., 287
Mosoni, L., 259
Mosser, J., 363
Mottola-Hartshorn, C., 133(8), 136
Moulder, J. E., 3
Mueller, B. M., 222, 223(12)
Mueller-Klieser, W., 13(45), 14
Mügge, A., 437
Muir, D., 286
Mulder, D. W., 495
Muller, M., 223
Muller-Berghaus, G., 225(43), 226
Muller-Ladner, U., 248
Mulliken, J. B., 422
Mullins, M. E., 348
Mulsch, A., 437, 508, 509
Munzel, T., 437, 439, 453
Murai, A., 439
Murohara, T., 409
Murphy, M. P., 348, 349(13)
Murphy, T. H., 183, 187, 188(2; 3; 9), 190(2), 191
Murphy, W. J., 423
Murrant, C. L., 310, 319(10), 320(10), 323(10)
Murray, I. V., 474
Musgrove, E. A., 156
Muta, M., 423
Muto, E., 495, 501(13)
Muzykantov, V., 430
Mwidau, A., 183

N

Naarala, J., 72
Nachman, R. L., 282
Nachury, M. V., 130
Nada, S., 192
Nadeau, R., 134
Nadler, J., 449
Nadler, S. G., 124
Nagano, T., 296, 297(9), 303(9), 309, 310, 321, 322, 323(13; 44–47), 324, 324(13), 351
Nagata, N., 324
Nagoshi, H., 296, 297(9), 303(9), 309, 321, 322, 323(44–47), 324, 351
Nagy, A., 464
Nagy, R., 464
Nakagawa, H., 192
Nakamoto, K., 519
Nakamura, H., 248, 253(3), 439
Nakamura, S. I., 225(44), 226
Nakata, N., 503
Nakatsu, K., 350
Nakatsubo, N., 296, 297(9), 303(9), 322, 323(45–47), 351
Nakielny, S., 123, 124(7), 132

Nalase, Y., 479
Nantermet, P. V., 211
Napoli, C., 346
Nararajan, R., 449
Narayanan, P. K., 64(9), 66, 69(9), 70(9), 312, 319(17)
Narla, R. K., 474
Naruse, M., 494, 496, 497(23), 503(23)
Nathan, C. F., 230, 451
Nathan, R., 483, 487(5)
Naughton, D., 507
Navab, M., 208
Navas, P., 268
Nawata, H., 495
Nawaz, D., 445
Nawroth, P. P., 222, 223(10)
Nayar, R., 312
Nazzal, D., 62
Near, J. A., 112(4), 113
Negishi, K., 191
Negre-Salvayre, A., 62, 67, 69
Nehls, M., 161
Neild, G. H., 360
Neish, A. S., 204, 209(18)
Nelson, D. L., 168
Nemerson, Y., 221, 222, 223
Nerem, R. M., 442, 445
Nervi, F. O., 365
Nethery, D., 310, 313(12)
Neuhof, S., 136
Neumann, H. W., 136
Neuringer, L. J., 4
Neve, R. L., 183, 189
Newman, P. J., 394
Nguyen, K., 63, 453
Nicholls, D. G., 313, 314(20)
Nichols, D. L., 66
Nichtberger, S., 223
Nielsen, E. O., 509
Nigg, E. A., 124
Niki, E., 172
Nilsen, S., 463
Nims, R. W., 350
Nishino, H., 350
Nishitoh, H., 54
Nixon, J. C., 292
Nobel, C. S. I., 442
Noda, K., 71, 310, 322, 323(13), 324(13)
Noell, W. K., 378
Nolan, J., 14, 15(57), 27(57)
Noll, L., 24
Nomura, K., 446
Nong, Z., 448
Nonner, D., 191
Nordsmark, M., 4
Noronha-Dutra, A., 360
Norris, P., 213
Novikov, K. N., 390
Novotny, D. B., 24
Nozaki, T., 496
Numagami, Y., 507, 509(7)
Numaguchi, Y., 449
Nuovo, G., 442

O

Obara, H., 500
Oberley, L. W., 92
Oberley, T. D., 373, 374, 374(7), 375(3; 4; 7), 377(3; 7), 461, 465
Obeso, J., 422, 424(4)
Obolenskaya, M. Y., 508
O'Brien, K. D., 346
O'Conner, M., 483, 487
O'Connor, P. M., 262, 263, 264, 265(16)
O'Donnell, V. B., 423, 451
O'Donovan, K., 189
Oesch, F., 92
Oeth, P., 222
Oettgen, P., 208
Ogata, N., 225(44), 226
Ogata, T., 333, 500
Ogino, T., 134
Ognibene, F. P., 347, 348(8), 350(8), 351(8)
Oh, B.-H., 394
Ohashi, N., 444
Ohnishi, S. T., 507, 509, 509(7), 520
Ohnishi, T., 507, 509, 509(7), 520
Ohya-Nishiguchi, H., 500
Oi, T., 494, 496, 497(23), 503(23)
Oikawa, K., 509
Ojimba, J., 163
Oka, A., 191
Oka, H., 225(44), 226
Oka, M., 310, 322, 323(13), 324(13)
Okada, M., 192
Okada, S., 134
Okada, Y., 415
Okamoto, T., 248, 348, 350

Okumura, K., 449
Okunieff, P., 4
Okura, Y., 341
Olea, A. F., 41
Olesen, S.-P., 309
Oliveira, C. R., 191
Oliver, C. N., 455
Ollivier, V., 225(39), 226
Olson, E. C., 72
Olson, N. E., 444
Olson, S. J., 373
Oltavi, Z. N., 161
O'Malley, Y. Q., 64(10), 66, 307, 309, 316(2), 317(2), 318(2), 319(2), 325(2)
Onat, T., 448
O'Neill, P., 14
O'Neill, R. D., 112(3), 113
Ono, S., 394
Op den Kamp, J. A. F., 81, 89(20), 90, 159, 160(1), 162(29; 30), 163, 164(30)
Oppenheim, J. J., 423
O'Regan, J. P., 495
O'Reilly, M. S., 424
Organisciak, D. T., 390
Orie, N. N., 315(27), 316
Orrenius, S., 120, 151, 160, 161, 442
Orthner, C. L., 225(42), 226
Osanai, H., 449
Osawa, T., 374
Osborne, J., 347, 350(4), 351(4), 357(4), 358(4)
Oshino, N., 38
Oshino, R., 479
Osmanian, C., 5
Ostbye, K. M., 324
Ota, I. M., 123, 124(11)
Otagiri, M., 348, 350
Ouellet, M., 360
Overgaard, J., 3(6; 7), 4
Overgaard, M., 4
Owens, G. K., 223
Oyama, Y., 71
Ozand, P., 369
Ozawa, T., 500
Ozer, N. K., 226

P

Pacelli, R., 350
Pacifici, E. H., 161
Paciucci, R., 114
Pack, D., 333
Packer, L. J., 183, 191, 229, 326, 329(5), 331(5), 333, 336, 423, 428, 429, 432, 433(5)
Pagano, P. J., 442, 452, 453
Pahan, K., 369, 372
Paigen, B., 442
Paillous, N., 67
Palinski, W., 340, 342, 345, 345(9), 346
Palitzsch, K. D., 247
Pallàs, M., 71, 72, 73
Pallone, F., 237, 247(24)
Palmer, R. M. J., 507
Palubo, G., 346
Panayi, G. S., 248
Panda, A., 519
Pang, Z., 458
Pani, G., 91, 92, 100
Pannel, R., 354
Pantano, P., 112(7), 113
Panter, S. S., 80
Panza, J. A., 347, 348(8), 350(8), 351(8)
Paoletti, C., 307, 314(1)
Papapetropoulos, A., 280
Papassotiropoulos, A., 141
Papendrecht-van der Voort, E. A. M., 248, 253, 254, 255, 255(6), 256, 256(6), 257, 257(6)
Parce, J. W., 35, 315
Parenti, A., 408, 419(7), 421, 421(7; 8)
Parka, M. H., 153
Parmentier, G., 369
Parnaik, V. K., 123, 124(14)
Parola, M., 428
Parrick, J., 15
Parry, G. C., 222, 223
Parthasarathy, S., 340, 474
Pasqualucci, C. A., 434, 435, 449
Pasterkamp, G., 434, 435(14)
Patel, C. V., 224, 225(38)
Pathobiological Determinants of Atherosclerosis in Youth (PDAY) Research Group, 343, 346
Patterson, C., 221, 448, 453
Patterson, D., 495
Patterson, M., 394
Paul, R., 192
Paul, T., 508
Pawlita, M., 92
Payne, N. N., 360
Pearce, L. L., 358

Pearlstein, R. D., 487
Pease-Fye, M. E., 347, 348(8), 350(8), 351(8)
Pedersen, M., 3(6), 4
Pedersen, W. A., 474
Pedro, M. A., 432, 434, 435(10), 436(10), 438, 440, 442(10), 443, 443(10), 446(10), 447(10), 448, 450(36), 452, 452(10), 454(57; 60)
Pe'er, J., 392
Pei, H., 449
Pelech, S. L., 54
Pelle, E., 270
Pellieux, C., 54
Pemberton, L. F., 124
Penix, L., 455
Penn, J. S., 378
Penn, M. S., 224, 225(38; 45; 46), 226
Peppelembosh, M. P., 97(17), 98
Percival, M. D., 360
Pereira, A. C., 434, 436(11), 437(11), 441, 442(11), 444, 445(11), 446(11), 447(11), 448, 452(11), 454(11)
Pereira, C. M., 191
Perex-Tur, J., 465
Perez, G. M., 24
Perez-Terzic, C., 124
Perlman, H., 434
Perocak-Vance, M. A., 495
Perrone, A., 237, 247(24)
Perry, G., 455, 495
Pervaiz, S., 150, 151, 152, 152(5), 159(5; 11)
Peters, E. J., 348
Peters, K. G., 221
Peters, S., 379, 389(21)
Peters, S. P., 100
Peyton, K. J., 449
Pham, D. G., 457, 461(13), 473, 473(13)
Phan, S. H., 212
Philippens, K. M. H., 437
Phillips, T. L., 3(8; 9), 4, 24(9)
Phumala, N., 333, 495
Piantadosi, C. A., 347, 350(2)
Pichiule, P., 32
Pierce, G. N., 123, 124, 130(29), 131
Pierlot, C., 54
Piermattei, D., 259
Pietri, S., 440, 457, 467(10)
Pileggi, F., 434, 435, 449
Pincus, M. R., 54
Pintilie, M., 4
Pintucci, G., 444
Pirzer, U., 232
Pitt, B. R., 163, 164, 358, 359, 359(42)
Placidi, G., 333
Plaisancie, H., 224, 226(36)
Plate, K. H., 392
Ploegh, H. L., 101
Pober, J. S., 212, 223
Podda, M., 326, 329(5), 331(5)
Poderoso, J. J., 296
Polack, A., 92
Politz, J. C., 123
Pollard, V., 132
Pollman, M. J., 444, 445
Pontén, U., 476
Ponton, A., 151, 152(5), 159(5)
Pook, D. R., 17(81), 21, 26
Poon, M., 223
Poot, M., 296
Porasuphatana, S., 516, 517(37c)
Porro, D., 261
Porter, C. D., 153
Post, J. A., 162(30), 163, 164(30)
Post, M. J., 449
Post, S., 232
Postiglione, A., 346
Pothier, P., 69
Potter, B., 201
Potter, D., 261
Pou, S., 516, 517(37c)
Poulos, A., 363, 369
Poustka, A. M., 363
Powell, H. C., 346
Poyton, R. O., 31
Poznansky, M. J., 389
Prabhakar, N. R., 31
Prada, C. M., 464, 465
Pratt, S. E., 259
Pratt, W. B., 259
Preedy, V. R., 319
Prehn, J. H., 296, 472
Preissner, K. T., 222, 223(10)
Prendergast, F. G., 124
Presta, M., 421
Price, D. L., 464
Pries, C., 167
Pritchard, K. A., 437
Prosnitz, L. R., 4, 5(14)
Pryor, W. A., 80, 333
Przedborski, S., 503

Puceat, M., 124
Pushpa-Rehka, T. R., 167
Puttfarcken, P., 191
Pyle, J., 124
Pyles, J. M., 444

Q

Qian, S. Y., 317, 465
Qian, Y. Z., 394
Qiao, L., 232
Qiu, R.-G., 97(17), 98
Qiu, J. H., 81
Qu, Z. C., 268, 271, 275
Quehenberger, O., 160
Quehenberger, P., 223
Queral, A. E., 191
Quinn, M. T., 453
Quinn, P. J., 163, 171, 172(28), 358, 359(42)
Quinn, T., 232
Quinones, M. J., 394
Quistorff, B., 477

R

Rabinowitz, J. D., 113
Rabizadeh, S., 474
Radda, G. K., 278
Radi, R., 325, 437
Radin, N. S., 367
Raff, M. C., 93, 151
Ragni, M., 224, 225(37)
Rahma, L. O., 365
Rahmani, Z., 495
Raingeaud, J., 58
Rainville, A. M., 369
Rajagopalan, S., 453
Raleigh, J. A., 14, 15, 24
Ramadass, P., 201
Ramakrishnan, N., 321
Ramakrishnan, S., 392
Ramalingan, J., 151
Ramasamy, S., 442
Ramdev, P., 348
Ramos, L. E., 474
Rampling, R., 4
Randin, O., 360
Rao, G. N., 221
Rapp, L. M., 378, 390
Rasey, J. S., 24
Rashba-Step, J., 167
Ratan, R. R., 183, 187, 188(3; 9), 189
Ratcheson, R. A., 476
Ratovitsky, T., 464
Rauckman, E. J., 495
Rauhala, P., 350
Rauth, A. M., 10, 14
Ray, C. J., 445
Raynor, J. B., 509
Razzack, J., 358
Recknagel, R. O., 80
Reed, D. J., 261
Reed, J. C., 93
Reers, M., 133(8), 136
Refolo, L., 465
Regan, L. M., 222
Reichmann, H., 141
Reid, M. B., 310, 315, 318(24), 319, 319(10), 320(10), 323(10)
Reidy, M. A., 435, 444
Reilly, J., 348
Reinke, L. A., 510
Reis, E. D., 434
Reiss, U., 378, 389(11)
Reivich, M., 487
Remans, P. H. J., 248, 253, 254, 255, 255(6), 256, 256(6), 257, 257(6)
Remick, D. G., 212
Renauer, D., 92
Requen, J. R., 341
Reszka, K. J., 509
Rey, F. E., 442
Reynolds, I., 183
Reynolds, T. Y., 5
Rhee, S. G., 92
Rice-Evans, C., 287
Richardson, G., 437
Richter, C., 296
Richter, I., 367
Riely, C. A., 81
Riepe, M. W., 139, 141(13)
Riethmuller, G., 208
Rimarachin, J. A., 449
Rincon, M., 53
Risau, W., 392
Ritov, V. B., 160, 163(8), 164, 170(8), 172(8)
Rivard, A., 391

Rivier, C., 301
Roa, J. C., 341
Robakis, N. K., 189
Robbesyn, F., 62
Robert, A., 191
Roberts, J. T., 24
Roberts, L. J., 205
Roberts, S. M., 272
Robertus, L., 449
Robino, G., 428
Robinson, J. P., 64(9), 66, 69(9), 70(9), 312, 319(17)
Robinson, N., 457, 461(13), 473(13)
Rocha, T. R., 435
Rochenstein, E., 474
Rochette, L., 394
Rockman, H. A., 394
Rockwell, K. J., 24
Rockwell, S. C., 3, 5
Roder, J. C., 464
Rodgers, G. M., 225(42), 226
Rodgers, M. A. J., 41
Rodriguez, C. M., 150
Rodriguez, M. S., 123, 124(16)
Rodriguez-Aguilera, J. C., 268
Roe, J. A., 474
Roelofsen, B., 81, 89(20), 90, 162(29), 163
Roesen, R., 323
Rogers, J. S., 58
Rohan, R. M., 392
Rojas, J. D., 280, 281(9), 295(9)
Rokach, J., 199
Ronai, Z., 53, 54
Ronn, L. C., 296, 299(8), 302, 322, 323(49), 324(49)
Roque, M., 434
Rosario, L., 54
Rose, D. R., 495
Rose, R. C., 269, 270, 272(7)
Rosen, G. M., 495, 516, 517(37c)
Rosenberg, P. A., 191
Rosenblum, J. S., 4, 124
Rosenfeld, M. E., 340, 342, 345(9)
Rosenfield, C. L., 222, 223
Rosenthal, G. L., 392
Rosner, G. L., 4
Ross, J., 394
Ross, R., 432
Rossant, J., 464
Rossikhina, M., 222, 223
Rota, C., 307, 312(4), 316, 316(3; 4), 320(3), 321(3; 4; 28)
Rotenberg, S. A., 112, 115, 118(26), 119(26), 122
Roth, S., 237, 246(19; 20)
Rothe, G., 70
Rothman, R. B., 503
Roubaud, V., 153, 437
Rouleau, G. A., 495
Rousou, J. A., 392
Roy, S., 183, 191, 326, 422, 429
Royall, J. A., 62, 63(4; 5), 64(4), 66(4), 67(4), 69(4; 5), 70(5), 71(5), 92, 316, 317, 318, 318(26), 319(26), 320(26), 468
Rubbnicki, E., 165
Rubin, S., 21
Ruch, W., 92
Ruef, J., 221, 448, 449, 453
Ruf, W., 222, 223, 223(12)
Rugtveit, J., 237, 247(23)
Rumbaugh, R. C., 80(11), 81
Rumsey, S. C., 270, 272(9)
Runge, M. S., 221, 448, 449, 453
Russo, A., 495
Russo, L., 264, 265, 266
Russo, T., 258, 259, 261, 262, 262(13), 263, 264, 265, 265(15; 16), 266
Rutledge, B. J., 423
Ryan, U. S., 282
Rydel, R. E., 461, 462(26), 473
Ryu, H., 183, 189

S

Saadawi, S., 69
Sacchi, N., 227
Sadler, J. E., 223
Sadler, J. J., 249
Sadrzadeh, S. M. H., 80
Sagara, Y., 188, 191, 474
Saiki, I., 312, 313
Saji, H., 423
Sakurai, K., 309, 321, 323(44), 324
Salama, G., 163
Salcedo, R. P. M., 423
Salford, L. G., 476
Salgia, R., 80
Salvayre, R., 62, 67, 69
Salyapongse, A. N., 280

Samali, A., 160
Sambrook, J., 355, 357(37)
Samelson, L. E., 249, 257(7)
Samouilov, A., 339, 350, 351(30), 454, 506, 507, 509, 512, 515, 515(36), 517(36), 519(26), 521(37a)
Samuni, A., 495
Sanchez, E. R., 259
Sandberg, E., 3(6), 4
Sanders, S. P., 333, 519
Sandersen, A. P., 3(6), 4
Sandhir, R., 361, 364(7), 367(7), 369, 371(7; 45)
Sandstrom, P. A., 151, 161
Sang, H., 510
Sanghera, J. S., 54
Sano, H., 494, 496, 497(22; 23), 503(23)
Sano, T., 495
Sapkota, P. G., 507
Sapp, P., 495
Sardana, M., 54
Sardi, C. O., 363
Sarembock, I. J., 223
Sarkar, K. D., 442
Sarks, J. P., 379
Sarks, S. H., 379
Sarma, J. M., 449
Sarti, P., 360
Sartorelli, A. C., 3
Sarvazyan, N., 318, 319(38), 320(38)
Sasaki, H., 391
Sasson, I., 16, 28(71)
Sato, H., 415
Sato, S., 509
Saunders, M. I., 24
Savolainen, K. M., 72
Savvides, P., 81, 86(18), 88(18), 89(18)
Sawa, T., 348
Sayre, L. M., 455, 495
Scanlon, C. E. O., 346
Scarpati, E. M., 223
Schacter, E., 426
Schafer, A., 92, 449
Schalkwijk, C., 162(29), 163
Schapira, A. H., 474
Schau, M., 32
Schaur, R. J., 373
Schecheter, A. N., 347, 348(8), 350(8), 351(8)
Schecter, A. D., 222, 223
Scheel-Kruger, J., 509
Scheffer, R. C. H., 452, 454(57)
Scheidegger, K. J., 341
Schelling, M. E., 414
Schieffer, B., 453
Schild, L., 136
Schinder, A. F., 72
Schini-Kerth, V. B., 225(47), 226, 227(47), 230(47)
Schirmer, H., 236
Schlegel, J., 442
Schlenger, K., 4
Schlenstedt, G., 123
Schlidt, S. A., 319
Schliess, F., 54
Schlosser, R. J., 318
Schmidt, D. H., 393
Schmidt, J. V., 32
Schmitz, M. L., 53, 54(4)
Schnaar, R. L., 183, 188(2), 190(2), 191
Schneider, R. F., 29
Schoener, B., 479, 482
Schölmerich, J., 247
Scholz, R., 135
Schon, E. A., 458
Schor, N. F., 160, 163, 163(9), 171, 172(28)
Schraermeyer, U., 378, 379, 389(18–21)
Schröder, R., 141
Schroeder, T. J., 112(4), 113
Schroit, A. J., 159, 160(2)
Schubert, D., 183, 188, 188(4), 191
Schubert, W., 133(7), 136, 141, 141(7)
Schuett, W. H., 483
Schuette, W. H., 483, 487
Schulman, H., 123
Schultz, J. E., 394
Schulze, W., 136
Schulze-Osthoff, K., 53, 54(4)
Schumaker, P. T., 32, 309, 311, 319, 319(16)
Schürmann, G., 232
Schutgens, R. B. H., 363
Schutz, S., 92
Schwamborn, K., 123
Schwartz, D., 444
Schwartz, S. M., 221
Schweitzer, G., 437
Schweizer, M., 296
Scita, G., 229
Scognamiglio, A., 224, 225(37)
Scott, H., 237, 247(23)
Scott, K. F., 199
Scully, S. P., 4, 5(14)

Sedlov, A., 172
Seeds, M. C., 35, 315
Seeger, M., 464
Seekins, S., 464
Seel, C., 232
Seftel, H., 342
Segal, A. W., 453
Segal, B. H., 442
Seger, R., 59
Seghezzi, G., 444
Sehrt, E., 389
Seibel, P., 141
Seidner, S. R., 5, 16(27), 28(27)
Seikh, F., 372
Seiki, M. A., 415
Sekhsaria, S., 64(11), 66
Seki, T., 123
Sekimoto, T., 123
Sekine, M., 132
Sellinger, O. Z., 365
Sellinger, R., 365
Selvakumar, B., 442
Semenza, G. L., 31, 32, 189
Sen, C. K., 183, 191, 326, 422, 423, 429, 430, 432, 433(5)
Sengupta, T., 363
Seo, M. S., 92
Serbinova, E. A., 163
Serezhenkov, V. A., 348, 509, 520(32)
Serianni, A. S., 268
Sessa, W. C., 280
Sevanian, A., 161, 167
Seyed, M. A., 152
Shah, G. M., 152
Shah, S., 208
Shalit, M., 64(11), 66
Shan, S., 14
Shankar, R. L., 495
Shao, Z., 309, 311, 319(16)
Shapiro, I. M., 28
Sharbrough, F. W., 487, 488, 489
Sharikabad, M. N., 324
Sharma, S. K., 221
Sharma, V. K., 236, 237(12)
Sharma, Y. K., 453
Sharp, F. R., 16
Sharp, P., 369
Sharplin, J., 14
Shatwell, K. P., 453
Shaw, P. E., 58
Shawven, L. K., 392
Shears, L. L., 280
Sheikh, F., 363
Shelhamer, J. H., 347, 348(8), 350(8), 351(8)
Shergil, J. K., 509
Shertzer, H. G., 309, 317, 321(29)
Shibao, G. N., 259
Shibata, S. S., 389
Shibata, T., 503
Shibuki, K., 288
Shiino, A., 475, 478
Shima, D. T., 392
Shimakura, A., 495
Shimizu, S., 151
Shimomura, O., 439
Shinagawa, A., 415
Shing, Y., 424
Shingu, T., 446
Shin-Ichi, I., 333
Shinobu, L. A., 510
Shioi, J., 189
Shiue, C.-Y., 29, 30(86)
Shortal, B. P., 345
Showell, H. J., 212
Shriver, A. E., 443
Shubinsky, G., 53, 54(4)
Shuman, H., 309, 310, 311(7), 312(7), 313(7), 314(7), 315
Shvedova, A. A., 159, 163, 170(25), 171, 378, 390
Shweiki, D., 392
Siddique, T., 495
Siddon, N. A., 123
Sido, B., 232, 239, 247(25)
Sidorov, A. S., 390
Siebenlist, U., 446
Siemann, D. W., 16
Sies, H., 54, 206, 326, 373, 440
Siesjö, B. K., 476
Silber, R., 270
Silver, I. A., 4
Silver, M., 409
Silver, P. A., 123
Simon, D. I., 347, 350(4), 351(4), 357(4), 358(4)
Simon, M. C., 32
Simonian, N. A., 72
Simos, G., 123
Sims, N. R., 477
Sims, P. J., 160

Sindermann, I., 437
Singel, D. J., 347, 348, 350(4), 351(4), 357(4), 358(4)
Singer, V. L., 296
Singh, A. K., 362, 363, 364, 365(24), 369
Singh, G., 465
Singh, H., 363, 369
Singh, I., 361, 362, 362(5), 363, 363(5), 364, 364(5; 7; 23), 365(23; 24), 367(7), 369, 371(7; 45), 372
Singh, P. K., 288, 508
Singh, R. J., 347
Sinha, B., 453
Siomi, M. C., 132
Sisk, A., 474
Sisodia, S. S., 464
Skatchkov, M. P., 437, 439
Sketch, M. H., 221
Skjeldal, O. H., 369
Sklar, L. A., 64(8), 66
Slater, A. F. G., 442
Slater, T. F., 81
Slaughter, C., 58
Slevin, C. J., 113, 116(14)
Sloan-Lancaster, J., 249
Sloan-Stanley, G. H., 164, 173(32)
Sloop, G. D., 340, 341, 345, 346(12)
Slunt, H. H., 464
Slupsky, J. R., 225(43), 226
Smeal, T., 53, 58(8)
Smiley, S. T., 133(8), 136
Smith, B., 363
Smith, C., 160
Smith, C. D., 455
Smith, G. C. M., 451
Smith, J. N., 347, 349(3)
Smith, K. A., 15
Smith, K. M., 221
Smith, M. A., 455, 495
Smith, R. C., 434
Smyth, C., 495
Sneller, M., 232
Snyder, S. H., 92, 417
Sobreira, C., 318
Somers, M. J., 423, 440, 452(35)
Sondhi, D., 192
Sonenshein, G. E., 446
Sonmezgoz, E., 351
Sopher, B. L., 457, 461(13), 473, 473(13)
Sorentino, J. D., 201
Sorescu, D., 221, 443, 447, 448, 452(52), 453(52)
Soriano, E., 495
Sotgiu, A., 333
Souza, H. P., 434, 435(10), 436(10; 11), 437(11), 438, 441, 442(10; 11), 443, 443(10), 444, 445(11), 446(10; 11), 447(10; 11), 448, 452, 452(10; 11), 454, 454(11; 59; 60)
Souza, J. M., 325, 474
Souza, L. C., 434, 435, 435(10), 436(10; 11), 437(11), 438, 441, 442(10; 11), 443(10), 444, 445(11), 446(10; 11), 447(10; 11), 448, 449, 452, 452(10; 11), 454(11; 57)
Sozmen, E. Y., 448
Spector, A. A., 201
Speidel, C. M., 223
Sperling, D., 437, 439
Spiteller, G., 200
Spitzer, N. C., 72
Spokas, E. G., 199
Spotnitz, W. D., 348
Springer, J. E., 474
Sprirchev, V. B., 378
Srinivasan, M., 123
Srivastava, S., 449
St. Clair, D. K., 458, 461, 461(23), 465, 466(23), 468(23), 469(23), 472(23)
St. Croix, C. M., 359
Staal, F. J., 233, 248, 253(3)
Staal, G. E., 233
Stack, R. S., 221
Stade, B. G., 208
Stadtman, E. R., 259, 373, 374, 429, 455, 494
Stamenkovic, I., 151
Stamler, J. S., 347, 348, 349, 350, 350(1; 2; 4), 351(4), 357(4), 358(4)
Stanciu, M., 183
Stanley, W., 363, 364, 365(24)
Stanzl, K., 339
Stao, S., 520
Starke-Reed, P. E., 455
Starnes, S. L., 392
Stassen, J. M., 434
State, H. M., 270
States, S., 392
Stearns-Kurosawa, D. J., 222
Steed, D. L., 280
Steer, C. J., 150
Stehno-Bittel, L., 124
Steinberg, B. M., 444

Steinberg, D., 160, 340, 341
Steinbracher, U. P., 346
Steiner, S. M., 458, 461(23), 466(23), 468(23), 469(23), 472(23), 474
Stern, A., 92
Sterne-Marr, R., 124, 130(18)
Stevens, C., 21, 507, 509(3)
Stevens, R. C., 507
Stobbe, C. C., 29
Stocken, L. A., 102(9), 103
Stocker, R., 271
Stofan, D., 310, 313(12)
Stolk, J., 230
Stone, H. B., 4
Storrie, B., 171
Strand, O. A., 351, 357(34)
Stratford, M. R., 15
Strein, T. S., 112(5), 113
Strieter, R. M., 212
Striggow, F., 136
Stuart, B. H., 483, 487(5)
Stubauer, G., 360
Stuehr, D. J., 230, 451
Stukan, R. A., 509, 512, 515(36), 517(36), 520(32)
Sturrock, J., 7
Sturrock, R. D., 248
Stypinski, D., 29
Su, J. H., 455
Suarez, E., 437
Sud'ina, G. F., 100
Sugarbaker, D. J., 348
Sugars, K., 213
Sugden, P. H., 54
Sugimoto, K., 301, 323
Sugita, Y., 233, 236, 236(8), 237(8)
Sugiyama, N., 342
Sugiyama, S., 225(44), 226
Suh, Y. A., 449
Sukhova, G., 444
Summers, S., 201
Sun, E. E., 268
Sun, I. L., 268
Sun, X.-G., 115
Sun, Y., 92
Sundaresan, M., 92
Sundt, T. M., Jr., 486, 487, 488, 489, 490
Supinsky, G., 310, 313(12)
Sureda, F. X., 71, 72, 73, 74
Sustmann, R., 508
Sutcliffe, L. H., 333
Suter, C., 101, 102(6)
Suthanthiran, M., 236, 237(12)
Sutherland, R. M., 4, 13(45), 14
Suzuki, S., 348
Svardal, A. M., 237, 239(21)
Swaal, F. R. A., 160
Swallen, S. F., 324, 325(59)
Swanson, J. A., 324, 325(59)
Swartz, H. M., 333, 495
Swerlick, R. A., 201
Swift, L. M., 318, 319(38), 320(38)
Symes, J. F., 409
Symons, M. H., 97(17), 98, 509
Szabo, P., 449
Szedja, P., 315
Szejda, P., 35
Sziraki, I., 350
Szollosi, J., 101, 102(6)
Szscs, K., 443, 448
Sztejnberg, L., 66, 309, 311, 312(15), 314(15)
Szweda, L. I., 373, 374, 374(7), 375(3; 4; 7), 377(3; 7), 429

T

Tabrizi, S. J., 474
Taby, O., 222
Tachibana, T., 123
Tada, S., 123
Tagawa, T., 123
Tager, J. M., 363
Taha, S., 226
Tailor, P., 249
Tait, J., 160
Tak, P. P., 248, 253(3), 257(4)
Takahashi, H., 41
Takano, T., 346
Takashima, S., 392
Takayama, F., 509
Takei, H., 345, 346(12)
Takemasa, T., 301, 323
Takemura, S., 357, 358(38)
Takenouchi, T., 474
Takeshima, E., 346
Takeshita, K., 333, 495, 501, 503(20)
Takino, T., 415
Tamagami, H., 496
Tamaki, T., 516, 517(37b)

Tan, S., 188, 191
Tanaka, H., 432, 433(2), 434(2), 435(2), 444
Tanaka, K., 446
Tanaka, M., 373
Tang, C. S., 449
Tang, D., 123
Tanji, K., 458
Tannenbaum, S. R., 349
Tanzi, R. E., 495
Tao, J., 54
Tappel, A. L., 378, 389(11)
Tarpey, M. M., 437, 453
Tateishi, N., 237, 238(18)
Tatoyan, A., 167, 296
Tauber, A. I., 452
Taubman, M. B., 221, 222, 223, 434
Taylor, F. B., 222
Taylor, Y. C., 14
Teague, D., 350
Teicher, B. A., 3
Tekle, E., 92
Telo, J. P., 259
Ten Brink, H. J., 369
Tepel, M., 315(27), 316
Terada, L. S., 259
Teranishi, K., 439
Terao, J., 389
Terrand, J., 394
Tertoolen, L. G. J., 97(17), 98
Terzic, A., 124
Tew, K. D., 53, 54
Tewsbury, D., 354
Tewson, T. K., 24
Teyssier, J. R., 394
Thannickal, V. J., 150
Theofanidis, P., 208
Therriault, H., 207
Thews, G., 4
Thews, O., 14
Thiers, J.-C., 62
Thihes Sempoux, M., 365
Thinakaran, G., 464
Thiruvikraman, S. V., 221
Thistlethwaite, P. A., 393
Thom, S. R., 62, 63(5), 69(5), 70(5), 71(5)
Thomae, O., 58
Thomas, M., 35
Thomas, P. D., 389
Thompson, R. H. S., 102(9), 103
Thomsen, L. L., 409
Thomsen, M., 69
Thorpe, S. R., 341
Thrall, D. E., 24
Thumann, G., 378
Thurman, R. G., 135
Tickner, T. R., 81
Tienrungroj, W., 259
Tiertze, F., 207
Tijssen, K., 268, 274, 275(3), 277, 278(3), 279
Tobena, R., 213
Toborek, M., 198, 199, 200, 201, 203, 204, 205, 206, 207, 209, 209(18), 212, 217
Tocchi, M., 449
Toda, N., 310, 322, 323(13), 324(13)
Toda, S., 422
Todd, R. F., 99
Toi, M., 423
Toki, Y., 449
Tolentino, M., 392
Tomasi, A., 495
Tominaga, T., 509
Tominago, T., 520
Tomlinson, F. H., 483, 484(6), 486(6)
Tomlinson, M. G., 249
Tommaga, T., 520
Topol, E. J., 434, 435(12)
Tordo, P., 440, 441
Totter, J. R., 80
Toyokuni, S., 373, 374, 375(3), 377(3)
Traber, M. G., 326, 329(5), 331(5), 428
Tracy, M., 16, 24(68), 27(68)
Travis, J., 354
Tremethick, D. J., 123
Trempe, C. L., 41
Tresini, M., 224, 446
Trible, R. P., 249, 257(7)
Trimarco, B., 224, 225(37)
Tritschler, H. J., 191, 428
Trojanowski, J. Q., 474
Tronheim, K., 452
Trullas, R., 73
Trumbauer, M. E., 465
Trunkey, D., 357, 358(39)
Trush, M. A., 153, 230, 437, 452
Tsacmacidis, N., 54
Tsai, P., 516, 517(37c)
Tsang, A. W., 324, 325(59)
Tsao, J. Y., 313
Tseng, E., 280

Tsien, R. Y., 102(12), 103, 496
Tsimikas, S., 345
Tsionsky, M., 113, 116, 116(13)
Tso, M. O., 390
Tsoukas, C. D., 249
Tsuchihashi, N., 333
Tsuchiya, K., 515, 516, 517(37b), 520(37)
Tsujimoto, Y., 151
Tsukeda, H., 111
Tuccio, B., 440
Tufolo, P. C., 434, 435(15)
Tulis, D. A., 449
Turkoz, Y., 351
Turrens, J. F., 38
Tyurin, V. A., 159, 160, 160(9), 161(10), 163, 163(9–11; 25–28), 170, 170(11; 25), 171(39), 172, 172(28), 347, 358, 359, 359(42), 392
Tyurina, Y. Y., 159, 160, 160(9), 161(10), 163, 163(9–11; 25–28), 170, 170(11; 25), 171(39), 172, 172(28), 347, 358, 359, 359(42)

U

Uchida, I., 116
Uchida, K., 373, 374, 429, 457, 458, 469(19), 472(19)
Uckun, F. M., 474
Ueda, S., 508
Ueha, T., 71
Ueland, P. M., 237, 239(21)
Ueno, N., 41
Ueno, T., 423
Uint, L., 435
Ukai, Y., 310, 322, 323(13), 324(13)
Ulevitch, R. J., 58
Ullrich, A., 392
Ullrich, V., 100
Umeda, F., 495
Underwood, A., 101, 102(7)
Unsoeld, H., 53, 54(4)
Unwin, P. R., 113, 116(14)
Upchurch, R. G., 49
Urano, S., 500
Urano, Y., 322, 323, 323(47)
Uriel, J., 310, 312(11), 313(11)
Urtasun, R. C., 3(9), 4, 15, 24(9)
Ushio-Fukai, M., 221, 447, 452(52), 453(52)
Ustunes, L., 448
Utepbergenov, D., 508
Utsumi, H., 333, 494, 495, 496, 497(22; 23), 501, 501(13), 502(14), 503(20; 23)
Uysal, F., 448

V

Vacchino, J. F., 113
Valberg, P. A., 315, 318(24)
Valentine, J. S., 474
Valeri, C. R., 347, 350(4), 351(4), 357(4), 358(4)
Valet, G., 70, 315
Vallance, P., 507
Vallyathan, V., 333
Vamecq, J., 362
van Asseldonk, J. T., 452, 454(57)
Van Brockhoven, A., 363
Van Buren, J. M., 483
Van de Broek, G., 363
van den Berg, J. J. M., 81, 89(20), 90, 162(29), 163
van den Berg, R., 495
Van den Bosch, H., 363
Van Den Broek, P. J. A., 268, 274, 275(3), 276, 277, 278(3), 279
Van den Dobbelsteen, D. J., 442
Vanden Hoek, T. L., 309, 311, 319(16)
van den Oord, J. J., 434
Van de Ploeg, L. H., 465
van Der Helm, Y. J., 449
Van der Vliet, A., 167
van der Voort, E. A., 248, 253(3)
Van Der Zee, J., 268, 274, 275(3), 276, 277, 278(3), 279
VanDuijn, M. M., 268, 274, 275(3), 276, 277, 278(3), 279
Van Dyke, R. A., 486
Van Gent, C. M., 167
Van Grunsven, E. G., 362, 364(9)
Vanhoutte, F., 369
Vanin, A. F., 348, 506, 508, 509, 512, 515(36), 517(36), 520, 520(32)
Vannini, V., 495
Vanoni, M., 261, 264, 265, 266
Van Os-Corby, D. J., 14
van Pelt, N., 448
VanRaalte, G., 3(8), 4
Van Roermund, C. W. T., 363, 369

VanSteveninck, J., 268, 274, 275(3), 277, 278(3), 279
Van Veldhoven, P. P., 362, 363, 364(8), 369
van Vliet, A. I., 248, 253(3)
Varenne, O., 448
Varghese, A. J., 14
Varia, M., 24
Varon, S., 286
Vasilenko, I. A., 160
Vasios, G., 424
Vásquez-Vivar, J., 437
Vaupel, P., 4, 14
Velarde, V., 63
Velasco, A., 213
Venojarvi, M., 191, 326
Venuta, S., 261, 262(13)
Verdu, J., 114
Vergari, W. A., 409, 410(10)
Verheijen, J. H., 449
Verhoeven, A. J., 230
Vermeij, C. L., 248
Vern, B., 487
Versendaal, J., 248
Verweij, C. L., 248, 253, 253(3), 254, 255, 255(6), 256, 256(6), 257, 257(4; 6)
Via, D. P., 283
Vichitbandha, S., 465
Victorov, A. V., 160
Vieira, A. J. S. C., 259
Vielhaber, S., 139, 141, 141(13)
Villalba, J. M., 268
Vinters, H. V., 455
Vita, J., 347, 349, 350(4), 351(4), 357(4), 358(4)
Vitale, A., 487
Viveros, O. H., 112(4), 113
Vlessis, A. A., 357, 358(39)
Vodovotz, Y., 350
Voelker, D. R., 160
Voest, E. E., 392
Voevodskaya, N. V., 520
Vogt, P. K., 123
Vogt, W., 259
Volicer, L., 327
Volk, A., 364, 366(25)
Volpe, J. J., 191
von Bossanyi, P., 141, 143(14)
Vordran, B., 4
Vourinen, K. H., 135
Vowells, S. J., 64(11), 66
Voyta, J. C., 283
Vreman, H., 487
Vu, T. H., 458
Vuillard, L., 123, 124(16)
Vukomanovic, D. V., 350

W

Waclawiw, M. A., 347, 348(8), 350(8), 351(8)
Wada, K., 519
Wada, T., 132
Waeg, G., 462, 469(29)
Wagner, E. F., 422
Wagner, J. R., 207
Wahlbrink, T., 296, 472
Wakefield, L. M., 278
Waleh, N., 5, 16, 16(27), 28(27; 70)
Walker, V. S., 378
Wallace, M., 315
Wallich, R., 232
Walsh, K., 434
Walsh, M. L., 66
Walter, U., 448
Waltiaux, P., 365
Walton, M. I., 24
Wanders, R. J. A., 362, 363, 364(9), 369
Wang, H. M., 390
Wang, J. H., 192
Wang, L. H., 449
Wang, P., 507, 509, 519(26)
Wang, R., 464
Wang, S., 346
Wang, Y., 183, 270, 272(9)
Wanke, V., 261
Ward, J. M., 423
Ward, M. R., 434, 435(14)
Wardman, P. O., 14, 309, 325
Ware, C. B., 457
Warltier, D. C., 394
Warner, E. D., 422, 424
Warner, M. L., 160
Wasserman, K., 423
Wasserman, T. H., 3(8; 9), 4, 24(9)
Watanabe, H., 237
Watanabe, M., 392
Waters, C. R., 393
Watkins, B. A., 200, 201
Watkins, P. A., 369
Watkins, S., 183
Watkins, S. C., 280

Watson, A. D., 208
Watson, S. J., 296, 299(8), 301, 302, 322, 323(49), 324(49)
Weber, C., 326, 329(5), 331(5)
Weber, J., 422, 424(4)
Weber, R., 21
Wei, C., 113, 116(12)
Wei, G., 520
Wei, Z., 430
Weindruch, R., 373, 374, 374(7), 375(7), 377(7)
Weintraub, N. L., 443, 448
Weir, E. K., 31
Weis, K., 130
Weise, M. J., 363, 364, 365(24)
Weiss, A., 249, 257(8)
Weiss, H., 5, 16(27), 28(27)
Weksler, B. B., 449
Wells, K. S., 296
Wendehenne, D., 323
Wendling, P., 4
Wenger, R. H., 32
Wessels, D. A., 64(10), 66, 307, 309, 316(2), 317(2), 318(2), 319(2), 325(2)
West, M. S., 315, 318(24)
Wharram, B. L., 212
Whetsell, W., Jr., 373
Whitaker, C., 282
White, C. R., 437
Whitehouse, R., 423
Whitehouse, W. C., 483, 487
Whitesell, R. R., 275
Whitlow, P. L., 434, 435(16)
Wiebe, L. L., 29
Wiedau-Pazos, M., 474
Wiedemann, D. J., 112(2), 113
Wiedemann, F. R., 141
Wiedmer, T., 160
Wiegand, R. D., 378
Wieser, R. J., 92
Wiggins, R., 212
Wightman, R. M., 112(2; 4), 113
Wigland, M. J. A., 363
Wilcox, J. N., 221
Wilensky, R. L., 444
Wilkinson, F., 41
Williams, D. A., 319
Williams, J. A., 426
Williams, R. L., 422
Williams, T. P., 390
Williamson, J. R., 135
Willuweit, A., 225(43), 226
Wilson, D. F., 9(40), 10, 11(40)
Wink, D. A., 350
Winkler, K., 133(7), 135, 136, 139, 141, 141(7; 13), 143(14), 149
Winokur, A. L., 225(45), 226
Winyard, P. D., 507
Wissler, R. W., 346
Witt, E., 229
Witting, L. A., 378
Witz, G., 80
Witztum, J. L., 160, 340, 342, 345, 345(9), 346
Woldman, Y. Y., 508
Wolf, C. R., 54
Wolf, P. L., 393
Wolfauer, G., 345
Wolff, S. P., 80, 205
Wolin, M. S., 62, 230, 423
Wong, G., 54
Wong, J., 474
Wong, J. M. S., 54
Wong, P. Y., 199
Wong, R., 4
Wong-Staal, F., 123
Wood, J. D., 423
Wood, K. S., 507
Woods, M. R., 16
Workman, J., 474
Workman, P., 4, 24
Wright, J. R., 80(11), 81
Wright, R. M., 259
Wright, T. L., 349
Wright, V. P., 310, 312(14), 313(14), 314(14)
Wu, C., 439
Wu, G., 280, 281(9), 287, 289, 295, 295(9)
Wu, J. F., 467
Wu, Y., 221
Wyllie, A. H., 151

X

Xia, Y., 440, 506, 519
Xie, Z., 394
Xu, A., 349
Xu, Q., 444
Xu, Z., 474

Y

Yager, D., 464, 465
Yalowich, J. C., 164
Yam, P. T. W., 101, 102(5), 103(5), 112(5)
Yamada, H., 116
Yamada, Y., 519
Yamamoto, E., 415
Yamamoto, T., 192
Yamanashi, Y., 192
Yamasaki, K., 280
Yamashita, H., 192
Yamashita, K., 301, 323
Yamauchi, A., 236, 237(13)
Yan, L. J., 428
Yan, W., 295
Yan, Y., 135
Yang, F., 463
Yao, Z., 394
Yarwood, H., 213
Yasa, M., 448
Yasue, H., 225(44), 226
Yegudin, J., 349
Yen, H. C., 458, 461(23), 465, 466(23), 468(23), 469(23), 472(23)
Yeo, K. T., 392
Yeo, T. K., 392
Yetik, G., 448
Yeung, A. C., 434, 435(14)
Yi, C., 112(9), 113
Yin, D., 378, 379(12), 389(12)
Yin, L. Y., 221
Yin, X.-M., 161
Yin, Z., 53, 54
Ylä-Herttuala, S., 340, 342, 345(9)
Yodoi, J., 248
Yohoyama, H., 333
Yokoyama, H., 500, 509
Yokoyama, R., 503
Yologlu, S., 351
Yoneda, Y., 123
Yoritaka, A., 373
Yoshida, S., 248
Yoshida, Y., 346, 363, 364, 365(24)
Yoshimura, T., 348, 509
Yoshizumi, M., 515, 516, 517(37b), 520(37)
Young, A. E., 422
Young, H. A., 423
Young, R. W., 379
Young, S. D., 5
Younkin, S., 463, 464, 465
Yu, A. Y., 32
Yu, H., 465
Yu, X., 465
Yu, Z. X., 92
Yun, J., 444
Yurekli, M., 351

Z

Zafiriou, O. C., 289
Zainal, T. A., 373, 374, 374(7), 375(7), 377(7)
Zakar, J., 24
Zaman, A., 207
Zaman, K., 189
Zambrano, N., 262, 263, 264, 265(16)
Zamzami, N., 310, 312(11), 313(11)
Zaruba, M. E., 71
Zastrow, L., 339
Zehr, C., 465
Zeiher, A. M., 161
Zeitz, M., 232
Zeng, M., 349
Zenk, B., 465
Zetter, B. R., 282, 283
Zhang, H. T., 408, 411(5), 413(5), 415(5), 416(5), 421(5)
Zhang, P., 21
Zhang, P. J., 17(81), 26
Zhang, W., 249, 257(7)
Zhang, Y., 223, 296
Zhang, Z. G., 192, 474
Zhong, W., 373, 375(3; 4)
Zhou, J., 116
Zhu, G. Y., 449
Zhu, H., 153, 309, 317, 321(29), 437
Zhu, M., 249
Zhyang, W. X., 434
Ziche, M., 280, 407, 408, 409, 410, 410(10), 411(4; 5), 412(4), 413(4; 5), 415(4–6), 416(4; 5), 419(3; 4; 7), 420(3; 4), 421, 421(5–8)
Zidek, W., 315(27), 316
Ziegelstein, R. C., 452, 454(59)
Zierz, S., 141
Zieske, A., 345, 346(12)
Zilversmit, D. B., 200

Zimmer, G., 333, 336
Zimmerman, J. B., 112(2), 113
Zoldhelyi, P., 448
Zollinger, M., 347, 350(5)
Zollner, H., 373
Zoutman, D. E., 350
Zuo, L., 307, 310, 312(14), 313(14), 314(14)
Zwaal, R. F. A., 159, 160(2)
Zweier, J. L., 153, 333, 339, 350, 351(30), 437, 440, 452, 454, 454(59; 60), 495, 506, 507, 509, 509(7), 512, 515, 515(36), 517(36), 519, 519(26), 520, 521(37a)

Subject Index

A

AD, *see* Alzheimer's disease
Alzheimer's disease
  β-amyloid precursor protein
    cell exposure conditions, 461
    immunocytochemistry, 473–474
    mutations, 455
    oxidative stress induction, 455, 457–458
    transgenic mice expressing mutants, 463–464
    Western blot, 473
  calcium measurement in cytoplasm, 472–473
  fluorescence imaging of oxidative species and damage
    hydrogen peroxide, 470
    hydroxyl radicals, 468–469
    lipid peroxidation, 470–472
    peroxynitrite radicals, 468–469
    precautions, 467
    superoxide, 468
  hippocampal neurons from rodents
    apoptosis assay, 462
    primary culture, 459–460
  manganese superoxide dismutase
    immunocytochemistry, 473–474
    PC12 cell overexpression, 460–461
    transgenic mice overexpression, 465–466
    Western blot, 473
  mitochondrial measurements
    calcium, 472
    membrane potential, 472
  oxidative damage overview, 455, 457, 474
  presenelin-1
    immunocytochemistry, 473–474
    mutations, 455, 457
    oxidative stress induction, 458
    PC12 cell overexpression of mutant protein, 460–461
    transgenic mice expressing mutants, 464–465
    Western blot, 473
3-Amino, 4-aminomethyl-2′,7′-difluorofluorescein
  nitric oxide detection in tissues
    fluorescence detection, 323–324
    loading, 323
    reactivity, 322–323
  structure, 322
Aminophospholipid translocase, phosphatidylserine translocation in apoptosis, 160
β-Amyloid precursor protein, *see* Alzheimer's disease
Angiogenesis, *see also* Hemangioma
  angiogenic factors, 175, 392
  disease roles, 175, 407
  endothelial cell migration, *see* Endothelial cell
  hypoxia triggering, 392
  myocardial infarction, *see* Myocardial infarction
  nitric oxide stimulation
    cyclic GMP assay, 419
    endothelial cell characterization
      bromodeoxyuridine proliferation assay, 415
      chemotaxis and invasion, 414
      gelatinase zymography, 414–415
    fibroblast growth factor-2 induction, 421
    nitrate/nitrite assays
      colorimetric Griess reaction, 416–417
      fluorimetric assay, 417
    nitric oxide synthase assays
      cell monolayer assay, 418
      ion-exchange resin preparation, 419
      lysate assay, 417–418
      reverse transcription polymerase chain reaction of isoform expression, 419–420
    overview, 280, 407–409, 420
    pharmacological tools for nitric oxide level modification, 412–413, 415–416
    rabbit cornea assay
      advantages, 409–410
      cell and tissue implants, 410–411

considerations for success, 411–412
histological examination, 411
quantification of angiogenesis, 411
slow-release preparations for testing, 410
surgery, 410
AP-1, electrophoretic mobility shift assay, 208–209
Apoptosis
endothelial cells following myocardial infarction, 403–404
hydrogen peroxide induction studies
cell lines, 152
hydrogen peroxide treatment, 152
intracellular oxidant measurement
flow cytometry assay for hydrogen peroxide, 155
lucigenin-based assay for superoxide, 153–155
intracellular pH assay with 2′,7′-bis(2-carboxyethyl)-5,6-carboxyfluorescein, 156–157, 159
poly(ADP-ribose) polymerase cleavage detection by Western blot, 152–153
reductive stress-induced apoptosis, 152
neuron oxidative stress, 183–184
cis-parinic acid reporter of phospholipid oxidation
availability, 161–162
cell culture for membrane integration, 161–162
high-performance liquid chromatography of labeled phospholipids, 164–166
high-performance thin-layer chromatography of phospholipids during oxidation-induced apoptosis, 166–167
phosphatidylserine site-selective oxidation during oxidation-induced apoptosis, 168, 170–172
phospholipid integration in different cell types, 161–162, 164
positional distribution in phospholipids, 167–168
vitamin E protective effects, 172–173
phosphatidylserine translocation, 159–160
reactive oxygen species induction, overview, 150–152, 159, 160–161
ARNT, *see* Hypoxia inducible factor-1
Ascorbic acid
antioxidant defense, 268
assays
ascorbate free radical
electron spin resonance, 273–274
stability, 273
ascorbate
high-performance liquid chromatography, 270–271
spectrophotometric assay, 270
dehydroascorbic acid
derivatization, 272
nuclear magnetic resonance, 272–273
radiolabeling, 272
reduction, 272
materials, 269–270
tris-(ethylenediamide)-nickel(II) chloride 2-hydrate preparation, 270
free radical
ascorbate-dependent reduction assays
electron spin resonance, 278–279
ultraviolet spectroscopy, 277–278
generation, 275–276
intracellular level modification
dehydroascorbic acid incubation, 274–275
TEMPOL treatment, 275
plasma membrane redox system, 268–269
transport, 274–275
Atherosclerosis, lysine oxidation in plaques, *see* 4-Hydroxynonenal-lysine; Malondialdehyde-lysine

B

BrdU, *see* Bromodeoxyuridine
Breast cancer, *see* Scanning electrochemical microscopy
Bromodeoxyuridine, endothelial cell proliferation assays, 286–287, 402–403, 415

C

5-(and 6)-Carboxy-2,7′-dichlorodihydrofluorescein
reactive oxygen species sensitivity, 315–316, 318
structure, 315
tissue radical studies
detection, 320–321
loading, 318–320

Cercosporin
  Cercospora resistance, 48–49
  phosphorescence detection of singlet oxygen in fungus
    data acquisition, 52
    light activation, 47
    overview, 42
    quenchers in media, 50–51
    sample preparation, 49–50
  toxicity, 47–48
Chemiluminescence, vascular injury oxidative stress assays
  NAD(P)H oxidase, 450, 452
  lucigenin chemiluminescence, 436–438, 442–443
  coelenterazine chemiluminescence, 438–439, 443
Computed tomography, electron paramagnetic resonance dual studies, *see* Electron paramagnetic resonance
Confocal microscopy
  α-lipoamide dehydrogenase redox state changes in muscle fiber mitochondria, 136–139
  reactive oxygen species effects on nuclear import, 128, 130, 132
Cysteine, lamina propria T lymphocyte effects and assays
  acid-soluble thiol assay with 5,5′-dithiobis(2-nitrobenzoic acid), 238–239
  cystine versus cysteine effects on CD3 reactivity, 245–246
  high-performance liquid chromatography assay, 239–241
  lamina propria macrophages as source, 237, 247
  peripheral blood monocytes as source, 237, 244, 247

D

DAF-2, *see* 4,5-Diaminofluorescein
DAF-FM, *see* 3-Amino, 4-aminomethyl-2′,7′-difluorofluorescein
Dehydroascorbic acid, *see* Ascorbic acid
DEM, *see* Diethylmaleate
4,5-Diaminofluorescein
  nitric oxide detection in tissues
    fluorescence detection, 323–324
    loading, 323
    mitochondrial imaging, *see* Nitric oxide
    reactivity, 322–323
  S-nitrosylthiol assay
    advantages, 360
    fluorescence quenching by proteins, 353–354
    metallothionein analysis, 358–360
    plasma analysis, 354–357
    pregnancy studies, 357–358
    principles, 351–352
    sensitivity, 354
    validation, 352–353
  specificity of nitric oxide reactivity, 351
  structure, 322
2′,7′-Dichlorofluorescein diacetate
  fatty acid-induced oxidation assay, 203–205
  reactive nitrogen species sensitivity, 317
  reactive oxygen species assays
    cell density effects
      assay considerations, 92–93
      cell-to-volume versus cell-to-surface ratio effects, 96
      controls, 95–96
      dye loading, 93
      fibroblastoid cell line and culture, 93, 96
      flow cytometry, 93–94
    hydrogen peroxide assay, 155
    neuron flow cytometry
      cerebellar granule cell cultures, 78
      dissociated cerebellar cells, 76, 78
      dye loading, 74
      overview, 72–73
    spectrofluorometry
      cell lysis and fluorometry, 35–36
      dye uptake, 35
      materials, 35
      overview, 35
  reactive oxygen species sensitivity, 315–317
  structure, 64
  structure, 315
  tissue radical studies
    detection, 320–321
    loading, 318–320
Diethylmaleate
  advantages in oxidation studies, 267
  cell cycle regulation studies
    electrophoretic mobility shift assay
      glucocorticoid receptor, 262
      p53, 262

glutathione depletion conditions for eukaryotic cells, 261
materials, 260–261
p21 effects
Northern blot analysis, 262–263
Western blot analysis, 265
retinoblastoma protein effects, 265
reversibility of effects, 267
glutathione depletion mechanism, 260, 265, 267
Dihydroethidium, reactive oxygen species assay, 66
Dihydrofluorescein
hydrogen peroxide oxidation, 316
reactive oxygen species sensitivity, 315–317
structure, 315
tissue radical studies
detection, 320–321
loading, 318–320
Dihydrorhodamine 123
cell permeability, 314
reactive nitrogen species sensitivity, 318
reactive oxygen species
assay, 64, 66
sensitivity, 315–316, 318
structure, 315
tissue radical studies
detection, 320–321
loading, 318–320
3-(4,5-Dimethylthiazol-2-yl)-2,5-diphenyltetrazolium bromide
cortical neuron viability assay, 188–189
endothelial cell proliferation assay, 287
Dithiothreitol
lamina propria T lymphocyte proliferation effects, 236
oxidative damage reversal, 259
DTT, *see* Dithiothreitol

E

EF5
detection techniques, 15–16, 29
fluorescence-based immunohistochemistry of adducts
calibration, 20–21, 26–28, 30
CD31 colocalization, 16–17
dye conjugation with antibodies, 19–20
fixation, 20
monoclonal antibodies, 18
nonspecific binding of antibodies, 21, 23
quantitative analysis, 18, 25–28
oxygen effects on pharmacokinetics, 24
oxygen measurement
criteria for absolute measurement, 17–18
prospects, 28
rationale, 13–14
principles of 2-nitroimidazole-binding technique of oxygen measurement, 14–15
Electron paramagnetic resonance
ascorbate free radical, 273–274, 278–279
brain imaging of free radical reactions *in vivo*
computed tomography dual imaging
deconvolution, 495–496
probe distribution in brain, 503–504
time-resolved imaging, 504
mouse model, 500
nitroxyl probes
blood–brain barrier permeability, 496
esterase hydrolysis, 498–500
3-methoxycarbonyl-2,2,5,5-tetramethylpyrrolidine-1-oxyl synthesis and advantages, 496, 506
partition coefficients, 498
reactive oxygen species reactivity, 495
synthesis, 496–498
tissue distribution following injection, 501–503
types, 496–497
diethylmaleate effects on cell cycle proteins
glucocorticoid receptor, 262
p53, 262
nitric oxide detection in endothelial cells, 288–290
nitric oxide detection with iron dithiocarbamate traps
advantages, 507–508
ferrous and ferric complex interactions, 514–515, 517–521
nitric oxide and adduct preparation for studies, 510, 512
nitrite interference, 515, 519–521
organic solutions for traps, 518–519
oxygen effects on traps and adducts, 512–514, 516–517

rationale, 508–509
traps
comparative advantages and disadvantages, 521–522
types and structures, 509–510
redox-regulated transcription factors in endothelial cells
AP-1, 208–209
binding reaction, 211–212
gel electrophoresis, 212
materials, 209
nuclear extract preparation, 211
nuclear factor-κB, 208–209
principles, 209
probe labeling, 211
skin oxidative stress assay
antioxidative factor, 338–339
antioxidative potential, 338
applications, 339
instrumentation, 334, 336
nitroxide reduction, 333–334
principles, 334
S-band spectroscopy with TEMPO, 336, 338–339
vascular injury oxidative stress assays, 439–440, 442, 450, 452–453
ELISA, *see* Enzyme-linked immunosorbent assay
EMSA, *see* Electrophoretic mobility shift assay
Endothelial cell
fatty acid inflammation mediation studies
electrophoretic mobility shift assay of redox-regulated transcription factors
AP-1, 208–209
binding reaction, 211–212
gel electrophoresis, 212
materials, 209
nuclear extract preparation, 211
nuclear factor-κB, 208–209
principles, 209
probe labeling, 211
fatty acid-enriched media preparation, 203
flow cytometry detection of adhesion molecules, 216
glutathione assays
overview, 206–207
o-phthalaldehyde assay, 207
human umbilical vein endothelial cell preparation
exposure conditions, 201
isolation and culture, 202
overview, 200–201
solutions, 201–202
low-density lipoprotein uptake and measurement, 202–203
oxidative stress measurement
2′,7′-dichlorofluorescein diacetate assay, 203–205
lipid peroxidation assays, 205–206
thiobarbituric acid-reactive substances, 205
reverse transcription polymerase chain reaction of inflammatory gene induction
adhesion molecules, 212
amplification reaction, 215–216
gel electrophoresis, 216
materials, 213–214
monocyte chemoattractant protein-1, 212
overview, 213
reverse transcription, 215
RNA isolation, 214–215
tumor necrosis factor-α, 213
transient transfection
cell culture, 218
dual luciferase reporter gene assay following fatty acid treatment, 217–219
materials, 218
overview, 217
transfection, 218
migration
Boyden chamber assay, 175–176
high-throughput assay
advantages, 182
angiogenic factors, 178, 181–182
cell culture, 176–178
cell number effects, 181
fluorescence labeling of cells, 179–180
invasion assay, 178–179
Matrigel preparation, 176–177
migration assay, 179
principles, 175–176
trypsinization of cells, 177–178
myocardial infarction response
apoptosis and TUNEL assay, 403–404
bromodeoxyuridine-positive cells, 402–403
nitric oxide regulation of proliferation

cell isolation
coronary endothelial cell isolation from rat, 282–283
overview, 281–282
skeletal muscle endothelial cell isolation from rat, 283–284
vascular segment endothelial cell isolation from human, 284–285
nitric oxide production assays
electron paramagnetic resonance, 288–290
GTP cyclohydrolase I activity assay, 292–294
NADPH analysis, 294
nitric oxide synthase activity and Western blot, 290–291
tetrahydrobiopterin assay, 291–292
overview, 280–281
proliferation assays
bromodeoxyuridine incorporation, 286–287
cell counting, 285
MTT assay, 287
proliferating cell nuclear antigen assay, 287–288
tritiated thymidine incorporation, 285–286
tetrahydrobiopterin production manipulation, 295
nitric oxide response characterization
bromodeoxyuridine proliferation assay, 415
chemotaxis and invasion, 414
gelatinase zymography, 414–415
vascular injury, *see* Vascular injury
Enzyme-linked immunosorbent assay
hemangioma oxidative stress effects on monocyte chemoattractant protein-1 expression, 431–432
tissue factor, 227
Epifluorescence microscopy
EF5 adduct immunohistochemistry for oxygen determination
calibration, 20–21, 26–28, 30
CD31 colocalization, 16–17
dye conjugation with antibodies, 19–20
fixation, 20
monoclonal antibodies, 18
nonspecific binding of antibodies, 21, 23
quantitative analysis, 18, 25–28
lysine oxidation product immunohistochemistry, *see* 4-Hydroxynonenal-lysine; Malondialdehyde-lysine
nitric oxide imaging in mitochondria, *see* Nitric oxide
reactive oxygen species assays, 69
redox state changes in mitochondria
fibroblast NAD(P) redox state determination, 145
Fp/NAD(P)H ratio imaging in muscle fibers, 139, 141, 143, 145
instrumentation, 137
Src immunolocalization
cell culture, 196
fluorescence microscopy, 198
indirect immunofluorescence staining, 196–198
materials, 196
EPR, *see* Electron paramagnetic resonance
Etanidazole, oxygen effects on pharmacokinetics, 24
Extracellular signal-regulated kinase, *see* Mitogen-activated protein kinase

F

Fatty acid, inflammation mediation
dietary sources, 199–200
endothelial cell studies
electrophoretic mobility shift assay of redox-regulated transcription factors
AP-1, 208–209
binding reaction, 211–212
gel electrophoresis, 212
materials, 209
nuclear extract preparation, 211
nuclear factor-κB, 208–209
principles, 209
probe labeling, 211
exposure conditions, 201
fatty acid-enriched media preparation, 203
flow cytometry detection of adhesion molecules, 216
glutathione assays
overview, 206–207
o-phthalaldehyde assay, 207
human umbilical vein endothelial cell preparation

overview, 200–201
solutions, 201–202
isolation and culture, 202
low-density lipoprotein uptake and measurement, 202–203
oxidative stress measurement
2′,7′-dichlorofluorescein diacetate assay, 203–205
lipid peroxidation assays, 205–206
thiobarbituric acid-reactive substances, 205
reverse transcription polymerase chain reaction of inflammatory gene induction
adhesion molecules, 212
amplification reaction, 215–216
gel electrophoresis, 216
materials, 213–214
monocyte chemoattractant protein-1, 212
overview, 213
reverse transcription, 215
RNA isolation, 214–215
tumor necrosis factor-α, 213
transient transfection
cell culture, 218
dual luciferase reporter gene assay following fatty acid treatment, 217–219
materials, 218
overview, 217
transfection, 218
pathways, 198–199
Fatty acid oxidation, *see also* Lipid peroxidation
α-oxidation
assays for peroxisomes
cell studies, 371–372
cerebronic acid assay and carbon-14 labeling, 370–372
isolated organelle fatty acid oxidation, 370
phytanic acid assay, 370
defects in disease, 369
peroxisome pathway, 368–369
β-oxidation
assays for peroxisomes
cell studies, 366
isolated organelle fatty acid oxidation, 366–367
lignoceric acid carbon-14 labeling, 367
nervonic acid carbon-14 labeling, 367–368
peroxisome isolation by density gradient centrifugation, 364–365
peroxisome isolation by immunomagnetic sorting, 366
pathways
enzymatic organization, 362–364
mitochondria, 361–364
peroxisomes, 361–363
Flow cytometry
endothelial cell adhesion molecule detection, 216
lipid peroxidation assay with fluoresceinated phosphoethanolamine
advantages, 86, 90–91
benzoyl peroxide oxidation studies, 88–89
data acquisition, 86
materials, 82
principles, 81
probe stability, 85
red blood cells
isolation, 82
labeling, 82, 84–85
vitamin E effects, 89–90
mononuclear cell redox state changes in mitochondria, 137–138, 149–150
neuron studies
advantages, 71
cerebellar neurons
culture, 73–74
isolation from rat pup, 73
instrument settings, 74–76
mitochondrial membrane potential measurement
cerebellar granule cell cultures, 78
dissociated cerebellar cells, 76, 78
rationale, 72
rhodamine 123 loading, 74
propidium iodide viability assay, 74
prospects, 79
reactive oxygen species measurement
cerebellar granule cell cultures, 78
2′,7′-dichlorofluorescein diacetate loading, 74
dissociated cerebellar cells, 76, 78
overview, 72–73
reactive oxygen species
assay, 69–70
cell density effects on production, 93–94

Fluorescence microscopy, *see* Confocal microscopy; Epifluorescence microscopy
Fluoromisonidazole, oxygen effects on pharmacokinetics, 24

G

Glutamate
  cortical neuron oxidative death induction
    advantages of system, 183–184
    gene delivery into immature neurons, 189–190
    glutamate depletion of glutathione and antioxidant protection, 187–188, 190
    plating of cells for cytotoxicity studies, 186–187
    primary culture for glutathione depletion studies, 184–186
    viability assays
      lactate dehydrogenase assay, 188–189
      morphology, 188
      MTT reduction, 188–189
  c-Src activation in neurons
    immunolocalization
      cell culture, 196
      fluorescence microscopy, 198
      indirect immunofluorescence staining, 196–198
      materials, 196
    overview, 191–192
    Src kinase assay
      gel electrophoresis, 195
      immunoprecipitation, 195
      materials, 194–195
      substrates, 195
    tyrosine phosphorylation profile
      cell culture, 192–193
      materials, 192
      metabolic labeling, 193
      Western blot, 193–194
  toxicity pathways, 191
Glutathione
  assays
    overview, 206–207
    o-phthalaldehyde assay, 207
  depletion effects on cell cycle, *see* Diethylmaleate
  inhibitors of synthesis, 260
  intracellular concentrations, 101
  lamina propria T lymphocytes
    assay, 246
    enzyme inhibitor studies, 236
    metabolism, 232–233
  linker for activation of T cells, effects
    modulators, 250, 255–256
    phosphorylation, 257
    subcellular localization, 253, 255
  vascular injury oxidative stress assay, 440–443
4-[*N*-(*S*-Glutathionylacetyl)amino] phenylarsenoxide, *see* GSAO
GSAO
  5,5′-dithiobis(2-nitrobenzoic acid) reactivity, 108–109
  rationale for synthesis, 102–103
  synthesis
    4-[*N*-(bromoacetyl)amino]phenylarsonic acid, 104–105
    4-[*N*-(bromoacetyl)amino] phenylarsenoxide, 105–106
    4-[*N*-(*S*-glutathionylacetyl)amino] phenylarsenoxide, 106–107
GSAO-B
  5,5′-dithiobis(2-nitrobenzoic acid) reactivity, 108–109
  redox-active protein labeling
    cell surface labeling, 109–110
    Western blot detection of labeled proteins, 110–111
  structure, 103
  synthesis from GSAO, 107–108
GSH, *see* Glutathione
GTP cyclohydrolase I, endothelial cell activity assay, 292–294

H

HE, *see* Hydroethidine
Hemangioma
  clinical features, 422
  EOMA cell model, 422–424
  4-hydroxy-2-noneal protein adduct immunodetection, 428–429
  monocyte chemoattractant protein-1
    macrophage chemotaxis role, 423
    oxidative stress effects on expression, 431–432
  phorbol 12-myristate 13-acetate induction of oxidation, 429–432

Western blot of protein carbonyls
controls, 428
derivatization, 426, 428
EOMA cell culture, 425–426
extraction, 426
gel electrophoresis and staining, 426
Hemoglobin, fluorescence minimization in NADH imaging, 483, 487–489
HIF-1, *see* Hypoxia inducible factor-1
High-performance liquid chromatography
ascorbate assay, 270–271
cysteine assay, 239–241
cis-parinic acid-labeled phospholipids, 164–166
vitamin E assay with coulometric electrode array detection in biological samples
array features, 327–328
current–voltage response curve, 330
detection, 327
extraction, 331
instrumentation, 328
mobile phase, 328
rationale, 326–327
sensitivity, 332
standard curve, 329
tissue preparation, 331
HPLC, *see* High-performance liquid chromatography
Hydroethidine
detection with ethidium product, 313–314
loading of tissues
precautions, 313
solutions, 312
technique, 313
peroxynitrite sensitivity, 311–312
reactive oxygen species sensitivity, 311–312
Hydrogen peroxide
Alzheimer's disease, fluorescence imaging in models, 470
apoptosis induction studies
cell lines, 152
hydrogen peroxide treatment, 152
intracellular oxidant measurement
flow cytometry assay for hydrogen peroxide, 155
lucigenin-based assay for superoxide, 153–155
intracellular pH assay with 2′,7′-bis(2-carboxyethyl)-5,6-carboxyfluorescein, 156–157, 159
overview of reactive oxygen species induction, 150–152, 159
poly(ADP-ribose) polymerase cleavage detection by Western blot, 152–153
reductive stress-induced apoptosis, 152
lamina propria T lymphocyte proliferation effects, 235
nuclear import effects in cultured cells
advantages of assay, 132, 134
caveats of assay, 132, 134
confocal microscopy, 128, 130, 132
digitonin-permeabilized cell assay, 130, 132
hydrogen peroxide treatment, 128–129
import substrate preparation, 126–127
incubation conditions for import, 128
smooth muscle cell culture, 125–126
peroxisome production, 372
Hydroxyl radical, fluorescence imaging in Alzheimer's disease models, 468–469
4-Hydroxynonenal-lysine
disease accumulation in proteins, 340–341, 377
fluorescence microscopy immunohistochemistry
antibodies
dilution, 345
incubations and washes, 343–344
sources, 342
atherosclerotic lesions
animals, 345–346
humans, 346–347
controls, 344–345
counterstaining, 344
human tissue sample sources, 342–343
principles, 341–342
slide preparation, 343
hemangioma immunodetection, 428–429
ultrastructural localization of modified proteins
antibody specificity, 374–375
controls, 376–377
fixation, 375
immunogold electron microscopy, 375–376
light microscopy, 375
quantitative data analysis, 376
Hypoxia
angiogenesis triggering, 392
definition, 31

measurement in tissue, *see* EF5
mitochondrial response, 32
pharmacokinetic effects, 23–24
tumors, 3–5
vascular insufficiency disorders, 5
Hypoxia inducible factor-1
activation, 32
components, 32
detection of HIF-1α in cells
materials, 33
nuclear extract preparation, 33–34
Western blot, 34
mitochondria role in induction
mitochondrial genome-depleted cell lines
desferrioxamine response, 37–38
materials, 36, 37
polymerase chain reaction for genome depletion verification, 37
principles, 36
reactive oxygen species generation in response to hypoxia, 37–38
mitochondrial inhibitor studies
inhibitor incubations and responses, 39–40
materials, 38–39
overview, 38
overview, 32
oxygen sensing, 31–32
target genes, 32

I

Iron dithiocarbamate traps, *see* Electron paramagnetic resonance

J

Jun N-terminal kinase, *see* Mitogen-activated protein kinase

L

Lamina propria T lymphocyte
CD3 reactivity restoration by peripheral blood monocytes and lamina propria macrophages, 242–246
cysteine effects and assays
acid-soluble thiol assay with 5,5′-dithiobis(2-nitrobenzoic acid), 238–239
cystine versus cysteine effects on CD3 reactivity, 245–246
high-performance liquid chromatography assay, 239–241
lamina propria macrophages as source, 237, 247
peripheral blood monocytes as source, 237, 244, 247
function, 232
glutathione
assay, 246
enzyme inhibitor studies, 236
metabolism, 232–233
oxidative stress in disease, 247
preparation from human gut, 233–234
proliferation
dithiothreitol effects, 236
hydrogen peroxide effects, 235
limitations, 232
2-mercaptoethanol effects, 236–237
stimulation *in vitro*, 234–235
LAT, *see* Linker for activation of T cells
Linker for activation of T cells
functions, 248–249
phosphorylation
rheumatoid arthritis defects, 257
signaling, 249
redox studies of regulation
glutathione effects
modulators, 250, 255–256
phosphorylation, 257
subcellular localization, 253, 255
immunofluorescence microscopy, 251–252, 255–256
T lymphocyte isolation, 249–250
Western blot analysis
phosphorylated protein, 252–253, 257
subcellular localization, 250–251
whole-cell lysates, 252
structure, 249
Lipid peroxidation
Alzheimer's disease, fluorescence imaging in models, 470–472
cellular targets, 80
detection techniques, 80–81
fatty acid
inflammation mediation assays, 205–206
oxidation, *see* Fatty acid oxidation
flow cytometry assay with fluoresceinated phosphoethanolamine

advantages, 86, 90–91
benzoyl peroxide oxidation studies, 88–89
data acquisition, 86
materials, 82
principles, 81
probe stability, 85
red blood cells
isolation, 82
labeling, 82, 84–85
vitamin E effects, 89–90
lysine oxidation by products, *see* 4-Hydroxynonenal-lysine; Malondialdehyde-lysine
cis-parinic acid reporter of phospholipid oxidation
availability, 161–162
cell culture for membrane integration, 161–162
high performance liquid chromatography of labeled phospholipids, 164–166
high-performance thin-layer chromatography of phospholipids during oxidation-induced apoptosis, 166–167
phosphatidylserine site-selective oxidation during oxidation-induced apoptosis, 168, 170–172
phospholipid integration in different cell types, 161–162, 164
positional distribution in phospholipids, 167–168
vitamin E protective effects, 172–173
ultrastructural localization of lipid peroxides in retinal pigment epithelium
benzidine-reactive substance detection, 379, 381, 390
eye processing and fixation, 380–381
light induction of retinal damage
age-related macular degeneration, 378–379
exposure protocols, 379–380
rat eye effects
constant light exposure of 6,000 lux, 381, 385
constant light exposure of 6,000 lux followed by dark period, 385, 388
short light exposure of 150,000 lux, 385
tetramethylbenzidine reaction with lipid peroxides, 389–391
α-Lipoamide dehydrogenase
fluorescence, 480–481
redox state changes in muscle fiber mitochondria, 136–139
Lysine oxidation, *see* 4-Hydroxynonenal-lysine; Malondialdehyde-lysine

M

Magnetic resonance imaging, EF5 detection, 15, 29
N-[3-(*N*-Maleimidyl)proprionyl] biocytin
5,5′-dithiobis(2-nitrobenzoic acid) reactivity, 108–109
redox-active protein labeling
cell surface labeling, 109–110
Western blot detection of labeled proteins, 110–111
thiol reactivity on cell surface, 102
Malondialdehyde-lysine
disease accumulation, 340–341
fluorescence microscopy immunohistochemistry
antibodies
dilution, 345
incubations and washes, 343–344
sources, 342
atherosclerotic lesions
animals, 345–346
humans, 346–347
controls, 344–345
counterstaining, 344
human tissue sample sources, 342–343
principles, 341–342
slide preparation, 343
MAPK, *see* Mitogen-activated protein kinase
MCP-1, *see* Monocyte chemoattractant protein-1
MDA-lysine, *see* Malondialdehyde-lysine
Metallothionein, *S*-nitrosylation analysis, 358–360
3-Methoxycarbonyl-2,2,5,5-tetramethylpyrrolidine-1-oxyl, *see* Electron paramagnetic resonance
Mitochondria
Alzheimer's disease model measurements
calcium, 472
membrane potential, 472
fatty acid oxidation, *see* Fatty acid oxidation
flow cytometry, neuron mitochondrial membrane potential measurement

cerebellar granule cell cultures, 78
dissociated cerebellar cells, 76, 78
rationale, 72
rhodamine 123 loading, 74
nitric oxide imaging with fluorescent dyes
advantages, 301, 303
cell cultre, 298–299
controls, 299–300
4,5-diaminofluorescein diacetate
preparation and characteristics, 297
dye loading, 299
fluorescence microscopy and image acquisition, 300–301
limitations, 303
materials, 297
MitoTracker preparation and characteristics, 297–298
synthase in mitochondria, 296
reactive oxygen species production in response to hypoxia
mitochondrial genome-depleted cell lines and hypoxia inducible factor-1 induction
desferrioxamine response, 37–38
materials, 36, 37
polymerase chain reaction for genome depletion verification, 37
principles, 36
reactive oxygen species generation in response to hypoxia, 37–38
mitochondrial inhibitor studies of hypoxia inducible factor-1 induction
inhibitor incubations and responses, 39–40
materials, 38–39
overview, 38
overview, 32
redox state imaging
confocal microscopy of α-lipoamide dehydrogenase redox state changes in muscle fibers, 136–139
epifluorescence microscopy
fibroblast NAD(P) redox state determination, 145
Fp/NAD(P)H ratio imaging in muscle fibers, 139, 141, 143, 145
instrumentation, 137
fibroblast culture, 136
mononuclear cell isolation and flow cytometry, 137–138, 149–150
muscle fiber preparation, 135–136
nicotinamide adenine dinucleotide system, 135
Mitogen-activated protein kinase
phosphorylation cascades, 53
reactive oxygen species activation
extracellular signal-regulated kinases, 55
immune complex kinase assays, 57–58
in-gel kinase assays, 58–59
Jun N-terminal kinase, 53–54
kinase inhibitor analysis of specific signaling cascades, 59, 61
p38, 54
Western blot analysis of phosphoproteins, 55–57
types and functions, 53
Monocyte chemoattractant protein-1
hemangioma studies
macrophage chemotaxis role, 423
oxidative stress effects on expression, 431–432
reverse transcription polymerase chain reaction, 212
MPB, *see* *N*-[3-(*N*-Maleimidyl)proprionyl] biocytin
MRI, *see* Magnetic resonance imaging
MTT, *see* 3-(4,5-Dimethylthiazol-2-yl)-2,5-diphenyltetrazolium bromide
Myocardial infarction, *see* Angiogenesis
chronic model in rat
blood flow measurement by neutron microsphere technique, 398–399
dobutamine stress testing, 396–398
endothelial cell response
apoptosis and TUNEL assay, 403–404
bromodeoxyuridine-positive cells, 402–403
immunohistochemistry of angiogenic factors, 405–407
left ventricular infarct zone measurement, 395–396
postoperative monitoring, 394–395
pressure recordings, 396
surgery, 393–394
tissue processing, 399
vessel density measurement, 399–400
Western blotting of angiogenic factors, 404–405
ischemia–reperfusion injury and angiogenesis, 392–393

N

NAD(P)H
  endothelial cell producion assay, 294
  frozen-quenched tissue three-dimensional NADH redox imaging
    brain freezing optimization, 475–476
    cutting and grinding of tissue, 476
    flavoprotein/PN ratio, 479–481
    fluorescence measurements, 476–477
    forebrain ischemia model, 478
    freeze trapping, 481–482
    redox ratio calibration, 477–478
    seizure model, 478–479
  redox ratio, 475
  redox state changes in mitochondria
    epifluorescence microscopy instrumentation, 137
    fibroblast NAD(P) redox state determination, 145
    Fp/NAD(P)H ratio imaging in muscle fibers, 139, 141, 143, 145
  in vivo flourescence imaging of NADH redox state in brain
    focal cerebral ischemia studies at different glycemia states, 490, 494
    hemoglobin interference, 483, 487–489
    image analysis, 486
    instrumentation, 481–486
    photodecomposition, 486–487
NAD(P)H oxidase
  antisense inhibition in smooth muscle cells, 230
  tissue factor induction role, 230–231
  vascular injury activation
    assays
      chemiluminescence assays, 450, 452
      electron paramagnetic resonance, 450, 452–453
      particular fraction preparation, 447, 450
    evidence, 447
    inhibitor study limitations, 451–452
    number of oxidases, 454
    phagocyte oxidase system comparison, 453–454
Neuron
  apoptosis induction by oxidative stress, 183–184
  flow cytometry studies
    advantages, 71
    cerebellar neurons
      culture, 73–74
      isolation from rat pup, 73
    instrument settings, 74–76
    mitochondrial membrane potential measurement
      cerebellar granule cell cultures, 78
      dissociated cerebellar cells, 76, 78
      rationale, 72
      rhodamine 123 loading, 74
    propidium iodide viability assay, 74
    prospects, 79
    reactive oxygen species measurement
      cerebellar granule cell cultures, 78
      2′,7′-dichlorofluorescein diacetate loading, 74
      dissociated cerebellar cells, 76, 78
      overview, 72–73
  glutamate-induced oxidative death in cortical neurons
    advantages of system, 183–184
    gene delivery into immature neurons, 189–190
    glutamate depletion of glutathione and antioxidant protection, 187–188, 190
    plating of cells for cytotoxicity studies, 186–187
    primary culture for glutathione depletion studies, 184–186
    viability assays
      lactate dehydrogenase assay, 188–189
      morphology, 188
      MTT reduction, 188–189
  reactive oxygen species pathogenesis, 72, 183
  c-Src, glutamate activation studies
    immunolocalization
      cell culture, 196
      fluorescence microscopy, 198
      indirect immunofluorescence staining, 196–198
      materials, 196
    overview, 191–192
    Src kinase assay
      gel electrophoresis, 195
      immunoprecipitation, 195
      materials, 194–195
      substrates, 195
    tyrosine phosphorylation profile
      cell culture, 192–193

materials, 192
metabolic labeling, 193
Western blot, 193–194
NF-κB, *see* Nuclear factor-κB
Nitric oxide
angiogenesiss stimulation
cyclic GMP assay, 419
endothelial cell characterization
bromodeoxyuridine proliferation assay, 415
chemotaxis and invasion, 414
gelatinase zymography, 414–415
fibroblast growth factor-2 induction, 421
nitrate/nitrite assays
colorimetric Griess reaction, 416–417
fluorimetric assay, 417
nitric oxide synthase assays
cell monolayer assay, 418
ion-exchange resin preparation, 419
lysate assay, 417–418
reverse transcription polymerase chain reaction of isoform expression, 419–420
overview, 280, 407–409, 420
pharmacological tools for nitric oxide level modification, 412–413, 415–416
rabbit cornea assay
advantages, 409–410
cell and tissue implants, 410–411
considerations for success, 411–412
histological examination, 411
quantification of angiogenesis, 411
slow-release preparations for testing, 410
surgery, 410
electron paramagnetic resonance detection with iron dithiocarbamate traps
advantages, 507–508
ferrous and ferric complex interactions, 514–515, 517–521
nitric oxide and adduct preparation for studies, 510, 512
nitrite interference, 515, 519–521
organic solutions for traps, 518–519
oxygen effects on traps and adducts, 512–514, 516–517
rationale, 508–509
traps
comparative advantages and disadvantages, 521–522
types and structures, 509–510
endothelial cell proliferation regulation
cell isolation
coronary endothelial cell isolation from rat, 282–283
overview, 281–282
skeletal muscle endothelial cell isolation from rat, 283–284
vascular segment endothelial cell isolation from human, 284–285
nitric oxide production assays
electron paramagnetic resonance, 288–290
GTP cyclohydrolase I activity assay, 292–294
NADPH analysis, 294
nitric oxide synthase activity and Western blot, 290–291
tetrahydrobiopterin assay, 291–292
overview, 280–281
proliferation assays
bromodeoxyuridine incorporation, 286–287
cell counting, 285
MTT assay, 287
proliferating cell nuclear antigen assay, 287–288
tritiated thymidine incorporation, 285–286
tetrahydrobiopterin production manipulation, 295
fluorescent probes for detection in tissues, *see also* 3-Amino, 4-aminomethyl-2′,7′-difluorofluorescein; 4,5-Diaminofluorescein; 2,7′-Dichlorodihydrofluorescein
dye loading, 310–311
overview, 307–310, 324–325
precautions and controls, 307, 310
units for data reporting, 310
heme protein complexes, 508
mitochondrial imaging with fluorescent dyes
advantages, 301, 303
cell cultre, 298–299
controls, 299–300
4,5-diaminofluorescein diacetate preparation and characteristics, 297
dye loading, 299
fluorescent microscopy and image acquisition, 300–301
limitations, 303

materials, 297
MitoTracker preparation and characteristics, 297–298
synthase in mitochondria, 296
nitrate formation, 507
nitrite formation, 507
solubility, 506
sources in cells, 506–507
vascular effects, 280
[2-(2-Nitro-1*H*-imidazol-1-yl)-*N*-(2,2,3,3,3-pentafluoropropyl)acetamide], *see* EF5
S-Nitrosylation
albumin, 348, 357–358
assays
4,5-diaminofluorescein assay
advantages, 360
fluorescence quenching by proteins, 353–354
metallothionein analysis, 358–360
plasma analysis, 354–357
pregnancy studies, 357–358
principles, 351–352
sensitivity, 354
validation, 352–353
overview of approaches, 350–351
biological functions, 349–350
decomposition, 348–349
formation, 348
glutathione, 347–348
tissue distribution, 347–348
NO, *see* Nitric oxide
Northern blot, p21, 262–263
Nuclear factor-κB
electrophoretic mobility shift assay, 208–209
vascular injury activation assay, 445–446
Nuclear import
nuclear localization signals, 123–124
overview, 123–124
reactive oxygen species effects in cultured cells
advantages of assay, 132, 134
caveats of assay, 132, 134
confocal microscopy, 128, 130, 132
digitonin-permeabilized cell assay, 130, 132
hydrogen peroxide treatment, 128–129
import substrate preparation, 126–127
incubation conditions for import, 128
smooth muscle cell culture, 125–126
regulation, 123–125

O

Oxidation-specific epitopes, *see* 4-Hydroxynonenal-lysine; Malondialdehyde-lysine
Oxygen
cell sensing, *see* Hypoxia inducible factor-1
iron dithiocarbamates, effects on traps and nitric oxide adducts, 512–514, 516–517
measurement, *see also* EF5
Controls Katharobic sensor, 7, 9
difficulty in tissues, 5
electrode calibration with ascorbic oxidase, 10
gas–liquid phase equilibrium in spinner cultures, 11, 13
sensor development, 5–7
spinner-vial system, 9–10
steep oxygen gradient measurement approaches, 13–14
tubing in measurement systems, 10–11
radiation damage modification, 5
singlet oxygen, *see* Singlet oxygen
tumor oxygenation, 3–5

P

p21, diethylmaleate effects
Northern blot analysis, 262–263
Western blot analysis, 265
p38, *see* Mitogen-activated protein kinase
cis-Parinic acid, *see* Lipid peroxidation
Peroxisome
functions, 361
α-oxidation of fatty acids
assays
cell studies, 371–372
cerebronic acid assay and carbon-14 labeling, 370–372
isolated organelle fatty acid oxidation, 370
phytanic acid assay, 370
defects in disease, 369
pathway, 368–369
β-oxidation of fatty acids
assays
cell studies, 366
isolated organelle fatty acid oxidation, 366–367
lignoceric acid carbon-14 labeling, 367

nervonic acid carbon-14 labeling, 367–368
peroxisome isolation by density gradient centrifugation, 364–365
peroxisome isolation by immunomagnetic sorting, 366
pathway and enzymatic organization, 361–364
oxygen consumption, 372
reactive oxygen species generation, 372
Peroxynitrite
Alzheimer's disease, fluorescence imaging in models, 468–469
fluorescent probes for detection in tissues, *see also* 2,7′-Dichlorodihydrofluorescein; Dihydrorhodamine 123; Hydroethidine
dye loading, 310–311
overview, 307–310
precautions and controls, 307, 310
units for data reporting, 310
PET, *see* Positron emission tomography
Phosphatidylserine
cis-parinic acid site-selective oxidation during oxidation-induced apoptosis, 168, 170–172
translocation in apoptosis, 159–160
Phosphorescence, *see* Singlet oxygen
Pimonidazole, oxygen effects on pharmacokinetics, 24
PNCA, *see* Proliferating cell nuclear antigen
Positron emission tomography, EF5 detection, 15, 29
Presenelin-1, *see* Alzheimer's disease
Proliferating cell nuclear antigen, endothelial cell proliferation assay, 287–288

R

RA, *see* Rheumatoid arthritis
Rb, *see* Retinoblastoma protein
Reactive oxygen species, *see also individual species*
antioxidant defenses, 220, 258
apoptosis mediation, *see* Hydrogen peroxide
brain imaging *in vivo, see* Electron paramagnetic resonance
cell density effects on production
assay considerations, 92–93
cell-to-volume versus cell-to-surface ratio effects, 96
contact inhibition role of reactive oxygen species, 100
controls, 95–96
2′,7′-dichlorofluorescein diacetate loading, 93
fibroblastoid cell line and culture, 93, 96
flow cytometry, 93–94
mechanism of cell density effects, 98–99
oxygen radical source identification with inhibitors, 97–98
damage mechanisms, 258
fatty acid induction, *see* Fatty acid, inflammation mediation
flow cytometry, neuron measurements
cerebellar granule cell cultures, 78
2′,7′-dichlorofluorescein diacetate loading, 74
dissociated cerebellar cells, 76, 78
overview, 72–73
fluorescent probe assays
2′,7′-dichlorofluorescein diacetate assay
cell lysis and fluorometry, 35–36
dye uptake, 35
materials, 35
overview, 35
flow cytometry, 69–70
fluorescence microscopy, 69
principles, 63
probe selection, 63–64, 66, 70–71
spectrofluorometry, 66–69
tissue assay probes, *see* 5-(and 6)-Carboxy-2,7′-dichlorodihydrofluorescein; 2,7′-Dichlorodihydrofluorescein; Dihydrofluorescein; Dihydrorhodamine 123
generation systems, 259–260
immune system function, 91, 259
ischemia–reperfusion injury, 392–393
lipid peroxidation, *see* Lipid peroxidation
mitochondrial production in response to hypoxia
mitochondrial genome-depleted cell lines and hypoxia inducible factor-1 induction
desferrioxamine response, 37–38
materials, 36, 37
polymerase chain reaction for genome depletion verification, 37
principles, 36

reactive oxygen species generation in response to hypoxia, 37–38
mitochondrial inhibitor studies of hypoxia inducible factor-1 induction
inhibitor incubations and responses, 39–40
materials, 38–39
overview, 32
mitogen-activated protein kinase induction, *see* Mitogen-activated protein kinase
neuropathogenesis, 72, 455, 457–459, 494–495
nuclear transport effects, *see* Nuclear import
signal transduction, 62, 91–92, 220, 373
sources in cells, 62, 72, 220, 258
tissue factor modulation, *see* Tissue factor
vascular injury, *see* Vascular injury
Redox-active protein probes, *see* GSAO, GSAO-B; *N*-[3-(*N*-Maleimidyl)proprionyl] biocytin
Retinal pigment epithelium, light-induced lipid peroxidation, *see* Lipid peroxidation
Retinoblastoma protein, diethylmaleate effects, 265
Reverse transcription polymerase chain reaction
inflammatory gene induction in endothelial cells
adhesion molecules, 212
amplification reaction, 215–216
gel electrophoresis, 216
materials, 213–214
monocyte chemoattractant protein-1, 212
overview, 213
reverse transcription, 215
RNA isolation, 214–215
tumor necrosis factor-α, 213
nitric oxide synthase isoforms, 419–420
Rheumatoid arthritis
linker for activation of T cells, defects, *see* Linker for activation of T cells
oxidative stress, 248
T cell hyporesponsiveness in synovial fluid, 248
Rhodamine 123
flow cytometry of neuron mitochondrial membrane potential
cerebellar granule cell cultures, 78
dissociated cerebellar cells, 76, 78
dye loading, 74
rationale, 72
reduced form, *see* Dihydrorhodamine 123
ROS, *see* Reactive oxygen species
RT-PCR, *see* Reverse transcription polymerase chain reaction

S

Scanning electrochemical microscopy
breast cancer cells
cell culture, 115
current–distance curves, 120–121
data acquisition, 116
instrumentation, 116
materials, 115
protein kinase Cα expression, 114–115, 120–121
redox activity mapping, 117, 119–122
topographic imaging with hydrophilic mediators, 116–117, 122
commercial instruments, 122
microelectrode, 113–114
principles, 112–113
SECM, *see* Scanning electrochemical microscopy
Single cell voltammetry, *see* Scanning electrochemical microscopy
Singlet oxygen
phosphorescence detection
cercosporin-generating fungus studies
data acquisition, 52
light activation, 47
overview, 42
quenchers in media, 50–51
sample preparation, 49–50
instrumentation, 42–44
keratinocytes stained with Rose Bengal
artifacts, 45
cell suspensions, 45–46
data acquisition, 45–47
model system advantages, 44
sample preparation, 44–45
resonance energy transfer, 41
spectrum, 41
sources in cells, 51
Skin oxidative stress, *see* Electron paramagnetic resonance
Smooth muscle cell, *see* Tissue factor
SOD, *see* Superoxide dismutase
c-Src

glutamate activation studies in neurons
immunolocalization
cell culture, 196
fluorescence microscopy, 198
indirect immunofluorescence staining, 196–198
materials, 196
overview, 191–192
Src kinase assay
gel electrophoresis, 195
immunoprecipitation, 195
materials, 194–195
substrates, 195
tyrosine phosphorylation profile
cell culture, 192–193
materials, 192
metabolic labeling, 193
Western blot, 193–194
neuron expression, 191
phosphorylation, 192
Superoxide
Alzheimer's disease, fluorescence imaging in models, 468
antioxidant defenses, 220
lucigenin-based assay, 153–155
Superoxide dismutase
manganese superoxide dismutase, Alzheimer's disease models
immunocytochemistry, 473–474
PC12 cell overexpression, 460–461
transgenic mice overexpression, 465–466
Western blot, 473
vascular injury oxidative stress assay, 442–443

T

T cell
activation adaptor protein, *see* Linker for activation of T cells
redox regulation, *see* Lamina propria T lymphocyte
Tetrahydrobiopterin, endothelial cells
assay, 291–292
production manipulation, 295
Tetramethylbenzidine, lipid peroxide reaction and ultrastructural localization, 389–391
TF, *see* Tissue factor
Tissue factor
activation, 221–222
blood coagulation role, 221
disease roles, 222–223
redox-regulated expression
antioxidant inhibition studies, 224–226, 228–229
reactive oxygen species source determinaation, 229–231
smooth muscle cell studies
enzyme-linked immunosorbent assay, 227
Northern blot, 227
procoagulant activity, 228
tissue culture, 227
summary of studies, 224–225
transactivation assays, 230–21
signal transduction and transcriptional regulation, 223–224
structure, 222
TNF-α, *see* Tumor necrosis factor-α
Tocopherols, *see* Vitamin E
Tocotrienols, *see* Vitamin E
Tumor
hypoxia, *see* Hypoxia
redox activity, *see* Scanning electrochemical microscopy
Tumor necrosis factor-α, reverse transcription polymerase chain reaction, 213

V

Vascular endothelial growth factor
angiogenesis studies in myocardial infarction
immunohistochemistry, 405–406
Western blot, 404–405
angiogenic factor, 175, 392–393
gene therapy, 393
Vascular injury
antioxidant protection studies, 446
NAD(P)H oxidase activation
assays
chemiluminescence assays, 450, 452
electron paramagnetic resonance, 450, 452–453
particular fraction preparation, 447, 450
evidence, 447
inhibitor study limitations, 451–452
number of oxidases, 454
phagocyte oxidase system comparison, 453–454

oxidative stress characterization in models
coelenterazine chemiluminescence, 438–439, 443
control arteries, 436
electron paramagnetic resonance, 439–440, 442
glutathione assay, 440–443
lucigenin chemiluminescence, 436–438, 442–443
superoxide dismutase assay, 442–443
phases of response, 432–433
rabbit balloon injury model, 433–435
reactive oxygen species sources, 447–449
redox-dependent signaling
nuclear factor-κB activation assay, 445–446
overview of events, 443–445
VEGF, *see* Vascular endothelial growth factor
Vitamin C, *see* Ascorbic acid
Vitamin E
compounds with activity, 326
high-performance liquid chromatography with coulometric electrode array detection in biological samples
array features, 327–328
current–voltage response curve, 330
detection, 327
extraction, 331
instrumentation, 328
mobile phase, 328
rationale, 326–327
sensitivity, 332
standard curve, 329
tissue preparation, 331
oxidation protection in fluoresceinated phosphoethanolamine flow cytometry assay, 89–90
phospholipid oxidation protective effects, 172–173
Western blot
Alzheimer's disease protein expression
β-amyloid precursor protein, 473
manganese superoxide dismutase, 473
presenelin-1, 473
angiogenic factors, 404–405
c-Src, 193–194
GSAO-B-labeled proteins, 110–111
hemangioma protein carbonyls
controls, 428
derivatization, 426, 428
EOMA cell culture, 425–426
extraction, 426
gel electrophoresis and staining, 426
hypoxia inducible factor-1, 34
linker for activation of T cells
phosphorylated protein, 252–253, 257
subcellular localization, 250–251
whole-cell lysates, 252
mitogen-activated protein kinase substrates, 55–57
nitric oxide synthase isoforms, 290–291
p21, 265
poly(ADP-ribose) polymerase cleavage detection, 152–153
trivalent arsenical-labeled protein detection, 110–111
tyrosine phosphoproteins in glutamate activation, 193–194
vascular endothelial growth factor, 404–405

ISBN 0-12-182255-9

90051

9 780121 822552